Choices in Relationships

About the Authors

David Knox, Ph.D., is Professor of Sociology at East Carolina University, where he teaches Courtship and Marriage, Marriage and the Family, and Sociology of Human Sexuality. He is a marriage and family therapist and the author or coauthor of 10 books and 60 professional articles. He and Caroline Schacht are married.

Caroline Schacht, M.A. in Sociology and M.A. in Family Relations, is Instructor of Sociology at East Carolina University and teaches Courtship and Marriage, Introduction to Sociology, and The Sociology of Food. Her clinical work includes marriage and family relationships. She is also a divorce mediator and the coauthor of several books, including *Understanding Social Problems* (Wadsworth, 2008).

Choices in Relationships

An Introduction to Marriage and the Family · NINTH EDITION

David Knox

East Carolina University

Caroline Schacht

East Carolina University

THOMSON
WADSWORTH

Australia • Canada • Mexico • Singapore • Spain • United Kingdom • United States

THOMSON

™

WADSWORTH

Choices in Relationships: An Introduction to Marriage and the Family, **Ninth Edition**

David Knox and Caroline Schacht

Acquisitions Editor: Chris Caldeira

Development Editor: Kristin Marrs

Assistant Editor: Christina Ho

Editorial Assistant: Tali Beesley

Technology Project Manager: David Lionetti

Marketing Manager: Michelle Williams

Marketing Assistant: Jaren Boland

Marketing Communications Manager: Linda Yip

Project Manager, Editorial Production: Matt Ballantyne

Creative Director: Rob Hugel

Art Director: Tippy McIntosh

Print Buyer: Doreen Suruki

Permissions Editor: Sarah D'Stair

Production Service: Graphic World Inc.

Text Designer: Lou Ann Thesing

Photo Researcher: Terri Wright

Copy Editor: Graphic World Inc.

Illustrator: Graphic World Inc.

Cover Designer: Yvo Riezebos

Cover Image: Christian Hoehn/Taxi/Getty Images, Inc.

Compositor: Graphic World Inc.

Text and Cover Printer: Transcontinental Printing

Thomson Higher Education
10 Davis Drive
Belmont, CA 94002-3098
USA

Printed in Canada

1 2 3 4 5 6 7 11 10 09 08 07

Library of Congress Control Number: 2006940530

Student Edition: ISBN-13: 978-0-495-09185-1
ISBN-10: 0-495-09185-5

Instructor's Edition: ISBN-13: 978-0-495-09952-9
ISBN-10: 0-495-09952-X

For more information about our products, contact us at:
Thomson Learning Academic Resource Center
1-800-423-0563

For permission to use material from this text or product, submit a request online at **http://www.thomsonrights.com**.

Any additional questions about permissions can be submitted by e-mail to **thomsonrights@thomson.com**.

To Cameron Emmott,
who will someday make choices in relationships for herself.

Contents in Brief

Chapters

Contents

CHAPTER *5*

Sexuality in Relationships *130*

CHAPTER *6*

Same-Sex Couples and Families *161*

CHAPTER *11*

Parenting *315*

CHAPTER *12*

Balancing Work and Family Life *347*

Preface

The title of this text is *Choices in Relationships* because of an important, endurable fact: the choices we make in regard to our relationships have a long-term impact on the well being of us as individuals, our partners, our marriage, our parents, and our children. By being attentive to the importance of making conscious, deliberate, research-based relationship choices (we also recognize the importance of emotional factors in U.S. culture), everyone (including society) wins. Not to take our relationship choices seriously is to limit our ability to have, to share, and to enjoy a fulfilling emotional life—the only game in town that matters.

New to the Ninth Edition: Chapter-by-Chapter Changes

In addition to new research reflected in every chapter, we have added new content to this edition:

Chapter 1 Choices in Relationships: An Introduction

New conceptions of family—friends, pets

Explanation of traditional, modern, postmodern families

Biosocial framework

Social class differences

Evaluation of research in marriage and the family

Families amid a context of terrorism

Amish/Mennonite marriage/family life—unique cultural examples

Chapter 2 Gender in Relationships

Brain anatomy as the cause of intersexuality

Gender roles in Caribbean, Greek, South African Families

Research Application: Self-made man: One woman's journey into manhood

Personality types that may be problematic

Cold feet—Jennifer Wilbanks

Chapter 8 Marriage Relationships

Research Application: *How college students view weddings*

Muslim American marriage

Military marriage

Marital happiness across the family life cycle

The value of marriage education programs

Healthy marriage resource center

Financial changes after marriage

Age-discrepant relationships where the woman is older

New data on desire for children and time commitment to role

Chapter 9 Communication in Relationships

New focus on fair fighting

Research Application: *Marital quality—keep it high for good health*

Expansion of section on self-disclosure, dishonesty

Expansion of section on gender differences in communication

New section on forgiveness

Chapter 10 Planning Children and Contraception

New data on having children at later ages

Reasons for remaining childfree

Research Application: *Baby Think It Over*

The latest information on contraceptive technology

Self-Assessment: *Attitudes toward parenthood scale*

New data on prevalence of infertility

New data on women more likely to blame themselves if infertile

Self-Assessment: *Abortion attitude scale*

Chapter 11 Parenting

Social Policy focused on whether government or parents should control Internet content for kids

Single parenting moved to this chapter

New Self-Assessment: Spanking versus time-out scale

Avoiding overindulgence

New data on effect of children on marital satisfaction

Family health promotion as an important role of parenting

Features of the Text

To emphasize the importance of using research-based information to understand interpersonal relationships/choices, Research Application is a new feature added to the text in which we detail a research study. Examples of the eight research applications in this edition include how marital quality affects one's health, how being responsible for a baby changes one's life, and how some aspects of mate selection remain traditional (women seeking traditional husbands). Other pedagogical features of previous editions (identified by adopters and reviewers as valuable) have been retained in this ninth edition and include the following.

Self-Assessment Scales

One of the most popular features of the text is the Self-Assessment Scale. Six new scales, including the "Open-Mindedness Scale," "Attitudes toward Premarital Sex," and "Attitudes toward Infidelity Scale," have been added to the text. Students can complete the scales to measure a particular aspect of themselves and/or their relationship with their partner and compare their scores with other students'.

Personal Choices

Consistent with our theme of Choices, embedded in each chapter are detailed discussions of one or two personal choices relevant to the content of the chapter. Examples include "Who Is the Best Person for You to Marry?" "Should I Get Involved in a Long-Distance Dating Relationship?" and "Are the Benefits of 'Coming Out' Worth the Risks?"

Social Policy

These identify social policy issues related to marriage and the family with which U.S. society is confronted. Topics discussed include abstinence sex education in the public schools, adoption by same-sex couples, and increasing the requirements for a marriage license.

National Data and International Data

To replace speculation with facts, we present data from national samples as well as data from around the world.

Diversity in the United States

These emphasize racial, age-related, religious, economic, and educational differences in regard to marriage and family in the United States.

Diversity in Other Countries

These provide a glimpse of gender roles, love, courtship, and marriage and family patterns in other societies and cultures. For example, over a thousand (1368 aged 16 or older) high school students in the United Kingdom revealed their expectations regarding oral sex. Of sexually experienced males, almost half (48.9%) expected oral sex; 46.5% of the sexually experienced females expected oral sex. Only 13.7 percent of the males and 24.7% of the females reported that they believed it was important to use a condom during fellatio (Stone et al., 2006).

Summary

The format for the summary is questions and answers, with each question highlighting a major point in the chapter.

Key Terms

Boldface type indicates key terms in the text, listed at the end of each chapter, and defined in the glossary at the end of the text.

Web Links

Internet addresses are provided at the end of each chapter. We have checked each of these links to ensure that each site was operative at the time the text went to press.

Supplements and Resources

Choices in Relationships, Ninth Edition, is accompanied by a wide array of supplements prepared for both the instructor and the student. Some new resources have been created specifically to accompany the ninth edition, and all of the continuing supplements have been thoroughly revised and updated.

Supplements for the Instructor

Instructor's Edition of *Choices in Relationships: An Introduction to Marriage and the Family*, Ninth Edition   The Instructor's Edition contains a Visual Preface, a walk-through of the text that provides an overview of the text's key features, themes, and supplements.

Instructor's Resource Manual with Test Bank   This manual provides instructors with learning objectives, a list of major concepts and terms (with page references), detailed lecture outlines, extensive student projects and classroom activities, InfoTrac® College Edition and Internet exercises, and self-assessment handouts for each chapter. Also included is a concise user guide for InfoTrac College Edition and a table of contents for the *ABC News* Marriage and Family Video Series. The Test Bank contains 40 to 50 multiple-choice, 10 to 15 true/false, 10 short answer/discussion, and 3 to 5 essay questions per chapter and Special Topic. The Test Bank items are also available electronically on the ExamView Computerized Testing CD-ROM and on the Multimedia Manager CD-ROM.

Multimedia Manager Instructor Resource CD: A Microsoft® PowerPoint® Link Tool
The Multimedia Manager Instructor Resource CD-ROM includes book-specific PowerPoint Lecture Slides, graphics from the book itself, the IRM/TB Word documents, ABC Video Clips, and links to many of Wadsworth's important sociology resources in addition to the Test Bank items.

Marriage & Family DVD   Enhance your classroom lecture with unique video clips on topics typically taught in the marriage and family course, including gender, relationships, and sexuality.

NEW! Classroom Activities and Lecture Ideas for Marriage and Family   Made up of contributions by instructors who teach the course, this book will add new life to your lectures. Includes group exercises, lecture ideas, and homework assignments.

***ExamView* Computerized Testing**   Create, deliver, and customize tests and study guides (both print and online) in minutes with this easy-to-use assessment and tutorial system. *ExamView* offers both a Quick Test Wizard and an Online Test Wizard that guide you step by step through the process of creating tests. The test appears on screen exactly as it will print or display online. Using *ExamView's* complete word-processing capabilities, you can enter an unlimited number of new questions or edit existing questions included with *ExamView*.

JoinIn™ on TurningPoint®   Transform your lecture into an interactive student experience with *JoinIn*. Combined with your choice of keypad systems, *JoinIn* turns your PowerPoint® application into audience-response software. With a click on a handheld device, students can respond to multiple-choice questions, short polls, interactive exercises, and peer-review questions. You can also take attendance, check student comprehension of concepts, collect student demographics to better assess student needs, and even administer quizzes. In addition, there are interactive text-specific slide sets that you can modify and merge with any of your own PowerPoint lecture slides. This tool is available to qualified adopters (http://turningpoint.thomsonlearningconnections.com).

Wadsworth's Marriage and Family 2008 Transparency Masters   A selection of one-color masters consisting of tables and figures from Wadsworth's marriage and family texts is available to help prepare lecture presentations. Free to qualified adopters.

***ABC News* Marriage and Family Video Series, Volumes I–II**   Illustrate how the principles that students learn in the classroom apply to the stories they see on television with the *ABC News* Marriage and Family Video Series, an exclusive series jointly created by Wadsworth and ABC. Each volume consists of approximately 45 minutes of footage originally broadcast on ABC and selected specifically to illustrate concepts relevant to the marriage and the family course. The videos are broken into short 2- to 5-minute segments, perfect for classroom use as lecture launchers or to illustrate key concepts. An annotated table of contents accompanies each video, with descriptions of the segments. Special adoption conditions apply. Examples include: Living Together vs. Marriage, Sperm Donors: The Aftermath, and Effect of War on Marriage.

Wadsworth Sociology Video Library   This large selection of thought-provoking films, including some from the Films for the Humanities collection, is available to adopters based on adoption size.

Supplements for the Student

Study Guide   Each chapter of the Study Guide includes learning objectives, key terms (with page references), a detailed chapter outline, InfoTrac College Edition exercises, Internet exercises, and a personal application section. Students can test and apply their knowledge of concepts with chapter practice tests including 15 to 20 multiple-choice, 10 true/false, 10 to 15 completion, and 5 to 10 short answer questions, all with page references, as well as 5 to 10 essay questions. The Study Guide includes practice tests for not only each of the chapters but also the Special Topics sections.

Marriage and Family: Using Microcase® ExplorIt, Third Edition   Written by Kevin Demmitt of Clayton College, this software-based workbook is an exciting way to get students to view marriage and the family from the sociological perspective.

With this workbook and accompanying ExplorIt software and data sets, your students will use national and cross-national surveys to examine and actively learn marriage and family topics. This inexpensive workbook will add an exciting dimension to your marriage and family course.

***Families and Society,* First Edition**   Written by Scott L. Coltrane of The University of California, Riverside, this reader is designed to promote a sociological understanding of families while demonstrating the diversity and complexity of contemporary family life. The different parts or sections of the reader are designed to "map" onto most sociology of the family textbooks and course syllabi.

The articles emphasize social constructionist and sociological views of families. The reader is thus designed to dispel the myth that families are separate from society. Virtually every selection illustrates the myriad links that exist between families and their various social, cultural, economic, and political contexts. The first reading in each section provides a classic theoretical overview of the topic. These classic articles provide students with a historical understanding of the development of the field and offer insight into some of the enduring sociological facts about families.

***Readings in the Marriage and Family Experience,* Third Edition** Written by Bryan Strong, Christine DeVault, and Barbara W. Sayad, this *free* reader contains articles, essays, and excerpts from books, journals, magazines, and newspapers on the topic of marriage and the family. They have all been carefully selected for their diverse and stimulating content. Critical-thinking questions are included for each reading in order to encourage classroom discussion and student participation.

Marriage and Family Case Studies CD-ROM This unique student CD-ROM includes a series of 10 interactive case study videos that provide a dramatic enactment illustrating key topics and concepts from the text. Students watch each video and answer critical-thinking questions, applying marriage and family theories to the case-study videos. Students then compare their analysis with that of a marriage and family expert. Also on the CD-ROM is a direct link to InfoTrac College Edition, where students can search for related articles from sociology periodicals, and a link to Wadsworth's Virtual Society, where the companion website for *Choices in Relationships,* Ninth Edition, offers a wide range of book-specific study tools.

Online Resources

Wadsworth's Virtual Society: The Wadsworth Sociology Resource Center http://www .wadsworth.com/sociology Here you will find a wealth of resources for students and instructors, including the Marriage and Family Resource Center. This fun, interactive site is filled with additional data, exercises, and enriching resources:

- GSS Activities—Students can compare their attitudes with GSS data in this interactive exercise.
- Marriage and Family Activities—Fun self-quizzes for students to learn more about themselves.
- Marriage and Family Links—Online resources.
- Resources and Organizations—A library of useful resources.

The Companion Website for *Choices in Relationships: An Introduction to Marriage and the Family,* Ninth Edition www.thomsonedu.com/sociology/knox
Access useful learning resources for each chapter of the book. Some of these resources include:

- Tutorial practice quizzes that can be scored and emailed to the instructor
- Flash cards
- Census 2000 information
- And much more!

InfoTrac College Edition Give your students anytime, anywhere access to reliable resources with InfoTrac College Edition, the online library. This fully searchable database offers 20 years' worth of full-text articles from almost 5000 diverse sources, such as academic journals, newsletters, and up-to-the-minute periodicals, including *Time, Newsweek, Science, Forbes,* and *USA Today.* The incredible depth and breadth of material—available 24 hours a day from any computer with Internet access—makes conducting research so easy, your students will want to use it to enhance their work in every course! Through InfoTrac College Edition's InfoWrite, students now also have instant access to critical-thinking and paper-writing tools. Both adopters and their students receive unlimited access for 4 months.

Opposing Viewpoints Resource Center (OVRC) This online center allows you to expose your students to all sides of today's most compelling issues! The Opposing Viewpoints Resource Center draws on Greenhaven Press's acclaimed social issues series, as well as core reference content from other Gale and Macmillan Reference USA sources. The result is a dynamic online library of current event topics—the facts as well as the arguments of each topic's proponents and detractors. Special sections focus on critical thinking (walks student through how to critically evaluate point-counterpoint arguments) and on researching and writing papers.

Acknowledgments

Texts are always collaborative efforts. This text reflects the commitment and vision of the new editor for the ninth edition, Chris Caldeira. We thank Chris for her content guidance through this new edition. We also thank Kristin Marrs, our developmental editor, who was always prompt in moving the revision forward; Christina Ho, the assistant editor; Michelle Williams, the marketing manager; Dave Lionetti, the technology project manager; Matt Ballantyne, the production project manager; and Joohee Lee, the permissions editor. All were superb, and we appreciate their professionalism and attention to detail.

We also thank Sarah Raji, Britt Sholar, and Angela DeCuzzi for researching various topics; Beth Credle for updating the information on contraception and sexually transmitted infections; and Kelly Lewis and E. Fred Johnson, Jr., who provided several stunning photos for the text.

Reviewers for the Ninth Edition

Terry Hatkoff, California State University
Tina Mougouris, San Jacinto College
Lloyd Pickering, University of Montevallo
Janice Purk, Mansfield University
Teresa Tsushima, Iowa State University

Reviewers for the Previous Editions

Grace Auyang, University of Cincinnati
Rosemary Bahr, Eastern New Mexico University
Von Bakanic, College of Charleston
Mary Beaubien, Youngstown State University
Sampson Lee Blair, Arizona State University
Mary Blair-Loy, Washington State University
David Daniel Bogumil, Wright State University
Elisabeth O. Burgess, Georgia State University
Craig Campbell, Weber State University
Michael Capece, University of South Florida
Lynn Christie, Baldwin-Wallace College
Laura Cobb, Purdue University and Illinois State University
Jean Cobbs, Virginia State University
Donna Crossman, Ohio State University
Karen Dawes, Wake Technical Community College
Susan Brown Donahue, Pearl River Community College
Doug Dowell, Heartland Community College

John Engel, University of Hawaii
Mary Ann Gallagher, El Camino College
Shawn Gardner, Genesee Community College
Heidi Goar, St. Cloud State University
Norman Goodman, State University of New York at Stony Brook
Ted Greenstein, North Carolina State University
Jerry Ann Harrel-Smith, California State University, Northridge
Gerald Harris, University of Houston
Rudy Harris, Des Moines Area Community College
Terry Hatkoff, California State University, Northridge
Sheldon Helfing, College of the Canyons
Tonya Hilligoss, Sacramento City College
Rick Jenks, Indiana University
Richard Jolliff, El Camino College
Diane Keithly, Louisiana State University
Steve Long, Northern Iowa Area Community College
Patricia B. Maxwell, University of Hawaii
Carol May, Illinois Central College
Scott Potter, Marion Technical College
Cherylon Robinson, University of Texas at San Antonio
Cynthia Schmiege, University of Idaho
Eileen Shiff, Paradise Valley Community College
Scott Smith, Stanly Community College
Tommy Smith, Auburn University
Beverly Stiles, Midwestern State University
Dawood H. Sultan, Louisiana State University
Elsie Takeguchi, Sacramento City College
Myrna Thompson, Southside Virginia Community College
Janice Weber-Breaux, University of Southwestern Louisiana
Loreen Wolfer, University of Scranton

We love the study, writing, and teaching of Marriage and the Family and recognize that no one has a corner on relationships. We welcome your insights, stories, and suggestions for improvement in the tenth edition of this text. We check our email frequently and invite you to write us.

David Knox, email: *Knoxd@ecu.edu*
Caroline Schacht, email: *Schachtc@ecu.edu*

Marriage has become more joyful, more loving, and more satisfying for many couples than ever before in history. At the same time it has become optional and more brittle.

Stephanie Coontz, *family historian*

Choices in Relationships: An Introduction

Authors

Contents

True or False?

1. Marriage is losing its allure—based on the new trend toward singlehood, it is estimated that less than 60 percent of adults will eventually marry.

2. The high rate of interracial marriages reflects the fact that in today's society, people are free to marry whomever they please.

3. An example of a postmodern family is a gay family unit or a mother who is single by choice.

4. The most enduring relationships in the family are those between siblings, particularly between sisters.

5. The social exchange framework is the most common framework used in empirical studies of marriage and the family.

Answers: **1.** F   **2.** F   **3.** T   **4.** T   **5.** T

You are everything you choose to be.

Shad Helmstedder, self-talk guru

An aging marriage and family professor revealed a pattern of always giving the same tests each year. Colleagues thought the teacher both lazy and unfair to students since the tests would get out. But the teacher justified giving the same exams every year on the premise that the answers kept changing. One answer that has not changed is the consistently high value that college students place on marriage and the family. In a national survey of over 263,000 first-year undergraduates at over 400 two- and four-year universities in the United States, "raising a family" was listed as the number one value (75.9%) with "being well off financially" a close second (74.5%) (American Council on Education and University of California, 2005–2006). In a sample of 1027 undergraduate university students, 98 percent responded "yes" to the statement, "Someday, I want to marry" (96% also responded "yes" to "Someday, I want to have children") (Knox and Zusman, 2006).

This text emphasizes a proactive take-charge-of-your-life/make-wise-relationship-choices theme. The World Health Organization defines *health* as a state of complete physical, mental, and social well being, and not merely the absence of disease or infirmity. This definition underscores the importance of social relationships as an important element in our individual and national health. Making the right choices in our marriage and family relationships is critical to our health, happiness, and sense of well being. Our times of greatest elation and sadness are in reference to our love relationships. The choices we make in these relationships affect these outcomes. We begin by discussing in detail this theme for the text.

Your choice of a spouse is one of the most important choices you will ever make.

Authors

Choices in Relationships— The View of This Text

The central theme of this text is choices in relationships. Although there are over a hundred such choices, five major choices include whether to marry, whether to have

Chapter 1 Choices in Relationships: An Introduction

Will You Make Choices Deliberately or by Default?
It is a myth that we can avoid making relationship decisions, because not to make a decision is to make a decision by default. Some examples follow:

- If we don't make a decision to pursue a relationship with a particular person, we have made a decision (by default) to let that person drift out of our life.
- If we don't decide to do the things that are necessary to keep or improve the relationships we have, we have made a decision to let these relationships slowly disintegrate.
- If we don't make a decision to be faithful to our partner or spouse, we have made a decision to be open to situations and relationships in which we are vulnerable to being unfaithful.
- If we don't make a decision to avoid having intercourse early in a new relationship, we have made a decision to let intercourse occur at any time.
- If we are sexually active and don't make a decision to use birth control or a condom, we have made a decision to expose ourselves to risk for pregnancy or a sexually transmitted infection.
- If we don't make a decision to break up with an abusive partner or spouse, we have made a decision to continue the relationship with him or her.

Throughout the text, as we discuss various relationship choices, consider that you automatically make a choice by being inactive—that not to make a choice is to make one. We encourage a proactive style whereby you make deliberate relationship choices.

The three biggest mistakes in life are failing to make wise choices, failing to recognize bad choices, and failing to correct bad choices (if possible) once they have been made.

Marty E. Zusman, Sociologist

children, whether to have one or two earners in one marriage, whether to remain emotionally and sexually faithful to one's partner, and whether to use a condom. Though structural and cultural influences are operative, a choices framework emphasizes that individuals have some control over their relationship destiny by making deliberate choices to initiate, respond to, nurture, or terminate intimate relationships.

Facts about Choices in Relationships
The facts to keep in mind when making relationship choices include the following.

Not To Decide Is To Decide   Not making a decision is a decision by default. If you are sexually active and decide not to use a condom, you have made a decision to increase your risk for contracting a sexually transmissible infection, including HIV. If you don't make a deliberate choice to end a relationship that is going nowhere, you will continue in that relationship and have little chance of getting into a more positive and satisfying relationship. If you don't make a decision to be faithful to your partner, you are vulnerable to cheating. See the Personal Choices section at the top of this page for more examples.

Some Choices Require Corrections   Some of our choices, while appearing to be correct at the time that we make them, turn out to be disasters. Once it is evident that a choice is having consistently negative consequences, it is important to stop defending the old choice, reverse positions, make new choices, and move forward. Otherwise, one remains consistently locked into the continued negative outcomes of a "bad" choice. For example, the choice of a partner who was once loving and kind but who has turned out to be abusive and dangerous requires correcting that choice. To stay in the abusive relationship will have predictable disastrous consequences—to make the decision to disengage and to move on opens the opportunity for a better relationship with another partner. Other examples of making a correction may involve ending a dead relationship (we are

The bottom line is that I am responsible for my own well-being, my own happiness. The choices and decisions I make regarding my life directly influence the quality of my days.

Kathleen Andrus, actress

not suggesting that this be done without investing time and effort to improve one's relationship), changing majors, or changing one's job or career.

Choices Involve Trade-offs By making one choice you relinquish others. Every relationship choice you make will have a downside and an upside. If you decide to get married, you will give up your freedom to pursue other emotional/ sexual relationships, and you will also give up some of your control over how

The Relationship Involvement Scale

This scale is designed to assess the level of your involvement in a current relationship. Please read each statement carefully, and write the number next to the statement that reflects your level of disagreement to agreement, using the following scale.

1	2	3	4	5	6	7
Strongly Disagree						Strongly Agree

_____ **1.** I have told my friends that I love my partner.

_____ **2.** My partner and I have discussed our future together.

_____ **3.** I have told my partner that I want to marry him/her.

_____ **4.** I feel happier when I am with my partner.

_____ **5.** Being together is very important to me.

_____ **6.** I cannot imagine a future with anyone other than my partner.

_____ **7.** I feel that no one else can meet my needs as well as my partner.

_____ **8.** When talking about my partner and me, I tend to use the words "us," "we," and "our."

_____ **9.** I depend on my partner to help me with many things in life.

_____ **10.** I want to stay in this relationship no matter how hard times become in the future.

Scoring

Add the numbers you assigned to each item. A 1 reflects the least involvement and a 7 reflects the most involvement. The lower your total score (10 is the lowest possible score), the lower your level of involvement; the higher your total score (70), the greater your level of involvement. A score of 40 places you at the midpoint between a very uninvolved and very involved relationship.

Other Students Who Completed the Scale

Valdosta State University. The participants were 31 male and 86 female undergraduate psychology students haphazardly selected from Valdosta State University. They received course credit for their participation. These participants ranged in age from 18 to 59 with a mean age of 20.25 (SD = 4.52). The ethnic background of the sample included 70.9% white, 23.9% Black, 1.7% Hispanic, and 3.4% from other ethnic backgrounds. The college classification level of the sample included 46.2% freshmen, 36.8% sophomores, 14.5% juniors, and 2.6% seniors.

East Carolina University. Also included in the sample were 60 male and 129 female undergraduate students haphazardly selected from East Carolina University. These participants ranged in age from 18 to 43 with a mean age of 20.40 (SD 5 3.58). The ethnic background of the sample included 76.2% white, 14.3% Black, 0.5% Hispanic, 1.6% Asian, 2.6% American Indian, and 4.8% from other ethnic backgrounds. The college classification level of the sample included 40.7% freshmen, 19.0% sophomores, 19.6% juniors, and 20.6% seniors. All participants were treated in accordance with the ethical guidelines of the American Psychological Association (1992).

Scores of Participants

When students from both universities were combined, the average score of the men was 50.06 (SD = 14.07) and the average score of the women was 52.93 (SD = 15.53), reflecting moderate involvement for both sexes. There was no significant difference between men and women in level of involvement. However, there was a significant difference ($p < .05$) between whites and non-whites, with whites reporting greater relationship involvement (M = 53.37; SD = 14.97) than non-whites (M = 48.33; SD = 15.14).

In addition, there was a significant difference between the level of relationship involvement of seniors compared with juniors ($p < .05$) and freshmen ($p < .01$). Seniors reported more relationship involvement (M = 57.74; SD = 12.70) than did juniors (M = 51.57; SD 5= 15.37) or freshmen (M = 50.31; SD = 15.33).

Source

"The Relationship Involvement Scale" 2004 by Mark Whatley, Ph.D. Department of Psychology, Valdosta State University, Valdosta, Georgia 31698. Information on validity and reliability may be obtained from Dr. Whatley. The scale is used by permission of Dr. Whatley. Other uses of this scale by written permission only; e-mail *mwhatley@valdosta.edu* you spend your money—but you may get a wonderful companion for life if you marry. If you decide to have children, you will give up more of your freedom and will have even less money to spend on yourself; if you don't have children, you will miss what many regard as wonderful relationship with your own child. If you maintain a long-distance relationship with someone, you continue involvement in a relationship that is obviously important to you. But you may spend many weekends alone when you could be discovering new relationships. The Relationships Involvement Scale (see the Self-Assessment section on the previous page) will help you to identify the level of involvement in your current relationship.

Choices Include Selecting a Positive or Negative View Thomas Edison said as he progressed toward inventing the light bulb, "I have not failed. I have found ten thousand ways that won't work." In spite of an unfortunate event in your life, you can choose to see the bright side. Regardless of your circumstances, you can choose to view a situation in positive terms. A breakup or divorce can be viewed as the end of a disastrous relationship and an opportunity to become involved in a new, more fulfilling relationship. Being overwhelmed with debt can be viewed as an opportunity to learn how to manage money and to live on less and as an opportunity to reevaluate the importance of relationships versus money as the ultimate source of personal contentment. The discovery of infidelity on the part of one's partner can be viewed as an opportunity to open up communication channels with one's partner and develop a stronger relationship. Finally, a debilitating illness (e.g., need for kidney transplant) can be viewed as a challenge to face adversity together. One's point of view does make a difference—it is the one thing we have control over.

Choices Produce Ambivalence Choosing among options and trade-offs often creates ambivalence—conflicting feelings that produce uncertainty or indecisiveness as to what course of action to take. There are two forms of ambivalence: sequential and simultaneous. In sequential ambivalence, the individual experiences one wish and then another. For example, a person may vacillate between wanting to stay in a less-than-fulfilling relationship or end it. In simultaneous ambivalence, the person experiences two conflicting wishes at the same time. For example, the individual may at the same time feel both the desire to stay with the partner and the desire to break up. The latter dilemma is reflected in the saying, "You can't live with them and you can't live without them."

The woods are lovely, dark and deep,
But I have promises to keep,
And miles to go before I sleep,
And miles to go before I sleep.
Robert Frost

Most Choices Are Revocable; Some Are Not Most choices are revocable; that is, they can be changed. For example, a person who has chosen to be sexually active with multiple partners can later decide to be monogamous or to abstain from sexual relations. Or individuals who in the past have chosen to emphasize career (money and advancement) over marriage and family can choose to emphasize relationships over economic and career-climbing behaviors.

Other choices are less revocable. For example, backing out of the role of spouse is much easier than backing out of the role of parent. Whereas the law permits disengagement from the role of spouse (formal divorce decree), the law ties parents to dependent offspring (e.g., through child support). Hence the decision to have a child is usually irrevocable.

Choices Are Influenced by the Stage in the Family Life Cycle The choices a person makes tend to be individualistic or familistic, depending on which stage in the family life cycle the person is in. Before marriage, individualism characterizes most thinking and decisions. People are delaying marriage in favor of completing school, becoming established in a career, and enjoying the freedom of singlehood.

The hardest thing to learn in life is which bridge to cross and which to burn.
David Russell, guitarist

Once married, and particularly when they have children, their values and choices become more familistic. Evidence for familism in this regard is the fact that the divorce rate has decreased in recent years (*Statistical Abstract of the United States, 2006*). Also, couples with children have much lower divorce rates than those without children.

Making Choices Is Facilitated with Decision-Making Skills Choices occur at the individual, couple, and family level. Deciding to transfer to another school or take a job out of state may involve all three levels, whereas the decision to lose weight is more likely to be an individual decision. Regardless of the level, the steps in decision making include setting aside enough time to evaluate the issues involved in making a choice, identifying alternative courses of action, carefully weighing the consequences for each choice, and being attentive to your own inner state and motivations. The goal of most people is to make relationship choices that result in the most positive and least negative consequences. Choices are also difficult.

We asked our students to identify their "most difficult" choices. Their answers included "continuing a relationship with someone I was intensely in love with but who was abusive," "leaving my partner to come to college . . . I thought we had a solid relationship . . . I was wrong," and "forgiving my partner for cheating on me."

Structural and Cultural Influences on Choices

Choices are influenced by structural and cultural factors. This section reviews the ways in which social structure and culture impact choices in relationships. Although a major theme of this book is the importance of taking active control of your life in making relationship choices, it is important to be aware that such choices are restricted and channeled by the social world in which you live. For example, enormous social disapproval for marrying someone of another race is part of the reason 95 percent of all individuals in the United States marry someone of the same race.

Social Structure The social structure of a society consists of institutions, social groups, statuses, and roles.

1. Institutions. The largest elements of society are social institutions, which may be defined as established and enduring patterns of social relationships. Major institutions include the family, economy, education, and religion. Institutions affect individual decision-making. For example, the Protestant, Catholic, and Jewish faiths encourage their members to restrict premarital sexual behavior and to marry someone of the same faith. You live in a capitalistic society where economic security is valued (recall it is the number two value held by college students). In effect, the more time you spend focused on obtaining money, the less time you have for relationships. The family is a universal institution. Spouses who "believe in the institution of the family" are highly committed to maintaining their marriage and do not regard divorce as an option. Some evidence suggests that wives married to husbands with high institutional commitment report higher marital quality than wives married to husbands without this view/value (Wilcox and Noch, 2006).

2. Social groups. Institutions are made up of social groups, defined as two or more people who have a common identity, interact, and form a social relationship. Most individuals spend their day going between social groups. You may awaken in the context (social group) of a roommate, partner, or spouse. From there you go to class with classmates, lunch with friends, job with boss, and talk on the phone to your parents. So, within 24 hours you have been in at least five social groups. These social groups have varying influences on your choices. Your roommate affects who can be in the room for how long, your friends may want to lunch at a particular place, your boss will assign you certain duties, and your parents may want you to come home for the weekend.

Your interpersonal choices are influenced mostly by your partner and peer (e.g., your sexual values, use of a condom, the amount of alcohol you consume Thus the importance of selecting your partner and peers carefully. Falstaff, of Shakespeare's characters, said, "Company, villainous company, hath been spoil of me."

Our lives are shaped by the choices and actions of others around us.

Tepperman and Susannah

Social groups may be categorized as primary or secondary. **Primary groups,** which tend to involve small numbers of individuals, are characterized by interaction that is intimate and informal. Parents are members of one's primary group and may exercise enormous influence over one's mate choice. Grace Kelly was a movie star of the 1950s who married (with her parents approval) Prince Rainier of Monaco. Previously, she had been in love with and wanted to marry various men. But, as her biographer expressed it:

Her life, as it differed from family strictures, was simply not her own . . . no matter how many important magazines ran profiles of her. It was a depressing realization. Her choice of a marriage partner, she feared, could not be based upon her own needs and desires, but rather on what was best for the Kelly family, what advanced their position in the eyes of their neighbors and the world. As Cassini (a lover) put it, 'Her family regarded her as a prize possession, a property, like a racehorse that must be handled, above all invested, wisely—not wasted" (Spada, 1987, 105).

Other parents may be less direct in how they influence the mate choice of their son or daughter. By living in certain neighborhoods and sending their children to certain schools/universities, parents are influencing the context in which their children are likely to meet and select others as partners.

In contrast to primary groups, **secondary groups** may involve small or large numbers of individuals and are characterized by interaction that is impersonal and formal. Being in a context of classmates, coworkers, or fellow students in the library are examples of secondary groups. Members of secondary groups have much less influence over one's relationship choices than members of one's primary groups.

Most people regard primary groups as crucial for their personal happiness and feel adrift if they have only secondary relationships. Indeed, in the absence of close primary ties, they may seek meaning in secondary group relationships. Comedian George Carlin said that his "fans were his family" since he was in a different town performing 36 of 52 weekends a year, implying he had no "real" family.

3. Statuses. Just as institutions consist of social groups, social groups consist of statuses. A **status** is a position a person occupies within a social group. The statuses we occupy largely define our social identity. The statuses in a family may consist of mother, father, child, sibling, stepparent, and so on. In discussing family issues we refer to statuses such as teenager, cohabitant, and spouse. Statuses are relevant to choices in that many choices are those that significantly change one's status. Making decisions that change one's status from single person to spouse to divorcee can influence how people feel about themselves and how others treat them.

4. Roles. Every status is associated with many roles, or sets of rights, obligations, and expectations associated with a status. Our social statuses identify who we are; our **roles** identify what we are expected to do. Roles guide our behavior and allow us to predict the behavior of others. Spouses adopt a set of obligations and expectations associated with their status. By doing so, they are better able to influence and predict each other's behavior.

Because individuals occupy a number of statuses and roles simultaneously, they may experience role conflict. For example, the role of the parent may conflict with the role of the spouse, employee, or student. If your child needs to be driven to the math tutor and your spouse needs to be picked up at the airport

*All the world's a stage,
And all the men and women merely players.
They have their exits and entrances;
Each man in his time plays many parts.*

William Shakespeare

and your employer wants you to work late and you have a final exam all at the same time, you are experiencing role conflict. Researchers have found that having multiple roles and balancing these roles actually has a positive effect on one's health and ability to handle stress (Stuart and Garrison, 2002). Hence, while having multiple roles may cause stress since there is a lot to do, multiple roles also provide diversion and distraction if one role becomes problematic.

Culture Just as social structure refers to the parts of society, culture refers to the meanings and ways of living that characterize persons in a society. Two central elements of culture are beliefs and values.

1. Beliefs. **Beliefs** refer to definitions and explanations about what is true. The beliefs of an individual or couple influence the choices they make. Dual-career couples who believe that young children flourish best with a full-time parent in the home make different child-care decisions than do couples who believe that day care offers opportunities for enrichment. The belief that children are best served by being reared with two parents will influence a decision regarding what to do about a premarital pregnancy or an unhappy marriage that is different from a decision where there is a belief that single-parent families can provide an enriching context for rearing children.

2. Values. **Values** are standards regarding what is good and bad, right and wrong, desirable and undesirable. Values influence choices. Individualism, in contrast to familism, reflects the values of the society for the individual or the family. **Individualism** involves making decisions that are more often based on what serves the individual's rather than the family's interests (**familism**). Those who remain single, who live together, who seek a childfree lifestyle, and who divorce are more likely to be operating from an individualistic philosophical perspective than those who marry, do not live together before marriage, rear children, and stay married.

The elements of social structure and culture that we have just discussed play a central role in making interpersonal choices/decisions. One of the goals of this text is to encourage awareness of how powerful social structure and culture are in influencing decision making. Sociologists refer to this awareness as the **sociological imagination** (or sociological mindfulness). For example, though most people in the United States assume that they are free to select their own sex partner, this choice (or lack of it) is in fact heavily influenced by structural and cultural factors. Most people date, have sex with, and marry a person of the same racial background. Structural forces influencing race relations include segregation in housing, religion, and education. The fact that blacks and whites live in different neighborhoods, worship in different churches, and often attend different schools makes meeting a person of another race less likely. And when such encounters occur, they are so influenced by the prejudices and bias brought into the interaction that individuals are hardly "free" to act as they choose. Hence, cultural values (transmitted by and through parents and peers) generally do not promote mixed racial interaction, relationship formation, and marriage. DeCuzzi et al. (2006) found in their study of college students that both whites and blacks tended to view their respective groups more positively. Consider the last three relationships in which you were involved, the level of racial similarity, and the structural and cultural influences on those relationships.

It's not what we don't know that hurts, it's what we know that ain't so.

Will Rogers, humorist

Diversity in the United States

Gay and lesbian relationships are developed and maintained in the context of disapproving institutions (most religions disapprove of homosexuality) and primary groups (family members often react with shock and grief when their child "comes out"). In addition, while the status terms in marriage are husband and wife, gay and lesbian individuals struggle with which term to use (lover, partner, significant other, companion, spouse, life mate, etc.), and their roles as partners in the relationship are not socially scripted.

Other Influences on Relationship Choices

Aside from structural and cultural influences on relationship choices, other influences include your **family of origin** (the family in which you were reared), unconscious motivations, habit patterns, individual personality, and previous experiences. Your family of origin is a major influence on your subsequent choices and relationships. Over half (55.9%) of 658 (a random sample phone interview) undergraduates at a southeastern university reported that they were "very close" to their family and almost another third (31.2%) reported that they were "somewhat close" (Bristol and Farmer, 2005). Such closeness may translate into the desire for parental approval for one's choice of partner. McNeely et al. (2002) also found that adolescent girls (ages 14 to 15) who had close relationships with their mothers reported a later sexual debut (McNeely et al., 2002). Busby et al. (2005) analyzed data on 6,744 individuals in regard to the impact of their family of orientation on their current relationship functioning and found that the quality of the parents' relationship was strongly related to the quality of their own relationship. "Coming from intact families where parents loved each other provided a core foundation for subsequent relationships for children growing up in that context."

Being reared in an economically impoverished family/neighborhood is also associated with deviant peers and behavior (Barrera et al., 2002). Living apart from both parents, whether as a result of divorce or being born out of wedlock, is associated with an increased risk of divorce (Teachman, 2002). Humor in one's family of origin (among women) has also been associated with one's emotional and physical health (Nelson, 2002).

Unconscious processes have been demonstrated in to be operative in our reactions and choices (Winkielman et al., 2005; Mahaffey et al., 2005). A person reared in a lower-class home without adequate food and shelter may become overly concerned about the accumulation of money and make all decisions in reference to obtaining/holding/hoarding economic resources.

Habit patterns also influence choices. A person who is accustomed to and enjoys spending a great deal of time alone may be reluctant to make a commitment to live with a person who makes demands on his or her time. A person who is a workaholic is unlikely to allocate enough time to a relationship to make it flourish. Frequent alcohol consumption is associated with a higher number of sexual partners and not using condoms to avoid pregnancy and contraction of sexually transmitted infections (LaBrie et al., 2002).

Personalities (e.g., introvert, extrovert; passive, assertive) also influence choices. For example, a person who is assertive is more likely than someone who is passive to initiate a conversation with someone he or she is attracted to at a party. A person who is very quiet and withdrawn may never choose to initiate a conversation even though he or she is attracted to someone. A divorced woman who is very individualistic in her thinking is less likely to remarry (DeGraaf and Kalmijn, 2003).

Finally, a person with a bipolar disorder who is manic one part of the semester (or relationship) and depressed the other part is likely to make different choices when each process is operative (Young 2003).

Current and past relationship experiences also influence one's perceptions and choices. Individuals in a current relationship are more likely to hold relativistic sexual values (choose intercourse over abstinence). And individuals who have had prior sexual experience are more likely to engage in sexual activity in the future (Gillmore et al., 2002).

The life of Ann Landers illustrates how life experience changes one's view and choices. After Landers's death, her only child, Margo Howard (2003), noted that her mother was once against intercourse before marriage, divorce, and involvement with a married man. But when Margo told her that unmarried youth were having intercourse, Landers shifted her focus to the use of contraception. When Margo divorced, Landers began to say that ending an unfulfilling marriage is an option. And when Landers was divorced and fell in love with a married man, she said that you can't control who you fall in love with.

Having emphasized that choices is the theme of this text and reviewed the mechanisms operative in those choices, we turn to a discussion of the meaning of marriage and the family in our society.

Some people change when they see the light, others when they feel the heat.

Caroline Schoeder

Marriage

Traditionally, **marriage** has been viewed as a legal relationship that binds a man and a woman together for reproduction and the subsequent care (physical/ emotional) and socialization of children. Attempts to legalize gay marriage throughout the United States continue. Massachusetts recognizes a marriage between two men or two women. We will discuss the issue of gay marriage in detail later in this chapter.

Each society works out its own details of what marriage is. In the United States, marriage is a legal contract between a heterosexual couple and the state in which they reside that specifies economic cooperation and encourages sexual fidelity. The fine print/implied factors implicit during a wedding ceremony include the following elements:

National Data

Over 2.3 million marriages occur annually (Sutton 2003). Over 95 percent of U.S. adult women (96.3%) and men (95.9%) aged 65 and older have married at least once (*Statistical Abstract of the United States: 2006*, Table 34).

Elements of Marriage

Several elements comprise the meaning of marriage in the United States.

Legal Contract   Marriage in our society is a legal contract that may be entered into only by two people of different sexes of legal age (usually 18 or older) who are not already married to someone else. The marriage license certifies that the individuals were married by a legally empowered representative of the state, often with two witnesses present. Though common-law marriages exist in ten states and the District of Columbia, they are the exception.

Under the laws of the state, the license means that all future property acquired by the spouses will be jointly owned and that each will share in the estate of the other. In most states, whatever the deceased spouse owns is legally transferred to the surviving spouse at the time of death. In the event of divorce and unless there was a prenuptial agreement, the property is usually divided equally regardless of the contribution of each partner. The license also implies the expectation of sexual fidelity in the marriage. Through less frequent because of no-fault divorce, infidelity is a legal ground for both divorce and alimony in some states.

The marriage license is also an economic license that entitles a spouse to receive payment by a health insurance company for medical bills if the partner is insured, to collect Social Security benefits at the death of the other spouse, and to inherit from the estate of the deceased. One of the goals of gay-rights advocates who seek the legalization of marriage between homosexuals is that the couple will have the same rights and benefits as heterosexuals.

Though the definition of what constitutes a "family" is being reconsidered by the courts, the law is currently designed to protect spouses, not lovers or cohabitants. An exception is **common-law marriage,** which means that if a heterosexual couple cohabit and present themselves as married, they will be regarded as legally married in those states that recognize such marriages.

Emotional Relationship   "Being in love" was a top life goal (second only to being happy in life) identified by a random sample of students at Bucknell University (Abowitz and Knox, 2003). Indeed, most people in the United States regard being in love with the person they marry as an important reason for staying married. Over two thirds (67%) of 1027 undergraduates reported that they would divorce someone they did not love (Knox and Zusman, 2006). Our emphasis on love is not shared throughout the world. Individuals in other cultures (e.g., India, Iran) do not require love feelings to marry—love is expected to follow, not precede, marriage. And, love is not necessary to continue a marriage, as family considerations (keeping the marriage together for the children) are more important.

Sexual Monogamy   Marital partners expect sexual fidelity. Almost three fourths (73%) of 1027 undergraduates agreed, "I would divorce a spouse who had an affair" and a similar percentage (74%) agreed that they would end a relationship with a partner who cheated on them (Knox and Zusman, 2006).

Legal Responsibility for Children   Although individuals marry for love and companionship, one of the most important reasons for the existence of marriage from the viewpoint of society is to legally bond a male and a female for the nurture and support of any children they may have. In our society, childrearing is the primary responsibility of the family, not the state.

Children are the living messages we send to a time we will never see.
John W. Whitehead, attorney/author

Marriage is a relatively stable relationship that helps to ensure that children will have adequate care and protection, will be socialized for productive roles in society, and will not become the burden of those who did not conceive them. Thus, while it is less normative among African Americans to be married at the time they have children, there is tremendous social pressure to birth children as spouses. Even at divorce, the legal obligation of the father and mother to the child is theoretically maintained through child-support payments.

Announcement/Ceremony   The legal bonding of a couple is often preceded by an announcement in the local newspaper and a formal ceremony in a church or synagogue.

Such a ceremony reflects the cultural importance of the event. Telling parents, siblings, and friends about wedding plans helps to verify the commitment of the partners and also helps to marshal the social and economic support to launch the couple into marital orbit. Most people in our society decide to marry, and the benefits of doing so are enormous (Knox and Zusman, 2006; Wilmoth and Koso, 2002; Kim and McKenry, 2002). When married persons are compared with singles, the differences are striking (Table 1.1.).

Table 1.1   Benefits of Marriage and the Liabilities of Singlehood

	Benefits of Marriage	**Liabilities of Singlehood**
Health	Spouses have fewer hospital admissions, see a physician more regularly, are "sick" less often	Singles are hospitalized more often, have fewer medical checkups, and are "sick" more often.
Longevity	Spouses live longer than singles.	Singles die sooner than marrieds.
Happiness	Spouses report being happier than singles.	Singles report less happiness than marrieds.
Sexual satisfaction	Spouses report being more satisfied with their sex lives, both physically and emotionally.	Singles report being less satisfied with their sex lives, both physically and emotionally.
Money	Spouses have more economic resources than singles.	Singles have fewer economic resources than marrieds.
Lower expenses	Two can live more cheaply together than separately.	Cost is greater for two singles than one couple.
Drug use	Spouses have lower rates of drug use and abuse.	Singles have higher rates of drug use and abuse.
Connectedness	Spouses are connected to more individuals who provide a support system—partner, in-laws, etc.	Singles have fewer individuals upon whom they can rely for help.
Children	Rates of high school dropouts, teen pregnancies, and poverty are lower among children reared in two-parent homes.	Rates of high school dropouts, teen pregnancies, and poverty are higher among children reared by single parents.
History	Spouses develop a shared history across time with significant others.	Singles may lack continuity and commitment across time with significant others.
Crime	Spouses are less likely to be involved in crime.	Singles are more likely to be involved in crime.
Loneliness	Spouses are less likely to report loneliness.	Singles are more likely to report being lonely.

The "one size fits all" model of relationships and marriage is nonexistent. Individuals may be described as existing on a continuum from heterosexuality to homosexuality, from rural to urban dwellers, and from being single and living alone to being married and living in communes. Emotional relationships range from being close and loving to being distant and violent. Family diversity includes two parents (other or same sex), single-parent families, blended families, families with adopted children, multigenerational families, extended families, and families representing different racial, religious, and ethnic backgrounds. Diversity is the term that accurately describes marriage and family relationships today.

Polygyny, whereby a man can have multiple wives, is the marriage form found in more places and at more times than any other.

Stephanie Coontz, family historian

If you charge one, where do you stop? You start prosecuting 10,000 people and have 20,000 kids go into the welfare system.

Mark Shurtleff, Utah Attorney General

Types of Marriage

While we think of marriage in the United States as involving one man and one woman, other societies view marriage differently. **Polygamy** is a form of marriage in which there are more than two spouses. Polygamy occurs in societies or subcultures whose norms sanction multiple partners. Polygamous marriages/families have been associated with lower levels of education, socioeconomic status, and family functioning (Al-Krenawi et al., 2002). There are three forms of polygamy: polygyny, polyandry, and pantogamy.

Polygyny Polygyny or plural marriage has been given renewed cultural visibility with *Big Love*, a Home Box Office series. **Polygyny** involves one husband and two or more wives and is practiced illegally in the United States by some religious fundamentalist groups. These groups are primarily in Arizona, New Mexico, and Utah, and have splintered off from the Church of Jesus Christ of Latter-day Saints (commonly known as the Mormon Church) into what is called the Fundamentalist Church of the Latter-day Saints (FLDS). It is estimated that there are over 60,000 Mormon Fundamentalists in the state of Utah (Colorado City is the town with the largest concentration). This sect represents only about five percent of all Mormons in Utah. While polygyny is against the law, individuals are rarely prosecuted for several reasons:

1. Living together—some men live with several wives and have children with them (living together is not against the law).

2. Population—Mormons outnumber non-Mormons, representing 70 percent of the population in Utah (there is limited public momentum for prosecuting Mormons).

3. Prosecution witness—the absence of finding someone willing to testify for the prosecution makes it difficult to mount a case.

4. Prosecution priorities—local prosecutors elect to spend available resources on organized crime, and drug trafficking rather than on illegal civil relationships.

5. Jail space—there is no jail space for housing individuals if convicted.

In spite of the tolerance toward polygyny in Utah, there are some abuses. Some former wives of plural marriages report living in poverty (it is difficult for one man to financially support 15 wives), nonconsent (some existing wives resent their husbands taking new wives), and child sex abuse (some children as young as 14 are made "new wives"). Warren Steed Jeffs, one of the leaders of the Fundamentalist sects, has been charged in Arizona with child sex abuse for arranging the marriage of a teenager to an older married man. He was on the FBI wanted list, where $60,000 was offered in bounty money (Florio and Passey, 2006). He was found and jailed in 2006.

Since the wives have limited education, job skills, and contact with the outside community, they are disadvantaged in extricating themselves from their plight. Tapestry Against Polygamy is an organization that has helped women break free from bigamous marriages.

Polygyny among fundamentalist Mormons serves a religious function in that large earthly families are believed to result in large heavenly families. Notice that polygynous sex is only a means to accomplish another goal—large families. See http://www.absalom.com/mormon/polygamy/faq.htm for more information.

It is often assumed that polygyny exists to satisfy the sexual desires of the man, that the women are treated like slaves, and that jealousy among the wives is common. In most polygynous societies, however, polygyny has a political and economic rather than a sexual function. Polygyny is a means of providing many

The HBO program Big Love gave visibility to the issues involved with multiple wives/plural marriages among some polygynous Mormon families.

male heirs to continue the family line. In addition, when a man has many wives, a greater number of children for domestic/farm labor can be produced. Wives are not treated like slaves (although women have less status than men in general), as all household work is evenly distributed among the wives and each wife is given her own house or own sleeping quarters. Jealousy is minimal because the husband often has a rotational system for conjugal visits, which ensures that each wife has equal access to sexual encounters.

Polyandry   The Buddhist Tibetans foster yet another brand of polygamy, referred to as **polyandry,** in which one wife has two or more (up to five) husbands. These husbands, who may be brothers, pool their resources to support one wife. Polyandry is a much less common form of polygamy than polygyny. The major reason for polyandry is economic. A family that cannot afford wives or marriages for each of its sons may find a wife for the eldest son only. Polyandry allows the younger brothers to also have sexual access to the one wife or marriage that the family is able to afford.

Polyamory   Polyamory (also known as **open relationships**) is a lifestyle in which two lovers do not forbid each other from having other lovers. By agreement, each partner may have numerous other emotional and sexual relationships. Some (about 5%) of the 100 members of Twin Oaks Intentional Community in Louisa, Virginia, are polyamorous in that each partner may have several emotional/physical relationships with others at the same time. While not legally married, these adults view themselves as emotionally bonded to each other and may even rear children together. Polyamory is not swinging, as polyamorous lovers are concerned about enduring, intimate relationships that include sex (Adamson and Star, 2005; McCullough and Hall, 2003). We discuss polyamory in greater detail in the chapter on love.

It's complicated.
Pax Starr, polyamorist

Pantogamy   This describes a group marriage in which each member of the group is married to the others. Pantogamy is a more formal arrangement than polyamory and is reflected in communes (e.g., Oneida) of the nineteenth and twentieth centuries.

Our culture emphasizes monogamous marriage and values stable marriages. One expression of this value is the concern for marriage education (see the Social Policy feature).

Marriage Education in Public Schools

SOCIAL POLICY

Social policies are purposive courses of action taken by individuals or groups of individuals about a particular issue or problem of concern (Moen and Coltrane, 2005). Public schools are a major source of socialization of values and skills in our youth. With almost half of new marriages ending in divorce, politicians ask whether social policies designed to educate youth about the realities of marriage might be beneficial in promoting marital quality and stability. Might students profit from education about marriage, before they get married, if such relationship skill training is made mandatory in the school curriculum? One-hundred percent of students enrolled in a marriage education course noted that they wished to have happy marriages/relationships, 59 percent feared divorce, and 47 percent reported they were having actual trouble in current or past relationships (Nielsen et al., 2004).

Sociologist Marline Pearson emphasized that teenagers "need to hear about relationships before they hit college" (and she has developed a curriculum for high school students) (Madison, 2003). The philosophy behind premarital education is that it is better to build a fence at the top of a cliff than to put an ambulance at the bottom. Over 2000 public schools nationwide offer such a course. In Florida, all public school high school seniors are required to take a marriage and relationship-skills course. Utah budgeted over $600,000 for marriage-promotion education and trains teachers on marriage issues (Serafini and Zeller, 2002). McGeorge and Carlson (2006) compared couples exposed to an 8-week premarital education course with those not exposed and found greater readiness for marriage among those who had premarital education. Kirby (2006) also found in a pre and post test of over 1000 married couples, those who took a marriage education course scored significantly higher on marital satisfaction after completing the course.

The Federal government has a stake in marriage education programs. One motivation is economic, since divorce often leads to poverty. According to Dr. Steven Noch "for every three divorces, one woman and her children end up below the poverty line. Now that states have more responsibility for welfare, they've hopped on the marriage bandwagon" (Serafini and Zeller, 2002).

Another motivation is control of content. The School Textbook Marriage Protection Act, considered in the House education committee in Arkansas, would require that marriage textbooks define marriage as between one man and one woman (House bill, 2005).

There is also opposition to marriage education programs. Opponents question using school time for relationship courses. Teachers are already seen as overworked, and an additional course on marriage seems to press the system to the breaking point. In addition, some teachers lack the training to provide relationship courses. Although training teachers would stretch already thin budgets, many schools already have programs in family and consumer sciences and teachers in these programs are trained in teaching about marriage and the family. A related concern with teaching about marriage and the family in high school is the fear on the part of some parents that the course content may be too liberal. Some parents who oppose teaching sex education in the public schools fear that such courses lead to increased sexual activity.

Your Opinion?

1. To what degree do you believe marriage education belongs in the public school system?
2. How should marriage be defined?
3. Should marriage be encouraged by the Federal government?

Sources

House bill to define marriage in textbooks heads back to education committee. 2005. *Education Daily* 38(16): 4.

Kirby, J. S. 2005. A study of the marital satisfaction levels of participants in a marriage education course. Dissertation Abstracts, International A: *The Humanities and Social Sciences,* 66, 4, Oct, 1513-A

Madison, R. C. 2003. Relationships 101. *Time,* 24 November: 63.

McGeorge, C. and T. Carlson. 2006. Premarital education: An assessment of program efficacy. *Contemporary Family therapy: An International Journal* 28: 165-190

Nielsen, A., W. Pinsof, C. Rampage, A. H. Solomon, and S. Goldstein. 2004. Marriage 101: An integrated academic and experiential undergraduate marriage education course. *Family Relations* 53: 485–494

Serafini, M. W., and S. Zeller. 2002. Get hitched, stay hitched. *National Journal* 34:694–99.

Family

Other things may change, but we all start and end with the family.

Anthony Brandt, historian

Most people who marry choose to have children and become a family. But the definition of what constitutes a family is sometimes unclear. This section examines how families are defined, their numerous types, and how marriages and families have changed in the past 50 years.

Definitions of Family

The U.S. Census defines **family** as a group of two or more persons related by blood, marriage, or adoption. This definition has been challenged since it does not include foster families or long-term couples (heterosexual or homosexual) that live together. The answer to "who is family?" is important since access to resources such as health care, social security, and retirement benefits is involved. Cohabitants are typically not viewed as "family" and are not accorded health benefits, social security, and retirement benefits of the partner. Indeed, the "live-in partner" or a partner who is gay, although a long-term significant other, may not be allowed to see the beloved in the hospital, which limits visitation to "family only." Nevertheless, the definition of who counts as family is being challenged. In some cases families are being defined by function rather than by structure—what is the level of emotional and financial commitment and interdependence? How long have they lived together? Do the partners view themselves as a family?

Friends sometimes become family. Due to mobility, spouses may live several states away from their respective families. While they may visit their families for holidays, they often develop close friendships with others on whom they rely locally for emotional and physical support on a daily basis. Persons with close friendship relationships are healthier (physically and mentally) and live longer (Lawson, 2005).

Sociologically, a family is defined as a kinship system of all relatives living together or recognized as a social unit, including adopted persons. The family is regarded as the basic social institution because of its important functions of procreation and socialization, and because it is found, in some form, in all societies. Henley-Walters et al., (2002) noted that in many respects,

> the experience of living in a family is the same in all cultures. For example, relationships between spouses and parents and children are negotiated; most relationships within families are hierarchical; the work of the home is primarily the responsibility of the wife; . . . destructive conflict between spouses or parents and children is damaging to children . . . . (p.449)

Same-sex couples (e.g., Rosie O'Donnell, her partner, and their children) certainly define themselves as family. Massachusetts recognizes marriages between same-sex individuals. Short of marriage, Vermont, California, and Connecticut recognize committed gay relationships as **civil unions.**

While other states typically do not recognize the same-sex marriages of Massachusetts or the civil unions of Vermont, California, or Connecticut (and thus persons moving from these states to another state lose the privileges associated with marriage), over 24 cities and counties (including Canada) recognize some form of domestic partnership. In addition, some corporations, such as Disney, are recognizing the legitimacy of such relationships by providing medical coverage for partners of employees. **Domestic partnerships** are considered an alternative to marriage by some individuals and tend to reflect more egalitarian relationships than those between traditional husbands and wives.

As an aside, some individuals view their pets as part of their family. Some victims of hurricane Katrina refused to get on a bus to take them to safety since their pets were not allowed on the bus. Some who lost their animals experienced grief not unlike the loss of another family member (Donohue, 2005; Morley and Fook, 2005). Custody battles have been fought and financial trusts have been set up for pets, and lawsuits against veterinarians have been filed over pets that did not make it out of surgery (Parker, 2005).

National Data

Based on a survey of 1000 registered voters age 18 and older, almost 70% (69%) considered family pets as a "member of the family"; 19% considered them property and 8% considered them both family members and property (Best Friends Survey, 2006).

Call it a clan, call it a network, call it a tribe, call it a family. Whatever you call it, whoever you are, you need one.

Jane Howard, anthropologist

What we've shown is a process of suffusion—family becomes more friend-like and friends become more family like.

Ray Pahl, sociologist

This man views the family dog as a member of the family.

Authors

Families are like fudge—mostly sweet, with a few nuts.

Barbara Johnson

Types of Families

There are various types of families.

Family of Origin Also referred to as the **family of orientation,** this is the family into which you were born or the family in which you were reared. It involves you, your parents, and your siblings. When you go to your parents' home for the holidays, you return to your **family of origin.** Your experiences in your family of origin have an impact on subsequent outcome behavior (Shaw et al., 2003). For example, if you are born into a family on welfare, you are less likely to attend college than if your parents are both educated and affluent.

Novilla et al. (2006) emphasized the important role the family of orientation plays in health promotion. Children learn the importance of a healthy diet, exercise, moderate alcohol use, and so on, from their family of orientation and may duplicate these patterns in their own families.

Siblings represent an important part of one's family of origin. Indeed, a team of researchers (Meinhold et al., 2006) noted that the relationship with one's siblings, particularly sister–sister relationships, represent the most enduring relationship in a person's lifetime. Sisters who lived near each other and who did not have children reported the greatest amount of intimacy and contact.

Family of Procreation The **family of procreation** represents the family that you will begin when you marry and have children. Over 95 percent of U.S. citizens living in the United States marry and establish their own family of procreation (*Statistical Abstract of the United States: 2006*). Across the life cycle, individuals move from the family of orientation to the family of procreation.

Nuclear Family The **nuclear family** refers to either a family of origin or a family of procreation. In practice, this means that your nuclear family consists of you, your parents, and your siblings; or you, your spouse, and your children. Generally, one-parent households are not referred to as nuclear families. They are binuclear families if both parents are involved in the child's life, or single-parent families if only one parent is involved in the child's life and the other parent is totally out of the picture.

Sociologist George Peter Murdock (1949) emphasized that the nuclear family is a "universal social grouping" found in all of the 250 societies he studied. Not only does it channel sexual energy between two adult partners who reproduce, but also these partners cooperate in the care for and socialization of offspring to be productive members of society. "This universal social structure, produced through cultural evolution in every human society, as presumably the only feasible adjustment to a series of basic needs, forms a crucial part of the environment in which every individual grows to maturity" (p. 11). Neyer and Lang (2003) emphasized that closeness to one's kinship members continues throughout one's life and found some evidence for "blood being thicker than water" (p. 310).

Traditional, Modern, and Postmodern Family   Silverstein and Auerbach (2005) distinguished between three central concepts of the family. The **traditional family** is the two-parent nuclear family, with the husband as breadwinner and wife as homemaker. The **modern family** is the dual-earner family, where both spouses work outside the home. **Postmodern families** represent a departure from these models such as lesbian or gay couples and mothers who are single by choice, which emphasizes that a healthy family need not be heterosexual or include two parents.

Binuclear Family   A **binuclear family** is a family in which the members live in two separate households. It is created when the parents of the children divorce and live separately, setting up two separate units, with the children remaining a part of each unit. Each of these units may also change again when the parents remarry and bring additional children into the respective units (**blended family**). Hence, the children go from a nuclear family with both parents to a binuclear unit with parents living in separate homes to a blended family when parents remarry and bring additional children into the respective units.

Extended Family   The **extended family** includes not only your nuclear family but other relatives as well. These relatives include your grandparents, aunts, uncles, and cousins. An example of an extended family living together would be a husband and wife, their children, and the husband's parents (the children's grandparents).

Families are diverse. In the next section we look at unique courtship, marriage, and family patterns among the Amish/Mennonites.

Amish/Mennonite Marriage/Families: Unique Cultural Examples*

In the sixteenth century Martin Luther broke from the Catholic Church in what is known as the Protestant Reformation. Around the same time the Anabaptist Movement emphasized a return to the simplicity of faith and practice as seen in the early Christian church. They believed that adult, not infant, baptism was preferable (the adult could make decisions; the child could not). The name *Mennonite* came from Menno Simons of Holland, who was a leader of the Anabaptist movement in 1536, and the group became known as the Mennonites. But in 1693, a Mennonite elder named Jacob Amish felt the church was losing its purity and broke off to form the Amish sect. The Amish (the smaller of the

Friendships come and go, but families are forever.
Kenny Rogers, singer

---

*Based on author visits to the Amish community in Lancaster County, Pennsylvania as well as: Denlinger, A. M. 1993. *Real People: Amish and Mennonites in Lancaster County, Pennsylvania.* 4th ed. Scottdale, PA: Herald Press; Good, M. and Good, P. 1995. *Twenty most asked questions about the Amish and Mennonites.* Intercourse, PA.: Good Books; and Koehn, R. 1996. *A threefold cord.* Moundridge, KS: Gospel Publishers. Appreciation is also expressed to Merlin and Edith Nichols, members of The Church of God in Christ Mennonites in Ayden, North Carolina, for their assistance in the development of this section.

two groups) are known as the right wing—the more conservative—of the two groups. While there are great variations in the values and behaviors of specific Amish or Mennonite families or groups, the Amish are less likely to use electricity and cars. However, both groups see themselves as attempting to be separate from the world of individualism, materialism, and secularism in favor of being more familistic, simplistic, and spiritual.

Today there are over 180,000 Amish living in 22 states and Ontario, Canada. Between 16,000 and 18,000 Old Order Amish live in Lancaster County, Pennsylvania. There are over 1 million Mennonite church members in 60 countries. The Mennonite Church, USA, is one of nearly 20 formally organized groups of Mennonites in North America. For example, The Church of God in Christ Mennonites is one such group of 20,000. Some general beliefs of this group regarding courtship, marriage, and the family include:

Endogamy   Church members are expected to marry someone of their own faith. Among the Amish, marriage to a member outside the faith is forbidden. Since the community is "closed" to outsiders and since "families" know each other, eligible offspring are known to everyone. Mennonites are also encouraged to marry someone of their faith and, in practice, typically meet someone within their church/community.

Courtship   The Amish couple typically meets at one of the youth sings at church. The boy will then take the girl home, meet her parents (whom he no doubt knows because of the small community) and ask if he can return to see her again. Such meetings typically occur in the living room of the girl's home with other family members nearby. When the couple is ready to marry their intent is published by the deacon about 2 weeks before the wedding in November (harvest is over). The ceremony takes place in the bride's home and is followed by an array of food for the wedding guests. The couple does not take a honeymoon but visit extended family throughout the winter and set up their own place in the spring.

Among the Church of God in Christ Mennonites, a boy who is interested in a particular girl he has seen at church is expected to pray to God and seek spiritual guidance that she is "the one." If he feels God is leading him to this "sister" as his mate, he approaches his minister who further prays with him for God's direction. If the two feel that God encourages this union, the minister will approach the girl's parents and alert them of the boy's interest. If the parents approve, the minister takes the proposal to the girl. If she is interested and feels that God is leading her to marry this man, the wedding, involving a very holy ceremony of commitment whereby the individual lives are bound together into one strong cord, will soon follow in the church. The church membership brings gifts to the couple. After the wedding ceremony, food is served on the grounds and the couple opens their gifts. Notice that no "courtship" precedes the wedding and there is very little time between the mutual acknowledgement of interest and the wedding.

Marital roles   Amish and Mennonite roles are clear and distinct, with the man being the spiritual leader of the family and his wife complementing (i.e., submitting to) to his role/leadership. The role of the man is to work the farm/be the provider; the role of the woman is to take care of the home, cook, sew, and so on. Women are not encouraged to work outside the home. Indeed, Mennonite women who feel that outside work would be more interesting than taking care of the home, "should reject all such thoughts at once" and recognize that homemaking is a "privilege" (Koehn, 1996, p. 30).

Children   Both groups emphasize large families, with family life the focus of the adults. Children are typically home-schooled and do not attend school beyond the 8th grade (public schools are seen as secular and encouraging impure values) but encouraged to work on the farm.

Elderly Both the Amish and Mennonite treat their elderly with respect and regard aging as honorable. The elderly are rarely put in nursing homes, and are instead cared for by their children in their (offspring's) own home. It is not unusual to see a new rooms added to a home in order to accommodate the needs of the parents.

Divorce Divorce is strongly discouraged by both Amish and Mennonites, resulting in very low divorce rates. Marital conflict is seen as an individual problem that can be resolved by getting one's spiritual life in order, which will have a positive impact on the marriage. Hence, when a couple have a disagreement, the resolution is to work on one's own relationship with God, and when this is "fixed," the marriage problem will disappear.

Differences between Marriage and Family

The concepts of marriage and the family are often used in tandem. Marriage can be thought of as a set of social processes that lead to the establishment of family. Indeed, every society/culture has mechanisms (from "free" dating to arranged marriages) of guiding their youth into permanent emotionally/legally/socially bonded heterosexual relationships that are designed to lead to reproduction and care of offspring (children). Although the concepts of marriage and the family are closely related, they are distinct. Some of these differences are identified by sociologist Dr. Lee Axelson in Table 1.2.

Diversity in Other Countries

Asians are also more likely than Anglo-Americans to live with their extended families. Among Asians, the status of the elderly in the extended family derives from religion. Confucian philosophy, for example, prescribes that all relationships are of the subordinate-super ordinate type—husband-wife, parent-child, and teacher-pupil. For traditional Asians to abandon their elderly rather than include them in larger family units would be unthinkable. However, commitment to the elderly may be changing as a result of the Westernization of Asian countries such as China, Japan, and Korea.

In addition to being concerned for the elderly, Asians are socialized to subordinate themselves to the group. Familism (what is best for the group) and group identity are valued over individualism (what is best for the individual) and independence. Divorce is not prevalent because Asians are discouraged from bringing negative social attention to the family.

Table 1.2 Differences between Marriage and the Family in the United States	
Marriage	**Family**
Usually initiated by a formal ceremony.	Formal ceremony not essential.
Involves two people.	Usually involves more than two people.
Ages of the individuals tend to be similar.	Individuals represent more than one generation.
Individuals usually choose each other.	Members are born or adopted into the family.
Ends when spouse dies or is divorced.	Continues beyond the life of the individual.
Sex between spouses is expected and approved.	Sex between near kin is neither expected nor approved.
Requires a license.	No license needed to become a parent.
Procreation expected.	Consequence of procreation.
Spouses are focused on each other.	Focus changes with addition of children.
Spouses can voluntarily withdraw from marriage with approval of state.	Spouses/parents cannot easily withdraw voluntarily from obligations to children.
Money in unit is spent on the couple.	Money is used for the needs of children.
Recreation revolves around adults.	Recreation revolves around children.

Reprinted by permission of Dr. Lee Axelson.

Changes in Marriage and the Family

Whatever family we experience today was different previously and will change yet again. A look back at some changes in marriage and the family follow.

The Industrial Revolution and Family Change

The Industrial Revolution refers to the social and economic changes that occurred when machines and factories, rather than human labor, became the dominant mode for the production of goods. Industrialization occurred in the United States during the early- and mid-1800s and represents one of the most profound influences on the family.

Before industrialization, families functioned as an economic unit that produced goods and services for its own consumption. Parents and children worked together in or near the home to meet the survival needs of the family. As the United States became industrialized, more men and women left the home to sell their labor for wages. The family was no longer a self-sufficient unit that determined its work hours. Rather, employers determined where and when family members would work. Whereas children in preindustrialized America worked on farms and contributed to the economic survival of the family, children in industrialized America became economic liabilities rather than assets. Child labor laws and mandatory education removed children from the labor force and lengthened their dependence on parental support. Eventually, both parents had to work away from the home to support their children. The dual-income family had begun.

During the Industrial Revolution, urbanization occurred as cities were built around factories and families moved to the city to work in the factories. Living space in cities was crowded and expensive, which contributed to a decline in the birthrate and thus smaller families. The development of transportation systems during the Industrial Revolution made it possible for family members to travel to work sites away from the home and to move away from extended kin. With increased mobility, many extended families became separated into smaller nuclear family units consisting of parents and their children. As a result of parents' leaving the home to earn wages and the absence of extended kin in or near the family household, children had less adult supervision and moral guidance. Unsupervised children roamed the streets, increasing the potential for crime and delinquency.

Industrialization also affected the role of the father in the family. Employment outside the home removed men from playing a primary role in childcare and in other domestic activities. The contribution men made to the household became primarily economic.

Finally, the advent of industrialization, urbanization, and mobility is associated with the demise of familism and the rise of individualism. When family members functioned together as an economic unit, they were dependent on one another for survival and were concerned about what was good for the family. This familistic focus on the needs of the family has since shifted to a focus on self-fulfillment—individualism. Families from familistic cultures such as China who immigrate to the United States soon discover that their norms, roles, and values begin to alter in reference to the industrialized, urbanized, individualistic patterns and thinking. Individualism and the quest for personal fulfillment are thought to have contributed to high divorce rates, absent fathers, and parents' spending less time with their children.

Hence, although the family is sometimes blamed for juvenile delinquency, violence, and divorce, it is more accurate to emphasize changing social norms and conditions of which the family is a part. When industrialization takes parents out of the home so that they can no longer be constant nurturers and supervisors, the likelihood of aberrant acts by children and adolescents increases. One explana-

Other things may change us, but we start and end with family.

Anthony Brandt, author

tion for school violence is that absent, career-focused parents have failed to provide close supervision for their children.

Changes in the Last Half-Century

There have been enormous changes in marriage and the family in the last fifty years. Changes include divorce replacing death as the endpoint for the majority of marriages, marriage and intimate relations as legitimate objects of scientific study, and the rise of feminism/changes in gender roles in marriage (Pinsof, 2002). Other changes include a delay in age at marriage, increased acceptance of singlehood, cohabitation, and child-free marriages. Even the definition of what constitutes a family is being revised with some emphasizing that durable emotional bonds between individuals is the core of "family" while others insist on a more legalistic view, emphasizing connections by blood marriage or adoption mechanisms. Table 1.3 reflects some of the changes as we near the year 2010.

Families amid a Context of Terrorism

Prior to September 11, 2001, when terrorists attacked the World Trade Center and the Pentagon, which resulted in the loss of life of an estimated three to five thousand individuals, terrorism was thought to exist mostly in foreign countries (exception—Timothy McVeigh bombed the federal building in Oklahoma in 1995). Today, with "threat levels," "airport security measures," and "new terrorist bombing reports" as part of the evening news, husbands, wives, parents, and children live under a new cloud of terror anxiety. The ever-present awareness of the possibility of a terrorist attack testifies to the new reality. Indeed, almost one fourth (23.6%) of 1027 under-graduates agreed, "I am afraid that a terrorist attack will personally hit me or someone that I love" (Knox and Zusman, 2006). Half of U.S. citizens report that they have cried because of the war in Iraq (Page, 2006). Military families, who have always lived with an increased sense of vulnerability, are a first line of defense against terrorism. These families are featured in the chapter on Marriage Relationships.

As citizens, we are constantly reminded that we live in a context of terrorism.

Authors

Theoretical Frameworks for Viewing Marriage and the Family

Although we emphasize choices in relationships as the framework for viewing marriage and the family, other conceptual theoretical frameworks are helpful in understanding the context of relationship decisions. All **theoretical frameworks** are the same in that they provide a set of interrelated principles designed to explain a particular phenomenon and provide a point of view. In essence, theories are explanations (Moen and Coltrane, 2005).

Social Exchange Framework

In a review of 673 empirical articles, of those using a theoretical perspective, exchange theory was most common (Taylor and Bagd, 2005). The social exchange framework views interaction in terms of giving and receiving. Two researchers (Van de Rijt and Macy, 2006) studied the concept of sexual effort and found that "the most important determinant of sexual effort is effort received" (p. 1464).

Make all your relationships win-win.

Jack Turner, psychologist

Table 1.3 Changes in Marriages and Families from 1950 toward 2010

	1950	Moving toward 2010
Family Relationship Values	Strong values for marriage and the family. Individuals who wanted to remain single or childless were considered deviant, even pathological. Husband and wife should not be separated by jobs or careers.	Individuals who remain single or child-free experience social understanding and sometimes encouragement. The single and child-free are no longer considered deviant or pathological but are seen as self-actuating individuals with strong job or career commitments. Husband and wife can be separated by job or career reasons and live in a commuter marriage. Married women in large numbers have left the role of full-time mother and housewife to join the labor market.
Gender Roles	Rigid gender roles, with men earning income iand wives staying at home, taking care of children.	Fluid gender roles, with most wives in workforce, even after birth of children.
Sexual Values	Marriage was regarded as the only appropriate context for intercourse in middle-class America. Living together was unacceptable, and a child born out of wedlock was stigmatized. Virginity was sometimes exchanged for marital commitment.	For many, concerns about safer sex have taken precedence over the marital context for sex. Virginity is no longer exchanged for anything. Living together is regarded as not only acceptable but sometimes preferable to marriage. For some, unmarried single parenthood is regarded as a lifestyle option. It is certainly less stigmatized.
Homogamous Mating	Strong social pressure existed to date and marry within one's own racial, ethnic, religious, and social class group. Emotional and legal attachments were heavily influenced by obligation to parents and kin.	Dating and mating have become more heterogamous, with more freedom to select a partner outside one's own racial, ethnic, religious, and social class group. Attachments are more often by choice.
Cultural Silence on Intimate Relationships	Intimate relationships were not an appropriate subject for the media.	Talk shows, interviews, and magazine surveys are open about sexuality and relationships behind closed doors.
Divorce	Society strongly disapproved of divorce. Familistic values encouraged spouses to stay married for the children. Strong legal constraints kept couples together. Marriage was forever.	Divorce has replaced death as the endpoint of a majority of marriages. Less stigma is associated with divorce. Individualistic values lead spouses to seek personal happiness. No-fault divorce allows for easy divorce. Marriage is tenuous. Increasing numbers of children are being reared in single-parent households apart from other relatives.
Familism versus Individualism	Families were focused on the needs of children. Mothers stayed home to ensure that the needs of their children were met. Adult concerns were less important.	Adult agenda of work and recreation has taken on increased importance, with less attention being given to children. Children are viewed as more sophisticated and capable of thinking as adults, which frees adults to pursue their own interests. Day care used regularly.
Homosexuality	Same-sex emotional and sexual relationships were a culturally hidden phenomenon. Gay relationships were not socially recognized.	Gay relationships are increasingly a culturally open phenomenon. Some definitions of the family include same-sex partners. Domestic partnerships are increasingly given legal status in some states. Same-sex marriage is a hot social/political issue.
Scientific Scrutiny	Aside from Kinsey, few studies on intimate relationships.	Acceptance of scientific study of marriage and intimate relationships.

The social exchange framework also operates from a premise of **utilitarianism**—that individuals rationally weigh the rewards and costs associated with behavioral choices. Each interaction between spouses, parents, and children can be understood in terms of each individual's seeking the most benefits at the least cost so as to have the highest "profit" and avoid a "loss" (White and Klein, 2002). Both men and women marry because they perceive more benefits than costs for doing so. Similarly, those who remain single or divorce perceive fewer benefits and more costs for marriage. We examine how the social exchange framework is operative in mate selection later in the text.

A social exchange view of marital roles emphasizes that spouses negotiate the division of labor on the basis of exchange. For example, he participates in childcare in exchange for her earning an income, which relieves him of the total financial responsibility. Social exchange theorists also emphasize that power in relationships is the ability to influence, and avoid being influenced by, the partner.

The various bases of power, such as money, the need for a partner, and brute force, may be expressed in various ways, including withholding resources, decreasing investment in the relationship, and violence.

Family Life Course Development Framework

The **family life course development** framework is the second most frequently used theory in empirical family studies (Taylor and Bagd, 2005) and emphasizes the process of how families—both heterosexual and homosexual (Cohler, 2005)—change over time. Family development is a process that follows distinct, norm-related stages (e.g., parenthood). As families move systematically through these stages, they change. The family's life course, or **family career,** is comprised of all the stages and events that have occurred within the family (White and Klein, 2002). At different stages there may be developmental tasks or expectations of what typically occurs. The timing of the movement across stages and the completion of developmental tasks are important.

It's never too late to become what you might have been.

George Eliot, Victorian novelist

If developmental tasks at one stage are not accomplished, functioning in subsequent stages will be impaired. For example, one of the developmental tasks of early marriage is to emotionally and financially separate from one's family of origin. If such separation does not take place, independence as individuals and as a couple is impaired.

Tasks are sometimes completed out of order or simply at the wrong time. Becoming a grandparent at age 45 may seem "off-time" and may make it difficult to accept this new role. Social timing also states that at a certain age, particular events should follow. Off-time, or out-of-sequence, events also can lead to disruptions later in life. For example, having a baby typically follows marriage. This sequence typically occurs after graduating from high school and perhaps from college. If a young teenager has a baby prior to marriage, her education may well be disrupted.

The family life course development framework may also help to identify the choices with which many individuals are confronted throughout life. Each family stage presents choices. For example, the never-married are choosing partners, the newly married are making choices about careers and when to begin their family, the soon-to-be-divorced are making decisions about custody/child support/division of property, and the remarried are making choices with regard to stepchildren and ex-spouses. Grandparents are making choices about how much childcare they want to commit to, and widows/widowers are concerned with where to live (children, retirement home, with a friend, alone).

Structural–Functional Framework

Just as the human body is made up of different parts that work together for the good of the individual, society is made up of different institutions (family, education, economics, etc.) that work together for the good of society. **Functionalists** view the family as an institution with values, norms, and activities meant to provide stability for the larger society. Such stability is dependent on families' performing various functions for society.

First, families serve to replenish society with socialized members. Since our society cannot continue to exist without new members, we must have some way of ensuring a continuing supply. But just having new members is not enough. We need socialized members—those who can speak our language and know the norms and roles of our society. So-called **feral** (meaning wild, not domesticated) children are those who are thought to have been reared by animals. Newton (2002) details nine such children, the most famous of which was Peter the Wild Boy found in the Germanic woods at the age of 12 and brought to London in 1726. He could not speak; growling and howling were his modes of expression. He lived until the age of 70 and never learned to talk. Feral children emphasize that it is social interaction and family context that make us human. The legal bond of marriage and the obligation to nurture and socialize offspring help to ensure that this socialization will occur.

Second, marriage and the family promote the emotional stability of the respective spouses. Society cannot provide enough counselors to help us whenever we have problems. Marriage ideally provides an in-residence counselor who is a loving and caring partner with whom a person shares his or her most difficult experiences.

Children also need people to love them and to give them a sense of belonging. This need can be fulfilled in a variety of family contexts (two-parent family, single-parent family, extended family). The affective function of the family is one of its major offerings. No other institution focuses so completely on meeting the emotional needs of its members as marriage and the family.

Third, families provide economic support for their members. Although modern families are no longer self-sufficient economic units, they provide food, shelter, and clothing for their members. One need only consider the homeless in our society to be reminded of this important function of the family.

In addition to the primary functions of replacement, emotional stability, and economic support, other functions of the family include the following:

- Physical care—families provide the primary care for their infants, children, and aging parents. Other agencies (neonatal units, day care, assisted-living residences) may help, but the family remains the primary and recurring caretaker. Spouses are also concerned about the physical health of each other by encouraging the partner to take medications and to see the doctor.
- Regulation of sexual behavior—spouses are expected to confine their sexual behavior to each other, which reduces the risk of having children who do not have socially and legally bonded parents, and of contracting/spreading HIV or other sexually transmitted infections.
- Status placement—being born in a family provides social placement of the individual in society. One's social class, religious affiliation, and future occupation are largely determined by one's family of origin. Prince William, the son of Prince Charles and the late Diana, by virtue of being born into a political family, was automatically in the upper class and destined to be in politics.
- Social control—spouses in high-quality, durable marriages provide social control for each other that results in less criminal behavior. Parole boards often note that the best guarantee that a person released from prison will not return to prison is a spouse who expects the partner to get a job and avoid criminal behavior and who reinforces these goals.

Conflict Framework

Conflict theorists recognize that family members have different goals and values that result in conflict. Conflict is inevitable between social groups (e.g., parents and children). Conflict theory provides a lens through which to view these dif-

The family is one of nature's masterpieces.

George Santayana, philosopher

ferences. Whereas functionalists look at family practices as good for the whole, conflict theorists recognize that not all family decisions are good for every member of the family. Indeed, some activities that are good for one member are not good for others. For example, a woman who has devoted her life to staying home and taking care of the children may decide to return to school or to seek full-time employment. This may be a good decision for her personally, but her husband and children may not like it. Similarly, divorce may have a positive outcome for spouses in turmoil but a negative outcome for children, whose standard of living and access to the noncustodial parent are likely to decrease.

Conflict theorists also view conflict not as good or bad but as a natural and normal part of relationships. They regard conflict as necessary for change and growth of individuals, marriages, and families. Cohabitation relationships, marriages, and families all have the potential for conflict. Cohabitants are in conflict about commitment to marry, spouses are in conflict about the division of labor, and parents are in conflict with their children over rules such as curfew, chores, and homework. These three units may also be in conflict with other systems. For example, cohabitants are in conflict with the economic institution for health benefits for their partners. Similarly, employed parents are in conflict with their employers for flexible work hours, maternity/paternity benefits, and day-care or eldercare facilities.

Karl Marx emphasized that conflict emanates from struggles over scarce resources and for power. Though Marxist theorists viewed these sources in terms of the conflict between the owners of the means of production (bourgeoisie) and the workers (proletariat), they are also relevant to conflicts within relationships. The first of these concepts, conflict over scarce resources, reflects the fact that spouses, parents, and children compete for scarce resources such as time, affection, and space. Spouses may fight with each other over how much time should be allocated to one's job, friends, or hobbies. Parents are sometimes in conflict with each other over who does what housework/child care. Children are in conflict with their parents and with each other over time, affection, what programs to watch on television, and money.

Conflict theory is also helpful in understanding choices in relationships with regard to mate selection and jealousy. Unmarried individuals in search of a partner are in competition with other unmarried individuals for the scarce resources of a desirable mate. Such conflict is particularly evident in the case of older women in competition for men. At age 85 and older, there are twice as many women as there are men (*Statistical Abstract of the United States: 2006* Table 14). Jealousy is also sometimes about scarce resources. People fear that their "one and only" will be stolen by someone else who has no partner.

Conflict theorists also emphasize conflict over power in relationships. Premarital partners, spouses, parents, and teenagers also use power to control each other. The reluctance of some courtship partners to make a marital commitment is an expression of wanting to maintain their autonomy, since marriage implies a relinquishment of power from each partner to the other. Spouse abuse is sometimes the expression of one partner trying to control the other through fear, intimidation, or force. Divorce may also illustrate control. The person who executes the divorce is often the person with the least interest in the relationship. Having the least interest gives that person power in the relationship. Parents and adolescents are also in a continuous struggle over power. Parents attempt to use privileges and resources as power tactics to bring compliance in their adolescent. But adolescents may use the threat of suicide as their ultimate power ploy to bring their parents under control.

Feminist Framework

Feminist thought "originated with a social movement for change" (White and Klein 2002, 171) and is embodied by no one feminist framework. Indeed there are eleven perspectives, including lesbian feminism (oppressive heterosexuality

When one ceases from conflict, whether because he has won, because he has lost, or because he cares no more for the game, the virtue passes out of him.

George Horton, sociologist

Though I have worked many years to make marriage more equal, I never expected to take advantage of marriage myself.

Gloria Steinem, feminist (married at age 66)

and men's domination of social spaces), psychoanalytic feminism (cultural domination of men's phallic-oriented ideas and repressed emotions), and standpoint feminism (neglect of women's perspective and experiences in the production of knowledge) (Lorber, 1998). Regardless of which feminist framework is being discussed, all feminist frameworks have the themes of inequality and oppression. According to feminist theory, gender structures our experiences (i.e., women and men will experience life differently because there are different expectations for the respective genders) (White and Klein, 2002).

Symbolic Interaction Framework

Marriages and families represent symbolic worlds in which the various members give meaning to each other's behavior. Human behavior can be understood only by the meaning attributed to behavior (White and Klein, 2002). Herbert Blumer (1969) used the term *symbolic interaction* to refer to the process of interpersonal interaction. Concepts inherent in this framework include the definition of the situation, the looking-glass self, and the self-fulfilling prophecy.

Definition of the Situation   Two people who have just spotted each other at a party are constantly defining the situation and responding to those definitions. Is the glance from the other person (1) an invitation to approach, (2) an approach, or (3) a misinterpretation—the other person was looking at someone behind the person? The definition a person arrives at will affect subsequent interaction.

Looking-Glass Self   The image people have of themselves is a reflection of what other people tell them about themselves (Cooley, 1964). The people at the party previously referred to develop an idea of who they are by the way others act toward them. If no one looks at or speaks to them, according to Cooley, they will begin to feel unsettled. Similarly, family members constantly hold up social mirrors for one another into which the respective members look for definitions of self.

The importance of the family (and other caregivers) as an influence on the development and maintenance of a positive self-concept cannot be overemphasized. Orson Welles, known especially for his film *Citizen Kane,* once said that he was taught that he was wonderful and that everything he did was perfect. He never suffered from a negative self-concept. Cole Porter, known for creating such memorable songs as "I Get a Kick Out of You," had a mother who held up social mirrors offering nothing but praise and adoration. Since children spend their formative years surrounded by their family, the self-concept they develop in that setting is important to their feelings about themselves and their positive interaction with others.

G. H. Mead (1934) believed that people are not passive sponges but evaluate the perceived appraisals of others, accepting some opinions and not others. Although some children are taught by their parents that they are worthless, they may eventually overcome the definition.

Self-Fulfilling Prophecy   Once people define situations and the behaviors they are expected to engage in, they are able to behave toward one another in predictable ways. Such predictability of behavior also tends to exert influence on subsequent behavior. If you feel that your partner expects you to be faithful to him or her, your behavior is likely to conform to these expectations. The expectations thus create a self-fulfilling prophecy.

Symbolic interactionism as a theoretical framework helps to explain various choices in relationships. Individuals who decide to marry have defined their situation as a committed reciprocal love relationship. This choice is supported by the belief that the partners will view each other positively (looking-glass self) and be faithful spouses and cooperative parents (self-fulfilling prophecies).

If we define things as real, they are real in their consequences.

W. I. Thomas, sociologist

Chapter 1 Choices in Relationships: An Introduction

Family Systems Framework

Systems theory is the most recent of all the theories for understanding family interaction (White and Klein, 2002). Its basic premise is that each member of the family is part of a system and the family as a unit develops norms of interacting, which may be explicit (e.g., parents specify chores for the children) or implied (e.g., spouses expect fidelity from each other). These rules serve various functions such as allocating the resources (e.g., money for vacation), specifying the division of power (e.g., who decides how money is spent), and defining closeness and distance between systems (e.g., seeing or avoiding parents/grandparents).

Rules are most efficient if they are flexible. For example, they should be adjusted over time in response to children's growing competence. A rule about not leaving the yard when playing may be appropriate for a 4-year-old but inappropriate for a 15-year-old. The rules and individuals can be understood only by recognizing that "all parts of the system are interconnected" (White and Klein, 2002, 122).

Family members also develop boundaries that define the individual and the group and separate one system or subsystem from another. A boundary is a "border between the system and its environment that affects the flow of information and energy between the environment and the system" (White and Klein, 2002, p. 124). A boundary may be physical, such as a closed bedroom door, or social, such as expectations that family problems will not be aired in public. Boundaries may also be emotional, such as communication, which maintains closeness or distance in a relationship. Some family systems are cold and abusive; others are warm and nurturing.

The nuclear family, rather than the individual, is the emotional unit.

Murray Bowan, theorist

In addition to rules and boundaries, family systems have roles (leader, follower, scapegoat) for the respective family members. These roles may be shared by more than one person or may shift from person to person during an interaction or across time. In healthy families, individuals are allowed to alternate roles rather than being locked into one role. In problem families, one family member is often allocated the role of scapegoat, or the cause of all the family's problems (e.g., an alcoholic spouse).

Family systems may be open, in that they are open to information and interaction with the outside world, or closed, in that they feel threatened by such contact. The Amish have closed family systems and minimize contact with the outside world. Some communes also encourage minimal outside exposure. Twin Oaks Intentional Community of Louisa, Virginia, does not permit any of its almost one hundred members to own or keep a television in their rooms. Exposure to the negative drumbeat of the evening news is seen as harmful.

Human Ecological Framework

Human ecology is the study of **ecosystems,** or the interaction of families with their environment (Bubolz and Sontag, 1993). Individuals, couples, and families are dependent on the environment for air, food, and water, and on other human beings for social interaction (White and Klein, 2002). The well being of individuals and families cannot be considered apart from the well being of the ecosystem. This framework emphasizes the importance of examining families within multiple contexts. For example, nutrition and housing are important to the functioning of families. If a family does not have enough to eat or adequate housing, it will not be able to function at an optimal level.

Human beings and their environment also are interdependent in that they depend on each other and they influence each other. Individuals and families do not operate in isolation but interact with multiple environments—schools, workplace, and neighborhoods (Bronfenbrenner,1979; Jenkins et al., 2003). These interactions are reciprocal. For example, stress at work will influence one's family life and vice versa. Judkins and Presser (2005) noted that wives engage in more environmentally sustainable domestic labor than did their husbands (e.g., wives would spend more time sorting cans and bottles to be thrown away in different containers).

Biosocial Framework

The **biosocial framework** emphasizes the interaction of one's biological/genetic inheritance with one's social environment to explain and predict human behavior. Borgerhoff Mulder and McCabe (2006) noted that while sociobiology is sometimes dismissed as purely genetic determinism, sociobiology is not merely genetics but primarily interested in the environmental and social context of variability. Hence, while human behavior can be explained as having an evolutionary function, it operates in a current social context.

Caspi et al. (2003) found that the presence of certain genes affect the degree to which a person responds to stressful life events with depression. Borrowing from evolutionary psychology, sociobiology, and psychobiology, biosocial theory uses the concepts of adaptation, fitness, and natural selection to explain such phenomena as mate selection, gender roles, and the incest taboo (Ingoldsby et. al., 2004).

Adaptation is the changing of the individual to fit in and maximize one's potential. Fitness is the blending in to solidify one's position. Natural selection emphasizes that it is "natural" for the individual to want to survive by adapting and fitting in. That men tend to seek young women with whom to procreate is related to the biological fact that young women are more fertile and produce more healthy offspring than older women. Women, on the other hand, tend to seek men who are older and more economically stable who can provide the economic resources for their offspring. Hence, both biological (youth/fertility) and social (economic stability) factors combine to explain the mate selection process.

Understanding of gender roles may also benefit from a biosocial perspective. That women carry their babies to term and provide milk for their sustenance reflects a biological background for intense maternal bonding. That men do not have such a biological link but are rewarded for economic productivity in the workplace may help to explain the discrepancy in male/female child-care involvement.

The incest taboo may also be viewed from a biosocial perspective. While mating between parents and children may produce defective offspring, there are also social reasons for the incest taboo. Individuals who look outside the family unit for new partners add to the range of resources available. When one marries outside one's own family, the mate and the mate's parents offer a new set of economic and emotional resources that would otherwise be missed.

Beck and Shaw (2005) suggested other uses of the biosocial model. They found that perinatal (at or around the time of birth) difficulties in combination with family adversity predicted antisocial behavior in boys in middle school. Their finding was based on an analysis of birth records of 210 boys whom they followed in a longitudinal study for ten years. Self-reports for youths in middle childhood also confirmed their findings. Hence while counselors/parents may look to environmental/social factors (e.g. peers) for explanations of antisocial behavior, the biosocial or early biological experiences may be instructive.

Stratification/Race Framework

Though not formal theoretical frameworks, stratification and race provide ways of viewing and understanding choices in relationships.

Stratification   **Stratification** refers to the ranking of people into layers or strata according to their socioeconomic status or social class, usually indexed according to income, occupation, and educational attainment. Passengers on the *Titanic* were stratified and assigned to different decks, which influenced who

Our minds have been built by selfish genes, but they have been built to be social, trustworthy, and cooperative.

Matt Ridley, "The Origins of Virtue"

If you have to ask the price, you can't afford it.

J. P. Morgan, Billionnaire

Chapter 1 Choices in Relationships: An Introduction

they would interact with. Individuals who occupy different socioeconomic statuses are less likely to meet and marry. Katrina victims varied by social class. The wealthy hired cab drivers to drive them to Atlanta. The impoverished were relegated to the Superdome for weeks with limited food and water.

Marriages and families are also stratified into different social classes, such as the upper, middle, working, or lower social class (see Table 1.4). Families in these various social classes reflect dramatic differences in their attitudes, values, and behavior. For example, individuals from the lower class are more likely to divorce than individuals from the higher social classes. Parents in lower socioeconomic classes are also more likely to discuss personal and financial problems with their children than parents in higher socioeconomic groups. The former feel that the sooner their children become aware of the harsh realities of life, the better. Middle-class parents, on the other hand, believe that they should protect their children from the realities that lie ahead. One's social class and, by implication, occupation also influence the time available for leisure activities. Persons in low-skilled occupations have long hours and low pay with little leisure time (Salmon

Diversity in Other Countries

The social class of one's family may have an effect on one's selection of a mate. Middle- and upper-class individuals in Ecuador and Latin America typically have longer courtships, engagements, and wedding ceremonies than those from lower socioeconomic backgrounds—hence the respective families of the bride and groom want considerable involvement. In contrast, individuals from lower socioeconomic families are more likely to pair bond in their teens, to cohabit, and to experience an out-of-marriage pregnancy/childbirth (Schvaneveldt, 2003).

Table 1.4   105 Million Family Households by Social Class*

Class Identification	Percentage of Population	Household Income	Education/Occupation	Lifestyle	Example
Upper Class (3%)					
Upper-upper class (old money capitalists)	1%	$500,000+	Prestigious schools/ wealth passed down	Large, spacious homes in lush residential areas	Kennedys
Lower-upper class (nouveau riche)	2%	$200,000+	Prestigious schools/ investors in or owners of large corporations	(same as above)	Bill Gates, Donald Trump
Middle Class (50%)					
Upper-middle class	23%	$75,000–$200,000	Post-graduate degrees/ physicians, lawyers, managers of large corporations	Nice homes, nice neighborhoods, send children to state universities	Your physician
Lower-middle class	27%	$40,000–$75,000	College degrees/nurses, elementary- or high-school teachers	Modest homes, older cars	High-school English teacher
Working Class	25%	$20,000–$40,000	High school diploma/ Waitresses, mechanics,	Home in lower-income suburb; children get job after high school	Employee at fast-food restaurant
Working Poor	15%	Below poverty line of $14,680 for family of 3	Some high school/ service jobs	Live in poorest of housing; barely able to pay rent/buy food	Janitor
Underclass	10%	—	Unemployed/ unemployable; survive via public assistance, begging, hustling/ illegal behavior (e.g., selling drugs)	Homeless; contact with mainstream society is via criminal justice system	Bag lady

*Appreciation is expressed to Arunas Juska, Ph.D. for his assistance in the development of this table. Estimates of percent in each class are in reference to household incomes as published in *Statistical Abstract of the United States: 2006,* Table 695.

et al., 2000). Pahl and Pevalin (2005) also noted that one's social class affects one's choice of friends.

Race Racial heritage also influences choices in relationships. For example, blacks are more likely than whites to report having had sexual intercourse earlier. In a national longitudinal study, almost half (45%) of blacks reported having had sexual intercourse by age 15 to 16, as compared with 31 percent for other racial groups (Cooksey et al. 2002).

The term **race** is a social construct the meaning of which has less to do with biological differences than with social, cultural, political, economic, behavioral, fertility, and health differences. With regard to the latter, the life expectancy of white men born in 2005 is projected to be 75.4 compared with 69.9 for black men; the figures for white women and black women are projected to be 81.1 and 76.8, respectively (*Statistical Abstract of the United States: 2006*, Table 96).

A person's membership in a particular racial or ethnic group is sometimes not as clear-cut as might be expected. For example, not all black people are African Americans. Black immigrants from the Caribbean have a strong ethnic identity that is not African American. In addition, differences that appear to be racial may actually be social-class differences. For example, individuals in the lower class (whether white or black) have higher rates of unemployment, premarital pregnancies, divorce, and crime. In looking at the comparisons between blacks and whites throughout this text, it is important to keep in mind that many presumed racial differences are really those of social class. However, racism still exists in the form of discrimination against minorities in education, employment, and housing, which affects spouses, parents, and children. The major theories are summarized in Table 1.5.

Evaluating Research in Marriage and the Family

"New Research Study" is a frequent headline in popular magazines (e.g., *Cosmopolitan, Glamour, Redbook*) promising accurate information about "hooking up," "what women want," "what men want," or other relationship/marriage/

After all, there is but one race—humanity

George Moore, Irish novelist

My latest survey shows that people don't believe in surveys.

Laurence J. Peter, Canadian humorist

New research in marriage and the family is presented at regional and national conferences.

Authors

Chapter 1 Choices in Relationships: An Introduction

Table 1.5 105 Million Family Households by Social Class*

Theory	Description	Concepts	Level of Analysis	Strengths	Weaknesses
Social Exchange	In their relationships, individuals seek to maximize their benefits and minimize their costs.	Benefits Costs Profit Loss	Individual Couple Family	Provides explanations of human behavior based on outcome.	Assumes that people always act rationally and all behavior is calculated.
Family Life Course Development	All families have a life course or family career that is composed of all the stages and events that have occurred within the family.	Family career Stages Transitions Timing	Institution Individual Couple Family	Families are seen as dynamic rather than static. Useful in working with families who are facing transitions in their life courses.	Difficult to adequately test the theory through research.
Structural-Functional	The family has several important functions within society; within the family, individual members have certain functions.	Structure Function	Institution	Emphasizes the relation of family to society; noting how families affect and are affected by the larger society.	Families with non-traditional structures (single-parent, same-sex couples) are seen as dysfunctional.
Conflict	Conflict in relationships is inevitable, due to competition over resources and power.	Conflict Resources Power	Institution	Views conflict as a normal part of relationships and as necessary for change and growth.	Sees all relationships as conflictual, and does not acknowledge cooperation.
Feminism	Women's experience is central and different from man's experience of social reality.	Inequality Power Oppression	Institution Individual Couple Family	Exposes inequality and oppression as explanations for frustrations experienced by women.	Multiple branches of feminism may inhibit central accomplishment of increased equality.
Symbolic Interaction	People communicate through symbols and interpret the words and actions of others.	Definition of the situation Looking-glass self Self-fulfilling prophecy	Couple	Emphasizes the perceptions of individuals, not just objective reality or the viewpoint of outsiders.	Ignores the larger social context and minimizes the influence of external forces.
Family Systems	The family is a system of interrelated parts that function together to maintain the unit.	Subsystem Roles Rules Boundaries Open system Closed System	Couple Family	Very useful in working with families who are having serious problems (violence, alcoholism). Describes the effect family members have on each other.	Based on work with troubled families and may not apply to non-problem families.
Human Ecological	Families interact with and are interdependent with their environment.	Ecosystem Interdependence Environment	Institution Individual Couple Family	Can be applied to families of different structures and ethnic or racial backgrounds.	Scope of the theory may be too broad.
Biosocial	Individual behavior is influenced by biological/genetic inheritance that is constrained by others/the environment.	Biosocial Adaptation Natural Selection	Individual Couple Family	Emphasizes interaction of biological and social factors.	Unclear whether biological or social influences are more important. Some predictions of biosocial theory are not borne out in the real world.

family issues. As you read such articles, as well as the research in such texts as this, be alert to their potential flaws. Following are specific issues to keep in mind when evaluating research in marriage and the family.

Sample

Some of the research on marriage and the family is based on random samples. In a **random sample,** each individual in the population has an equal chance of being included in the sample. Random sampling involves selecting individuals at random from an identified population. That population often does not include the homeless or persons living on military bases.

Studies that use random samples are based on the assumption that the individuals studied are similar to and therefore representative of the population that the researcher is interested in. For example, suppose you want to know the percentage of unmarried seniors (USs) on your campus who are living together. Although the most accurate way to get this information is to secure an anonymous yes or no response from every US, doing so is not practical. To save yourself time, you could ask a few USs to complete your questionnaire and assume that the rest of them would say yes or no in the same proportion as those who answered the questions. To decide who those few USs would be, you could put the name of every US on campus on a separate note card, stir these cards in your empty bathtub, put on a blindfold, and draw one hundred cards. Because each US would have an equal chance of having his or her card drawn from the tub, you would obtain a random sample. After administering the questionnaire to this sample and adding the yes and no answers, you would have a fairly accurate idea of the percentage of USs on your campus that are living together.

The term *random sample,* however, may not always mean "random." For example, in the preceding study of unmarried seniors, not all the names you put in the bathtub to select from would have addresses and phone numbers. Hence, even if you drew the person's name, finding her or him to complete a questionnaire could be difficult. In addition, some people refuse to complete a questionnaire.

Because of the trouble and expense of obtaining random samples, most researchers study subjects to whom they have convenient access. This often means students in the researchers' classes. The result is an overabundance of research on "convenience" samples consisting of white, Protestant, middle-class college students. Because college students cannot be assumed to be similar to their non-college peers or older adults in their attitudes, feelings, and behaviors, research based on college students cannot be generalized beyond the base population. To provide a balance, this text included data that reflected people of different ages, marital statuses, racial backgrounds, lifestyles, religions, and social classes. When only data on college samples are presented, it is important not to generalize the findings too broadly.

In addition to having a random sample, it is important to have a large sample. The study of over 230,000 first-semester undergraduates at U.S. colleges and universities referred to earlier in this chapter represented a large national sample, which provides credible data on first-year college students (American Council on Education and University of California, 2005–2006). If only 50 college students had been in the sample, the results would have been very unreliable in terms of generalizing beyond that sample. Similarly, Corra et al., (2006) analyzed national data collected over a 30-year period from the 1972–2002 General Social Survey to discover the influence of sex (male or female) and race (white or black) on the level of reported marital happiness. This is an enormous set of random samples and gives validity to the findings.

Be alert to the sample size of the research you read. Most studies are based on small samples. Other researchers have emphasized the limitations of norms based on populations of college students only (Meyers and Shurts, 2002).

Control Groups

Any study that concludes that an abortion (or any independent variable) is associated with negative outcomes (or any dependent variable) must necessarily include two groups: in this example, the groups would be (1) women who have had an abortion and (2) women who have not had an abortion. The latter would serve as a **control group**—the group not exposed to the independent variable you are studying (**experimental group**). Hence, if you find that women in both groups in your study develop negative attitudes toward sex, you know that abortion cannot be the cause.

Be alert to the existence of a control group, which is usually *not* included in research studies. An exception is Bradshaw and Slade (2005), who compared 98 women (experimental group) who had a first-trimester abortion with a control group of 51 never-pregnant women to evaluate the effect of an abortion on sexuality. The researchers found that having an abortion was followed by *more* positive attitudes and feelings towards sexual matters. Similarly, Kelly et al. (2006) wanted to find out if poor communication in marriage is related to the inability of a woman to have an orgasm. He studied women in relationships of both good (control group) and bad communication (experimental group) and found that women in the latter were more likely to report difficulty achieving an orgasm.

Age and Cohort Effects

In some research designs, different cohorts or age groups are observed and/or tested at one point in time. One problem that plagues such research is the difficulty—even impossibility—of discerning whether observed differences between the subjects studied are due to the research variable of interest, cohort differences, or some variable associated with the passage of time (e.g., biological aging).

A good illustration of this problem is found in research on changes in marital satisfaction over the course of the family life cycle. In such studies, researchers may compare the levels of marital happiness reported by couples that have been married for different lengths of time. For example, a researcher may compare the marital happiness of two groups of people—those who have been married for fifty years and those who have been married for five years. But differences between these two groups may be due to (1) differences in age (age effect), (2) the different historical time period that the two groups have lived through (cohort effect), or (3) being married different lengths of time (research variable). It is helpful to keep these issues in mind when you read studies on marital satisfaction over time.

Prior to the 2002 National Survey of Family Growth, which provided sexuality data on 12,571 respondents—male and female, between the ages of 15 and 44—we had no national data on sexual behavior since 1994 (Michael et al., 1994). The need to continually provide new data is clear.

Terminology

In addition to being alert to potential shortcomings in sampling and control groups, you should consider how the phenomenon being researched is defined. For example, in a preceding illustration of unmarried seniors living together, how would you define *living together*? How many people, of what sex, spending what amount of time, in what place, engaging in what behaviors will constitute your definition? Indeed, researchers have used more than twenty definitions of what constitutes living together.

What about other terms? Considerable research has been conducted on *marital success*, but how is the term to be defined (DeOllos, 2005)? What is meant by *marital satisfaction, commitment, interpersonal violence,* and *sexual fulfillment*? Even the term *romantic behavior* has a number of referents. Quiles (2003) specified fifteen such behaviors, including kissing, making love, sending flowers, saying "I love you," hugging, candlelight dinner, slow dancing, cuddling, love cards or letters, holding

What's in a name? That which we call a rose by any other name would smell as sweet.

William Shakespeare

hands, and so on. Before reading too far in a research study, be alert to the definitions of the terms being used. Exactly what is the researcher trying to measure?

Researcher Bias

Although one of the goals of scientific studies is to gather data objectively, it may be impossible for researchers to be totally objective. McGraw et al. (2000) emphasized that marriage and family research is inherently political in content and method. Researchers are human and have values, attitudes, and beliefs that may influence their research methods and findings. It may be important to know what the researcher's bias is in order to evaluate his or her findings. For example, a researcher who does not support abortion rights may conduct research that focuses only on the negative effects of abortion.

In addition, some researchers present an interpretation of what other researchers have done. Two layers of bias may be operative here: (1) bias affecting the original collection and interpretation of the data, and (2) bias affecting the second researcher's reading and interpretation of the data. Much of this text was based on interpretations of other researchers' studies. As a consumer you should be alert to the potential bias in reading such secondary sources. To help control for this bias, we have provided references to the original sources for your own reading.

Even the particular topics selected by researchers can reflect a gender bias. For example, as a result of male bias in the scientific community, research on women's issues has not been a major focus. Thus research on male contraception is almost nonexistent, since male researchers focused on the "female pill." With regard to sexual dysfunctions, women are often presented as dysfunctional because of their difficulty in achieving an orgasm, whereas in fact the lack of adequate stimulation provided by the male might be a more accurate emphasis.

A more gender-neutral approach to research in marriage and the family would include more qualitative studies based on women's experiences, recognition that gender is a socially created category, and commitment to design research with the aim of eliminating bias and improving the lives of women.

Time Lag

Typically, a 2-year lag exists between the time a study is completed and the study's appearance in a professional journal. Because textbook production takes even longer than getting an article printed in a professional journal, they do not always present the most current research, especially on topics in flux. In addition, even though a study may have been published recently, the data on which the study was based may be old. Be aware that the research you read in this or any other text may not reflect the most current, cutting-edge research.

Distortion and Deception

Our society is no stranger to distortion and deception—weapons of mass destruction? Writers at the prestigious *New York Times* have been fired after being discovered fabricating articles. South Korean researcher Hwang Woo Suk resigned at Seoul National University over his fabrication of published research claiming that he had cloned a human embryo and extracted stem cells from it, and that he had cloned the first dog. South Korean newspapers called the affair "Hwang-gate" (Weise and Vergano, 2006). The Hwang affair gave visibility to other fraudulent research. In 1996, chemists at the University of Utah reported that they had discovered "cold fusion." "They hadn't, it turned out. . . ." (Lemonick, 2006, p. 43).

Distortion and deception, deliberate or not, also exist in marriage and family research. Marriage is a very private relationship that happens behind closed doors; individual interviewees and respondents to questionnaires have been socialized not to reveal to strangers the intimate details of their lives. Hence, they are prone to distort, omit, or exaggerate information, perhaps unconsciously, to cover up what they may feel is no one else's business. Thus, the researcher sometimes obtains inaccurate information.

When I was younger I could remember anything, whether it happened or not.

Mark Twain, author and humorist

Marriage and family researchers know more about what people say they do than about what they actually do. An unintentional and probably more common form of distortion is inaccurate recall. Sometimes researchers ask respondents to recall details of their relationships that occurred years ago. Time tends to blur some memories, and respondents may relate not what actually happened, but rather what they remember to have happened, or, worse, what they wish had happened.

Other Research Problems

Nonresponse on surveys and the discrepancy between attitudes and behaviors are other research problems. With regard to nonresponse, not all individuals who complete questionnaires or agree to participate in an interview are willing to provide information about such personal issues as date rape and partner abuse. Such individuals leave the questionnaire blank or tell the interviewer they would rather not respond. Others respond but give only socially desirable answers. The implications for research are that data gatherers do not know the nature or extent to which something may be a problem because people are reluctant to provide accurate information. Computer-administered self-interviewing (CASI) has been suggested as a way of collecting sensitive (e.g., sexual) information. But a study by Testa et. al. (2005) comparing CASI and self-administered mailed questionnaires found that the latter had a higher response rate with greater disclosure of sensitive information.

The discrepancy between people's attitudes and their behavior is another cause for concern about the validity of research data. It is sometimes assumed that if a person has a certain attitude (for example, a belief that extramarital sex is wrong), then his or her behavior will be consistent with that attitude (avoidance of extramarital sex). However, this assumption is not always accurate. People do indeed say one thing and do another. This potential discrepancy should be kept in mind when reading research on various attitudes.

Finally, most research reflects information provided by volunteers. But volunteers may not represent nonvolunteers when they are completing surveys. In view of the research cautions identified here, you might ask, "Why bother to report the findings?" The quality of some family science research is excellent. For example, articles published in *Journal of Marriage and the Family* (among other journals) reflect the high level of methodologically sound articles that are being published. Even less sophisticated journals provide useful information on marital, family, and other relationship data. Particularly when there are multiple replications of a study to assess consistency of results, there is increasing confidence in the research process (Riniolo and Schmidt, 2000). Table 1.6 summarizes potential inadequacies of any research study.

Table 1.6   Potential Research Problems in Marriage/Family

Weakness	Consequences	Example
Sample not random	Cannot generalize findings	Opinions of college students do not reflect opinions of other adults.
No control group	Inaccurate conclusions	Study on the effect of divorce on children needs control group of children whose parents are still together.
Age differences between groups of respondents	Inaccurate conclusions	Effect may be due to passage of time or to cohort differences.
Unclear terminology	Inability to measure what is not clearly defined	What is living together, marital happiness, sexual fulfillment, good communication, quality time?
Researcher bias	Slanted conclusions	Male researcher may assume that since men usually ejaculate each time they have intercourse, women should have an orgasm each time they have intercourse.
Time lag	Outdated conclusions	Often-quoted Kinsey sex research is over fifty years old.
Distortion	Invalid conclusions	Research subjects exaggerate, omit information, and/or recall facts or events inaccurately. Respondents may remember what they wish had happened.

What is the theme of this text?

A central theme of this text is to encourage you to be proactive—to make conscious, deliberate relationship choices to enhance your own well-being and the well-being of those in your intimate groups. Five major choices include whether to marry, whether to have children, whether to have one or two earners in one marriage, whether to remain emotionally and sexually faithful to one's partner, and whether to use a condom. Though structural and cultural influences are operative, a choices framework emphasizes that individuals have some control over their relationship destiny by making considered choices to initiate, respond to, nurture, or terminate intimate relationships. Important issues to keep in mind about a choices framework for viewing marriage and the family are that not to decide is to decide, some choices require correcting, all choices involve trade-offs, choices include selecting a positive or negative view, making choices produces ambivalence, and some choices are not revocable.

What is marriage?

Marriage is a system of binding a man and a woman together for the reproduction, care (physical/emotional), and socialization of offspring. Marriage in the United States is a legal contract between a heterosexual couple and the state in which they reside that regulates their economic and sexual relationship. Other elements of marriage involve emotion, sexual monogamy, and a formal ceremony. The federal government supports marriage education in the public school system with the intention of reducing divorce (which is costly to both individuals and society).

What is family?

The Census Bureau defines family as a group of two or more persons related by blood, marriage, or adoption. In recognition of the diversity of families, the definition of family is increasingly becoming two adult partners whose interdependent relationship is long-term and characterized by an emotional and financial commitment. The traditional family is the two-parent nuclear family with the husband as breadwinner and wife as homemaker. The modern family is the dual-earner family where both spouses work outside the home. Postmodern families represent a departure from these models, such as lesbian and gay couples and single mothers by choice, which emphasizes that a healthy family need not be heterosexual or include two parents. The Amish/Mennonites represent a system where courtship, marriage, and family patterns are tightly controlled. The result is a very low divorce rate.

How do the concepts of marriage and the family differ?

Marriage involves a license; having a child does not. Marriage ends when one person dies or the couple divorce; relationships between the family members of parent and child continue. Marriages involve spouses who are expected to have sex with each other; families involve parents and children, and sex between them is prohibited.

How have marriage and the family changed?

The advent of industrialization, urbanization, and mobility involved the demise of familism and the rise of individualism. When family members functioned together as an economic unit, they were dependent on one another for survival and were concerned about what was good for the family. This familistic focus on the needs of the family has since shifted to a focus on self-fulfillment—individualism.

Other changes in the last 50 years include divorce replacing death as the endpoint for the majority of marriages, marriage and intimate relations emerging as legitimate objects of scientific study, the rise of feminism/changes in gender roles in marriage, and an increase in age at marriage, and increased acceptance of singlehood, cohabitation, and childfree marriages. Families today function in social context of terrorism. Approximately one fourth of college students in one study reported fear that a terrorist attack would hit them personally or someone they loved. Half of U.S. citizens report having cried over the War in Iraq.

What are the theoretical frameworks for viewing marriage and the family?

Theoretical frameworks provide a set of interrelated principles designed to explain a particular phenomenon and provide a point of view. Those used to study the family are the social exchange framework (spouses exchange resources, and decisions are made on the basis of perceived profit and loss), family life course development framework (the process of how families change over time), structural–functional framework (how the family functions to serve society), conflict framework (family members are in conflict over scarce resources of time and money), feminist framework (inequality and oppression), symbolic interaction framework (symbolic worlds in which the various family members give meaning to each other's behavior), systems framework (each member of the family is part of a system and the family as a unit develops norms of interaction), the human ecological framework (individuals, couples, and families are dependent on the environment for air, food, and water, and on other human beings for social interaction), biosocial framework (the interaction of one's biological/genetic inheritance with one's social environment to explain and predict human behavior), and the stratification/race framework(families operate in different social classes and reflect divergent racial identities/histories). The framework used in most empirical family studies is the social exchange framework.

What factors should be kept in mind when evaluating research in marriage and the family?

A random sample (the respondents providing the data reflect those who were not in the sample), control group (the group not subjected to the experimental design so there is a basis of comparison), terminology (the phenomenon being studied should be objectively defined), researcher bias (present in all studies but be alert to the bias issue), time lag (takes two years from study to print), and distortion/deception (while rare, some researchers distort their data) are factors to be used in evaluating research. Few studies avoid all research problems.

KEY TERMS

beliefs	familism	marriage	random sample
binuclear family	family	modern family	role
biosocial framework	family career	nuclear family	secondary group
blended family	family life course development	open relationship	sociological imagination
civil union	family of orientation	polyamory	status
common-law marriage	family of origin	polyandry	stratification
control group	family of procreation	polygamy	theoretical framework
domestic partnership	feral	polygyny	traditional family
ecosystem	functionalists	postmodern family	utilitarianism
experimental group	human ecology	primary group	values
extended family	individualism	race	

The Companion Website for *Choices in Relationships: An Introduction to Marriage and the Family,* Ninth Edition
http://www.thomsonedu.com/sociology/knox

Supplement your review of this chapter by going to the companion website to take one of the Tutorial Quizzes, use the flash cards to master key terms, and check out the many other study aids you'll find there. You'll also find special features such as the Marriage and Family Resource Center, Census 2000 information, and other data and resources at your fingertips to help you with that special project or to do some research on your own.

WEBLINKS

Gilder Lehrman Institute of American History—History of the Family
http://www.digitalhistory.uh.edu/historyonline/familyhistory.cfm

National Center for Health Statistics
http://www.cdc.gov/nchs/

National Council on Family Relations
http://www.ncfr.org/

National Marriage Project
http://marriage.rutgers.edu/about.htm

National Survey of Families and Households
http://www.ssc.wisc.edu/nsfh/home.htm

U.S. Census Bureau
http://www.census.gov/

REFERENCES

Abowitz, D. A. and D. Knox. 2003. College student life: Gender, gender ideology, and the effects of Greek status. Paper presented at the 73rd Annual Meeting of the Eastern Sociological Society, Philadelphia, February 28.

Adamson, K. and P. Star. 2005. Polyamory. Presentation to Sociology of Human Sexuality class, Department of Sociology, East Carolina University, Greenville, NC. October.

Al-Krenawi, A., J. R. Graham, and V. Slonim-Nevo. 2002. Mental health aspects of Arab-Israeli adolescents from polygamous versus monogamous families. *Journal of Social Psychology* 142:446–60.

American Council on Education and University of California. 2005–2006. *The American freshman: National norms for Fall, 2005.* Los Angeles: Higher Education Research Institute. U.C.L.A. Graduate School of Education and Information Studies.

Barrera, M., H. M. Prelow, L. E. Dumka, N. A. Gonzales, G. P. Knight, M. L. Michaels, M. W. Roosa, and J. Tien. 2002. Pathways from family economic conditions to adolescents' distress: Supportive parenting, stressors outside the family, and deviant peers. *Journal of Community Psychology* 30:135–52.

Beck, J. E. and D. S. Shaw. 2005. The influence of perinatal complications and environmental adversity on boy's antisocial behavior. *Journal of Child Psychology & Psychiatry and Allied Disciplines.* 46:35–46.

Best Friends Survey (2006) All in the famiy. *USA Today,* June 21, A 1.

Blumer, H. G. 1969. The methodological position of symbolic interaction. In *Symbolic interactionism: Perspective and method.* Englewood Cliffs, NJ: Prentice-Hall.

Borgerhoff Mulder, M. and C. McCabe. 2006. Whatever happened to human sociobiology? *Anthropology* 22: 21–22.

Bradshaw, Z. and P. Slade. 2005. The relationships between induced abortion, attitudes toward sexuality and sexuality problems. *Sexual and Relationship Therapy* 20:391–406.

Bronfenbrenner, U. 1979. *The ecology of human development: Experiments by nature and design.* Cambridge, MA: Harvard University Press.

Bubolz, M. M., and M. S. Sontag. 1993. Human ecology theory. In *Sourcebook of family theory and methods: A contextual approach,* edited by P.G. Boss, W. J. Doherty, R. LaRossa, W. R. Schumm, and S. K. Steinmetz. New York: Plenum.

Busby, D. M., B. Gardner, C. Brandt and N. Taniguchi 2005. The family of origin parachute model: Landing safely in adult romantic relationships. *Family Relations* 54:254–264.

Caspi, A., K. Sugden, T. Moffitt, A. Taylor, I. W. Craig, H. L. Harrington, J. McClay, J. Mill, J. Martin, A. Braithwaite, and R. Poulton. 2003. Influence of life stress on depression: Moderation by a polymorphism in the 5-htt gene. *Science* 301:386–389.

Cohler, B. J. 2005. Life course social science perspectives on the GLBT family. *Journal of GLBT Family Studies* 1:69–95.

Cooksey, E. C., F. L. Mott, and S. A. Neubauer. 2002. Friendships and early relationships: Links to sexual initiation among American adolescents born to young mothers. *Perspectives on Sexual and Reproductive Health* 34:118–126.

Cooley, C. H. 1964. *Human nature and the social order.* New York: Schocken.

Coontz, S. 2005. *Marriage, A History: How love conquered marriage.* New York: Penguin Books.

Corra, M., J. S. Carter, and D. Knox. 2006. Marital happiness by sex and race: A second look. Paper, American Sociological Association, Annual Meeting, New York, August.

DeCuzzi, A., D. Knox, and M. Zusman. 2006. Racial differences in perceptions of women and men. *College Student Journal* 40: 343–349

DeGraaf, P. M., and M. Kalmijn. 2003. Alternative routes in the remarriage market: Competing risk analysis of union formation after divorce. *Social Forces* 81:1459–98.

DeOllos, I. Y. 2005. Predicting marital success or failure: Burgess and beyond. In *Sourcebook of family theory & research,* edited by Vern L. Bengtson, Alan C. Acock, Katherine R. Allen, Peggye Dilworth-Anderson, and David M. Klein. Thousand Oaks, CA: Sage Publications, 134–136.

Donohue, K. M. 2005. Pet loss: Implications for social work practice. *Social Work* 50:187–190.

Bristol, K. and B. Farmer 2005. *Sexuality Among Southeastern University Students: A Survey.* Unpublished data, East Carolina University, Greenville, NC.

Florio, G. and B. Passey. 2006. Some members of polygamy set fleeing as law closes in. *USA Today* April 13, p. 3A.

Gillmore, M. R., M. E. Archibald, D. M. Morrison, A. Wilsdon, E. A. Wells, M. J. Hoppe, D. Nahom, and E. Murowchick. 2002. Teen sexual behavior: Applicability of the theory of reasoned action. *Journal of Marriage and the Family* 64:885–97.

Henley-Walters, L., W. Warzywoda-Kruszynska, and T. Gurko. 2002. Cross-cultural studies of families: Hidden differences. *Journal of Comparative Family Studies* 33:433–50.

Howard, Margo. 2003. *A life in letters: Ann Landers's letters to her only child.* New York: Warner.

Ingoldsby, B. B., S. R. Smith, and J. E. Miller. 2004. *Exploring family theories.* Los Angeles, CA: Roxbury Publishing Co.,

Jenkins, J. M., J. Rasbash, and T. G. O'Connor. 2003. The role of the shared family context in differential parenting. *Developmental Psychology* 39:99–113.

Judkins, B. and L. Presser. 2005. Gender asymmetry in the division of eco-friendly domestic labor. Paper, Society of Social Problems, Philadelphia, PA, August.

Kelly, M., D. Strassberg, and C. Turner. 2006. Behavioral assessment of couple's communication in female orgastic disorder. *Journal of Sex and Marital Therapy* 32:81–95

Kim, H. K., and P. C. McKenry. 2002. The relationship between marriage and psychological well-being. *Journal of Family Issues* 23:885–911.

Knox, D. and M. E. Zusman. 2006. Relationship and sexual behaviors of a sample of 1027 university students. Unpublished data collected for this text. Department of Sociology, East Carolina University, Greenville, NC.

Koehn, R. 1996. *A threefold cord.* Moundridge, KS: Gospel Publishers.

LaBrie, J. W., J. Schiffman, and M. Earleywine. 2002. Expectancies specific to condom use mediate the alcohol and sexual risk relationship. *Journal of Sex Research* 39:145–52.

Lawson, W. 2005. Good friends, long life. *Psychology Today* 38:32–32.

Lemonick, M. D. 2006. The rise and fall of the cloning king. *Time,* January 6, pp. 40–43.

Lorber, J. 1998. *Gender inequality: Feminist theories and politics.* Los Angeles, CA: Roxbury.

Mahaffey, A. L., A. Bryan, and K. E. Hutchison. 2005. Sex differences in affective responses to homoerotic stimuli: evidence for an unconscious bias among heterosexual men, but not heterosexual women. *Archives of Sexual Behavior* 34: 537–546.

McCullough, D. and D. S. Hall. 2003. Polyamory—What it is and what it isn't. *Electronic Journal of Human Sexuality* 6: Feb 27. Retrieved August 8, 2006 from http://www.ejhs.org/volume6/polyamory.htm

McGraw, A., M. Zvonkovic, and A. J. Walker. 2000. Studying postmodern families: A feminist analysis of ethical tensions in work and family research. *Journal of Marriage and the Family* 62:68–77.

McNeely, C., M. L. Shew, T. Beuhrign, R. Sieving, B. C. Miller, and R. W. Blum. 2002. Mothers' influence on the timing of first sex among 14- and 15-year-olds. *Journal of Adolescent Health* 31:256–65.

Mead, G. H. 1934. *Mind, self, and society.* Chicago: University of Chicago Press.

Meinhold, J. L., A. Acock, and A. Walker. 2006. The influence of life transition statuses on sibling intimacy and contact in early adulthood. An earlier version was presented at the Annual Meeting of the National Council on Family Relations in Orlando in November 2005. Submitted for publication.

Meyers, J. F., and M. Shurts. 2002 Measuring positive emotionality: A review of instruments assessing love. *Measurement and Evaluation in Counseling and Development* 34:238–54.

Michael, R. T., J. H. Gagnon, E. O. Laumann and G. Kolata. 1994. *Sex in America: A definitive survey.* Boston, MA: Little, Brown.

Moen, P. and S. Coltrane. 2005. Families, theories, and social policy. In *Sourcebook of family theory & research,* edited by Vern L. Bengtson, Alan C. Acock, Katherine R. Allen, Peggye Dilworth-Anderson, and David M. Klein. Thousand Oaks, CA: Sage Publications, 543–556.

Morley, C. and J. Fook. 2005. The importance of pet loss and some implications for services. *Mortality* 10:127–143.

Murdock, G. P. 1949. *Social structure.* New York: Free Press.

Nelson, S. 2002. Sense of humor and women's psychological health. *Progress: Family Systems Research and Therapy* 11:117–24.

Newton, N. 2002. *Savage girls and wild boys: A history of feral children.* New York: Thomas Dunne Books/St. Martin's Press.

Neyer, F. J., and F. R. Lang. 2003. Blood is thicker than water: Kinship orientation across adulthood. *Journal of Personality and Social Psychology* 84:310–21.

Novilla, M., B. Lelinneth, M. D. Barnes, N. G. De La Cruz, P. N. Williams and J. Rogers. 2006. Public health perspectives on the family. *Family and Community Health* 29:28–42.

Odero, D. R. 2004. Families in Kenya. *Family Focus* 49:f14–f15.

Page, S. 2006. War has hurt USA. *USA Today* March 17, page A1.

Pahl, R. and D. J. Pevalin. 2005. Between family and friends: a longitudinal study of friendship choice. *The British Journal of Sociology* 56:433–451.

Parker, L. 2005. When pets die at the vet, grieving owners call lawyers. *USA Today*, March 15, p. A1.

Pinsof, W. 2002. Introduction to the special issue on marriage in the 20th century in Western civilization: Trends, research, therapy, and perspectives. *Family Process* 41:133–34.

Quiles, J. A. 2003. Romantic behaviors of university students: A cross-cultural and gender analysis in Puerto Rico and the United States. *College Student Journal* 37:354–66.

Riniolo, T. C., and L. A. Schmidt. 2000. Searching for reliable relationships with statistical packages: An empirical example of the potential problems. *Journal of Psychology* 13:143–51.

Salmon, J., N. Owen, A. Bauman, M. K. H. Schmitz, and M. Booth. 2000. Leisure-time, occupational, and household physical activity among professional, skilled, and less-skilled workers and homemakers. *Prevention Medicine* 30:191–99.

Schvaneveldt, P. L. 2003. Mate selection preferences and practices in Ecuador and Latin America. In *Mate Selection Across Cultures* edited by R. R. Hamon and B. B. Ingoldsby. Thousand Oaks, California: Sage Publications, pp. 43–59.

Shaw, D. S., M. Gilliom, E. M. Ingoldsby, and D. S. Nagin. 2003. Trajectories leading to school-age conduct problems. *Developmental Psychology* 39:189–200.

Silverstein, L. B. and C. F. Auerbach. 2005. (Post) modern families. In *Families in global perspective.*, edited by Jaipaul L. Roopnarine and U. P. Gielen. Boston, MA. Pearson Education, 33–48.

Spada, J. 1987. *Grace.* Garden City, New York: Doubleday & Company.

Statistical Abstract of the United States: 2006. 125th ed. Washington, DC: U.S. Bureau of the Census.

Stuart, T. D., and M. E. B. Garrison. 2002. The influence of daily hassles and role balance on health status: A study of mothers of grade school children. *Women and Health* 36:1–11.

Sutton, P. D. 2003. Births, marriages, divorces, and deaths: Provisional data for January–March 2002. *National Vital Statistics Report* 51, no. 6. Hyattsville, MD: National Center for Health Statistics.

Taylor, A. C. and A. Bagd. 2005. The lack of explicit theory in family research: The case analysis of the *Journal of Marriage and the Family 1990–1999.* In *Sourcebook of family theory & research,* edited by Vern L. Bengtson, Alan C. Acock, Katherine R. Allen, Peggye Dilworth-Anderson, and David M. Klein. Thousand Oaks, CA: Sage Publications, pp. 22–25.

Teachman, J. D. 2002. Childhood living arrangements and the intergenerational transmission of divorce. *Journal of Marriage and Family* 64:717–29.

Testa, M., J. A. Livingston, and C. VanZile-Tamsen. 2005. The impact of questionnaire administration mode on response rate and reporting of consensual and nonconsensual sexual behavior. *Psychology of Women Quarterly* 29:345–352.

Van de Rijt, A. and M. W. Macy. 2006. Power and dependence in intimate exchange. *Social Forces* 84:1456–1460.

Weise, W. and D. Vergano 2006. South Korean stem cell scandal could sink deeper. *USA Today* January 3, p. 8D.

Whatley, M., G. M. Little, D. Knox. 2006. A scale to measure college student relationship involvement. *College Student Journal* 40:55–62.

White, J. M., and D. M. Klein. 2002. *Family theories,* 2d ed. Thousand Oaks, Calif.: Sage Publications.

Wilcox, W. B. and S. L. Nock. 2006. What's love got to do with it? Equality, equity, commitment and marital quality. *Social Forces* 84:1321–1345.

Wilmoth, J., and G. Koso. 2002. Does marital history matter? Marital status and wealth outcomes among preretirement adults. *Journal of Marriage and the Family* 64:254–68.

Winkielman, P., K. C. Berridge, and J. L. Wilbarger. 2005. Unconscious affective reactions to masked happy versus angry faces influence consumption behavior and judgments of value. *Personality & Social Psychology Bulletin* 31:121–136.

Yee, B. W. K. 1992. Gender and family issues in minority groups. In *Cultural diversity and families,* edited by K. G. Arms, J. K. Davidson, Sr., and N. B. Moore. Dubuque, IA: Bench Press, pp. 5–10.

Young, J. R. 2003. Prozac campus. *Chronicle of Higher Education* 69:A37–38.

chapter 2

Relations between women and men have changed more in the past thirty years than they have in the past three thousand years.

Stephanie Coontz, Marriage, A History

Gender in Relationships

Contents

True or False?

1. The case of David Reimer, who was reared as a girl, provides convincing evidence of the profound influence of biological wiring on one's gender identity.

2. One's biological sex may most accurately be described as being on a continuum rather than being dichotomously described as being female or male.

3. Up until age 6 or 7 years, children think that they can change their gender and become someone of the other sex.

4. How undergraduate women view themselves is very similar to the way undergraduate men view them.

5. Women earn fewer than 20 percent of the Ph.D.s granted annually, and this percentage has remained stable in recent years.

Answers: **1.** T   **2.** T   **3.** T   **4.** F   **5.** F

*T*he Sopranos and *Desperate Housewives* captured the attention of the American viewing public in 2006. While these stories emphasized the mob in New Jersey and suburban life on Wisteria Lane, what went virtually unnoticed was that traditional and nontraditional gender roles in relationships were a staple of each television show. Tony and Carmella (*Sopranos*) represented the "man provides/woman stays at home and rears two children" American family. In contrast, *Desperate Housewives* featured Lynette Scavo as the breadwinning wife and Tom Scavo as the husband who stays at home and rears three sons. The theme was clear: diversity of role relationships in families. In this chapter we examine variations in gender roles and the way they express themselves in various relationships. We begin by looking at the terms used to discuss gender issues.

Terminology of Gender Roles

I don't think boys in general watch the emotional world of relationships as closely as girls do. Girls track that world all day long, like watching the weather.

Carol Gilligan, Gender researcher

In common usage the terms *sex* and *gender* are often interchangeable, but to sociologists, family/consumer science educators, human development specialists, and health educators, these terms are not synonymous. After clarifying the distinction between sex and gender, we discuss other relevant terminology, including *gender identity, gender role,* and *gender role ideology.*

Sex

Sex refers to the biological distinction between females and males. Hence, to be assigned as a female or male, several factors are used to determine the biological sex of an individual:

- *Chromosomes:* XX for female; XY for male
- *Gonads:* Ovaries for female; testes for male
- *Hormones:* Greater proportion of estrogen and progesterone than testosterone in the female; greater proportion of testosterone than estrogen and progesterone in the male
- *Internal sex organs:* Fallopian tubes, uterus, and vagina for female; epididymis, vas deferens, and seminal vesicles for male
- *External genitals:* Vulva for female; penis and scrotum for male

Even though we commonly think of biological sex as consisting of two dichotomous categories (female and male), biological sex exists on a continuum. Sometimes not all of the items identified are found neatly in one person (who would be labeled as a female or a male). Rather, items typically associated with

Chapter 2 Gender in Relationships

females or males might be found together in one person, resulting in mixed or ambiguous genitals; such persons are called **hermaphrodites** or **intersexed individuals.** Indeed, the genitals in these individuals are not clearly male or female. Looy and Bouma (2005) discussed the confusion intersexed individuals feel in that they struggle with whether they are female or male. To be uncertain and to be shifting between the two can be very disconcerting in a world that emphasizes dichotomies. Meyer-Bahlburg (2005) suggested that the genesis of the ambiguity may be the brain anatomy, whereby the wiring is different for those with gender identity disorder (GID). Genetics, hormones, and brain mechanisms might all underlie neuroanatomic changes inducing intersexuality.

There is no unisex brain. It follows that these two brain models can produce quite different behaviors.
Louann Brizendine, *The Female Brain*

Gender

Gender refers to the social and psychological characteristics associated with being female or male. For example, women see themselves (and men agree) as moody and easily embarrassed; men see themselves (and women agree) as competitive, sarcastic, and sexual (Knox et al., 2004).

In popular usage, gender is dichotomized as an either/or concept (feminine or masculine). Each gender has some characteristics of the other. But gender may also be viewed as existing along a continuum of femininity and masculinity.

There is an ongoing controversy about whether gender differences are innate as opposed to learned or socially determined. Just as sexual orientation may be best explained as an interaction of biological and social/psychological variables, gender differences may also be seen as a consequence of both biological and social/psychological influences. For example, Irvolino et al., (2005) studied the genetic and environmental effects on the sex-typed behavior of 3999 three- to four-year-old twin and non-twin sibling pairs and concluded that their gender role behavior was a function of both genetic inheritance (e.g., chromosomes, hormones) and social factors (e.g., male/female models such as parents, siblings, peers).

Whereas some researchers emphasize an interaction of the biological and social, others emphasize a biological imperative as the basis of gender role behavior. As evidence for the latter, John Money, psychologist and former director of the now defunct Gender Identity Clinic at Johns Hopkins University School of Medicine, encouraged the parents of a boy (Bruce) to rear him as a girl (Brenda) because of a botched circumcision that rendered the infant without a penis. Money argued that social mirrors dictate one's gender identity, and thus if the parents treated the child as a girl (name, dress, toys, etc.), the child would adopt the role of a girl and later that of a woman. The child was castrated and sex reassignment began.

But the experiment failed miserably; the child as an adult (David Reimer—his real name) reported that he never felt comfortable in the role of a girl and had always viewed himself as a boy. He later married and adopted the two children of his wife. In the book *As Nature Made Him: The Boy Who Was Raised as a Girl* (Colapinto, 2000), David worked with a writer to tell his story. His courageous decision to make his poignant personal story public has shed light on scientific debate

The gender role of hunter is traditionally associated with men. Its origin is that men were responsible for hunting and bringing meat back to the social unit for food. Women, who were nursing/caring for young children, were less mobile.

Authors

on the "nature/nurture" question. In the past, David's situation was used as a textbook example of how "nurture" is the more important influence in gender identity, if a reassignment is done early enough. Today, his case makes the point that one's biological wiring dictates gender outcome (Colapinto, 2000). Indeed, David Reimer noted in a television interview, "I was scammed," referring to the absurdity of trying to rear him as a girl. Distraught with the ordeal of his upbringing and beset with financial difficulties, he committed suicide in May 2004 via a gunshot to the head.

Although the story of David Reimer is a landmark in terms of the power of biology in determining gender identity, other research supports the critical role of biology. Cohen-Kettenis (2005) emphasized that biological influences in the form of androgens in the prenatal brain are very much at work in creating one's gender identity.

Nevertheless, **socialization** (the process through which we learn attitudes, values, beliefs, and behaviors appropriate to the social positions we occupy) does impact gender role behaviors, and social scientists tend to emphasize the role of social influences in gender differences. Although her research is controversial, Margaret Mead (1935) focused on the role of social learning in the development of gender roles in her study of three cultures.

She visited three New Guinea tribes in the early 1930s and observed that the Arapesh socialized both men and women to be feminine, by Western standards. The Arapesh people were taught to be cooperative and responsive to the needs of others. In contrast, the Tchambuli were known for dominant women and submissive men—just the opposite of our society. And both of these societies were unlike the Mundugumor, which socialized only ruthless, aggressive, "masculine" personalities. The inescapable conclusion of this cross-cultural study is that human beings are products of their social and cultural environment and that gender roles are learned. As Peoples (2001, 18) observed, "cultures construct gender in different ways."

Gender Identity

Gender identity is the psychological state of viewing oneself as a girl or a boy, and later as a woman or a man. Such identity is largely learned and is a reflection of society's conceptions of femininity and masculinity. Some individuals experience **gender dysphoria,** a condition in which one's gender identity does not match one's biological sex. An example of gender dysphoria is transsexualism (discussed in the next section).

Transgenderism

The word **transgender** is a generic term for a person of one biological sex who displays characteristics of the other sex. **Cross-dresser** is a broad term for individuals who may dress or present themselves in the gender of the other sex. Some cross-dressers are heterosexual adult males who enjoy dressing and presenting themselves as women.

Cross-dressers may also be women who dress as men and present themselves as men. Some cross-dressers are bisexual or homosexual. Another term for cross-dresser is **transvestite,** although the latter term is commonly associated with homosexual men who dress provocatively as women to attract men—sometimes as sexual customers.

This heterosexual, married father of two children enjoys dressing as a woman.

Authors

Transsexuals are persons with the biological/anatomical sex of one gender (e.g., male) but the self-concept of the other sex (e.g., female). "I am a woman trapped in a man's body" reflects the feelings of the male-to-female transsexual, who may take hormones to develop breasts and reduce facial hair and may have surgery to artificially construct a vagina. This person lives full-time as a woman.

The female-to-male transsexual is one who is a biological/anatomical female but feels "I am a man trapped in a female's body." This person may take male hormones to grow facial hair and deepen her voice and may have surgery to create an artificial penis. This person lives full-time as a man.

Individuals need not take hormones or have surgery to be regarded as transsexuals. The distinguishing variable is living full-time in the role of the gender opposite one's biological sex. A man or woman who presents himself or herself full-time as a woman or man is a transsexual by definition. Table 2.1 may help to keep the categories clear.

Some transsexuals prefer the term **transgenderist,** which means an individual who lives in a gender role that does not match his or her biological sex but has no desire to surgically alter his or her genitalia (as does a transsexual). Another variation is the she/male, who looks like a woman and has the breasts of a woman yet has the genitalia and reproductive system of a male.

Colleges and universities are beginning to consider the needs of transgender students, but they are uncertain what their needs are and how to support them (Beemyn et al., 2005). One such need concerns the increased vulnerability of the transgender community to human immunodeficiency virus (HIV) infection. Kenagy and Hsieh (2005) compared the HIV risk of male-to-female transsexuals and female-to-male transsexuals and found the latter were significantly less likely to have used protection the last time they had sex. What is instructive is that this segment of college students is being recognized and that their needs are being considered.

Gender Roles

Gender roles are the social norms that dictate what is socially regarded as appropriate female and male behavior. All societies have expectations of how boys and girls, men and women "should" behave. Gender roles influence women and men in virtually every sphere of life, including family and occupation.

For example, traditional gender role expectations have influenced women to be homemakers, day-care workers, and nurses. Martin (2003) noted that women during childbirth are expected to be "nice," "polite," "kind," and "selfless" (even when they are hurt, bleeding, or dying). Her study emphasized how women are compelled to act in gender-normative ways. Rodkin et al., (2006) studied 526 fourth- to sixth-graders and found that aggressive boys were more likely to be labeled "cool" by girls who were themselves aggressive.

How Undergraduate Women View Men

McNeely et al. (2004) assessed how a sample of 326 undergraduates viewed men. She compared women's and men's views and found that women were significantly more likely to believe that all men cheat on their partners at least once, that a man will not call when he says he will, that men prefer to cohabit with a woman rather than marry her, that they think about sex more than women, and

Few women admit their age. Fewer men act it.

Bumper sticker

Table 2.1	Transgender Categories		
Category	**Biological Sex**	**Sexual Orientation**	**Most Usual Case**
Cross-dresser	Either	Either	Male heterosexual; dresses as woman
Transvestite	Male	Gay	Gay male; dresses as woman
Transsexual	Either	Either	Heterosexual male in woman's body; wants surgery

Attitudes toward Infidelity Scale

Infidelity can be defined as a person being unfaithful in a committed monogamous relationship. The purpose of this scale is to gain a better understanding of what people think and feel about issues associated with infidelity. There are no right or wrong answers to any of these statements; we are interested in your honest reactions and opinions. Please read each statement carefully, and respond by using the following scale:

1	2	3	4	5	6	7
Strongly Disagree						Strongly Agree

_____ 1. Being unfaithful never hurt anyone

_____ 2. Infidelity in a marital relationship is grounds for divorce.

_____ 3. Infidelity is acceptable for retaliation of infidelity.

_____ 4. It is natural for people to be unfaithful.

_____ 5. Online/internet behavior (e.g., sex chatrooms, porn sites) is an act of infidelity.

_____ 6. Infidelity is morally wrong in all circumstances regardless of the situation.

_____ 7. Being unfaithful in a relationship is one of the most dishonorable things a person can do.

_____ 8. Infidelity is unacceptable under any circumstances if the couple is married.

_____ 9. I would not mind if my significant other had an affair as long as I did not know about it.

_____10. It would be acceptable for me to have an affair, but not my significant other.

_____11. I would have an affair if I knew my significant other would never find out.

_____12. If I knew my significant other was guilty of infidelity, I would confront him/her.

Scoring

Selecting a 1 reflects the least acceptance of infidelity; selecting a 7 reflects the greatest acceptance of infidelity. Before adding the numbers you selected, reverse score items #2, #5, #6, #7, #8, and #12 (i.e., 1 = 7; 2 = 6; 3 = 5; 4 = 4; 5 = 3; 6 = 2; 7 = 1). For example, if you responded to question #2 with a "6," change this number to a "2." If you responded to question #12 with a "7," change this number "1." After making these changes, add the numbers. The lower your total score (12 is the lowest possible score) the less accepting you are of infidelity; the higher your total score (84 is the highest possible score) the greater your acceptance of infidelity. A score of 48 places you at the midpoint between being very disapproving of infidelity and very accepting of infidelity.

Scores of Other Students Who Completed the Scale

The scale was completed by 150 male and 136 female student volunteers at Valdosta State University. Their ages ranged from 18 to 49 with a mean age of 23.36 (SD = 5.13). The average score on the scale was 27.85 (SD = 12.02) suggesting a generally negative view of infidelity. In regard to sex of the participants, male participants reported more positive attitudes toward infidelity (M = 31.53, SD = 11.86) than did female participants (M = 23.78, SD = 10.86) ($p < .05$).

The ethnic background of the sample included 60.8% White, 28.3% African American, 2.4% Hispanic, 3.8% Asian, 0.3% American Indian, and 4.2% from other ethnic backgrounds. White participants had more negative attitudes toward infidelity (M = 25.36, SD = 11.17) than did non-white participants (M = 31.71, SD = 12.32) ($p < .05$). The college classification level of the sample included 11.5% Freshman, 18.2% Sophomore, 20.6% Junior, 37.8% Senior, 7.7% graduate student, and 4.2% post-baccalaureate. There were no significant differences in regard to college classification and views of infidelity.

Source

"Attitudes toward Infidelity Scale" 2006 by Mark Whatley, Ph.D. Department of Psychology, Valdosta State University, Valdosta, Georgia 31698-0100. Used by permission. Other uses of this scale by written permission of Dr. Whatley only. His email is mwhatley@valdosta.edu. Information on the reliability and validity of this scale is available from Dr. Whatley.

that they have poorer communication skills than women. These views include that men are more sex focused than women and less likely to be faithful. The Self-Assessment on Attitudes toward Infidelity (above) was taken by both undergraduate men and women. Men were significantly more likely than women to have positive views of infidelity.

How Undergraduate Men View Women

McNeely et al., (2005) also assessed how 326 undergraduates viewed women. They compared men's and women's agreement with various beliefs about women and found several significant differences. Table 2.2 reflects the percentage differences and suggests that, indeed, what men think about women is very different from what women think about themselves.

Table 2.1 Gender Differences in Beliefs About Women (N = 326)

Beliefs about Women	% Men Believing	% Women Believing
Unmarried women aged 30+ years are unhappy/depressed	16.3%	6.2%
Women assume men are mind–readers	55.4%	40.8%
Women are controlling	58.2%	36.4%
Red-haired women are fiery and saucy	23.7%	9.2%
Women want marriage, not cohabitation	84.0%	68.2%
Women love money	16.7%	3.5%
Women are possessive	52.1%	32.9%
Women are manipulative	58.3%	33.3%

McNeely, A., D. Knox, and M.E. Zusman. 2005. College student beliefs about women: Some gender differences. *College Student Journal* 39:769–74. Used by permission of *College Student Journal*.

National Data

Sixteen percent of a random sample of all first-year college undergraduate women and 26 percent of a random sample of all first-year undergraduate men in colleges and universities in the United States agreed that the "activities of married women are best confined to the home" (American Council on Education and the University of California, 2005–2006).

The term **sex roles** is often confused with and used interchangeably with the term *gender roles*. However, whereas gender roles are socially defined and can be enacted by either women or men, sex roles are defined by biological constraints and can be enacted by members of one biological sex only—for example, wet nurse, sperm donor, child-bearer.

Gender Role Ideology

Gender role ideology refers to beliefs about the proper role relationships between women and men in any given society. In spite of the rhetoric regarding the entrenchment of egalitarian interaction between women and men in the United States, there is evidence of traditional gender roles. When 692 undergraduate females at a large southeastern university were asked if they had ever asked a new guy to go out, more than 60 percent (61.1%) responded "no" (Knox and Zusman, 2006). Thirty percent reported a preference of marrying a traditional man (one who saw his role as provider and who was supportive of his wife staying home to rear the children) (McGinty et al., 2006). Such a preference for traditional roles is also found among undergraduate men. In a national sample of first-year undergraduates, over a fourth (26%) preferred having a traditional family in which the woman stays home to rear the children (American Council on Education and University of California, 2005–2006). However, inside the family, egalitarian wives are most happy with their marriages if their husbands share both the work and the emotions of managing the home and caring for the children (Wilcox and Nock, 2006).

Traditional American gender role ideology has perpetuated and reflected patriarchal male dominance and male bias in almost every sphere of life. Even our language reflects this male bias. For example, the words *man* and *mankind* have traditionally been used to refer to all humans. There has been a growing trend away from using male-biased language. Dictionaries have begun to replace *chairman* with *chairperson* and *mankind* with *humankind*.

Women always worry about the things men forget. Men always worry about the things women remember.

Unknown

Theories of Gender Role Development

Various theories attempt to explain why women and men exhibit different characteristics and behaviors.

Biosocial

We noted in the discussion of gender, at the beginning of the chapter, the profound influence of biology on one's gender. In 2005, Harvard University president Larry Summers also suggested that biological factors could explain why fewer females than males are top-tiered, tenured science professors. Included among his explanations was the suggestion that men may have more "intrinsic aptitude" than women (Ripley, 2005). In effect, he suggested a biosocial explanation—that men are biologically wired with more aptitude in science—causing a firestorm of criticism. (Summers subsequently lost his job.)

Biosocial theory emphasizes that social behaviors (e.g., gender roles) are biologically based and have an evolutionary survival function. For example, women tend to select and mate with men whom they deem will provide the maximum parental investment in their offspring. The term **parental investment** refers to any investment by a parent that increases the offspring's chance of surviving and thus increases reproductive success. Parental investments require time and energy. Women have a great deal of parental investment in their offspring (including 9 months' gestation) and tend to mate with men who have high status, economic resources, and a willingness to share those economic resources.

The biosocial explanation (also referred to as **sociobiology**) for mate selection is extremely controversial. Critics argue that women may show concern for the earning capacity of a potential mate because they have been systematically denied access to similar economic resources, and selecting a mate with these resources is one of their remaining options. In addition, it is argued that both women and men, when selecting a mate, think more about their partners as companions than as future parents of their offspring.

Social Learning

In contrast to the biological explanation for gender roles is that of social learning. Derived from the school of behavioral psychology, the social learning theory emphasizes the roles of reward and punishment in explaining how a child learns gender role behavior. For example, two young brothers enjoyed playing "lady." Each of them would put on a dress, wear high-heeled shoes, and carry a pocketbook. Their father came home early one day and angrily demanded, "Take those clothes off and never put them on again. Those things are for women." The boys were punished for playing lady but rewarded with their father's approval for playing cowboys, with plastic guns and "Bang! You're dead!" dialogue.

Reward and punishment alone are not sufficient to account for the way in which children learn gender roles. Direct instruction ("girls wear dresses," "a man stands up and shakes hands") by parents or peers is another way children learn. In addition, many of society's gender rules are learned through modeling. In modeling, the child observes another's behavior and imitates that behavior. Gender role models include parents, peers, siblings, and characters portrayed in the media.

The impact of modeling on the development of gender role behavior is controversial. For example, a modeling perspective implies that children will tend to imitate the parent of the same sex, but children in all cultures are usually reared mainly by women. Yet this persistent female model does not seem to interfere with

Diversity in Other Countries

In his classic research, David Buss (1989) found that the pattern of men seeking physically attractive young women and women seeking economically ambitious men was true in 37 groups of women and men in 33 different societies.

the male's development of the behavior that is considered appropriate for his gender. One explanation suggests that boys learn early that our society generally grants boys and men more status and privileges than girls and women. Therefore, boys devalue the feminine and emphasize the masculine aspects of themselves.

Identification

Freud was one of the first researchers to study gender role development. He suggested that children acquire the characteristics and behaviors of their same-sex parent through a process of identification. Girls identify with their mothers; boys identify with their fathers. For example, girls are more likely to become involved in taking care of children because they see women as the primary caregivers of young children. In effect, they identify with their mothers and will see their own primary identity and role as those of a mother. Likewise, boys will observe their fathers and engage in similar behaviors to lock in their own gender identity. The classic example is the son who observes his father shaving and wants to do likewise (be a man too).

Cognitive-Developmental Theory

The cognitive-developmental theory of gender role development reflects a blend of biological and social learning views. According to this theory, the biological readiness, in terms of cognitive development, of the child influences how the child responds to gender cues in the environment (Kohlberg, 1966). For example, gender discrimination (the ability to identify social and psychological characteristics associated with being female or male) begins at about age 30 months. However, at this age, children do not view gender as a permanent characteristic. Thus, even though young children may define people who wear long hair as girls and those who never wear dresses as boys, they also believe they can change their gender by altering their hair or changing clothes.

Not until age 6 or 7 does the child view gender as permanent (Kohlberg, 1966, 1969). In Kohlberg's view, this cognitive understanding involves the development of a specific mental ability to grasp the idea that certain basic characteristics of people do not change. Once children learn the concept of gender permanence, they seek to become competent and proper members of their gender group. For example, a child standing on the edge of a school playground may observe one group of children jumping rope while another group is playing football. That child's gender identity as either a girl or a boy connects with the observed gender-typed behavior, and the child joins one of the two groups. Once in the group, the child seeks to develop the behaviors that are socially defined as appropriate for her or his gender.

Agents of Socialization

Three of the four theories discussed in the preceding section emphasize that gender roles are learned through interaction with the environment. Indeed, though biology may provide a basis for one's gender identity, cultural influences in the form of various socialization agents (parents, peers, religion, and the media) shape the individual toward various gender roles. These powerful influences in large part dictate what a person thinks, feels, and does in his or her role as a man or a woman. In the next section we look at the different sources influencing your gender socialization.

Family

The family is a gendered institution with female and male roles highly structured by gender. Lorber (2001, p. 23) noted that parents "create a gendered world for

You are the bows from which your children as living arrows are sent forth.

Kahlil Gibran, poet

their newborns by naming, birth announcements, and dress." Parents may also relate differently to their children on the basis of gender. Lindsey and Mize (2001) observed videotapes of parent-child play behavior among 33 preschool children (18 boys, European-American, from middle- and upper-middle-class families) and observed that during the physical play session, fathers and sons engaged in more physical play than did fathers and daughters.

The importance of the father in the family was noted in Pollack's 2001 study of adolescent boys. "America's boys are crying out for a new gender revolution that does for them what the last forty years of feminism has tried to do for girls and women," he stated (p. 18). This new revolution will depend on fathers who teach their sons that feelings and relationships are important. How equipped do you feel today's fathers are to provide these new models for their sons?

Siblings also influence gender role learning. As noted in Chapter 1, the relationship with one's sibling (particularly in sister-sister relationships) is likely to be the most enduring of all relationships (Meinhold et al., 2006). Also, growing up in a family of all sisters or all brothers intensifies social learning experiences toward femininity or masculinity. A male reared with five sisters and a single-parent mother is likely to reflect more feminine characteristics than a male reared in a home with six brothers and a stay-at-home dad.

Race/Ethnicity

The race and ethnicity of one's family also influence gender roles. Although African-American families are often stereotyped as being matriarchal, the more common pattern of authority in these families is egalitarian (Taylor, 2002). However, the fact that Black women have increased economic independence provides a powerful role model for young Black women. A similar situation exists among Hispanics, who represent the fastest-growing segment of the U.S. population. There is great variability among Mexican-American marriages, but the employment opportunities of Mexican-American women provide them with both resources and autonomy (Baca Zinn and Pok, 2002). The result is a less-than-docile model of the emerging Hispanic female (e.g., Jennifer Lopez).

Peers

One's friends are that part of the human race with which one can be human.

Santayana

Though parents are usually the first socializing agents that influence a child's gender role development, peers become increasingly important during the school years. Haynie and Osgood (2005) analyzed data from the National Longitudinal Study of Adolescents reflecting responses from adolescents in grades 7 through 12 at 132 schools over an 11-year period and confirmed the influence of peers in delinquent behavior. If friends drank, smoked cigarettes, skipped school without an excuse, and became involved in serious fights, then there was an increased likelihood that an individual would also engage in delinquent acts. Regarding gender, the gender role messages from adolescent peers are primarily traditional. Boys are expected to play sports and be career-oriented. Female adolescents are under tremendous pressure to be physically attractive and thin, popular, and achievement-oriented. Female achievements may be traditional (cheerleading) or nontraditional (sports or academics). Adolescent females are sometimes in conflict because high academic success may be viewed as being less than feminine.

Peers also influence gender roles throughout the family life cycle. In Chapter 1 we discussed the family life cycle and noted the various developmental tasks throughout the cycle. With each new stage, role changes are made and one's peers influence those role changes. For example, when a couple moves from being childfree to being parents, peers quickly socialize them into the role of parent and the attendant responsibilities. This process of peer socialization continues throughout life. Senator Orin Hatch was seen on *60 Minutes* reporting that he had told Senator Edward Kennedy to "grow up" (translation: stop the wild drinking and partying). The seasoned Senator Kennedy noted that the chiding was warranted and that he had indeed changed his behavior (e.g., allowed his peer to influence his role behavior).

Religion

Religion remains a potentially important influence in the lives of most individuals. Almost two-thirds (65.2%) of a sample of 1027 undergraduates at a large southeastern university viewed themselves as "religious" (Knox and Zusman, 2006). More than 85 percent of adults acknowledge a religious preference (Hout and Fischer, 2002). Because women (particularly white women) are "socialized to be submissive, passive, and nurturing," they may be predisposed to greater levels of religion and religious influence (Miller and Stark, 2002). Such exposure includes a traditional framing of gender roles. Male dominance is indisputable in the hierarchy of religious organizations, where power and status have been accorded mostly to men.

The Roman Catholic Church does not have female clergy, and men dominate the 19 top positions in the U.S. dioceses. Popular books marketed to the Christian Right also emphasize traditional gender roles. Denton (2004) found that conservative Protestants are committed to the ideology that the husband is the head of the family. However, this implies "taking spiritual leadership" and may not imply dominance in marital decision-making (p. 1174).

Education

The educational institution serves as an additional socialization agent for gender role ideology. But such an effect must be considered in the context of the society/culture in which the "school" exists and of the school itself. Schools are basic cultures of transmission in that they make deliberate efforts to reproduce the culture from one generation to the next. Blair (2002) emphasized how schools reflect the broader U.S. culture "and its patriarchal gender roles in their structure, organization, curriculum, and interaction" (p. 22). Sumsion (2005) noted that even having male teachers in the lower grades does not seem to disrupt traditional gender stereotypes that young children have.

To be truly ignorant, be content with your own knowledge.
Chuang Tzu, 300 B.C.

Economy

The economy of the society influences the roles of the individuals in the society. The economy is a very gendered institution. **Occupational sex segregation** denotes the fact that women and men are employed in gender-segregated occupations, that is, occupations in which workers are either primarily male or primarily female. Examples include a preponderance of men as mechanical and electrical engineers and of females as flight attendants and interior designers (Lippa, 2002). Female-dominated occupations tend to require less education, have lower status, and pay lower salaries than male-dominated occupations. If men typically occupy the role, it tends to pay more. For example, the job of child-care attendant requires more education than the job of animal shelter attendant. However, animal shelter attendants are more likely to be male and earn more than child-care attendants, who are more likely to be female.

Mass Media

Mass media, such as movies, television, magazines, newspapers, books, music, and computer games, both reflect and shape gender roles. Media images of women and men typically conform to traditional gender stereotypes, and media portrayals depicting the exploitation, victimization, and sexual objectification of women are common. *Sex in the City,* a television program now in reruns, portrayed "Mr. Big" (the on-again-off-again love interest of Carrie) as cool and in control, whereas Carrie and her three girlfriends (Miranda, Samantha, and

Charlotte) were often frazzled about their relationships with men. However, one of the characters, Samantha, prided herself on "being like a man" in that she was sexually aggressive and able to separate sex and love. Her role provided a model for females who wanted to feel in more control of their sexuality.

Self-help parenting books are also biased toward traditional gender roles. A team of researchers conducted a content analysis of six of the best-selling self-help books for parents and found that 82% were wrought with stereotypical gender role messages (Krafchick et al., 2005).

The cumulative effects of family, peers, religion, education, the economy, and mass media perpetuate gender stereotypes. Each agent of socialization reinforces gender roles that are learned from other agents of socialization, thereby creating a gender role system that is deeply embedded in our culture.

PERSONAL CHOICES

Do You Want a Nontraditional Occupational Role?

The concentration of women in certain occupations and men in others is referred to as **occupational sex segregation.** But some jobs traditionally occupied by one gender are now open to the other. Men may become nurses and librarians, and women may become construction workers and lawyers. Increasingly, occupations are becoming less segregated on the basis of gender, and social acceptance of nontraditional career choices has increased. The U.S. government is committed to opening occupations to both genders and confirms that women are making inroads into traditional male jobs (2005).

The trend continues; women now fly jet aircraft on military combat missions and study at the previously all-male West Point and Virginia Military Institute. However, only 15 percent and 3 percent, respectively, of the cadets at West Point and Virginia Military Institute are female. Diamond (2003) noted that the military remains a less viable option for many women. Even when women enter male-dominated professions like law, they tend to remain at the lower levels of practice due to the priority they give to their families (Bacik and Drew, 2006).

Choosing nontraditional occupational roles may have benefits both for the individual and for society. On the individual level, women and men can make career choices on the basis of their personal talents and interests rather than on the basis of arbitrary social restrictions regarding who can and cannot have a particular job or career. Because traditionally male occupations are generally higher paying than traditionally female occupations, women who make nontraditional career choices can gain access to higher-paying and higher-status jobs.

On the societal level, an increase in nontraditional career choices reduces gender-based occupational segregation, thereby contributing to social equality among women and men. In addition, women and men who enter nontraditional occupations may contribute greatly to the field that they enter. For example, such traditionally male-dominated fields as politics, science and technology, and medicine may benefit greatly from the increased involvement of women. Similarly, among preschool/kindergarten teachers (98% of whom are female), there are not enough male role models for their students. Some parents now ask for male nannies as role models for their male children (Cullen, 2003).

Sources

Cullen, L. T. 2003. I want your job, lady! *Time* 12 May, 52 et passim.

Diamond, D. A. 2003. Breaking down the barricades: The admission of women at Virginia Military Institute and the United States Military Academy at West Point. Paper, Annual Meeting of the Eastern Sociological Society, Philadelphia, February.

Mastracci, S. H. 2005. Persistent problems demand consistent solutions: Evaluating policies to mitigate occupational segregation by gender. *Review of Radical Political Economics* 37:23–49.

Gender Roles in Other Societies

Because gender roles are largely influenced by culture, individuals reared in different societies typically display the gender role patterns of those societies. The following subsections discuss how gender roles differ in Afghanistan, the Caribbean, Greece, Sweden, and East and South Africa.

Gender Roles of Women in Afghanistan under the Taliban

Because of the war on terrorism and the war in Iraq, Afghanistan is very much in the news. It is a country about the size of Texas with an estimated population of 26 million. The Taliban reached the peak of their dominance there in 1996. The Revolutionary Association of the Women of Afghanistan (RAWA) compiled an "abbreviated" list of restrictions against women (including "creating noise when they walk" and "wearing white socks"). The life for many women and children was often cruel, demeaning, and fatal. Some women drank household bleach rather than continue to endure their plight. They were not allowed to go to school or work and thus were completely dependent economically. Indeed, they were required to stay in the house, to paint the windows black, and to leave the house only if they were fully clothed (wearing a burqa) and accompanied by a male relative. Some could not afford burqas and had no living male relatives. According to Skaine, "There are two places for women: one is the husband's bed and the other is the graveyard" (2002, p. 64). One mother reported that the Taliban came to her house, dragged her 19-year-old daughter from it, and drove away with her. "They sell them," she lamented (p. 116).

Although the plight of Afghan women to an outsider may seem horrible, some Afghan women who are thoroughly socialized in the culture and tradition may not feel oppressed but accepting of their role. They may feel love, protection, and security inside the context of their marriage and family. Some may relish tradition and be against those who wish to change their sacred traditions.

Nevertheless, subsequent to 9/11, the United States attacked the Taliban in Afghanistan with the goal of removing them, and in doing so it improved the lives of Afghan women. But the role of women in Afghanistan, particularly in the rural areas, has typically been one of submissiveness. "Most of the women in rural areas (which comprise over 80% of Afghanistan) have never had the opportunity to get out of their own little house, little village, little province" (Consolatore, 2002, p.13).

Hence, outside of Kabul, Afghan women go uneducated, become child brides, produce children, and rarely expect their daughters' lives to be different. The patriarchal social structure and the absence of a centralized and modernized state in Afghanistan predict that changes will be limited for Afghan women (Moghadam, 2002).

Gender Roles in Caribbean Families

For spring break, college students sometimes go to the Bahamas, Jamaica, or other English-speaking islands in the Caribbean (e.g., Barbados, Trinidad, Guyana) and may wonder about the family patterns and role relationships of the people they encounter. The natives of the Caribbean represent more than 30 million, with a majority being of African ancestry. Their family patterns are diverse but are often characterized by women and their children as the primary family unit—the fathers of these children rarely live in the home (at a rate as low as 29% in St. Kitts and as high as 55% in Jamaica) (Roopnarine et al., 2005). Hence, men may have children with different women and be psychologically and physically absent from their children's lives. When they do live with a woman, traditional division of labor prevails, with women taking care of domestic and child-care tasks.

Barbara Walters asked Afghani women, "Why do you now seem happy with the old custom of walking 5 paces behind your husband?" "Land mines" was the answer.

About half of all female household heads have never been married. Women view motherhood, not marriage, as the symbol of their womanhood. Hence, their focus is on taking care of their children and the children of others. Caribbean fathers vary from showing negligible levels of involvement with their children (e.g., in Belize) to showing levels comparable to those in other societies (e.g., in Trinidad, Guyana, and Jamaica) (Roopnarine et al., 2005).

Gender Roles in Greek Families

Greek families reflect predictable roles for the father, mother, children, and married couple. Although adherence to traditional values is changing, it is doing so slowly. The father is regarded as the head of the household, economic provider, and disciplinarian of the children. The mother takes care of the children, is the go-between for them and the father, and supervises them carefully. Indeed, the "first goal in life is to be a good mother" (Georgas et al., 2005, p. 213). Greek parents are to teach their children to behave properly and to support them financially. In addition, they are to "be involved in the private lives of their married children" (p. 213).

The role of the children is to do as they are told, to be seen and not heard, to respect their parents, and to take care of them when they are old. They are to "keep no secrets" from their parents and to respect their grandparents as well.

Role relationships within the marriage are also traditional, with the wife being responsible for food preparation and cleaning. As an aside, the first author was in the home of a Greek couple. After dinner he got up and started taking dinner plates to the kitchen. The husband politely said, "Sit down, that is woman's work." The author looked at the wife, who smiled and said, "We are Greek," emphasizing they both were very firm in their traditional roles.

Gender Roles in Swedish Families

The Swedish government is strongly concerned with equality between women and men. In 1974 Sweden became the first country in the world to introduce a system that enables mothers and fathers to share parental leave (paid by the government) from their jobs in any way they choose. Furthermore, Swedish law states that employers may not penalize the career of a working parent because he or she has used parental rights. By encouraging fathers to participate more in child-rearing, the government aims to provide more opportunities for women to pursue other roles. Women hold about one-quarter of the seats in the Swedish Parliament.

However, few Swedish women are in high-status positions in business, and governmental efforts to reduce gender inequality are weak compared with the power of tradition. Nevertheless, a comparison of Swedish and U.S. students revealed greater acceptance of gender egalitarianism (Weinberg et al., 2000).

Gender Roles in East and South African Families

Maasai men may have several wives who cooperate to take care of them and their children.

Shaabon Mgonza, Guide, Arusha, Tanzania, East Africa

Africa is a diverse continent with more than 50 nations. The cultures range from Islamic/Arab cultures of Northern Africa to industrial and European influences in South Africa. In some parts of East Africa (e.g., Kenya), gender roles are in flux.

Meredith Kennedy (2000) has lived in East Africa and makes the following observations of gender roles:

> The roles of men and women in most African societies tend to be very separate and proscribed, with most authority and power in the men's domain. For instance, Maasai wives of East Africa do not travel much, since when a husband comes home he expects to find his wife (or wives) waiting for him with a gourd of sour milk. If she is not, he has the right to beat her when she shows up. As attempts are made to soften these boundaries and equalize the roles, the impacts are very visible and cause

a lot of reverberations throughout these communal societies. Many African women who believe in and desire better lives will not call themselves "feminists" for fear of social censure. Change for people whose lives are based on tradition and "fitting in" can be very traumatic.

In South Africa, where more than 75 percent of the racial population is African and around 10 percent are white, the African family is also known for its traditional role relationships and patriarchy. African men were socialized by the Dutch with firm patriarchal norms and adopted this style in their own families. In addition, women were subordinate to men "within a wider kinship system, with the chief as the controlling male. An unequal division of labor according to age and sex prevailed, and this system was exacerbated by the absence of men from the home because of migrant labor" (Pretorius 2005, 372). Gold had been discovered in 1870, and young men were recruited to work the mines while living in single quarters (separated from their families).

Consequences of Traditional Gender Role Socialization

This section discusses different consequences, both negative and positive, for women and men, of traditional female and male socialization in the United States. On page 56 we look at what it is like to be a man, through the eyes of a woman who inhabited the identity of a male.

The woman most in need of liberation is the woman in every man and the man in every woman.

William Sloan Coffin, clergyman and political activist

Consequences of Traditional Female Role Socialization

Table 2.3 summarizes some of the negative and positive consequences of being socialized as a woman in U.S. society. Each consequence may or may not be true for a specific woman. For example, though women in general have less education and income, a particular woman may have more education and a higher income than a particular man.

Negative Consequences of Traditional Female Role Socialization   There are several consequences of being socialized as a woman in our society.

1. Less Income. While women now earn more Ph.D.s than men (Aissen and Houvouras, 2006), they have lower academic rank (Probert, 2005) and earn less money. The lower academic rank is because women give priority to the care of their children/family (Aissen and Houvouras, 2006). In addition, women tend to be more concerned about the nonmonetary aspects of work. In a study of 102 seniors and 504 alumni from a mid-sized midwestern public university who rated 48 job characteristics, women gave significantly higher ratings to family life accommodations, pleasant working conditions, travel, and interpersonal relationships. Men were significantly more concerned about the pay and promotion aspects than the women (Heckert et al., 2002).

Table 2.3 Consequences of Traditional Female Role Socialization	
Negative Consequences	**Positive Consequences**
Less education/income (more dependent)	Longer life
Feminization of poverty	Stronger relationship focus
Higher STD/HIV infection risk	Keep relationships on track
Negative body image	Bonding with children
Less marital satisfaction	Identity not tied to job

Self-Made Man: One Woman's Journey into Manhood

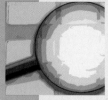

Journalist Norah Vincent wanted to know what it is like to be a man. So, for 18 months, she observed and participated in the world of men, not as Norah Vincent but as "Ned"—the male persona she created for the purpose of her research. Norah Vincent describes her journey into manhood in her book *Self-Made Man* (2006).

Methods

Norah Vincent utilized a method of research known as *participant observation*—a research method in which the researcher participates in the phenomenon being studied in order to obtain an insider's perspective of the people and/or behavior being observed. The main advantage of participant observation research is that it provides detailed qualitative information about the behavior, norms,

Norah posed as Ned and revealed the "man's" world of bowling, strip clubs, and a men's group meeting (from Norah Vincent, 2006. *Self-Made Man.* New York: Viking Penguin).

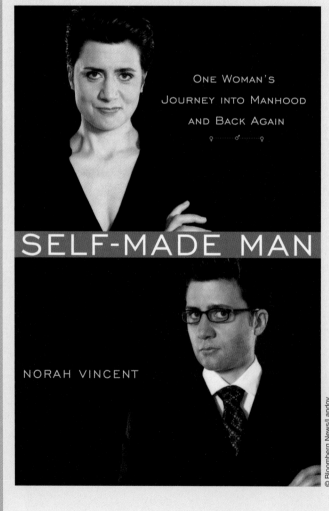

ONE WOMAN'S JOURNEY INTO MANHOOD AND BACK AGAIN ♀ ⎯ ♂ ⎯ ♀

SELF-MADE MAN

NORAH VINCENT

© Bloomberg News/Landov

values, rituals, beliefs, and emotions of those being studied. However, participant observation research is based on small samples, so the findings may not be generalizable. That particular researcher's values and perspectives may also bias the researcher's observations. Norah Vincent admits that her book is not a scientific or objective study; rather, it is "one person's observations about her own experience" (p. 17). She notes, "my observations are full of my own prejudices and preconceptions," and she describes her book as "a travelogue as much as anything else . . . a six-city tour of an entire continent, a woman's-eye view of one guy's approximated life, not an authoritative guide to the whole vast and variegated terrain of manhood in America" (p. 17).

To participate in and observe the world of men, Norah sought the help of a make-up artist to transform her into "Ned." She learned how to apply an artificial beard stubble, wore masculine-looking glasses, had her hair cut into a man's hairstyle, wore men's clothing, and stuffed her pants with a prosthetic penis known as a "packable softie." Norah bulked up her upper body muscle through 6 months of weight lifting and increased protein consumption, and she flattened her breasts by wearing a flat-front sports bra two sizes too small. Norah, who already had a deep voice, hired a voice coach to help her speak like a man.

Having achieved a male physical identity, "Ned" spent 18 months exploring what it means socially to be a man by (1) joining a men's bowling team; (2) going to men's strip clubs; (3) dating women; (4) living at a monastery; (5) working at a sales job; and (6) participating in a men's movement group.

Findings and Discussion

Examples of "Ned's" experiences and observations include the following.

Bowling with the Guys

Our evenings together always started out slowly with a few grunted hellos that among women would have been interpreted as rude. This made my female antennae twitch a little. Were they pissed off at me about something? But among these guys no interpretation was necessary. Everything was out and aboveboard, never more, never less, than what was on anyone's mind. If they were pissed at you, you'd know it. These gruff greetings were indicative of nothing so much as fatigue and appropriate male distance. . . . They were coming from long, wearing workdays, usually filled with hard physical labor and the slow, soul-deadening deprecation that comes of being told what to do all day. . . . They didn't have the energy for pretense. (pp. 29–30)

Strip Clubs

"Ned" went to strip clubs to get a glimpse of "a substratum of the male sexual psyche that most women either don't know about, don't want to know about, or both" (p. 65). After weeks of observing, and interacting with various strippers and other patrons, Ned offers the following reflections:

> I'd been inside a part of the male world that most women and even a lot of men never see, and I'd seen it as just another one of the boys. In those places male sexuality felt like something you weren't supposed to feel but did, like something heavy you were carrying around and had nowhere to unload except in the lap of some damaged stranger, and then only for five minutes. . . . It wasn't nearly so simple as men objectifying women. . . . Nobody won . . . nobody was more or less victimized than anyone else. The girls got money. The men got an approximation of sex and flirtation. But in the end everyone was equally debased by the experience. (pp. 90–91)

Ned recounted an interaction he had with Phil, a male acquaintance who went with Ned to a strip club. Phil said, "I go to some of these bars . . . and I say to myself, these girls were somebody's daughter. Somebody put them to bed, somebody kissed them and hugged them and gave them love and now they're in this pit." Ned responded, "Or, maybe someone didn't." "Yeah," Phil nodded, "I've thought about that too." (p. 91).

Dating

As a lesbian in real life, Norah Vincent had dated women. But dating women as Ned provided unique insights into male/female relationships. Ned found his dates at bars and on the Internet.

Ned found that most of the women he dated "were carrying the baggage of previous hurts at the hands of men, which in many cases had prejudiced them unfairly against the male sex" (p. 100). Ned described that as a "man" dating women,

> . . . I often felt attacked, judged, on the defensive. Whereas with the men I met and befriended as Ned there was a presumption of innocence—that is, you're a good guy until you prove otherwise—with women there was quite often a presumption of guilt: you're a cad like every other guy until you prove otherwise. (p. 101)

Ned experienced the difficulty that heterosexual men have in living up to the ideals of women, observing that "women wanted a take-control man; at the same time, they wanted a man who was vulnerable to them, a man who would show his colors and open his doors, someone expressive, intuitive, attuned" (p. 111). Ned noted, " . . . a man is expected to be modern . . . to support feminism . . .

to see and treat women as equals in every respect . . ." but, Ned continued, "he is on the other hand often still expected to be traditional at the same time, to treat a lady like a lady, to lead the way and pick up the check" (p. 112).

A Catholic Monastery

In the weeks Ned spent at a Catholic monastery, he observed men in a context where women and sexual expression were, at least in principle, excluded from the male experience. Even in this context, however, male gender role expectations were evident, as in the following excerpt:

> . . . I made the mistake once of referring to one of the other monks as cute—the kind of thing that women say all the time about sweet elderly gentleman like the one to whom I was referring. . . . But as soon as the offending remark was out of my mouth, Father Jerome pounced on it, sneering. "He's not cute. You don't call other men cute." (p. 144)

On another occasion, Ned was at the dinner table with the monks and he told Father Richard that he looked very good for his age.

> As soon as the remark came out of my mouth, everyone at the table stopped eating mid-forkful and looked at me as if I had three heads. Father Richard . . . said a very suspicious, squint-eyed "thank you," and looked away, clearly embarrassed. But the implication from other quarters was clear: "What the hell's wrong with you, kid? Don't you know that properly socialized males don't behave that way with each other?" (pp. 144–45)

Work

Ned interviewed for several jobs and landed a job as a salesman. One of his co-workers advised him, "You're a man. . . . You gotta pitch like a man." Ned learned that in sales work, women and men had different strategies.

> Girls pitched differently. They flirted. They cajoled and smiled and eased their way into the sales underhandedly, which was exactly how I'd started out trying to do it. I'd tried initially to ask for the sale the way I asked for food in a restaurant as a woman, or the way I asked for help at a gas station—pleadingly. But coming from a man this was off-color. It didn't work. It bred contempt in both men and women. . . . People see weakness in a woman and they want to help. They see weakness in a man and they want to stamp it out. (p. 213)

Men's Movement Group

Ned participated in a men's movement group in which 25 to 30 guys met once a month. He noted that hugging was a central part of the group.

Most men don't tend to share much physical affection with their male friends, so here the guys made a point of hugging each other long and hard at every possible opportunity as a way of offsetting what the world had long deprived them of, and what they in turn had been socialized to disallow themselves. (p. 232)

Based on his experiences in the men's group, Ned arrived at the following conclusion:

. . . these guys were trying to find the love that their fathers had been unable to give them, or possibly the love that the entire culture had conspired to keep men from giving each other. . . . They had a profound need for other men's love. . . . They needed a man's affection and respect, a man's approval, and a man's shared perspective on their feelings. Having a mother's or a woman's love just wasn't and could never be the same. It couldn't fill the hole. (p. 241)

Concluding Thoughts

In the concluding chapter of *Self-Made Man*, Norah explains that the experience of being a guy was not what she had expected.

I had thought that by being a guy I would get to do all the things I didn't get to do as a woman, things I'd always envied about boyhood when I was a child: the perceived freedoms of being unafraid in the world. . . .

But when it actually came to the business of being Ned I rarely felt free at all. (pp. 275–76)

As Ned, Norah found that "somebody is always evaluating your manhood. Whether it's other men, other women, even children. And everybody is always on the lookout for your weakness or your inadequacy" (p. 276). Norah described the male role as a "straightjacket . . . that is no less constrictive than its feminine counterpart. You're not allowed to be a complete human being. Instead you get to be a coached jumble of stoic poses. You get to be what's expected of you" (p. 276).

In conclusion, Norah Vincent's study challenges the notion that white men have all the advantages. "White manhood in America isn't the standard anymore by which women and all other minorities are being measured and found wanting . . . It's just another set of marching orders, another stereotype to inhabit." Although the Woman's Movement has largely liberated U.S. women from the stereotypical traditional female role, Ms. Vincent notes that "men haven't had their movement yet . . . And they're due for it, as are the women who live with, fight with, take care of and love them" (p. 287).

Source

"Self-Made Man: One Woman's Journey into Manhood" by Norah Vincent. Copyright © 2006 by Viking Penguin.

Women still earn about two-thirds of what men earn, even when the level of educational achievement is identical (see Table 2.4).

Women also seem to be invisible in high-paid executive jobs. Of *Fortune* 500 companies, 393 have no women among their top executives. Even in those companies that have women CEOs, only three are among the best paid in those firms (Jones, 2003). As noted on page 55, women are less likely to climb to the rank of full professor than male academicians (Probert, 2005). The reason: they give priority to their children/family life (Aissen and Houvouras, 2006).

With divorce being a close to 45 percent probability for marriages begun in the 2000s, the likelihood of being a widow for 7 or more years, and the almost certain loss of her parenting role midway through her life, a woman without education and employment skills is often left high and dry. As one widowed mother of four said, "The shock of realizing you have children to support and no skills to do it is a worse shock than learning that your husband is dead." In the words of a divorced, 40-year-old mother of three, "If young women think it can't happen to them, they are foolish."

Table 2.4 Women's and Men's Median Income with Similar Education			
	Bachelor's	**Master's**	**Doctoral Degree**
Men	$50,916	$61,698	$73,853
Women	$31,309	$41,334	$53,003

Source: *Statistical Abstract of the United States: 2006*. 125th ed. Washington, D.C.: U.S. Bureau of the Census, Table 685.

Chapter 2 Gender in Relationships

2. Feminization of Poverty. Another reason many women are relegated to a lower income status is the **feminization of poverty.** This term refers to the disproportionate percentage of poverty experienced by women living alone or with their children. Single mothers are particularly associated with poverty.

When head-of-household women are compared with married-couple households, the median income is $26,550 versus $62,281 (*Statistical Abstract of the United States: 2006*, Table 682). The process is cyclical—poverty contributes to teenage pregnancy, because teens have limited supervision and few alternatives to parenthood.

Such early childbearing interferes with educational advancement and restricts their earning capacity, which keeps them in poverty. Their offspring are born into poverty, and the cycle begins anew.

Even if they get a job, women tend to be employed fewer hours than men, and even when they work full-time, they earn less money. Not only is discrimination in the labor force operating against women, but women usually make their families a priority over their employment, which translates into less income. Such prioritization is based on the patriarchal family, which ensures that women stay economically dependent on men and are relegated to domestic roles. Such dependence limits the choices of many women.

Low pay for women is also related to the fact that they tend to work in occupations that pay relatively low incomes. Indeed, women's lack of economic power stems from the relative dispensability of women's labor (it is easy to replace) and how work is organized (men control positions of power) (Wermuth and Ma'At-Ka-Re Monges, 2002). Women also live longer than men, and poverty is associated with being elderly (Lipsitz, 2005).

When women move into certain occupations, such as teaching, there is a tendency in the marketplace to segregate these occupations from men's, and the result is a concentration of women in lower-paid occupations. The salaries of women in these occupational roles increase at slower rates. For example, salaries in the elementary and secondary teaching profession, which is predominately female, have not kept pace with inflation.

Conflict theorists assert that men are in more powerful roles than women and use this power to dictate incomes and salaries of women and "female professions." Functionalists also note that keeping salaries low for women keeps women dependent and in child-care roles so as to keep equilibrium in the family. Hence, for both conflict and structural reasons, poverty is primarily a feminine issue. One of the consequences of being a woman is to have an increased chance of feeling economic strain throughout life.

I make the most of all that comes and the least of all that goes.

Sara Teasdale, poet

3. Higher Risk for Sexually Transmitted Infections. Gender roles influence a woman's vulnerability to sexually transmitted infection and HIV infection, not only because women receive more bodily fluids from men, who have a greater number of partners (and are therefore more likely to be infected) but also because some women feel limited power to influence their partners to wear condoms. Moreover, when a woman possesses a condom, she is viewed as sexually willing, a fact that could undermine any claim of sexual assault (Hynie et al., 2003). In developing countries, other negative health outcomes of being socialized as a woman are evident—unsafe abortion, maternal mortality, depression, and psychosomatic symptoms. The progress toward relieving these issues is "uneven and slow" (Murphy, 2003).

4. Negative Body Image. There are more than 3800 beauty pageants annually. The effect for many women who do not match the cultural ideal is to have a negative body image. Such a negative view of one's body begins early. A team of researchers studied 136 girls aged 11 to 16 years who had exposure to trim and average-size models via magazines and noted their negative body images (Clay et al., 2005). The researchers noted the need for early education to help young girls deconstruct advertising and body images.

Female Genital Alteration

Worldwide, 140 million girls and women have undergone some form of female genital alteration (FGA, also referred to as female genital cutting, female genital mutilation, female genital operation, and female circumcision). Roughly 90 percent of all female genital alterations take place in Egypt, Somalia, and Mali (Wallis, 2005). Although FGAs occur primarily in African and some Middle Eastern and Asian countries, the practice also occurs in the United States among immigrant families who bring their cultural traditions with them.

Although FGA has been a federal crime in the United States since 1997 (Nour, 2005), each year, about 7000 immigrants to the United States (from countries that practice various forms of female genital operations) undergo the procedure either in the United States or during a visit to their homeland (Bosch, 2001). The reason for the practice is cultural: parents believe that female circumcision makes their daughters marketable for marriage. Many daughters view it as a rite of passage, which improves their chance for marriage. To be circumcised as a female is to be marriageable.

Changing a country's deeply held beliefs and values concerning this practice cannot be achieved by denigration. More effective approaches to discouraging the practice include the following (James and Robertson, 2002; Yount, 2002):

1. Respect the beliefs and values of countries that practice female genital operations. Calling the practice "genital mutilation" and "a barbaric practice" and referring to it as a form of "child abuse" and "torture" convey disregard for the beliefs and values of the cultures where it is practiced. In essence, we might adopt a culturally relativistic point of view (without moral acceptance of the practice). However, Kennedy (2002) traveled extensively in Africa and emphasized that the term *female circumcision* should not be used as the equivalent of female genital operations in general, or infibulation (stitching together of the labia majora so as to leave a tiny hole for menses and urine) specifically. Female genital operations, she noted, are not equivalent to the removal of foreskin. FGAs can cause recurrent urinary tract infection, painful sexual intercourse and menstruation, and difficulties with childbirth (Wallis, 2005).

2. Remember that genital operations are arranged and paid for by loving parents who deeply believe that the surgeries are for their daughters' welfare.

3. It is important to be culturally sensitive to the meaning of being a woman. Indeed, genital cutting is mixed up with how a woman sees herself; thus Westerners are becoming involved in her identity when attempting to alter long-held historical practices.

4. Raising the access of a female to education is related to a decline in female genital alteration. Indeed, educated women more often "oppose circumcision" and give prevention of sexual satisfaction as their reason (Yount, 2002, p. 352).

Your Opinion?

1. To what degree do you feel the United States should become involved in the practice of female genital alterations of U.S. citizens?

2. To what degree can you regard the practice from the view of traditional parents and daughters?

3. How could not having the operation be a liability and a benefit for the Middle Eastern woman whose culture supports the practice?

Sources

Bosch, X. 2001. Female genital mutilation in developed countries. *The Lancet* 358:1177–82.

James, S. M. and C. C. Robertson, eds. 2002. Genital cutting and transnational sisterhood: Disputing U.S. polemics. Urbana, IL: University of Illinois Press.

Kennedy, M. 2002. Presentation on infibulation. Department of Sociology, East Carolina University, Greenville, N.C., January.

Nour, N. M. 2005. Female genital cutting. *Family Practice News* 35:12–13.

Wallis, L. 2005. When rites are wrong: Many women suffer long-term consequences from female genital mutilation. *Nursing Standard* 20:24–27.

Yount, K. M. 2002. Like mother, like daughter? Female genital cutting in Minia, Egypt. *Journal of Health and Social Behavior* 43:336–58.

Women also live in a society that devalues them in a larger sense. Their lives and experiences are not taken as seriously. **Sexism** is defined as an attitude, action, or institutional structure that subordinates or discriminates against an individual or group because of their sex. Sexism against women reflects the tradition of male dominance and presumed male superiority in our society. It is reflected in the fact that women are rarely found in power positions in our society. In the 108th Congress, of the 435 members of the House of Representatives, only 62 are women. There are 14 women and 86 men serving as senators.

5. Less Marital Satisfaction? Although there is disagreement over whether wives are less happy than husbands (Kurdek, 2005), Corra et al. (2006) analyzed General Social Survey data over a 30-year period (1972–2002), controlled for socioeconomic factors such as income and education, and found that women reported less marital satisfaction than men. Similarly, twice as many husbands as wives among 105 late-life couples (average age, 69 years) reported that they had "no disappointments in the marriage" (15 percent vs. 7 percent), suggesting greater dissatisfaction among wives (Henry et al., 2005). The lower marital satisfaction of wives is attributed to power differentials in the marriage. Traditional husbands expect to be dominant, which translates into their earning an income and the expectation that the wife not only will earn an income but also will take care of the house and children. The latter expectation results in a feeling of unfairness. Analysis of other large national samples has yielded the same finding of lower marital satisfaction among wives (Faulkner et al., 2005).

Before leaving this section on negative consequences of being socialized as a woman, look on page 60 at the issue of **female genital alteration** in this chapter's Social Policy feature. This is more of an issue for females born in some African, Middle Eastern, and Asian countries than for women in the United States. But the practice continues even here.

Positive Consequences of Traditional Female Role Socialization   We have discussed the negative consequences of being born and socialized as a woman. But there are also decided benefits.

National Data

Females born in the year 2005 are expected to live to the age of 81, in contrast to men, who are expected to live to the age of 75 (*Statistical Abstract of the United States: 2006*, Table 96).

1. Longer Life Expectancy. Women have a longer life expectancy than men. Greater life expectancy may be related more to biological than to social factors.

2. Stronger Relationship Focus. Crossley and Langdridge (2005) compared women and men on sources of happiness and found that women ranked "helping others," having a "close family," and being "loved by loved ones" significantly higher than men. Previous research supports the idea that women value close friends and relatives more than men do (Abowitz and Knox, 2003).

3. Keep Relationships on Track. Since women evidence more concern for relationships, they are more likely to be motivated to keep them on track and to initiate conversation when there is a problem. In a study of 203 undergraduates, two-thirds of the women, in contrast to 60 percent of the men, reported that they were likely to start a discussion about a problem in their relationship (Knox et al., 1998).

4. Bonding with Children. Another advantage of being socialized as a woman is the potential to have a closer bond with children. In general, women tend to be more emotionally bonded with their children than men do. Although the new cultural image of the father is of one who engages emotionally with his children, many fathers continue to be content for their wives to take care of their children, with the result that mothers, not fathers, become more bonded with their children.

As you become more clear about who you really are, you'll be better able to decide what is best for you—the first time around.

Oprah Winfrey

Consequences of Traditional Male Role Socialization

Male socialization in our society is associated with its own set of consequences. Both the negative and positive consequences are summarized in Table 2.5. As with women, each consequence may or may not be true for a specific man.

Table 2.5 Consequences of Traditional Male Role Socialization

Negative Consequences	Positive Consequences
Identity tied to work role	Higher income, occupational status
Limited emotionality	More positive self-concept
Fear of intimacy, more lonely	Less job discrimination
Disadvantaged in getting custody	Freedom of movement, more partners to select from, more normative to initiate relationships
Shorter life	Happier marriage

We need a new definition of masculinity in this new century: a definition that is more about the character of men's hearts and the depths of their souls than about the size of their biceps, wallets, or penises.

Michael S. Kimmel, Manhood in America

Negative Consequences of Traditional Male Role Socialization   There are several negative consequences associated with being socialized as a man in U.S. society.

National Data

Seventy-three percent of men, compared with 59 percent of women, were in the civilian work force in 2004 (*Statistical Abstract of the United States: 2006*, Table 578).

1. Identity Synonymous with Occupation. Ask men who they are, and many will tell you what they do. Society tends to equate a man's identity with his occupational role. Male socialization toward greater involvement in the labor force is evident in governmental statistics.

Maume (2006) analyzed national data on taking vacation time and found that men were much less likely to do so. They cited fear that doing so would affect their job/career performance evaluation. Women, on the other hand, were much more likely to use all of their vacation time. However, the "work equals identity" equation for men may be changing. Increasingly, there are more stay-at-home dads and more fathers seeking full custody in divorce litigation. These changes challenge cultural notions of masculinity.

That men work more and play less may translate into fewer friendships/relationships. In a study of 377 university students, over a quarter (25.9%) of the men compared to 16.7% of the women reported feeling a "deep sense of loneliness" (Vail-Smith et al., 2006). Similarly, Grief (2006) reported that a quarter of 386 adult men reported that they did not have enough friends. Grief also suggested some possible reasons for having few friends—homophobia, lack of role models, fear of being vulnerable, and competition between men. McPherson et al. (2006) also found that men reported fewer confidentes than women.

2. Limited Expression of Emotions. Some men feel caught between society's expectations that they be competitive, aggressive, and unemotional and their own desire to be more cooperative, passive, and emotional. Indeed, men are pressured to disavow any expression that could be interpreted as feminine (e.g., be emotional). Cordova et al. (2005) studied a sample of husbands and wives and confirmed that the men were less able to express their emotions than the women. Notice that men are repeatedly told to "prove their manhood" (which implies not being emotional), whereas there is no dictum in our culture for women "to 'prove their womanhood'—the phrase itself sounds ridiculous" (Kimmel, 2001, p. 33). Indeed, today men are encouraged to shed their traditional masculinity, with the result that they will "live longer, happier, and healthier lives, lives characterized by close and caring relationships with children, with women, and with other men" (Kimmel, 2006, p. 187).

3. Fear of Intimacy. Men may be socialized to withhold information about themselves that encourages the development of intimacy. Giordano et al., (2005) analyzed data from the National Longitudinal Study of Adolescent Health, consisting of more than 9000 interviews, and found that adolescent boys reported less willingness to disclose than adolescent girls.

4. Custody Disadvantages. Courts are sometimes biased against divorced men who want custody of their children. Because divorced fathers are typically regarded as career-focused and uninvolved in child care, some are relegated to seeing their children on a limited basis, such as every other weekend or four evenings a month.

5. Shorter Life Expectancy. Men typically die 6 years sooner (at age 75) than women (*Statistical Abstract of the United States: 2006,* Table 96). Although biological factors may play a role in greater longevity for women than for men, traditional gender roles play a major part. For example, the traditional male role emphasizes achievement, competition, and suppression of feelings, all of which may produce stress. Not only is stress itself harmful to physical health, but it may lead to compensatory behaviors such as smoking, alcohol and other drug abuse, and dangerous risk-taking behavior.

In sum, the traditional male gender role is hazardous to men's physical health. However, as women have begun to experience many of the same stresses and behaviors as men, their susceptibility to stress-related diseases has increased. For example, since the 1950s, male smoking has declined while female smoking has increased, resulting in an increased incidence of lung cancer in women.

Benefits of Traditional Male Socialization As a result of higher status and power in society, men tend to have a more positive self-concept and greater confidence in themselves. They also enjoy higher incomes and an easier climb up the good-old-boy corporate ladder; they are rarely stalked or targets of sexual harassment. Other benefits are the following.

1. Freedom of Movement. Men typically have no fear of going anywhere, any time. Their freedom of movement is unlimited. Unlike women, who are taught to fear rape and to be aware of their surroundings, walk in well-lit places, and not walk alone after dark, men are oblivious to these fears and perceptions. They can be alone in public and be anxiety-free about something ominous happening to them.

2. Greater Available Pool of Potential Partners. Because of the mating gradient, whereby men tend to pair with younger women, the pool of eligible women for a 35-year-old man includes women who are in their 20s as well as those in their 40s. In contrast, the pool of eligible men for 35-year-old women is more likely to be men her age and older.

3. Norm of Initiating a Relationship. Men are advantaged because traditional norms allow men to be aggressive in initiating relationships with women. In contrast, women are less often aggressive in initiating a relationship. In a study of 1027 undergraduates, 61.1 percent of the female respondents reported that they had not "asked a guy to go out with me." (Knox and Zusman, 2006).

We have been discussing the respective ways in which women and men are affected by traditional gender role socialization. Table 2.6 on page 64 summarizes 12 implications that traditional gender role socialization has for the relationships of women and men.

Changing Gender Roles

Imagine a society in which women and men each develop characteristics, lifestyles, and values that are independent of gender role stereotypes. Characteristics such as strength, independence, logical thinking, and aggressiveness are no longer associated with maleness, just as passivity, dependence, showing emotions, intuitiveness, and nurturing are no longer associated with femaleness. Both sexes are considered equal, and women and men may pursue the

Therapists report that the most common complaint of women in distressed marriages is that their husbands are too withdrawn and don't share openly enough.

Howard Markman, marriage therapist

Table 2.6 Effects of Gender Role Socialization on Relationship Choices

Women

1. A woman who is not socialized to pursue advanced education (which often translates into less income) may feel pressure to stay in an unhappy relationship with someone on whom she is economically dependent.

2. Women who are socialized to play a passive role and not initiate relationships are limiting interactions that could develop into valued relationships.

3. Women who are socialized to accept that they are less valuable and important than men are less likely to seek or achieve egalitarian relationships with men.

4. Women who internalize society's standards of beauty and view their worth in terms of their age and appearance are likely to feel bad about themselves as they age. Their negative self-concept, more than their age or appearance, may interfere with their relationships.

5. Women who are socialized to accept that they are solely responsible for taking care of their parents, children, and husband are likely to experience role overload. Potentially, this could result in feelings of resentment in their relationships.

6. Women who are socialized to emphasize the importance of relationships in their lives will continue to seek relationships that are emotionally satisfying.

Men

1. Men who are socialized to define themselves more in terms of their occupational success and income and less in terms of positive individual qualities leave their self-esteem and masculinity vulnerable should they become unemployed or work in a low-status job.

2. Men who are socialized to restrict their experience and expression of emotions are denied the opportunity to discover the rewards of emotional interpersonal sharing.

3. Men who are socialized to believe it is not their role to participate in domestic activities (child-rearing, food preparation, house-cleaning) will not develop competencies in these life skills. Potential partners often view domestic skills as desirable qualities.

4. Heterosexual men who focus on cultural definitions of female beauty overlook potential partners who might not fit the cultural beauty ideal but who would nevertheless be good life companions.

5. Men who are socialized to view women who initiate relationships in negative ways are restricted in their relationship opportunities.

6. Men who are socialized to be in control of relationship encounters may alienate their partners, who may desire equal influence in relationships.

If there were a ritual dance of the androgyne, T'ai Chi as performed by this master could be that dance. It is neither a masculine nor a feminine dance. It has the strength and grace of both.

June Singer

same occupational, political, and domestic roles. Some gender scholars have suggested that persons in such a society would be neither feminine nor masculine but would be described as androgynous. The next subsections discuss androgyny, gender role transcendence, and gender postmodernism.

Androgyny

Androgyny typically refers to being neither male nor female but a blend of both traits. Two forms of androgyny are described here.

1. Physiological androgyny: refers to intersexed individuals, discussed earlier in the chapter. The genitals are neither clearly male nor female, and there is a mixing of "female" and "male" chromosomes and hormones.

2. Behavioral androgyny refers to the blending or reversal of traditional male and female behavior, so that a biological male may be very passive, gentle, and nurturing and a biological female may be very assertive, rough, and selfish. When looking at an androgynous person it may be difficult to identify the individual as male or female.

Androgyny may also imply flexibility of traits; for example, an androgynous individual may be emotional in one situation, logical in another, assertive in another, and so forth. Ward (2001) classified 311 (159 male/152 female) undergraduates at the National University of Singapore as androgynous (33.8% men and 16.0% women), feminine (11.0% men and 39.6% women), masculine

(35.7% men and 13.9% women), and undifferentiated (19.5% men and 30.6% women). Peters (2005) emphasized that the blending of genders is inevitable.

Cheng (2005) found that androgynous individuals have a broad coping repertoire and are much more able to cope with stress. As evidence, Moore et al. (2005) found that androgynous individuals with Parkinson's disease not only were better able to cope with their disease but also reported having a better quality of life than those with the same disease who expressed the characteristics of one gender only. Similarly, androgynous individuals reported much less likelihood of having an eating disorder (Hepp et al., 2005).

Woodhill and Samuels (2003) emphasized the need to differentiate between positive and negative androgyny. **Positive androgyny** is devoid of the negative traits associated with masculinity (aggression, hardheartedness, indifference, selfishness, showing off, and vindictiveness). Antisocial behavior has also been associated with masculinity (Ma, 2005). Negative aspects of femininity include being passive, submissive, temperamental, and fragile. The researchers also found that positive androgyny is associated with psychological health and well-being.

This androgynous person has physical features of both a woman and a man.

Authors

Gender Role Transcendence

Beyond the concept of androgyny is that of gender role transcendence. We associate many aspects of our world, including colors, foods, social/occupational roles, and personality traits, with either masculinity or femininity. The concept of **gender role transcendence** involves abandoning gender schema (i.e., becoming "gender aschematic" [Bem, 1983]) so that personality traits, social/occupational roles, and other aspects of our lives become divorced from gender categories. But such transcendence is not equal for women and men. Although females are becoming more masculine, in part because our society values whatever is masculine, men are not becoming more feminine. Indeed, adolescent boys may be described as very gender-entrenched.

Beyond gender role transcendence is gender postmodernism.

Gender Postmodernism

Mirchandani (2005) emphasized that empirical postmodernism can help us see into the future. Such a view would abandon the notion that the genders are natural and focus on the social construction of individuals in a gender-fluid society. Monro (2000) previously noted that people would no longer be categorized as male or female but be recognized as capable of many identities—"a third sex" (p. 37). A new conceptualization of "trans" people calls for new social structures "based on the principles of equality, diversity and the right to self determination" (p. 42). No longer would our society telegraph transphobia but embrace pluralization "as an indication of social evolution, allowing greater choice and means of self-expression concerning gender" (p. 42).

No longer is the female destined solely for the home and the rearing of the family nor the male for the marketplace and the world of ideas.

William J. Brennan, Supreme Court Justice

What are the important terms related to gender?

Sex refers to the biological distinction between females and males. One's biological sex is identified on the basis of one's chromosomes, gonads, hormones, internal sex organs, and external genitals and exists on a continuum rather than being a dichotomy. *Gender* refers to the social and psychological characteristics often associated with being female or male. Other terms related to gender include *gender identity* (one's self-concept as a girl or boy), *gender role* (social norms of what a girl or boy "should" do), *gender role ideology* (how women and men "should" interact), *transgendered* (expressing characteristics different from one's biological sex), and *transgenderist* (person who lives in a role other than his or her biological sex).

What theories explain gender role development?

Biosocial theory emphasizes that social behaviors (e.g., gender roles) are biologically based and have an evolutionary survival function. Women stayed in the nest or gathered food nearby, while men could go afar to find food. Such a conceptualization focuses on the division of labor between women and men as functional for the survival of the species. Social learning theory emphasizes the roles of reward and punishment in explaining how a child learns gender role behavior. Identification theory says that children acquire the characteristics and behaviors of their same-sex parent through a process of identification. Boys identify with their fathers; girls identify with their mothers. Cognitive-developmental theory emphasizes biological readiness, in terms of cognitive development, of the child's responses to gender cues in the environment. Once children learn the concept of gender permanence, they seek to become competent and proper members of their gender group.

What are the various agents of socialization?

Various socialization influences include parents/siblings (representing different races and ethnicities), peers, religion, the economy, education, and mass media. These shape the individual toward various gender roles and influence what a person thinks, feels, and does in his or her role as a woman or a man.

How are gender roles expressed in other societies?

The conditions imposed upon women and their children (what happens to women affects their children) in Afghanistan under the Taliban were cruel, demeaning, and often fatal. Some women drank household bleach rather than continue to endure their plight. They were not allowed to go to school or work and thus were completely dependent economically. Some women of the Taliban do not regard their role negatively, however.

The natives of the Caribbean represent over 30 million people, with a majority being of African ancestry. Their family patterns are diverse but are often characterized by women and their children as the primary family unit, with men often not living in the home (at a rate as low as 29% in St. Kitts and as high as 55% in Jamaica). Men may have children with different women and be psychologically and physically absent from their children's lives. Women view motherhood, not marriage, as the symbol of their womanhood.

Greek families reflect predictable roles for the father, mother, children, and married couple. The father is regarded as the head of the household, economic provider, and disciplinarian of the children. The mother takes care of the children, is the go-between for them and the father, and supervises them carefully. In contrast, Swedish families are considerably more egalitarian. Swedish women hold about one-quarter of the seats in the Swedish Parliament. Gender roles in Africa reflect a diverse continent with more than 50 nations. The cultures range from the Islamic/Arab cultures of Northern Africa to industrial and European influences in South Africa. In some parts of East Africa (e.g., Kenya), gender roles are in flux.

What are the consequences of traditional gender role socialization?

Traditional female role socialization may result in negative outcomes such as less education, less income, negative body image, and lower marital satisfaction but positive outcomes such as a longer life, a stronger relationship focus, keeping relationships on track, and a closer emotional bond with children. Traditional male role socialization may result in the fusion of self and occupation, a more limited expression of emotion, disadvantages in child custody disputes, and a shorter life but higher income, greater freedom of movement, a greater available pool of potential partners, and greater acceptance in initiating relationships.

How are gender roles changing?

Androgyny refers to a blend of traits that are stereotypically associated with both masculinity and femininity. It may also imply flexibility of traits; for example, an androgynous individual may be emotional in one situation, logical in another, assertive in another, and so forth. The concept of gender role transcendence involves abandoning gender schema (i.e., becoming "gender aschematic") so that personality traits, social/occupational roles, and other aspects of our lives become divorced from gender categories. But such transcendence is not equal for women and men. Although females are becoming more masculine, in part because our society values whatever is masculine, men are not becoming more feminine.

Indeed, adolescent boys may be described as very gender-entrenched. A new era of gender postmodernism would involve a dissolution of male and female categories as currently conceptualized in Western capitalist society. In essence, people would no longer be categorized as male or female but be recognized as capable of many identities—"a third sex."

KEY TERMS

androgyny	gender role ideology	positive androgyny	transgenderism
cross-dresser	gender role transcendence	sex	transgenderist
female genital alteration	gender roles	sex roles	transsexual
feminization of poverty	hermaphrodites	sexism	transvestite
gender	intersexed individuals	socialization	
gender dysphoria	occupational sex segregation	sociobiology	
gender identity	parental investment	transgender	

The Companion Website for *Choices in Relationships: An Introduction to Marriage and the Family*, Ninth Edition
www.thomson.edu/sociology/knox

Supplement your review of this chapter by going to the companion website to take one of the Tutorial Quizzes, use the flash cards to master key terms, and check out the many other study aids you'll find there. You'll also find special features such as the Marriage and Family Resource Center, Census 2000 information, and other data and resources at your fingertips to help you with that special project or to do some research on your own.

Androgyny

http://www.lilithgallery.com/feminist/males_crying.html

Equal Employment Opportunity Commission

http://www.eeoc.gov/

Men's Studies Association

http://www.mensstudies.org/

National Organization for Women (NOW)

http://www.now.org/

Transgender Forum

http://www.tgforum.com/

REFERENCES

Abowitz, D. A. and D. Knox. 2003. College student life goals: Gender, gender ideology, and the effects of Greek status. Paper, 73rd Annual Meeting of the Eastern Sociological Society, Philadelphia, February 28.

Aissen, K. and S. Houvouras. 2006. Family first: Negotiating motherhood and doctoral studies. Southern Sociological Society, New Orleans, March 24.

American Council on Education and University of California. 2005-2006. The American freshman: National norms for fall, 2005. Los Angeles: Higher Education Research Institute.

Baca Zinn, M. and A. Y. H. Pok. 2002. Tradition and transition in Mexican-origin families. In *Minority families in the United States: A Multicultural perspective*, edited by Ronald L. Taylor. Upper Saddle River, NJ: Prentice-Hall, 79–100.

Bacik, I. and E. Drew. 2006. Struggling with juggling: Gender and work/life balance in the legal professions. *Women's Studies in International Forum* 29: 136–146

Beemyn, B., B. Curtis, M. Davis, and N. J. Tubbs. 2005. Transgender issues on college campuses. *New Directions for Student Services,* 111:49–60.

Bem, S. L. 1983. Gender schema theory and its implications for child development: Raising gender- aschematic children in a gender-schematic society. *Signs* 8:596–616.

Blair, K. D. 2002. School social work, the transmission of culture, and gender roles in schools. *School Social Work* 24:21–33.

Buss, D. M. 1989. Sex differences in human mate preferences: Evolutionary hypotheses tested in 37 cultures. *Behavioral and Brain Sciences* 12:1–13.

Cheng, C. 2005. Processes underlying gender-role flexibility: Do androgynous individuals know more or know how to cope? *Journal of Personality*, 73:645–74.

Clay, D., V. L. Vignoles, and H. Dittmar. 2005. Body image and self-esteem among adolescent girls: Testing the influence of sociocultural factors. *Journal of Research on Adolescence* 15:451–77.

Cohen-Kettenis, P. T. 2005. Gender change in 46,XY persons with 5[alpha]-reductase-2 deficiency and 17[beta]-hydroxysteroid dehydrogenase-3 deficiency. *Archives of Sexual Behavior* 34:399–411.

Colapinto, J. 2000. *As nature made him: The boy who was raised as a girl.* New York: Harper Collins.

Consolatore, D. 2002. What next for the women of Afghanistan? *The Humanist* 62:10–15.

Cordova, J. V., C. B. Gee, and L. Z. Warren. 2005. Emotional skillfulness in marriage: Intimacy as a mediator of the relationship between emotional skillfulness and marital satisfaction. *Journal of Social & Clinical Psychology* 24:218–35.

Corra, M., J. S. Carter, and D. Knox. 2006. Marital happiness by sex and race: A second look. Paper, Annual Meeting of the American Sociological Association, New York, August.

Crossley, A. and D. Langdridge. 2005. Perceived sources of happiness: A network analysis. *Journal of Happiness* 6:107–35.

Cullen, L. T. 2003. I want your job, lady! *Time.* 12 May, 52 et passim.

Denton, M. L. 2004. Gender and marital decision making: Negotiating religious ideology and practice. *Social Forces* 82:1151–80

Diamond, D. A. 2003. Breaking down the barricades: The admission of women at Virginia Military Institute and the United States Military Academy at West Point. Paper, Annual Meeting of the Eastern Sociological Society, Philadelphia, February.

Faulkner, R., A. M. Davey, and A. Davey. 2005. Gender-related predictors of change in marital satisfaction and marital conflict. *American Journal of Family Therapy* 33:61–83.

Georgas, J., T. Bafiti, K. Mylonas, and L. Papademou. 2005. Families in Greece. In *Families in global perspective,* edited by J. L. Roopnaraine and U. P. Gielen. Boston: Pearson, Allyn & Bacon, 207–24.

Giordano, P. C., W. D. Manning, and M. A. Longmore. 2005. The romantic relationships of African-American and white adolescents. *The Sociological Quarterly* 46:545–68.

Greif, G. L. 2006. Male friendships: Implications from research for family therapy *Family Therapy, 33,* 1–15.

Haynie, D. L. and D. W. Osgood. 2005. Reconsidering peers and delinquency: How do peers matters? *Social Forces* 84:1109–30.

Heckert, T. M., H. E. Dorste, G. W. Farmer, P. J. Adams, J. C. Bradley, and B. M. Bonness. 2002. Effect of gender and work experience on importance of job characteristics when considering job offers. *College Student Journal* 36:344–51.

Henry, R. G., R. B. Miller, and R. Giarrusso. 2005. Difficulties, disagreements, and disappointments in late-life marriages. *International Journal of Aging & Human Development* 61:243–65.

Hepp, U., A. Spindler, and G. Milos. 2005. Eating disorder symptomatology and gender role orientation. *International Journal of Eating Disorders* 37:227–33.

Hout, M. and C. S. Fischer. 2002. Why more Americans have no religious preference: Politics and generations. *American Sociological Review* 67:165–90.

Hynie, M., R. A. Schuller, and L. Couperthwaite. 2003. Perceptions of sexual intent: The impact of condom possession. *Psychology of Women Quarterly* 27:75–79.

Irvolino, A. C., M. Hines, S. E. Golombok, J. Rust, and R. Plomin. 2005. Genetic and environmental influences on sex-typed behavior during the preschool years. *Child Development* 76:826–40.Jones, D. 2003. Few women hold top executive jobs, even when CEO's are female. *USA Today* 27 January, B1.

Kenagy, G. P. and C. M. Hsieh. 2005. The risk less known: female-to-male transgender persons' vulnerability to HIV infection. *AIDS Care* 17:195–207.

Kennedy, M. 2000. Gender role observations of East Africa. Written exclusively for this text.

Kimmel, M. S. 2006. *Manhood in America: A cultural history*. New York: Oxford University Press.

Kimmel, M. S. 2001. Masculinity as homophobia: Fear, shame, and silence in the construction of gender identity. In *Men and masculinity: A text reader*, edited by T. F. Cohen. Belmont, CA: Wadsworth, 29–41.

Knox, D. and Zusman, M. E. 2006. Relationship and sexual behaviors of a sample of 1027 university students. Unpublished data collected for this text. Department of Sociology, East Carolina University, Greenville, NC.

Knox, D., M. E. Zusman, and H. R. Thompson. 2004. Emotional perceptions of self and others: Stereotypes and data. *College Student Journal* 38:130–42.

Knox, D., S. Hatfield, and M. E. Zusman. 1998. College student discussion of relationship problems. *College Student Journal* 32:19–21.

Kohlberg, L. 1966. A cognitive-developmental analysis of children's sex-role concepts and attitudes. In *The development of sex differences*, edited by E. E. Macoby. Stanford, CA: Stanford University Press.

———. 1969. State and sequence: The cognitive developmental approach to socialization. In *Handbook of socialization theory and research*, edited by D. A. Goslin. Chicago: Rand McNally, 347– 480.

Krafchick, J. L., T. S. Zimmerman, S. A. Haddock, and J. H. Banning. 2005. Best-selling books advising parents about gender: A feminist analysis. *Family Relations* 54:84–101.

Kurdek, L. A. 2005. Gender and marital satisfaction early in marriage: A growth curve approach. *Journal of Marriage and the Family* 67:68–95.

Lindsey, E. W., and J. Mize. 2001. Contextual differences in parent-child play: Implications for children's gender role development. *Sex Roles: A Journal of Research* 44:155–76.

Lippa, R. A. 2002. Gender-related traits of heterosexual and homosexual men and women. *Archives of Sexual Behavior* 31:83–98.

Lipsitz, L. A. 2005. The elderly people of post-soviet Ukraine: Medical, social, and economic challenges. *Journal of the American Geriatrics Society* 53:2216–20.

Looy, H. and H. Bouma III. 2005. The nature of gender: gender identity in persons who are intersexed or transgendered. *Journal of Psychology and Theology* 33:166–80.

Lorber, J. 2001. "Night to his day": The social construction of gender. In *Men and masculinity: A text reader*, edited by T. F. Cohen. Belmont, CA: Wadsworth, 19–28.

Ma, M.K. 2005. The relation of gender-role classifications to the prosocial and antisocial behavior of Chinese adolescents. *Journal of Genetic Psychology* 166:189–201.

Martin, K. A. 2003. Giving birth like a girl. *Gender and Society* 17:54–72.

Maume, D. J. 2006. Gender differences in taking vacation time. *Work and Occupations* 33:161–90.

McGinty, K., D. Knox, and M. E. Zusman. 2006. Research report on undergraduate women who prefer a traditional man. Created for this text.

McNeely, A., D. Knox, and M. E. Zusman. 2004. Beliefs about men: Gender differences among college students. Poster, Annual Meeting of the Southern Sociological Society, Atlanta, April 16–17.

McNeely, A., D. Knox, and M.E. Zusman. 2005. College student beliefs about women: Some gender differences. *College Student Journal* 39:769–74.

McPherson, M., L. Smith-Lovin, and M. E. Brashears. 2006. Social isolation in America, 1985-2004. *American Sociological Review* 71: 353–375.

Mead, M. 1935. *Sex and temperament in three primitive societies*. New York: William Morrow.

Meinhold, J. L., A. Acock, and A. Walker. 2006. The influence of life transition statuses on sibling intimacy and contact in early adulthood. Previously presented at the National Council on Family Relations Annual meeting in Orlando in 2005 and submitted for publication.

Meyer-Bahlburg, H. F. L. 2005. Introduction: Gender dysphoria and gender change in persons with intersexuality. *Archives of Sexual Behavior* 34:371–74.

Miller, A. S., and R. Stark 2002. Gender and religiousness: Can socialization explanations be saved? *American Journal of Sociology* 107:1399–423.

Mirchandani, R. 2005. Postmodernism and sociology: From the epistemological to the empirical. *Sociological Theory* 23:86–115.

Moghadam, V. M. 2002. Patriarchy, the Taliban, and the politics of public space in Afghanistan. *Women's Studies International Forum* 25:19–31.

Monro, S. 2000. Theorizing transgender diversity: Towards a social model of health. *Sexual and Relationship Therapy* 15:33–42.

Moore, O., S. Kreitler, M. Ehrenfeld, and N. Giladi. 2005. Quality of life and gender identity in Parkinson's disease. *Journal of Neural Transmission* 112:1511–22.

Murphy, E. M. 2003. Being born female is dangerous to your health. *American Psychologist* 58:205–10.

Peoples, J. G. 2001.The cultural construction of gender and manhood. In *Men and masculinity: A text reader,* edited by T. F. Cohen. Belmont, CA: Wadsworth, 9–18.

Peters, J. K. 2005 Gender remembered: The ghost of "unisex" past, present, and future. *Women's Studies* 34:67–83.

Pollack, W. S. (with Shuster, T.). 2001. *Real boys' voices.* New York: Penguin Books.

Pretorius, E. 2005. Family life in South Africa. In *Families in global perspective,* edited by J. L. Roopnaraine and U. P. Gielen. Boston: Pearson, Allyn & Bacon, 363–80.

Probert, B. 2005. "I Just Couldn't Fit It In": Gender and unequal outcomes in academic careers. *Gender, Work & Organization* 12:50–73.

Ripley, A. 2005. Who says a woman can't be Einstein? *Time* 51–60.

Rodkin, P. C., T. W. Farmer, R. Pearl, and R. Van Acker. 2006. They're cool: Social status and peer group supports for aggressive boys and girls. *Social Development* 15:175–204.

Roopnarine, J. L., P. Bynoe, R. Singh, and R. Simon. 2005. Caribbean families in English-speaking countries. In *Families in global perspective,* edited by J. L. Roopnaraine and U. P. Gielen. Boston: Pearson, Allyn & Bacon, 311–29.

Sax, L. 2005. *Why gender matters.* New York: Doubleday.

Skaine, R. 2002. *The women of Afghanistan under the Taliban.* Jefferson, NC: McFarland.

Sumsion, J. 2005. Male teachers in early childhood education: Issues and case study. *Early Childhood Research Quarterly* 20:109–23.

Statistical Abstract of the United States: 2006. 125th ed. Washington, D.C.: U.S. Bureau of the Census.

Taylor, R. L. 2002. Black American families. In *Minority families in the United States: A multicultural perspective,* edited by Ronald L. Taylor. Upper Saddle River, NJ: Prentice Hall, 19–47.

Vail-Smith, K., D. Knox, and M. Zusman 2006. The lonely college male. *International Journal of Men's Health.* In press.

Ward, C. A. 2001. Models and measurement of psychological androgyny: A cross-cultural extension of theory and research. *Sex Roles: A Journal of Research* 43: 529–552.

Weinber, M. S., I. Lottes, and F. M. Shaver. 2000. Sociocultural correlates of permissive sexual attitudes: A test of Reiss's hypothesis about Sweden and the United States. *Journal of Sex Research* 37:44–52.

Wermuth, L. and M. Ma'At-Ka-Re Monges. 2002. Gender stratification: A structural model for examining case examples of women in less developed countries. *Frontiers* 23:1–22.

Wilcox, W. B. and S. L. Nock. 2006. What's love got to do with it? Equality, equity, commitment and marital quality. *Social Forces* 84:1321–45.

Woodhill, B. M., and C. A. Samuels. 2003. Positive and negative androgyny and their relationship with psychological health and well-being. *Sex Roles* 48:555–65.

Say you love me every waking moment,
turn my head with talk of summertime.

Christy to her young lover, *Phantom of the Opera*

Love in Relationships

Contents

Kelly Lewis

True or False?

1. Most college students report that they would seek a divorce if they fell out of love.

2. The most common love style of college students is that of passion and romance (eros).

3. College students report that they are more likely to make relationship decisions with their heart than their head.

4. Undergraduate women are more likely than men to believe that "jealousy shows love."

5. Heavy women who lose weight are more likely to become involved in a romantic relationship.

Answers: **1.** T **2.** T **3.** T **4.** F **5.** T

. . . sometimes when I write about love I think of a flying squirrel. There's got to be a moment when that squirrel looks from the end of one branch to the tree six feet away and thinks twice about making a leap. Falling in love is no different; it's the moment that we close our eyes and throw away everything that seems reasonable and hope to God there's someone or something waiting to catch us on the other side. (Picoult, 2001, p. 439)

We leap for good reason. Being in love is related to our personal happiness, and having a strong, intimacy-focused relationship is one of our best chances for happiness (Sanderson and Karetsky, 2002). Indeed, falling out of love paves the way for considering divorce. More than two-thirds (67%) of 1027 university students reported that they would divorce someone they did not love (Knox and Zusman, 2006). Love is very much a part of student life. More than half (51.8%) of the sample of 1027 undergraduates reported that they were emotionally involved with one person, engaged, or married (most [46.1] were not engaged or married).

This chapter is concerned with the nature of love (both ancient and modern views), various theories of the origin of love, how love develops in a new relationship, and problems associated with love. Because jealousy in love relationships is common, we examine its causes and consequences.

For love is such a mystery,
I cannot find it out.
For when I think I'm best resolved,
I then am most in doubt.

John O'Brien, *Happy Marriage*

Descriptions of Love

Love is elusive and incapable of being defined by those caught in its spell. Watts and Stenner (2005) wrote that the various definitions of love "struggle to capture our actual experiences of love" (p. 85). Nevertheless, researchers have conceptualized love as a dichotomy (i.e., "romantic or conjugal"), as a vague construct involving relationship dynamics (i.e., "meeting of needs") (Meyers and Shurts, 2002), and as a phenomenon ranging from the mundane to the transcendental (Watts and Stenner, 2005). Love is often confused with lust and infatuation (Jefson, 2006). Love is about deep, abiding feelings, lust is about sexual desire, and infatuation is about emotional feelings based on little actual exposure to the love object.

Chapter 3 Love in Relationships

Romantic versus Realistic Love

Love may also be described as being on a continuum from romanticism to realism. For some people, love is romantic; for others, it is realistic. **Romantic love** is characterized by such beliefs as "there is only one true love," "love at first sight," and "love conquers all." Regarding these beliefs, more than a quarter (26%) of 1027 undergraduates believed that true love comes only once, almost a third (29.8%) reported they had experienced love at first sight, and more than 80 percent (81.7%) agreed that "deep love can get a couple through any difficulty or difference" (Knox and Zusman, 2006).

The symptoms of romantic love include drastic mood swings, palpitations of the heart, and intrusive thoughts about the partner. F. Scott Fitzgerald immortalized the concept of romantic obsession in *The Great Gatsby*. Of Daisy Buchanan he wrote, "She was the first girl I ever loved and I have faithfully avoided seeing her . . . to keep that illusion perfect." He actually was writing about a real-life true love, Ginevra King, whom he had met when she was 16; she eventually married another man (West, 2005).

Whether men or women are more romantic varies by study. Sharp and Ganong (2000) found that men were more likely than women to fall in love quickly. However, Medora et al. (2002) found that American, Turkish, and Indian women tended to be more romantic than men.

In a study of 197 emotional perceptions of college students, women saw themselves (and men agreed) as more romantic than men (Knox et al., 2004). Huston et al. (2001) found that after two years of marriage, the couples that had fallen in love more slowly were just as happy as couples that fell in love at first sight.

Infatuation is sometimes regarded as synonymous with romantic love. **Infatuation** comes from the same root word as *fatuous*, meaning "silly" or "foolish," and refers to a state of passion or attraction that is not based on reason. Infatuation is characterized by the tendency to idealize the love partner. People who are infatuated magnify their lover's positive qualities ("My partner is always happy") and overlook or minimize their negative qualities ("My partner doesn't have a problem with alcohol; he just likes to have a good time").

In contrast to romantic love is realistic love. Realistic love is also known as conjugal love. **Conjugal (married) love** is less emotional, passionate, and exciting than romantic love and is characterized by companionship, calmness, comfort, and security. *Companionate love* is a term often used for conjugal love. Sprecher and Regan (1998) found that companionate love was more satisfying than passionate love. However, individuals may experience both companionate and passionate love for the same partner.

The Love Attitudes Scale (p. 76) offers a way for you to assess the degree to which you tend to be romantic or realistic in your view of love. When you determine your love attitudes score, be aware that your tendency to be a romantic or a realist is neither good nor bad. Both romantics and realists can be happy individuals and successful relationship partners.

This couple reflects conjugal love. They are happily married, have reared two children, and enjoy their lives together.

Authors

Do You Make Relationship Choices with Your Heart or Head?

Lovers are frequently confronted with the need to make decisions about their relationships, but they are divided on whether to let their heart or head rule in such decisions. Some evidence suggests that the heart rules. More than a third (36.6%) of 1027 undergraduates disagreed with the statement "I make relationship decisions more with my head than my heart" (Knox and Zusman, 2006), suggesting that the heart tends to rule in relationship matters. You will notice examples from both women and men that give precedence to the heart. We asked students in our Marriage and Family classes to fill in the details about deciding with their heart or head. Some of their answers follow.

Heart

Those who relied on their heart (women more than men) for making decisions felt that emotions were more important than logic and that listening to your heart made you happier. One woman said:

> In deciding on a mate, my heart would rule because my heart has reasons to cry and my head doesn't. My heart knows what I want, what would make me most happy. My head tells me what is best for me. But I would rather have something that makes me happy than something that is good for me.

Some men also agreed that your heart should rule. One said:

> I went with my heart in a situation, and I'm glad I did. I had been dating a girl for two years when I decided she was not the one I wanted and that my present girlfriend was. My heart was saying to go for the one I loved, but my head was telling me not to because if I broke up with the first girl, it would hurt her, her parents, and my parents. But I decided I had to make myself happy and went with the feelings in my heart and started dating the girl who is now my fiancée.

Relying on one's emotions does not always have a positive outcome, as the following experience illustrates:

> Last semester, I was dating a guy I felt more for than he did for me. Despite that, I wanted to spend any opportunity I could with him when he asked me to go somewhere with him. One day he had no classes, and he asked me to go to the park by the river for a picnic. I had four classes that day and exams in two of them. I let my heart rule and went with him. Nothing ever came of the relationship and I didn't do well in those classes.

Head

In contrast to making relationship decisions with one's heart, more than a fourth (26.5%) of 1027 university students reported that they made such decisions with their head (Knox and Zusman, 2006). Some comments follow from our Marriage and Family students about making relationship decisions rationally.

> In deciding on a mate, I feel my head should rule because you have to choose someone that you can get along with after the new wears off. If you follow your heart solely, you may not look deep enough into a person to see what it is that you really like. Is it just a pretty face or a nice body? Or is it deeper than that, such as common interests and values? After the new wears off, it's the person inside the body that you're going to have to live with. The "heart" sometimes can fog up this picture of the true person and distort reality into a fairy tale.

Another student said:

> Love is blind and can play tricks on you. Two years ago, I fell in love with a man who I later found out was married. Although my heart had learned to love this man, my mind knew the

Half our mistakes in life arise from feeling where we ought to think, and thinking where we out to feel.

John Churton, English literary critic (1848–1908)

consequences and told me to stop seeing him. My heart said, "Maybe he'll leave her for me," but my mind said, "If he cheated on her, he'll cheat on you." I got out and am glad that I listened to my head.

Some individuals feel that both the head and the heart should rule when making relationship decisions.

When you really love someone, your heart rules in most of the situations. But if you don't keep your head in some matters, then you risk losing the love that you feel in your heart. I think that we should find a way to let our heads and hearts work together.

There is an old saying, "Don't wait until you find the person you can live with; wait and find the person that you can't live without!" One individual hearing this quote said, "I think both are important. I want my head to let me know it 'feels' right" (authors' files).

Figure 3.1
Triangular Theory of Love

Triangular View of Love

Sternberg (1986) developed the "triangular" view of love, consisting of three basic elements: intimacy, passion, and commitment. The presence or absence of these three elements creates various types of love experienced between individuals regardless of their sexual orientation.

1. Nonlove—the absence of intimacy, passion, and commitment. Two strangers looking at each other from afar have a nonlove.

2. Liking—intimacy without passion or commitment. A new friendship may be described in these terms of the partners liking each other.

3. Infatuation—passion without intimacy or commitment. Two persons flirting with each other in a bar may be infatuated with each other.

4. Romantic love—intimacy and passion without commitment. Love at first sight reflects this type of love.

5. Companionate love—intimacy and commitment without passion. A couple that has been married for 50 years is said to have companionate love.

6. Fatuous love—passion and commitment without intimacy. Couples who are passionately wild about each other and talk of the future but do not have an intimate connection with each other have a fatuous love.

7. Empty love—commitment without passion or intimacy. A couple who stay together for social and legal reasons but who have no spark or emotional sharing between them have an empty love.

8. Consummate love—combination of intimacy, passion, and commitment; Sternberg's view of the ultimate, all-consuming love.

Individuals bring different combinations of the elements of intimacy, passion, and commitment (the triangle) to the table of love. One lover may bring a predominance of passion, with some intimacy but no commitment (romantic love), while the other person brings commitment but no passion or intimacy (empty love). The triangular theory of love allows lovers to see the degree to which they are matched in terms of passion, intimacy, and commitment in their relationship.

A common class exercise among professors who teach about marriage and the family is to randomly ask class members to identify one word they most closely

Diversity in the United States

Sixty-two percent of Americans buy a valentine. Hallmark researchers found that of the 2000 different valentines for sale each year, the one consistently chosen most in the United States has these words on the front: "For the One I Love." Inside, the words are "Each time I see you, hold you, think of you, here's what I do....I fall deeply, madly, happily in love with you. Happy Valentine's Day" (Sedensky, 2006).

Diversity in Other Countries

We have already noted the importance of romantic love in the United States. But what about other societies? Jankowiak and Fischer (1992) found evidence of passionate love in 147 of 166 (88.5%) of the societies they studied. Passionate love was defined by the presence of at least one of the following: accounts depicting personal anguish and longing, love songs, elopement due to mutual affection, or native accounts affirming the existence of passionate love. The researchers' study stands "in direct contradiction to the popular idea that romantic love is essentially limited to or the product of Western culture. Moreover, it suggests that romantic love constitutes a human universal, or at the least a near-universal" (p. 154).

The Love Attitudes Scale

This scale is designed to assess the degree to which you are romantic or realistic in your attitudes toward love. There are no right or wrong answers.

Directions

After reading each sentence carefully, circle the number that best represents the degree to which you agree or disagree with the sentence.

1	2	3	4	5
Strongly agree	Mildly agree	Undecided	Mildly disagree	Strongly disagree

	SA	MA	U	MD	SD
1. Love doesn't make sense. It just is.	1	2	3	4	5
2. When you fall "head over heels" in love, it's sure to be the real thing.	1	2	3	4	5
3. To be in love with someone you would like to marry but can't is a tragedy.	1	2	3	4	5
4. When love hits, you know it.	1	2	3	4	5
5. Common interests are really unimportant; as long as each of you is truly in love, you will adjust.	1	2	3	4	5
6. It doesn't matter if you marry after you have known your partner for only a short time as long as you know you are in love.	1	2	3	4	5
7. If you are going to love a person, you will "know" after a short time.	1	2	3	4	5
8. As long as two people love each other, the educational differences they have really do not matter.	1	2	3	4	5
9. You can love someone even though you do not like any of that person's friends.	1	2	3	4	5
10. When you are in love, you are usually in a daze.	1	2	3	4	5
11. Love "at first sight" is often the deepest and most enduring type of love.	1	2	3	4	5
12. When you are in love, it really does not matter what your partner does because you will love him or her anyway.	1	2	3	4	5
13. As long as you really love a person, you will be able to solve the problems you have with the person.	1	2	3	4	5
14. Usually you can really love and be happy with only one or two people in the world.	1	2	3	4	5
15. Regardless of other factors, if you truly love another person, that is a good enough reason to marry that person.	1	2	3	4	5
16. It is necessary to be in love with the one you marry to be happy.	1	2	3	4	5
17. Love is more of a feeling than a relationship.	1	2	3	4	5
18. People should not get married unless they are in love.	1	2	3	4	5
19. Most people truly love only once during their lives.	1	2	3	4	5
20. Somewhere there is an ideal mate for most people.	1	2	3	4	5
21. In most cases, you will "know it" when you meet the right partner.	1	2	3	4	5
22. Jealousy usually varies directly with love; that is, the more you are in love, the greater your tendency to become jealous will be.	1	2	3	4	5
23. When you are in love, you are motivated by what you feel rather than by what you think.	1	2	3	4	5
24. Love is best described as an exciting rather than a calm thing.	1	2	3	4	5
25. Most divorces probably result from falling out of love rather than failing to adjust.	1	2	3	4	5
26. When you are in love, your judgment is usually not too clear.	1	2	3	4	5
27. Love comes only once in a lifetime	1	2	3	4	5
28. Love is often a violent and uncontrollable emotion.	1	2	3	4	5
29. When selecting a marriage partner, differences in social class and religion are of small importance compared with love.	1	2	3	4	5
30. No matter what anyone says, love cannot be understood.	1	2	3	4	5

Scoring

Add the numbers you circled. 1 (strongly agree) is the most romantic response and 5 (strongly disagree) is the most realistic response. The lower your total score (30 is the lowest possible score), the more romantic your attitudes toward love. The higher your total score (150 is the highest possible score), the more realistic your attitudes toward love. A score of 90 places you at the midpoint between being an extreme romantic and an extreme realist. Both men and women undergraduates typically score above 90, with men scoring closer to 90 than women.

A team of researchers (Medora et al., 2002) gave the scale to 641 young adults at three international universities in America, Turkey, and India. Female respondents in all three cultures had higher romanticism scores than male respondents (reflecting their higher value for, desire for, and thoughts about marriage). When the scores were compared by culture, American young adults were the most romantic, followed by Turkish students, with Indians having the lowest romanticism scores.

Reference

Medora, N. P., J. H. Larson, N. Hortacsu, and P. Dave. 2002. Perceived attitudes towards romanticism: A cross-cultural study of American, Asian-Indian, and Turkish young adults. *Journal of Comparative Family Studies* 33:155–78.

associate with love. Invariably, students identify different words (commitment, feeling, trust, altruism, etc.), which suggests that there is great variability in the way we think about love. Indeed, just the words "I love you" have different meanings, depending on whether they are said by a man or a woman. In a study of 147 undergraduates (72% female, 28% male), men (more than women) reported that saying "I love you" was a ploy to get the partner to have sex, whereas women (more than men) reported that saying "I love you" was a reflection of their feelings, independent of a specific motive (Brantley et al., 2002).

Love Styles

Theorist John Lee (1973, 1988) identified a number of styles of love that describe the way lovers relate to each other. Keep in mind that the same individual may view love in more than one way at a time or may view love in different ways at different times. These love styles are also independent of one's sexual orientation—no one love style is characteristic of heterosexuals or homosexuals.

1. Ludus. Country-western singer George Strait's song "She'll Leave You with a Smile" reflects involvement with a ludic lover: "You're gonna give her all your heart/ Then she'll tear your world apart. / You're gonna cry a little while/ Still she'll leave you with a smile." The ludic lover views love as a game, refuses to become dependent on any one person, and does not encourage another's intimacy. Two essential skills of the ludic lover are to juggle several partners at the same time and to manage each relationship so that no one partner is seen too often.

These strategies help to ensure that the relationship does not deepen into an all-consuming love. Don Juan represented the classic ludic lover. "Love 'em and leave 'em" is the motto of the ludic lover. Tzeng et al. (2003) found that whereas men were more likely than women to be ludic lovers, ludic love characterized the love style of college students the least.

In a study (Paul et al., 2000) of "hookups" between college students, certain love styles were characteristic of students who hooked up. Distinguishing features of those who had noncoital hookups were a ludic love style and high concern for personal safety. These individuals may have been participating in collegiate cultural expectations by engaging in "playful" sexual exploration but refraining from intercourse out of their concern for personal safety. Indeed, those who engaged in coital hookups were also characterized by ludic love styles, along with symptoms of alcohol intoxication. The researchers worried that the combination of ludic orientation (motivated by the thrill of the game) and alcohol intoxication could be a precursor to sexual experiences that were forced or unwanted by a partner.

The game of love cannot be played with the cards on the table.

Unknown

The **ludic love style** is sometimes characterized as manipulative and noncaring. But ludic lovers may also be compassionate and very protective of another's feelings. For example, some uninvolved soon-to-graduate seniors avoid involvement with anyone new and become ludic lovers so as not to encourage anyone.

2. Pragma. The **pragma love style** is the love of the pragmatic—that which is logical and rational. The pragma lover assesses his or her partner on the basis of assets and liabilities. Economic security may be regarded as very important. The pragma lover does not become involved in interracial, long-distance, or age-discrepant partners, because logic argues against doing so. Bulcroft et al. (2000) noted that, increasingly, individuals are becoming more pragmatic about their love choices.

The main character in the movie *Alfie* was a ludic lover, juggling different women and committing to none of them.

This university couple reflects the eros style of love, which is one of passion and romance.

Authors

3. Eros. Just the opposite of the pragmatic love style, the **eros love style** is one of passion and romance. Intensity of both emotional and sexual feelings dictates one's love involvements. Tzeng et al. (2003) assessed the love styles of more than 700 college students and found that eros was the most common love style of women and men. Hendrick et al. (1988) found that couples who were more romantically and passionately in love were more likely to remain together than couples who avoided intimacy by playing games with each other.

4. Mania. The person with **mania love style** feels intense emotion and sexual passion but is out of control. The person is possessive and dependent and "must have" the beloved. Persons who are extremely jealous and controlling reflect manic love. "If I can't have you, no one else will" is sometimes the mantra of the manic lover. Stalking is an expression of love gone wild (discussed on page 89).

5. Storge. The **storge love style** is a calm, soothing, nonsexual love devoid of intense passion. Respect, friendship, commitment, and familiarity are characteristics that help to define the relationship. The partners care deeply about each other but not in a romantic or lustful sense. Their love is also more likely to endure than fleeting romance. One's grandparents who have been married 50 years and who still love and enjoy each other are likely to have a storge type of love.

6. Agape. One of the forms of love identified by the ancient Greeks, the **agape love style** involves selflessness and giving, without expecting anything in return. These nurturing and caring partners are concerned only about the welfare and growth of each other. The love parents have for their children is often described as agape love. Inman-Amos et al. (1994) observed that the predominant love style of married couples is agape.

International Data

Data from a survey of 641 young adults at three international universities indicated that American young adults are the most romantic, followed by Turkish students, with Indians having the lowest romanticism scores (Medora et al., 2002).

Love in Social and Historical Context

Though we think of love as an individual experience, the society in which we live exercises considerable control over our love object/choice and conceptualizes it in various ways.

Diversity in Other Countries

Love among Chinese couples is of the storge variety. Pimentel (2000) studied a large representative sample of married couples in urban China and found that "Chinese couples have what Westerners might characterize as a relatively unromantic vision of love, more like companionship. The words most often accompanying remarks about love were "respect," "mutual understanding," and "support." Expressions of passion, of "sparks flying," or similar phrases, were not noted.

Social Control of Love

Love may be blind, but it knows what color a person's skin is. Indeed, one might ask, "Is love color-blind, or is love blinded by color?" The data are clear—love seems to see only people of similar color, as more than 95 percent of people marry someone of their own racial background (*Statistical Abstract of the United States: 2006*, Table 54). Hence, parents and peers may approve of

Love in Black and White

RESEARCH APPLICATION

To what degree are African-American and white adolescent experiences with love similar or different?

Sample and Methods

To find out, a team of researchers analyzed data collected during interviews with adolescents in grades 7 through 11 at more than 80 high schools. The sample consisted of 575 African-American girls, 379 African-American boys, 1,528 white girls, and 985 white boys. Each of these respondents reported having a "current" or "recent" relationship. The data are part of Add Health, a longitudinal study of a nationally representative sample of adolescents

Selected Findings and Conclusions

1. *Whites valued romantic love relationships more than African-Americans.* When the respondents were asked, "How much would you like to have a romantic relationship in the next year?" they selected a number on a continuum from 1 to 7 (the higher the number, the greater the importance):

   White respondents, compared with African-American respondents, rated having a romantic relationship in the next year as significantly more important (mean = 3.47 and 3.24, respectively). Interestingly, a similar percentage (36% of whites and 34% of African-Americans) reported current involvement in a love relationship.

2. *Whites engaged in more romantic behaviors than African-Americans.* When "romantic behaviors" were defined as "told other people that we were a couple," "went out together alone," "kissed," "held hands," "gave each other presents," "told each other we loved each other," and "thought of ourselves as a couple," white adolescents, on average, reported significantly more romantic behaviors with the current partner than did African-American teens.

3. *African-Americans were less likely to report involvement in an exclusive relationship.* Although more than 90 percent of both whites and African-Americans reported current involvement in only one relationship, African-American respondents were less likely than white respondents to report exclusive involvement in their current relationship (94% vs. 98%).

4. *African-Americans were less likely to report intimate self-disclosure than whites.* Self-disclosure was identified in terms of telling a partner about a problem. Females, older individuals, and those who had been in a relationship for a considerable amount of time were also more likely to self-disclose than males, younger adolescents, and those who had known each other for only a short period of time.

5. *African-Americans reported longer current relationships and more inclusion of sexual intercourse.* The duration of African-American relationships was about a month longer than those of white respondents. African-Americans also more often reported the inclusion of sexual intercourse in their current relationships, a circumstance which might be related to the fact that sexual involvement increases with relationship duration.

What are the implications of this study? One, love remains an important experience for adolescents, even as young as the seventh grade. Two, while there are statistically "significant" differences in some of the variables studied, the actual experienced differences may be irrelevant. For example, the mean scores of 3.47 and 3.24, respectively, for whites and African-Americans on a scale of 1 to 7 (reflecting the importance of wanting to be involved in a love relationship) may actually reflect more similarities than differences between the races. Other findings above reflect the similarity of the races, rather than the differences (e.g., more than 90 percent of both races reported exclusive involvement in a relationship).

Source

Giordan, P.C., W. D. Manning, and M. A. Longmore. 2005. The romantic relationships of African-American and white adolescents. *The Sociological Quarterly* 46:545–68.

their offspring's and friend's love choice when it is a same-race partner and disapprove of the selection when it does not.

National Data

Fewer than 1 percent of the almost 60 million married couples in the United States include a black spouse and a white spouse (*Statistical Abstract of the United States: 2006,* Table 54).

Our society also encourages love relationships because pair-bonded individuals provide mutual aid and are less of a drain on human-services resources. Our society cannot provide counselors for the mental health roller coaster of life. But a love partner who is caring and empathetic is just what the doctor ordered.

Love requires respect and friendship as well as passion. Because there comes a time when you have to get out of bed.

Erica Jong, novelist

Another example of the social control of love is that individuals attracted to someone of the same sex quickly feel the social and cultural disapproval of this attraction. Ennis Del Mar and Jack Twist of *Brokeback Mountain* (nominated in 2006 for 8 Academy Awards, including Best Picture) lamented that they could be open about their love relationship only when in the wilderness of Wyoming. They acutely felt the social constraints on their relationship (and one was murdered because he was homosexual). To "fit in," they each married and had children but remained attached and in love with each other. Diamond (2003) emphasized that individuals are biologically wired and capable of falling in love with and establishing intense emotional bonds with members of their own or opposite sex (hence, one's love-desire and sexual-desire partners can be different).

Romantic love is such a powerful emotion and marriage such an important relationship, connecting an outsider into an existing family and peer network, that mate selection is not left to chance. Parents inadvertently influence the mate choice of their children by moving to certain neighborhoods, joining certain churches, and enrolling their children in certain schools. Doing so increases the chance that their offspring will "hang out" with, fall in love with, and marry persons who are similar in race, education, and social class. Although twenty-first-century parents normally do not have large estates and are not concerned about the transfer of wealth, they usually want their offspring to meet someone who will "fit in" and with whom they will feel comfortable.

Peers exert a similar influence on homogamous mating by approving of certain partners and disapproving of others. Their motive is similar to that of parents—they want to feel comfortable around the people their peers bring with them to social encounters. Both parents and peers are influential, as most offspring and friends end up falling in love with and marrying persons of the same race, education, and social class.

Social approval of one's partner is normally important for a love relationship to proceed on course. Even the engagement of Prince Charles to Camilla Parker Bowles received the "blessing" of Queen Elizabeth, and 70 percent of Britons either approved of it or did not care (Soriano, 2005).

Love is also used by partners to control each other. Fehr and Harasymchuk (2005) noted that a comment expressing dissatisfaction by a romantic partner has considerably more negative emotional impact than a similar comment by a friend. In effect, we give considerable credence to what our love partner thinks of us, and we get upset when they criticize us.

Love is also an issue of control in the workplace (see Social Policy on page 81).

Ancient Views of Love

Many of our present-day notions of love stem from early Buddhist, Greek, and Hebrew writings.

Buddhist Conception of Love The Buddhists conceived of two types of love—an "unfortunate" kind of love (self-love) and a "good" kind of love (creative spiritual attainment). Love that represents creative spiritual attainment was described as "love of detachment," not in the sense of withdrawal from the emotional concerns of others but in the sense of accepting people as they are and not requiring them to be different from their present selves as the price of friendly affection. To a Buddhist, the best love is one in which you accept others as they are without requiring them to be like you.

Love in the Workplace

In 2005, Harry Stonecipher (a 68-year-old married man and head of Boeing) was forced to resign because he was having an affair with a female employee. He was fired not because of the affair (there were no company rules) but because he caused embarrassment to Boeing by, for example, sending steamy company e-mails to his lover ("The End of the Office Affair?" *The Economist*, 2005). These types of love relationships are sometimes problematic in the workplace.

With an increase of women in the workforce, an increase in the age at first marriage, and longer work hours, the workplace has become a common place for romantic relationships to develop. More future spouses may meet at work than in school, social, or neighborhood settings.

Pros and Cons of Office Romances

The energy that both fuels and results from intense love feelings can also fuel productivity on the job. And if the co-workers eventually marry or enter a nonmarital but committed, long-term relationship, they may be more satisfied with and committed to their jobs than spouses whose partners work elsewhere. Working at the same location enables married couples to commute together, go to company-sponsored events together, and talk shop together.

Recognizing the potential benefits of increased job satisfaction, morale, productivity, creativity, and commitment, some companies even look favorably upon love relationships among employees. Prior to the economic downturn following 9/11, Apple Computer, in Cupertino, California, encouraged socializing among employees by sponsoring get-togethers every Friday afternoon with beer, wine, food, and, on occasion, live bands. The company also had ski clubs, volleyball clubs, and Frisbee clubs, providing employees with opportunities to meet and interact socially. Some companies hire two employees who are married, reflecting a focus on the value of each employee to the firm rather than on their love relationship outside work.

However, workplace romances can also be problematic for the individuals involved as well as for their employers. When a workplace romance involves a supervisor/subordinate relationship, other employees might make claims of favoritism or differential treatment. In a typical differential-treatment allegation, an employee (usually a woman) claims that the company denied her a job benefit because her supervisor favored a female co-worker—who happens to be the supervisor's girlfriend.

If a workplace relationship breaks up, it may be difficult to continue to work in the same environment (and others at work may experience the fallout). A breakup that is less than amicable may result in efforts by partners to sabotage each other's work relationships and performance, incidents of workplace violence, harassment, and/or allegations of sexual harassment. In a survey of 1221 human-resource managers conducted by the Society for Human Resource Management and CareerJournal.com, 81 percent of human resource professionals and 76 percent of executives saw office romances as "dangerous" (Franklin, 2002).

Workplace Policies on Intimate Relationships

Some companies such as Disney, Universal, and Columbia have "anti-fraternization" clauses that impose a cap on workers' talking about private issues or sending personal e-mails. Some British firms have "love contracts" that require workers to tell their managers if they are involved with anyone from the office. Although these restrictions are rare, they seem to have the desired effect of curtailing office romances (Cooper, 2003).

Most companies (Wal-Mart is an example) do not prohibit romantic relationships among employees. However, the company may have a policy prohibiting open displays of affection between employees in the workplace and romantic relationships between supervisor and subordinate. Most companies have no policy regarding love relationships at work and generally regard romances between co-workers as "none of their business." There are some exceptions to the general permissive policies regarding workplace romances. Many companies have written policies prohibiting intimate relationships when one member of the couple is in a direct supervisory position over the other. These policies may be enforced by transferring or dismissing employees who are discovered in romantic relationships.

Your Opinion?

1. To what degree do you believe corporations should develop policies in regard to workplace romances?
2. What are the advantages and disadvantages of a workplace romance for a business?
3. What are the advantages and disadvantages for the individuals involved in a workplace romance?

Sources

Cooper, C. 2003. Office affairs are hard work. *The Australian* 5 March.

Franklin, R. 2002. Office romances: Conduct unbecoming? *Business Week Online* 14 February:1.

The end of the office affair? 2005. *The Economist* 10 March, 374:64.

Greek and Hebrew Conceptions of Love   Three concepts of love introduced by the Greeks and reflected in the New Testament are phileo, agape, and eros. *Phileo* refers to love based on friendship and can exist between family members, friends, and lovers. The city of Philadelphia was named after this phileo type of love. Another variation of phileo love is *philanthropia,* the Greek word meaning "love of humankind."

Agape refers to a love based on a concern for the well-being of others. Agape is spiritual, not sexual, in nature. This type of love is altruistic and requires nothing in return. "Whatever I can do to make your life happy" is the motto of the agape lover, even if this means giving up the beloved to someone else. Such love is not always reciprocal.

Eros refers to sexual love. This type of love seeks self-gratification and sexual expression. In Greek mythology, Eros was the god of love and the son of Aphrodite. Plato described "true" eros as sexual love that existed between two men. According to Plato's conception of eros, homosexual love was the highest form of love because it existed independent of the procreative instinct and free from the bonds of matrimony. Also, women had low status and were uneducated and were therefore not considered ideal partners for men. By implication, love and marriage were separate.

Love in Medieval Europe—from Economics to Romance

Love in the 1100s was a concept influenced by economic, political, and family structure. In medieval Europe, land and wealth were owned by kings controlling geographical regions—kingdoms. When so much wealth and power were at stake, love was not to be trusted as the mechanism for choosing spouses for royal offspring. Rather, marriages of the sons and daughters of the aristocracy were arranged with the heirs of other states with whom an alliance was sought. Love was not tied to marriage but was conceptualized, even between people not married or of the same sex, as an adoration of physical beauty (often between a knight and his beloved) and as spiritual and romantic. Hence, romantic love had its origin in extramarital love and was not expected between spouses (Trachman and Bluestone, 2005).

The presence of kingdoms and estates and the patrimonial households declined with the English revolutions of 1642 and 1688 and the French Revolution of 1789. No longer did aristocratic families hold power; it was transferred to individuals through parliaments or other national bodies. Even today, English monarchs are figureheads, with the real business of international diplomacy being handled by parliament. Because wealth and power were no longer in the hands of individual aristocrats, the need to control mate selection decreased and the role of love changed. Marriage became less of a political and business arrangement and more of a mutually desired emotional union. Just as partners in medieval society were held together by bureaucratic structure, a new mechanism—love—would now provide the emotional and social bonding.

Hence, love in medieval times changed from a feeling irrelevant to marriage—since individuals (representing aristocratic families) were to marry even though they were not in love—to a feeling that bonded a woman and a man together for marriage.

Love in Colonial America

Love in Colonial America was similar to that in medieval times. Marriage was regarded as a business arrangement between the fathers of the respective families (Dugan, 2005). An interested suitor would approach the father of the girl to express his desire to court his daughter. The fathers would generally confer on the amount of the **dowry,** which was the money and/or valuables the girl's father would pay the boy's father. Because unmarried women were stigmatized, getting them married off was desirable; thus, the dowry was an added inducement for the boy to marry the girl. Fathers could deny their daughters a dowry if they were unwilling to marry the choice of the father. Love was not totally absent, however; sometimes a girl could persuade her father to tell the suitor she was not interested.

Theories on the Origins of Love

Various theories have been suggested with regard to the origins of love.

Evolutionary Theory

Love has an evolutionary purpose by providing a bonding mechanism between the parents during the time their offspring are dependent infants. Love's strongest bonding lasts about four years, the time when children are most dependent and two parents can cooperate in handling their new infant. "If a woman was carrying the equivalent of a 12-lb bowling ball in one arm and a pile of sticks in the other, it was ecologically critical to pair up with a mate to rear the young," observes anthropologist Helen Fisher (Toufexis, 1993). The "four-year itch" is Fisher's term for the time at which parents with one child are most likely to divorce—the time when the woman can more easily survive without parenting help. If the couple has a second child, doing so resets the clock, and "the seven-year itch" is the next most vulnerable time.

Learning Theory

Unlike evolutionary theory, which views the experience of love as innate, learning theory emphasizes that love feelings develop in response to certain behaviors engaged in by a partner. Individuals in a new relationship who look at each other, smile at each other, compliment each other, touch each other endearingly, do things for each other, and do enjoyable things together are engaging in behaviors that make it easy for love feelings to develop. In effect, love can be viewed as a feeling that results from a high frequency of positive behavior and a low frequency of negative behavior. The reverse is also true. Persons who "fall out of love" may note the high frequency of negative behavior on the part of their partner and the low frequency of positive behavior. Persons who say "this is not the person I married" are saying the ratio of positives to negatives has changed dramatically.

The magic of first love is our ignorance that it can ever end.
Benjamin Disraeli, British politician

Cunningham et al. (2005) used the term **social allergy** to refer to being annoyed and disgusted by a repeated behavior on the part of the partner. Examples are uncouth habits (e.g., picking one's teeth), inconsiderate acts (e.g., not offering to get something from the kitchen when going one's self), intrusive behaviors (e.g., opening one's mail/email) and norm violations (e.g., drinking out of someone else's glass). The researchers found that these types of behaviors increased over time and were associated with both decreased relationship satisfaction and termination of the relationship.

Sociological Theory

Forty-five years ago, Ira Reiss (1960) suggested the wheel model as an explanation for how love develops. Basically, there are four stages of the wheel—rapport, self-revelation, mutual dependency, and personality need fulfillment. In the rapport stage, each partner has the feeling of having known the partner before, feels comfortable with the partner, and wants to deepen the relationship.

Such desire leads to self-revelation or self-disclosure, whereby each reveals intimate thoughts to the other about one's self, the partner, and the relationship. Such revelations deepen the relationship because it is assumed that the confidences are shared only with special people, and each partner feels special when listening to the revelations of the other.

As the level of self-disclosure becomes more intimate, a feeling of mutual dependency develops. Each partner is happiest in the presence of the other and begins to depend on the other for creating the context of these euphoric feelings. "I am happiest when I am with you" is the theme of this stage.

The feeling of mutual dependency involves the fulfillment of personality needs. The desires to love and be loved, to trust and be trusted, and to support and be supported are met in the developing love relationship.

Psychosexual Theory

According to psychosexual theory, love results from blocked biological sexual desires. In the sexually repressive mood of his time, Sigmund Freud ([1905] 1938), referred to love as "aim-inhibited sex." Love was viewed as a function of the sexual desire a person was not allowed to express because of social restraints. In Freud's era, people would meet, fall in love, get married, and have sex. Freud felt that the socially required delay from first meeting to having sex resulted in the development of "love feelings." By extrapolation, Freud's theory of love suggests that love dies with marriage (access to one's sexual partner).

Biochemical Theory

There may be a biochemical basis for love feelings. **Oxytocin** is a hormone that encourages contractions during childbirth and endears the mother to the suckling infant. It has been referred to as the "cuddle chemical" because of its significance in bonding. Later in life, oxytocin seems operative in the development of love feelings between lovers during sexual arousal. Oxytocin may be responsible for the fact that more women than men prefer to continue cuddling after intercourse.

Phenylethylamine (PEA) is a natural, amphetamine-like substance that makes lovers feel euphoric and energized. The high that they report feeling just by being with each other is from the PEA in their bloodstream that has been released by the brain. The natural-chemical high associated with love may explain why the intensity of passionate love decreases over time. As with any amphetamine, the body builds up a tolerance to PEA, and it takes more and more to produce the special kick. Hence, lovers develop a tolerance for each other. "Love junkies" are those who go from one love affair to the next in rapid succession to maintain the high. Alternatively, some lovers break up and get back together frequently as a way of making the relationship new again and keeping the high going.

Attachment Theory

The attachment theory of love emphasizes that a primary motivation in life is to be connected with other people. Monteoliva et al. (2005) confirmed that the attachment style an individual has with one's parents is associated with the quality of one's later romantic relationships. Specifically, a secure emotional attachment with loving adults as a child is associated with later involvement in a satisfying, loving, communicative relationship. This finding was true regardless of ethnic or racial background. Persons who evidence a secure attachment to a love partner also report higher levels of commitment/dedication (Pistole and Vocaturo, 2000). Hence, the benefits of a secure love attachment are enormous. One form of family therapy is "emotionally focused family therapy" (EFFT), which emphasizes intensifying the emotional bonds between family members on the premise that such emotional connectedness creates a context for resolving family problems (Furrow et al., 2005).

Each of the theories of love presented in this section has its critics (Table 3.1).

Unless we practice loving feelings toward everyone we meet, day in, day out, we're missing out on the most joyous part of life. If we can actually open our hearts, there's no difficulty in being happy.

Ayya Khema, *Be an Island*

Chapter 3 Love in Relationships

Table 3.1 Love Theories and Criticisms

Theory	Criticism
Evolutionary	Assumption that women and children need men for survival is not necessarily true today. Women can have and rear children without male partners.
Learning	Does not account for (1) why some people will share positive experiences yet will not fall in love and (2) why some people stay in love despite negative behavior.
Psychosexual	Does not account for people who report intense love feelings yet are having sex regularly.
Ego-Ideal	Does not account for the fact that people of similar characteristics fall in love.
Ontological	The focus of an ontological view of love is the separation of women and men from each other as love objects. Homosexual love is unaccounted for.
Biochemical	Does not specify how much of what chemicals result in the feeling of love. Chemicals alone cannot create the state of love; cognitions are also important.
Attachment	Not all people feel the need to be emotionally attached to others. Some prefer to be detached.

How Love Develops in a New Relationship

The development of love relationships is affected by various social, physical, psychological, physiological, and cognitive conditions.

Social Conditions for Love

Love is a label given to an internal feeling. Our society promotes love through popular music, movies, television, and novels. These media convey the message that love is an experience to enjoy and to pursue and that you are missing something if you are not in love.

More traditional societies attempt to directly influence one's love choice. In India and other countries (e.g., Iran) where marriages have been arranged, the development of romantic love relationships is tightly controlled. For example, parents select the mate for their child in an effort to prevent any potential love relationship from forming with the "wrong" person and to ensure that the child marries the "right" person. Such a person must belong to the desired social class and have the economic resources desired by the parents. Marriage is regarded as the linking of two families; the love feelings of the respective partners are irrelevant. Love is expected to follow marriage, not precede it.

Body Type Condition for Love

The probability of being involved in a love relationship is influenced by approximating the cultural ideal of physical appearance. Halpern et al. (2005) analyzed data on a nationally representative sample of 5487 black, white, and Hispanic adolescent females and found that for each one point increase in body mass index (BMI), the probability of involvement in a romantic relationship dropped by 6%. Hence, to the degree that a woman approximates the cultural ideal of being trim and "not being fat," there is an increased chance of attracting a partner and becoming involved in a romantic love relationship. A former student of the authors dropped from 225 to 125 pounds and noted, "You wouldn't believe the dramatic difference in the way guys noticed and talked to me [between] when I was beefed up and when I lost weight. I am now in a love relationship and we are talking marriage."

Men as a rule love with their eyes, but women with their ears.

Oscar Wilde

Psychological Conditions for Love

Two psychological conditions associated with the development of healthy love relationships are high self-esteem and self-disclosure.

Self-Esteem High self-esteem is important for defining success (Bianchi and Povilavicus, 2006). It is also important for developing healthy love relationships because it enables an individual to feel worthy of being loved. Feeling good about yourself allows you to believe that others are capable of loving you. Individuals with low self-esteem doubt that someone else can love and accept them (DeHart et al., 2002). Having self-esteem provides other benefits:

1. It allows one to be open and honest with others, about both strengths and weaknesses.

2. It allows one to feel generally equal to others.

3. It allows one to take responsibility for one's own feelings, ideas, mistakes, and failings.

4. It allows for the acceptance of both strengths and weaknesses in one's self and others.

5. It allows one to validate one's self and not to expect the partner to do this.

6. It permits one to feel empathy—a very important skill in relationships.

7. It allows separateness and interdependence, as opposed to fusion and dependence.

Positive physiological outcomes also follow from high self-esteem. People who feel good about themselves are less likely to develop ulcers and are likely to cope with anxiety better than those who don't. In contrast, low self-esteem has devastating consequences for individuals and the relationships in which they become involved. Not feeling loved as a child and, worse, feeling rejected and abandoned creates the context for the development of a negative self-concept and mistrust of others. People who have never felt loved and wanted may require constant affirmation from their partner as to their worth and may cling desperately to that person out of fear of being abandoned. Such dependence (the modern term is *codependency*) may also encourage staying in unhealthy (for example, abusive and alcoholic) relationships, because the person may feel "this is all I deserve." Fuller and Warner (2000) studied 257 college students and observed that women had higher codependency scores than men. Codependency was also associated with being reared in families that were stressful and alcoholic.

One characteristic of individuals with low self-esteem is that they may love too much and be addicted to unhealthy love relationships. Petrie et al. (1992) studied 52 women who reported that they were involved in unhealthy love relationships in which they had selected men with problems (such as alcohol or other drug addiction) that they attempted to solve at the expense of neglecting themselves. "Their preoccupation with correcting the problems of others may be an attempt to achieve self-esteem," the researchers noted (p. 17).

Although it helps to have positive feelings about one's self going into a love relationship, sometimes these develop after one becomes involved in the relationship. "I've always felt like an ugly duckling," said one woman. "But once I fell in love with him and he with me, I felt very different. I felt very good about myself then because I knew that I was somebody that someone else loved." High self-esteem, then, is not necessarily a prerequisite for falling in love. People who have low self-esteem may fall in love with someone else as a result of feeling deficient (ego-ideal theory of love). The love they perceive the other person having for them may compensate for the deficiency and improve their self-esteem.

Self-Disclosure Disclosing one's self is necessary if one is to love—to feel invested in another (Radmacher and Azmitia, 2006). Ross (2006) identified eight dimensions of self-disclosure (background/history, feelings toward the partner, self feelings, feelings about one's body, social issue attitudes, tastes/interests,

Chapter 3 Love in Relationships

money/work, and feelings about friends). Disclosing feelings about the partner included disclosing "how much I like the partner," "my feelings about our sexual relationship," "how much I trust my partner," "things I dislike about my partner," and "my thoughts about the future of our relationship," all of which were associated with relationship satisfaction. Of interest in Ross's findings is that disclosing one's tastes and interests was negatively associated with relationship satisfaction. By telling a partner too much detail about what one likes, the partner may discover something that turns him or her off and lowers relationship satisfaction.

Kito (2005) examined the self-disclosure patterns of 145 college students (both American and Japanese) and found that self-disclosure was higher in romantic relationships than in friendships and that Americans were more disclosing than the Japanese. The researcher also found that disclosure was higher in same-sex friendships than in cross-sex friendships.

It is not easy for some people to let others know who they are, what they feel, or what they think. They may fear that if others really know them, they will be rejected as a friend or lover. To guard against this possibility, they may protect themselves and their relationships by allowing only limited information about their past behaviors and present thoughts and feelings.

Trust is the condition under which people are most willing to disclose themselves. When people trust someone, they tend to feel that whatever feelings or information they share will not be judged and will be kept safe with that person. If trust is betrayed, a person may become bitterly resentful and vow never to disclose herself or himself again. One woman said: "After I told my partner that I had had an abortion, he told me that I was a murderer and he never wanted to see me again. I was devastated and felt I had made a mistake telling him about my past. You can bet I'll be careful before I disclose myself to someone else" (authors' files).

Gallmeier et al. (1997) studied the communication patterns of 360 undergraduates at two universities and found that women were significantly more likely to disclose information about themselves. Specific areas of disclosure included previous love relationships, what they wanted for the future of the relationship, and what their partners did that they did not like.

Love usually begins by deceiving oneself, and usually ends by deceiving the other person.

Unknown

Physiological and Cognitive Conditions for Love

Physiological and cognitive variables are also operative in the development of love. The individual must be physiologically aroused and interpret this stirred-up state as love (Walster and Walster, 1978).

> *Suppose, for example, that Dan is afraid of flying, but his fear is not particularly extreme and he doesn't like to admit it to himself. This fear, however, does cause him to be physiologically aroused. Suppose further that Dan takes a flight and finds himself sitting next to Judy on the plane. With heart racing, palms sweating, and breathing labored, Dan chats with Judy as the plane takes off. Suddenly, Dan discovers that he finds Judy terribly attractive, and he begins to try to figure out ways that he can continue seeing her after the flight is over. What accounts for Dan's sudden surge of interest in Judy? Is Judy really that appealing to him, or has he taken the physiological arousal of fear and mislabeled it as attraction? (Brehm, 1992, p. 44)*

Although most people who develop love feelings are not aroused in this way, they may be aroused or anxious about other issues (being excited at a party, feeling apprehensive about meeting someone) and mislabel these feelings as those of attraction when they meet someone.

In the absence of one's cognitive functioning, love feelings are impossible. Individuals with brain cancer who have had the front part of their brain (between the eyebrows) removed are incapable of love. Indeed, emotions are not present in them at all (Ackerman, 1994). The social, physical, psychological, physiological, and cognitive conditions are not the only factors important for the development of love feelings. The timing must also be right. There are only certain times in your life when you are seeking a love relationship. When those times occur, you are likely to fall in love with the person who is there and who is also seeking a love relationship. Hence, many love pairings exist because each of the individuals is available to the other at the right time—not because they are particularly suited for each other.

Love as a Context for Problems

Though love may bring great joy, it also creates a context for problems. Four such problems are simultaneous loves, involvement in an abusive relationship, making risky or dangerous choices, and the emergence of stalking.

Destruction of Existing Relationships

Sometimes the development of one love relationship is at the expense of another. A student in the authors' classes noted that when she was 16 she fell in love with a person at work who was 25. Her parents were adamant in their disapproval and threatened to terminate the relationship with their daughter if she continued to see this man. The student noted that she initially continued to see her lover and to keep their relationship hidden. But eventually she decided that giving up her family was not worth the relationship, so she stopped seeing him permanently. There are others in the same situation who would end the relationship with their parents and continue the relationship with their beloved. Choosing to end a relationship with one's parents is one of the downsides of love.

Simultaneous Loves

For all the wonder of love, it can create heartbreak to be aware that one's partner is in love with or having sex with someone else. As we will discuss in the section on jealousy later in the chapter, multiple involvements are not a problem for some individuals/couples (e.g., in compersion/polyamory). However, most people are not comfortable knowing their partner has other emotional/sexual relationships. Only 2.7% of 1027 undergraduates agreed "I can feel good about my partner having an emotional/sexual relationship with someone else" (Knox and Zusman, 2006). In a study on "undesirable marriage forms," 91.2% of 111 undergraduates reported that they would "never participate" in a "group marriage" (Billingham et al., 2005). Hence, for most individuals, simultaneous lovers are viewed as a problem.

Abusive Relationships

Another problem associated with love is being in love with someone who may be emotionally or physically abusive (see *When "I Love You" Turns Violent*, Johnson, 2005). Almost a third (29.5%) of 1027 undergraduates reported that they had been involved in an emotionally abusive relationship with a partner (10.4% reported previous involvement in a physically abusive relationship) (Knox and Zusman, 2006). Someone who criticizes you ("you're ugly, stupid, pitiful"), is dishonest with you (sexually unfaithful, for instance), or physically harms you will create a context of interpersonal misery for you. Nevertheless, you might love that person and feel emotionally drawn to him or her.

Most marriage therapists suggest that you examine why you love and continue to stay with such a person. Do you feel that you deserve this treatment because you are "no good" or feel you would not be able to find a better alternative (low self-concept)? Do you feel you would rather be with a person who treats you badly than be alone (fear of the unknown)? Or do you hang on because he looks forward to a better tomorrow (hope)?

Another explanation for why some people who are abused by their partners continue to be in love with them is that the abuse is only one part of the relationship. When the partner is not being abusive, he or she may be kind, loving, and passionate. Shackelford et al. (2005) analyzed data on 1461 men and found that they often used "mate retention behaviors" such as giving flowers or gifts to entice the partner to stay in the relationship, which becomes even more abusive.

It is the presence of these mate retention behaviors, which happen every now and then (a periodic reinforcement), that keeps the love feelings alive. Love stops when the extent of the abuse is so great that there are insufficient positive behaviors to counteract the abusive behavior. One abused partner said that "when he started abusing my kids, that was it."

Love relationships that do not involve emotional or physical abuse may be unfulfilling for other reasons. Partners in love relationships may experience a lack of fulfillment if they have radically different values, religious beliefs, role expectations, recreational or occupational interests, sexual needs, or desires concerning family size. These intensely frustrated/unhappy partners may be in a relentless struggle to make their relationship work because "we love each other."

Stalking: When Loves Goes Mad

In the name of love, people have stalked their beloved. **Stalking** is defined as a repeated malicious pursuit that threatens the safety of the victim. It may involve following a victim; threats of physical harm to the victim, one's self, or another person; or restricting the behavior of the victim, including kidnapping or home invasion. The most common stalking behavior (which is also prohibited by stalking laws) is unwanted "obsessional following" (Meloy and Fisher, 2005). Stalking behaviors typically cause great distress or fear and impact the emotional well being and social and work activities of the victim.

Who are the stalkers? Men are most often stalkers (85%) and women are their victims. Two primary reasons for stalking are rejection by a sexual intimate (hence the male has been in a previous emotional or sexual relationship with the woman and obsessively tries to win her back) or rejection by a stranger the stalker is infatuated with, who fails to return the stalker's romantic overtures. The stalking of celebrity females (e.g., Jodie Foster) sometimes becomes visible in the media. The two most common emotions of the stalker are anger (over being rejected) and jealousy (at being replaced) (Meloy and Fischer, 2005). Exercising a great deal of control in an existing relationship is predictive that the controlling partner will become a stalker when the other partner ends the relationship (King, 2003). For the 15 percent of stalkers who are women, the most common victim is another woman. These may be partners in lesbian relationships that have ended. The stalker feels rejected and wants to renew the relationship.

Stalkers are obsessional and very controlling. Obsessional thinking is their most common cognitive trait (Meloy and Fisher, 2005). They are typically mentally ill and have one or more personality disorders involving paranoid, antisocial, or obsessive-compulsive behaviors. Reid (2005) and Meloy and Fischer (2005) noted a neurobiological component in the stalker. Meloy and Fischer (2005) identified three primary brain systems involved in stalking: sex drive, whereby the individual is motivated to achieve sexual gratification; attraction, whereby the individual is driven to emotionally connect with a specific mating partner; and attachment, whereby the individual is motivated to experience a secure relationship with a long-term partner. In effect, the stalker feels a barrier in access to the beloved (physically and emotionally), which intensifies the drive to be with the rejecting partner (referred to as abandonment rage). Indeed, brain activity can be observed with magnetic resonance imaging (MRI) that reveals differences between a person who is happily in love and a person who is the "spurned or unrequited stalker" (Meloy and Fischer, 2005, p. 1475).

To those of us who knew the pain of Valentines that never came.
Janis Ian, singer

My love for you has driven me insane.
Rumi, Persian mystic and poet

Although various coping strategies have been identified, additional research is needed on how to manage unwanted attention (Regan, 2000). A survey (Spitzberg and Cupach, 1998) of young adults identified several general coping categories:

1. Make a direct statement to the person ("I am not interested in dating you, my feelings about you will not change, and I know that you will respect my decision and direct your attention elsewhere.") (Regan, 2000, 266).

2. Seek protection through formal channels (police, court restraining order).

3. Avoid the perpetrator (ignore, don't walk with or talk to, hang up if the person calls).

4. Use informal coping methods (use telephone caller identification; seek advice from others).

Direct statements and actions that unequivocally communicate lack of interest are probably the most effective types of intervention.

Unrequited/Unfulfilling Love Relationships

It is not unusual that lovers vary in the intensity of their love for each other. Interesting indeed is the question of whether it is better to be the person who loves more or less in a relationship. Certainly, the person who loves less may suffer more anguish. Such was the case of Jack Twist, who was hurt that his love interest, Ennis Del Mar, would not make time for them to continue their clandestine meetings on *Brokeback Mountain*.

Context for Risky/Dangerous/Questionable Choices

Plato said that "love is a grave mental illness," and some research suggests that individuals in love make risky, dangerous, or questionable decisions. In a study on "what I did for love," college students reported that "driving drunk," "dropping out of school to be with my partner," and "having sex without protection" were among the more dubious choices they had made while they were under the spell of love (Knox et al., 1998). Similarly, a team of researchers examined the relationship between having a romantic love partner and engaging in minor acts of delinquency (smoking cigarettes, getting drunk, skipping school) and found that females were particularly influenced by their "delinquent" boyfriends (Haynie et al., 2005). Their data source was the National Longitudinal Study of Adolescent Health.

National Data

Anderson et al. (2000) analyzed national data on U.S. adults and noted that only 19 percent of those in an ongoing (love) relationship used a condom, whereas 62 percent used a condom if their partner was not a person they were seeing regularly.

Jealousy in Relationships

Jealousy can be defined as an emotional response to a perceived or real threat to an important or valued relationship. People experiencing jealousy fear being abandoned and feel anger toward the partner or the perceived competition (Guerrero et al., 2005). As Buss (2000) emphasized, "Jealousy is an adaptive emotion, forged over millions of years. . . . It evolved as a primary defense against threats of infidelity and abandonment" (p. 56). Persons become jealous when they fear replacement. Although jealousy does not occur in all cultures (polyandrous societies value cooperation, not sexual exclusivity (Cassidy and Lee, 1989), it does occur in our society and among both heterosexuals and homosexuals.

More than 40 percent (42.1%) of 1027 university students reported, "I am a jealous person" (Knox and Zusman, 2006). One hundred eighty-five students

Living with someone you love can be lonelier—than living entirely alone!—if the one that y'love doesn't love you back.

Maggie to her husband Brick in Tennessee William's, *Cat on a Hot Tin Roof*

When a young man complains that a young lady has no heart, it is a pretty certain sign that she has his.

George D. Prentice, U.S. newspaperman, editor, poet (1802–1870)

provided information about their experience with jealousy (Knox et al., 1999). On a continuum of 0 ("no jealousy") to 10 ("extreme jealousy"), with 5 representing "average jealousy," these students reported feeling jealous at a mean level of 5.3 in their current or last relationship. Students who had been dating their partner a year or less were significantly more likely to report higher levels of jealousy (mean = 4.7) than those who had dated 13 months or more (mean = 3.3). Hence, jealously is more likely to occur early in a couple's relationship.

Causes of Jealousy

Jealousy can be triggered by external or internal factors.

External Causes   External factors refer to behaviors the partner engages in that are interpreted as (1) an emotional and/or sexual interest in someone (or something) else or (2) a lack of emotional and/or sexual interest in the primary partner. In the study of 185 students referred to above, the respondents identified "actually talking to a previous partner" (34%) and "talking about a previous partner" (19%) as the most common sources of their jealousy. Also, men were more likely than women to report feeling jealous when their partner talked to a previous partner, whereas women were more likely than men to report feeling jealous when their partner danced with someone else.

Next to the atom bomb, the greatest explosion is set off by an old flame.
Evan Esar, humorist

Internal Causes   Jealousy may also exist even when there is no external behavior that indicates the partner is involved or interested in an **extradyadic relationship**—an emotional/sexual involvement between a member of a pair and someone other than the partner. Internal causes of jealousy refer to characteristics of individuals that predispose them to jealous feelings, independent of their partner's behavior. Examples include being mistrustful, having low self-esteem, being highly involved in and dependent on the relationship, and having no perceived alternative partners available (Pines, 1992). These internal causes of jealousy are explained below.

1. *Mistrust.* If an individual has been deceived or cheated on in a previous relationship, that individual may learn to be mistrustful in subsequent relationships. Such mistrust may manifest itself in jealousy. Mistrust and jealousy may be intertwined. Tilley and Brackley (2005) examined the factors involved for men convicted of assaulting a female and suggested that both jealousy and mistrust may have been involved in aggression against a female.

2. *Low self-esteem.* Individuals who have low self-esteem tend to be jealous because they lack a sense of self-worth and hence find it difficult to believe anyone can value and love them (Khanchandani, 2005). Feelings of worthlessness may contribute to suspicions that someone else is valued more.

3. *Anxiety.* In general, individuals who experience higher levels of anxiety also display more jealousy (Khanchandani, 2005).

4. *Lack of perceived alternatives.* Individuals who have no alternative person or who feel inadequate in attracting others may be particularly vulnerable to jealousy. They feel that if they do not keep the person they have, they will be alone.

5. *Insecurity.* Individuals who feel insecure in the relationship with their partner may experience higher levels of jealousy. Khanchandani (2005) found that individuals who had been in relationships for a shorter time, who were in less committed relationships, and who were less satisfied with their relationships were more likely to be jealous.

Consequences of Jealousy

Jealousy can have both desirable and undesirable consequences.

Desirable Outcomes   Jealousy may be functional if it occurs at a low level and results in open and honest discussion about the relationship. Not only may jealousy keep the partner aware that he or she is cared for (the implied message is "I love you and don't want to lose you to someone else"), but also the partner may learn that the development of other romantic and sexual relationships is unacceptable.

One wife said:

> *When I started spending extra time with this guy at the office my husband got jealous and told me he thought I was getting in over my head and asked me to cut back on the relationship because it was "tearing him up." I felt he really loved me when he told me this and I chose to stop having lunch with the guy at work. (Authors' files)*

According to Buss (2000), the evoking of jealousy also has the positive functions of assessing the partner's commitment and of alerting the partner that one could leave for greener mating pastures. Hence, one partner may deliberately evoke jealousy to solidify commitment and ward off being taken for granted. In addition, sexual passion may be reignited if one partner perceives that another would take his or her love object away. That people want what others want is an adage that may underlie the evocation of jealousy.

Undesirable Outcomes Shakespeare referred to jealousy as the "green-eyed monster," suggesting that it sometimes leads to undesirable outcomes for relationships.

The emotional torment for one's self and one's partner when feelings of jealousy are obsessive is evident. In addition, one's partner can tire of unwarranted jealous accusations and end the relationship.

In its extreme form, jealousy may have devastating consequences. In the name of love, people have stalked the beloved, shot the beloved, and killed themselves in reaction to rejected love. The next section details the different ways women and men cope with jealousy.

Gender Differences in Coping with Jealousy

Jealousy is a theme of popular movies and part of our cultural language. Defined as "one's emotional reaction to the perception that one's love relationship may end because of a third party," jealousy was the topic of a 44-item questionnaire completed anonymously by 291 undergraduates at a large southeastern university. More than half (51.9%) of respondents "agreed" or "strongly agreed" that "jealousy is normal," and more than a third (35%) reported, "I am a jealous person."

Analysis of the data on women's and men's reactions to jealousy revealed four significant differences.

1. *Women eat when they feel jealous.* Women were significantly more likely than men to report that they turned to food when they felt jealous: 30.3 percent of women, in contrast to 22 percent of men, said that they "always, often, or sometimes" looked to food when they felt jealous.

2. *Men drink alcohol when they feel jealous.* Men were significantly more likely than women to report that they drank alcohol or used drugs when they felt jealous: 46.9% of men, in contrast to 27.1 percent of women, said that they "always, often, or sometimes" would drink or use drugs to make the pain of jealousy go away.

3. *Women more often confide with friends when they feel jealous.* Women were significantly more likely than men to report that they turned to friends when they felt jealous: 37.9 percent of women, in contrast to 13.5 percent of men, said that they "always" turned to friends for support when feeling jealous.

*This section is abridged and adapted from a presentation by Ronda Breed, David Knox, and Marty Zusman, at the Annual Research Symposium at East Carolina University, 2004. A related version of this paper was published as "Men and Jealousy" by David Knox, Marty Zusman, and Rhonda Breed. *College Student Journal* 2007, in press.

The way to hold a husband is to keep him a little jealous; the way to lose him is to keep him a little more jealous.

H. L. Mencken

4. *Women reject the belief that "jealousy shows love."* Women were significantly more likely than men to disagree or to strongly disagree that "jealousy shows how much your partner loves you": 63.2 percent of the women, in contrast to 42.6 percent of the men, disagreed with the statement. This difference may be related to the fact that women are more often victimized by jealous and abusive males.

Where there is no jealousy, there is no love.

Elizabeth Taylor, actress

Coping Strategies   Implications of the data may be relevant to women who may be alert to the "extra urge" to eat in reaction to jealousy, and turn instead to vigorous exercise as a way of reducing stress. Not only might exercise better reduce the stress; it will do so without adding pounds, which could lead to further self-deprecation and depression. Similarly, men might consider talking with a buddy rather than turning to the bottle; strengthening friendships would be more productive than risking a hangover or a fatal car wreck.

Compersion and Polyamory

Compersion denotes a situation in which an individual feels positive about a partner's emotional and sexual enjoyment with another person, and it is sometimes thought of as the opposite of jealousy. **Polyamory** means multiple loves (poly = many, amorous = love); polyamorous relationships may be heterosexual or homosexual (Bettinger, 2005). Persons in polyamorous relationships agree that they will have emotional/sexual relationships with others and seek to rid themselves of jealous feelings and to increase their level of compersion. To feel happy for a partner who delights in the attention and affection of—and sexual involvement with—another person is the goal of polyamorous couples.

This couple has a polyamorous relationship—an emotional and sexual relationship with each other. They also have similar relationships with others. They note that no one person can be all things to another.

There are both advantages and disadvantages to embracing polyamory and compersion (Adamson and Starr, 2005). Advantages of polyamory include greater variety in one's emotional/sexual life; the avoidance of hidden affairs and the attendant feelings of deception, mistrust, or betrayal; and the opportunity to have different needs met by different people. Of the latter advantage, one polyamorous partner said,

I have one partner with whom I enjoy movies/books, another with whom I fish, another with whom I cook, and still another with whom I play the guitar. Each of these partners does not have the other three interests, which I dearly love and enjoy, so polyamory allows one to enjoy different things with different people.

Sheff (2006) noted, "The vast majority of polyamorists espouse gender equality." One polyamorous woman said:

Women with multiple lovers are usually called sluts, bitches, very derogatory, very demeaning in sexual context. Whereas men who have multiple lovers—they're studs, they're playboys, they're glorified names, where with a woman it's very demeaning. So to be a woman and have multiple partners, it's been very empowering and claiming some of that back, saying I have just as much right to be a sexual person with many lovers as men do . . . without the shame and the guilt . . .

The disadvantages of polyamory involve having to manage one's feelings of jealousy, greater exposure of

Authors

Jealousy in Relationships   93

one's self and partners to human immunodeficiency virus and other sexually transmitted infections, and limited time with each partner. Of the latter, one polyamorous partner said, "With three relationships and a full-time job, I just don't have much time to spend with each partner so I'm frustrated about who I'll be with next. And managing the feelings of the other partners who want to spend time with me is a challenge." More information about the various nuances of polyamory is presented in the following section. We are discussing polyamory in the chapter on love rather than sexuality because polyamory is as much about emotional intimacy as sexuality.

A Short Primer on Polyamory*

This very brief and incomplete discussion focuses on polyamory, a relationship style in which people openly conduct sexual relationships with multiple partners. Polyamory is more emotionally intimate than swinging and offers the possibility of greater gender equality than polygyny because both men and women can have more than one partner. These relationships have a number of different elements that include (but are not limited to) levels of sexual exclusivity, numbers of people involved, and various degrees of emotional intimacy between partners.

It is better to love two too many than one too few.

Sir John Harrington

Sexual Exclusivity   Many community members use the term polyamory, or more commonly "poly," as an umbrella term to encompass both polyamory and polyfidelity. Those in *polyamorous* relationships generally have sexually and (ideally) emotionally intimate extradyadic relationships, with no promise of sexual exclusivity. *Polyfidelity* differs from polyamory in that polyfideles (the term for someone who practices polyfidelity) expect their partners to remain sexually exclusive within a group that is larger than two people, though some polyfidelitous groups have members who do not have sex with each other. Almost all polyfideles see each other as family members, regardless of the degree of sexual contact within their relationships. Not all polys in a relationship have sex with each other, and I call those who are emotionally intimate but not sexually connected *polyaffective.*

Polygeometry   The number of people involved in poly relationships varies and can include open couples, vees, triads, quads, and moresomes (all described below). As the number of people involved in a relationship rises, the relationships become rarer and potentially less stable. The most common form is the *open couple,* usually composed of two people who often are in a long-term relationship, cohabitate (some married, others unmarried), and have extradyadic sexual relationships. *Vees* are three-person relationships in which one member is sexually connected to each of the two others. The relationship between the two nonlovers can range from strangers (who are aware of and cordial with each other), to casual friends, to enemies. A *triad,* commonly understood as a *ménage à trois,* generally includes three sexually involved adults. Sometimes triads begin as threesomes, but more often they form when a single joins an open couple or a larger group loses member(s). *Quads,* as the name implies, are groups of four adults most commonly formed when two couples join, although sometimes they develop when a triad adds a fourth or a moresome loses member(s). Quads are notoriously unstable, frequently losing someone to poly-style divorce. *Moresomes,* groups with five or more adult members, are larger, more fragile, and more complicated than quads.

Emotional Intimacy   Polys frequently use the terms *primary, secondary,* and *tertiary* to describe their varied levels of intimacy. *Primary* partners—sometimes corre-

*This section was written for this text by Dr. Elizabeth Sheff, Department of Sociology, Georgia State University, Atlanta, Georgia. Dr. Sheff conducted more than 40 interviews with persons involved in polyamory.

sponding to the larger cultural conception of a spouse—usually have long-term relationships, have joint finances, cohabitate, and make major life decisions together, and sometimes they have children. *Secondary* partners tend to keep their lives more separate than primary partners, frequently maintain separate finances and residences, may have less intense emotional connections than primaries, and usually discuss major life decisions, though they generally do not make those decisions jointly. *Tertiary* relationships are often less emotionally intimate, sometimes with long distance or more casual partners. Some tertiary relationships closely resemble swinging. Some poly families have *spice,* the poly word for more than one spouse.

SUMMARY

What are some ways love has been described?

Love remains an elusive and variable phenomenon. Researchers have conceptualized love as a continuum from romanticism to realism, as a triangle consisting of three basic elements (intimacy, passion, and commitment), and as a style (from playful ludic love to obsessive and dangerous/manic).

How has love expressed itself in various social and historical contexts?

The society in which we live exercises considerable control over our love object/choice and conceptualizes it in various ways. Love may be blind, but it knows what color a person's skin is. Romantic love is such a powerful emotion and marriage such an important relationship, connecting an outsider with an existing family and peer network, that mate selection is not left to chance. Parents inadvertently influence the mate choice of their children by moving to certain neighborhoods, joining certain churches, and enrolling their children in certain schools. Doing so increases the chance that their offspring will "hang out" with, fall in love with, and marry persons who are similar in race, education, and social class.

In the 1100s in Europe, love was not expected between spouses but existed primarily between the unmarried, such as a knight and his beloved. Marriage was an economic and political arrangement that linked two families. As aristocratic families declined after the French Revolution, love became an emotion to bind a woman and man together to bear and rear children. Previously, Buddhists, Greeks, and Hebrews had their own views of love. Love in colonial America was also tightly controlled.

What are the various theories of love?

Theories of love include evolutionary (love provides the social glue needed to bond parents with their dependent children and spouses with each other to care for their dependent offspring), social learning (positive experiences create love feelings), sociological (Reiss's "wheel" theory), psychosexual (love results from a blocked biological drive), biochemical (love involves feelings produced by biochemical events), and attachment (a primary motivation in life is to be connected with other people).

How does love develop in a new relationship?

Love occurs under certain conditions. Social conditions include a society that promotes the pursuit of love, peers who enjoy it, and a set of norms that link love and marriage. Psychological conditions involve high self-esteem and a willingness to disclose one's self to others. Physiological and cognitive conditions imply that the individual experiences a stirred-up state and labels it "love."

How is love a context for problems?

Love sometimes provides a context for problems in that a young person in love will lie to parents and become distant from them so as to be with the lover. Also, lovers experience problems such as being in love with two people at the same time, being in love with someone who is abusive, and making risky/dangerous/ questionable choices while in love (e.g., not using a condom) or reacting to a former lover who has become a stalker.

How do jealousy and love interface?

Jealousy is an emotional response to a perceived or real threat to a valued relationship. Jealous feelings may have both internal and external causes and may have both positive and negative consequences for a couple's relationship. Compersion is the opposite of jealousy and involves feeling positive about a partner's emotional and physical relationship with another person. Polyamory (many loves) is an arrangement whereby lovers agree to have numerous emotional/sexual relationships with others at the same time, and each person is aware of every relationship. Polyamorous lovers note both the difficulties of this arrangement (coping with feelings of jealousy) and the advantages (not having to lie about being interested in others).

KEY TERMS

agape love style	extradyadic relationship	mania love style	romantic love
compersion	infatuation	oxytocin	social allergy
conjugal love	jealousy	polyamory	stalking
dowry	ludic love style	pragma love style	storge love style
eros love style			

The Companion Website for *Choices in Relationships: An Introduction to Marriage and the Family,* Ninth Edition
www.thomson.edu/sociology/knox

Supplement your review of this chapter by going to the companion website to take one of the Tutorial Quizzes, use the flash cards to master key terms, and check out the many other study aids you'll find there. You'll also find special features such as the Marriage and Family Resource Center, Census 2000 information, and other data and resources at your fingertips to help you with that special project or to do some research on your own.

WEBLINKS

Love Quotes
http://library.lovingyou.com/quotes/

Marriage Builders
http://marriagebuilders.com/

Polyamory
http://www.polyamorysociety.org/
http://lovemore.com/home

REFERENCES

Ackerman, D. 1994. *A natural history of love.* New York: Random House.

Adamson, K., and P. Starr. 2005. Polyamory and Compersion. Presentation, Courtship and Marriage class, Department of Sociology, East Carolina University, Fall.

Anderson, J. E., R. Wilson, L. Doll, T. S. Jones, and P. Barker. 2000. Condom use and HIV risk behaviors among U.S. adults: Data from a national survey. *Family Planning Perspectives* 31:24–28.

Bettinger, M. 2005. Polyamory and gay men: A family systems approach. *Journal of GLBT Family Studies* 1:97–116

Bianchi, A. J., and L. Povilavicius 2006. Cultural meanings versus academic meaning of self-esteem. Paper, Southern Sociological Society, New Orleans, LA, March 24.

Billingham, R. E., P. B. Perera, and N. A. Ehlers. 2005. College women's rankings of the most undesirable marriage and family forms. *College Student Journal* 39:749–753

Brantley, A., D. Knox, and M. E. Zusman. 2002. When and why gender differences in saying "I love you" among college students. *College Student Journal* 36:614–15.

Brehm, S. S. 1992. *Intimate relationships.* 2nd ed. New York: McGraw-Hill.

Bulcroft, R., K. Bulcroft, K. Bradley, and C. Simpson. 2000. The management and production of risk in romantic relationships: A postmodern paradox. *Journal of Family History* 25:63–92.

Buss, D. M. 2000. Prescription for passion. *Psychology Today* May/June, 54–61.

Cassidy, M. L., and G. Lee. 1989. The study of polyandry: A critique and synthesis. *Journal of Comparative Family Studies* 20:1–11.

Coontz, S. 2005. *Marriage, A History: How love conquered marriage.* New York: Penguin Books.

Cunningham, M. R., S. R. Shamblen, A. P. Barbee, and L. K. Ault. 2005. Social allergies in romantic relationships: Behavioral repetition, emotional sensitization, and dissatisfaction in dating couples. *Personal Relationships* 12:273–95.

DeHart, T., S. L. Murray, B. W. Pelham, and P. Rose. 2002. The regulation of dependency in parent-child relationships. *Journal of Experimental Social Psychology* 39:59–67.

Diamond, L. M. 2003. What does sexual orientation orient? A biobehavioral model distinguishing romantic love and sexual desire. *Psychological Review* 110:173–92.

Dugan, J. 2005. Colonial America. Retrieved from Suite.com at http://www.suite101.com/article.cfm/colonial_america_retired/61531/1 on December 12.

Fehr, B., and C. Harasymchuk. 2005. The experience of emotion in close relationships: Toward and integration of the emotion-in-relationships and interpersonal script models. *Personal Relationships* 12:181–96.

Freud, S. 1938. Three contributions to the theory of sex. In *The basic writings of Sigmund Freud,* edited by A. A. Brill. New York: Random House (article originally published in 1905).

Fuller, J. A., and R. M. Warner. 2000. Family stressors as predictors of codependency. *Genetic, Social, and General Psychology Monographs* 126:5–22.

Furrow, J. L., B. Bradley, and S. M. Johnson. 2005. Emotionally focused family therapy with stepfamilies. In *Sourcebook of family theory & research,* edited by Vern L. Bengtson, Alan C. Acock, Katherine R. Allen, Peggye Dilworth-Anderson, and David M. Klein. Thousand Oaks, CA: Sage Publications, 220–22.

Gallmeier, C. P., M. E. Zusman, D. Knox, and L. Gibson. 1997. Can we talk? Gender differences in disclosure patterns and expectations. *Free Inquiry in Creative Sociology* 25:219–25.

Giordano, P.C., W. D. Manning, and M. A. Longmore. 2005. The romantic relationships of African-American and White adolescents. *The Sociological Quarterly* 46:545–68.

Guerrero, L. K., M. R. Trost, and S. M. Yoshimura. 2005. Romantic jealousy: Emotions and communicative responses. *Personal Relationships* 12: 233–52.

Halpern, C. T., R. B. King, S. G. Oslak, and J. R. Udry. 2005. Body mass index, dieting, romance, and sexual activity in adolescent girls: Relationships over time. *Journal of Research on Adolescence* 15:535–59.

Haynie, D. L., P. C. Giordano, W. D. Manning, and M. A. Longmore. 2005. Adolescent romantic relationships and delinquency involvement. *Criminology* 43:177–210.

Hendrick, S. S., C. Hendrick, and N. L. Adler. 1988. Romantic relationships: Love, satisfaction, and staying together. *Journal of Personality and Social Psychology* 54:980–88.

Huston, T. L., J. P. Caughlin, R. M. Houts, S. E. Smith, and L. J. George. 2001. The connubial crucible: Newlywed years as predictors of marital delight, distress, and divorce. *Journal of Personality and Social Psychology* 80:237–52.

Inman-Amos, J., S. S. Hendrick, and C. Hendrick. 1994. Love attitudes: Similarities between parents and between parents and children. *Family Relations* 43:456–61.

Jankowiak, W. R., and E. F. Fischer. 1992. A crosscultural perspective on romantic love. *Ethnology* 31:149–55.

Jefson, C. 2006. Candy hearts: Messages about love, lust, and infatuation. *Journal of School Health.* 76:117–22.

Johnson, S. 2005. *When "I love you" turns violent: Recognizing and confronting dangerous relationships.* New York: New Horizon Press.

Jones, D. 2006. One of USA's exports: Love, American style. *USA Today* February 14:1B.

Khanchandani, L. 2005. Jealousy during dating among college women. Paper, Third Annual East Carolina University Undergraduate Research and Creative Activities Symposium, Greenville, NC, April 8.

King, P. A. 2003. Stalking: A control factor. Paper, 73rd Annual Meeting of the Eastern Sociological Society, Philadelphia, February 28.

Kito, M. 2005. Self-disclosure in romantic relationships and friendships among American and Japanese college students. *Journal of Social Psychology* 145:127–40.

Knox, D., and M. E. Zusman. 2006. Relationship and sexual behaviors of a sample of 1027 university students. Unpublished data collected for this text. Department of Sociology, East Carolina University, Greenville, NC.

Knox, D., M. E. Zusman, L. Mabon, and L. Shivar. 1999. Jealousy in college student relationships. *College Student Journal* 33:328–29.

Knox, D., M. Zusman, and W. Nieves. 1998. What I did for love: Risky behavior of college students in love. *College Student Journal* 32:203–05.

Knox, D., M. E. Zusman, and H. R. Thompson. 2004. Emotional perceptions of self and others: Stereotypes and data. *College Student Journal* 38:130–35.

Lee, J. A. 1973. *The colors of love: An exploration of the ways of loving.* Don Mills, Ontario: New Press.

———. 1988. Love-styles. In *The Psychology of Love,* edited by R. Sternberg and M. Barnes. New Haven, CN: Yale University Press, 38–67.

Medora, N. P., J. H. Larson, N. Hortacsu, and P. Dave. 2002. Perceived attitudes towards romanticism: A cross-cultural study of American, Asian-Indian, and Turkish young adults. *Journal of Comparative Family Studies* 33:155–78.

Meloy, J. R. and H. Fisher. 2005. Some thoughts on the neurobiology of stalking. *Journal of Forensic Science* 50:1472–80.

Meyers, J. E., and M. Shurts. 2002. Measuring positive emotionality: A review of instruments assessing love. *Measurement and Evaluation in Counseling and Development* 34:238–54.

Monteoliva, A., J. Garcia-Martinez, and A. Miguel. 2005. Adult attachment style and it's effect on the quality of romantic relationships in Spanish students. *Journal of Social Psychology* 145:745–47.

Paul, E. L., B. McManus, and A. Hayes. 2000. "Hookups": Characteristics and correlates of college students' spontaneous and anonymous sexual experiences. *Journal of Sex Research* 37:76–88.

Petrie, J., J. A. Giordano, and C. S. Roberts. 1992. Characteristics of women who love too much. *Affilia: Journal of Women and Social Work* 7:7–20.

Picoult, J. 2001. *Salem Falls*. New York: Washington Square Press.

Pimentel, E. E. 2000. Just how do I love thee? Marital relations in urban China. *Journal of Marriage and the Family* 62:32–47.

Pines, A. M. 1992. *Romantic jealousy: Understanding and conquering the shadow of love.* New York: St. Martin's Press.

Pistole, M. C., and L. C. Vocaturo. 2000. Attachment and commitment in college students' romantic relationships. *Journal of College Student Development* 40:710–20.

Radmacher, K. and M. Azmitia. 2006. Are there gendered pathways to intimacy in early adolescents' and emerging adults' friendships? *Journal of Adolescent Research* 21: 415–448

Regan, P. 2000. Love relationships. In *Psychological perspectives on human sexuality,* edited by L. T. Szuchman and F. Muscarella. New York: Wiley, 232–82.

Reid, M. J. 2005. Some thoughts on the neurobiology of stalking. *Journal of Forensic Sciences* 50:1472–80.

Reiss, I. L. 1960. Toward a sociology of the heterosexual love relationship. *Journal of Marriage and Family Living* 22:139–45.

Ross, C. B. 2006. An exploration of eight dimensions of self-disclosure on relationship. Paper, Southern Sociological Society, New Orleans, LA. March 24.

Sanderson, C. A., and K. H. Karetsky. 2002. Intimacy goals and strategies of conflict resolution in dating relationships: A mediational analysis. *Journal of Social and Personal Relationships* 19:317–37.

Sedensky, M. 2006. Research helps Hallmark find ways to express love. *The Daily Reflector,* February 14:A10.

Shackelford, T. K., A. T. Goetz, D. M. Buss, H. A. Euler, and S. Hoier. 2005. When we hurt the ones we love: Predicting violence against women from men's mate retention. *Personal Relationships* 12:447–63.

Sharp, E. A., and L. H. Ganong. 2000. Raising awareness about marital expectations: An unrealistic beliefs change by integrative teaching? *Family Relations* 49:71–76.

Sheff, E. 2006. The reluctant polyamorist: Conducting auto-ethnographic research in a sexualized setting. In *Sex Matters: The Sexuality and Society Reader,* edited by M. Stombler, D. Baunach, E. Burgess, D. Donnelly, and W. Simonds. New York: Pearson, Allyn, and Bacon.

Soriano, C. G. 2005. Prince Charles and Camilla to wed. *USA Today* A1.

Spitzberg, B. H., and W. R. Cupach, eds. 1998. *The dark side of close relationships.* Mahway, N.J.: Erlbaum.

Sprecher, S., and P. C. Regan. 1998. Passionate and companionate love in courting and young married couples. *Sociological Inquiry* 68:163–85.

Statistical Abstract of the United States: 2006. 125th ed. Washington, D.C.: U.S. Bureau of the Census.

Sternberg, R. J. 1986. A triangular theory of love. *Psychological Review* 93:119–35.

Tilley, D. S., and M. Brackley. 2005. Men who batter intimate partners: A grounded theory study of the development of male violence in intimate partner relationships. *Issues in Mental Health Nursing* 26:281–97.

Toufexis, A. 1993. The right chemistry. *Time* 15 February, 49–51.

Trachman, M. and C. Bluestone. 2005. What's love got to do with it? *College Teaching* 53:131–36.

Tzeng, O. C. S., K. Wooldridge, and K. Campbell. 2003. Faith love: A psychological construct in intimate relations. *Journal of the Indiana Academy of the Social Sciences* 7:11–20.

Walster, E., and G. W. Walster. 1978. *A new look at love.* Reading, MA: Addison-Wesley.

Watts, S., and P. Stenner. 2005. The subjective experience of partnership love: A Q methodological study. *British Journal of Social Psychology* 44:85–107.

West III, J. L. W. 2005. *The perfect hour.* New York: Random House.

Authors

Everywhere, marriage is becoming more optional.

Stephanie Coontz, A History of Marriage

Being Single, Hanging Out, Hooking Up, and Living Together

Contents

True or False?

1. Never-married single women report higher life satisfaction than never-married single men.

2. Black adults and white adults of all ages (>18 years) are about equal in terms of the percentage who are unmarried.

3. Singlehood is becoming normative in that 60 percent of individuals will choose not to marry.

4. Persons who live together and then marry are less likely to get divorced than those who don't live together before marriage.

5. Friendship, not romance or sex, is the goal of university students who go on the Internet looking for someone.

Answers: **1.** T **2.** F **3.** F **4.** F **5.** T

C oncerns about launching one's career, paying off debts, and enjoying the freedom of singlehood (which implies avoiding expectations of either a spouse or a child) have propelled a new generation of individuals to delay marriage into their late twenties. The pattern adopted by today's youth is basically a ten-year float from the late teens to the late 20s. And U.S. young adults are not alone: individuals in France, Germany, and Italy are engaging in a similar pattern of delaying marriage. In the meantime, the process of courtship has evolved, with various labels and patterns, including "hanging out" (undergraduates rarely use the term **dating**), "hooking up" (the new term for "one night stand"), and "pairing off," which may include cohabitation as a prelude to marriage. We begin with examining singlehood versus marriage.

Singlehood

In this section we discuss the effect of social movements on the acceptance of singlehood, the various categories of singles, the choice to be permanently unmarried, and the human immunodeficiency virus (HIV) infection risk associated with this choice.

Social Movements and the Acceptance of Singlehood

Though more than 95 percent of U.S. adults eventually marry (*Statistical Abstract of the United States,* 2006; Table 50), more people are delaying marriage and enjoying singlehood. The acceptance of singlehood as a lifestyle can be attributed to social movements—the Sexual Revolution, the Women's Movement, and the Gay Liberation movement.

The Sexual Revolution involved openness about sexuality and permitted intercourse outside the context of marriage. No longer did people feel compelled to wait until marriage for involvement in a sexual relationship. Hence, the sequence changed from dating, love, maybe intercourse with a future spouse, and then marriage and parenthood to "hanging out," "hooking up" with numerous partners, maybe living together (in one or more relationships), marriage, and children.

The Women's Movement emphasized equality in education, employment, and income for women. As a result, rather than get married and depend on a husband for income, women earned higher degrees, sought career opportunities, and earned their own income. This economic independence brought with it independence of choice. Women could afford to remain single or to leave an unfulfilling or abusive relationship.

The Gay Liberation movement, with its push for recognition of same-sex marriage, has increased the visibility of gay people/relationships. Though some gays still marry heterosexuals to provide a traditional social front, the gay liberation movement has provided support for a lifestyle consistent with one's sexual orientation. This includes rejecting traditional heterosexual marriage. Today, some gay pair-bonded couples regard themselves as married even though they are not legally wed. Some gay couples have formal wedding ceremonies in which they exchange rings and vows of love and commitment.

In effect, there is a new wave of youth who feel that their commitment is to themselves in early adulthood and to marriage in their late 20s and 30s, if at all.

The increased acceptance of singlehood translates into staying in school or getting a job, establishing oneself in a career, and becoming economically and emotionally independent from one's parents. The old pattern was to leap from high school into marriage. The new pattern is to wait until after college or after becoming established in a career—in effect, to enjoy youthhood (see next section). A few (less than five percent of U.S. adults) opt for remaining single forever.

The value young adults attach to singlehood may vary by race. Brewster (2006) interviewed 40 African-American men 18 to 25 years of age in New York City. These men dated numerous women and some maintained serious relationships with multiple women simultaneously. These men "have redefined family for themselves, and marriage is not included in the definition" (p. 32).

Youthhood

Sociologist James Cote identified *youthhood* as that period of time between adolescence and adulthood, which is characterized by lower percentages of youth finishing school, leaving home, getting married, having a child, and reaching financial independence by age 30 than in previous years (Jayson, 2004). In effect, today's youth are in no hurry to rush into marriages that they see may end in divorce or lodge themselves in a career they feel unsure about. They are moving back home to save money, traveling, and keeping their options for the future open.

Categories of Singles

The term *singlehood* is most often associated with young unmarried individuals. But there are three categories of singles: the never-married, the divorced, and the widowed.

Never-Married Singles   Kevin Eubanks (*Tonight Show* music director), Oprah Winfrey, Diane Keaton, and Drew Carey are examples of heterosexuals who have never married. Nevertheless, it is rare for a person to remain single his or her entire life. One reason is stigma. DePaulo (2006) asked 950 undergraduate college students to describe singles. In contrast to marrieds, who were described as "happy, loving, stable,"singles were described as "lonely, unhappy, and insecure."

National Data

By age 75, only 3.6 percent of U.S. women and 3.8 percent of U.S. men have never married (*Statistical Abstract of the United States: 2006*, Table 51)

Though the never-married singles consist mostly of those who want to marry someday, increasingly these individuals are comfortable delaying marriage to pursue educational and career opportunities. Others, such as African-American women who have never married, note a lack of potential marriage partners. Educated black women report a particularly difficult time finding eligibles from which to choose.

There's only one thing worse than the old who think nothing matters but the past, and that's the young who think nothing matters but the present.

Evan Esar, humorist

Most women find themselves never-married by accident, not design. Almost all are surprised by the fact that they're not married. It's not something they dreamed of.

Carol Anderson, psychologist

This woman chose to never marry. She noted, "I've had a wonderful life with many lovers. I'm sure I've had more loves and adventures than if I had married."

The 25 women I interviewed ages 36 to 83, who had never married, were happy, with satisfying life work and strong attachments to their family, friends, and community.

Judith Sills, psychologist

National Data

Among adults 18 years and older, about 38 percent of black women and 42 percent of black men have never married, in contrast to about 17 percent of white women and 26 percent of white men (*Statistical Abstract of the United States: 2006,* Table 50).

As noted in the national data, there is a great racial divide in terms of remaining single. In her article on "Marriage is for White People," Jones (2006, p. B5) notes:

Sex, love, and childbearing have become a la carte *choices rather than a package deal that comes with marriage. Moreover, in an era of brothers on the "down low," the spread of sexually transmitted diseases and the decline of the stable blue-collar jobs that black men used to hold, linking one's fate to a man makes marriage a risky business for a black woman.*

In spite of the viability of singlehood as a lifestyle, stereotypes remain, as the never married are viewed as desperate or swingers. Jessica Donn (2005) emphasized that since mature "adulthood" implies that one is married, singles are left to negotiate a positive identity outside of marriage. She studied the subjective well being of 171 self-identified heterosexual, never-married singles (40 men and 131 women), aged 35 to 45 years, who were not currently living with a romantic partner. Results revealed that women participants reported higher life satisfaction and positive affect than men. Donn hypothesized that having social connections/close friendships (greater frequency of social contact, more close friends, someone to turn to in times of distress, and greater reciprocity with a confidant) was the variable associated with higher subjective well being for both women and men (but women evidenced greater connectedness). Other findings included that, for never-married men, relationships, career, older age, and avoiding thoughts of old age contributed to their sense of well being. For women, financial security, relationships, achievement, and control over their environment contributed to a sense of well being. See Table 4.1 on page 103 for the issues involved in being a single woman.

When singles and marrieds are compared, marrieds report being happier. Lucas et al. (2003) analyzed data from a 15-year longitudinal study of more than 24,000 individuals and found that married people were happier than single people and hypothesized that marriage may draw persons who are already more satisfied than average (e.g., unhappy people may be less in demand as marriage partners).

Divorced Singles   People who are divorced are also among the single. For many of the divorced, the return to singlehood is not an easy transition. The separated and divorced are the least likely to say that they are "very happy" with their life: only 18 percent of divorced men, compared with 36 percent of married men, said they were very happy. Likewise, only 19 percent of divorced women, compared with 40 percent of married women, said they were very happy (Glenn and Weaver, 1988).

National Data

There were 12,804,000 divorced females and 8,956,000 divorced males in the United States in 2004 (*Statistical Abstract of the United States: 2006,* Table 51).

Table 4.1 A Never-Married Single Woman's View of Singlehood

A never-married woman, 40 years of age, spoke to the authors' Marriage and Family class about her experience as a single woman. The following is from the outline she developed and the points she made about each topic.

Stereotypes about Never-Married Women

Various assumptions are made about the never-married woman and why she is single. These include the following:

Unattractive—she's either overweight or homely, or else she would have a man.

Lesbian—she has no real interest in men and marriage because she is homosexual.

Workaholic—she's career-driven and doesn't make time for relationships.

Poor interpersonal skills—she has no social skills and she embarrasses men.

History of abuse—she has been turned off to men by the sexual abuse of, for example, her father, a relative, or a date.

Negative previous relationships—she's been rejected again and again and can't hold a man.

Man-hater—deep down, she hates men.

Frigid—she hates sex and avoids men and intimacy.

Promiscuous—she is indiscriminate in her sexuality so that no man respects or wants her.

Too picky—she always finds something wrong with each partner and is never satisfied.

Too weird—she would win the Miss Weird contest, and no man wants her.

Positive Aspects of Being Single

1. Freedom to define self in reference to own accomplishments, not in terms of attachments (e.g., spouse).
2. Freedom to pursue own personal and career goals and advance without the time restrictions posed by a spouse and children.
3. Freedom to come and go as you please and to do what you want, when you want.
4. Freedom to establish relationships with members of both sexes at desired level of intensity.
5. Freedom to travel and explore new cultures, ideas, values.

Negative Aspects of Being Single

1. Increased extended-family responsibilities. The unmarried sibling is assumed to have the time to care for elderly parents.
2. Increased job expectations. The single employee does not have marital or family obligations and consequently can be expected to work at night and on weekends.
3. Isolation. Too much time alone does not allow others to give feedback such as "Are you drinking too much?" "Have you had a checkup lately?" or "Are you working too much?"
4. Decreased privacy. Others assume the single person is always at home and always available. They may call late at night or drop in whenever they feel like it. They tend to ask personal questions freely.
5. Less safety. A single woman living alone is more vulnerable than a married woman with a man in the house.
6. Feeling different. Many work-related events are for couples, husbands, and wives. A single woman sticks out.
7. Lower income. Single women have much lower incomes than married couples.
8. Less psychological intimacy. The single woman may have no one supportive waiting for her at the end of the workday.
9. Negotiation skills lie dormant. Since there is no one with whom the single woman must negotiate issues on a regular basis, she may become deficient in compromise and negotiation skills.
10. Patterns become entrenched. Since there is no other person around to express preferences, the single person may establish a very repetitive lifestyle.

Maximizing One's Life as a Single

1. Frank discussion. Talk with parents about your commitment to and enjoyment of the single lifestyle and request that they drop marriage references. Talk with siblings about joint responsibility for aging parents and your willingness to do your part. Talk with employers about spreading workload among all workers, not just those who are unmarried and childfree.
2. Relationships. Develop and nurture close relationships with parents, siblings, extended family, and friends to have a strong and continuing support system.
3. Participate in social activities. Go to social events with or without a friend. Avoid becoming a social isolate.
4. Be cautious. Be selective in sharing personal information such as your name, address, and phone number.
5. Money. Pursue education to maximize income; set up retirement plan.
6. Health. Exercise, have regular checkups, and eat healthy food. Take care of yourself.

Divorced persons tend not to live as long as marrieds or are more likely to die at a younger age. On the basis of a study of 44,000 deaths, Hemstrom (1996) observed that "on the whole, marriage protects both men and women from the higher mortality rates experienced by unmarried groups" (p. 376). One explanation is the protective aspect of marriage. "The protection against diseases and mortality that marriage provides may take the form of easier access to social support, social control, and integration, which leads to risk avoidance, healthier lifestyles, and reduced vulnerability" (p. 375).

Widowed Singles   Whereas some divorced people choose to be single rather than remain in an unhappy marriage, the widowed are forced into singlehood. As a group, the widowed are happier than the divorced but not as happy as the married. Twenty-one percent of widowed men, in contrast to 18 percent of divorced men and 36 percent of married men, reported that they were very happy. Among women, 29 percent of the widowed, compared with 19 percent of the divorced and 40 percent of the married, reported that they were very happy (Glenn and Weaver, 1988). Lucas et al. (2003) also found that widows are struggling to regain their previous level of satisfaction.

National Data

There were 11,141,000 widowed females and 2,641,000 widowed males in the United States in 2004 (*Statistical Abstract of the United States: 2006,* Table 51).

Of all singles, consisting of the never married, divorced, and widowed, the divorced and widowed represent the majority of singles. These individuals are in their middle years. But being single does not mean being uninvolved. In a study of single women between the ages of 40 and 69 conducted by the American Association of Retired Persons (AARP), most were attached or involved in a relationship. Almost a third (31%) were in an exclusive relationship; another third (32%) were dating nonexclusively. Of the remaining 37 percent, only 13 percent reported that they were actively looking. Ten percent said that they had no desire to date; the rest said they were open to meeting someone but not obsessive about it (Mahoney, 2006).

Are More People Choosing Singlehood?

As noted earlier, about 96 percent of both women and men eventually marry. Whether today's individuals are embracing singlehood forever or just delaying marriage is unknown. We won't know until they reach their 70s, which is the age by which more than 95 percent marry. Indeed, Gloria Steinem, the ardent feminist, who once spoke of marriage as a prison, married when she was 66. We do know, however, that people are opting for singlehood longer than in the past. Although the median age for marriage is 25 for women and 27 for men, today about a quarter of women and a third of men aged 30 to 34 are single. This percentage is three to four times higher than in the 1970s (Vanderkam, 2006). Table 4.2 lists the standard reasons people give for remaining single. The primary advantage of remaining single is freedom and control over one's life. Once a decision has been made to involve another in one's life, it follows that one's choices become vulnerable to the influence of that other person. The person who chooses to remain single may view the needs and influence of another person as things to avoid.

Some people do not set out to be single but drift into being single longer than they anticipated—and discover that they like it. Meredith Kennedy is a never-married veterinarian who has found that singlehood works very well for her. She writes:

> *As a little girl, I always assumed I'd grow up to be swept off my feet, get married, and live happily ever after. Then I hit my 20s, became acquainted with reality, and*

Table 4.2 Reasons to Remain Single

Benefits of Singlehood	Limitations of Marriage
Freedom to do as one wishes	Restricted by spouse or children
Variety of lovers	One sexual partner
Spontaneous lifestyle	Routine, predictable lifestyle
Close friends of both sexes	Pressure to avoid close other-sex friendships
Responsible for one person only	Responsible for spouse and children
Spend money as one wishes	Expenditures influenced by needs of spouse/children
Freedom to move as career dictates	Restrictions on career mobility
Avoid being controlled by spouse	Potential to be controlled by spouse
Avoid emotional/financial stress of divorce	Possibility of divorce

discovered that I had a lot of growing up to do. I'm still working on it now, in my late 30s.

Growing up for me has become an ongoing adventure in individuality and creating my own path. This is a very personal journey, and remaining single has helped me envision who I am and who I'm evolving into. Marriage for a lot of people (it seems to me) brings an end, or a plateau, to individual development, leading to boredom and disenchantment. Life is not a dress rehearsal, so I want to continue to learn and grow as much as I can, and not settle for any situation that hinders this. A marriage in which both partners could really continue to evolve as individuals would be ideal, but this takes a lot of maturity and self-awareness, and I'm still working on growing up.

I've gotten a lot out of my relationships with men over the years, some serious and some not so serious, and they've all left their impression on me. But gradually I've moved away from considering myself "between boyfriends" to getting very comfortable with being alone, and finding myself good company. The thought of remaining single for the rest of my life doesn't bother me, and the freedom that comes with it is very precious. I've worked and traveled all over the world, and my schedule is my own. I've lived in West Africa (learning to play drums), East Africa (teaching marine biology), and southern Africa (wildlife biology), and now I'm taking six months off to write a book. Currently I have two professions and I'm working on a transition to a third, all of which fill my life with passion, intensity, and creativity. I don't think this could have come about with the responsibilities of marriage and a family, and the time and space I have as a single woman have allowed me to really explore who I am in this life.

I love children, and I'm fortunate to have a number of friends with families, so I'm "auntie" to several wonderful little people. Parenting is a very difficult job, and I think there's a need for parents and their children to have relationships with single adults who can add their love, time, and perspective.

Traditionally the demands of parenthood would have been spread out among an extended family, but that doesn't happen much anymore, so I see myself (and other single people) filling an important role in helping kids grow up. This then becomes my contribution to society, and I don't need to have my own children to feel complete.

It's not always easy to explain why I'm single in a culture that expects women to get married and have children, but the freedom and independence I have allow me to lead a unique and interesting life.

Diversity in the United States

Ferguson (2000) studied 62 never-married Chinese-American and Japanese-American women and found that 40 percent expressed some regret at not being married. However, for most of these women, their regret was primarily about not having children and not about never marrying. Instead, most of these never-married women were happy with the decisions they made and were living rich and fulfilling lives. Most were economically successful, immersed in a community of friends and family, and actively involved in their work or community projects (p. 155).

Is Singlehood for You?

Singlehood is not a one-dimensional concept. Whereas some are committed to single-hood, others enjoy it for now but intend to eventually marry, and still others are conflicted about it. There are many styles of singlehood from which to choose. As a single person, you may devote your time and energy to career, travel, privacy, heterosexual or homosexual relationships, living together, communal living, or a combination of these experiences over time. An essential difference between traditional marriage and single-hood is the personal and legal and social freedom to do as you wish. Although single-hood offers freedom, single people are sometimes challenged by such issues as loneliness, less money, and establishing an identity.

1. Loneliness. For some singles, being alone is a desirable and enjoyable experience. "The major advantage of being single," said one 29-year-old man, "is that I don't have to deal with another person all the time. I like my privacy." Ninety-three percent of the single women in the AARP survey noted that their "independence" was important for their quality of life (and that this overshadowed the occasional feelings of loneliness) (Mahoney, 2006). Henry David Thoreau, who never married, spent two years alone on 14 acres bordering Walden Pond in Massachusetts. He said of his experience, "I love to be alone. I never found the companion that was so companionable as solitude."

Nevertheless, some singles are lonely. In the AARP study of single women aged 40 to 69, 28 percent reported that in the past two weeks they had felt lonely occasionally or most of the time (13% of married women reported these same feelings) (Mahoney, 2006). One researcher (Pinquart, 2003) noted that unmarried single men report more loneliness than single women.

2. Less money. Married couples who combine their incomes usually have more income than single people living alone. The median income of a married couple is $62,281, compared with $38,032 for a male householder with no wife and $26,550 for a female householder with no husband (*Statistical Abstract of the United States:* 2006, Table 682). Lichter et al. (2003) also confirmed higher rates of poverty among the never married and divorced. In addition, debt can be a serious issue. The average female aged 45 to 59 carries $11,414 in revolving debt; divorced women and those between the ages of 45 and 49 are the least likely to pay off their credit cards (Mahoney, 2006).

3. Social Identity. Single people must establish a social identity—a role—that helps to define who they are and what they do, independent of the role of spouse. Couples eat together, sleep together, party together, and cooperate economically. They mesh their lives into a cooperative relationship that gives them the respective identity of being a spouse. On the basis of their spousal roles, we can predict what they will be doing most of the time. For example, at noon on Sunday, they are most likely to be having lunch together. Not only can we predict what they will be doing, but their roles as spouses tell them what they will be doing—interacting with each other.

The single person finds other roles and avenues to identity. A meaningful career is the avenue most singles pursue. A career provides structure, relationships with others, and a strong sense of identity ("I am a veterinarian and love my work," said one woman).

To the degree that singles find meaning in their work, they are successful in establishing autonomous identities independent of the marital role. In evaluating the single lifestyle, to what degree, if any, do you feel that loneliness is or would be a problem for you? What are your economic needs, and to what degree might education help to meet those needs? The old idea that you can't be happy unless you are married is no longer credible. Whereas marriage will be the first option for some, it will be the last option for others. As one 76-year-old single-by-choice said, "A spouse would have to be very special to be better than no spouse at all."

A woman is destined to remain single when she starts buying her shoes for comfort and her sweaters for warmth.

Evan Esar

Sources

Lichter, D. T., D. R. Graefe, and J. B. Brown. 2003. Is marriage a panacea? Union formation among economically disadvantaged unwed mothers. *Social Problems* 50:60–86.

Mahoney, S. 2006. The secret lives of single women—lifestyles, dating and romance: A study of midlife singles. *AARP: The Magazine* May/June 62–69.

Statistical Abstract of the United States: 2006. 125th ed. Washington, D.C.: U.S. Bureau of the Census.

Singlehood and Human Immunodeficiency Virus Infection Risk

Individuals who are not married or not living with someone are at greater risk for contracting human immunodeficiency virus (HIV) and other sexually transmitted infections (STIs). In a study comparing single and married white women, the former reported feeling more vulnerable to HIV and STI than the latter, and they are more likely to engage in risky sexual behavior (Wayment et al., 2003). Though women typically report having had fewer sexual partners than men, the men they had sex with usually had had multiple sexual partners. Hence, women are more likely to get infected from men than men are from women. In addition to the social reason for increased risk of infection, there is a biological reason—sperm that may be HIV-infected is deposited into the woman's body.

Intentional Communities

Although singles often live in apartments, condominiums, or single-family houses, living in an intentional community (previously called a commune) is an alternative. More than one-third (35.3%) of 1027 undergraduates at a large southeastern university agreed, "Someday, I would like to live in an intentional community [a commune]" (Knox and Zusman, 2006). Intentional communities are not just for the single or the young. Marrieds and seniors also embrace communal living. Blue Heron Farm, for example, is home to a collective of individuals from their 20s to 70s who live in ten houses (Yeoman, 2006). The range of intentional community alternatives may be explored via the Cohousing Association of the United States (www.cohousing.org) and the Federation of Egalitarian Communities (www.thefec.org). In the next section we describe one such intentional community.

Twin Oaks: An Alternative Context for Living

Twin Oaks is a community of 90 adults and 15 children living together on 450 acres of land in Louisa, Virginia (about 45 minutes east of Charlottesville and one hour west of Richmond). Known as an **intentional community** (comprising a group of people who choose to live together on the basis of a set of shared values), the **commune** (an older term now replaced by the newer term) was founded in 1967 and is one of the oldest nonreligious intentional communities in the United States.

We embrace diversity. Our two rules are no TV and no guns.

Pat Starr, Twin Oaks member of eight years

The membership is 55 percent male and 45 percent female. Most of the members are white, but there is a wide range of racial, ethnic, and social class backgrounds. Members include gay, straight, bisexual, and transgender people; most are single never-married adults, but there are married couples and families. Sexual values range from celibate to monogamous to polyamorous (involvement in more than one emotional/sexual relationship at the same time). The age range of the members is newborn to 78, with an average age of 40.

The average length of stay for current members is over seven years. There are no officially sanctioned religious beliefs at Twin Oaks—it is not a "spiritual" community, although its members represent various religious values (Jewish, Christian, pagan, atheist, etc.). The core values of the community are nonviolence (no guns, low tolerance for violence of any kind in the community, no parental use of violence against children), egalitarianism (no leader, with everyone having equal political power and the same access to resources), and environmental sustainability (the community endeavors to live off the land, growing its own food and heating its buildings with wood from the forest).

Another core value is income sharing. All the members work together to support everyone in the community instead of working individually to support themselves. Money earned from the three community businesses (weaving hammocks, making tofu, and writing indexes for books) is used to provide food and other basic needs for all members. No one needs to work outside the community. Each member works in the community in a combination of income-producing and domestic jobs. An hour of cooking or gardening receives the same credit as an hour of fixing computers or business management. No matter what work one chooses to do, each member's commitment to the community is 42 hours of work a week. This includes preparing meals, taking care of children, cleaning bathrooms, and maintaining buildings—activities not considered in the typical mainstream 40-hour workweek.

The community places high value on actively creating a "homegrown" culture. Members provide a large amount of their own entertainment, products, and services. They create homemade furniture, present theatrical and musical performances, and enjoy innovative community holidays such as Validation Day (a distant relative of Valentine's Day, minus the commercialism). The community also encourages participation in activism outside the community. Many members are activists for peace-and-justice, feminist, and ecological organizations. (This section is based on information provided by Kate Adamson and Ezra Freeman, members of Twin Oaks, and is used with their permission.)

Undergraduate Interest in Finding a Partner

A significant portion of college students are not involved in an emotional relationship but are looking for a partner. In a sample of 1027 undergraduates, more than one quarter (27.6%) reported that they were not dating and not involved with anyone (Knox and Zusman, 2006). In a sample of 377 first-year students at the same university, more than half of the men (57%) and more than a third of the women (43%) reported that finding a girlfriend/boyfriend was important (Vail-Smith et al., 2006). Indeed, all societies have a way of moving women and men from same-sex groups into pair-bonded legal relationships for the reproduction and socialization of children.

Ways of Finding a Partner

The meaning of the word dating has become ambiguous, and new phrases such as "hanging out," "hooking up," and "just talking" have emerged as descriptors of intimate relationships with varying degrees of commitment.

Tracy Luff and Kristi Hoffman, sociologists

One of the unique qualities of the college/university environment is that it provides a context in which to meet thousands of potential partners of similar age, education, and social class. This context will likely never recur following graduation. Although people often meet through friends or on their own in school, work, or recreation contexts, an increasing number are open to a range of alternatives, from hanging out to the Internet.

Hanging Out

The term *hanging out* has made its way into the professional literature (Harcourt, 2005; Thomas, 2005). **Hanging out,** also referred to as getting together, refers to going out in groups where the agenda is to meet others and have fun. The individuals may watch television, rent a video, go to a club or party, and/or eat out. Hanging out may be considered "testing the waters," as a possible prelude to more serious sexual involvement and/or commitment (Luff and Hoffman, 2006). More than 90 percent of 1027 undergraduates reported that "hanging out for me is basically about meeting people and having fun" (Knox and Zusman, 2006). Hanging out may occur in group settings such as at a bar, a sorority or fraternity party, or a small gathering of friends that keeps expanding. Individuals may be introduced by friends or meet someone "cold," as in initiating a conversation. There is usually no agenda beyond meeting and having fun. Fewer than five percent of the 1027 respondents said that hanging out was about beginning a relationship that may lead to marriage (3.7%) or meeting someone to hook up with (2.8%) (Knox and Zusman, 2006).

Hooking Up

Hooking up is also a term that has entered the social science literature and become the focus of research. **Hooking up** is defined as a one-time sexual encounter in which there is generally no expectation of seeing each other again. The nature of the sexual expression may be making out, oral sex, and/or sexual intercourse. The term is also used to denote getting together periodically for a sexual encounter, with no strings attached (Luff and Hoffman, 2006). Renshaw (2005) wrote his dissertation on hooking up and noted that hooking up involves playful/nonserious interaction where one's body and the context of alcohol move the encounter to a sexual ending. The Institute for American Values (a conservative think tank) collected data from 1000 undergraduates, 40 percent of whom reported that they had "hooked up" at least once (Glenn and Marquardt, 2001). In a sample of 615 university students, a little more than 20 percent reported never having hooked up. Of those who had hooked up, 20 percent had done so five to ten times; more than a third had hooked up more than ten times (England and Thomas, 2006).

Bogle (2003) interviewed 57 college students and alumni at two universities in the eastern United States about their experiences with dating and sex. She found that hooking up had become the *primary* means for heterosexuals to get together on campus. About half (47%) of hook ups start at a party and involve alcohol, with men averaging 5 drinks and women 3 drinks (England and Thomas, 2006). The sexual behaviors that were reported to occur during a hook up included kissing and nongenital touching (34%), hand stimulation of genitals (19%), oral sex (22%), and intercourse (23%) (England and Thomas, 2006).

Researchers (Bogle, 2003; Glenn and Marquardt, 2001) note that although hooking up may be an exciting sexual adventure, it is fraught with feelings of regret. Some of the women in their studies were particularly disheartened to discover that hooking up usually did not result in the development of a relationship that went beyond a one-night encounter.

Indeed, Bogle (2002) noted three outcomes of hooking up. In the first, described above, nothing results from the first-night sexual encounter. In the second, the college students will repeatedly hook up with each other on subsequent "hanging out" occasions. However, a low level of commitment characterizes this type of relationship, in that each person is still open to hooking up with someone else. A third outcome of hooking up, and the least likely, is that the two people begin going out or spending time together in an exclusive relationship. Hence, hooking up is most often a sexual adventure that rarely results in the development of a relationship. However, there are exceptions. When 1027 undergraduates were presented with the statement "People who 'hook up' and have sex the first night don't end up in a stable relationship," 43.3% agreed, 20.6% disagreed, and 33.1% neither agreed nor disagreed (Knox and Zusman, 2006).

Renshaw (2005) noted that hooking up is becoming normative and replacing contemporary patterns of dating. He suggested that it is "shrouded in deception," "contains individual health risks associated with sex," and "may also threaten traditional conceptions of marriage and family."

The Internet—Meeting Online

According to Jupiter Research, more than 17 million individuals browse the Internet annually to find a partner. About one-third of them are satisfied with their online experience; most do not anticipate the time it takes to go through

I look for love, not hook-ups, alas. I hate to admit it.

Erica Jong, feminist

This Internet is a quick way to meet a variety of new people—but be cautious.

Kelly Lewis

I've found Mr. Cheap, Mr. Crude, and Mr. Cocky, so where's Mr. Right?

Personal ad on Match.com

profiles and e-mail potential partners (CNN, 2006). More than ten percent of 1027 undergraduates reported that they had used the Internet to look for a partner (Knox and Zusman, 2006). Internet dating is used more often by persons older than 30 who are busy with their career and have no special place to hang out and meet. "There are only six guys in my office," noted one Internet user in her 30s. "Four are married and the other two are alcoholics."

There are more than 200 Web sites designed for the purpose of meeting a partner online (RightMate at Heartchoice.com is one of them and offers not only a way to meet others but a free "RightMate Checkup" to evaluate whether the person is right for you). Some Web sites are specific to certain communities, such as BlackPlanet.com (black community), Jdate.com (Jewish community), and Gay.com (gay community).

To provide more information about Internet dating, 191 never-married undergraduate university students completed an anonymous 28-item questionnaire designed to assess their attitudes toward and involvement in use of the Internet to find a mate (Knox et al., 2001). The data revealed that among these college students, friendship, not romance or sex, was the primary interest when going on the Internet to meet new people. The survey also found that more than 60 percent of these respondents were successful in establishing an online friendship and almost half felt more comfortable meeting a person online than in person; 40 percent reported that they had lied online.

The advantages of such online meetings include the ability to sift through large numbers of potential partners quickly to identify persons who have the qualities you require (e.g., nonsmoking), the potential to exchange information quickly to assess a person's sense of humor, and the opportunity to develop a relationship with another on the basis of content independent of visual distraction. The Internet also allows one to avoid noisy, smoky bars and to try on new identities. Defining identify as an understanding of who one is, Yurchisin et al. (2005) interviewed individuals who had used an Internet dating service in the past year and found that some posted a profile of who they wanted to be rather than who they were. For example, a respondent reported wanting to be more athletic so she checked various recreational activities she didn't currently engage in but wanted to. The Internet is also a place for persons to "try out" a gay identity if they are very uncomfortable doing so in real life, around persons they know.

Chapter 4 Being Single, Hanging Out, Hooking Up, and Living Together

The disadvantages of online meeting include deception; the potential to fall in love quickly as a result of intense mutual disclosure; not being able to assess "chemistry" or to observe nonverbal cues/gestures or how a person interacts with your friends or family; and the tendency to move too quickly (from e-mail to phone to meeting to first date) to marriage, without spending the requisite two years to get to know each other. Forty percent of the respondents in the Internet study referred to above reported that they had lied online. Men tend to lie about their economic status and women tend to lie about their weight or age. Gibbs et al. (2006) studied online daters at Match.com and found deceptions about physical appearance, age, income, and marital status.

McGinty (2006) noted the importance of using online Internet etiquette, such as not going overboard with information (phone numbers and address), posting photos that are too revealing (these can be copied and posted elsewhere), complaining, or deceiving. "Let them know who you are and who you are looking for," she suggests.

Although some relationships begun online result in lifetime marriages, one should be cautious of meeting someone online. See the Web site WildXAngel (the address is in the Weblinks at the end of this chapter) for horror stories of online dating. Although the Internet is a good place to meet new people, it also allows someone you rejected or an old lover to monitor your online behavior. Most sites note when you have been online last, so if you reject someone online by saying "I'm really not ready for a relationship," that same person can log on and see that you are still looking.

If you are looking for love on the Internet, you better look here first.
Lisa, founder of WildXAngel.com

Video Chatting

Video chatting moves beyond communicating by typewritten words and allows potential partners to see each other while chatting online. iSpQ ("Eye Speak," http://www.ispq.com/) is one of the largest online communities with downloadable software that enables people to visually meet with others all over the world. Hodge (2003) noted, "Unlike conventional chat rooms, iSpQ does not have a running dialogue or conversation for anyone to view. Video chatting allows users to have personal or private conversations with another user. Hence an individual can not only write information but see the person with whom he or she is interacting online." Half of the respondents in Hodge's study of video chat users reported "meeting people and having fun" as their motivation for video chatting.

Speed Dating: The 8-Minute Date

Dating innovations that involve the concept of speed include the 8-minute date. The Web site http://www.8minutedating.com/ identifies these "8 Minute Dating Events" throughout the country, where a person has eight one-on-one dates that last 8 minutes each. If both parties are interested in seeing each other again, the organizer provides contact information so that the individuals can set up another date. Speed dating is cost-effective because it allows the daters to meet face to face without burning up a whole evening. Wilson et. al. (2006) collected data on 19 young men who had 3-minute social exchanges with 19 young women and found that those partners who wanted to see each other again had more in common than those who did not want to see each other again. Common interests were assessed using the Compatibility Quotient (CQ).

Diversity in the United States

Matchmakers and astrologers were the precursors to Match.com and eHarmony. Prior to 1950, Chinese parents with the help of a professional matchmaker arranged for the marriage of their children. The matchmaker was "usually an elderly woman who knew the birthday, temperament, and appearance of every unmarried man and woman in her community" (Xia and Zhou, 2003, p. 231). This woman would visit parents with children who were ready for marriage and propose specific individuals, usually of similar social and economic status. If the parents liked a man the matchmaker was proposing for their daughter, the matchmaker would meet with the man's parents and alert them of the family's interest in their son marrying the daughter. If the parents agreed, a Chinese astrologer would be consulted to see if the signs of the zodiac were compatible. If the signs were off, the marriage would be too, and there would be no further contact between the families.

Speed dating allows individuals to "interview" a series of new people in a single evening.

I got a whole gob of 'em married off.

Ivan Thompson, "Cupid Cowboy"

A variation of speed dating is the baseball blind date. In Richmond, Virginia (where the event originated), 300 singles are given free tickets to a Richmond Braves baseball game, where all they do is take their seat in the stands and meet their blind date. The event initially was advertised as a "Shot in the Dark" and was designed to break the world record for Largest Blind Date.

International Dating

Go to google.com and type in INTERNATIONAL BRIDES, and you will see the array of sites dedicated to finding foreign women for Americans. Not listed is Ivan Thompson, who specializes in finding Mexican women for his American clients. As documented in the movie Cupid Cowboy (Ohayon, 2005), for $3000 Ivan takes males (one at a time) to Mexico (Torreon is his favorite place), places an ad in a local newspaper for a young (age 20–35), trim (<130 pounds) single woman "interested in meeting an American male for romance and eventual marriage" and waits in a hotel for the phone to ring. They then meet/interview an array of "candidates" in the hotel lobby. Ivan says his work is done when his client finds a woman he likes.

Advertising for a partner is not unusual. Jagger (2005) conducted a content analysis of 1094 advertisements and found that young men and older women were the most likely to advertise for a partner. The researcher also noted a trend in women seeking younger men.

Functions of Involvement with a Partner

Meeting and becoming involved with someone has at least six functions: confirmation of a social self, recreation, companionship/intimacy/sex, anticipatory socialization, status achievement, and mate selection.

1. Confirmation of a social self. In Chapter 1 we noted that symbolic interactionists emphasize the development of the self. Parents are usually the first social mirrors in which we see ourselves and receive feedback about who we are; new partners continue the process. When you are hanging out with a person, you are continually trying to assess how that person sees you: Does the person like me? Will the person want to be with me again? When the person gives you positive feedback through speech and gesture, you feel good about yourself and tend to view yourself in positive terms. Hanging out provides a context for the confirmation of a strong self-concept in terms of how you perceive your effect on other people.

2. Recreation. The focus of hanging out and pairing off is fun. Reality television programs such as *Blind Date, Elimidate,* and *The Bachelor* always have recreational activities to help the participants interact; being a fun person seems to be a criterion for being selected. The couples may make only small talk and learn very little about each other—what seems important is not that they have common interests, values, or goals but that they "have fun."

3. *Companionship/intimacy/sex.* Beyond fun, major motivations for finding a new person and pairing off are companionship, intimacy, and sex. The impersonal environment of a large university makes a secure relationship very appealing. "My last two years have been the happiest ever," remarked a senior in interior design. "But it's because of the involvement with my partner. During my freshman and sophomore years, I felt alone. Now I feel loved, needed, and secure."

4. *Anticipatory socialization.* Before puberty, boys and girls interact primarily with same-sex peers. A fifth-grade boy or girl may be laughed at if he or she shows an interest in someone of the other sex. Even when boy-girl interaction becomes the norm at puberty, neither sex may know what is expected of the other. Meeting a new partner and hanging out provides the first opportunity for individuals to learn how to interact with other-sex partners. Though the manifest function of hanging out is to teach partners how to negotiate differences (e.g., how much sex and how soon), the latent function is to help them learn the skills necessary to maintain long-term relationships (empathy, communication, negotiation). In effect, pairing off involves a form of socialization that anticipates a more permanent union in one's life. Individuals may also try out different role patterns, like dominance or passivity, and try to assess the feel and comfort level of each.

5. *Status achievement.* Being involved with someone is usually associated with more status than being unattached and alone. Some may seek such involvement because of the associated higher status. Others may become involved for peer acceptance and conformity to gender roles, not for emotional reasons. Though the practice is becoming less common, a gay person may pair off with someone of the other sex so as to provide a heterosexual cover for his or her sexual orientation.

6. *Mate selection.* Finally, pairing off may eventually lead to marriage, which remains a major goal in our society (American Council on Education and University of California, 2005–2006). Selecting a mate has become big business. B. Dalton, one of the largest bookstore chains in the United States, carries about 200 titles on relationships, about 50 of which are specifically geared toward finding a mate.

PERSONAL CHOICES

Should I Get Involved in a Long-Distance Dating Relationship?

As a result of individuals' meeting online, the delay of marriage, and the respective desires of women and men to finish their educations or launch their careers (which may take them to different states), or separations due to military service, long-distance dating relationships (LDDRs) are increasing. Indeed, there are more than 1.5 million such relationships (and another million marriages separated by careers, not counting those separated by military service [Guldner, 2003]). Although the advantages (Mietzner and Lin, 2005) of these relationships include positive labeling ("even though we are separated, we care about each other enough to maintain our relationship"), staying "high" on a relationship despite not having regular access to each other depends on the following: developing autonomy/independence for the times when you are apart, developing the nonphysical aspect of the relationship (since frequent sex is not an option), developing communication skills so that you can tell your partner how you think and feel, learning how to trust (because you can't always be with your partner, this is especially important), being patient, and permitting each other to have personal time and space.

Knox et al. (2002) analyzed a sample of 438 undergraduates at a large southeastern university on their attitudes and involvement in an LDDR—defined as being sepa-

The effect of separation on a love relationship is like water on fire—a little water makes the fire blaze while a lot of water puts it out.

Unknown

rated from a love partner by at least 200 miles for a period of not less than three months. The median number of miles these LDDR respondents had been separated was 300 to 399 (about a six-hour drive), and the median length of time they were separated was five months. Of the total sample, 20 percent were currently involved in an LDDR, and 37 percent had been previously.

Being separated was a strain on the couple's relationship. One in five (21.5%) broke up, and another one in five (20%) said that the separation made their relationship worse. Only 18 percent reported that the separation improved their relationship (other responses included a mixed effect for 33% and no effect for 9%). Does absence make the heart grow fonder for the beloved? Most of those who have not been separated seem to think so. But 40 percent of those who had experienced an LDDR believed that "out of sight, out of mind" was a more accurate characterization (Knox et al., 2002). One respondent said, "I got tired of being lonely, and the women around me started looking good." However, Guldner (2003) noted that LDDRs are no more likely to end because of infidelity than those relationships in which the partners live in the same town. In this regard, he noted that the quality of the relationship and personality of the individuals are more important factors than distance.

Being frustrated over not being able to be with the partner, loneliness, feeling as though one is missing out on other activities and relationships, missing physical intimacy, and spending a lot of money on phone calls or travel are among the disadvantages of being involved in an LDDR. Another is that your partner may actually look better from afar than up close. One respondent noted that he and his partner could not wait to live together after they had been separated—but "when we did, I found out I liked her better when she wasn't there."

Nevertheless, here are some issues to consider in making an LDDR manageable and keeping the relationship together:

1. Maintain daily contact. In the Knox et al. (2002) study referred to above, actual contact between the lovers during the period of separation was limited. Only 11 percent reported seeing each other weekly, and 16 percent reported that they never saw each other. However, more than three-fourths (77%) reported talking with each other by phone several times each week (22% daily), and more than half (53%) e-mailed the partner several times each week (18% daily). Some partners maintain daily contact by Web cams. One student reported:

> We get to see each other every day, in real time, whenever we want to. Because the connection doesn't interfere with the telephone, we stay connected 24 hours a day. It has been a big help in keeping our relationship going strong. If we need to talk about an important issue, we can do it face to face without worrying about time or money. Also, because we can't physically be together, this device has helped our personal lives as well. We can see each other whenever we want in whatever way that we want. We have been together for over a year, and during that year we have been connected by the Web cams for over eleven months. Technology has certainly helped our relationship last!

2. Enjoy/use the time when apart. Because being separated is often for a period of months (Guldner's [2003] sample expected to be separated for 26 months), it is important to be involved in worthwhile activities with study, friends, work, sports, and personal projects when apart. Otherwise, resentments may spill over into the interaction with your partner.

3. Avoid conflictual phone conversations. Talking on the phone should involve the typical sharing of events. When the need to discuss a difficult topic arises, the phone is not the best place for such a discussion. Rather, it may be wiser to wait and have the discussion face to face. If you decide to settle a disagreement over the phone, stick to it until you have a solution acceptable to both of you.

A woman should leave her husband long enough to increase his appreciation, but not long enough for him to seek consolation.

Unknown

4. Stay monogamous. Agreeing not to be open to other relationships is crucial to maintaining a long-distance relationship. This translates into not dating others while apart. Individuals who say "Let's date others to see if we are really meant to be together" often discover that they are capable of being attracted to and becoming involved with others. Such other involvements usually predict the end of the LDDR. Lydon et al. (1997) studied 69 undergraduates who were involved in LDDRs and found that "moral commitment" predicted the survival of the relationships. Individuals committed to maintaining their relationships are often successful in doing so.

5. Other strategies. Researchers at the University of Pittsburgh revealed that partners who were separated from each other reported preserving, smelling, and wearing the clothes of a sexual partner. Over half the men and almost 90% of the women had deliberately smelled their partner's blouse or shirt to feel a sense of closeness with the partner from whom they were separated (Gardiner, 2005).

Sources

Gardiner, D. 2005. A sniff of your sweetie. *Psychology Today* 38:31–2.

Guldner, G. T. 2003. *Long Distance Relationships: The Complete Guide.* Corona, CA: JFMilne Publications.

Knox, D., M. Zusman, V. Daniels, and A. Brantley. 2002. Absence makes the heart grow fonder? Long-distance dating relationships among college students. *College Student Journal* 36:365–67.

Lydon, J., T. Pierce, and S. O'Regan. 1997. Coping with moral commitment to long-distance dating relationships. *Journal of Personality and Social Psychology* 73:104–13.

Mietzner, S. and L. Li-Wen. 2005. Would you do it again? Relationship skills gained in a long-distance relationship. *College Student Journal* 39:192–200.

Dating after Divorce

Over 2 million Americans get divorced each year. As evidenced by the fact that more than three-quarters of the divorced remarry within five years, most of the divorced are open to a new relationship. But there are differences between this single-again population and those becoming involved for the first time.

1. Older population. Divorced individuals are, on the average, ten years older than persons in the marriage market who have never been married before. Hence the divorced are in their mid- to late thirties. Widows and widowers are usually 40 and 30 years older, respectively (hence, around ages 65 and 55), when they begin to date the second time around. Most divorced persons date and marry others who are divorced.

2. Fewer potential partners. Most men and women who are dating the second time around find fewer partners from whom to choose than when they were dating before their first marriage. The large pool of never-marrieds (24% of the population) and currently marrieds (59% of the population) is usually not considered an option (*Statistical Abstract of the United States: 2006,* Table 50). Most divorced persons (10% of the population) date and marry others who have been married before.

3. Increased HIV risk. The older an unmarried person, the greater the likelihood that he or she has had multiple sexual partners, which is associated with increased risk of contracting HIV and other STIs (Mosher et al., 2005). Therefore, individuals entering the dating market for the second time are advised to be more selective in choosing their sexual partners because of this higher likelihood.

4. Children. Anderson and Greene (2005) emphasized the challenge of forming new romantic relationships and incorporating them into existing relationships with their children. More than half of the divorced persons who are dating again have children from a previous marriage. How these children feel about their parents' dating, how the partners feel about each other's children,

Nothing grows again more easily than love.

Seneca, Roman philosopher

and how the partners' children feel about each other are complex issues. Deciding whether to have intercourse when one's children are in the house, when a new partner should be introduced to the children, and what the children should call the new partner are other issues familiar to parents dating for the second time. Cohen and Rinzi-Dottan (2005) studied 49 divorced couples who were beginning new relationships and found that mothers who reported making their children their priority were more satisfied with their parenting role. Smyth (2005) noted that nonresident fathers often feel that their time with their children is stilted, shallow, artificial, and brief and that these feelings impacted the way the fathers felt about their role.

5. *Ex-spouse issues.* Ties to an ex-spouse, in the form of child support or alimony, and phone calls will have an influence on the new dating relationship. Some individuals remain psychologically and sexually involved with their exes. In other cases, if the divorce was bitter, the partner may be preoccupied or frustrated in his or her attempts to cope with a harassing ex-spouse (and feel emotionally distant).

6. *Brief courtship.* Divorced people who are dating again tend to have a shorter courtship period than first-marrieds. In a study of 248 individuals who remarried, the median length of courtship was nine months, as opposed to 17 months the first time around (O'Flaherty and Eells, 1988). A shorter courtship may mean that sexual decisions are confronted more quickly—timing of first intercourse, discussing the use of condoms and contraceptives, and clarifying whether the relationship is to be monogamous.

Cultural and Historical Background of Dating

Any consideration of current pairing-off patterns must take into account a historical view. Contemporary hanging-out patterns in the United States today are radically different from courtship and dating in other cultures and times.

Traditional Chinese "Dating" Norms

The freedom with which U.S. partners today select each other on the basis of love is a relatively recent phenomenon. At most times and in most cultures, marriages were more often arranged by parents. Love feelings between the partners, if they existed, were given either no or limited consideration. In traditional China, **blind marriages,** wherein the bride and groom were prevented from seeing each other for the first time until their wedding day, were the norm.

The marriage of two individuals was seen as the linking of two families. Though the influence of parents in the mate selection of their offspring is decreasing (such arranged marriages are no longer the norm), Chinese parents continue to be involved. Indeed, China's dating culture is very different from that of the United States, where individuals meet and date new partners. In a large representative study of couples in urban China, more than three-fourths (77%) of the women and two-thirds (66%) of the men reported dating no one or only their spouse prior to marriage (Pimentel, 2000).

Dating during the Puritan Era in the United States

To the Puritans, all things were impure.

Unknown

Although less strict than traditional courtship norms among the Chinese, the European marriage patterns brought to America were conservative. The Puritans who settled on the coast of New England in the seventeenth century were radical Protestants who had seceded from the Church of England. They valued marriage and fidelity, as reflected in a very rigid pattern of courtship.

Bundling, also called *tarrying,* was a courtship custom commonly practiced among the Puritans. It involved the would-be groom's sleeping in the girl's bed in her parents' home. But there were rules to restrict sexual contact. Both partners had to be fully clothed, and a board was placed between them. In addition, the young girl might be encased in a type of long laundry bag up to her armpits, her clothes might be sewn together at strategic points, and her parents might be sleeping in the same room.

The justifications for bundling were convenience and economics. Aside from meeting at church, bundling was one of the few opportunities a couple had to get together to talk and learn about each other. Since daylight hours were consumed by heavy work demands, night became the only time for courtship. But how did bed become the courtship arena? New England winters were cold. Firewood, oil for lamps, and candles were in short supply. By talking in bed, the young couple could come to know each other without wasting valuable sources of energy. Although bundling flourished in the middle of the eighteenth century, it provoked a great deal of controversy. By about 1800, the custom had virtually disappeared.

Effects of the Industrial Revolution on Dating

The transition from a courtship system controlled by parents to the relative freedom of mate selection experienced today occurred in response to a number of social changes. The most basic change was the Industrial Revolution, which began in England in the middle of the eighteenth century. No longer were women needed exclusively in the home to spin yarn, make clothes, and process food from garden to table. Commercial industries had developed to provide these services, and women transferred their activities in these areas from the home to the factory. The result was that women had more frequent contact with men.

Women's involvement in factory work decreased parental control, since parents were unable to dictate the extent to which their offspring could interact with those they met at work. Hence, values in mate selection shifted from the parents to the children. In the past, the "good wife" was valued for her domestic aptitude—her ability to spin yarn, make clothes, cook meals, preserve food, and care for children. The "good husband" was evaluated primarily in terms of being an economic provider. Though these issues may still be important, contemporary mates are more likely to be selected on the basis of personal qualities, particularly for love and companionship, than for either utilitarian or economic reasons.

Changes in Dating in the Past 50 Years

The Industrial Revolution had a profound effect on courtship patterns, but these patterns have continued to change in the past 50 years. The changes include an increase in the age at marriage, which has been accompanied by each person's having a longer period of time during which he or she becomes involved with more people. Marrying at age 29 rather than 24 provides more time and opportunity to date more people.

Things do not change—we do.

Henry David Thoreau, social activist

The dating pool today also includes an increasing number of individuals in their 30s who have been married before. These individuals often have children, which changes the nature of a date from two adults going out alone to a movie to renting a movie and baby-sitting in the apartment or home of one of the partners.

As we will note later in this chapter, cohabitation has become more normative. For some couples, the sequence of date, fall in love, and get married has been replaced by date, fall in love, and live together. Such a sequence results in the marriage of couples that are more relationship-savvy than those who dated and married out of high school.

Not only do individuals now date more partners and more often live together, but also gender role relationships have become more egalitarian. Though the double standard still exists, women today are more likely than

Relationships Dynamics Scale

Please answer each of the following questions in terms of your relationship with your "mate" if married, or your "partner" if dating or engaged. We recommend that you answer these questions by yourself (not with your partner), using the ranges following for your own reflection.

Use the following 3-point scale to rate how often you and your mate or partner experience the following:

1 = almost never
2 = once in a while
3 = frequently

1 2 3   Little arguments escalate into ugly fights with accusations, criticisms, name calling, or bringing up past hurts.

1 2 3   My partner criticizes or belittles my opinions, feelings, or desires.

1 2 3   My partner seems to view my words or actions more negatively that I mean them to be.

1 2 3   When we have a problem to solve, it is like we are on opposite teams.

1 2 3   I hold back from telling my partner what I really think and feel.

1 2 3   I think seriously about what it would be like to date or marry someone else.

1 2 3   I feel lonely in this relationship.

1 2 3   When we argue, one of us withdraws...that is, doesn't want to talk about it anymore; or leaves the scene.

       Who tends to withdraw more when there is an argument?
         Male
         Female
         Both Equally
         Neither Tend to Withdraw

Stanley S.M., and H. J. Markman. 1997. *Marriage and Family: A Brief Introduction.* Reprinted with permission of PREP, Inc.

Where Are You in Your Marriage

We devised these questions based on seventeen years of research at the University of Denver on the kinds of communication and conflict management patterns that predict if a relationship is headed for trouble. We have recently completed a nationwide, random phone survey using these questions. The average score was 11 on this scale. While you should not take a higher score to mean that your relationship is somehow destined to fail, higher scores can mean that your relationship may be in greater danger unless changes are made. (These ranges are based only on your individual ratings—not a couple total.)

8 to 12 "Green Light"

If you scored in the 8–12 range, your relationship is probably in good or even great shape at *this time,* but we emphasize "*at this time"* because relationships don't stand still. In the next twelve months, you'll either have a stronger, happier relationship, or you could head in the other direction.

To think about it another way, it's like you are traveling along and have come to a green light. There is no need to stop, but it is probably a great time to work on making your relationship all it can be.

13 to 17 "Yellow Light"

If you scored in the 13–17 range, it's like you are coming to a "yellow light." You need to be cautious. While you may be happy now in your relationship, your score reveals warning signs of patterns you don't want to let get worse. You'll want to be taking action to protect and improve what you have. Spending time to strengthen your relationship now could be the best thing you could do for your future together.

18 to 24 "Red Light"

Finally, if you scored in the 18–24 range, it's like approaching a red light. Stop, and think about where the two of you are headed. Your score indicates the presence of patterns that could put your relationship at significant risk. You may be heading for trouble—or already be there. But there is *good news.* You can stop and learn ways to improve your relationship now!

For more information on danger signs and constructive tools for strong marriages, see: Markman, H. J., Stanley, S. M., & Blumberg, S. L. (1994). *Fighting for Your Marriage: Positive Steps for a Loving and Lasting Relationship.* San Francisco: Jossey Bass, Inc. (PREP 1-800-366-0611).

To: Those interested in using the Relationships Dynamics Scale Form: PREP, Inc.

1. We wrote these items based on understanding of many key studies in the field. The content or themes behind the questions are based on numerous in-depth studies on how people think and act in their marriages. These kinds of dynamics have been compared with patterns on many other key variables, such as satisfaction, commitment, problem intensity, etc. Because the kinds of methods researchers can use in their laboratories are quite complex, this actual measure is far simpler than many of the methods we and others use to study marriages over time. But the themes are based on many solid studies. Caution is warranted in interpreting scores.

2. The discussion of the Relationships Dynamics Scale gives rough guidelines for interpreting the meaning of the scores. The ranges we suggest for the measure are based on results from a nationwide, random phone survey of 947 people (85% married) in January 1996. These ranges are meant as a rough guideline for helping couples assess the degree to which they are experiencing key danger signs in their marriages. The measure as you have it here powerfully discriminated between those doing well in their marriages/relationships and those who were not doing well on a host of other dimensions (thoughts of divorce, low satisfaction, low sense of friendship in the relationship, lower dedication, etc.). Couples scoring more highly on these items are truly more likely to be experiencing problems (or, based on other research, are more likely to experience problems in the future).

3. This measure in and of itself should not be taken as a predictor of couples who are going to fail in their marraiges. No couple should be told they will "not make it" based on a higher score. That would not be in keeping with our intention in developing this scale or with the meaning one could take from it for any one couple. While the items are based on studies that assess such things as the likelihood of a marriage working out, we would hate for any one person to take this and assume the worst about their future based on a high score. Rather, we believe that the measure can be used to motivate high and moderately high scoring people to take a serious look at where their marriages are heading—and take steps to turn such negative patterns around for the better.

For more information on constructive tools for strong marriages, see: Markman, H.J., Stanley, S. M., & Blumberg, S. L. (1994). *Fighting for Your Marriage: Positive Steps for a Loving and Lasting Relationship.* If you have questions about the measure and the meaning of it, please write to us at:

PREP, Inc.
P. O. Box 102530
Denver, Colorado 80250-2530
E-mail: PREPinc@aol.com
To order books, audio or videotapes: Call 1-800-366-0166
Scott M. Stanely, Ph.D.
Howard J. Markman, Ph.D.

women in the 1950s to ask men to go out, to have sex with them without requiring a commitment, and to postpone marriage until their own educational and career goals have been met. Women no longer feel desperate to marry but consider marriage one of many goals they have for themselves.

Unlike during the fifties, both sexes today are aware of and somewhat cautious of becoming HIV-infected. Sex has become potentially deadly, and condoms are being used more frequently. The 1950s fear of asking a druggist for a condom has been replaced by the confidence and mundaneness of buying condoms along with one's groceries.

Finally, couples of today are more aware of the impermanence of marriage. However, most couples continue to feel that divorce will not happen to them, and they remain committed to domestic goals. Over three-quarters (75.9%) of all first-year college students in the United States reported that "raising a family" was "an essential or very important goal" for them (American Council on Education and University of California, 2005–2006).

To assess the relationship with your partner at this time, complete the Relationships Dynamics Scale on page 118.

Cohabitation

Cohabitation, also known as living together, is becoming more normative. Almost three-quarters (72.3%) of 1027 undergraduates at a large southeastern university reported that they would live with a partner they were not married to. Eighteen percent were or had already done so (Knox and Zusman, 2006). Reasons for the increase in cohabitation include career or educational commitments; increased tolerance of society, parents, and peers; improved birth-control technology; desire for a stable emotional and sexual relationship without legal ties; and greater disregard for convention. Twenge (2006) surveyed university students and found that 62 percent paid little attention to social conventions. Cohabitants also regard living together as a vaccination against divorce. Later, we will review studies emphasizing that this hope is more often an illusion.

Cohabitation appears not to be very helpful and may be harmful as a try-out for marriage.

David Popenoe, Barbara Dafoe Whitehead, sociologists

National Data

There are 5.6 million unmarried-couple households in the United States (*Statistical Abstract of the United States, 2006*, Table 57).

Persons who live together before marriage are more likely to be high school dropouts than college graduates (60% vs. 37%), to have been married, to be less religious/traditional, and to be supportive of egalitarian gender roles (Baxter, 2005). Teens who are not close to their parents are more likely to cohabit and to have a child as an unmarried couple (Houseknecht and Lewis 2005). Cohabitants are also less likely to be exchanging support with their parents and to turn to parents in an emergency (Eggebeen, 2005).

Same-Sex Cohabitation and Race

Although U.S. Census surveys do not ask about sexual orientation or gender identity, same-sex cohabiting couples may identify themselves as "unmarried partners." Those couples in which both partners are men or both are women are considered to be same-sex couples or households for purposes of research. According to the 2000 Census, there are 600,000 same-sex couples of all races; about 14 percent of these, or 85,000, are black couples. These families must cope with both racism and homophobia.

Diversity in Other Countries

Dolbik-Vorobei (2005) noted that university student attitudes in Russia are increasingly accepting of cohabitation and of intercourse out of marriage as long as "close spiritual relations" have been established between the partners.

This couple moved in together after knowing each other for only a short time.

Authors

Definitions of Cohabitation

In research, more than 20 definitions of cohabitation (also referred to as **living together**) have been used. These various definitions involve variables such as duration of the relationship, frequency of overnight visits, emotional/sexual nature of the relationship, and sex of the partners. Most research on cohabitation has been conducted on heterosexual live-in couples. Even the partners in the relationship may view the meaning of their cohabitation differently, with women viewing it more as a sign of a committed relationship moving toward marriage and men viewing it as an alternative to marriage or as a test to see whether future commitment is something to pursue. We define cohabitation as two unrelated adults involved in an emotional and sexual relationship who sleep overnight in the same residence on a regular basis. The terms used to describe live-ins include cohabitants and **POSSLQs** (people of the opposite sex sharing living quarters), the latter term used by the U.S. Bureau of the Census.

Diversity in Other Countries

Iceland is a homogeneous country of 250,000 descendants of the Vikings. Their sexual norms include early (age 14) protected intercourse, nonmarital parenthood, and living together before marriage. Indeed, a wedding photo often includes not only the couple but the children they have already had. One American woman who was involved with an Icelander noted, "My parents were upset with me because Ollie and I were thinking about living together, but his parents were upset that we were not already living together" (authors' files).

The engagement that is too short is usually followed by a marriage that is too long.

Unknown

Eight Types of Cohabitation Relationships

There are various types of cohabitation:

1. Here and now. The new partners have an affectionate relationship and are focused on the here and now, not the future of the relationship. Only a small proportion of persons living together report that the "here and now" type characterizes their relationship (Jamieson et al., 2002).

2. Testers. The couples are involved in a relationship and want to assess whether they have a future together. As in the case of here-and-now cohabitants, only a small proportion of cohabitants characterize themselves as "testers" (Jamieson et al., 2002).

3. Engaged. These couples are in love and are planning to marry. While not all cohabitants consider marriage their goal, most view themselves as committed to each other (Jamieson et al., 2002). Oppenheimer (2003) studied a national sample of cohabitants and found that cohabiting whites were much more likely to have married the partner than cohabiting blacks—51 percent versus 22 percent. Dush et al. (2005) compared persons in marriage; persons in cohabiting, steady dating, and casual dating relationships; and persons who dated infrequently or not at all. They found that individuals in a happy relationship, independent of the nature of the relationship, reported higher subjective well being. In addition, the more committed the relationship, the higher the subjective well being. Hence, for cohabiting persons who define their relationships as involved, committed, or engaged, we would expect higher levels of subjective well being.

4. Money savers. The couples live together primarily out of economic convenience. They are open to the possibility of a future together but regard such a possibility as unlikely.

5. Pension partners. A variation of the money savers category is that of pension partners. These individuals are older, have been married before, still derive benefits from their previous relationships, and are living with someone new. Getting married would mean giving up their pension benefits from the previous marriage. An example is a widow from the war with Iraq who was given military benefits due to her husband's death. If she remarries, she forfeits both health and pension benefits; she is now living with a new partner but getting the benefits from the previous marriage.

   6. Security blanket cohabiters. Also known as "Linus blanket" cohabiters, some of the individuals in these relationships are drawn to each other out of a need for security rather than mutual attraction.

   7. Rebellious cohabiters. Some couples use cohabitation as a way of making a statement to their parents that they are independent and can make their own choices. Their cohabitation is more about rebelling from parents than being drawn to each other.

   8. Marriage never. These couples feel that a real relationship is a commitment of the heart, not a legal document. Living together provides both companionship and sex without the responsibilities of marriage. Skinner et al. (2002) found that individuals in long-term cohabiting relationships scored the lowest in terms of relationship satisfaction when compared with married and remarried couples. The "marriage never" couples are rare (celebrities Johnny Depp/Vanessa Paradis and Goldie Hawn/Kurt Russell are examples of couples who live together, have children, and have opted not to marry).

   There are various reasons and motivations for living together as a permanent alternative to marriage. Some may have been married before and don't want the entanglements of another marriage. Others feel that the real bond between two people is (or should be) emotional. They contend that many couples stay together because of the legal contract, even though they do not love each other any longer. "If you're staying married because of the contract," said one partner, "you're staying for the wrong reason." Some couples feel that they are "married" in their hearts and souls and don't need or want the law to interfere with what they feel is a private act of commitment. For most couples, living together is a short-lived experience. About 55 percent will marry and 40 percent will break up within five years of beginning cohabitation (Smock, 2000). Some couples who view their living together as "permanent" seek to have it defined as a **domestic partnership** (see the Social Policy box in this section).

Some people never marry because they do not believe in divorce.

Evan Esar

Consequences of Cohabitation

McGinnis (2003) noted that one of the effects of cohabitation was to increase the chance that a couple would end up getting married. Although living together before marriage does not ensure a happy, stable marriage, it has some potential advantages.

Advantages of Cohabitation   Many unmarried couples who live together report that it is an enjoyable, maturing experience. Other potential benefits of living together include the following:

   1. Sense of well being. Cohabitants are likely to report a sense of well being. They are in love, the relationship is new, and the disenchantment that frequently occurs in long-term relationships has not had time to surface. One student reported, "We have had to make some adjustments in terms of moving all our stuff into one place, but we very much enjoy our life together."

   2. Delayed marriage. Another advantage of living together is remaining unmarried—and the longer one waits to marry, the better. Being older at the time of marriage is predictive of marital happiness and stability, just as being young (particularly 18 years and below) is associated with marital unhappiness and divorce. Hence, if a young couple who have known each other for a short time is faced with the choice of living together or getting married, their delaying marriage while they live together seems to be the better choice. Also, if they break up, the split will not go on their "record" as would a divorce.

Diversity in Other Countries

Over 90 percent of first marriages in Sweden are preceded by cohabitation; however, only 12 percent of first marriages in Italy are preceded by cohabitation (Kiernan, 2000).

Domestic Partnerships

Domestic partnerships involve two adults who have chosen to share each other's lives in an intimate and committed relationship of mutual caring. Such cohabitants, both heterosexual and homosexual, want their employers, whether governmental or corporate, to afford them the same rights as spouses. Specifically, an employed person who pays for health insurance would like his or her domestic partner to be covered in the same way that one's spouse would be. Employers have been reluctant to legitimize domestic partners as qualifying for benefits because of the additional expense. One reason for the reluctance is the fear that a higher proportion of partners may be HIV infected, which would involve considerable medical costs.

Aside from the economic issue, domestic partner benefits have been criticized by fundamentalist religious groups as eroding family values by giving nonmarital couples the same rights as married couples. California and New Jersey lead the way in domestic partner benefits (Asancheyev, 2005), with the law providing rights and responsibilities in areas as varied as child custody, legal claims, housing protections, bereavement leave, and state government benefits.

To receive benefits, domestic partners must register, which involves signing an affidavit of domestic partnership verifying that they are a nonmarried cohabiting couple 18 years of age or older and unrelated by blood close enough to bar marriage in the state of residence. Other criteria typically used to define a domestic partnership include that the individuals must be jointly responsible for debts to third parties, they must live in the same residence, they must be financially interdependent, and they must intend to remain in the intimate committed relationship indefinitely. Should they terminate their domestic partnership, they are required to file notice of such termination.

Domestic partnerships offer a middle ground between those states that want to give full legal recognition to same-sex marriages and those that deny any legitimacy to same-sex unions. Such relationships also include long-term-committed heterosexuals who are not married (Bowman, 2004). Glenn (2004) noted that noncommitted individuals who pose as partners just to get the benefits have sometimes abused the law.

Your Opinion?

1. To what degree do you believe benefits should be given to domestic partners?
2. What criteria should be required in order for a couple to be regarded as domestic partners?
3. How can abuses of those claiming to be domestic partners be eliminated?

Sources

Asancheyev, N. 2005. Same-sex couples: Marriage, civil unions, and domestic partnerships. *Georgetown Journal of Gender and the Law* 6:731–60.

Bowman, C. G. 2004. Legal treatment of cohabitation in the United States. *Law and Policy* 26:119–24.

Glenn, N. D. 2004. The struggle for same-sex marriage. *Society* 24:25–28.

3. Learning about self and partner. Living with an intimate partner provides couples with an opportunity for learning more about themselves and their partner. For example, individuals in living-together relationships may find that their role expectations are more (or less) traditional than they had previously thought. Learning more about one's partner is a major advantage of living together. A person's values (calling parents daily), habits (leaving the lights on), and relationship expectations (how emotionally close or distant) are sometimes more fully revealed in a living-together context than in a traditional dating context.

Disadvantages of Cohabitation   There is a downside for individuals and couples who live together.

1. Feeling used or tricked. We have mentioned that women are more prone than men to view cohabitation as reflective of a more committed relationship. When expectations differ, the more invested partner may feel used or tricked if the relationship does not progress toward marriage. One partner said, "I always felt we would be getting married, but it turns out that he never saw a future for us."

2. Problems with parents. Some cohabiting couples must contend with parents who disapprove of or do not fully accept their living arrangement. For example, cohabitants sometimes report that when visiting their parents' homes, they are required to sleep in separate beds in separate rooms. Some cohabitants who have parents with traditional values respect these values, and sleeping in separate

Will Living Together Ensure a Happy, Durable Marriage?

Couples who live together before getting married assume that doing so will increase their chances of having a happy and durable marriage relationship. But will it? In a word, no. Researchers refer to the **cohabitation effect** as the tendency for couples who cohabit to end up in less happy and shorter-lived marriages (more likely to divorce). Cohabitants are more likely not only to divorce but to report more disagreements, more violence, lower levels of happiness, and lower levels of ability to negotiate conflict (Cohan and Kleinbaum, 2002).

One explanation for the higher divorce rate among persons who cohabit before marriage is that cohabitants tend to be people who are willing to violate social norms by living together before marriage. Once they marry, they may be more willing to break another social norm and divorce if they are unhappy than are unhappily married persons who tend to conform to social norms and have no history of unconventional behavior. A second explanation is that since cohabitants are less committed to the relationship than marrieds, this may translate into withdrawing from conflict by terminating the relationship rather than communicating about the problems and resolving them since the stakes are higher (White et al., 2004). A third explanation is that it is not the types of people that cohabitation attracts but the experience of cohabitation that is associated with subsequent divorce. "According to this perspective, cohabitation changes people and their relationships in ways that undermine later marital quality and commitment" (Kamp Dush et al., 2003, p. 545). For example, being in a relationship with an uncertain commitment to the future may make people less invested and committed and therefore most vulnerable to divorce. A fourth explanation is the selection hypothesis, that cohabitation may select or draw people with nontraditional values (who are not ready to commit to each other or to the institution of marriage) and who may have poor relationship skills. Hence, cohabitation does not change them but collects the noncommitted with poor relationship skills (Smock, 2000). Whatever the reason, cohabitants should not assume that cohabitation will make them happier spouses or insulate them from divorce.

Not all researchers have found that there are negative effects of cohabitation on relationships. Skinner et al. (2002) compared those who had cohabited and married and those who married but did not cohabit and found no distinguishing characteristics. They concluded that "cohabiting couples may not be stigmatized if there is an expectation that marriage will occur." In addition, Musick (2005) examined national longitudinal data on marrieds and cohabitants and found few differences between the two groups on the variable of well being the first three years. But after three years, marrieds reported higher levels of well being. The researcher suggested that "institutional commitment adds value to relationships" (p.104).

Courtship is a short period of long kisses followed by a long period of short kisses.

Unknown

Sources

Kamp Dush, C. M. K., C. L. Cohan, and P. R. Amato. 2003. The relationship between cohabitation and marital quality and stability: Change across cohorts? *Journal of Marriage and the Family* 65:539–49.

Musick, K. 2005. Does marriage make people happier? Marriage, cohabitation, and trajectories in well-being. In *Sourcebook of Family Theory and Research*, edited by Vern L. Bengtson, Alan C. Acock, Katherine R. Allen, Peggye Dilworth-Anderson, and David M. Klein. Thousand Oaks, CA: Sage Publications, 103–04.

Skinner, K. B., S. J. Bahr, D. R. Crane, and V. R. A. Call. 2002. Cohabitation, marriage, and remarriage. *Journal of Family Issues* 23:74–90.

Smock, P. J. 2000. Cohabitation in the United States: An appraisal of research themes, findings, and implications. *Annual Review of Sociology* 26:1–20.

White, A. M, F. S. Christopher, and T. K. Poop. 2004. Cohabitation and the early years of marriage. Poster session, National Council on Family Relations, November, Orlando, Fla.

rooms is not a problem. Other cohabitants feel resentful of parents who require them to sleep separately. Some parents express their disapproval of their child's cohabiting by cutting off communication, as well as economic support, from their child. Other parents display lack of acceptance of cohabitation in more subtle ways. One woman who had lived with her partner for two years said that her partner's parents would not include her in the family's annual photo portrait. Emotionally, she felt very much a part of her partner's family and was deeply hurt that she was not included in the family portrait (authors' files). Still other parents are completely supportive of their children's cohabiting and support their doing so. "I'd rather my kid live together than get married, and besides it is safer for her and she's happier," said one father.

3. Economic disadvantages. Some economic liabilities exist for those who live together instead of getting married. In the Social Policy section on domestic partnerships, we noted that cohabitants typically do not benefit from their partner's health insurance, Social Security, or retirement benefits. In most cases only spouses qualify for such payoffs. We have already noted that cohabitants tend to be from lower socioeconomic strata than noncohabitants (Smock, 2000).

Given that most living-together relationships are not long-term and that breaking up is not uncommon, cohabitants might develop a written and signed legal agreement should they purchase a house, car, or other costly items together. The written agreement should include a description of the item, to whom it belongs, how it will be paid for, and what will happen to the item if the relationship terminates. Purchasing real estate together may require a separate agreement, which should include how the mortgage, property taxes, and repairs will be shared. The agreement should also specify who gets the house if the partners break up and how the value of the departing partner's share will be determined.

If the couple have children, another agreement may be helpful in defining custody, visitation, and support issues in the event the couple terminates their relationship. Such an arrangement may take some of the romance out of the cohabitation relationship, but it can save a great deal of frustration should the partners decide to go their separate ways.

In addition, couples who live together instead of marrying can protect themselves from some of the economic disadvantages of living together by specifying their wishes in wills; otherwise, their belongings will go to next of kin or to the state. They should also own property through joint tenancy with rights of survivorship.

This means that ownership of the entire property will revert to one partner if the other partner dies. In addition, the couple should save for retirement, since Social Security benefits may not be accessed by live-in companions, and some company pension plans bar employees from naming anyone other than a spouse as the beneficiary.

4. Effects on children. About 40 percent of children will spend some time in a home where the adults are cohabiting. In addition to being disadvantaged in terms of parental income and education, they are likely to experience more disruptions in family structure. Raley et al. (2005) analyzed data on children who lived with cohabiting mothers (from the National Survey of Families and Households) and found that these children fared exceptionally poorly and sometimes were significantly worse off than were children who lived with divorced or remarried mothers. The authors reasoned that the instability associated with cohabitation may account for why these children do less well.

5. Other issues. Over a million cohabitants are over the age of 50. When compared to marrieds, they report more depressive symptoms independent of their economic resources, social support, and physical health (Brown et al., 2005). It is possible that these middle-aged individuals would prefer to be married and their unhappiness reflects that preference.

Legal Aspects of Living Together

In recent years, the courts and legal system have become increasingly involved in living-together relationships. Some of the legal issues concerning cohabiting partners include common-law marriage, palimony, child support, and child inheritance. Lesbian and gay couples also confront legal issues when they live together.

Technically, cohabitation is against the law in some states. For example, in North Carolina, cohabitation is a misdemeanor punishable by a fine not to exceed $500, imprisonment for not more than six months, or both. Most law enforcement officials view cohabitation as a victimless crime and feel that the general public can be better served by concentrating upon the crimes that do real damage to citizens and their property.

Common-Law Marriage The concept of **common-law marriage** dates to a time when couples who wanted to be married did not have easy or convenient access to legal authorities (who could formally sanction their relationship so that they would have the benefits of legal marriage). Thus, if the couple lived together, defined themselves as husband and wife, and wanted other people to view them as a married couple, they would be considered married in the eyes of the law.

Despite the assumption by some that heterosexual couples who live together a long time have a common-law marriage, only eleven jurisdictions recognize such marriages. In ten states (Alabama, Colorado, Idaho, Iowa, Kansas, Rhode Island, South Carolina, Montana, Pennsylvania, and Texas) and the District of Columbia, a heterosexual couple may be considered married if they are legally competent to marry, if there is an agreement between the partners that they are married, and if they present themselves to the public as a married couple. A ceremony or compliance with legal formalities is not required.

In common-law states, individuals who live together and who prove that they were married "by common law" may inherit from each other or receive alimony and property in the case of "divorce." They may also receive health and Social Security benefits, as would other spouses who have a marriage license. In states not recognizing common-law marriages, the individuals who live together are not entitled to benefits traditionally afforded married individuals. More than three-quarters of the states have passed laws prohibiting the recognition of common-law marriages within their borders.

Palimony A take-off on the word *alimony,* **palimony** refers to the amount of money one "pal" who lives with another "pal" may have to pay if the partners end their relationship. In 2005, for instance, comedian Bill Maher was the target of a $9 million palimony suit by ex-girlfriend Coco Johnsen (*Forbes,* 2005).

Avellar and Smock (2005) compared the economic well being of cohabitants who ended their relationship. Whereas the economic standing of the cohabitant man declined moderately, that of the former cohabitant woman declined steeply, leaving a substantial proportion of women in poverty (and even more so for African-American and Hispanic women).

Child Support Heterosexual individuals who conceive children are responsible for those children whether they are living together or married. In most cases, the custody of young children will be given to the mother, and the father will be required to pay child support. In effect, living together is irrelevant with regard to parental obligations. However, a woman who agrees to have a child with her lesbian partner cannot be forced to pay child support if the couple breaks up. In 2005, the Massachusetts Supreme Judicial Court ruled that their informal agreement to have a child together did not constitute an enforceable contract.

Couples who live together or who have children together should be aware that laws traditionally applying only to married couples are now being applied to

Palimony is like paying installments on a car after it is wrecked.

Unknown

many unwed relationships. Palimony, distribution of property, and child support payments are all possibilities once two people cohabit or parent a child.

Child Inheritance   Children born to cohabitants who view themselves as spouses and who live in common-law states are regarded as legitimate and can inherit from their parents. However, children born to cohabitants who do not present themselves as married or who do not live in common-law states are also able to inherit. A biological link between the parent and the offspring is all that needs to be established.

SUMMARY

What are the attractions of singlehood and the social movements that created it, and what is Twin Oaks?

The primary attraction of singlehood is the freedom to do as one chooses. As a result of the Sexual Revolution, the Women's Movement, and the Gay Liberation movement, there is increased social approval of being unmarried. Singles include not only the never-married but also the divorced and widowed.

Twin Oaks is a community of 90 adults and 15 children living together on 450 acres of land in Louisa, Virginia. Known as an intentional community (a group of people who choose to live together on the basis of a set of shared values), the commune (an older term now replaced by the newer term) was founded in 1967 and is one of the oldest nonreligious intentional communities in the United States. Members include gay, straight, bisexual, and transgender people; most are single never-married adults, but there are married couples and families. Sexual values range from celibate to monogamous to polyamorous (involvement in more than one emotional/sexual relationship at the same time).

The age range of the members is newborn to 78, with an average age of 40. The average length of stay for current members is about eight years. The core values of the community are nonviolence (no guns, low tolerance for violence of any kind in the community, no parental use of violence against children), egalitarianism (no leader, with everyone having equal political power and the same access to resources), and environmental sustainability (the community endeavors to live off the land, growing its own food and heating its buildings with wood from the forest). The community also encourages participation in activism outside the community. Many members are activists for peace-and-justice, feminist, and ecological organizations.

To what degree are undergraduates interested in finding a partner?

About a quarter of university students in one study reported that they were not involved with anyone. More than half of the men and about four in ten women reported that they were looking for a partner.

How do university students go about finding a partner?

Besides the traditional way of meeting people at work/school or through friends and going out on a date, couples today may also "hang out," which may lead to "hooking up." Internet dating, video dating, and speed dating are new forms for finding each other.

What issues do divorced persons face when they start dating again?

The divorced are older and select from an older, more limited population with a higher chance of having a sexually transmitted infection (STI). Most have children from a previous marriage and are dealing with an ex-spouse.

What is the cultural and historical background of dating?

The freedom with which U.S. partners today select each other on the basis of love is a relatively recent phenomenon. At most times and in most cultures, marriages were more often arranged by parents. Unlike during the 1950s, both sexes today are aware of and somewhat cautious of becoming HIV/STI-infected. Sex has become potentially deadly, and condoms are being used more frequently. Couples of today are also more aware of the impermanence of marriage.

What is cohabitation like among today's youth?

Cohabitation, also known as living together, is becoming a "normative life experience," with almost 60 percent of U.S. women reporting that they had cohabited before marriage. Reasons for an increase in living together since 1970 include a delay of marriage for educational or career commitments, fear of marriage, increased tolerance of society for living together, and a desire to avoid the legal entanglements of marriage. Types of living-together relationships include the here-and-now, testers (testing the relationship), engaged couples (planning to marry), and cohabitants forever (never planning to marry). Most people who live together eventually get married but not necessarily to each other.

Domestic partners are two adults who have chosen to share each other's lives in an intimate and committed relationship of mutual caring. Such cohabitants, both heterosexual and homosexual, want their employers, whether governmental or corporate, to afford them the same rights as spouses. Only about 10 percent of firms recognize domestic partners and offer them benefits.

Although living together before marriage does not ensure a happy, stable marriage, it has some potential advantages. These include a sense of well being, delayed marriage, learning about yourself and your partner, and being able to disengage with minimal legal hassle. Disadvantages include feeling exploited, feeling guilty about lying to parents, and not having the same economic benefits as those who are married. Social Security and retirement benefits are paid to spouses, not live-in partners.

KEY TERMS

blind marriage	common-law marriage	hanging out	palimony
bundling	commune	hooking up	POSSLQ
cohabitation	dating	intentional community	
cohabitation effect	domestic partnership	living together	

The Companion Website for *Choices in Relationships: An Introduction to Marriage and the Family,* Ninth Edition
http://www.thomson.edu/sociology/knox

Supplement your review of this chapter by going to the companion website to take one of the Tutorial Quizzes, use the flash cards to master key terms, and check out the many other study aids you'll find there. You'll also find special features such as the Marriage and Family Resource Center, Census 2000 information, and other data and resources at your fingertips to help you with that special project or to do some research on your own.

WEBLINKS

Right Mate at Heartchoice
http://www.heartchoice.com/rightmate/

Speed-Dating: The 8-Minute Date
http://www.8minutedating.com/

WildXAngel
http://www.wildxangel.com/

Independent Women's Forum
http://www.happinessonline.org/BeFaithfulToYourSexualPartner/p17.htm

REFERENCES

American Council on Education and University of California. 2005–2006. The American freshman: National norms for fall, 2005. Los Angeles, CA: Los Angeles Higher Education Research Institute.

Anderson, E. R. and S. M. Greene. 2005. Transitions in parental repartnering after divorce. *Journal of Divorce and Remarriage* 43:47–62.

Avellar, S. and P. J. Smock. 2005. The economic consequences of the dissolution of cohabiting unions. *Journal of Marriage and the Family* 67:315–327.

Baxter, J. 2005. To marry or not to marry: Marital status and the household division of labor. *Journal of Family Issues* 26:300–321.

Bogle, K. A. 2002. From dating to hooking up: Sexual behavior on the college campus. Paper, Annual Meeting of the Society for the Study of Social Problems, Summer.

———. 2003. Sex and "dating" in college and after: A look at perception and behavior. Paper, 73rd Annual Meeting of the Eastern Sociological Society, Philadelphia, February 28.

Brewster, C. D. D. 2006. African American male perspective on sex, dating, and marriage. *The Journal of Sex Research* 43:32–33.

Brown, S. L., J.R. Bulanda, and G. R. Lee. 2005. The significance of nonmarital cohabitation: Marital status and mental health benefits among middle-aged and older adults. *Journals of Gerontology Series B—Psychological Sciences and Social Sciences*, 60(1):S21–S29.

Cohan, C. L., and S. Kleinbaum. 2002. Toward a greater understanding of the cohabitation effect: Premarital cohabitation and marital communication. *Journal of Marriage and the Family* 64:180–92.

Cohen, O. and R. Rinzi-Dottan. 2005. Parent-child relationships during the divorce process: From attachment theory and intergenerational perspective. *Contemporary Family Therapy: An International Journal* 27:81–99.

CNN. 2006. Online Internet dating. http://onlinepersonalswatch.typepad.com/news/jupiter_research/ Retrieved Feb 13.

DePaulo, B. 2006. *Singled out: How singles are stereotyped, stigmatized, and ignored, and still live happily ever after*. New York: St. Martin's Press.

Dolbik-Vorobei, T. A. 2005. What college students think about problems of marriage and having children. *Russian Education and Society* 47:47–58.

Donn, J. 2005. Adult development and well-being of midlife never-married singles. Unpublished dissertation. Miami University, Oxford, Ohio. For more information on the measures developed in this study, contact Jessica Donn at etoiles24@yahoo.com.

Dush, C., M. Kamp, and P. R. Amato. 2005. Consequences of relationship status and quality for subjective well-being. *Journal of Social and Personal Relationships* 22:607–27.

Eggebeen, D. J. 2005. Cohabitation and exchanges of support. *Social Forces* 83:1097–110.

England, P. and R. J. Thomas. 2006. The decline of the date and the rise of the college hook up. In *Family in transition: 14th ed.*, edited by A. S. Skolnick and J. H. Skolnick. Boston: Pearson/Allyn Bacon, 151–62.

Ferguson, S. J. 2000. Challenging traditional marriage: Never married Chinese American and Japanese American women. *Gender and Society* 14:136–59.

Gibbs, J. L., N. B. Ellison, and R. D. Heino. 2006. Self-presentation in online personals: The role of anticipated future interaction, self-disclosure, and perceived success in Internet dating. *Communication Research* 33:152–77.

Glenn, N. and E. Marquardt. 2001. *Hooking up, hanging out, and hoping for Mr. Right: College women on dating and mating today*. New York: Institute for American Values.

Glenn, N. D. and C. N. Weaver. 1988. The changing relationship of marital status to reported happiness. *Journal of Marriage and the Family* 50:317–24.

Guldner, G. T. 2003. *Long distance relationships: The complete guide*. Corona, CA: JFMilne Publications.

Harcourt, W. 2005. Gender and community in the social construction of the Internet/Hanging out in the virtual pub: Masculinities and relationships online. *Signs: Journal of Women in Culture and Society* 30: 1981–84.

Hemstrom, O. 1996. Is marriage dissolution linked to differences in mortality risks for men and women? *Journal of Marriage and the Family* 58:366–78.

Hodge, A. 2003. Video chatting and the males who do it. Paper, 73rd Annual Meeting of the Eastern Sociological Association, Philadelphia, February 28.

Houseknecht, S. K. and S. K. Lewis. 2005. Explaining teen childbearing and cohabitation: Community embeddedness and primary ties. *Family Relations* 54:607–20.

Jagger, E. 2005. Is thirty the new sixty? Dating, age and gender in postmodern consumer society. *Sociology* 9:89–106.

Jamieson, L., M. Anderson, D. McCrone, F. Bechhofer, R. Stewart, and Y. Li. 2002. Cohabitation and commitment: Partnership plans of young men and women. *The Sociological Review* 50:356–77.

Jayson, S. 2004. It's time to grow up—later. *USA Today* Sept 30, D1.

Jones, J. 2006. Marriage is for white people. *The Washington Post* 26 March, B3–B4.

Kamp Dush, C. M. K., C. L. Cohan, and P. R. Amato. 2003. The relationship between cohabitation and marital quality and stability. Change across cohorts? *Journal of Marriage and the Family* 65:539–49.

Kellner, T. The O'Reilly defense. 2005. *Forbes* 14 Feb, 175(3):52.

Kiernan, K. 2000. European perspectives on union formation. In *The ties that bind*, edited by L. J. Waite. New York: Aldine de Gruyter, 40–58.

Knox, D., V. Daniels, L. Sturdivant, and M. E. Zusman. 2001. College student use of the Internet for mate selection. *College Student Journal* 35:158–60.

Knox, D. and Zusman, M. E. 2006. Relationship and sexual behaviors of a sample of 1027 university students. Unpublished data collected for this text. Department of Sociology, East Carolina University, Greenville, NC.

Knox, D., M. Zusman, V. Daniels, and A. Brantley. 2002. Absence makes the heart grow fonder? Long-distance dating relationships among college students. *College Student Journal* 36:365–67.

Lichter, D. T., D. R. Graefe, and J. B. Brown. 2003. Is marriage a panacea? Union formation among economically disadvantaged unwed mothers. *Social Problems* 50:60–86.

Lucas, R. E., A. E. Clark, Y. Georgellis, and E. Diener. 2003. Reexamining adaptation and the set point model of happiness: Reactions to changes in marital status. *Journal of Personality and Social Psychology* 84:527–39.

Luff, T. and K. Hoffman. 2006. College dating patterns: Cultural and structural influences. Roundtable presentation, Southern Sociological Society, New Orleans, LA, March 24.

Lydon, J., T. Pierce, and S. O'Regan. 1997. Coping with moral commitment to long-distance dating relationships. *Journal of Personality and Social Psychology* 73:104–13.

Mahoney, S. 2006. The secret lives of single women—lifestyles, dating and romance: A study of midlife singles. *AARP: The Magazine* May/June 62–69.

Maldonado, S. 2005. Beyond economic fatherhood: Encouraging divorced fathers to parent. *University of Pennsylvania Law Review* 153:921–1009.

McGinnis, S. L. 2003. Cohabiting, dating, and perceived costs of marriage: A model of marriage entry. *Journal of Marriage and Family* 65:105–16.

McGinty, C. 2006. Internet dating. Presentation to Courtship and Marriage class, East Carolina University, Greenville, NC.

Mosher, W., A. Chandra, and J. Jones. 2005. Sexual behavior and selected health measures: Men and women 15–44 years of age, United States, 2002. *Advanced Data from Vital and Health Statistics* Number 362, September 15. Centers for Disease Control and Prevention. Retrieved January 14, 2006, from http://www.cdc.gov/nchs/data/ad/ad362.pdf

O'Flaherty, K. M., and L. W. Eells. 1988. Courtship behavior of the remarried. *Journal of Marriage and the Family* 50:499–506.

Ohayon, M. 2005. *Cowboy Amor* (video released in 2005).

Oppenheimer, V. K. 2003. Cohabiting and marriage during young men's career-development process. *Demography* 40:127–49.

Pimentel, E. E. 2000. Just how do I love thee? Marital relations in urban China. *Journal of Marriage and the Family* 62:32–47.

Pinquart, M. 2003. Loneliness in married, widowed, divorced, and never married older adults. *Journal of Social and Personal Relationships* 20: 31–53

Raley, R. K., M. L. Frisco, and E. Wildsmith. 2005. Maternal cohabitation and educational success. *Sociology of Education* 78:144–164.

Renshaw, S. W. 2005. "Swing Dance" and "Closing Time": Two ethnographies in popular culture. Dissertation Abstracts International. A: *The Humanities and Social Sciences* 25:4355A–56A.

Skinner, K. B., S. J. Bahr, D. R. Crane, and V. R. A. Call. 2002. Cohabitation, marriage, and remarriage. *Journal of Family Issues* 23:74–90.

Smyth, B. 2005. Time to rethink time: The experience of time with children after divorce. *Family Matters* 71:4–10

Smock, P. J. 2000. Cohabitation in the United States: An appraisal of research themes, findings, and implications. *Annual Review of Sociology* 26:1–20.

Statistical Abstract of the United States: 2006. 125th ed. Washington, D.C.: U.S. Bureau of the Census.

Thomas, M. E. 2005. Girls, consumption space and the contradictions of hanging out in the city. *Social and Cultural Geography* 6:587–605.

Twenge, J. 2006. *Generation me.* New York: Free Press.

Vail-Smith, K., D. Knox, and M. Zusman. 2006. The lonely college male. *International Journal of Men's Health* In press.

Vanderkam, L. 2006. Love (or not) in an iPod world. *USA Today* 14 Feb, 13A.

Wayment, H. A., G. E. Wyatt, M. B. Tucker, G. J. Romero, J. V. Carmona, M. Newcomb, B. M. B. M. Solis, M. Riederle, and C. Mitchell-Kernan. 2003. Predictors of risky and precautionary sexual behaviors among single and married white women. *Journal of Applied Social Psychology* 33: 791–816

Wilson, G. D., J. M. Cousins, and B. Fink. 2006. The CQ as a predictor of speed-date outcomes. *Sexual & Relationship Therapy* 21: 163–169

Xia, Y. R. and Z. G. Zhou 2003. The transition of courtship, mate selection, and marriage in China. In *Mate Selection Across Cultures,* edited by R. R. Hamon and B. B. Ingoldsby. Thousand Oaks, CA: Sage Publications, 231–46.

Yeoman, B. 2006. Rethinking the commune. *AARP: The Magazine* March/April:88–97.

Yurchisin, J., K. Watchravesrighkan, and D. M. Brown. 2005. An exploration of identity re-creation in the context of internet dating. *Social Behavior and Personality: An International Journal* 33:735–50.

Authors

chapter 5

My sweetheart
You have aroused my passion
Your touch has filled me with
desire
I am no longer separate from
you.

Rumi, Persian poet

Sexuality in Relationships

Contents

True or False?

1. Students who pledge to remain a virgin are as likely to contract a sexually transmitted infection as those who do not take the "virginity pledge."

2. Male undergraduates are almost twice as likely as female undergraduates to report having been involved in a "friends with benefits" relationship.

3. There is no longer a double standard in regard to hooking up: women who hook up are not regarded more negatively than men who do the same.

4. Having sex to avoid conflict in a relationship has a positive outcome for the individual, the couple, and their relationship.

5. Greater equality between women and men is associated with higher sexual satisfaction between the partners.

Answers: **1.** T **2.** F **3.** F **4.** F **5.** T

"It is not sex that gives the pleasure but the lover," wrote Marge Piercy. Her quote reveals that it is the relationship with the partner that is the source of physical pleasure. It is not the technical skills of the lover (though these should not be underestimated) but the feelings the lovers have for each other, that make the sexual experience enjoyable and memorable. Sexuality in relationships is the focus of this chapter. We begin by discussing the sexual values lovers bring to the encounter.

Sexual Values

The following are some examples of choices (reflecting sexual values) with which individuals in a new relationship are confronted:

How much sex and how soon in a relationship are appropriate?
Require a condom for vaginal or anal intercourse?
Require a condom and/or dental dam (vaginal barrier) for oral sex?
Require testing for sexually transmitted infection (STI) and human immunodeficiency virus (HIV) infection before becoming sexually active with a new partner?
Tell partner *actual* number of previous sexual partners?
Tell partner of sexual fantasies?
Reveal previous/current same-sex behavior or interests?

But what are sexual values? **Sexual values** are moral guidelines for making sexual choices in nonmarital, marital, heterosexual, and homosexual relationships. Attitudes and values predict sexual behavior (Meier, 2003). One's sexual values may be identical to one's sexual choices. For example, a person who values abstinence until marriage may choose to remain a virgin until marriage. But one's behavior does not always correspond with one's values. Some who express a value of waiting until marriage have intercourse before marriage. One explanation for the discrepancy between values and behavior is that a person may engage in a sexual behavior, then decide the behavior was wrong, and adopt a sexual value against it.

The self-assessment in this section allows you to examine your own premarital sexual values.

Unless we gratify our sex desire, the race is lost; unless we restrain it, we destroy ourselves.

Bernard Shaw

Attitudes Toward Premarital Sex Scale

Premarital sex is defined as engaging in sexual intercourse prior to marriage. The purpose of this survey is to assess your thoughts and feelings about intercourse before marriage. Read each item carefully and consider how you feel about each statement. There are no right or wrong answers to any of these statements, so please give your honest reactions and opinions. Please respond by using the following scale:

1	2	3	4	5	6	7
Strongly Disagree						Strongly Agree

_____ 1. I believe that premarital sex is healthy.

_____ 2. There in nothing wrong with premarital sex.

_____ 3. People who have premarital sex develop happier marriages.

_____ 4. Premarital sex is acceptable in a long-term relationship.

_____ 5. Having sexual partners before marriage is natural.

_____ 6. Premarital sex can serve as a stress reliever.

_____ 7. Premarital sex has nothing to do with morals.

_____ 8. Premarital sex is acceptable if you are engaged to the person.

_____ 9. Premarital sex is a problem among young adults.

_____10. Premarital sex puts unnecessary stress on relationships.

Scoring

Selecting a 1 reflects the most negative attitude toward premarital sex; selecting a 7 reflects the most positive attitude toward premarital sex. Before adding the numbers you assigned to each item, change the scores for items #9 and #10 as follows: replace a score of 1 with a 7; 2 with a 6; 3 with a 5; 4 with a 4; 5 with a 3; 6 with a 2; and 7 with a 1. After changing these numbers, add your ten scores. The lower your total score (10 is the lowest possible score), the less accepting you are of premarital sex; the higher your total score (70 is the highest possible score), the greater your acceptance of premarital sex. A score of 40 places you at the midpoint between being very disapproving of premarital sex and very accepting of premarital sex.

Scores of Other Students Who Completed the Scale

The scale was completed by 252 student volunteers at Valdosta State University, in Valdosta, Georgia. The mean score of the students was 40.81 (standard deviation [SD] = 13.20), reflecting that the students were virtually at the midpoint between a very negative and a very positive attitude toward premarital sex. For the 124 males and 128 females in the total sample, the mean scores were 42.06 (SD = 12.93) and 39.60 (SD = 13.39), respectively (not statistically significant). In regard to race, 59.5 percent of the sample was white and 40.5 percent was nonwhite (35.3% black, 2.4% Hispanic, 1.6% Asian, 0.4% American Indian, and 0.8% other). The mean scores of whites, blacks, and nonwhites were 41.64 (SD = 13.38), 38.46 (SD = 13.19), and 39.59 (SD = 12.90) (not statistically significant). Finally, regarding year in college, 8.3 percent were freshmen, 17.1 percent sophomores, 28.6 percent juniors, 43.3 percent seniors, and 2.8 percent graduate students. Freshman and sophomores reported more positive attitudes toward premarital sex (mean = 44.81; SD = 13.39) than did juniors (mean = 40.32; SD = 12.58) or seniors and graduate students (mean = 38.91; SD = 13.10) ($p < .05$).

Source

M. Whatley. Attitudes Toward Premarital Sex Scale. 2006. Valdosta, Georgia: Department of Psychology, Valdosta State University. Used by permission. Other uses of this scale by written permission of Dr. Whatley only. Information on the reliability and validity of this scale is available from Dr. Whatley (mwhatley@valdosta.edu).

Alternative Sexual Values

> Remember that sex is not out there, but in here, in the deepest layer of your own being. There is not only a morning after—there are lots of days and years afterwards.
>
> Jacob Heusner, Words of Wisdom

There are at least three sexual value perspectives that guide choices in sexual behavior: absolutism, relativism, and hedonism. People sometimes have different sexual values at different stages of the family life cycle. For example, elderly individuals are more likely to be absolutist, whereas those in the middle years are more likely to be relativistic. Young unmarried adults are more likely than the elderly to be hedonistic.

Absolutism

Absolutism refers to a belief system based on unconditional allegiance to the authority of science, law, tradition, or religion. A religious absolutist makes sexual choices on the basis of moral considerations. To make the correct moral choice is to comply with God's will, and to not comply is a sin. A legalistic absolutist makes sexual decisions on the basis of a set of laws. People who are guided by absolutism in their sexual choices have a clear notion of what is right and wrong.

The official creeds of fundamentalist Christian and Islamic religions encourage absolutist sexual values. Intercourse is solely for procreation, and any sexual acts (masturbation, oral sex, homosexuality) that do not lead to procreation are immoral and regarded as sins against God, Allah, self, and community. Waiting until marriage to have intercourse is also an absolutist sexual value.

"True Love Waits" is an international campaign designed to challenge teenagers and college students to remain sexually abstinent until marriage. Under this program, created and sponsored by the Baptist Sunday School Board, young people are asked to agree to the absolutist position and sign a commitment to the following: "Believing that true love waits, I make a commitment to God, myself, my family, my friends, my future mate, and my future children to be sexually abstinent from this day until the day I enter a biblical marriage relationship" (True Love Waits Web site, 2006, http://www.lifeway.com/tlw/students/join.asp).

A team of researchers (Rostosky et al., 2003) found that adolescents who had taken the "virginity pledge" were more likely to delay having first intercourse. However, recent analysis of data from the National Longitudinal Study of Adolescent Health reveals more complex long-term consequences of making such a pledge (Brucker and Bearman, 2005). Although youth who took the pledge were more likely than other youth to experience a later "sexual debut," had fewer partners, and married earlier, most eventually engaged in premarital sex, were less likely to use a condom when they first had intercourse, and were more likely to substitute oral and/or anal sex in the place of vaginal sex. There was no significant difference in STIs between "pledgers" and "nonpledgers." The researchers speculated that the emphasis on virginity may have encouraged the pledgers to engage in noncoital (nonintercourse) sexual activities, which still exposed them to STIs and to be less likely to seek testing and treatment for STIs.

In regard to college students, 14 percent of 1019 undergraduates at a southeastern university selected absolutism as their primary sexual value. Demographic factors associated with absolutism among this university sample included being black (25%), being religious (48%), and being engaged/married (65%) (O'Reilly et al., 2006a). These percentages reflect the influence of race, religion, and level of relationship commitment. The fact that significantly more black than white students reported "I view myself as a religious person" (76% vs. 66%) helps to account for the fact that blacks adhere to a more absolutist position. It is not surprising that persons who are engaged or married are more absolutist in their sexual values than those with less relationship involvement.

Some individuals still define themselves as virgins even though they have engaged in oral sex. Of 1027 university students, 61.6 percent agreed that "having sex was having sexual intercourse, not having oral sex." Hence, in their opinion, having oral sex with someone is not really having sex (Knox and Zusman, 2006).

Carpenter (2003) discussed in her paper "Like a Virgin . . . Again?" the concept of **secondary virginity**—the conscious decision of a sexually active person to refrain from intimate encounters for a specified period of time. Secondary virginity closely resembles a pattern scholars have called "regretful" nonvirginity, the chief difference being the adoption of the label *virgin* by the nonvirgin in question. Secondary virginity may be a result of physically painful, emotionally distressing, or romantically disappointing sexual encounters. Of the 61 young adults Carpenter interviewed, more than half (women more than men) believed that a person could, under some circumstances, be a virgin more than once. Fifteen people contended

Give me chastity and continence, but not quite yet.

St. Augustine, theological fountainhead of the Reformation

Diversity in Other Countries

Absolutism is still a viable sexual value in the Hai Duong Province, Vietnam. In a survey of 800 married respondents, 40 percent of the husbands and 70 percent of the wives reported that premarital intercourse was not acceptable. Persons most likely to have absolutist values were female, had less education, and were from rural areas (Ghuman, 2005).

that a person could resume her or his virginity in an emotional, psychological, or spiritual sense. Terence Deluca (27 years old, heterosexual, white, Roman Catholic), explained:

> There is a different feeling when you love somebody and when you just care about somebody. So I would have to say if you feel that way, then I guess you could be a virgin again. Christians get born all the time again, so. . . . When there's true love involved, yes, I believe that.

A subcategory of absolutism is **asceticism.** The ascetic believes that giving in to carnal lust is unnecessary and attempts to rise above the pursuit of sensual pleasure into a life of self-discipline and self-denial. Accordingly, spiritual life is viewed as the highest good, and self-denial helps one to achieve it. Catholic priests, monks, nuns, and some other celibates have adopted the sexual value of asceticism.

Relativism

Relativism is a value system emphasizing that sexual decisions should be made in the context of a particular situation. Whereas an absolutist might feel that it is wrong for unmarried people to have intercourse, a relativist might feel that the moral correctness of sex outside marriage depends on the particular situation. For example, a relativist might feel that in some situations, sex between casual dating partners is wrong (such as when one individual pressures the other into having sex or lies in order to persuade the other to have sex). But in other cases—when there is no deception or coercion and the dating partners are practicing "safer sex"— intercourse between casual dating partners may be viewed as acceptable.

Sixty-six percent of 1019 undergraduates selected "relativism" as their prevailing sexual value. Most of these undergraduates felt that if they were in a secure, mutual love relationship, then sexual intercourse was justified (O'Reilly et al., 2006a). Women were more likely to be relativists than men (72% vs. 52%). In addition, whites were more likely to be relativists than blacks (69% vs. 55%).

Sexual values and choices that are based on relativism often consider the degree of love, commitment, and relationship involvement as important factors. In a study designed to assess "turn ons" and "turn offs" in sexual arousal, women spoke of "feeling desired versus feeling used" by the partner. "Many women talked about how their arousal was increased with partners who seemed particularly interested in them as individual women, rather than someone that they just wanted to have sex with" (Graham et al., 2004).

A disadvantage of relativism as a sexual value is the difficulty of making sexual decisions on a relativistic case-by-case basis. The statement "I don't know what's right anymore" reflects the uncertainty of a relativistic view. Once a person decides that mutual love is the context justifying intercourse, how often and how soon is it appropriate for the person to fall in love? Can love develop after two hours of conversation? How does one know that love feelings are genuine? The freedom that relativism brings to sexual decision-making requires responsibility, maturity, and judgment. In some cases, individuals may convince themselves that they are in love so that they will not feel guilty about having intercourse. Though one may feel "in love," "secure," and "committed" at the time first intercourse occurs, of all first intercourse experiences reported by women, only 17 percent of these are with the person they eventually marry (Raley, 2000).

Whether or not two unmarried people should have intercourse would be viewed differently by absolutists and relativists. Whereas an absolutist would say that it is wrong for unmarried people to have intercourse and right for married people to do so, a relativist would say, "It depends on the situation." Suppose, for example, that a married couple do not love each other and intercourse is an abusive, exploitative act. Suppose also that an unmarried couple

Sex is the proof that it is easier to get two bodies together than two souls.
Unknown

Diversity in Other Countries

In a study of sexual values among university students in Russia, Dolbik-Vorobei (2005) noted that 60 percent of the undergraduates in his sample reported that sexual relations outside of marriage was acceptable as long as "close spiritual relations" had been established between the partners.

Chapter 5 Sexuality in Relationships

Abstinence Sex Education in the Public Schools

Sexuality education was introduced in the American public school system in the late nineteenth century with the goal of combating STIs (referred to then as venereal diseases) and instilling sexual morality (typically understood as abstinence till marriage). Over time the abstinence agenda became more formalized. As Rose (2005, p. 1208) noted, "Since 1996, nearly a billion in state and federal funding has been allocated for abstinence-only education despite the lack of evidence supporting the effectiveness of the approach." To qualify for these funds, only sex education programs that emphasized or promoted abstinence were eligible. Those that also discussed contraception and other means of pregnancy protection were not eligible. However, abstinence education may include giving credence to nonsexual antecedents of sexual behavior, including skills (goal setting, decision making, and assertiveness), deals (fidelity, friendships), and psychological factors such as self-esteem (Wilson et al., 2005).

To what degree does exposure to abstinence-only sex education programs result in delay of first intercourse? Bristol (2006) conducted a phone survey of 472 university students and found that the type of sex education program they experienced in high school (abstinence vs. comprehensive) was unrelated to age at first intercourse. Zanis (2005) provided data on 31 students aged 12 to 16 who were exposed to a ten-session abstinence-only program. Students who had not had intercourse before the program began praised the program for supporting their abstinence values and said that the program made it easier to tell their boyfriends "no." All of the students who had had intercourse before the program began continued to do so after the program ended. They expressed support for the sex education program but were hopeful that it would provide information about contraception.

Santelli et al. (2006) noted that although federal support of abstinence-only programs has grown rapidly since 1996, "the evaluations of such programs find little evidence of efficacy in delaying initiation of sexual intercourse." In addition, the American Civil Liberties Union charged that more than a million in federal funds had been used to fund a "road show to convert teens to Christianity" under the guise of "abstinence education." As a result of the suit, the United States agreed to stop funding the Silver Ring Thing program as of January 31, 2006 (Lindsay, 2006).

Your Opinion?

1. To what degree do you support abstinence education in public schools?
2. Should condoms be made available for students already having sex?
3. Should parents control the content of sex education in public schools?

Sources

Bristol, K. 2006. The influence of community factors on sex education program effectiveness. Honors Thesis Sociology, Sociology 4551, East Carolina University, Greenville, NC.

Lindsay, J. 2006. U.S. agrees not to fund abstinence program. Associated Press. http://news.yahoo.com/s/ap/20060224/ap_on_re_us/abstinence_suit (Retrieved February 24).

Rose, S. 2005. Going too far? Sex, sin and social policy. *Social Forces* 84:1207–32.

Santelli, J., M.A. Ott, M. Lyon, J. Rogers, and D. Summers. 2006. Abstinence-only education policies and programs: A position paper of the Society for Adolescent Medicine. *Journal of Adolescent Health* 38:83–87.

Zanis, D. A. 2005. Use of sexual abstinence only curriculum with sexually active youths. *Children and Schools* 27:59–63.

love each other and their intercourse experience is an expression of mutual affection and respect. A relativist might conclude that in this particular situation, it is "more right" for the unmarried couple than the married couple to have intercourse. A specific expression of relativism is reflected by students who become involved in a "friends with benefits" relationship.

Friends with Benefits

A new trend is emerging in relational/sexual behavior. **Friends with benefits** (FWB) is a relationship of nonromantic friends who also have a sexual relationship. In a sample of 1027 undergraduates at a large southeastern university, more than half (50.1%) reported that they had been in an FWB relationship (Knox and Zusman, 2006). In a smaller sample of 170 undergraduates at the same university, almost 60 percent of these undergraduates (57.3%) reported that they were or had been involved in an FWB relationship. There were no significant differences between the percentages of women and men reporting involvement in an FWB relationship. This is one of the few studies finding no difference in sexual behavior between women and men (e.g., one would expect men to have more FWB relationships than women). However, the percentages of women and men in our sample were very similar in their reported rates of FWB involvement—57.1 percent and 57.9 percent. Is a new sexual equality operative in FWB relationships? (McGinty et al., 2007). Continued analysis of the data revealed other significant differences between women and men college students in regard to various aspects of the FWB relationship.

To err is human, but it feels divine.

Mae West, actress

1. Women were more emotionally involved. Women were significantly more likely than men (62.5% vs. 38.1%) to view their current FWB relationship as an emotional relationship. In addition, women were significantly more likely than men to be perceived as being more emotionally involved in the FWB relationship. More than 40 percent (43.5%) of the men, compared with 13.6 percent of the women, reported "my partner is more emotionally involved than I am."

2. Men were more sexually focused. As might be expected from the first finding, men were significantly more likely than women to agree with the statement, "I wish we had sex more often than we do" (43.5% vs. 13.6%).

3. Men were more polyamorous. With polyamory defined as desiring to be involved in more than one emotional/sexual relationship at the same time, men were significantly more likely than women to agree that "I would like to have more than one FWB relationship going on at the same time" (34.8% vs. 4.5%). Serial FWB relationships may already be occurring. More than half of the men (52.2%), compared with almost a quarter of the women (24.6%), reported that they had been involved in more than one FWB relationship.

University students who consider involvement in an FWB relationship might be aware that women and men bring different perceptions, expectations, and definitions to the relationship table. These data suggest that women tend to view such a relationship as emotional, with the emphasis on friends, whereas men tend to view the relationship as more casual, with an emphasis on benefits (as in sexual benefits). Indeed, when the women and men who were involved in an FWB relationship were asked if they were more friends than lovers, almost 85 percent (84.4%) of the women compared with less than 15 percent (14.8%) of the men reported that they were more friends than lovers. Hughes et al. (2005) studied 143 undergraduates in FWB relationships and noted that a ludic, playful, noncommittal love characterized them. Hughes et Al. (2005) did not provide data on gender differences.

Hedonism

Pleasure is the only thing one should live for.

Oscar Wilde, playwright, novelist

Hedonism is the belief that the ultimate value and motivation for human actions lie in the pursuit of pleasure and the avoidance of pain. The hedonistic value is reflected in the statement "If it feels good, do it." Hedonism assumes that sexual desire, like hunger and thirst, is an appropriate appetite and its expression is legitimate.

More than one in five (20.3%) of 1019 undergraduates selected "hedonism" as their primary sexual value. Persons who were dating casually were three times more likely than those who were involved in a love relationship (38.5% vs. 12.2%) to be hedonists. Undergraduates who were willing to live together were almost twice as likely to have a hedonistic view of sexuality than those who were against cohabitation (23.3% vs. 12.9%). Finally, those who had been in an FWB relationship were four times as likely to report hedonistic sexual values than those who had not been in an FWB relationship (32.1% vs. 8.3%) (O'Reilly et al., 2006a). Clearly, hedonism is a liberal sexual value that casual daters, those who are open to cohabitation, and those who have experienced an FWB relationship are more likely to support (O'Reilly et al., 2006a).

Hedonism may also emerge in reference to social context. Milhausen et al., (2006) studied 180 men and 120 women who experienced Mardi Gras in New Orleans and noted that sexual behavior could be predicted by perceptions of peer sexual activity, intentions, and previous sexual experience. If one's peers were perceived as engaging in sexual behavior, if the individual intended to have sex with a new partner, and if the individual had a history of previous sexual partners, hedonism was likely to emerge. Indeed, among the males, almost half expected to have oral or vaginal intercourse with a new partner at Mardi Gras (20% expected to have anal intercourse with a new partner).

Sexual Double Standard

The **sexual double standard**—the view that encourages and accepts sexual expression of men more than women—is reflected in Table 5.1. Indeed, one study showed that men were about three times more hedonistic than women (O'Reilly et al., 2006a). Acceptance of the double standard is evident in that hedonistic men are thought of as "studs" but hedonistic women as "sluts."

Beware of the girl with the baby stare—a man is safer in the electric chair.

Evan Esar, Humorist

Table 5.1 Sexual Value by Sex of Respondent			
Respondents	**Absolutism**	**Relativism**	**Hedonism**
Male students	13.0% (43)	52.2% (173)	34.8% (115)
Female students	14.4% (43)	72.2% (497)	13.4% (92)

Source: O'Reilly, S., D. Knox, and M. Zusman. 2006. Correlates of sexual values: Data on 1019 undergraduates. Poster, Southern Sociological Society, New Orleans, March 24.

The sexual double standard is also evident in that there is lower disapproval of men having higher numbers of sexual partners but high disapproval of women for having the same number of sexual partners as men. Welles (2005) conceptualized the sexual double standard as allowing men to seek and enjoy sexual pleasure while women are relegated to the role of being attractive and being objects of sexual pleasure; hence, their own development of and right to sexual desire are secondary.

In a recent study, England and Thomas (2006) noted that the double standard was operative in hooking up. Women who hooked up too often with too many men and had sex too easily were vulnerable to getting a bad reputation. Men who did the same thing got a bad reputation among women, but there was less stigma. In addition, men gained status among other men for their exploits; women were more quiet.

Evidence of the focus on women in terms of youth and beauty was revealed by Lauzen and Dozier (2005). They reviewed the 100 top-grossing domestic films of 2002 and found that the majority of the men were in their 30s and 40s, whereas the majority of women were in their 20s and 30s. Finally, Greene and Faulkner (2005) studied 689 heterosexual couples and found that women were disadvantaged in negotiating sexual issues with their partners, particularly when their traditional gender roles were operative.

National Data

Fifty-eight percent of undergraduate men versus 34 percent of undergraduate women in a random sample of more than 263,000 students at more than 460 colleges and universities agreed that "if two people really like each other, it's all right for them to have sex if they've known each other for only a very short time" (American Council on Education and University of California, 2005–2006).

Diversity in Other Countries

The double standard in other countries may be particularly pronounced. In Turkey women are "significantly disrespected and penalized for having sex before marriage" (Sakalh-Ugurlu and Glick, 2003).

Sources of Sexual Values

The sources of one's sexual values are numerous and include one's school, religion, family, and peers, as well as technology, television, social movements, and the Internet. Halstead (2005) emphasized that schools play a powerful role in shaping a child's sexual values. Previously we have noted that public schools in the United States promote absolutist sexual values through abstinence education and that the effectiveness of these programs has been questioned.

Religion is also an important influence. More than 45 percent of 657 undergraduates at a large southeastern university (assessed via random digit dialing) reported that religion had been influential on their sexual choices: for 26.9 percent, "very influential," and for 18.4 percent, "somewhat influential" (Bristol and Farmer, 2005). A team of researchers (Weinberg et al., 2000) compared undergraduates at an American and a Swedish university and found that the religiosity among Americans was higher and contributed to their more restrictive views toward sexuality. Buddhism, Hinduism, and Islam all encourage waiting until marriage for sexual intercourse.

Sexual attitudes of parents may provide a model for the sexual values of their children. More than 40 percent of 657 undergraduates at a large southeastern university (assessed via random digit dialing) reported that their parents had been influential in their sexual choices: for 17.5 percent, "very influential," and for 24.7 percent, "somewhat influential" (Bristol and Farmer, 2005). Similarly, among 918 university students, "mom" was identified as the most influential source of sexual information.

Siblings are also influential. Kornreich et al. (2003) found that girls who had older brothers held more conservative sexual values. "Those with older brothers in the home may be socialized more strongly to adhere to these traditional standards in line with power dynamics believed to shape and reinforce more submissive gender roles for girls and women" (p. 197).

Peers are an important source of sex education and are important influences on one's sexual values. Silver and Bauman (2005) studied 422 sexually inexperienced adolescents and found that they believed that their peers were also virgins. Somers and Gleason (2001) evaluated the comparative contribution that multiple sources of education about sexual topics (family, peers, media, school, and professionals) made on teen sexual knowledge, attitudes, and behavior and found that, in general, teens tended to get less of their sex education from schools and more of their sex education from nonsibling family.

Reproductive technologies such as birth-control pills, the morning-after pill, and condoms influence sexual values by affecting the consequences of behavior. Being able to reduce the risk of pregnancy and HIV infection with the pill and condoms allows one to consider a different value system than if these methods of protection did not exist.

This French couple is in Paris, where kissing in public is normative. One's sexual values are learned from one's society.

Authors

Sexuality is a major theme of television in the United States. A television advertisement shows an affectionate couple with minimal clothes on in a context where sex could occur. "Be ready for the moment" is the phrase of the announcer, and Levitra, the new quick-start Viagra, is the product for sale. The advertiser uses sex to get the attention of the viewer and punches in the product.

However, as a source of sexual values and responsible treatments of contraception, condom usage, abstinence, and consequences of sexual behavior, television is woefully inadequate. Indeed, the viewer learns that sex is romantic and exciting but learns nothing about discussing the need for contraception or HIV and STI protection. With few exceptions, viewers are inundated with role models who engage in casual sex without protection.

Social movements such as the Women's Movement affect sexual values by empowering women with an egalitarian view of sexuality. This translates into encouraging women to be more assertive about their own sexual needs and giving them the option to experience sex in a variety of contexts (e.g., without love or commitment) without self-deprecation. The net effect is a potential increase in the frequency of recreational, hedonistic sex. The Gay Liberation movement has also been influential in encouraging values that are accepting of sexual diversity.

Another influence on sexual values is the Internet; its sexual content is extensive. The Internet features erotic photos, videos, and stripping/"live" sex acts by Web-cam sex artists. Individuals can exchange nude photos, have explicit sex dialogue, arrange to have "phone sex" or meet in person, or find a prostitute. Individuals who "previously would never have ventured into a porn shop or ordered sex products through the mail now feel comfortable exploring explicit material from the privacy of their personal computers" (Cooper et al., 2002). Annually, an estimated 32 million people visit one or more of the 260 million adult Web pages; thus, a quarter of all people who use the Internet have visited an adult site (Reavill, 2005). However, there is concern that the Internet has gone too far in its sexual content. Dean (2005) notes, "We have been assimilated, stimulated, and simulated. I'm glad I can still remember the real human touch of desire." Some suggest that the government should intervene and control the sexual content on the Internet.

The Internet is a virtual store open 24 hours a day. . . . with millions of users on-line... someone who knows how to navigate the World Wide Web can, at any moment, find a kindred spirit with a similar sexual interest or desire.

Al Cooper, Sylvain Boies, Marlene Maheu, David Greenfield, researchers

Sexual Behaviors

We have been discussing the various sources of sexual values. We now focus on what people report that they do sexually in terms of masturbation, oral sex, vaginal intercourse, and anal sex. Our discussion of sexual behavior ends with an examination of gender differences in sexual behavior.

Masturbation

Masturbation involves stimulating one's own body with the goal of experiencing pleasurable sexual sensations. Of almost a thousand (973) undergraduates at a large southeastern university, more than two-thirds (68.9%) reported having masturbated. Those more likely to report having masturbated were male, had hedonistic sexual values, were nonreligious, and were cohabitants (O'Reilly et al., 2006b). Alternative terms for masturbation include *autoeroticism, self-pleasuring, solo sex,* and *sex without a partner.* Replacing various myths about masturbation (e.g., it causes blindness) is now an appreciation of its benefits. Most health care providers and therapists today regard masturbation as a normal and healthy sexual behavior. Furthermore, masturbation has become known as a form of safe sex in that it involves no risk of transmitting diseases (such as HIV) or produc-

Masturbation: the primary sexual activity of mankind. In the nineteenth century it was a disease; in the twentieth, it's a cure.

Thomas Szasz, psychiatrist

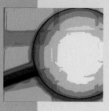

Sex Education/Attitudes Among North American Women

RESEARCH APPLICATION

Most individuals, couples, and parents want to avoid an unplanned pregnancy. How to provide sex education to youth remains a primary concern of sex educators. This study sought to identify those sources of sex education associated with reducing unplanned pregnancies.

Sample and Methods
To find out, Williams and Bonner (2006) analyzed data provided via an Internet survey completed by 1400 North American women. The average age of the respondents was 19.5 years. Almost a quarter (24.7%) had experienced an unplanned pregnancy; 74.4 percent had not done so. When sex education was defined as some type of birth control information, 82 percent received such sex education in school, 66.5 percent from parents, 47.7 percent from a doctor, and 40.2 percent from a clinic/agency. Others found information about sex from friends (77.7%), 43.3 percent from books, 36.8 percent from the Internet, and 19.2 percent from other sources. Less than 3 percent of the sample had not received sex education from any source.

Selected Findings and Conclusions
A number of findings emerged from the analysis.

1. Satisfaction with source of sex education. Women rated their level of satisfaction on a scale from 1 to 7. Women were more satisfied with sex education from friends, books, and the Internet than from parents, schools, clinics, and doctors. Indeed, women who reported that they received no information about either contraception or abstinence were least satisfied with the sex education from their school.

2. Source of sex education and unplanned pregnancy. Students receiving abstinence or equal parts abstinence and contraceptive sex education were 1.68 times less likely to experience unplanned pregnancy than those receiving no sex education. In addition, those receiving contraceptive education were 1.45 times less likely to experience unplanned pregnancy than those receiving no sex education.

3. School sex education and abortion. Students who received both abstinence and a combination of contraceptive and abstinence education were significantly more likely to have fewer abortions than those with no sex education.

4. Religion and unplanned pregnancy. Persons who self-identified as having a religious affiliation were less likely to report having had an unplanned pregnancy than persons reporting no religious affiliation.

5. Religion and type of sex education. Parents who identified as Christian were more likely to teach abstinence. More than a third of non-Christians had not received any birth control information from their parents.

There are two implications from this study: (1) some sex education from school is better than none in terms of reducing the frequency of an unplanned pregnancy, and (2) women seem to learn more from their friends than from parents or teachers.

Source
Williams, M. T. and L. Bonner. 2006. Sex education attitudes and outcomes among North American women. *Adolescence* 41:1–14.

---

ing unintended pregnancy. Masturbation is also associated with orgasm during intercourse. In a study by Thomsen and Chang (2000), 292 university undergraduates reported whether they had ever masturbated and whether they had an orgasm during their first intercourse experience. The researchers found that the strongest single predictor of orgasm and emotional satisfaction with first intercourse was previous masturbation.

Diversity in Other Countries

More than a thousand high school students (1368, aged years and older) in the United Kingdom revealed their expectations regarding oral sex. Of the sexually experienced males, almost half (48.9%) expected oral sex during a sexual experience; 46.5 percent of the sexually experienced females expected oral sex. However, only 14 percent of the males and 29 percent of the females reported that they believed it was important to use a condom during fellatio (Stone et al., 2006).

Oral Sex

In a sample of 1027 undergraduates, more than three-quarters reported that they had received (78.5%) or given (75.9%) oral sex (Knox and Zusman, 2006). **Fellatio** is oral stimulation of the man's genitals by his partner. In many states, legal statutes regard fellatio as a "crime against nature." "Nature" in this case refers to reproduction, and the "crime" is sex that does not produce babies. Nevertheless, most men have experienced fellatio.

Sexuality has become a legitimate research topic.

Cunnilingus is oral stimulation of the woman's genitals by her partner. With regard to their most recent sexual event, 20 percent of women reported that the last time they had sex their partner performed cunnilingus on them. Women who were neither married nor living with a partner reported the highest frequency of cunnilingus as their most recent sexual event (26%, noncohabiting; 22%, cohabiting; 17%, married) (Michael et al., 1994).

African-Americans typically have lower rates of oral sex than whites. Gagnon (2004) noted:

> *I think that these differences are mostly the result of education and religion—but there may be other factors relating to the symbolic meanings of oral sex in Western cultures. As a result of the economic and racial oppression of African-Americans, potential threats to men's power and masculinity are often more present when African-American men have sex. There is always the threat of symbolic subordination when men perform oral sex, particularly if it is not identified by the woman as masculinity enhancing and it is not reciprocated. This is a parallel to the symbolic subordination of women when they perform unreciprocated oral sex. However, if the men will not do it, then the women will not either, since they view fellatio without reciprocation as simply servicing the man.*

We noted earlier that, increasingly, youth who have oral sex regard themselves as virgins, believing that only sexual intercourse constitutes "having sex." Of 1027 university students, 61.6 percent agreed that "having sex was having sexual intercourse, not having oral sex" (Knox and Zusman, 2006). One researcher found that among her respondents, "overwhelmingly, oral sex is seen as foreplay, rather than sex, even among those abstaining from intercourse" (Hertzog, 2004).

There is the mistaken belief that only intercourse carries the risk of contracting a STI. But STIs as well as HIV can be contracted orally. A team of researchers confirmed that a person may contract not only the herpes virus from oral sex but also HIV (Mbopi-Keou et al., 2005). Use of a condom or dental dam, a flat latex device that is held over the vaginal area, is recommended.

The average number of times per week 483 French men reported having had intercourse was 2; of 519 French women, 1.6. The decision preceded the act by a few seconds or minutes for 82.7 percent of subjects (Colson et al., 2006).

Vaginal Intercourse

Vaginal intercourse, or **coitus,** refers to the sexual union of a man and woman by insertion of the penis into the vagina. In a study of 1027 undergraduates, more than 85 percent (86.1) reported that they were open to having intercourse with someone they were not married to (Knox and Zusman, 2006). Kaestle et al. (2002) noted that females who are involved with an older partner are more likely to report having had intercourse. Data from 1975 females revealed that if a 17-year-old female was romantically involved with a partner six years older, the female was twice as likely to report having had intercourse as was a female with a partner the same age.

PERSONAL CHOICES

Deciding to Have Intercourse with a New Partner

A team of researchers (Michels et al., 2005) interviewed 42 adolescents and found that decisions about engaging in sexual activities are made after considering a variety of issues, including risks, benefits, and alternatives. The following are issues you might consider in making the decision to have sex with a new partner.

1. Personal consequences. How do you predict you will feel about yourself after you have had intercourse? An increasing percentage of college students are relativists and feel that if they are in love and have considered their decision carefully, the outcome will be positive. (The quotes in this section are from students in the authors' classes.)

> I believe intercourse before marriage is OK under certain circumstances. I believe that when a person falls in love with another, it is then appropriate. This should be thought about very carefully for a long time, so as not to regret engaging in intercourse.

Those who are not in love and have sex in a casual context sometimes feel bad about their decision:

> I viewed sex as a new toy—something to try as frequently as possible. I did my share of sleeping around, and all it did for me was to give me a total loss of self-respect and a bad reputation. Besides, guys talk. I have heard rumors that I have slept with guys that I have never slept with.

The effect intercourse will have on you personally will be influenced by your personal values, your religious values, and the emotional involvement with your partner. Some people prefer to wait until they are married to have intercourse and feel that this is the best course for future marital stability and happiness. There is often, but not necessarily, a religious basis for this value.

Strong personal and religious values against nonmarital intercourse may result in guilt and regret following an intercourse experience. In a sample of 270 undergraduates who had had intercourse, 71.9 percent regretted their decision to engage in sexual activity at least once. The most cited reason for these students' regret was going against their own morals (37%) (Oswalt et al., 2005). Similarly, in a sample of first-year students at a large southeastern university, one-third regretted their sexual activity in high school, and almost half reported that they had made changes in their sexual behavior: engaging in less sexual behavior, with fewer partners (Doub, 2006).

2. Partner consequences. Because a basic moral principle is to do no harm to others, it may be important to consider the effect of intercourse on your partner. Whereas intercourse may be a pleasurable experience with positive consequences for you, your partner may react differently. What is your partner's religious background and what are your partner's sexual values? A highly religious person with absolutist sexual values will typically have a very different reaction to sexual intercourse than a person with low religiosity and relativistic or hedonistic sexual values. In the study of 270 undergraduates referred to above almost a fourth (23%; more women than men) reported regret due to pressure from the partner (Oswalt et al., 2005). Anderson et al. (2005) also noted that university women might use varying tactics (sexy dancing, verbal intimidation, alcohol) to pressure a man to have sex.

3. **Relationship consequences.** What is the effect of intercourse on a couple's relationship? One's personal reaction to having intercourse may spill over into the relationship. When 163 unmarried undergraduates were asked to identify the effect of sexual intercourse on their relationship, 63 percent reported "no change" and 30 percent reported an "improved relationship" (Knox and Brigman, 1996).

4. **Contraception.** Another potential consequence of intercourse is pregnancy. Once a couple decides to have intercourse, a separate decision must be made as to whether intercourse should result in pregnancy. Most sexually active undergraduates do not want children. Persons who want to avoid pregnancy must choose and plan to use a contraceptive method. But many do not. In a study of almost 73,000 pregnancies, 45 percent were unintended (Naimi et al., 2003).

5. **HIV and other sexually transmissible infections.** Engaging in casual sex has potentially fatal consequences. Avoiding HIV infection and other STIs is an important consideration in deciding whether to have intercourse in a new relationship. The increase in the number of people having more partners results in the rapid spread of the bacteria and viruses responsible for numerous varieties of STIs. However, in a sample of 1027 undergraduates, less than half (49.5%) reported that they had used a condom the last time they had intercourse (Knox and Zusman, 2006).

Serovich and Mosack (2003) reported that only 37 percent of men with HIV infection told a casual partner of their positive HIV status. Those who did so reported, "I thought [my partner] had a right to know." Although depending on a person's integrity is laudable, it may not be wise. Indeed, a condom should be used routinely to help protect one from contracting a STI.

6. **Influence of alcohol and other drugs.** A final consideration with regard to the decision to have intercourse in a new relationship is to be aware of the influence of alcohol and other drugs on such a decision. Of students who reported regretting having had intercourse, 31.9 percent noted that alcohol was involved in their decision to have intercourse (Oswalt et al., 2005). A new term has emerged to describe these pregnancies: **alcohol-exposed pregnancy** (AEP) (Ingersoll et al., 2005).

Sources

Anderson, P. B., A. P. Kontos, H. Tanigoshi, and C. Struckman-Johnson. 2005. An examination of sexual strategies used by urban southern and rural Midwestern university women. *Journal of Sex Research* 42:335–41.

Ingersoll, K. S., S. D. Ceperich, M. D. Nettleman, K. Karanda, S. Brocksen, and B. A. Johnson. 2005. Reducing alcohol-exposed pregnancy risk in college women: Initial outcomes of a clinical trial of a motivational intervention. *Journal of Substance Abuse Treatment* 29:173–80.

Knox, D. and M. E. Zusman. 2006. Relationship and sexual behaviors of a sample of 1027 university students. Unpublished data collected for this text. Department of Sociology, East Carolina University, Greenville, NC.

Knox, D. and B. Brigman. 1996. University students' reactions to intercourse. *College Student Journal* 30:547–48.

Michels, T. M., R. Y. Kropp, S. L. Eyre, and B. L. Halpern-Felsher. 2005. Initiating sexual experiences: How do young adolescents make decisions regarding early sexual activity? *Journal of Research on Adolescence* 15:583–607.

Naimi, T. S., L. E. Lipscomb, R. D. Brewer, and B. C. Gilbert. 2003. Binge drinking in the preconception period and the risk of unintended pregnancy: Implications for women and their children. *Pediatrics* 111:1136–41.

Oswalt, S. B., K. A. Cameron, and J. J. Koob. 2005. Sexual regret in college students. *Archives of Sexual Behavior* 34:663–69.

Serovich, J. M., and K. E. Mosack. 2003. Reasons for HIV disclosure or nondisclosure to casual sexual partners. *AIDS Education and Prevention* 15:70–81.

First Intercourse

A team of researchers (Else-Quest et al., 2005) examined data from the National Health and Social Life Survey on the outcome of the respondent's first vaginal intercourse. The first experience was premarital for 82.9 percent of the respondents, at an average age of 17.7 years. The relationship status of the respondent was not associated with later psychological or physical health outcomes. However, if the first was "prepubertal, forced, with a blood relative or stranger, or the result of peer pressure, drugs, or alcohol, poorer psychological and physical outcomes were consistently reported in later life" (p. 102). The take-away message of this research is to delay first intercourse, which might best occur with a partner in an emotional relationship where no pressure, alcohol, or drugs are involved.

Anal Sex

Almost 40 percent (39.1%) of 1027 undergraduates (95.5% heterosexual) reported that they had engaged in anal sex (Knox and Zusman, 2006). The greatest danger of anal sex is that the rectum might tear, in which case blood contact can occur; STIs (including HIV infection) may then be transmitted. Gay men are particularly vulnerable. Wolitski (2005) emphasized that *barebacking* (intentional unprotected anal sex) is a significant threat to the health of gay, bisexual, and other men who have sex with men (MSM). Partners who use a condom during anal intercourse reduce their risk of not only HIV infection but also other STIs. Pain (physical as well as psychological) may also occur; 14 percent of 404 men who have sex with men reported experiencing *anodyspareunia*—frequent, severe pain during receptive anal sex (Damon and Simon Rosser, 2005).

Gender Differences in Sexual Behavior

What man sees in love is woman; what woman sees in man is love.

Arsène Houssaye, novelist

Do differences exist in the reported sexual behaviors of women and men? Yes. In national data based on interviews with 3432 adults, women reported thinking about sex less often than men (19% vs. 54% reported thinking about sex several times a day), reported having fewer sexual partners than men (2% vs. 5% reported having had five or more sexual partners in the previous year), and reported having orgasm during intercourse less often (29% vs. 75%) (Michael et al., 1994, pp. 102, 128, 156).

Men and women also differ in their motivations for sexual intercourse, with men viewing sex more casually (Lenton and Bryan, 2005). In a study of 99 undergraduates, men were significantly more likely than women to report that they were willing to have intercourse with someone they had known for three hours, to have intercourse with two different people within a six-hour period, to have intercourse with someone they did not love, and to have intercourse with someone with whom they did not have a good relationship (Knox et al., 2001). Earlier, we noted that undergraduate men (in comparison with undergraduate women) were three times more likely to report being hedonistic in their sexual values (O'Reilly et al., 2006a).

Sociobiologists explain males' more casual attitude toward sex, engaging in sex with multiple partners, and being hedonistic as biologically based (i.e., due to higher testosterone levels). Social learning theorists, on the other hand, emphasize that men are socialized by the media and peers to think about and to seek sexual experiences. Men are also accorded social approval and called "studs" for their sexual exploits. Women, on the other hand, are more often punished and labeled "sluts" if they have many sexual partners. Because sexual behavior is guided by **social scripts,** what individuals think, do, and experience is a reflection of what they have learned (Simon and Gagnon, 1998).

There are also differences in the perceptions of foreplay and intercourse. Miller and Byers (2004) compared the reported duration of actual foreplay (men = 13 minutes; women = 11 minutes) and desired foreplay (men = 18 minutes; women = 19 minutes) and the reported duration of actual intercourse (men = 8 minutes; women = 7 minutes) and desired intercourse (men = 18 minutes; women = 14 minutes) of heterosexual men and women in long-term relationships with each other. These findings suggested that both men and women underestimated their partner's desires for the duration of both foreplay and intercourse; also, they had similar preferences for duration of foreplay, but men wanted longer intercourse than women.

Sexual satisfaction reported by men and women seems to be equal, at least among the French. In a representative sample of 1002 French respondents (483 men and 519 women) aged 35 years, 83 percent reported relative or full satisfaction with their sex life (Colson et al., 2006).

Mood states typically affect both men and women equally (Lykins et al., 2006). In a study of 663 female college students and 399 college men, the researchers found that individuals who were depressed were less likely to be interested in engaging in sexual behavior. However, this was not always the case as about 10 percent of women and a higher percentage of men reported that they were interested in engaging in sexual behavior in spite of a negative mood state.

Finally, gender differences may also be influenced by ethnic background. Eisenman and Dantzker (2006) surveyed 128 men and 199 women at a Texas-Mexico border university and found that there were statistically significant gender differences on 26 of 38 items. For example, women were less permissive and more sex negative than men in regard to oral sex, premarital intercourse, and masturbation.

Pheromones and Sexual Behavior

The term *pheromone* comes from the Greek words *pherein,* meaning "to carry," and *hormon,* meaning to "excite." Pheromones are "chemical messengers that are emitted into the environment from the body, where they can then activate specific physiological or behavioral responses in other individuals of the same species" (Grammer et al., 2005, p.136). Pheromones are produced primarily by the apocrine glands located in the armpits and pubic region. The functions of pheromones include opposite-sex attractants, same-sex repellents, and mother-infant bonding.

Pheromones typically operate without the person's awareness; researchers disagree about whether pheromones do in fact influence human sociosexual behaviors. Although Levin (2004) reviewed the literature on chemical messengers in attraction, the strongest evidence for the effect of hormones on sexual behavior is that 38 male volunteers who applied a male hormone to their aftershave lotion reported significant increases in sexual intercourse and sleeping next to a partner when compared with men who had a placebo in their aftershave lotion (Cutler et al., 1998).

Sexuality in Relationships

Sexuality occurs in a social context that influences its frequency and perceived quality.

Sexual Relationships among the Never-Married

The never-married and those not living together report more sexual partners than those who are married or living together. In one study, 9 percent of the never-married and those not living together reported having had five or more sexual partners in the previous 12 months; 1 percent of the marrieds and 5 percent of cohabitants reported the same. However, the unmarried, when compared with marrieds and cohabitants, reported the lowest level of sexual satisfaction. One-third of a national sample of persons who were not married and not living with anyone reported that they were emotionally satisfied with their sexual relationships. In contrast, 85 percent of the married and pair-bonded reported emotional satisfaction in their sexual relationships. Hence, although the never-married have more sexual partners, they are less emotionally satisfied (Michael et al., 1994).

Sexual Relationships among the Married

Marital sex is distinctive for its social legitimacy, declining frequency, and satisfaction (both physical and emotional).

1. Social legitimacy. In our society, marital intercourse is the most legitimate form of sexual behavior. Homosexual, premarital, and extramarital intercourse do not have as high a level of social approval as does marital sex. It is not only okay to have intercourse when married, it is expected. People assume that married couples make love and that something is wrong if they do not.

2. Declining frequency. Sexual intercourse between spouses occurs about six times a month, which declines in frequency as the spouses age (Liu, 2003). Pregnancy also decreases the frequency of sexual intercourse (Gokyildiz and Beji, 2005). In addition to biological changes due to aging and pregnancy, satiation also contributes to the declining frequency of intercourse between spouses and partners in long-term relationships. Psychologists use the term **satiation** to mean that repeated exposure to a stimulus results in the loss of its ability to reinforce. For example, the first time you listen to a new CD, you derive considerable enjoyment and satisfaction from it. You may play it over and over during the first few days. But after a week or so, listening to the same music is no longer new and does not give you the same level of enjoyment that it first did. So it is with intercourse. The thousandth time that a person has intercourse with the same partner is not as new and exciting as the first few times.

3. Satisfaction (emotional and physical). Despite declining frequency and less satisfaction over time (Liu, 2003), marital sex remains a richly satisfying experience. Contrary to the popular belief that unattached singles have the best sex, it

is the married and pair-bonded adults who enjoy the most satisfying sexual relationships. In a national sample, 88 percent of married people said they received great physical pleasure from their sexual lives, and almost 85 percent said they received great emotional satisfaction (Michael et al., 1994). Individuals least likely to report being physically and emotionally pleased in their sexual relationships are those who are not married, not living with anyone, or not in a stable relationship with one person (Michael et al., 1994).

Diversity in Other Countries

Chineko et al. (2006) compared 722 Japanese spouses with 162 sexually active Japanese unmarried individuals and found that the latter reported longer and more varied foreplay (e.g., kissing the lips and genitals). Whereas men in the two groups were equally satisfied, the unmarried Japanese women reported more sexual satisfaction.

Sexual Relationships among the Divorced

Of the almost 2 million people getting divorced, most will have intercourse within one year of being separated from their spouses. The meanings of intercourse for the separated or divorced vary. For many, intercourse is a way to reestablish—indeed, repair—their crippled self-esteem. Questions like "What did I do wrong?" "Am I a failure?" and "Is there anybody out there who will love me again?" loom in the minds of the divorced. One way to feel loved, at least temporarily, is through sex. Being held by another and being told that it feels good give people some evidence that they are desirable. Because divorced people may be particularly vulnerable, they may reach for sexual encounters as if for a lifeboat. "I felt that as long as someone was having sex with me, I wasn't dead and I did matter," said one recently divorced person.

National Data

When cohabiting, married, widowed, single, and divorced persons were asked whether intercourse was occurring less frequently than they desired, the respective proportions who answered yes were 38 percent, 49 percent, 60 percent, 65 percent, and 74 percent. Hence, those most dissatisfied with frequency of intercourse are the divorced, and those most satisfied are cohabitants (Dunn et al., 2000, p. 145).

Because the divorced are usually in their 30s or older, they may not be as sensitized to the danger of contracting HIV as persons in their 20s. Yet acquired immunodeficiency syndrome (AIDS) does not affect only youths. The largest number of AIDS cases has occurred among persons aged 40 to 44 years, accounting for 21 percent of cases in 2004 (Centers for Disease Control and Prevention, 2005). Divorced individuals should always use a condom to lessen the risk of STI, including HIV infection, and AIDS.

Safe Sex: Avoiding Sexually Transmitted Infections, Including HIV Infection

The Student Sexual Risks Scale (see page 148) allows you to assess the degree to which you are at risk for contracting an STI, including HIV infection.

One of the negative consequences of sexual behavior is the risk of contracting an STI. Also known as sexually transmitted disease, or *STD*, **STI** refers to the general category of sexually transmitted diseases such as chlamydia, genital herpes, gonorrhea, and syphilis. The most lethal of all STIs is that due to **HIV,** which attacks the immune system and can lead to **AIDS.** Because the consequences of contracting HIV are the most severe, we focus on HIV here. In the Special Topics section at the end of the text, we discuss other STIs.

Some students have a hard time understanding that the consequence of one unprotected sexual encounter may not be reversible.

American College Health Association

Transmission of HIV and High-Risk Behaviors

HIV can be transmitted in several ways.

Student Sexual Risks Scale

The following self-assessment allows you to evaluate the degree to which you may be at risk for engaging in behavior that exposes you to HIV. Safer sex means sexual activity that reduces the risk of transmitting the AIDS virus. Using condoms is an example of safer sex. Unsafe, risky, or unprotected sex refers to sex without a condom, or to other sexual activity that might increase the risk of AIDS virus transmission. For each of the following items, check the response that best characterizes your option.

A = Agree
U = Undecided
D = Disagree

A U D

1. If my partner wanted me to have unprotected sex, I would probably give in.

2. The proper use of a condom could enhance sexual pleasure.

3. I may have had sex with someone who was at risk for HIV/AIDS.

4. If I were going to have sex, I would take precautions to reduce my risk for HIV/AIDS.

5. Condoms ruin the natural sex act.

6. When I think that one of my friends might have sex on a date, I ask him/her if he/she has a condom.

7. I am at risk for HIV/AIDS.

8. I would try to use a condom when I had sex.

9. Condoms interfere with romance.

10. My friends talk a lot about safer sex.

11. If my partner wanted me to participate in risky sex and I said that we needed to be safer, we would still probably end up having unsafe sex.

12. Generally, I am in favor of using condoms.

13. I would avoid using condoms if at all possible.

14. If a friend knew that I might have sex on a date, he/she would ask me whether I was carrying a condom.

15. There is a possibility that I have HIV/AIDS.

16. If I had a date, I would probably not drink alcohol or use drugs.

17. Safer sex reduces the mental pleasure of sex.

18. If I thought that one of my friends had sex on a date, I would ask him/her if he/she used a condom.

19. The idea of using a condom doesn't appeal to me.

20. Safer sex is a habit for me.

21. If a friend knew that I had sex on a date, he/she wouldn't care whether I had used a condom or not.

22. If my partner wanted me to participate in risky sex and I suggested a lower-risk alternative, we would have the safer sex instead.

23. The sensory aspects (smell, touch, etc.) of condoms make them unpleasant.

24. I intend to follow "safer sex" guidelines within the next year.

25. With condoms, you can't really give yourself over to your partner.

26. I am determined to practice safer sex.

27. If my partner wanted me to have unprotected sex and I made some excuse to use a condom, we would still end up having unprotected sex.

1. Sexual contact. HIV is found in several body fluids of infected individuals, including blood, semen, and vaginal secretions. During sexual contact with an infected individual, the virus enters a person's bloodstream through the rectum, vagina, penis (an uncircumcised penis is at greater risk because of the greater retention of the partner's fluids), and possibly the mouth during oral sex. Saliva, sweat, and tears are not body fluids through which HIV is transmitted.

_____ 28. If I had sex and I told my friends that I did not use condoms, they would be angry or disappointed. — — —

_____ 29. I think safer sex would get boring fast. — — —

_____ 30. My sexual experiences do not put me at risk for HIV/AIDS. — — —

_____ 31. Condoms are irritating. — — —

_____ 32. My friends and I encourage each other before dates to practice safer sex. — — —

_____ 33. When I socialize, I usually drink alcohol or use drugs. — — —

_____ 34. If I were going to have sex in the next year, I would use condoms. — — —

_____ 35. If a sexual partner didn't want to use condoms, we would have sex without using condoms. — — —

_____ 36. People can get the same pleasure from from safer sex as from unprotected sex. — — —

_____ 37. Using condoms interrupts sex play. — — —

_____ 38. It is a hassle to use condoms. — — —

(To be read after completing the scale)

Scoring

Begin by giving yourself eighty points. Subtract one point for every undecided response. Subtract two points every time that you disagreed with odd-numbered items or with item number 38. Subtract two points every time you agreed with even-numbered items 2 through 36.

Interpreting Your Score

Research shows that students who make higher scores on the SSRS are more likely to engage in risky sexual activities, such as having multiple sex partners and failing to consistently use condoms during sex. In contrast, students who practice safer sex tend to endorse more positive attitudes toward safer sex, and tend to have peer networks that encourage safer sexual practices. These students usually plan on making sexual activity safer, and they feel confident in their ability to negotiate safer sex even when a dating partner may press for

riskier sex. Students who practice safer sex often refrain from using alcohol or drugs, which may impede negotiation of safer sex, and often report having engaged in lower-risk activities in the past. How do you measure up?

(Below 15) Lower Risk

(Of 200 students surveyed by DeHart and Berkimer, 16 percent were in this category.) Congratulations! Your score on the SSRS indicates that, relative to other students, your thoughts and behaviors are more supportive of safer sex. Is there any room for improvement in your score? If so, you may want to examine items for which you lost points and try to build safer sexual strengths in those areas. You can help protect others from HIV by educating your peers about making sexual activity safer.

(15 to 37) Average Risk

(Of 200 students surveyed by DeHart and Berkimer, 68 percent were in this category.) Your score on the SSRS is about average in comparison with those of other college students. Though it is good that you don't fall into the higher-risk category, be aware that "average" people can get HIV, too. In fact, a recent study indicated that the rate of HIV among college students is 10 times that in the general heterosexual population. Thus, you may want to enhance your sexual safety by figuring out where you lost points and work toward safer sexual strengths in those areas.

(38 and Above) Higher Risk

(Of 200 students surveyed by DeHart and Berkimer, 16 percent were in this category.) Relative to other students, your score on the SSRS indicates that your thoughts and behaviors are less supportive of safer sex. Such high scores tend to be associated with greater HIV-risk behavior. Rather than simply giving in to riskier attitudes and behaviors, you may want to empower yourself and reduce your risk by critically examining areas for improvement. On which items did you lose points? Think about how you can strengthen your sexual safety in these areas. Reading more about safer sex can help, and sometimes colleges and health clinics offer courses or workshops on safer sex. You can get more information about resources in your area by contacting the CDC's HIV/AIDS Information Line at 1-800-342-2437.

Source

DeHart, D. D. and J. C. Birkimer. 1997. The Student Sexual Risks Scale (modification of SRS for popular use; facilitates student self-administration, scoring, and normative interpretation). Developed specifically for this text by Dana D. DeHart, College of Social Work at the University of South Carolina; John C. Birkimer, University of Louisville. Used by permission of Dana DeHart.

2. Intravenous drug use. Drug users who are infected with HIV can transmit the virus to other drug users with whom they share needles, syringes, and other drug-related implements.

3. Blood transfusions. HIV can be acquired by receiving HIV-infected blood or blood products. Currently, all blood donors are screened, and blood is not accepted from high-risk individuals. Blood that is accepted from donors is tested for the presence of HIV. However, prior to 1985, donor blood was not tested for

HIV. Individuals who received blood or blood products prior to 1985 may have been infected with HIV.

4. Mother-child transmission. A pregnant woman infected with HIV has a 40 percent chance of transmitting the virus through the placenta to her unborn child. These babies will initially test positive for HIV as a consequence of having the antibodies from their mother's bloodstream. However, azidothymidine (AZT, alternatively called zidovudine or ZVD) taken by the mother 12 weeks before birth seems to reduce by two-thirds the chance of transmission of HIV to her baby. HIV may also be transmitted, although rarely, from mother to infant through breast-feeding.

5. Organ or tissue transplants and donor semen. Receiving transplant organs and tissues, as well as receiving semen for artificial insemination, could involve risk of contracting HIV if the donors have not been HIV-tested. Such testing is essential, and recipients should insist on knowing the HIV status of the organ, tissue, or semen donor.

6. Other methods of transmission. For health care professionals, HIV can also be transmitted through contact with amniotic fluid surrounding a fetus, synovial fluid surrounding bone joints, and cerebrospinal fluid surrounding the brain and spinal cord.

Prevention of HIV and STI Transmission

The best way to avoid getting an STI is to avoid sexual contact or to have contact only with partners who are not infected. This means restricting your sexual contacts to those who limit their relationships to one person. The person most likely to get an STI has sexual relations with a number of partners or with a partner who has a variety of partners. Even if you are in a mutually monogamous relationship, you may be at risk for acquiring or transmitting an STI. This is because health officials suggest that when you have sex with someone, you are having sex (in a sense) with everyone that person has had sexual contact with in the past ten years.

Partners may believe that they are in a mutually monogamous relationship when they are not. It is not uncommon for partners in "monogamous" relationships to have extradyadic sexual encounters that are not revealed to the primary partner. Partners may also lie about how many sexual partners they have had and whether or not they have been tested for STIs.

Condoms should be used for vaginal, anal, and oral sex and should never be reused. However, in a sample of 1027 undergraduates, slightly less than half (49.5%) reported that they used a condom the last time they had intercourse (Knox and Zusman, 2006). Feeling that the partner is disease-free, having had too much alcohol, and believing that "getting an STI won't happen to me" are reasons individuals do not use a condom. Some partners are also forced to have sex or do not feel free to negotiate the use of a condom in their sexual relationship (Heintz and Melendez, 2006).

Using the condom properly is also important. Putting on a latex or polyurethane condom (natural membrane condoms do not block the transmission of STIs) before the penis touches the partner's body makes it difficult for STIs to be passed from one person to another. Care should also be taken to withdraw the penis while it is erect to prevent fluid from leaking from the base of the condom into the partner's genital area. If a woman is receiving oral sex, she should wear a dental dam, which will prevent direct contact between the genital area and her partner's mouth.

Sexuality in an age of HIV and STIs demands talking about safer sex issues with a new potential sexual partner. Bringing up the issue of condom use should be perceived as caring for oneself, the partner, and the relationship rather than as a sign of distrust. Some individuals routinely have a condom available, and it is a "given" in any sexual encounter. Figure 5.1 on page 151 illustrates that one is more likely to contract an STI through high alcohol use, low condom use, and having sex with multiple partners.

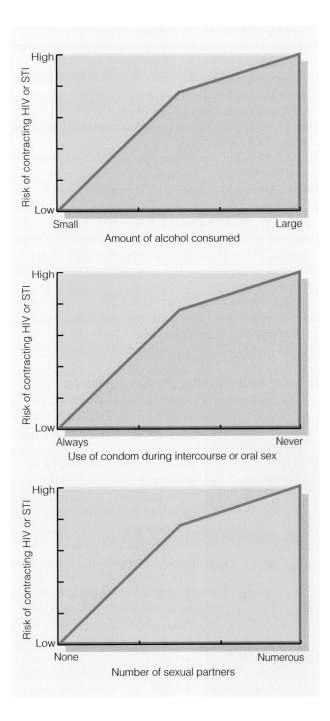

Figure 5.1
Risk of Contracting an STI, as Related to Alcohol, Condom Use, and Number of Partners

Sexual Fulfillment: Some Prerequisites

There are several prerequisites for having a good sexual relationship.

Self-Knowledge, Self-Esteem, Health

Sexual fulfillment involves knowledge about yourself and your body. Such information not only makes it easier for you to experience pleasure but also allows you to give accurate information to a partner about pleasing you. It is not possible to teach a partner what you don't know about yourself.

Sexual fulfillment also implies having a positive self-concept. To the degree that you have positive feelings about yourself and your body, you will regard your-

The secret of a good sexual relationship is making love with your partner, not to your partner.
Dianna Lowe and Kenneth Lowe, married couple

self as a person someone else would enjoy touching, being close to, and making love with. If you do not like yourself or your body, you might wonder why anyone else would.

Effective sexual functioning also requires good physical and mental health. This means regular exercise, good nutrition, lack of disease, and lack of fatigue. Performance in all areas of life does not have to diminish with age—particularly if people take care of themselves physically (see Chapter on Aging in Marriage and Family Relationships).

Good health also implies being aware that some drugs may interfere with sexual performance. Alcohol is the drug most frequently used by American adults. Although a moderate amount of alcohol can help a person become aroused through a lowering of inhibitions, too much alcohol can slow the physiological processes and deaden the senses. Shakespeare may have said it best: "It [alcohol] provokes the desire, but it takes away the performance" (*Macbeth,* act 2, scene 3). The result of an excessive intake of alcohol for women is a reduced chance of orgasm; for men, overindulgence results in a reduced chance of attaining an erection.

The reactions to marijuana are less predictable than the reactions to alcohol. Though some individuals report a short-term enhancement effect, others say that marijuana just makes them sleepy. In men, chronic use may decrease sex drive because marijuana may lower testosterone levels.

A kiss is something which you cannot give without taking, and cannot take without giving.

Anonymous

This couple is in a stable, loving, monogamous relationship.

Authors

A Good Relationship, Positive Motives

A guideline among therapists who work with couples who have sexual problems is to treat the relationship before focusing on the sexual issue. The sexual relationship is part of the larger relationship between the partners, and what happens outside the bedroom in day-to-day interaction has a tremendous influence on what happens inside the bedroom. The statement "I can't fight with you all day and want to have sex with you at night" illustrates the social context of the sexual experience.

McGuirl and Wiederman (2000) asked 185 men and 244 women to identify their preferences in an ideal sex partner. Men most valued a physically attractive partner, one who was open to discussing sex, who communicated her desires clearly, who was easily aroused, and who was uninhibited. Paying him compliments during sex and experiencing orgasm easily were also important.

Women most valued a partner who was open to discussing sex, who was knowledgeable about sex, who clearly communicated his desires, who was physically attractive, and who paid her compliments during sex. Being easily sexually aroused and being uninhibited were also important.

Sexual interaction communicates how the partners are feeling and acts as a barometer for the relationship. Each partner brings to a sexual encounter, sometimes unconsciously, a motive (pleasure, reconciliation, procreation, duty), a psychological state (love, hostility, boredom, excitement), and a physical state (tense, exhausted, relaxed, turned on). The combination of these factors will change from one encounter to another. Tonight the wife may feel aroused and loving and seek pleasure, but her husband may feel exhausted and hostile and have sex only out of a sense of duty. Tomorrow night, both partners may feel relaxed and have sex as a means of expressing their love for each other.

One's motives for a sexual encounter are related to the outcome. Impett et al. (2005) found that when individuals have intercourse out of the desire to enhance personal and interpersonal/relationship pleasure, the personal and interpersonal effect on well being is very positive. However, when sexual motives were to avoid conflict, the personal and interpersonal effects did not result in similar positive outcomes. In a study of 1002 French adults, sexuality was more synonymous with pleasure (44.0%) and love (42.1%) than with procreation, children, or motherhood (7.8%) (Colson et al., 2006).

An Equal Relationship

Laumann et al. (2006) surveyed 27,500 individuals in 29 countries and found that reported sexual satisfaction was higher where men and women were considered equal. Austria topped the list, with 71 percent reporting sexual satisfaction; only 25.7 percent of those surveyed in Japan reported sexual satisfaction. The United States was among those countries in which a high percentage of the respondents reported sexual satisfaction.

Open Sexual Communication, Feedback

Sexually fulfilled partners are comfortable expressing what they enjoy and do not enjoy in the sexual experience. Unless both partners communicate their needs, preferences, and expectations to each other, neither is ever sure what the other wants. In essence, the Golden Rule ("Do unto others as you would have them do unto you") is *not* helpful, because what you like may not be the same as what your partner wants. A classic example of the uncertain lover is the man who picks up a copy of *The Erotic Lover* in a bookstore and leafs through the pages until the topic on how to please a woman catches his eye. He reads that women enjoy having their breasts stimulated by their partner's tongue and teeth. Later that night in bed, he rolls over and begins to nibble on his partner's breasts. Meanwhile, she wonders what has possessed him and is unsure what to make of this new (possibly unpleasant) behavior.

Sexually fulfilled partners take the guesswork out of their relationship by communicating preferences and giving feedback. This means using what some therapists call the touch-and-ask rule. Each touch and caress may include the question "How does that feel?" It is then the partner's responsibility to give feedback. If the caress does not feel good, the partner should say what does feel good.

Guiding and moving the partner's hand or body are also ways of giving feedback. What women and men want each other to know about sexuality is presented in Table 5.2 on page 154.

I want a little sugar in my bowl.
Nina Simone, blues singer

Having Realistic Expectations

To achieve sexual fulfillment, expectations must be realistic. A couple's sexual needs, preferences, and expectations may not coincide. It is unrealistic to assume that your partner will want to have sex with the same frequency and in the same way that you do on all occasions. It may also be unrealistic to expect the level of sexual interest and frequency of sexual interaction in long-term relationships to remain consistently high.

Sexual fulfillment means not asking things of the sexual relationship that it cannot deliver. Failure to develop realistic expectations will result in frustration and resentment. One's health, feelings about the partner, age, previous sexual experiences (including child sexual abuse, rape, etc.) will have an affect on one's sexuality and one's sexual relationship.

Table 5.2 What Women and Men Want Each Other to Know about Sexuality

What Women Want Men to Know about Sex

Women tend to like a loving, gentle, patient, tender, and understanding partner. Rough sexual play can hurt and be a turnoff.

It does not impress women to hear about other women in the man's past.

If men knew what it is like to be pregnant, they would not be so apathetic about birth control.

Most women want more caressing, gentleness, kissing, and talking before and after intercourse.

Some women are sexually attracted to other women, not to men.

Sometimes the woman wants sex even if the man does not. Sometimes she wants to be aggressive without being made to feel that she shouldn't be.

Intercourse can be enjoyable without orgasm.

Many women do not have an orgasm from penetration only; they need direct stimulation of their clitoris by their partner's tongue or finger.

Men should be interested in fulfilling their partner's sexual needs.

Most women prefer to have sex in a monogamous love relationship.

When a woman says "no," she means it. Women do not want men to expect sex every time they are alone with their partner.

Many women enjoy sex in the morning, not just at night.

Sex is *not* everything.

Women need to be lubricated before penetration.

Men should know more about menstruation.

Many women are no more inhibited about sex than men are.

Women do not like men to roll over, go to sleep, or leave right after orgasm.

Intercourse is more of a love relationship than a sex act for some women.

The woman should not always be expected to supply a method of contraception. It is also the man's responsibility.

Men should always have a condom with them and initiate putting it on.

What Men Want Women to Know about Sex

Men do not always want to be the dominant partner; women should be aggressive.

Men want women to enjoy sex totally and not be inhibited.

Men enjoy tender and passionate kissing.

Men really enjoy fellatio and want women to initiate it.

Women need to know a man's erogenous zones.

Men enjoy giving oral sex; it is not bad and unpleasant.

Many men enjoy a lot of romantic foreplay and slow, aggressive sex.

Men cannot keep up intercourse forever. Most men tire more easily than women.

Looks are not everything.

Women should know how to enjoy sex in different ways and different positions.

Women should not expect a man to get a second erection right away.

Many men enjoy sex in the morning.

Pulling the hair on a man's body can hurt.

Many men enjoy sex in a caring, loving, exclusive relationship.

It is frustrating to stop sex play once it has started.

Women should know that not all men are out to have intercourse with them. Some men like to talk and become friends.

Avoiding Spectatoring

One of the obstacles to sexual functioning (we discuss sexual dysfunctions in detail at the end of this chapter) is **spectatoring,** which involves mentally observing your sexual performance and that of your partner. When the researchers in one extensive study observed how individuals actually behave during sexual intercourse, they reported a tendency for sexually dysfunctional partners to act as spectators by mentally observing their own and their partners' sexual performance. For example, the man would focus on whether he was having an erection, how complete it was, and whether it would last. He might also watch to see whether his partner was having an orgasm (Masters and Johnson, 1970).

Spectatoring, as Masters and Johnson conceived it, interferes with each partner's sexual enjoyment because it creates anxiety about performance, and anxiety blocks performance. A man who worries about getting an erection reduces his chance of doing so. A woman who is anxious about achieving an orgasm probably will not. The desirable alternative to spectatoring is to relax, focus on and enjoy your own pleasure, and permit yourself to be sexually responsive.

Spectatoring is not limited to sexually dysfunctional couples and is not necessarily associated with psychopathology. It is a reaction to the concern that the performance of one's sexual partner is consistent with his or her expectations. We all probably have engaged in spectatoring to some degree. It is when spectatoring is continuous that performance is impaired.

Debunking Sexual Myths

Sexual fulfillment also means not being victim to sexual myths. Some of the more common myths include that sex equals intercourse and orgasm, that women who love sex don't have values, and that the double standard is dead. Another common sexual myth is that the elderly have no interest in sex. Beckman et al. (2006) examined the sexual interests and needs of 563 70-year-olds and found that 95 percent reported the continuation of such interests and needs as they aged. Almost 70 percent (69%) of the married men and 60 percent (57%) of the married women reported continued sexual behavior. Nobre et al. (2006) noted that belief in sexual myths makes one vulnerable to sexual dysfunctions. Table 5.3 presents some other sexual myths.

Table 5.3   Common Sexual Myths
Masturbation is sick.
Women who love sex are sluts.
Sex education makes children promiscuous.
Sexual behavior usually ends after age 60.
People who enjoy pornography end up committing sexual crimes.
Most "normal" women have orgasms from penile thrusting alone.
Extramarital sex always destroys a marriage.
Extramarital sex will strengthen a marriage.
Simultaneous orgasm with one's partner is the ultimate sexual experience.
My partner should enjoy the same things that I do sexually.
A man cannot have an orgasm unless he has an erection.
Most people know a lot of accurate information about sex.
Using a condom ensures that you won't get HIV.
Most women prefer a partner with a large penis.
Few women masturbate.
Women secretly want to be raped.
An erection is necessary for good sex.
An orgasm is necessary for good sex.

What are sexual values?

Sexual values are moral guidelines for making sexual choices in nonmarital, marital, heterosexual, and homosexual relationships.

What are alternative sexual values?

Three sexual values are absolutism (rightness is defined by official code of morality), relativism (rightness depends on the situation—who does what, with whom, in what context), and hedonism ("if it feels good, do it"). Relativism is the sexual value held by most college students, with women being more relativistic than men and men being more hedonistic than women. About 60 percent of men and women college students in one study reported having had a "friends with benefits" relationship.

What is the sexual double standard?

The sexual double standard is the view that encourages and accepts sexual expression of men more than women. For example, men may have more sexual partners than women.

What are sources of sexual values?

The sources of sexual values include one's school, family, religion, and peers, as well as technology, television, social movements, and the Internet.

What are various sexual behaviors?

Masturbation involves stimulating one's own body with the goal of experiencing pleasurable sexual sensations. Fellatio is oral stimulation of the man's genitals by his partner. In many states, legal statutes regard fellatio as a "crime against nature." "Nature" in this case refers to reproduction, and the "crime" is sex that does not produce babies. Cunnilingus is oral stimulation of the woman's genitals by her partner. Increasingly, youth who have oral sex regard themselves as virgins, believing that only sexual intercourse constitutes "having sex." Vaginal intercourse, or coitus, refers to the sexual union of a man and woman by insertion of the penis into the vagina. Anal (not vaginal) intercourse is the sexual behavior associated with the highest risk of HIV infection. The potential for the rectum to tear and blood contact to occur presents the greatest danger. AIDS is lethal. Partners who use a condom during anal intercourse reduce their risk of not only HIV infection but also other STIs.

What are gender differences in sexual behavior?

In national data from interviews with 3432 adults, women reported thinking about sex less often than men (19% vs. 54% reported thinking about sex several times a day), reported having fewer sexual partners than men (2% vs. 5% reported having had five or more sexual partners in the past year), and reported having orgasm during intercourse less often (29% vs. 75%). Men are more casual and more hedonistic in their view of sex than women.

How do pheromones affect sexual behavior?

Pheromones are chemical messengers emitted from the body that activate physiological and behavioral responses. The functions of pheromones include opposite-sex attractants, same-sex repellents, and mother-infant bonding. Researchers disagree about whether pheromones do in fact influence human sociosexual behaviors. In one study men who applied a male hormone to their aftershave lotion reported significant increases in sexual intercourse and sleeping next to a partner in comparison with men who had a placebo in their aftershave lotion.

What are the sexual relationships of never-marrieds, married, and divorced persons?

The never-married and noncohabiting report more sexual partners than those who are married or living with a partner. Marital sex is distinctive for its social legitimacy, declining frequency, and satisfaction (both physical and emotional). The divorced have a lot of sexual partners but are the least sexually fulfilled.

How does one avoid contracting or transmitting STIs?

The best way to avoid getting an STI is to avoid sexual contact or to have contact only with partners who are not infected. This means restricting your sexual contacts to those who limit their relationships to one person. The person most likely to get an STI has sexual relations with a number of partners or with a partner who has a variety of partners. Even if you are in a mutually monogamous relationship, you may be at risk for acquiring or transmitting an STI. This is because health officials suggest that when you have sex with someone, you are having sex (in a sense) with everyone that person has had sexual contact with in the past ten years.

What are the prerequisites of sexual fulfillment?

Fulfilling sexual relationships involve self-knowledge, self-esteem, health, a good nonsexual relationship, open sexual communication, safer sex practices, and making love with, not to, one's partner. Other variables include realistic expectations ("my partner will not always want what I want") and not buying into sexual myths ("masturbation is sick").

KEY TERMS

absolutism	double standard	masturbation	sexual values
AIDS	fellatio	relativism	social script
asceticism	friends with benefits	satiation	spectatoring
coitus	hedonism	secondary virginity	STI
cunnilingus	HIV		

The Companion Website for *Choices in Relationships: An Introduction to Marriage and the Family,* Ninth Edition
www.thomson.edu/sociology/knox

Supplement your review of this chapter by going to the companion website to take one of the Tutorial Quizzes, use the flash cards to master key terms, and check out the many other study aids you'll find there.

WEBLINKS

Body Health: A Multimedia AIDS and HIV Information Resource
http://www.thebody.com

Centers for Disease Control and Prevention (CDC)
http://www.cdc.gov

Go Ask Alice: Sexuality
http://www.goaskalice.columbia.edu/

Sexual Intimacy
http://www.heartchoice.com/sex_intimacy/

Sexual Health
http://www.sexualhealth.com/

Sexuality Information and Education Council (SIECUS)
http://www.siecus.org

Sex Therapist On Call
www.therapyoncall.org

REFERENCES

American Council on Education and University of California. 2005–2006. *The American freshman: National norms for fall, 2005*. Los Angeles, CA: Los Angeles Higher Education Research Institute.

Beckman, N. M., M. Waern, I. Skoog, and The Sahlgrenska Academy at Goteborg University, Sweden. 2006. Determinants of sexuality in 70 year olds. *The Journal of Sex Research* 43:2–3.

Bristol, K. and B. Farmer. 2005. *Sexuality Among Southeastern university students: A survey*. Unpublished data. Greenville, NC: East Carolina University.

Brucker, H. and P. Bearman 2005. After the promise: The STD consequences of adolescent virginity pledges. *Journal of Adolescent Health* 36:271–78.

Carpenter, L. M. 2003. Like a virgin . . . again? Understanding secondary virginity in context. Paper, 73rd Annual Meeting of the Eastern Sociological Society, Philadelphia, February 28.

Chineko, A., I. Masami, and O. Reiko. 2006. Sexuality of the middle aged and elderly: A comparison between married and unmarried people. *The Journal of Sex Research* 43:2–3.

Centers for Disease Control and Prevention. *HIV/AIDS Surveillance Report 2004*. Vol 16. Atlanta: U.S. Department of Health and Human Services, Centers for Disease Control and Prevention.

Colson, M., A. Lemaire, P. Pinton, K. Hamidi, and P. Klein. 2006. Sexual behaviors and mental perception, satisfaction, and expectations of sex life in men and women in France. *The Journal of Sexual Medicine* 3:121–31.

Cooper, A., J. Morahan-Martin, R. M. Mathy, R. M. Maheu, and M. Maheu. 2002. Toward an increased understanding of user demographics in online sexual activities. *Journal of Sex & Marital Therapy*, 28:105–29.

Cutler, W. B., E. Friedmann, and N. L. McCoy. 1998. Pheromonal influences on sociosexual behavior in men. *Archives of Sexual Behavior* 27:1–13.

Damon, W. and B. R. Simon Rosser. 2005. Anodyspareunia in men who have sex with men: Prevalence, predictors, consequences and the development of DSM diagnostic criteria. *Journal of Sex & Marital Therapy* 31:129–41.

Dean, W. 2005. Web sluts of the webosphere: Watch me for only pennies a day. Cleansheets.com Web site: http://www.cleansheets.com/coverstories/dean_05.18.05.shtml (Retrieved January 15, 2006).

Dolbik-Vorobei, T. A. 2005. What college students think about problems of marriage and having children. *Russian Education and Society* 47:47–58.

Doub, L. J. 2006. Adolescent sexual behaviors and attitudes: A retrospective report. 4th Annual ECU Research and Creative Activities Symposium, April 21, East Carolina University, Greenville, NC.

Dunn, K. M., P. R. Croft, and G. I. Hackett. 2000. Satisfaction in the sex life of a general population sample. *Journal of Sex and Marital Therapy* 26:141–51.

Eisenman, R. and M. L. Dantzker. 2006. Gender and ethnic differences in sexual attitudes at a Hispanic-serving university. *The Journal of General Psychology* 133: 153–163,

Else-Quest, N. M., J. S. Hyde, and J. D. DeLamater. 2005. Context counts: Long-term sequelae of premarital intercourse of abstinence. *Journal of Sex Research* 42:102–12.

England, P. and R. J. Thomas. 2006. The decline of the date and the rise of the college hook up. In *Family in transition*, 14th edition, edited by A.S. Skolnick and J. H. Skolnick. Boston: Pearson/Allyn Bacon, 151–62.

Ghuman, S. 2005. Attitudes about sex and marital sexual behavior in Hai Duong Province, Vietnam. *Studies in Family Planning* 36:95–106.

Gokyildiz, S. and N. K. Beji 2005. The effects of pregnancy on sexual life. *Journal of Sex & Marital Therapy* 31:201–15.

Graham, C. A., S. A. Sanders, R. R. Milhausen, and K. R. McBride 2004. Turning on and turning off: A focus group study of the factors that affect women's sexual arousal. *Archives of Sexual Behavior*, 33, 527–38.

Grammer, K., F. Bernard, and N. Neave. 2005. Human pheromones and sexual attraction. *European Journal of Obstetrics & Gynecology and Reproductive Biology* 118:135–42.

Greene, K. and S. Faulkner. 2005. Gender, belief in the sexual double standard, and sexual talk in heterosexual dating relationships. *Sex Roles* 53:239–51.

Halstead, M. J. 2005. Teaching about love. *British Journal of Educational Studies* 53:290–305.

Heintz, A. J., and R. M. Melendez. 2006. Intimate partner violence and HIV/STD risk among lesbian, gay, bisexual and transgender individuals. *Journal of Interpersonal Violence* 21:193–208.

Hertzog, 2004. Negotiating the gray area: Women reflecting on the "sex talk" and abstinence. Poster session, National Council on Family Relations, Orlando, Florida, November.

Hughes, M., K. Morrison, and K. J. Asada. 2005. What's love got to do with it? Exploring the impact of maintenance rules, love attitudes, and network support on friends with benefits relationships. *Western Journal of Communication* 69:49–66.

Impett, E. A., L. A. Peplau, and S. L. Gable. 2005. Approach and avoidance sexual motives: Implications for personal and interpersonal well-being. *Personal Relationships* 12:465–82.

Kaestle, C. E., D. E. Morisky, and D. J. Wiley. 2002. Sexual intercourse and the age difference between adolescent females and their romantic partners. *Perspectives on Sexual and Reproductive Health* 34:304–30.

Knox, D. and Zusman, M. E. 2006. Relationship and sexual behaviors of a sample of 1027 university students. Unpublished data collected for this text. Department of Sociology, East Carolina University, Greenville, NC.

Knox, D., V. Daniels, L. Sturdivant, and M. E. Zusman. 2001. College student use of the Internet for mate selection. *College Student Journal* 35:158–60.

Knox, D., L. Sturdivant, and M.E. Zusman. 2001. College student attitudes toward sexual intimacy *College Student Journal* 35:241–43.

Kornreich, J. L., K. D. Hern, G. Rodriguez, and L. F. O'Sullivan. 2003 Sibling influence, gender roles, and the sexual socialization of urban early adolescent girls. *Journal of Sex Research* 40:101–10.

Laumann, E. O., A. Nicolosi, D. B. Glasser, A. Paik, C. Gingell, E. Moreira, and T. Wang. 2005. Sexual problems among women and men aged 40–80: Prevalence and correlates identified in the Global Study of Sexual Attitudes and Behaviors. *International Journal of Impotence Research* 17:39–57.

Laumann, E. O., A. Paik, D. B. Glasser, J-H. Kang, T. Wang, B. Levinson, E. D. Moreira, Jr., A. Nicolosi, and C. Gingell. 2006. A cross-national study of subjective sexual well-being among older women and men: Findings from the global study of sexual attitudes and behaviors. *Archives of Sexual Behavior* April.

Lauzen, M. M. and D. M. Dozier. 2005. Maintaining the double standard: Portrayals of age and gender in popular films. *Sex Roles* 437–46.

Lenton, A. P. and A. Bryan. 2005. An affair to remember: The role of sexual scripts in perceptions of sexual intent. *Personal Relationships* 12:483–98.

Levin, R. 2004. Smells and tastes: their putative influence on sexual activity in humans. *Sexual & Relationship Therapy* 19:451–62.

Lykins, A. D., E. Janssen, and C. A. Graham. 2006. The relationship between negative mood and sexuality in heterosexual college women and men. *The Journal of Sex Research* 43: 136–144.

Liu, C. 2003. Does quality of marital sex decline with duration? *Archives of Sexual Behavior* 32:55–60.

Masters, W. H., and V. E. Johnson. 1970. *Human sexual inadequacy*. Boston: Little, Brown.

Milhausen, R. R., M. Reece, and B. Perera. 2006. A theory-based approach to understanding sexual behavior at Mardi Gras. *The Journal of Sex Behavior* 43: 97–107

Mbopi-Keou, F. X, R. E. Mbu, H. Gonsu Kamga, G. C. M. Kalla, M. Monny Lobe, C. G. Teo, R. J. Leke, P. M. Ndumbe, and L. Belec. 2005. Interactions between human immunodeficiency virus and herpes viruses within the oral mucosa. *Clinical Microbiology and Infection* 11:83–85.

McGinty, K. D. Knox, and M. Zusman. 2007. Friends with benefits: Women want "friends," men want "benefits." *College Student Journal*, in press.

McGuirl, K. E., and M. W. Wiederman. 2000. Characteristics of the ideal sex partner: Gender differences and perceptions of the preferences of the other gender. *Journal of Sex and Marital Therapy* 26:153–59.

Meier, A. M. 2003. Adolescents' transition to first intercourse, religiosity, and attitudes about sex. *Social Forces* 81:1031–52.

Michael, R. T., J. H. Gagnon, E. O. Laumann, and G. Kolata. 1994. *Sex in America*. Boston: Little, Brown.

Miller, S. A., and E. S. Byers. 2004. Actual and desired duration of foreplay and intercourse: Discordant and misperceptions within heterosexual couples. *The Journal of Sex Research,* 41:301–09.

Naimi, T. S., L. E. Lipscomb, R. D. Brewer, and B. C. Gilbert. 2003. Binge drinking in the preconception period and the risk of unintended pregnancy: Implications for women and their children. *Pediatrics* 111:1136–41.

Nobre, P. J. and J. Pinto-Gouveia. 2006. Dysfunctional sexual beliefs as vulnerability factors for sexual dysfunction. *The Journal of Sex Research* 43:68–74.

O'Reilly, S., D. Knox, and M. Zusman. 2006a. Correlates of sexual values: Data on 1019 undergraduates. Poster, Southern Sociological Society, New Orleans, March 24.

O'Reilly, S., D. Knox, and M. Zusman. 2006b. "I have never masturbated": 973 college students who said "yes" or "no." Paper, Annual Meeting Southern Sociological Society, New Orleans, March.

Raley, R. K. 2000. Recent trends and differentials in marriage and cohabitation: The United States. In *The ties that bind*, edited by L. J. Waite. New York: Aldine de Gruyter, 19–39.

Reavill, G. 2005. *Smut: A sex industry insider (and concerned father) says enough is enough*. New York: Penguin Press.

Rose, S. 2005. Going too far? Sex, sin and social policy. *Social Forces* 84:1207–1232.

Rostosky, S. S., M. Regnerus, and M. L. C. Wright. 2003. The role of religiosity and sex attitudes in Add Health Survey. *Journal of Sex Research* 40:358–67.

Sakalh-Ugurlu, N., and P. Glick. 2003. Ambivalent sexism and attitudes toward women who engage in premarital sex in Turkey. *Journal of Sex Research* 40:296–302.

Simon, W., and J. Gagnon. 1998. Psychosexual development. *Society* 35:60–68.

Silver, E. J. and L. J. Bauman. 2005. The association of sexual experience with attitudes, beliefs, and risk behaviors of inner-city adolescents. *Journal of Research on Adolescence* 16:29–45

Somers, C. L., and J. H. Gleason. 2001. Does source of sex education predict adolescents' sexual knowledge, attitudes and behaviors? *Education* 121:674–82.

Stone, N., B. Hatherall, R. Ingham, and J. McEachron. 2006. Oral sex and condom use among young people in the United Kingdom. *Perspectives on Sexual and Reproductive Health* 38:6–13.

Thomsen, D., and I. J. Chang. 2000. Predictors of satisfaction with first intercourse: A new perspective for sexuality education. Poster, 62nd Annual Conference of the National Council on Family Relations, Minneapolis, November.

True Love Waits. 2006. Web site: http://www.lifeway.com/tlw/students/join.asp (Retrieved January 14, 2006).

Weinberg, M. S., Lottes, I., and Shaver, F. M. 2000. Sociocultural correlates of permissive sexual attitudes: A test of Reiss's hypothesis about Sweden and the United States. *Journal of Sex Research* 37:44–52.

Welles, C. E. 2005. Breaking the silence surrounding female adolescent sexual desire. *Women and Therapy* 28:31–45.

Williams, M. T. and L. Bonner. Sex education attitudes and outcomes among North American women. *Adolescence* 41:1–14.

Wilson, K. L., P. Goodson, B. E. Pruitt, E. Buhi, and E. Davis-Gunnels. 2005. A review of 21 curricula for abstinence-only-until marriage programs. *Journal of School Health* 75:90–99.

Wolitski, R. J. 2005. The emergence of barebacking among gay and bisexual men in the United States: A public health perspective. *Journal of Gay & Lesbian Psychotherapy* 9:9–34.

chapter 6

Homophobia alienates mothers and fathers from sons and daughters, friend from friend, neighbor from neighbor, Americans from one another.

Byrne Fone, Homophobia: A History

Same-Sex Couples and Families

Contents

True or False?

1. In general, heterosexuals have more favorable attitudes toward gay men and lesbian women if they have had prior contact with or know someone who is gay or lesbian.

2. According to the American Psychiatric Association and the American Psychological Association, homosexuality is a mental disorder that can be cured by reparative therapy.

3. Compared with children reared by heterosexual parents, children reared by lesbians have less emotional well being and are more likely to be homosexual as adults.

4. In over half the states it is legal to fire a person because of his or her sexual orientation.

5. Nearly one-fourth of U.S. adults say that someone they are related to is gay or lesbian.

Answers: **1.** T **2.** F **3.** F **4.** T **5.** T

When I was in the military they gave me a medal for killing two men and a discharge for loving one.

Epitaph of Leonard P. Matlovich

Rosie O'Donnell, talk show host, actress, and mother of four, organized the first-ever cruise for gay families in 2004 (see chapter-opening photo). She and her life partner, Kelli Carpenter O'Donnell, left New York City with more than 1500 passengers—gay and straight—for a 7-day cruise to the Caribbean. The families on board included same-sex couples and their children, and gay teens and adults with their straight family members. Rosie's gay family cruise became the subject of a 2006 HBO documentary, "All Aboard! Rosie's Family Cruise." Rosie said, "The illusion about gay people is that it's all about sex. What unites us on the cruise is not that we're gay, but that we're parents." Kelli explained, "As a gay person who brought my straight family members on board, it was a great thing to be able to show them what my community looks like. It's not what you see on television. . . . it's very simple and basic. It's more about love and happy families and happy children" (quoted in HBO Interview, no date).

In this chapter we discuss same-sex couples and families—relationships that are, in many ways, similar to heterosexual ones. A major difference, however, is that gay and lesbian couples and families are subjected to prejudice and discrimination. Although other minority groups also experience prejudice and discrimination, only sexual-orientation minorities are denied legal marital status and the benefits and responsibilities that go along with marriage, which we discuss later in this chapter. Also, gay couples are sometimes rejected by their own parents, siblings, and other family members.

If it wasn't for gay men, I wouldn't talk to men at all.

Margaret Cho, comedian

Homosexual behavior has existed throughout human history and in most (perhaps all) human societies (Kirkpatrick, 2000). In this chapter we focus on Western views of sexual diversity that define **sexual orientation** as the classification of individuals as heterosexual, bisexual, or homosexual, based on their emotional/cognitive/sexual attractions and self-identity. **Heterosexuality** refers to the predominance of emotional and sexual attraction to individuals of the other sex. **Homosexuality** refers to the predominance of emotional and sexual attraction to individuals of the same sex, and **bisexuality** is emotional and sexual attraction to members of both sexes. The term **lesbian** refers to homosexual women; **gay** can refer to either homosexual women or homosexual men. Lesbians, gays, and bisexuals, sometimes referred to collectively as the **lesbigay** population, are considered part of a larger population referred to as the transgendered community. **Transgendered** individuals include "a range of people whose gender identities do not conform to traditional notions of masculinity and femininity" (Cahill et al., 2002, p. 10). Transgendered individuals include not only homosexuals and bisexuals but also cross-dressers, transvestites, and transsexuals (see Chapter 2, Gender in Relationships). Because much of the current literature on the lesbigay population includes other members of the transgendered community, the term **LGBT** or **GLBT** is often used to refer collectively to lesbians, gays, bisexuals, and transgendered individuals.

Prevalence of Homosexuality, Bisexuality, and Same-Sex Couples

Before looking at prevalence data concerning homosexuality and bisexuality in the United States, it is important to understand the ways in which identifying or classifying individuals as "heterosexual," "homosexual," "gay," "lesbian," and "bisexual" is problematic.

Problems Associated with Identifying and Classifying Sexual Orientation

The classification of individuals into sexual orientation categories such as those listed above is problematic for a number of reasons (Savin-Williams, 2006). First, because of the social stigma associated with nonheterosexual identities, many individuals conceal or falsely portray their sexual-orientation identities to avoid prejudice and discrimination. In addition, distinctions among sexual-orientation categories are simply not as clear-cut as many people would believe.

Second, not all people who are sexually attracted to or have had sexual relations with individuals of the same sex view themselves as homosexual or bisexual. A final difficulty in labeling a person's sexual orientation is that an individual's sexual attractions, behavior, and identity may change across time. For example, in a longitudinal study of 156 lesbian, gay, and bisexual youth, 57 percent consistently identified as gay/lesbian and 15 percent consistently identified as bisexual over a one-year period, but 18 percent transitioned from bisexual to lesbian or gay (Rosario et al., 2006).

Early research on sexual behavior by Kinsey and his colleagues (1948, 1953) found that although 37 percent of men and 13 percent of women had had at least one same-sex sexual experience since adolescence, few of the individuals reported exclusive homosexual behavior. These data led Kinsey to conclude that most people are not exclusively heterosexual or homosexual. Rather, Kinsey suggested an individual's sexual orientation may have both heterosexual and homosexual elements. In other words, Kinsey suggested that heterosexuality and homosexuality represent two ends of a sexual-orientation continuum and that most individuals are neither entirely homosexual nor entirely heterosexual, but fall somewhere along this continuum. In other words, most people are, to some degree, bisexual. On the Heterosexual-Homosexual Rating Scale developed by Kinsey and his colleagues (1953), individuals with ratings of 0 or 1 are entirely or largely heterosexual; 2, 3, or 4 are more bisexual; and 5 or 6 are largely or entirely homosexual (Figure 6.1).

Sexual-orientation classification is also complicated by the fact that sexual behavior, attraction, love, desire, and sexual-orientation identity do not always match. For example, "research conducted across different cultures and historical periods (including present-day Western culture) has found that many individuals develop passionate infatuations with same-gender partners in the absence of same-gender sexual desires . . . whereas others experience same-gender sexual desires that never manifest themselves in romantic passion or attachment" (Diamond, 2003, p. 173).

The prejudice against gays and lesbians is meaner, nastier—a vindictiveness that's rooted in hatred, not ignorance.

Betty DeGeneres, mother of comedienne Ellen DeGeneres

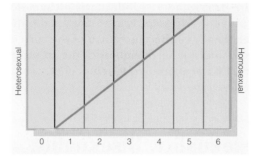

Based on both psychologic reactions and overt experience, individuals rate as follows:

0. Exclusively heterosexual with no homosexual
1. Predominantly heterosexual, only incidentallly homosexual
2. Predominantly heterosexual, but more than incidentally homosexual
3. Equally heterosexual and homosexual
4. Predominantly homosexual, but more than incidentally heterosexual
5. Predominantly homosexual, but incidentally heterosexual
6. Exclusively homosexual

Figure 6.1
The Heterosexual-Homosexual Rating Scale
Source: *Sexual Behavior in the Human Male,* W.B. Saunders, 1948. Reprinted by permission of the Kinsey Institute for Research in Sex, Gender, and Reproduction, Inc.

Consider the findings of a national study of U.S. adults that investigated (1) sexual attraction to individuals of the same sex, (2) sexual behavior with people of the same sex, and (3) homosexual self-identification (Michael et al., 1994). This survey found that 4 percent of women and 6 percent of men said that they are sexually attracted to individuals of the same sex, and 4 percent of women and 5 percent of men reported that they had had sexual relations with a same-sex partner after age 18. What these data tell us is that "those who acknowledge homosexual sexual desires may be far more numerous than those who actually act on those desires" (Black et al., 2000, p. 140).

Prevalence of Homosexuality, Heterosexuality, and Bisexuality

Despite the difficulties inherent in categorizing individuals' sexual-orientation, recent data reveal the prevalence of individuals in the United States who identify as lesbian, gay, or bisexual. In the national survey by Michael et al. (1994), fewer than 2 percent of women and 3 percent of the men identified themselves as homosexual or bisexual. More recently, Smith and Gates (2001) estimated that there are more than 10 million gay and lesbian adults in the United States, which represents between 4 percent and 5 percent of the total U.S. adult population.

These are two of the almost half a million U.S. college students who are gay.

Authors

Chapter 6 Same-Sex Couples and Families

Berg and Lein (2006) estimated that 7 percent of males and 4 percent of females were not heterosexual. A 2004 national poll showed that about 5 percent of U.S. high school students identified themselves as lesbian or gay (Curtis, 2004). Finally, among Americans older than 65 years, an estimated 1 to 3 million are gay, lesbian, bisexual, or transgender (National Gay and Lesbian Task Force, 2005a).

National Data
According to research cited above, estimates of the U.S. lesbigay population range from about two to seven percent of the U.S. adult population. An easy-to-remember percentage is five percent, which translates into 15 million gay individuals.

Prevalence of Same-Sex Couple Households
The 2000 census showed that about 1 in 9 (or 594,000) unmarried-partner households in the United States involve partners of the same sex (Simmons and O'Connell, 2003). Female same-sex couples and male same-sex couples were nearly equal in number (51% of same-sex couples were male).

Census 2000 data also revealed that 99.3 percent of U.S. counties reported same-sex cohabiting partners, compared with 52 percent of counties in 1990 (Bradford et al., 2002). Same-sex couples are more likely to live in metropolitan areas than in rural areas. However, the largest proportional increases in the number of same-sex couples self-reporting in 2000 versus 1990 came in rural, sparsely populated states.

Why are data on the numbers of GLBT individuals and couples in the United States relevant? The primary reason is that census numbers on the prevalence of GLBT individuals and couples can influence laws and policies that affect gay individuals and their families. In anticipation of the 2000 census, the National Gay and Lesbian Task Force Policy Institute and the Institute for Gay and Lesbian Strategic Studies conducted a public education campaign urging people to "out" themselves on the 2000 census. The slogan was, "The more we are counted, the more we count" (Bradford et al., 2002, p 3). "The fact that the Census documents the actual presence of same-sex couples in nearly every state legislative and U.S. Congressional district means anti-gay legislators can no longer assert that they have no gay and lesbian constituents" (Bradford et al., 2002, p. 8).

Origins of Sexual-Orientation Diversity

Much of the biomedical and psychological research on sexual orientation attempts to identify one or more "causes" of sexual-orientation diversity. The driving question behind this research is, "Is sexual orientation inborn or is it learned or acquired from environmental influences?" Although a number of factors have been correlated with sexual orientation, including genetics, gender role behavior in childhood, and fraternal birth order, there is no single theory that can explain diversity in sexual orientation.

"God made 'Adam and Eve' not 'Adam and Steve'? . . . hmmmm . . . then who made 'Steve'?"

Anonymous

Beliefs about What "Causes" Homosexuality
Aside from what "causes" homosexuality, social scientists are interested in what people *believe* about the "causes" of homosexuality. Most gays believe that homosexuality is an inherited, inborn trait. In a national study of homosexual men, 90 percent reported that they believed that they were born with their homosexual orientation; only 4 percent believed that environmental factors were the sole cause (Lever, 1994). The percentage of Americans who believe that homosexu-

ality is something a person is born with increased from 13 percent in 1977 to 40 percent in 2002 (Newport, 2002).

Individuals who believe that homosexuality is genetically determined tend to be more accepting of homosexuality and are more likely to be in favor of equal rights for lesbians and gays (Tyagart, 2002). In contrast, "those who believe homosexuals choose their sexual orientation are far less tolerant of gays and lesbians and more likely to feel that homosexuality should be illegal than those who think sexual orientation is not a matter of personal choice" (Rosin and Morin, 1999, p. 8).

Although the terms *sexual preference* and *sexual orientation* are often used interchangeably, the term *sexual orientation* avoids the implication that homosexuality, heterosexuality, and bisexuality are determined voluntarily. Hence, sexual orientation is used more often by those who believe that sexual orientation is inborn, and sexual preference is used more often by those who think that individuals choose their sexual orientation.

Can Homosexuals Change Their Sexual Orientation?

Individuals who believe that homosexuals choose their sexual orientation tend to think that homosexuals can and should change their sexual orientation. Various forms of reparative therapy or conversion therapy are dedicated to changing homosexuals' sexual orientation. Some religious organizations sponsor "ex-gay ministries," which claim to "cure" homosexuals and transform them into heterosexuals by encouraging them to ask for "forgiveness for their sinful lifestyle," through prayer and other forms of "therapy." Consider the following examples:

- One church counselor told a gay woman that God could heal her and make her "whole again." The counselor laid hands on her to rid her of the demonic "spirit of homosexuality," gave her Bible verses to memorize, and worked on her "femininity" by teaching her how to apply makeup, encouraging her to grow her hair long, and instructing her to replace her old jeans, sweatpants, and gym shorts with skirts and dresses (Human Rights Campaign, 2000).
- Sixteen-year-old Zach was enrolled by his parents in a Christian camp program to change him into a heterosexual. The program, called Refuge, discourages homosexual behavior by imposing the following rules on its participants: no secular music, no more than 15 minutes per day behind a closed bathroom door, no contact with any practicing homosexual, no masturbation, and (no joke) no Calvin Klein underwear (Buhl, 2005).

Critics of reparative therapy and ex-gay ministries take a different approach: "It is not gay men and lesbians who need to change . . . but negative attitudes and discrimination against gay people that need to be abolished" (Besen, 2000, p. 7). The American Psychiatric Association, the American Psychological Association, the American Academy of Pediatrics, the American Counseling Association, the National Association of School Psychologists, the National Association of Social Workers, and the American Medical Association agree that homosexuality is not a mental disorder and needs no cure—that efforts to change sexual orientation do not work and may, in fact, be harmful (Human Rights Campaign, 2000; Potok, 2005). An extensive review of the ex-gay movement concludes, "There is a growing body of evidence that conversion therapy not only does not work, but also can be extremely harmful, resulting in depression, social isolation from family and friends, low self-esteem, internalized homophobia, and even attempted suicide" (Cianciotto and Cahill, 2006, p. 77). According to the American Psychiatric Association, "clinical experience suggests that any person who seeks conversion therapy may be doing so because of social bias that has resulted in internalized homophobia, and that gay men and lesbians who have accepted their sexual orientation are better adjusted than those who have not done so" (quoted by Holthouse, 2005, p. 14).

We struggled against apartheid because we were being blamed and made to suffer for something we could do nothing about. It is the same with homosexuality. The orientation is a given, not a matter of choice. It would be crazy for someone to choose to be gay, given the homophobia that is present

Desmond Tutu, South African Anglican archbishop

Chapter 6 Same-Sex Couples and Families

Close scrutiny of reports of "successful" reparative therapy reveal that (1) many claims come from organizations with an ideological perspective on sexual orientation rather than from unbiased researchers, (2) the treatments and their outcomes are poorly documented, and (3) the length of time that clients are followed after treatment is too short for definitive claims to be made about treatment success (Human Rights Campaign, 2000). Indeed, at least 13 ministries of Exodus International—the largest ex-gay ministry network—have closed because their directors reverted to homosexuality (Fone, 2000). Michael Bussy, who in 1976 helped start Exodus International, said, "After dealing with hundreds of people, I have not met one who went from gay to straight. Even if you manage to alter someone's sexual behavior, you cannot change their true sexual orientation" (quoted by Holthouse, 2005, p. 14). Michael Bussy worked to help "convert" gay people for three years, until he and another male Exodus employee fell in love and left the organization.

Heterosexism, Homonegativity, Homophobia, and Biphobia

The United States, along with many other countries throughout the world, is predominantly heterosexist. **Heterosexism** refers to "the institutional and societal reinforcement of heterosexuality as the privileged and powerful norm" (Sex Information and Education Council of the United States, 2000). Heterosexism is based on the belief that heterosexuality is superior to homosexuality; it results in prejudice and discrimination against homosexuals and bisexuals. **Prejudice** refers to negative attitudes, whereas **discrimination** refers to behavior that denies individuals or groups equality of treatment. Before reading further, you may wish to complete this chapter's Self-Assessment feature on page 168, which assesses your behaviors toward individuals you perceive to be homosexual.

It always seemed to me a bit pointless to disapprove of homosexuality. It's like disapproving of rain.

Francis Maude, Member of Parliament for Horsham since 1997

Homonegativity and Homophobia

The term **homophobia** is commonly used to refer to negative attitudes and emotions toward homosexuality and those who engage in it. Homophobia is not necessarily a clinical phobia (i.e., one involving a compelling desire to avoid the feared object despite recognizing that the fear is unreasonable). Other terms that refer to negative attitudes and emotions toward homosexuality include **homonegativity** and **antigay bias.**

SIECUS, the Sex Information and Education Council of the United States, "strongly supports the right of each individual to accept, acknowledge, and live in accordance with his or her sexual orientation . . . and deplores all forms of prejudice and discrimination against people based on sexual orientation" (SIECUS, 2000, p. 1). Nevertheless, negative attitudes toward homosexuality are reflected in the high percentage of the U.S. population who disapprove of homosexuality. According to national surveys by the Gallup Organization, the percentage of Americans saying that homosexuality should be considered an acceptable alternative lifestyle is just over half (51%) (Saad, 2005). In a national sample of first-year college students, 27 percent agreed "It is important to have laws prohibiting homosexual relationships" (American Council on Education and University of California, 2005-2006).

In general, individuals who are more likely to have negative attitudes toward homosexuality and to oppose gay rights are men, older, and less educated; attend religious services; live in the South or Midwest; reside in small rural towns; and have had limited contact with someone who is gay or lesbian (Herek, 2002a; Curtis, 2003; Loftus, 2001; Page, 2003; Mohipp and Morry, 2004). Blacks are also

The Self-Report of Behavior Scale (Revised)

This questionnaire is designed to examine which of the following statements most closely describes your behavior during past encounters with people you thought were homosexuals. Rate each of the following self-statements as honestly as possible by choosing the frequency that best describes your behavior.

1. I have spread negative talk about someone because I suspected that he or she was gay.

   A. Never   B. Rarely   C. Occasionally   D. Frequently   E. Always

2. I have participated in playing jokes on someone because I suspected that he or she was gay.

   A. Never   B. Rarely   C. Occasionally   D. Frequently   E. Always

3. I have changed roommates and/or rooms because I suspected my roommate was gay.

   A. Never   B. Rarely   C. Occasionally   D. Frequently   E. Always

4. I have warned people who I thought were gay and who were a little too friendly with me to keep away from me.

   A. Never   B. Rarely   C. Occasionally   D. Frequently   E. Always

5. I have attended anti-gay protests.

   A. Never   B. Rarely   C. Occasionally   D. Frequently   E. Always

6. I have been rude to someone because I thought that he or she was gay.

   A. Never   B. Rarely   C. Occasionally   D. Frequently   E. Always

7. I have changed seat locations because I suspected the person sitting next to me was gay.

   A. Never   B. Rarely   C. Occasionally   D. Frequently   E. Always

8. I have had to force myself to keep from hitting someone because he or she was gay and very near me.

   A. Never   B. Rarely   C. Occasionally   D. Frequently   E. Always

9. When someone I thought to be gay has walked toward me as if to start a conversation, I have deliberately changed directions and walked away to avoid him or her.

   A. Never   B. Rarely   C. Occasionally   D. Frequently   E. Always

10. I have stared at a gay person in such a manner as to convey to him or her my disapproval of his or her being too close to me.

    A. Never   B. Rarely   C. Occasionally   D. Frequently   E. Always

11. I have been with a group in which one (or more) person(s) yelled insulting comments to a gay person or group of gay people.

    A. Never   B. Rarely   C. Occasionally   D. Frequently   E. Always

12. I have changed my normal behavior in a restroom because a person I believed to be gay was in there at the same time.

    A. Never   B. Rarely   C. Occasionally   D. Frequently   E. Always

13. When a gay person has checked me out, I have verbally threatened him or her.

    A. Never   B. Rarely   C. Occasionally   D. Frequently   E. Always

14. I have participated in damaging someone's property because he or she was gay.

    A. Never   B. Rarely   C. Occasionally   D. Frequently   E. Always

15. I have physically hit or pushed someone I thought was gay because he or she brushed his or her body against me when passing by.

    A. Never   B. Rarely   C. Occasionally   D. Frequently   E. Always

16. Within the past few months, I have told a joke that made fun of gay people.

    A. Never   B. Rarely   C. Occasionally   D. Frequently   E. Always

more likely than whites to view homosexual relations as "always wrong" (Lewis, 2003).

Negative social meanings associated with homosexuality can affect the self-concepts of LGBT individuals. **Internalized homophobia**—a sense of personal failure and self-hatred among lesbians and gay men resulting from social rejection and stigmatization—has been linked to increased risk for depression, substance abuse and addiction, anxiety, and suicidal thoughts (Bobbe, 2002; Gilman et al., 2001).

Biphobia

Just as the term *homophobia* is used to refer to negative attitudes toward homosexuality, gay men, and lesbians, **biphobia** (also referred to as **binegativity**) refers to a parallel set of negative attitudes toward bisexuality and those identified as bisexual. Although both homosexual- and bisexual-identified individuals are often rejected by heterosexuals, bisexual-identified women and men also face rejec-

17. I have gotten into a physical fight with a gay person because I thought he or she had been making moves on me.

    A. Never   B. Rarely   C. Occasionally   D. Frequently   E. Always

18. I have refused to work on school and/or work projects with a partner I thought was gay.

    A. Never   B. Rarely   C. Occasionally   D. Frequently   E. Always

19. I have written graffiti about gay people or homosexuality.

    A. Never   B. Rarely   C. Occasionally   D. Frequently   E. Always

20. When a gay person has been near me, I have moved away to put more distance between us.

    A. Never   B. Rarely   C. Occasionally   D. Frequently   E. Always

The Self-Report of Behavior Scale (Revised) (SBS-R) is scored by totaling the number of points endorsed on all items (Never = 1; Rarely = 2; Occasionally = 3; Frequently = 4; Always = 5), yielding a range from 20 to 100 total points. The higher the score, the more negative the attitudes toward homosexuals.

Comparison Data

The SBS was originally developed by Sunita Patel (1989) in her thesis research in her clinical psychology master's program at East Carolina University. College men (from a university campus and from a military base) were the original participants (Patel et al., 1995). The scale was revised by Shartra Sylivant (1992), who used it with a coed high school student population, and by Tristan Roderick (1994), who involved college students to assess its psychometric properties. The scale was found to have high internal consistency. Two factors were identified: a passive avoidance of homosexuals and active or aggressive reactions.

In a study by Roderick et al. (1998) the mean score for 182 college women was 24.76. The mean score for 84 men was significantly higher, at 31.60. A similar sex difference, although with higher (more negative) scores, was found in Sylivant's high school sample (with a mean of 33.74 for the young women, and 44.40 for the young men).

The following table provides detail for the scores of the college students in Roderick's sample (from a mid-sized state university in the southeast):

	N	Mean	Standard Deviation
Women	182	24.76	7.68
Men	84	31.60	10.36
Total	266	26.91	9.16

Sources

Patel, S. 1989. Homophobia: Personality, emotional, and behavioral correlates. Master's thesis, East Carolina University.

Patel, S., T. E. Long, S. L. McCammon, and K. L. Wuensch. 1995. Personality and emotional correlates of self reported antigay behaviors. *Journal of Interpersonal Violence* 10: 354–66.

Roderick, T. 1994. Homonegativity: An analysis of the SBS-R. Master's thesis, East Carolina University.

Roderick, T., S. L. McCammon, T. E. Long, and L. J. Allred. 1998. Behavioral aspects of homonegativity. *Journal of Homosexuality* 36:79–88.

Sylivant, S. 1992. The cognitive, affective, and behavioral components of adolescent homonegativity. Master's thesis, East Carolina University.

The SBS-R is reprinted by the permission of the students and faculty who participated in its development: S. Patel, S. L. McCammon, T. E. Long, L. J. Allred, K. Wuensch, T. Roderick, & S. Sylivant.

tion from many homosexual individuals. Thus bisexuals experience "double discrimination."

Some negative attitudes toward bisexual individuals "are based on the belief that bisexual individuals are really lesbian or gay individuals who are in transition or in denial about their true sexual orientation" (Israel and Mohr, 2004, p. 121). According to this view, bisexuals lack the courage to come out as lesbian or gay, or they are trying to maintain heterosexual privilege. Negative attitudes toward bisexuality are also based on the negative stereotype of bisexuals as incapable of or unwilling to be monogamous. In a review of research on bisexuality, Israel and Mohr (2004) state that "although bisexual individuals are more likely to value nonmonogamy as an ideal compared to lesbian, gay, and heterosexual individuals, research clearly indicates that some bisexual-identified individuals prefer monogamous relationships" (p. 122).

The *Resource Guide to Coming Out* is available from the Human Rights Campaign—the nation's largest GLBT civil rights organization.

A RESOURCE GUIDE TO COMING OUT

HUMAN RIGHTS CAMPAIGN
COMING OUT PROJECT

Talk about it.

It's better to be black than gay because when you're black you don't have to tell your mother.

Charles Pierce

Are the Benefits of "Coming Out" Worth the Risks?

In a society where heterosexuality is expected and considered the norm, heterosexuals do not have to choose whether or not to tell others that they are heterosexual. But for gay, lesbian, and bisexual individuals, decisions about **"coming out,"** or being open and honest about one's sexual orientation and identity, are some of the most difficult and important choices they face. Choices about coming out include whether to come out to others, who to come out to, and when and how to come out.

Risks of Coming Out

Whether a GLBT individual comes out is influenced by the perceived risks and reactions of others (Evans and Broido, 2000). Some of the risks involved in coming out include disapproval and rejection by parents and other family members; harassment and discrimination at school; discrimination and harassment in the workplace; and hate crime victimization.

1. Parental and Family Members' Reactions. When GLBT individuals come out to their parents, parental reactions range from "I already knew you were gay and I'm glad that you feel ready to be open with me about it" to "get out of this house, you are no longer welcome here." Mary Cheney reported that when she told her father, Vice President Dick Cheney, that she is a lesbian, his response was "You're my daughter and I love you and I just want you to be happy" (quoted in Walsh, 2006, p. 27). When Reverend Mel White—a closeted gay Christian man who was nearly driven to suicide after two decades of struggling to save his marriage and his soul with "reparative therapies"—finally came out to his mother, her response was, "I'd rather see you at the bottom of that swimming pool, drowned, than to hear this" (White, 2005, p. 28).

According to *The Resource Guide to Coming Out* (Human Rights Campaign, 2004), "many parents are shocked when their children say they are gay, lesbian, or bisexual. Some parents react in ways that hurt. Some cry. Some get angry. Some ask where they went wrong as a parent. Some call it a sin. Some insist it's a phase. Others try to send their child to counselors or therapists who attempt to change gay people into heterosexuals . . ." (p. 24).

Because blacks are more likely than whites to view homosexual relations as "always wrong," African-Americans who are gay or lesbian are more likely to face disapproval from their families (and straight friends) than are white lesbians and gays (Lewis, 2003). The result is that African-Americans are more likely to stay closeted—not let their parents know of their homosexuality (Grov, 2006).

The *Resource Guide* notes, however, that "for many parents it's very hard to completely reject their children" and that it takes time for parents to adjust—sometimes months, sometimes years. In some families with a GLBT member, the "gay issue" is not openly discussed, even though family members may know or suspect that a loved one is gay, lesbian, or bisexual. One gay male student explained, "My parents know I live with my "friend" Christopher, and they have invited Christopher to our family holiday gatherings and family vacations. I am sure my parents know that I'm gay. . . . We just don't talk about it" (authors' files).

To help parents and other family members adjust, it may be helpful to refer them to a local chapter of Parents, Families, and Friends of Lesbians and Gays (PFLAG) and to provide them with books and online resources, such as those found at Human Rights Campaign's National Coming Out Project (www.hrc.org/ncop).

2. Harassment and Discrimination at School. In a national survey of students aged 13–18 years, two-thirds (65%) of LGBT students reported that they had been verbally harassed, 16 percent physically harassed, and 8 percent physically assaulted because of their sexual orientation (Harris Interactive and GLSEN, 2005). "Students who openly identify as lesbian, gay, bisexual, and transgender (LGBT) have a more acute problem with being harassed at school" (p. 4). This survey found that LGBT students are over three times more likely than non-LGBT students to report that they feel unsafe at school (20% vs. 6%). Only eight states and the District of Columbia have statewide policies that prohibit antigay harassment in public schools (Snorton, 2005).

A study of GLBT college students, faculty, and staff/administrators found that half (51%) had concealed their sexual orientation or gender identity to avoid intimidation (Rankin, 2003). The same study found that in the previous year more than a third (36%) of undergraduates experienced harassment in the form of derogatory remarks, verbal threats, anti-gay graffiti, threats of physical violence, denial of services, and physical assault.

3. Discrimination and Harassment at the Workplace. The 2005 Workplace Fairness Survey found that 39 percent of lesbian and gay employees reported experiencing some form of discrimination or harassment in the workplace (Lambda Legal and Deloitte Financial Advisory Services, LLP, 2006). Although the majority (88%) of U.S. adults support equal employment rights for gays and lesbians, as of March 2006 it was legal in 33 states to fire, decline to hire or promote, or otherwise discriminate against an employee because of his or her sexual orientation (National Gay and Lesbian Task Force, 2006; Saad, 2005).

4. Hate Crime Victimization. Another risk involved in coming out is that of being victimized by anti-gay hate crimes—which are crimes against individuals or their property that are based on bias against the victim because of his or her perceived sexual orientation. Such crimes include verbal threats and intimidation, vandalism, sexual assault and rape, physical assault, and murder.

Benefits of Coming Out

Given the risks of coming out, it is not surprising that some GBLT individuals never come out, live a life of repressed feelings, and deny who they are to others and (sometimes) to themselves. But, according to the Human Rights Campaign, "most people come out because, sooner or later, they can't stand hiding who they are any more. Once they've come out, most people acknowledge that it feels much better to be open and honest than to conceal such an integral part of themselves" (2004, p. 16).

You could move.

Abigail Van Buren, "Dear Abby," in response to a reader who complained that a gay couple was moving in across the street and wanted to know what he could do to improve the quality of the neighborhood

In addition to the benefits for the individual who chooses to come out, there are also benefits for the entire GLBT population. Research has found that, in general, heterosexuals have more positive attitudes toward gays and lesbians if they have had prior contact with or know someone who is gay (Mohipp and Morry, 2004). One woman describes how her brother's coming out changed her views on homosexuality (Yvonne, 2004):

> I was raised in a devout born-again Christian family . . . to believe that homosexuality was evil and a perversion. When I was growing up, I used to wonder if any of the kids at my school could be gay. I couldn't imagine it could be so. As it happens, there was a gay individual even closer than I imagined. My brother Tommy came out of the closet in the early 1990s. After he came out, I had to confront my own denial about the fact that, in my heart of hearts, I had always known that Tommy was gay.

> Over the years, his partner Rod has come to be a loved and cherished member of our family, and we have all had to confront the prejudices and stereotypes we have held onto for so long about sexual orientation. It seems to me that coming out of the closet is the greatest weapon that gays and lesbians have. If my own brother had never come out, my family would never have been forced to confront the deep-seated prejudices we were raised with . . . I am still a Christian, but my husband (who also has a gay brother) and I attend a church that truly puts the teachings of Christ into practice—teachings about love, tolerance and inclusivity. As a Christian who grew up with a gay family member, I know that the propaganda put forth by the religious right on this issue is founded in fear, hatred and prejudice. None of these are values taught by Jesus!

Cheryl Jacques, a Massachusetts state senator who came out publicly in the Boston Globe, recognizes that coming out is a risk. But she suggests that:

> Coming out is a risk worth taking because it is one of the most powerful things any of us can do. I've yet to meet anyone who regretted the decision to live life truthfully. . . . That's why while coming out may be just one step in the life of a gay, lesbian, bisexual or transgender person, it contributes to a giant leap for all GLBT people (quoted in Human Rights Campaign, 2004, p. 4).

Heterosexism, Homonegativity, Homophobia, and Biphobia

Gay, Lesbian, Bisexual, and Mixed-Orientation Relationships

Research suggests that gay and lesbian couples tend to be more similar than different from heterosexual couples (Kurdek, 2005, 2006). However, there are some unique aspects of intimate relationships involving gay, lesbian, and bisexual individuals. In this section, we note the similarities as well as differences between heterosexual, gay male, and lesbian relationships in regard to relationship satisfaction, conflict and conflict resolution, and monogamy and sexuality. We also look at relationship issues involving bisexual individuals and mixed-orientation couples.

Relationship Satisfaction

For both heterosexual and LGBT partners, relationship satisfaction tends to be high in the beginning of the relationship and decreases over time. In a review of literature on lesbian and gay couples, Kurdek (1994) concluded, "The most striking finding regarding the factors linked to relationship satisfaction is that they seem to be the same for lesbian couples, gay couples, and heterosexual couples" (p. 251). These factors include having equal power and control, being emotionally expressive, perceiving many attractions and few alternatives to the relationship, placing a high value on attachment, and sharing decision-making.

Researchers who studied relationship quality among same-sex couples noted that "in trying to create satisfying and long-lasting intimate relationships, LGBT individuals face all of the same challenges faced by heterosexual couples, as well as a number of distinctive concerns" (Otis et al., 2006, p. 86). These concerns include if, when, and how to disclose their relationships to others and how to develop healthy intimate relationships in the absence of same-sex relationship models.

In one review of research on gay and lesbian relationships, the authors concluded that the main difference between heterosexual and nonheterosexual relationships is that, "Whereas heterosexuals enjoy many social and institutional supports for their relationships, gay and lesbian couples are the object of prejudice and discrimination" (Peplau et al., 1996, p. 268). Both gay male and lesbian couples must cope with the stress created by antigay prejudice and discrimination and by "internalized homophobia" or negative self-image and low self-esteem due to being a member of a stigmatized group. Not surprisingly, higher levels of such stress are associated with lower reported levels of relationship quality among LGBT couples (Otis et al., 2006).

Despite the stresses and lack of social and institutional support experienced by LGBT individuals, gay men and lesbians experience relationship satisfaction at a level that is at least equal to that reported by married heterosexual spouses (Kurdek, 2005). Partners of the same sex enjoy the comfort of having a shared gender perspective, which is often accompanied by a sense of equality in the relationship. For example, contrary to stereotypical beliefs, same-sex couples (male or female) typically do not assign "husband" and "wife" roles in the division of household labor and are more likely than heterosexual couples to achieve a fair distribution of household labor and at the same time accommodate the different interests, abilities, and work schedules of each partner (Kurdek, 2005). In contrast, division of household labor among heterosexual couples tends to be unequal, with wives doing the majority of such tasks.

Conflict and Conflict Resolution

All couples experience conflict in their relationships, and gay and lesbian couples tend to disagree about the same issues that heterosexual couples argue about. In one study, partners from same-sex and heterosexual couples identified

No government has a right to tell its citizens when or whom to love. The only queer people are those who don't love anybody.

Rita Mae Brown

the same sources of most conflict in their relationships: finances, affection, sex, being overly critical, driving style, and household tasks (Kurdek, 2004).

However, same-sex couples and heterosexual couples tend to differ in how they resolve conflict. In a study in which researchers videotaped gay, lesbian, and heterosexual couples discussing problems in their relationships, gay and lesbian partners began their discussions more positively and maintained a more positive tone throughout the discussion than did partners in heterosexual marriages (Gottman et al., 2003). Other research has found that compared with heterosexual married spouses, same-sex partners resolve conflict more positively, argue more effectively, and are more likely to suggest possible solutions and compromises (Kurdek, 2004). One explanation for the more positive conflict resolution among same-sex couples is that they value equality more and are more likely to have equal power and status in the relationship than are heterosexual couples (Gottman et al., 2003).

There is nothing wrong with going to bed with someone of your own sex. People should be very free with sex; they should draw the line at goats.

Elton John, singer

Monogamy and Sexuality

Like many heterosexual women, most gay women value stable, monogamous relationships that are emotionally as well as sexually satisfying. Women in U.S. society, gay and "straight," are taught that sexual expression should occur in the context of emotional or romantic involvement.

A common stereotype of gay men is that they prefer casual sexual relationships with multiple partners over monogamous long-term relationships. However, most gay men prefer long-term relationships, and when sex occurs outside of the primary relationship, it is usually infrequent and not emotionally involving (Green et al., 1996).

National Data

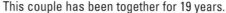

Between 8 percent and 21 percent of lesbian couples and between 18 percent and 28 percent of gay male couples have lived together for 10 or more years (Kurdek, 2004).

The degree to which gay males engage in casual sexual relationships is better explained by the fact that they are male than by the fact that they are gay. In this regard, gay men and straight men have a lot in common: they both tend to have fewer barriers to engaging in casual sex than do women (heterosexual or lesbian).

One unique aspect of gay male relationships in the United States involves coping with the high rate of human immunodeficiency virus (HIV) infection and acquired immunodeficiency syndrome (AIDS). Although most worldwide HIV infection occurs through heterosexual transmission, in the United States, male-to-male sexual contact is the most common mode of HIV transmission (Centers for Disease Control and Prevention, 2005). Women who have sex exclusively with other women have a much lower rate of HIV infection than do men (both gay and straight) and women who have sex with men. Many gay men have lost a love partner to HIV infection/AIDS; some have experienced multiple losses. Those still in relationships with partners who are HIV-positive experience profound changes such as developing a sense of urgency to "speed up" their relationship because they may not have much time left together (Palmer and Bor, 2001).

This couple has been together for 19 years.

Relationships of Bisexuals

Individuals who identify as bisexual have the ability to form intimate relationships with both sexes. However, research has found that the majority of bisexual women and men tend toward primary relationships with the other sex (McLean, 2004). Contrary to the common myth that bisexuals are, by definition, nonmonogamous, some bisexuals prefer monogamous relationships (especially in light of the widespread concern about HIV). In another study of 60 bisexual women and men, 25 percent of the men and 35 percent of the women were in exclusive relationships; 60 percent of the men and 53 percent of the women were in "open" relationships in which both partners agreed to allow each other to have sexual and or emotional relationships with others, often under specific conditions or rules about how this would occur (McLean, 2004). In these "open" relationships, nonmonogamy was not the same as infidelity, and the former did not imply dishonesty. The researcher concluded:

> Despite the stereotypes that claim that bisexuals are deceitful, unfaithful and untrustworthy in relationships, most of the bisexual men and women I interviewed demonstrated a significant commitment to the principles of trust, honesty and communication in their intimate relationships and made considerable effort to ensure both theirs and their partner's needs and desires were catered for within the relationship (McLean, 2004, p. 96).

Bisexuality immediately doubles your chance for a date on Saturday nights.

Woody Allen, director

Monogamous bisexual women and men find that their erotic attractions can be satisfied through fantasy and their affectional needs through nonsexual friendships (Paul, 1996). Even in a monogamous relationship, "the partner of a bisexual person may feel that a bisexual person's decision to continue to identify as bisexual . . . is somehow a withholding of full commitment to the relationship. The bisexual person may be perceived as holding onto the possibility of other relationships by maintaining a bisexual identity and, therefore, not fully committed to the relationship" (Ochs, 1996, p. 234). However, this perception overlooks the fact that one's identity is separate from one's choices about relationship involvement or monogamy. Ochs (1996) notes that "a heterosexual's ability to establish and maintain a committed relationship with one person is not assumed to falter, even though the person retains a sexual identity as 'heterosexual' and may even admit to feeling attractions to other people despite her or his committed status" (p. 234).

Mixed-Orientation Relationships

Mixed-orientation couples are those in which one partner is heterosexual and the other partner is gay, lesbian, or bisexual. Up to 2 million gay, lesbian, or bisexual persons in the United States have been in heterosexual marriages at some point (Buxton, 2004). Some lesbigay individuals do not develop same-sex attractions and feelings until after they have been married. Others deny, hide, or repress their same-sex desires.

In a study of 20 gay or bisexual men who had disclosed their sexual orientation to their wives, most of the men did not intentionally mislead or deceive their future wives with regard to their sexuality. Rather, they did not fully grasp their feelings toward men, although they had a vague sense of their same-sex attraction (Pearcey, 2004). The majority of the men in this study (14 of 20) attempted to stay married after disclosure of their sexual orientation to their wives, and nearly half (9 of 20) did stay married for at least three years.

Although gay and lesbian spouses in heterosexual marriages are not sexually attracted to their spouses, they may nevertheless love them. But that is little consolation to spouses, who upon learning that their husband or wife is gay, lesbian, or bisexual, often react with shock, disbelief, and anger. The Straight Spouse Network (http://www.ssnetwk.org) provides support to heterosexual spouses or partners, current or former, of GLBT mates.

Legal Recognition and Support of Same-Sex Couples and Families

As current divorce rates of heterosexuals suggest (4 in 10 marriages end in divorce), maintaining long-term relationships is challenging. But for same-sex couples who lack the many social supports and legal benefits of marriage, the challenge is even greater. A leading researcher and scholar on LGBT issues noted, "perhaps what is most impressive about gay and lesbian couples is . . . that they manage to endure without the benefits of institutionalized supports" (Kurdek, 2005, p. 253). In this section we discuss laws and policies designed to provide institutionalized support for same-sex couples and families.

Decriminalization of Sodomy

In the United States, a 2003 Supreme Court decision in *Lawrence v. Texas* invalidated state laws that criminalized **sodomy**—oral and anal sexual acts. The ruling, which found that sodomy laws were discriminatory and unconstitutional, removed the stigma and criminal branding that sodomy laws have long placed on GLBT individuals. Prior to this historic ruling, sodomy was illegal in 13 states. Sodomy laws, which carried penalties ranging from a $200 fine to 20 years' imprisonment, were usually not used against heterosexuals but were used primarily against gay men and lesbians. Same-sex sexual behavior is still considered a criminal act in many countries throughout the world.

Trust a nitwit society like this one to think that there are only two categories—fag and straight.

Gore Vidal, author

International Data

In more than 80 countries, sexual activity between consenting adults of the same sex is illegal; in 9 countries, individuals found guilty of engaging in same-sex sexual behavior may receive the death penalty (International Gay and Lesbian Human Rights Commission, 2003a).

Registered Partnerships, Civil Unions, and Domestic Partnerships

Aside from same-sex marriage (which we discuss later), other forms of legal recognition of same-sex couples exist in a number of countries throughout the world at the national, state, and/or local level. In addition, some workplaces recognize same-sex couples for the purposes of employee benefits. Legal recognition of same-sex couples, also referred to as registered partnerships, **civil unions,** or **domestic partnerships,** conveys most but not all the rights and responsibilities of marriage.

International Data

Federally recognized registered partnerships, civil unions, or domestic partnerships for same-sex couples are available in Croatia, Denmark, Finland, France, Germany, Iceland, Israel, New Zealand, Norway, Portugal, Slovenia, Sweden, Switzerland, and the United Kingdom (Human Rights Campaign, no date). A number of other countries recognize same-sex couples for the purposes of immigration policy.

State and Local Legal Recognition of Same-Sex Couples   There is no federal recognition of same-sex couples in the United States. However, a number of U.S. states allow same-sex couples legal status that entitles them to many of the same rights and responsibilities as married opposite-sex couples (see Table 6.1). For example, in Vermont and Connecticut, same-sex couples can apply for a civil union license, which entitles them to all the rights and responsibilities available under state law to married couples. Unlike marriage for other sex couples, the rights of partners in same-sex civil unions are not recognized by U.S. federal law, so they do not have the federal protections that go along with civil marriage, and their legal status is not recognized in other states.

Three states—Hawaii, California, and New Jersey—have enacted laws that provide varying degrees of protection for domestic partners. The rights and responsibilities granted to domestic partners vary from place to place but may include coverage under a partner's health and pension plan, rights of inheritance and community property, tax benefits, access to married-student housing, child custody and child and spousal support obligations, and mutual responsibility for debts. The California law (the Domestic Partner Rights and Responsibilities Act of 2003) provides the broadest array of protections, including eligibility for family leave, other employment and health benefits, the right to sue for wrongful death of partner or inherit from partner as next-of-kin, and access to the stepparent adoption process (National Gay and Lesbian Task Force, 2005–2006). Ten states and the District of Columbia, as well as several dozen U.S. municipalities, offer domestic partner benefits to the same-sex partners of public employees.

There is just one life for each of us: our own.

Euripides, philosopher

Recognition of Same-Sex Couples in the Workplace In 1991 the Lotus Development Corporation became the first major American firm to extend domestic partner recognition to gay and lesbian employees. By the end of 2004, the Human Rights Campaign (2005) identified 8250 employers that provided domestic partner health insurance benefits to their employees—an increase of 13 percent from the previous year. The percentage of Fortune 500 companies that offered health benefits to employees' domestic partners nearly doubled from 25 percent in 2000 to 49 percent in March 2006 (Luther, 2006). However, even when companies offer domestic partner benefits to same-sex partners of employees, these benefits are usually taxed as income by the federal government, whereas spousal benefits are not.

Table 6.1 States that Recognize Same-Sex Relationships

State	Same-Sex Relationship Recognition
California	California has a domestic partner registry that confers almost all the state-level spousal rights and responsibilities to registered domestic partners.
Connecticut	Same-sex couples in Connecticut may enter into civil unions that offer almost all the benefits of marriage under state law. Connecticut is the first state to establish civil unions voluntarily, without having been ordered to do so by a court.
Hawaii	Hawaii offers "reciprocal beneficiary" status to same-sex registered couples, certain rights and obligations associated with survivorship, inheritance, property ownership, and insurance.
Maine	A domestic partner registry provides registered same-sex couples with inheritance rights, next-of-kin status, victim's compensation, and priority in guardian and conservator rights.
Massachusetts	Massachusetts grants same-sex marriage licenses only to residents of Massachusetts. The license is not valid if the couple moves to another state.
New Jersey	New Jersey offers "domestic partner" status to same-sex and some opposite-sex couples, which allows the partner to be treated as a dependent for the purposes of administering certain retirement and health benefits.
Vermont	Vermont gives same-sex couples more than 300 state-level rights and responsibilities extended to opposite-sex spouses.

Source: Human Rights Campaign (2006); Luther (2006).

Same-Sex Marriage

In 2003, a Massachusetts Supreme Court ruling in *Goodridge* v. *Department of Public Health* ruled that same-sex couples have a constitutional right to be legally married and, effective May 2004, Massachusetts became the first U.S. state to offer civil marriage licenses to same-sex couples. In the first year after the court order went into effect, more than 5,000 same-sex couples were married in Massachusetts (Johnston, 2005). Unlike marriages between a man and a woman, same-sex marriages in Massachusetts are not recognized in other states, nor are they recognized by the federal government.

If the love of two people, committing themselves to each other exclusively for the rest of their lives, is not worthy of respect, then what is?

Andrew Sullivan, editor and gay rights advocate

Anti-Gay Marriage Legislation   In 1996 Congress passed and President Clinton signed the **Defense of Marriage Act (DOMA),** which states that marriage is a "legal union between one man and one woman" and denies federal recognition of same-sex marriage. In effect, this law allows states to either recognize or not recognize same-sex marriages performed in other states. As of March 2006, 36 states have banned gay marriage either through statute or a state constitutional amendment, and 17 states have passed broader antigay family measures that ban other forms of partner recognition in addition to marriage, such as domestic partnerships and civil unions. These broader measures, known as "Super DOMAs," potentially endanger employer-provided domestic partner benefits, joint and second-parent adoptions, health care decision-making proxies, or any policy or document that recognizes the existence of a same-sex partnership (Cahill and Slater, 2004). Some of these "Super DOMAs" ban partner recognition for unmarried heterosexual couples as well.

At the federal level, there are efforts to amend the U.S. Constitution to define marriage as being between a man and a woman. The Federal Marriage Amendment did not pass in 2004 in the Senate or the House, but supporters vowed to continue the fight. The Federal Marriage Amendment was reintroduced and voted on in 2006, and although it garnered more support than in 2004, it failed to reach the two-thirds majority vote necessary for proposal as an amendment. If it had passed, the constitutional amendment would have denied marriage and likely civil union and domestic partnership rights to same-sex couples (LAWbriefs, 2005). Such an amendment would also hurt the children in same-sex couple families. Dr. Kathleen Moltz, an assistant professor at Wayne State University School of Medicine,

Massachusetts permits same-sex couple marriages, but these are not recognized by the federal government or other states.

© AP/Wide World Photos

testified against the passage of an anti-gay constitutional amendment before a United States Senate Judiciary Committee, expressing her fears about how such an amendment would affect her family:

> I don't know what harm . . . a constitutional amendment might cause. I fear that families like mine, with young children, will lose health benefits; will be denied common decencies like hospital visitation when tragedy strikes; will lack the ability to provide support for one another in old-age. I fear that my loving, innocent children will face hatred and insults implicitly sanctioned by a law that brands their family as unequal. I know that these sweet children have already been shunned and excluded by people claiming to represent values of decency and compassion. I also know what such an amendment will not do. It will not help couples who are struggling to stay married. It will not assist any impoverished families struggling to make ends meet or to obtain health care for sick children. It will not keep children with their parents when their parents see divorce as their only option. It will not help any single American citizen to live life with more decency, compassion or morality (Moltz, 2005).

Disapproval of homosexuality cannot justify invading the houses, hearts and minds of citizens who choose to live their lives differently.

Harry A. Blackmun, Supreme Court Justice

Arguments in Favor of Same-Sex Marriage Advocates of same-sex marriage argue that banning or refusing to recognize same-sex marriages granted in other states is a violation of civil rights that denies same-sex couples the many legal and financial benefits that are granted to heterosexual married couples. Rights and benefits that married spouses have include the following:

- The right to inherit from a spouse who dies without a will;
- No inheritance taxes between spouses;
- The right to make crucial medical decisions for a partner and to take care of a seriously ill partner or parent of a partner under current provisions in the federal Family and Medical Leave Act;
- Social Security survivor benefits; and
- Health insurance coverage under a spouse's insurance plan.

Other rights bestowed on married (or once-married) partners include assumption of spouse's pension, bereavement leave, burial determination, domestic violence protection, reduced-rate memberships, divorce protections (such as equitable division of assets and visitation of partner's children), automatic housing lease transfer, and immunity from testifying against a spouse. As noted earlier, same-sex couples are taxed on employer-provided insurance benefits for domestic partners, whereas married spouses receive those benefits tax-free. Finally, unlike 17 other countries that recognize same-sex couples for immigration purposes, the United States does not recognize same-sex couples in granting immigration status because such couples are not considered "spouses." Another argument for same-sex marriage is that it would promote relationship stability among gay and lesbian couples. "To the extent that marriage provides status, institutional support, and legitimacy, gay and lesbian couples, if allowed to marry, would likely experience greater relationship stability" (Amato, 2004, p. 963). Indeed, same-sex relationships, like cohabitation relationships, end at a higher rate than marriage relationships (Wagner, 2006).

Recognized marriage, argues Amato, would be beneficial to the children of same-sex parents. Without legal recognition of same-sex families, children living in gay- and lesbian-headed households are denied a range of securities that protect children of heterosexual married couples. These include the right to get health insurance coverage and Social Security survivor benefits from a nonbiological parent. In some cases children in same-sex households lack the automatic right to continue living with their nonbiological parent should their biological mother or father die (Tobias and Cahill, 2003). It is ironic that the same pro-marriage groups that stress that children are better off in married-couple families disregard the benefits of same-sex marriage to children.

Finally, there are religious-based arguments in support of same-sex marriage. Although many religious leaders teach that homosexuality is sinful and prohibited by God, some religious groups, such as the Quakers and the United Church of Christ (UCC), are accepting of homosexuality, and other groups have made reforms toward increased acceptance of lesbians and gays. In 2005, the UCC became the largest Christian denomination to endorse same-sex marriages. In a sermon titled "The Christian Case for Gay Marriage," Jack McKinney (2004) interprets Luke 4: "Jesus is saying that one of the most fundamental religious tasks is to stand with those who have been excluded and marginalized. . . . [Jesus] is determined to stand with them, to name them beloved of God, and to dedicate his life to seeing them empowered." McKinney goes on to ask, "Since when has it been immoral for two people to commit themselves to a relationship of mutual love and caring? No, the true immorality around gay marriage rests with the heterosexual majority that denies gays and lesbians more than 1,000 federal rights that come with marriage."

Arguments Against Same-Sex Marriage   Whereas advocates of same-sex marriage argue that so long as same-sex couples cannot be legally married, they will not be regarded as legitimate families by the larger society, opponents do not want to legitimize same-sex couples and families. Opponents of same-sex marriage who view homosexuality as unnatural, sick, and/or immoral do not want their children to view homosexuality as socially acceptable.

Opponents of same-sex marriage commonly argue that such marriages would subvert the stability and integrity of the heterosexual family. However, Sullivan (1997) suggests that homosexuals are already part of heterosexual families:

> [Homosexuals] are sons and daughters, brothers and sisters, even mothers and fathers, of heterosexuals. The distinction between "families" and "homosexuals" is, to begin with, empirically false; and the stability of existing families is closely linked to how homosexuals are treated within them (p. 147).

Many opponents of same-sex marriage base their opposition on their religious views. In a Pew Research Center national poll, the majority of Catholics and Protestants opposed legalizing same-sex marriage, whereas the majority of secular respondents favored it (Green, 2004). However, churches have the right to say no to marriage for gays in their congregations. Legal marriage is a contract between the spouses and the state; marriage is a *civil* option that does not require religious sanctioning.

In previous years, opponents of gay marriage have pointed to public opinion polls that suggested that the majority of Americans are against same-sex marriage. But public opposition to same-sex marriage is decreasing. A 2006 Pew Research Center national poll found that about half (51%) of U.S. adults oppose legalizing gay marriage, down from 63 percent in 2004 (Pew Research Center, 2006). Support for gay marriage is higher among young adults. In a national survey of first-year college students, about two-thirds (64.3%) of female university students and half (50.1%) of male university students agree or strongly agree that same-sex couples should have the right to legal marital status (American Council on Education and University of California, 2005–2006).

If the last two days are any indication, the race for the White House will be pretty much decided by whether two women can open a joint checking account.

Jon Stewart

GLBT Parenting Issues

National Data

Nearly one quarter of all same-gender couples are raising children; 34.3% of lesbian couples are raising children, and 22.3% of gay male couples are raising children (compared with 45.6% of married heterosexual and 43.1% of unmarried heterosexual couples raising children) (Pawelski et al., 2006).

Of the more than 600,000 same-sex-couple households identified in the 2000 Census, 162,000 had one or more children living in the household. This is a low estimate of children who have gay or lesbian parents, as it does not count children in same-sex households who did not identify their relationship in the Census, those headed by gay or lesbian single parents, or those whose gay parent does not have physical custody but is still actively involved in the child's life.

National Data

Estimates of the number of U.S. children with gay or lesbian parents range from 1 to 14 million (Howard, 2006).

Many gay and lesbian individuals and couples have children from prior heterosexual relationships or marriages. Up to 2 million gay, lesbian, or bisexual persons in the United States have been in heterosexual marriages at some point (Buxton, 2005). Some of these individuals married as a "cover" for their homosexuality; others discovered their interest in same-sex relationships after they married. Children with mixed-orientation parents may be raised by a gay or lesbian parent, a gay or lesbian stepparent, a heterosexual parent, and a heterosexual stepparent.

A gay or lesbian individual or couple may have children through the use of assisted reproductive technology, including donor insemination, in vitro fertilization, and surrogate mothers. Others adopt or become foster parents.

Less commonly, some gay fathers are part of an emergent family form known as the *hetero-gay family*. In a hetero-gay family, a heterosexual mother and gay father conceive and raise a child together but reside separately.

Anti-gay views concerning gay parenting include the belief that homosexual individuals are unfit to be parents and that children of lesbians and gays will not develop normally and/or that they will become homosexual. As the following section suggests, research findings paint a more positive picture of the development and well being of children with gay or lesbian parents.

Development and Well Being of Children with Gay or Lesbian Parents

Anyone who wishes to examine the 20 years of peer-reviewed studies on the emotional, cognitive and behavioral outcomes of children of gay and lesbian parents will find not one shred of evidence that children are harmed by their parents' sexual orientation.

Carol Trust, executive director, National Association of Social Workers

A growing body of research on gay and lesbian parenting supports the conclusion that children of gay and lesbian parents are just as likely to flourish as are children of heterosexual parents. For example, one extensive review of research on gay and lesbian parenting concludes that "children raised by gay and lesbian parents adjust positively, and their families function well....There is no credible social science evidence to support that gay parenting (and by extension, gay adoptive parenting) negatively affects the well being of children" (Howard, 2006, p. 9). The American Psychological Association (2004) states that "results of research suggest that lesbian and gay parents are as likely as heterosexual parents to provide supportive and healthy environments for their children" and that "the adjustment, development, and psychological well being of children [are] unrelated to parental sexual orientation and that the children of lesbian and gay parents are as likely as those of heterosexual parents to flourish." Indeed, Pro-Family Pediatricians cheered when a proposal to ban gay marriage was defeated. "Our duty as pediatricians is to see that all children have the same security and protection regardless of the sexual orientation of their parents. Denying legal rights to same-sex couples injures their children" noted (Asch-Goodkin, 2006, 2)

For example, one study compared a national sample of 44 adolescents parented by same-sex couples with 44 adolescents parented by opposite-sex couples (Wainwright et al., 2004). On an array of assessments, the study showed that the personal, family, and school adjustment of adolescents living with same-sex parents did not differ from that of adolescents living with opposite-sex parents. Self-esteem, depressive symptoms and anxiety, academic achievement, trouble in school, quality of family relationships, and romantic relationships were similar in the two groups of adolescents. Regardless of family type, adolescents were more likely to show positive adjustment when they perceived more caring from adults and when parents described having close relationships with them. Thus, it was the qualities of adolescent-parent relationships rather than the sexual orientation of the parents that were significantly associated with adolescent adjustment.

In another study, researchers examined the quality of parent-child relationships and the socioemotional and gender development of a sample of 7-year-old children with lesbian parents, compared with 7-year-olds from two-parent heterosexual families and with single heterosexual mothers (Golombok et al., 2003). No significant differences between lesbian mothers and heterosexual mothers were found for most of the parenting variables assessed, although lesbian mothers reported smacking their children less and played more frequently with their children than did heterosexual mothers. No significant differences were found in psychiatric disorders or gender development of the children in lesbian families versus heterosexual families. The findings also suggest that having two parents is associated with more positive outcomes for children's psychological well being, but the gender of the parents is not relevant.

Stacey and Biblarz (2001) challenged the conclusion that there are no differences in developmental outcomes between children raised by lesbigay parents and those raised by heterosexual parents. Rather, they suggested that children with two same-gender parents develop in less gender-stereotypical ways than do children of opposite-sex parents and seem to be more open to homoerotic relationships (although the majority of children of lesbigay parents identify as heterosexual). Children with lesbian or gay parents must contend with the burdens of social stigma resulting from antigay attitudes and discrimination, but they also have more empathy for social diversity.

Discrimination in Child Custody, Visitation, Adoption, and Foster Care

A student in the authors' classes reported that after she divorced her husband, she became involved in a lesbian relationship. She explained that she would like to be open about her relationship to her family and friends, but she was afraid that if her ex-husband found out that she was in a lesbian relationship, he might take her to court and try to get custody of their children. Although several respected national organizations—including the American Academy of Pediatrics, the Child Welfare League of America, the American Bar Association, the American Medical Association, the American Psychological Association, the American Psychiatric Association, and the National Association of Social Workers—have gone on record in support of treating gays and lesbians without prejudice in parenting and adoption decisions (Howard, 2006; Landis, 1999), lesbian and gay parents are often discriminated against in child custody, visitation, adoption, and foster care.

Some court judges are biased against lesbian and gay parents in custody and visitation disputes. For example, in 1999 the Mississippi Supreme Court denied custody of a teenage boy to his gay father and instead awarded custody to his heterosexual mother who remarried into a home "wracked with domestic violence and excessive drinking" (*Custody and Visitation*, 2000, p. 1).

Gay and lesbian individuals and couples who want to adopt children can do so through adoption agencies or through the foster care system in at least 22 states and the District of Columbia. However, Florida and Mississippi forbid adoption by gay and lesbian people, Utah forbids adoption by any unmarried couple (which includes all same-sex couples), and Arkansas prohibits lesbians

The heterosexuals who hate us should stop having us.
Lynda Montgomery

Should We Prohibit Adoption by Lesbian and Gay Couples?

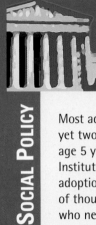

According to the U.S. Children's Bureau, in 2003 there were 119,000 children in the U.S. child welfare system waiting to be adopted, only 20,000 of whom were in pre-adoptive homes (Howard, 2006). Thousands of these children will never be adopted and will never have a stable, permanent home and family. Most adoptive parents want infants or young children, yet two-thirds of children waiting to be adopted are over age 5 years. A report by the Evan B. Donaldson Adoption Institute concludes that "laws and policies that preclude adoption by gay or lesbian parents disadvantage the tens of thousands of children mired in the foster care system who need permanent, loving homes" and that "adoption by gays and lesbians holds promise as an avenue for achieving permanency for many of the waiting children in foster care" (Howard, p. 2, 3).

A study of adoptive parents showed no significant differences between gay and lesbian adoptive parents and heterosexual adoptive parents on measures of family functioning and child behavior problems (Erich et al., 2005). Despite this and other research that finds positive outcomes for children raised by gay or lesbian parents, and despite the support for gay adoption by child advocacy organizations, placing children for adoption with gay or lesbian parents remains controversial. A 2006 poll reveals that 46 percent of U.S. adults support gay adoption, up from 38 percent in 1999 (Pew Research Center, 2006). Although public support for gay adoption has increased in recent years, fewer than half of U.S. adults reported support for gay adoption in 2006. At the time of this writing, efforts are under way in 16 states to ban gay adoption (Stone, 2006). Social policies that prohibit LGBT individuals and couples from adopting children result in fewer children being adopted. What happens to children who are not adopted? The thousands of children who "age out" of the foster care system annually experience high rates of homelessness, incarceration, early pregnancy, failure to graduate from high school, unemployment, and poverty (Howard, 2006). Essentially, social policies that prohibit gay adoption are policies that deny thousands of children the opportunity to have a nurturing family.

Your Opinion?

1. Do you believe children who grow up with same-sex parents are disadvantaged?
2. Do you believe that children without homes should be prohibited by law from being adopted by lesbian or gay individuals or couples?
3. If you were a 6-year-old child with no family, would you rather remain in an institutional setting or be adopted by a gay or lesbian couple?

and gay men from serving as foster parents (National Gay and Lesbian Task Force, 2004). This chapter's Social Policy section asks whether gay and lesbian individuals and couples should be prohibited from adopting.

Most adoptions by gay people are second-parent adoptions. A **second-parent adoption** (also called co-parent adoption) is a legal procedure that allows an individual to adopt his or her partner's biological or adoptive child without terminating the first parent's legal status as parent. Second-parent adoption gives children in same-sex families the security of having two legal parents. The potential benefits of second-parent adoption include the following (Clunis and Green, 2003):

- Places legal responsibility on the parent to support the child;
- Allows the child to live with the legal parent in the event that the biological (or original adoptive) parent dies or becomes incapacitated;
- Enables the child to inherit and receive Social Security benefits from the legal parent;
- Enables the child to receive health insurance benefits from the parent's employer; and
- Gives the legal parent standing to petition for custody or visitation in the event that the parents break up.

However, in four states (Colorado, Nebraska, Ohio, and Wisconsin), court rulings have decided that the state adoption law does not allow for second-parent adoption by members of same-sex couples, and in 22 states it is unclear whether the state adoption law permits second-parent adoption (National Gay and Lesbian Task Force, 2005b). Second-parent adoption is not possible when a parent in a same-sex relationship has a child from a previous heterosexual marriage or relationship, unless the former spouse or partner is willing to give up parental rights. Although "third-parent" adoptions have been granted in a small number of jurisdictions, this option is not widely available (National Center for Lesbian Rights, 2003).

Effects of Antigay Bias and Discrimination on Heterosexuals

As a junior and senior at Homewood-Flossmoor High School in the suburbs of Chicago, Myka Held played a key role in leading a campaign to promote tolerance of gay and lesbian students. The campaign involved selling gay-friendly t-shirts to students and teachers and having as many people as possible wear the t-shirts to school on a designated day. The t-shirts, made by Duke University, say, "gay? fine by me." "I think it's really important for gay people out there to know that there are straight people who support them," Ms. Held said (quoted in Puccinelli, 2005, p. 20). "I have always supported equal rights for every person and have been disgusted by discrimination and prejudice. As a young Jewish woman, I believe it is my duty to stand up and support minority groups. . . . In my mind, fighting for gay rights is a proxy for fighting for every person's rights" (Held, 2005). Myka Held's t-shirt campaign illustrates that fighting prejudice and discrimination against sexual-orientation minorities is an issue not just for lesbians, gays, and bisexuals but for all those who value fairness and respect for human beings in all their diversity.

The antigay and heterosexist social climate of our society is often viewed in terms of how it victimizes the gay population. However, heterosexuals are also victimized by heterosexism and antigay prejudice and discrimination. Some of these effects follow.

1. *Heterosexual victims of hate crimes.* As we discussed earlier in this chapter, extreme homophobia contributes to instances of violence against homosexuals—acts known as hate crimes. But hate crimes are crimes of perception, meaning that victims of antigay hate crimes may not be homosexual; they may just be *perceived* as being homosexual. The National Coalition of Anti-Violence Programs (2005) reported that in 2004, 192 heterosexual individuals in the United States were victims of antigay hate crimes, representing 9 percent of all antigay hate crime victims.

2. *Concern, fear, and grief over well being of gay or lesbian family members and friends.* Many heterosexual family members and friends of homosexuals experience concern, fear, and grief over the mistreatment of their gay or lesbian friends and/or family members. For example, heterosexual parents who have a gay or lesbian teenager often worry about how the harassment, ridicule, rejection, and violence their gay or lesbian child experiences at school might affect him or her. Will their child drop out of school, as one-fourth of gay youth do (Chase, 2000), to escape the harassment, violence, and alienation they endure there? Will their gay or lesbian child respond to the antigay victimization they experience by turning to drugs or alcohol or by committing suicide? Such fears are not unfounded: lesbian, gay, and bisexual youth who report high levels of victimization at school also have higher levels of substance use and suicidal thoughts than heterosexual peers who report high levels of at-school victimization (Bontempo and D'Augelli, 2002). A survey of youths' risk behavior conducted by the Massachusetts Department of Education in 1999 revealed that 30 percent of gay teens had attempted

Injustice anywhere is a threat to justice everywhere. We are caught in an inescapable network of mutuality, tied in a single garment of destiny. Whatever affects one directly affects all indirectly.

Martin Luther King, Jr., civil rights leader

Lucas Schaefer, shown in this photo, is one of several students at Duke University who began the "Gay? Fine By Me" t-shirt project in 2003, after the *Princeton Review* named Duke the most gay-unfriendly school in the United States. After graduating from Duke, Lucas became the Executive Director of Fine By Me, a non-profit organization based in Brooklyn, New York. Thousands of students from more than 100 colleges and high schools have participated in the Gay? Fine By Me movement since it began.

Authors

suicide in the previous year, compared with 7 percent of their straight peers (Platt, 2001).

Heterosexual individuals also worry about the ways in which their gay, lesbian, and bisexual family members and friends could be discriminated against in the workplace.

To heterosexuals who have lesbian and gay family members and friends, lack of family protections such as health insurance and rights of survivorship for same-sex couples can also be cause for concern. Finally, heterosexuals live with the painful awareness that their gay or lesbian family member or friend is a potential victim of anti-gay hate crime. Imagine the life-long grief experienced by heterosexual family members and friends of hate crime murder victims, such as Matthew Shepard, a 21-year-old college student who was brutally beaten to death in 1998 for no apparent reason other than he was gay.

National Data

As of March 2006, only 17 states had laws banning discrimination based on sexual orientation (National Gay and Lesbian Task Force, 2006).

3. *Restriction of intimacy and self-expression.* Because of the antigay social climate, heterosexuals, especially males, are hindered in their own self-expression and intimacy in same-sex relationships. "The threat of victimization (i.e., antigay violence) . . . causes many heterosexuals to conform to gender roles and to restrict their expressions of (nonsexual) physical affection for members of their own sex" (Garnets et al., 1990, p. 380). Homophobic epithets frighten youth who do not conform to gender role expectations, leading some youth to avoid activities that they might otherwise enjoy and benefit from (e.g., arts for boys, athletics for girls) (Gay, Lesbian, and Straight Education Network, 2000). A male student in the authors' class revealed that he always wanted to work with young children and had majored in early childhood education. His peers teased him relentlessly about his choice of majors, questioning both his masculinity and his heterosexuality. Eventually, this student changed his major to psychology, which his peers viewed as an acceptable major for a heterosexual male.

4. *Dysfunctional sexual behavior.* Some cases of rape and sexual assault are related to homophobia and compulsory heterosexuality. For example, college men who participate in gang rape, also known as "pulling train," entice each other into the act "by implying that those who do not participate are unmanly or homosexual" (Sanday, 1995, p. 399). Homonegativity also encourages early sexual activity among adolescent men. Adolescent male virgins are often teased by their male peers, who say things like "You mean you don't do it with girls yet? What are you, a fag or something?" Not wanting to be labeled and stigmatized as a "fag," some adolescent boys "prove" their heterosexuality by having sex with girls.

5. *School shootings.* Antigay harassment has also been a factor in many of the school shootings in recent years. In March 2001, 15-year-old Charles Andrew Williams fired more than 30 rounds in a San Diego suburban high school, killing two and injuring 13 others. A woman who knew Williams reported that the students had teased him and called him gay (Dozetos, 2001). According to the Gay, Lesbian, and Straight Education Network (GLSEN), Williams's story is not unusual. Referring to a study of harassment of U.S. students that was commissioned by the American Association of University Women, a GLSEN report concluded, "For boys, no other type of harassment provoked as strong a reaction on average; boys in this study would be less upset about physical abuse than they would be if someone called them gay" (Dozetos, 2001).

6. *Loss of rights for individuals in unmarried relationships.* The passage of state constitutional amendments that prohibit same-sex marriage can also result in denial of rights and protections to opposite-sex unmarried couples. For example, in 2005, Judge Stuart Friedman of Cuyahoga County (Ohio) agreed that a man

I was once involved in a same sex marriage. There was the same sex over and over and over.

David Letterman

who was charged with assaulting his girlfriend could not be charged with a domestic violence felony because the Ohio state constitutional amendment granted no such protections to unmarried couples (Human Rights Campaign, 2005). As we discussed earlier, some antigay marriage measures also threaten the provision of domestic partnership benefits to unmarried heterosexual couples.

Changing Attitudes toward Same-Sex Couples and Families

Over the past few decades, attitudes toward the morality of homosexuality have become more accepting, and support for protecting civil rights of gays and lesbians has increased (Loftus, 2001). Part of the explanation for these changing attitudes is the increasing levels of education in the U.S. population, because individuals with more education tend to be more liberal in their attitudes toward homosexuality. Increased contact between homosexuals and heterosexuals and positive depictions of gay and lesbian individuals and couples in the media also contribute to increased acceptance of homosexuality and same-sex relationships.

Increased Contact between Homosexuals and Heterosexuals

Greater acceptance of homosexuality may be due to increased personal contact between heterosexuals and openly gay individuals, as more gay and lesbian Americans are coming out to their family, friends, and co-workers. Psychologist Gordon Allport's (1954) "contact hypothesis" asserts that contact between groups is necessary for the reduction of prejudice. Recent research has found that, in general, heterosexuals have more favorable attitudes toward gay men and lesbian women if they have had prior contact with or know someone who is gay or lesbian (Mohipp and Morry, 2004). Contact with openly gay individuals reduces negative stereotypes and ignorance and increases support for gay and lesbian equality (Wilcox and Wolpert, 2000). Polls of U.S. adults find that more than half of respondents (56%) say they have a friend or acquaintance who is gay or lesbian, nearly one-third (32%) say they work with someone who is gay or lesbian, and nearly one-fourth (23%) say that someone in their family is gay or lesbian (Newport, 2002). A national poll of U.S. high school students showed that 16 percent of students have a gay or lesbian person in their family and 72 percent know someone who is gay or lesbian (Curtis, 2004).

The 2005 movie Brokeback Mountain depicts a love story between two cowboys and conveys the horror of antigay violence. If you have seen this movie, how did it affect your attitudes toward homosexuality?

© Focus Features/Photofest

Gays and Lesbians in the Media

Another explanation for the increasing acceptance of gays and lesbians is their positive depiction in the popular media. One study found that college students reported lower levels of antigay prejudice after watching television shows with prominent homosexual characters (e.g., *Six Feet Under* and *Queer Eye for the Straight Guy*) (Schiappa et al., 2005).

In 1998 Ellen DeGeneres came out on her sitcom *Ellen*, and by 2006 many television viewers had seen gay and lesbian characters in television shows such as *Will and Grace, Buffy the Vampire Slayer, The Sopranos,* and those mentioned earlier. The 2005

Oscar-nominated movie *Brokeback Mountain* portrayed the intimacy between two men in a story that left many viewers in tears. And the 2006 HBO documentary (see opening photo), *All Aboard! Rosie's Family Cruise* provided viewers with positive images of the love and care between same-sex couples and their children.

The Global Fight for Same-Sex Equality

The struggle that LGBT individuals, couples, and families face for acceptance and equal rights is worldwide. More than 20 countries have national laws that ban various forms of discrimination against gays, lesbians, and bisexuals (International Gay and Lesbian Human Rights Commission, 1999). In 1996 South Africa became the first country in the world to include in its constitution a clause banning discrimination based on sexual orientation. Fiji, Canada, Ecuador, and Portugal also have constitutions that ban discrimination based on sexual orientation.

Human rights treaties and transnational social movement organizations have increasingly asserted the rights of people to engage in same-sex relations. International organizations such as Amnesty International, which resolved in 1991 to defend those imprisoned for homosexuality, the International Lesbian and Gay Association (founded in 1978), and the International Gay and Lesbian Human Rights Commission (founded in 1990) continue to fight antigay prejudice and discrimination. The United Nations Human Rights Commission has proposed the Resolution on Sexual Orientation and Human Rights. This landmark resolution recognizes the existence of sexual orientation–based discrimination around the world, affirms that such discrimination is a violation of human rights, and calls on all governments to promote and protect the human rights of all people, regardless of their sexual orientation (International Gay and Lesbian Human Rights Commission, 2004).

In the United States, advancements in gay rights include the Supreme Court's 2003 decriminalization of sodomy, the growing adoption of nondiscrimination policies covering sexual orientation, the increased recognition of domestic partnerships, the state civil union rights and responsibilities offered to same-sex couples in Vermont, Connecticut, and California, and the Massachusetts ruling that same-sex marriage is allowable under the state constitution. But these victories for LGBT individuals and families fuel the backlash against them by groups who are determined to maintain traditional notions of family and gender. Often, this determination is rooted in and derives its strength from uncompromising religious ideology. Despite the worldwide movement toward increased acceptance and protection of homosexual individuals, the status and rights of lesbians and gays and their families in the United States continue to be some of the most divisive issues in U.S. society.

I would rather know fifty gays than one Jehovah's witness. You won't ever have a gay knocking on your front door trying to convert you.

Jeffrey Jena

SUMMARY

How prevalent are homosexuality/bisexuality and same-sex households and families?

Estimates of the lesbigay population in the United States range from about 2 percent to 5 percent of the total U.S. population. The 2000 census revealed that about 1 in 9 (or 594,000) unmarried-partner households in the United States involved partners of the same sex. About one-fifth (22.3%) of gay male couples and one-third (34.3%) of lesbian couples have children in the home.

Can homosexuals change their sexual orientation?

Individuals who believe that homosexuals choose their sexual orientation tend to think that homosexuals can and should change their sexual orientation. Various forms of reparative therapy or conversion therapy are dedicated to changing homosexuals' sexual orientation. However, the American Psychiatric Association, the American Psychological Association, the American Academy of Pediatrics, the American Counseling Association, the National Association of School Psychologists, the National Association of Social Workers, and the American Medical Association agree that homosexuality is not a mental disorder and needs no cure—that efforts to change sexual orientation do not work and may, in fact, be harmful.

Who tends to have more negative views toward homosexuality?

In general, individuals who are more likely to have negative attitudes toward homosexuality and to oppose gay rights are older, attend religious services, are less educated, live in the South or Midwest, and reside in small rural towns. In addition, blacks are more likely than whites to view homosexual relations as "always wrong."

How are gay and lesbian couples different from heterosexual couples?

Research suggests that gay and lesbian couples tend to be more similar to than different from heterosexual couples. Both gay and straight couples tend to value long-term monogamous relationships. For both heterosexual and same-sex partners, relationship satisfaction tends to be high in the beginning of the relationship and decreases over time, and the factors linked to relationship satisfaction (such as equal power and control and being emotionally expressive) are the same for same-sex and heterosexual couples. Gay and lesbian couples also tend to disagree about the same issues that heterosexual couples argue about.

Same-sex couples' relationships are different from heterosexuals' in that same-sex couples have more concern about when and how to disclose their relationships to others. Same-sex couples also are more likely than heterosexual couples to achieve a fair distribution of household labor and to argue more effectively and resolve conflict in a positive way. Most significantly, whereas heterosexuals enjoy many social and institutional supports for their relationships, gay and lesbian couples face prejudice and discrimination.

What forms of legal recognition of same-sex couples exist in the United States and in other countries?

In the United States, Massachusetts is the only state in which same-sex couples can be legally married (although these marriages are not recognized by the federal government or in other states). Same-sex couples can also be legally married in the Netherlands, Belgium, Spain, Canada, and South Africa.

Other forms of legal recognition of same-sex couples that exist in a number of countries throughout the world at the national, state, and/or local level include registered partnerships, civil unions, and domestic partnerships, which convey most but not all the rights and responsibilities of marriage. Some workplaces also recognize domestic partnerships of their employees for the purposes of benefits such as health insurance coverage and family leave.

What does research on gay and lesbian parenting conclude?

A growing body of credible, scientific research on gay and lesbian parenting concludes that children raised by gay and lesbian parents adjust positively and their families function well. Lesbian and gay parents are as likely as heterosexual parents to provide supportive and healthy environments for their children, and the children of lesbian and gay parents are as likely as those of heterosexual parents to flourish. Some research suggests that children raised by lesbigay parents develop in less-gender-stereotypical ways, are more open to homoerotic relationships, contend with the social stigma of having gay parents, and have more empathy for social diversity than children of opposite-sex parents. There is no credible social science evidence that gay parenting negatively affects the well being of children.

In what ways are heterosexuals victimized by heterosexism and antigay prejudice and discrimination?

Heterosexuals may be victims of antigay hate crimes if they are just perceived as gay. Many heterosexual family members and friends of homosexuals experience concern, fear, and grief over the mistreatment of and discrimination toward their gay or lesbian friends and/or family members. Because of the antigay social climate, heterosexuals, especially males, are hindered in their own self-expression and intimacy in same-sex relationships. Homophobia has also been linked to some cases of rape and sexual assault by males who are trying to prove they are not gay. Antigay harassment has been a factor in many of the school shootings in recent years. Finally, the passage of state constitutional amendments that prohibit same-sex marriage can also result in denial of rights and protections to opposite-sex unmarried couples.

How have attitudes toward homosexuality and same-sex relationships changed in recent decades?

Over the past few decades, attitudes toward the morality of homosexuality have become more accepting, and support for protecting civil rights of gays and lesbians has increased. In the United States, increased acceptance of homosexuality has resulted from increasing levels of education in the U.S. population (individuals with more education tend to be more liberal in their attitudes toward homosexuality), increased contact between homosexuals and heterosexuals (as a result of more gays and lesbians "coming out"), and positive depictions of gay and lesbian individuals and couples in the media.

KEY TERMS

antigay bias	discrimination	homonegativity	LGBT
binegativity	domestic partnerships	homophobia	prejudice
biphobia	gay	homosexuality	second-parent adoption
bisexuality	GLBT	internalized homophobia	sexual orientation
civil unions	heterosexism	lesbian	sodomy
coming out	heterosexuality	lesbigay population	transgendered
Defense of Marriage Act			

The Companion Website for *Choices in Relationships: An Introduction to Marriage and the Family,* Ninth Edition

http://sociology.wadsworth.com/knox_schacht/choices9e

Supplement your review of this chapter by going to the companion website to take one of the Tutorial Quizzes, use the flash cards to master key terms, and check out the many other study aids you'll find there. You'll also find special features such as the Marriage and Family Resource Center, Census 2000 information, and other data and resources at your fingertips to help you with that special project or to do some research on your own.

WEBLINKS

Advocate (Online Newspaper)
http://www.advocate.com/

APA (American Psychological Association). 2004. "Sexual Orientation, Parents, & Children." APA Online
http://www.apa.org

Bisexual Resource Center
http://www.biresource.org

Children of Lesbians & Gays Everywhere (COLAGE)
http://www.colage.org

Gay and Lesbian Support Groups for Parents
http://www.gayparentmag.com/29181.html

Human Rights Campaign
http://www.hrc.o:g

National Gay and Lesbian Task Force
http://www.ngltf.o:g

PFLAG (Parents, Families, and Friends of Lesbians and Gays)
http://www.pflag.org

PlanetOut (Online News Source)
http://www.planetout.com

Straight Spouse Network
http://www.ssnetwk.org

REFERENCES

Allport, G. W. 1954. *The nature of prejudice.* Cambridge, MA: Addison-Wesley.

Amato, Paul R. 2004. Tension between institutional and individual views of marriage. *Journal of Marriage and Family* 66:959–65.

American Council on Education and University of California. 2005–2006. *The American freshman: National norms for fall, 2005.* Los Angeles: Higher Education Research Institute, U.C.L.A. Graduate School of Education and Information Studies.

American Psychological Association. 2004. Sexual Orientation, Parents, & Children. APA Policy Statement on Sexual Orientation, Parents, & Children. APA Online. Available at http://www.apa.org.

Andersson, G., T. Noack, A. Seierstad, and H. Weedon-Fekjaer. 2006. The demographics of same-sex marriages in Norway and Sweden. *Demography* 43:79–98.

Asch-Goodkin, J. 2006. An unsuccessful attempt to adopt a constitutional amendment that bans gay marriage. *Contemporary Pediatrics* 23: 14–15.

Berg, N. and D. Lein. 2006. Same-sex behaviour: US frequency estimates from survey data with simultaneous misreporting and non-response. *Applied Economics* 39: 757–770

Besen, W. 2000. Introduction. In *Feeling free: Personal stories—How love and self-acceptance saved us from "ex-gay" ministries.* Washington, DC: Human Rights Campaign Foundation, 7.

Black, D., G. Gates, S. Sanders, and L. Taylor. 2000. Demographics of the gay and lesbian population in the United States: Evidence from available systematic data sources. *Demography* 37:139–54.

Bobbe, J. 2002. Treatment with lesbian alcoholics: Healing shame and internalized homophobia for ongoing sobriety. *Health and Social Work* 27:218–223.

Bontempo, D. E., and A. R. D'Augelli. 2002. Effects of at-school victimization and sexual orientation on lesbian, gay, or bisexual youths' health risk behavior. *Journal of Adolescent Health* 30:364–74.

Bradford, J., K. Barrett, and J. A. Honnold. 2002. *The 2000 census and same-sex households: A user's guide.* New York: National Gay and Lesbian Task Force Policy Institute, Survey and Evaluation Research Laboratory, and Fenway Institute. Available at http://www.thetaskforce.org.

Buhl, L. 2005 (June 16). Youth's blog stirs uproar over 'ex-gay' camp. *PlanetOut.* Available at http://www.planetout.com.

Buxton, A. P. 2004. Paths and pitfalls: How heterosexual spouses cope when their husbands or wives come out. *Journal of Couple and Relationship Therapy* 3.

Buxton, A. 2005. A family matter: When a spouse comes out as gay, lesbian, or bisexual. *Journal of GLBT Family Studies* 1:49–70.

Cahill, S., E. Mitra, and S. Tobias. 2002. *Family policy: Issues affecting gay, lesbian, bisexual, and transgender families.* National Gay and Lesbian Task Force Policy Institute. Available at http://www.thetaskforce.org.

Cahill, S., and S. Slater. 2004. *Marriage: Legal protections for families and children.* Policy brief. Washington, DC: National Gay and Lesbian Task Force Policy Institute.

Centers for Disease Control and Prevention. 2005. *HIV/AIDS Surveillance Report 2004,* Vol. 16. Available at http://www.cdc.gov.

Chase, B. 2000. *NEA president Bob Chase's historic speech from 2000 GLSEN Conference.* Available at http://www.glsen.org.

Cianciotto, J. 2005. *Hispanic and Latino same-sex couple households in the U.S.: A report from the 2000 Census.* National Gay and Lesbian Task Force Policy Institute. Available at http:www.ngltf.org.

Cianciotto, J. and S. Cahill. 2006. *Youth in the crosshairs: The third wave of ex-gay activism.* National Gay and Lesbian Task Force Policy Institute. Available at http:www.thetaskforce.org.

Clunis, D. M. and G. Dorsey Green. 2003. *The lesbian parenting book,* 2nd *edition.* Emeryville, CA: Seal Press.

Curtis, C. 2003 (October 7). Poll: U.S. public is 50–50 on gay marriage. *PlanetOut.* Available at http://www.planetout.com.

Curtis, C. 2004. Poll: 1 in 20 High school students is gay. *PlanetOut.* Available at http://www.planetout.com.

Custody and Visitation. 2000. Human Rights Campaign FamilyNet. Available at http://familynet.hrc.org.

Diamond, L. M. 2003. What does sexual orientation orient? A biobehavioral model distinguishing romantic love and sexual desire. *Psychological Review* 110:173–92.

Dozetos, B. 2001 (March 7). School shooter taunted as 'gay.' PlanetOut. Available at http:www.planetout.com.

Erich, S., P. Leung, and P. Kindle. 2005. A Comparative analysis of adoptive family functioning with gay, lesbian, and heterosexual parents and their children. *Journal of GLBT Family Studies* 1:43–60.

Evans, J. and E. M. Broido. 2000. Coming out in college residence halls: Negotiation, meaning making, challenges, supports. *Journal of College Student Development* 40:658–68.

Fone, B. 2000. *Homophobia: A history.* New York: Henry Holt.

Garnets, L., G. M. Herek, and B. Levy. 1990. Violence and victimization of lesbians and gay men: Mental health consequences. *Journal of Interpersonal Violence* 5:366–383.

Gay, Lesbian, and Straight Education Network. 2000. *Homophobia 101: Teaching respect for all.* Gay, Lesbian, and Straight Education Network. Available at http:www.glsen.org.

Gilman, S. E., S. D. Cochran, V. M. Mays, M. Hughes, D. Ostrow, and R. C. Kessler. 2001. Risk of psychiatric disorders among individuals reporting same-sex sexual partners in the National Comorbidity Survey. *American Journal of Public Health* 91:933–39.

Golombok, S., B. Perry, A. Burston, C. Murray, J. Money-Somers, M. Stevens, and J. Golding. 2003. Children with lesbian parents: A community study. *Developmental Psychology* 39:20–33.

Gottman, J. M., R. W. Levenson, C. Swanson, K. Swanson, R. Tyson, and D. Yoshimoto. 2003. Observing gay, lesbian, and heterosexual couples' relationships: Mathematical modeling of conflict interaction. *Journal of Homosexuality* 45:65–91.

Green, J. C. 2004. *The American religious landscape and political attitudes: a baseline for 2004.* Pew Forum on Religion and Public Life. Available at http://pewforum.org.

Green, R. J., J. Bettinger, and E. Sacks. 1996. Are lesbian couples fused and gay male couples disengaged? In *Lesbians and gays in couples and families,* edited by J. Laird and R. J. Green. San Francisco: Jossey-Bass, 185–230.

Grov, C., D. S. Bimbi, J. E. Nanin, and J. T. Parsons. 2006. Race, ethnicity, gender and generational factors associated with the coming-out process among gay, lesbian, and bisexual individuals. *The Journal of Sex Research* 43: 115–122

Harris Interactive and GLSEN. 2005. *From teasing to torment: School climate in America.* New York: GLSEN (Gay, Lesbian, and Straight Education Network).

Held, Myka. 2005 (March 16). *Mix it up: T-shirts and activism.* Available at http://www.tolerance.org/teens.

Herek, G. M. 2002a. Heterosexuals' attitudes toward bisexual men and women in the United States. *The Journal of Sex Research* 39:264–74.

Holthouse, D. 2005 (Spring). Curious cures. *Intelligence Report* 117:14.

Howard, J. 2006 (March). *Expanding resources for children: Is adoption by gays and lesbians part of the answer for boys and girls who need homes?* New York: Evan B. Donaldson Adoption Institute.

Human Rights Campaign. 2000. *Feeling free: Personal stories—How love and self-acceptance saved us from "ex-gay" ministries.* Washington, DC: Human Rights Campaign Foundation.

Human Rights Campaign. 2004. *Resource guide to coming out.* Washington, DC: Human Rights Campaign Foundation.

Human Rights Campaign. 2005. *The state of the workplace for lesbian, gay, bisexual, and transgendered Americans, 2004.* Washington, DC: Human Rights Campaign. Available at http://www.hrc.org.

Human Rights Campaign. 2006. *Massachusetts decision disappointing but progress continues.* Press release. Available at http://www.hrc.org.

Human Rights Campaign. *Marriage/relationship recognition laws: International.* Available at http://www.hrc.org.

International Gay and Lesbian Human Rights Commission. 1999. *Antidiscrimination Legislation.* Available at http://www.iglhrc.org/news/factsheets/990604-antidis.html.

International Gay and Lesbian Human Rights Commission. 2004 (January 30). *IGLHRC calls for global mobilization to help pass the United Nations Resolution on Sexual Orientation and Human Rights.* Available at http://www.iglhrc.org.

_____ 2003b. *Where You Can Marry: Global Summary of Registered Partnership, Domestic Partnership and Marriage Laws.* Available at http://www.iglhrc.org.

Israel, T. and J. J. Mohr. 2004. Attitudes toward bisexual women and men: Current research, future directions. In Ronald C. Fox (Ed.), *Current research on bisexuality,* edited by R. C. Fox. New York: Harrington Park Press, 117–34.

Johnston, E. 2005 (May 5). *Massachusetts releases data on same-sex marriages.* PlanetOut. Available at http://www.planetout.com.

Kinsey, A. C., W. B. Pomeroy, and C. E. Martin. 1948. *Sexual behavior in the human male.* Philadelphia: Saunders.

Kinsey, A. C., W. B. Pomeroy, C. E. Martin, and P. H. Gebhard. 1953. *Sexual behavior in the human female.* Philadelphia: Saunders.

Kirkpatrick, R. C. 2000. The evolution of human sexual behavior. *Current Anthropology* 41:385–414.

Kurdek, L. A. 1994. Conflict resolution styles in gay, lesbian, heterosexual nonparent, and heterosexual parent couples. *Journal of Marriage and the Family* 56:705–22.

_____. 2004. Gay men and lesbians: The family context. In *Handbook of contemporary families: Considering the past, contemplating the future,* edited by M. Coleman and L. H. Ganong. Thousand Oaks, CA: Sage, 96–115.

_____. 2005. What do we know about gay and lesbian couples? *Current Directions in Psychological Science* 14:251–54.

_____ 2006 Differences between partners from heterosexual, gay, and lesbian cohabiting couples. *Journal of Marriage and Family* 68: 509–528

Lambda Legal and Deloitte Financial Advisory Services LLP. 2006 (April). *2005 Workplace Fairness Survey*. Available at http:www.lambdalegal.org.

Landis, D. 1999 (February 17). Mississippi Supreme Court made a tragic mistake in denying custody to gay father, experts say. *American Civil Liberties Union News*. Available at http://www.aclu.org.

LAWbriefs. 2005 (April). Recent developments in sexual orientation and gender identity law. *LAWbriefs* 7(1).

Lever, J. 1994. The 1994 Advocate survey of sexuality and relationships: The men. *The Advocate* 23 August, 16–24.

Lewis, G. B. 2003. Black-white differences in attitudes toward homosexuality and gay rights. *Public Opinion Quarterly* 67:59–78.

Loftus, J. 2001. America's liberalization in attitudes toward homosexuality, 1973 to 1998. *American Sociological Review* 66:762–82.

Luther, S. 2006 (March). *Domestic partner benefits*. Human Rights Campaign. Available at http://www.hrc.org.

McKinney, J. 2004 (February 8). *The Christian case for gay marriage*. Pullen Memorial Baptist Church. Available at http://www.pullen.org.

McLean, K. 2004. Negotiating (non)monogamy: Bisexuality and intimate relationships. In *Current research on bisexuality*, edited by R. C. Fox. New York: Harrington Park Press, 82–97.

Michael, R. T., J. H. Gagnon, E. O. Laumann, and G. Kolata. 1994. *Sex in America: A definitive survey*. Boston: Little, Brown.

Mohipp, C. and M. M. Morry. 2004. Relationship of symbolic beliefs and prior contact to heterosexuals' attitudes toward gay men and lesbian women. *Canadian Journal of Behavioral Science* 36:36–44.

Moltz, K. 2005 (April 13). *Testimony of Kathleen Moltz*. Testimony given before the United States Senate Judiciary Committee, Subcommittee on the Constitution, Civil Rights and Property Rights. Human Rights Campaign. Available at http://www.hrc.org.

National Center for Lesbian Rights. 2003. *Second-parent adoptions: A snapshot of current law*. Available at http://www.nclrights.org.

National Coalition of Anti-Violence Programs. 2005. *2004 National hate crimes report: Anti-lesbian, gay, bisexual and transgender violence in 2004*. New York: National Coalition of Anti-Violence Programs.

National Gay and Lesbian Task Force. 2004 (June). *Anti-Gay parenting laws in the U.S.* National Gay and Lesbian Task Force. Available at http://www.thetaskforce.org .

National Gay and Lesbian Task Force. 2005a. *The Issues: Seniors*. National Gay and Lesbian Task Force. Available at http://www.thetaskforce.org.

National Gay and Lesbian Task Force. 2005b. *Second-parent adoption in the U.S.* National Gay and Lesbian Task Force. Available at http://www.thetaskforce.org.

National Gay and Lesbian Task Force. 2005–2006. *Marriage and partnership recognition*. Available at http://www.thetaskforce.org.

The National Gay and Lesbian Task Force. 2006. *State nondiscrimination laws in the U.S. (as of March 2006)*. Available at http:www.thetaskforce.com.

Newport, F. 2002 (September). *In-depth analysis: Homosexuality*. Gallup Organization. Available at http://www.gallup.com

Ochs, R. 1996. Biphobia: It goes more than two ways. In *Bisexuality: The psychology and politics of an invisible minority*, edited by B. A. Firestein. Thousand Oaks, CA: Sage, 217–39.

Otis, M. D., S. S. Rostosky, E. D. B. Riggle, and R. Hamrin. 2006. Stress and relationship quality in same-sex couples. *Journal of Social and Personal Relationships*. 23:81–99.

Page, S. 2003. Gay rights tough to sharpen into political "wedge issue." *USA Today* 28 July, 10A.

Palmer, R., and Bor, R. 2001. The challenges to intimacy and sexual relationships for gay men in HIV serodiscordant relationships: A pilot study. *Journal of Marital and Family Therapy* 27:419–31.

Paul, J. P. 1996. Bisexuality: Exploring/exploding the boundaries. In *The lives of lesbians, gays, and bisexuals: Children to adults*, edited by R. Savin-Williams and K. M. Cohen. Fort Worth, TX: Harcourt Brace, 436–61.

Pawelski, J. G., E. C. Perrin, J. M. Foy, C. E. Allen, J. E. Crawford, M. Del Monte, M. Kaufman, J. D. Klein, K. Smith, S. Springer, J. L. Tanner, and D. L. Vickers. 2006. The effects of marriage, civil union, and domestic partnership laws on the health and well-being of children *Pediatrics* 118: 349–355.

Pearcey, M. 2004. Gay and bisexual married men's attitudes and experiences: Homophobia, reasons for marriage, and self-identity. *Journal of GLBT Family Studies* 1:21–42.

Peplau, L. A., R. C. Veniegas, and S. N. Campbell. 1996. Gay and lesbian relationships. In *The lives of lesbians, gays, and bisexuals: Children to adults*, edited by R. C. Savin-Williams and K. M. Cohen. Fort Worth, TX: Harcourt Brace, 250–73.

Pew Research Center. 2006 (March 22). *Less opposition to gay marriage, adoption and military service*. Available at http://people-press.org.

Platt, L. 2001. Not your father's high school club. *American Prospect* 12:A37–39.

Potok, M. 2005 (Spring). Vilification and violence. *Intelligence Report* 117:1.

Puccinelli, M. 2005 (April 19). Students support, decry gays with t-shirts. *CBS 2 Chicago*. Available at http://cbs2chicago.com.

Rankin, S. R. 2003. *Campus climate for gay, lesbian, bisexual, and transgendered people: A national perspective*. New York: National Gay and Lesbian Task Force Policy Institute. Available at http://www.thetaskforce.org

Rosin, H., and R. Morin. 1999 (January 11). In one area, Americans still draw a line on acceptability. *Washington Post National Weekly Edition* 16:8.

Rosario, M., E. W. Schrimshaw, J. Hunter, and L. Braun. 2006. Sexual identity development among lesbian, gay, and bisexual youth: Consistency and change over time. *Journal of Sex Research* 43:46–58.

Saad, L. 2005 (May 20). *Gay rights attitudes a mixed bag.* Gallup Organization. Available at http://www.gallup.com.

Sanday, P. R. 1995. Pulling train. In *Race, class, and gender in the United States,* 3rd ed., edited by P. S. Rothenberg. New York: St. Martin's Press, 396–402.

Savin-Williams, R. C. 2006. Who's gay? Does it matter? *Current Directions in Psychological Science* 15:40–44.

Schiappa, E., P. B. Gregg, and D. E. Hewes. 2005. The parasocial contact hypothesis. *Communication Monographs* 72:92–115.

SIECUS (Sexuality Information and Education Council of the United States). 2000. *Fact sheets: Sexual orientation and identity.* New York: SIECUS.

Simmons, T. and M. O'Connell. 2003 (February). Married-couple and unmarried-partner households: 2000. Washington, DC: U.S. Census Bureau.

Smith, D. M. and G. J. Gates. 2001. *Gay and lesbian families in the United States: Same-sex unmarried partner households—Preliminary analysis of 2000 U.S. Census Data.* Washington, DC: Human Rights Campaign. Available at http://www.hrc.org.

Snorton, R. 2005 (April 1). *GLSEN's 2004 State of the States Report is the first objective analysis of statewide safe schools policies.* Available at http://www.glsen.org.

Stacey, J. and Biblarz, T. J. 2001. (How) does the sexual orientation of parents matter? *American Sociological Review* 66:159–83.

Stone, A. 2006. Drives to ban gay adoption up in 16 states. *USA Today* February 20, 2006. Available at http://www.usatoday.com.

Sullivan, A. 1997. The conservative case. In *Same sex marriage: Pro and con,* edited by A. Sullivan. New York: Vintage Books, 146–54.

Tobias, S. and S. Cahill. 2003. *School lunches, the Wright brothers, and gay families.* National Gay and Lesbian Task Force. Available at http://www.thetaskforce.org.

Tyagart, C. E. 2002. Legal rights to homosexuals in areas of domestic partnerships and marriages: Public support and genetic causation attribution. *Educational Research Quarterly* 25:20–29.

Yvonne. 2004 (December 1). *Devout Christian finds a reason to stand up for equality.* Human Rights Campaign. Available at http://www.hrc.org.

Wagner, C. G. 2006. Homosexual relationships. *Futurist* 40:6.

Wainwright, J., Russell, S. T., and Patterson, C. J. 2004. Psychosocial adjustment, school outcomes, and romantic relationships of adolescents with same-sex parents. *Child Development* 75:1886–98.

Walsh, Kenneth T. 2006 (May 15). And now it's her turn. *U.S. News & World Report* 27

White, M. 2005 (Spring). A thorn in their side. *Intelligence Report* 117:27–30.

Wilcox, C. and R. Wolpert. 2000. Gay rights in the public sphere: Public opinion on gay and lesbian equality. In *The Politics of Gay Rights,* edited by C. A. Rimmerman, K. D. Wald, and C. Wilcox. Chicago: University of Chicago Press, 409–32.

chapter 7

I know why I waited
Waited for so long,
I've been waiting for somebody,
Exactly like you.

Nina Simone, *Exactly Like You*

Mate Selection

Contents

True or False?

1. Persons who participate in a premarital education program show no benefits in relationship quality when compared with nonparticipants.

2. The more individuals have in common, the higher their reported relationship happiness and quality.

3. Persons who end up being in a happy, durable marriage knew each other at least two years before they married.

4. As of 2007, around ten percent of marriages are interracial.

5. Opposites not only attract but have more enduring and happy marriages because they keep a relationship exciting.

Answers: **1.** F   **2.** T   **3.** T   **4.** F   **5.** F

In spite of the fact that most people today delay marriage to complete their education, establish themselves in a career, pay off their debts, and/or enjoy their friends and freedom, more than 95 percent of U.S. adults end up getting married by age 75 (*Statistical Abstract of the United States: 2006*, Table 51). Indeed, 97.7 percent of 1027 undergraduates at a large southeastern university agreed, "Someday, I want to marry" (Knox and Zusman, 2006). No other choice in life is as important as your choice of a marriage partner. Your day-to-day happiness, health, and economic well being will be significantly influenced by the partner with whom you choose to share your life. Most have high hopes and even believe in the perfect mate. An anonymous comic said, "I married Miss Right. . . . I just didn't know her first name was Always." In this chapter we examine the cultural influences and pressures that influence the choice of one's mate, the tendency for individuals to select a partner with similar characteristics, and the effect of psychological factors on mate selection.

Since the heart of this text is about mate choice, we provide an inventory for those thinking about a life together to examine their relationship. We also submit for a couple's consideration the wisdom of calling off the wedding if several factors predictive of an unhappy and unstable relationship are present in their relationship.

Cultural Aspects of Mate Selection

It really doesn't matter who you select to marry as you are sure to find out you married someone else.

Jay Leno

Individuals are not free to marry whomever they please. Indeed, university students routinely assert, "I can marry whomever I want!" Hardly. Rather, their culture and society radically restrict and influence their choice. The best example of mate choice being culturally and socially controlled is the fact that *fewer than 1 percent* of persons marry someone outside their race (*Statistical Abstract of the United States: 2006*, Table 54). In addition, as emphasized in the movie *Brokeback Mountain*, homosexuals may not marry at all (federal law does permit same-sex marriages), and being open about a homosexual relationship is wrought with societal disapproval and danger.

Independent of the sexual orientation of the partners, two forms of cultural pressure operative in mate selection are endogamy and exogamy.

Endogamy

Endogamy is the cultural expectation to select a marriage partner within one's own social group, such as race, religion, and social class. Endogamous pressures involve social approval and encouragement to select a partner within your own group (e.g., someone of your own race and religion) and disapproval for selecting someone outside your own group. The pressure toward an endogamous mate choice is especially strong when race is concerned. Love may be blind, but it knows the color of one's partner.

Exogamy

In addition to the cultural pressure to marry within one's social group, there is also the cultural expectation that one will marry outside his or her family group. This expectation is known as **exogamy.**

Incest taboos are universal; in no society are children permitted to marry the parent of the other sex. In the United States, siblings and (in some states) first cousins are also prohibited from marrying each other. The reason for such restrictions is fear of genetic defects in children whose parents are too closely related.

Once cultural factors have identified the general pool of eligibles, individual mate choice becomes more operative. However, even when individuals feel that they are making their own choices, social influences are still operative.

Sociological Factors Operative in Mate Selection

Numerous sociological factors are at work in bringing two people together who eventually marry.

Homogamy

Whereas endogamy is a concept that refers to cultural pressure, **homogamy** refers to individual initiative toward sameness. The homogamy theory of mate selection states that we tend to be attracted to and become involved with those who are similar to ourselves in such characteristics as age, race, religion, and social class. In general, the more couples have in common, the higher the reported relationship satisfaction and the more durable the relationship (Wilson and Cousins, 2005).

Jepsen and Jepsen (2002) found homogamy operative in both same-sex and opposite-sex couples, whether married or unmarried. Some data suggest (and new

Film writer/director Woody Allen was the target of social disapproval when he violated the principle of exogamy by marrying the adopted daughter of his long-time partner, Mia Farrow.

© AP/Wide World Photos

Marry yourself.

Jack Wright, sociologist

Love is blind, but it sees the color of a person's skin.

Unknown

data confirm [Jepsen and Jepsen, 2002]) that homogamy is more important in selecting a mate than choosing a date (Knox et al., 1997). Two hundred seventy-eight undergraduates were asked to reveal the degree to which it was important to them to (1) date and (2) marry partners who had similar characteristics (race, religion, education, age, etc.), on a ten-point continuum from 0 (not important) to 10 (very important). The respondents averaged 7.4 per item when asked about the importance of a future mate (in contrast to a mean of 6.8, reflecting importance of similarity in a dating partner). Hence, the more involved the relationship, the greater the importance of finding a partner with whom they had a lot in common. Some of the more salient homogamous factors operative in mate selection include the following.

Race As noted above, racial homogamy operates strongly in selecting a living-together or marital partner (with greater homogamy for marital partners). Almost half (48.9%) of 1027 university students agreed that "it is important for me that I marry someone of my same race" (Knox and Zusman, 2006). Whites were more adamant than blacks about marrying someone of the same race (72.1% vs. 53.3%). Conversely, blacks were more willing to cross racial lines to marry (46.7% vs. 27.9%) (Knox and Zusman, 2006). Some explanations for this racial difference are that there are more benefits to blacks if they join the majority than vice versa, a greater number of whites are available to blacks than vice versa, and blacks have greater exposure to white culture than vice versa. Finally, black mothers and white fathers have different roles in the respective black and white communities in terms of setting the norms of interracial relationships. Hence, the black mother who approves of her son's or daughter's interracial relationship may be less likely to be overruled than the white mother.

Race may also affect one's perceptions. DeCuzzi et al (2006) found in their study of racial perceptions among college students, that while both races tended to view women and men of their own and the other race positively, there was a pronounced tendency to view women and men of their own race more positively and members of the other race more negatively.

Although most couples are open to interracial dating (this couple lives together), fewer are open to getting married.

Authors

Age   Most individuals select someone who is relatively close in age. Men tend to select women three to five years younger than themselves. The result is the "marriage squeeze," which is the imbalance of the ratio of marriageable-age men to marriageable-age women. In effect, women have fewer partners to select from since men choose from not only their same age group but also those younger than themselves. One 40-year-old recently divorced woman said, "What chance do I have with all these guys looking at these younger women?"

Education   Educational homogamy also operates strongly in selecting a living-together and marital partner (with greater homogamy for marital partners) (Kalmijn and Flap, 2001). Not only does college provide an opportunity to meet, date, live with, and marry another college student, but it also increases one's chance that only a college-educated partner becomes acceptable as a potential cohabitant or spouse. The very pursuit of education becomes a value to be shared. However, Lewis and Oppenheimer (2000) observed that when persons of similar education are not available, women are particularly likely to marry someone with less education. And the older the woman, the more likely she is to marry a partner with less education. In effect, the number of educated eligible males may decrease as she ages.

Open-Mindedness

People vary in the degree to which they are open-minded. The Self-Assessment on page 198 allows you to assess your open-mindedness. You might ask your partner to take the same assessment and compare your scores and views.

Social Class   You have been reared in a particular social class that reflects your parents' occupations, incomes, and educations as well as your residence, language, and values. If you were brought up in a home in which both parents were physicians, you probably lived in a large house in a nice residential area—summer vacations and a college education were givens. Alternatively, if your parents dropped out of high school and worked at Wal-Mart, your home would be smaller, in a less expensive part of town, and your opportunities (e.g., education) would be more limited. Social class affects one's comfort in interacting with others—we tend to feel more comfortable with others from our same social class.

The **mating gradient** refers to the tendency for husbands to be more advanced than their wives with regard to age, education, and occupational success. Indeed, husbands are typically older than their wives, have more advanced education, and earn higher incomes (*Statistical Abstract of the United States: 2006*).

Physical Appearance   It is commonly understood that physical appearance is important for feeling good about one's self (Delinsky, 2005). Marcus and Miller (2003) noted, "Beauty is quite clearly not entirely in the eye of the beholder. Instead, some of us are judged by almost everyone we meet as handsome or lovely, whereas others of us nearly always seem plain" (p. 333). In their study of physical attractiveness, they found that "people know they are pretty or handsome" (p. 325).

Homogamy is operative in regard to physical appearance in that people tend to become involved with those who are similar in physical attractiveness. However, a partner's attractiveness may be a more important consideration for men than for women. In a study of homogamous preferences in mate selection,

A thing of beauty is a joy for a while.

Hal Lee Luyah

Open-Mindedness Scale

The purpose of this scale is to assess the degree to which you are open-minded. Open-mindedness is one's receptiveness to arguments, ideas, suggestions, and opinions. As such, someone who is open-minded typically does not prejudge or have preconceptions of others and is receptive and tolerant of new information. After reading each statement, select the number that best reflects your answer, using the following scale:

1	2	3	4	5	6	7

Strongly
Disagree

Strongly
Agree

_____ 1. I am an open-minded person.

_____ 2. I like diversity of thoughts and ideas.

_____ 3. I think knowledge of different viewpoints is the only way to find the truth.

_____ 4. I have often changed my mind about something after reading more about it.

_____ 5. I consider more than one point of view before taking a stand on an issue.

_____ 6. It really bothers me if someone makes fun of another person's idea.

_____ 7. I am willing to explore a different point of view even if I do not agree with it.

_____ 8. I am willing to really listen to any point of view.

_____ 9. I evaluate information on the basis of its merit rather than my emotional reaction to it.

_____ 10. I try to read about both sides of an issue before I form an opinion.

Scoring

Selecting a 1 reflects the least open-mindedness; selecting a 7 reflects the greatest open-mindedness. Add the numbers you assigned to each item. The lower your total score (10 is the lowest possible score), the more closed-minded you are; the higher your total score (70 is the highest possible score), the greater your open-mindedness. A score of 40 places you at the midpoint between being very closed-minded and very open-minded.

Scores of Other Students Who Completed the Scale

The scale was completed by 44 male and 81 female students at Valdosta State University. They received course credit for their participation. Their ages ranged from 18 to 46 years, with a mean age of 20.50 (standard deviation [SD] = 4.00). The racial/ethnic background of the sample included 64.0 percent white, 29.6 percent black, 1.6 percent Hispanic, and 4.8 percent other. The college classification level of the sample included 61.6 percent freshmen, 27.2 percent sophomores, 8.0 percent juniors, 1.6 percent seniors, and 0.8 percent post-baccalaureate students. Male participants had higher open-mindedness scores (mean [M] = 57.55; SD = 6.85) than did female participants (M = 54.58; SD = 7.75; $p < .05$). Freshman had lower open-mindedness scores (M = 54.01; SD = 7.50) than did upper-classmen (M = 58.04; SD = 6.94; $p < .05$). There were no significant differences in regard to race.

Source

Open-Mindedness Scale. 2006. Mark Whatley, Ph.D., Department of Psychology, Valdosta State University, Valdosta, Georgia 31698. Used by permission. Other uses of this scale by written permission of Dr. Whatley only. Information on the reliability and validity of this scale is available from Dr. Whatley (mwhatley@valdosta.edu).

men and women rated physical appearance an average of 7.7 and 6.8 (out of 10) in importance, respectively (Knox et al., 1997).

Marital Status   The never-married tend to select as marriage partners the never-married, the divorced tend to select the divorced, and the widowed tend to select the widowed. Similar marital status may be more important to women than to men. In the study of homogamous preferences in mate selection, women and men rated similarity of marital status an average of 7.2 and 6.3 (out of 10) in importance, respectively (Knox et al., 1997).

Religion/Spirituality   Religion may be broadly defined as a specific fundamental set of beliefs (in reference to a supreme being, etc.) and practices generally agreed upon by a number of persons or sects. Similarly, some individuals view themselves as "not religious" but "spiritual," with spirituality defined as belief in the spirit as the seat of the moral or religious nature that guides one's decisions and behavior. Because religious or spiritual views reflect, in large part, who the person is, they have an enormous impact on one's attraction to a partner, the level of emotional engagement with that partner, and the durability of the relationship and marital happiness (Swenson et al., 2005).

Religious homogamy is operative in that persons of similar religion sometimes seek out each other. Almost a third (27%) of undergraduate women and 15 percent of undergraduate men in one study reported that they would marry only someone of their same religion (Knox, Zusman, and Daniels, 2002). Baptists were significantly more likely than Methodists or Catholics to believe that persons marrying outside their religion would eventually divorce.

Couples with religious homogamy may have greater marital stability and a lower chance of divorce because of the value of religion for resolving conflicts. Butler et al. (2002) found that couples who prayed about their marital conflicts reported a softening of their positions and a feeling of "healing." Indeed, the authors of the study suggested that marriage counselors should be aware of this spiritual source of healing.

The phrase "the couple that prays together, stays together" is more than just a cute cliché. However, it should also be pointed out that religion could serve as a divisive force. For example, when one partner becomes "born again" or "saved," unless the other partner shares the experience, the relationship can be dramatically altered and eventually terminated. A former rock-and-roll, hard-drinking, drug-taking wife noted that when her husband "got saved," it was the end of their marriage. "He gave our money to the church as the 'tithe' when we couldn't even pay the light bill," she said. "And when he told me I could no longer wear pants or lipstick, I left." Another example of how religious disharmony has a negative effect on relationships is the marriage of Ted Turner and Jane Fonda. Soon after she became a "Christian" and was "saved," she and Turner split up. In an interview, Fonda noted that she feared telling her husband about her religious conversion because she knew it would be the end of their 8-year marriage.

I don't believe in the afterlife, although I'm bringing a change of underwear.

Woody Allen

Status   When status is defined in terms of resources (education, income, power to influence others), Fu (2006) found that the higher a person's status, the more choices the person has in terms of potential partners. Not only does the high-status person have more choices for a marriage partner, high status gives an individual more freedom in dissolving a marriage. Hence, one's status is associated with power in making important decisions in family life.

Attachment   Individuals who report similar levels of attachment to each other report high levels of relationship satisfaction. This is the conclusion of Luo and Klohnen (2005), who studied 291 newlyweds to assess the degree to which similarity affected marital quality. Indeed, similarity of attachment was *the* variable most predictive of relationship quality.

Personality   Conservatives, liberals, and risk-takers tend to select each other as marital partners, and their doing so has positive consequences for them and their relationship. Partners who select others with similar personalities report high subjective well being (Arrindell and Luteijn, 2000).

Psychological Factors Operative in Mate Selection

Psychologists have focused on complementary needs, exchanges, parental characteristics, and personality types with regard to mate selection.

University Women Who Want a Traditional Husband

RESEARCH APPLICATION

A theme permeating the media in regard to relationships is that of equality. Women presumably seek a "modern" relationship in which the spouse is equal in terms of career status and division of labor. But do they? To what degree do women seek traditional husbands who view their primary role as provider and who are supportive of their wife staying at home to rear the children? This study sought to identify the degree to which undergraduate women at a large southeastern university seek a traditional husband and to identify the various background characteristics of these women.

Sample and Methods

To find out, researchers analyzed data from a sample of 692 undergraduate women at a large southeastern university who answered "yes" (30.9%) or "no" (69.1%) to the question, "As a female, I prefer to marry a traditional man who will be the provider and be supportive of my staying at home to rear the children." The sample was 80.8 percent white and 19.2 percent black, and the median age was 19 years. More than half (50.7%) were first-year students; 24.6 percent, sophomores; 14.6 percent, juniors, and 10.0 percent, seniors. In regard to current relationships, 43.4 percent were not dating anyone or were casually dating different people, whereas 56.6 percent were emotionally committed or involved.

Selected Findings and Conclusions

As noted, more than 30 percent (30.9%) of the 692 undergraduate women surveyed reported that they wanted to marry a traditional husband; more than two-thirds (69.1%) did not want to do so. Analysis of the data revealed five statistically significant findings in regard to the characteristics of the undergraduate women who wanted a traditional husband versus those who did not.

1. *Valued happy marriage over financial security or career.* Indicating their top three values, more than 40 percent (40.8%) of the women in this sample identified "having a happy marriage" in life, followed by "financial security" (27.1%) and "having a career I love" (17%). This finding came as no surprise, as we would expect women who prefer to marry a traditional man to value a happy marriage first and a career last. What did come as a surprise was that fewer than 20 percent (17%) of these women valued having a career as their top priority. Is the Women's Movement, with its emphasis on financial independence, losing support from women who re-evaluate being married to a man who makes a good income and takes care of his wife and children financially? Are feminists discovering that emphasizing a career leaves them empty? Did these undergraduate women have mothers who were devoted to their careers, did they feel cheated and abandoned by their mothers when they were growing up, and have they vowed to stay at home and take care of their own children?

2. *Believed children turn out better in traditional family.* One might expect women who preferred a traditional husband to be motivated by the desire to stay at home to rear their children and to believe that doing so would be beneficial to their children. Analysis of the data revealed this expectation to be true in that more than half (52.7%) of the women answered "yes" to the statement "Children turn out better when one parent stays home to take care of them"; in comparison, around ten percent (12.4%) answered "no."

3. *Against cohabitation.* A final characteristic of being conservative and having traditional values, as evidenced by the women in this sample, is their reluctance to cohabit. Almost 40 percent (38.3%) of those preferring a traditional husband reported that they would not live with a man before marriage, compared with the 27.6 percent who reported they would.

4. *White.* Almost exactly a third (33.4%) of the women who preferred a traditional husband were white, compared with the 18.6 percent who self-identified as black. Hence, white women were 14.8 percent more likely to prefer a traditional, provider husband than black women. This finding may say more about black women than white women. Because black women tend to have more education than black men and are more likely to come from matriarchal (female–dominated) homes, they may feel less need to have a man take care of them.

There are two implications of this research study for mate selection. (1) Although most undergraduate women in this sample preferred a "modern" man, 30 percent did not and were explicit about wanting a traditional partner to earn the income and be supportive of their desire to stay home and rear children. (2) Women who are looking for a traditional man have other traditional values: they are focused on a happy relationship rather than money, they feel that children flourish best with a stay-at-home mother, and they are against cohabitation.

Source

McGinty, K., D. Knox, and M. E. Zusman. 2006. *Traditional Husband? Traditional characteristics of university women who want one.* Research study conducted for this text. Greenville, NC: Department of Sociology, East Carolina University.

Complementary-Needs Theory

"In spite of the Women's Movement and a lot of assertive friends, I am a shy and dependent person," remarked a transfer student. "My need for dependency is met by Warren, who is the dominant, protective type." The tendency for a submissive person to become involved with a dominant person (one who likes to control the behavior of others) is an example of attraction based on complementary needs.

Complementary-needs theory states that we tend to select mates whose needs are opposite and complementary to our own. Partners can also be drawn to each other on the basis of nurturance versus receptivity. These complementary needs suggest that one person likes to give and take care of another, while the other likes to be the benefactor of such care. Other examples of complementary needs may involve responsibility versus irresponsibility, peacemaker versus troublemaker, and disorder versus order. The idea that mate selection is based on complementary needs was suggested by Winch (1955), who noted that needs can be complementary if they are different (e.g., dominant and submissive) or if the partners have the same need at different levels of intensity.

As an example of the latter, two individuals may have a complementary relationship if they both want to pursue graduate studies but want to earn different degrees. The partners will complement each other if, for instance, one is comfortable with aspiring to a master's degree and approves of the other's commitment to earning a Ph.D.

Winch's theory of complementary needs, commonly referred to as "opposites attract," is based on the observation of 25 undergraduate married couples at Northwestern University. The findings have been criticized by other researchers who have not been able to replicate Winch's study (Saint, 1994). Two researchers said, "It would now appear that Winch's findings may have been an artifact of either his methodology or his sample of married people" (Meyer and Pepper, 1977).

Three questions can be raised about the theory of complementary needs:

1. *Couldn't personality needs be met just as easily outside the couple's relationship as through mate selection?* For example, couldn't a person who has the need to be dominant find such fulfillment in a job that involved an authoritative role, such as being a supervisor?

2. *What is a complementary need as opposed to a similar value?* For example, is the desire to achieve at different levels a complementary need or a shared value?

3. *Don't people change as they age?* Could a dependent person grow and develop self-confidence so that he or she might no longer need to be involved with a dominant person? Indeed, the person might no longer enjoy interacting with a dominant person.

Exchange Theory

Exchange theory emphasizes that mate selection is based on assessing who offers the greatest rewards at the lowest cost. Five concepts help to explain the exchange process in mate selection.

1. *Rewards.* Rewards are the behaviors (your partner looking at you with the eyes of love), words (saying "I love you"), resources (being beautiful or handsome, having a car, condo, and money), and services (cooking for you, typing for you) your partner provides that you value and that influence you to continue the relationship. Increasingly, men are interested in women who offer "financial independence." In a study of Internet ads placed by women, the woman who described herself as "financially independent . . . successful and ambitious" produced 50 percent more responses than the next most popular ad, in which the woman described herself as "lovely . . . very attractive and slim" (Strassberg and Holty, 2003).

No matter how carefully one chooses a mate, there will always be qualities that the mate has that simply don't fit well.

Neil Jacobson and Andrew Christensen, psychologists

The best relationships are always win–win.

Jack Turner, psychologist

2. *Costs*. Costs are the unpleasant aspects of a relationship. A woman identified the costs associated with being involved with her partner: "He abuses drugs, doesn't have a job, and lives nine hours away." The costs her partner associated with being involved with this woman included "she nags me," "she doesn't like sex," and "she wants her mother to live with us if we marry." Ingoldsby et al. (2003) assessed the degree to which various characteristics were associated with reducing one's attractiveness on the marriage market. The most damaging traits were not being heterosexual, having alcohol/drug problems, having a sexually transmitted disease, and being lazy.

3. *Profit*. Profit occurs when the rewards exceed the costs. Unless the couple referred to above derive a profit from staying together, they are likely to end their relationship and seek someone else with whom there is a higher profit margin.

4. *Loss*. Loss occurs when the costs exceed the rewards.

5. *Alternative*. Is another person currently available who offers a higher profit margin?

Most people have definite ideas about what they are looking for in a mate. For example, Xie et al. (2003) found that men with good incomes were much more likely to marry than men with no or low incomes. The currency used in the marriage market consists of the socially valued characteristics of the persons involved, such as age, physical characteristics, and economic status. In our free choice system of mate selection, we typically get as much in return for our social attributes as we have to offer or trade. An unattractive, drug-abusing high school dropout with no job has little to offer an attractive, drug-free, 3.5 GPA college student who has just been accepted to graduate school.

Once you identify a person who offers you a good exchange for what you have to offer, other bargains are made about the conditions of your continued relationship. Waller and Hill (1951) observed that the person who has the least interest in continuing the relationship could control the relationship. This **principle of least interest** is illustrated by the woman who said, "He wants to date me more than I want to date him, so we end up going where I want to go and doing what I want to do." In this case, the woman trades her company for the man's acquiescence to her recreational choices.

Parental Characteristics

Whereas the complementary-needs and exchange theories of mate selection are relatively recent, Freud suggested that the choice of a love object in adulthood represents a shift in libidinal energy from the first love objects—the parents. Role theory and modeling theory emphasize that a son or daughter models after the parent of the same sex by selecting a partner similar to the one the parent selected.

This means that a man looks for a wife who has similar characteristics to those of his mother and that a woman looks for a husband who is very similar to her father.

Elimidate II: an In-Class Exercise

Based on the popular *Elimidate* television program, Elimidate II is a way for students to "see" the courtship process in class. Six students participate in the in-class exercise, whereby one man interviews five women, selects one, and goes out on a "real" date with his selection (Knox and McGinty, 2004). (Alternative versions of the *Elimidate II* include one women selecting from five men or same-sex partners selecting each other). Lecture/discussion of the "dating/mate selection process" precedes the exercise; *Elimidate* reflects traditional gender roles in society, the high probability of being rejected, and how homogamy operates in the selection of a partner.

The participants are recruited from the class. Participants must be able to take rejection (each female participant has an 80% chance of being rejected), uninvolved in a current relationship, and open to dating or involvement with persons of any racial/ethnic/religious background. No extra credit is awarded for playing the game. The first six people who e-mail their instructor (and specify agreement to the terms) will be included in the game. (The format involves a man selecting from among women because in past classes there were never enough volunteers for one woman to select from five men.)

Rules of the game include the following.

1. Participants become involved in a question-and-answer session in front of the class, during which the "dater" asks each of five potential dates two questions. One question is generated by the dater and another by class members or the instructor. Each date candidate may also ask the dater one question and may be rejected before he makes his selection. The dater ultimately selects the woman he would like to take out on a real date.

2. The couple goes out to dinner the following weekend (each pays his or her own way). They agree to no alcohol and no sex.

3. Both participants report to the class about their experience on the date. Each does so alone, while the other waits outside the classroom.

Advantages reported by students include "fun," "seeing guys answer questions asked by girls," and "learning how to ask questions." Disadvantages include the game doesn't last long enough, some of the questions can get too personal, and no one likes to get rejected.

Desired Personality Characteristics for a Potential Mate

In a study of 700 undergraduates, both men and women reported that the personality characteristics of being warm, kind, and open and having a sense of humor were very important to them in selecting a romantic/sexual partner. Indeed, these intrinsic personality characteristics were rated as more important than physical attractiveness or wealth (extrinsic characteristics) (Sprecher and Regan, 2002). Similarly, adolescents wanted intrinsic qualities such as intelligence and humor in a romantic partner but looked for physical appearance and high sex drive in a casual partner (no gender-related differences in responses were found) (Regan and Joshi, 2003).

In another study, women were significantly more likely than men to identify "having a good job" and "being well educated" as important attributes in a future mate, whereas significantly more men than women wanted their spouse to be physically attractive (Medora et al., 2002). Toro-Morn and Sprecher (2003) noted that both American and Chinese undergraduate women identified characteristics associated with status (earning potential, wealth) more than physical attractiveness.

The behavior that 60 percent of a national sample of adult single women reported as the most serious fault of a man was his being "too controlling" (Edwards, 2000). Women are also attracted to men who have good manners. In a study of 398 undergraduates, women were significantly more likely than men to report that they wanted to "date or be involved with [only] someone who had good manners," that "manners are very important," and that "the more well mannered the person, the more I like the person" (Zusman et al. 2003). Twenge (2006) noted that good manners are being displayed less often.

Personality Characteristics Predictive of Divorce

Researchers have defined several personality factors predictive of a divorce or unfulfilling relationship (Wilson and Cousins, 2005; Gattis et al., 2004).

An intelligent wife sees through her husband; an understanding wife sees him through.
Unknown

I've met a lot of hard-boiled eggs in my time, but you—you're 20 minutes.
Jan Sterling, *The Big Carnival*

1. *Disagreeable/low positives.* Gattis et al. (2004) studied 132 distressed couples seeking treatment and found that the personality characteristics of "low agreeableness" and "low positive expressions" were associated with their not getting along. Hence, partners who always find something to argue about and who find few opportunities to make positive observations or expressions should be considered with caution.

2. *Poor impulse control.* Persons who have poor impulse control have little self-restraint and may be prone to aggression and violence (Snyder and Regts, 1990). Lack of impulse control is also problematic in marriage because the person is less likely to consider the consequences of his or her actions. For example, to some people, having an affair might seem harmless but in most cases it will have devastating consequences for the spouses and their marriage.

3. *Hypersensitivity.* Hypersensitivity to perceived criticism involves getting hurt easily. Any negative statement or criticism is received with a greater impact than intended by the partner. The disadvantage of such hypersensitivity is that the partner may learn not to give feedback for fear of hurting the hypersensitive partner. Such lack of feedback to the hypersensitive partner blocks information about what the person does that upsets the other and what he or she could do to make things better. Hence, the hypersensitive one has no way of learning that something is wrong, and the partner has no way of alerting the hypersensitive partner. The result is a relationship in which the partners can't talk about what is wrong, so the potential for change is limited (Snyder and Regts, 1990).

4. *Inflated ego.* An exaggerated sense of oneself is another way of saying the person has a big ego and always wants things to be his or her way. A person with an inflated sense of self may be less likely to consider the other person's opinion in negotiating a conflict and prefer to dictate an outcome. Such disrespect for the partner can be damaging to the relationship (Snyder and Regts, 1990).

5. *Being neurotic.* Such individuals are perfectionists and require of themselves and others that they be perfect. This attitude is associated with relationship problems (Haring et al., 2003).

6. *Anxiety.* Husbands who report high levels of anxiety tend to report lower marital adjustment. No such association holds for wives. Although the explanation is not clear, the authors of one study postulated, "anxious husbands may have more negative interactions with their wives, which may also lower perceived marital quality" (Dehle and Weiss, 2002, p. 336).

7. *Insecurity.* Feelings of insecurity also compromise marital happiness. Researchers studied the personality trait of attachment and its effect on marriage in 157 couples at two time intervals and found that "insecure participants reported more difficulties in their relationships. . . [I]n contrast, secure participants reported greater feelings of intimacy in the relationship at both assessments (Crowell et al., 2002)

8. *Controlled.* Individuals who are controlled by their parents, grandparents, former partner, child, or whomever compromise the marriage relationship because their allegiance is external to the couple's relationship. Unless the person is able to break free of such control, the ability to make independent decisions will be thwarted, which will both frustrate the spouse and challenge the marriage.

Table 7.1 reflects some particularly troublesome personality types and how they may impact you negatively.

To summarize, a number of personality factors may be predictive of negative outcomes in marriage—being argumentative, poor impulse control, hypersensitivity, overly inflated ego, being neurotic, perfectionism, high anxiety, insecurity, and being controlled by others. Figure 7.1 summarizes the cultural, sociological, and psychological filters involved in mate selection.

Neurotics worry about things that didn't happen in the past instead of worrying like normal people about things that won't happen in the future.

Unknown

Table 7.1 Personality Types Problematic in a Potential Partner

Type	Characteristics	Impact on You
Paranoid	Suspicious, distrustful, thin-skinned, defensive	You may be accused of everything.
Schizoid	Cold, aloof, solitary, reclusive	You may feel that you can never "connect" and that this person is not capable of returning your love.
Borderline	Moody, unstable, volatile, unreliable	You will never know what your Jekyll-and-Hyde partner will be like.
Antisocial	Deceptive, untrustworthy, no conscience, remorseless	This person could cheat on you, lie, or steal from you and not feel guilty.
Narcissistic	Egotistical, demanding, greedy, selfish	This person views you only in terms of your value to him or her; don't expect him or her to see anything from your point of view.
Dependent	Helpless, weak, clingy, insecure	This person will demand your full time and attention, and other interests will incite jealousy.
Obsessive-compulsive	Rigid, inflexible	This person has rigid ideas about how you should think and behave and may try to impose them on you.

Cultural Filters

For two people to consider marriage to each other,

Endogamous factors   and   Exogamous factors
(same race, age)   ↓   (not blood-related)
                    must be met.
                         ↓

After the cultural prerequisites have been satisfied, sociological and psychological filters become operative.

Sociological Filters

Propinquity = the tendency to select a mate from among those who live, work, or go to school nearby.

Homogamy = the tendency to select a mate similar to oneself with regard to the following:

Race	Physical appearance
Education	Body clock compatibility
Social class	Religion
Age	Marital status
Intelligence	Interpersonal values

Psychological Filters

Complementary needs
Reward-cost ratio for profit
Parental characteristics
Desired personality characteristics

Figure 7.1
Cultural, Sociological, and Psychological Filters in Mate Selection

PERSONAL CHOICES

Who Is the Best Person for You to Marry?

In a study commissioned by *Time* magazine of 465 never-married adults, 78 percent of the women and 79 percent of the men thought they would find and marry their perfect mate (Edwards, 2000). Although there is no perfect mate, some individuals are more suited to you as a marriage partner than others. As we have seen in this chapter, persons who have a big ego, poor impulse control, and an oversensitivity to criticism and who are anxious and neurotic should be considered with great caution.

Equally as important as avoiding someone with problematic personality characteristics is selecting someone with whom you have a great deal in common. "Marry someone just like you" may be a worthy guideline in selecting a marriage partner. Homogamous matings with regard to race, education, age, values, religion, social class, and marital status (e.g., never-marrieds marry never-marrieds; divorced persons with children marry those with similar experience) tend to result in more durable, satisfying relationships. "Marry your best friend" is another worthy guideline for selecting the person you marry.

Finally, marrying someone with whom you have a relationship of equality and respect is associated with marital happiness. Relationships in which one partner is exploited or intimidated engender negative feelings of resentment and distance. One man said, "I want a co-chair, not a committee member, for a mate." He was saying that he wanted a partner to whom he related as an equal.

Reference

Edwards, T. M. 2000. Flying solo. *Time* August, 47–53.

Sociobiological Factors Operative in Mate Selection

In contrast to cultural, sociological, and psychological aspects of mate selection, which reflect a social learning assumption, the sociobiological perspective suggests that biological/genetic factors may be operative in mate selection.

Definition of Sociobiology

Sociobiology suggests a biological basis for all social behavior—including mate selection. Based on Charles Darwin's theory of natural selection, which states that the strongest of the species survive, sociobiology holds that men and women select each other as mates on the basis of their innate concern for producing offspring who are most capable of surviving.

According to sociobiologists, men look for a young, healthy, attractive, sexually conservative woman who will produce healthy children and who will invest in taking care of the children. Women, in contrast, look for an ambitious man with good economic capacity who will invest his resources in her children. Earlier in this chapter, we provided data supporting the idea that men seek attractive women and women seek ambitious, financially successful men.

Criticisms of the Sociobiological Perspective

The sociobiological explanation for mate selection is controversial. Critics argue that women may show concern for the earning capacity of men because women have been systematically denied access to similar economic resources, and selecting a mate with these resources is one of their remaining options. In addition, it is argued that both women and men, when selecting a mate, think about their partners more as companions than as future parents of their offspring.

Engagement

Engagement moves the relationship of a couple from a private love-focused experience to a public experience. Family and friends are invited to enjoy the happiness and commitment of the individuals to a future marriage. Unlike casual dating, engagement is a time in which the partners are emotionally committed, are sexually monogamous, and are focused on wedding preparations. The engagement period is your last opportunity before marriage to systematically examine your relationship, ask each other specific questions, find out about the partner's parents and family background, and participate in marriage education or counseling.

Asking Specific Questions

Because partners might hesitate to ask for or reveal information that they feel will be met with disapproval during casual dating, the engagement is a time to get specific about the other partner's thoughts, feelings, values, goals, and expectations. The Involved Couple's Inventory is designed to help individuals in committed relationships learn more about each other by asking specific questions.

Visiting Your Partner's Parents

Seize the opportunity to discover the family environment in which your partner was reared and consider the implications for your subsequent marriage. When visiting your partner's parents, observe their standard of living, the way they relate to each other, and the degree to which your partner is similar to the same-sex parent. How does their standard of living compare with that of your own family? How does the emotional closeness (or distance) of your partner's family compare with that of your family? Such comparisons are significant because both you and your partner will reflect your respective home environments to some degree.

Visiting in the home of one's intended spouse can be instructive. For example, the girl should focus on how her boyfriend's father treats his mother, as this is the way her boyfriend is likely to treat her. Likewise, the boy should look at how the girl's mother treats her father, as this is how the girlfriend is likely to treat him.

An engagement is a short period lacking in foresight, followed by a long period loaded with hindsight.
Unknown.

© Universal/Photofest

Involved Couple's Inventory

The following questions are designed to increase your knowledge of how you and your partner think and feel about a variety of issues. Assume that you and your partner have considered getting married. Each partner should ask the other the following questions.

Partner Feelings and Issues

1. If you could change one thing about me, what would it be?

2. On a scale of 0 to 10, how well do you feel I respond to criticism or suggestions for improvement?

3. What would you like me to say or not say that would make you happier?

4. What do you think of yourself? Describe yourself with three adjectives.

5. What do you think of me? Describe me with three adjectives.

6. What do you like best about me?

7. On a scale of 0 to 10, how jealous do you think I am? How do you feel about my level of jealousy?

8. How do you feel about me emotionally?

9. To what degree do you feel we each need to develop and maintain outside relationships so as not to focus all of our interpersonal expectations on each other? Does this include other-sex individuals?

10. Do you have any history of abuse or violence, either as an abused child or adult or as the abuser in an adult relationship?

11. If we could not get along, would you be willing to see a marriage counselor? Would you see a sex therapist if we were having sexual problems?

12. What is your feeling about prenuptial agreements?

13. Suppose I insisted on your signing a prenuptial agreement?

14. To what degree do you enjoy getting and giving a massage?

15. How important is it to you that we massage each other regularly?

16. On a scale of 0 to 10, how emotionally close do you want us to be?

17. How many intense love relationships have you had, and to what degree are these individuals still a part of your life in terms of seeing them or having e-mail or phone contact?

18. Have you been in a living-together relationship with anyone before? Are you open to our living together? What would be your understanding of the meaning of our living together: would we be "finding out more about each other" or would we be "committed to marriage"?

19. What do you want for the future of our relationship? Do you want us to marry? When?

20. On a ten-point scale (0 = very unhappy and 10 = very happy), how happy are you in general? How happy are you about us?

21. How depressed have you been? What made you feel depressed?

Feelings about Parents/Family

1. How do you feel about your mother? Your father? Your siblings?

2. On a ten-point scale, how close are you to your mom, dad, and each of your siblings?

3. How close were your family members to one another? On a ten-point scale, what value do you place on the opinions or values of your parents?

4. How often do you have contact with your father/mother? How often do you want to visit your parents and/or siblings? How often would you want them to visit us? Do you want to spend holidays alone or with your parents or mine?

5. What do you like and dislike most about each of your parents?

6. What do you like and dislike about my parents?

7. What is your feeling about living near our parents? How would you feel about my parents living with us? How do you feel about our parents living with us when they are old and cannot take care of themselves?

8. How do your parents get along? Rate their marriage on a scale of 0 to 10 (0 = unhappy, 10 = happy).

9. To what degree do your parents take vacations alone together? What are your expectations of our taking vacations alone or with others?

10. To what degree did members of your family consult one another on their decisions? To what degree do you expect me to consult you on the decisions that I make?

11. Who was the dominant person in your family? Who had more power? Who do you regard as the dominant partner in our relationship? How do you feel about this power distribution?

12. What "problems" has your family experienced? Is there any history of mental illness, alcoholism, drug abuse, suicide, or other such problems?

13. What did your mother and father do to earn an income? How were their role responsibilities divided in terms of having income, taking care of the children, and managing the household? To what degree do you want a job and role similar to that of the same-sex parent?

Social Issues, Religion, and Children

1. What are your political views? How do you feel about America being in Iraq?

2. What are your feelings about women's rights, racial equality, and homosexuality?

3. To what degree do you regard yourself as a religious/spiritual person? What do you think about religion, a Supreme Being, prayer, and life after death?

4. Do you go to religious services? Where? How often? Do you pray? How often? What do you pray about? When we are married, how often would you want to go to religious services? In what religion would you want our children to be reared? What responsibility would you take to ensure that our children had the religious training you wanted them to have?

5. How do you feel about abortion? Under what conditions, if any, do you feel abortion is justified?

6. How do you feel about children? How many do you want? When do you want the first child? At what intervals would you want to have additional children? What do you see as your responsibility in caring for the children—changing diapers, feeding, bathing, playing with them, and taking them to lessons and activities? To what degree do you regard these responsibilities as mine?

7. Suppose I did not want to have children or couldn't have them. How would you feel? How do you feel about artificial insemination, surrogate motherhood, in vitro fertilization, and adoption?

8. To your knowledge, can you have children? Are there any genetic problems in your family history that would prevent us from having normal children? How healthy (mentally and physically) are you? How often have you seen a physician in the last three years? What medications have you taken or do you currently take? What are these medications for? Have you seen a therapist/psychologist/psychiatrist? What for?

9. How should children be disciplined? Do you want our children to go to public or private schools?

10. How often do you think we should go out alone without our children? If we had to decide between the two of us going on a cruise to the Bahamas alone or taking the children camping for a week, what would you choose?

11. What are your expectations of me regarding religious participation with you and our children?

Sex

1. How much sexual intimacy do you feel is appropriate in casual dating, involved dating, and engagement?

2. Does "having sex" mean having sexual intercourse? If a couple has experienced oral sex only, have they "had sex"?

3. What sexual behaviors do you most and least enjoy? How often do you want to have intercourse? How do you want me to turn you down when I don't want to have sex? How do you want me to approach you for sex? How do you feel about just being physical together—hugging, rubbing, holding, but not having intercourse?

4. By what method of stimulation do you experience an orgasm most easily?

5. What do you think about masturbation, oral sex, homosexuality, sadism and masochism (S & M), and anal sex?

6. What type of contraception do you suggest? Why? If that method does not prove satisfactory, what method would you suggest next?

7. What are your values regarding extramarital sex? If I had an affair, would you want me to tell you? Why? If I told you about the affair, what would you do? Why?

8. How often do you view pornographic videos? Pornography on the Internet?

9. How important is our using a condom to you?

10. Do you want me to be tested for human immunodeficiency virus (HIV)? Are you willing to be tested?

11. What sexually transmitted infections (STIs) have you had?

12. How much do you want to know about my sexual behavior with previous partners?

13. How many "friends with benefits" relationships have you been in? What is your interest in our having such a relationship?

14. How much do you trust me in terms of my being faithful/monogamous with you?

15. How open do you want our relationship to be in terms of having emotional or sexual involvement with others, while keeping our relationship primary?

16. What things have you done that you are ashamed of?

17. What emotional/psychological/physical health problems do you have?

18. What are your feelings about your sexual adequacy? What sexual problems do you or have you had?

19. Give me an example of your favorite sexual fantasy.

Careers and Money

1. What kind of job or career will you have? What are your feelings about working in the evening versus being home with the family? Where will your work require that we live? How often do you feel we will be moving? How much travel will your job require?

2. What are your feelings about a joint versus a separate checking account? Which of us do you want to pay the bills? How much money do you think we will have left over each month? How much of this do you think we should save?

3. When we disagree over whether to buy something, how do you suggest we resolve our conflict?

4. What jobs or work experience have you had? If we end up having careers in different cities, how do you feel about being involved in a commuter marriage?

5. What is your preference for where we live? Do you want to live in an apartment or a house? What are your needs for a car, television, cable service, phone plan, entertainment devices, and so on?

6. How do you feel about my having a career? Do you expect me to earn an income? If so, how much annually? To what degree do you feel it is your responsibility to cook, clean, and take care of the children? How do you feel about putting young children or infants in day-care centers? When the children are sick and one of us has to stay home, who will that be?

7. To what degree do you want me to account to you for the money I spend? How much money, if any, do you feel each of us should have to spend each week as we wish without first checking with the other partner? What percentage of income, if any, do you think we should give to charity each year?

8. What assets or debts will you bring into the marriage?

9. How much child support/alimony do you get or pay each month? Tell me about your divorce.

10. May I read your divorce settlement agreement? When?

Recreation and Leisure

1. What is your idea of the kinds of parties or social gatherings you would like for us to go to together?

2. What is your preference in terms of us hanging out with others in a group versus being alone?

3. What is your favorite recreational interest? How much time do you spend enjoying this interest? How important is it for you that I share this recreational interest with you?

4. What do you like to watch on television? How often do you watch television and for what periods of time?

5. What are the amount and frequency of your current use of alcohol and other drugs (e.g., marijuana, cocaine, crack, speed)? What, if any, have been your previous alcohol and other drug behaviors and frequencies? What are your expectations of me regarding the use of alcohol and other drugs?

6. Where did you vacation with your parents? Where will you want us to go? How will we travel? How much money do you feel we should spend on vacations each year?

Relationships with Friends/Co-Workers

1. How do you feel about my three closest same-sex friends?

2. How do you feel about my spending time with my friends or co-workers, such as one evening a week?

3. How do you feel about my spending time with friends of the opposite sex?

4. What do you regard as appropriate and inappropriate affectional behaviors with opposite-sex friends?

Remarriage Questions

1. How and why did your first marriage end? What are your feelings about your former spouse now? What are the feelings of your former spouse toward you? How much "trouble" do you feel she or he will want to cause us? What relationship do you want with your former spouse?

2. Do you want your children from a previous marriage to live with us? What are your emotional and financial expectations of me in regard to your children? What are your feelings about my children living with us? Do you want us to have additional children? How many? When?

3. When your children are with us, who will be responsible for their food preparation, care, discipline, and driving them to activities?

4. Suppose your children do not like me and vice versa. How will you handle this? Suppose they are against our getting married?

5. Suppose our respective children do not like one another. How will you handle this?

It would be unusual if you agreed with each other on all of your answers to the previous questions. You might view the differences as challenges and then find out the degree to which the differences are important for your relationship. You might need to explore ways of minimizing the negative impact of those differences on your relationship. It is not possible to have a relationship with someone in which there is total agreement. Disagreement is inevitable; the issue becomes how you and your partner manage the differences.

Note: This self-assessment is intended to be thought-provoking and fun. It is not intended to be used as a clinical or diagnostic instrument.

If you want to know what your partner may be like in the future, look at his or her parent of the same sex. There is a tendency for a man to become like his father and a woman to become like her mother. And, if you want to know how your partner is likely to treat you in the future, observe the way your partner's parent of the same sex treats and interacts with his or her spouse. Their relationship is the model of a spousal relationship your partner is likely to duplicate in relating to you.

Premarital Education Programs

Various premarital education programs (also known as premarital prevention programs, premarital counseling, premarital therapy, and marriage preparation), both academic and religious, are designed to provide information to individuals and to couples about how to have a good relationship. Not only may couples assess the degree to which they are compatible but they also can learn communication skills to help resolve conflict. Carroll and Doherty (2003) conducted a review of the various outcome studies of these programs and found that the average participant in a premarital prevention program experienced about a 30 percent increase in measures of outcome success. Specifically, they were more likely than nonparticipants to experience immediate and short-term gains in interpersonal skills and overall relationship quality.

Prenuptial Agreement

Senator John Kerry and his wife Teresa Heinz (who is worth $500 million) have a prenuptial agreement. Other individuals include Donald Trump, Halle Berry (now divorced), and Britney Spears. Some couples, particularly those with considerable assets or those in subsequent marriages, might consider discussing and signing a prenuptial agreement. To reduce the chance that the agreement will later be challenged, each partner should hire an attorney (months before the wedding) to develop and/or review the agreement. Amy Irving and Stephen Spielberg did not get an attorney to monitor their prenuptial agreement and it was thrown out of court; Amy Irving was awarded $1 million for their four-year marriage.

One of the best things about any prenuptial contract is that it's a reality check.

Richard Dombrow, family law attorney

The primary purpose of a **prenuptial agreement** (also referred to as a premarital agreement, marriage contract, or antenuptial contract) is to specify how property will be divided if the marriage ends in divorce or when it ends by the death of one partner. In effect, the value of what you take into the marriage is the amount you are allowed to take out of the marriage. For example, if you bring $150,000 into the marriage and buy the marital home with this amount, at divorce, your ex-spouse is not automatically entitled to half the house. Some agreements may also contain clauses of no spousal support (no alimony) if the marriage ends in divorce. See Appendix for an example of a prenuptial agreement developed by a husband and wife who had both been married before and had assets and children.

Reasons for a prenuptial agreement include the following.

1. *Protecting assets for children from a prior relationship.* Persons who are in the middle or later years, who have considerable assets, who have been married before, and who have children are often concerned that money and property be kept separate in a second marriage so that the assets at divorce or death go to the children. Some children encourage their remarrying parent to draw up a prenuptial agreement with the new partner so that their (the offspring's) inheritance, house, or whatever will not automatically go to the new spouse upon the death of their parent.

2. *Protecting business associates.* A spouse's business associate may want a member of a firm or partnership to draw up a prenuptial agreement with a soon-to-

Increasing Requirements for a Marriage License

Should marriage licenses be obtained so easily? Should couples be required, or at least encouraged, to participate in premarital education before saying "I do"? Given the high rate of divorce today, policymakers and family scholars are considering this issue.

Although evidence of long-term effectiveness of premarital education remains elusive (Carroll and Doherty, 2003), some believe that "mandatory counseling will promote marital stability" (Licata, 2002, p. 518).

Several states have proposed legislation requiring premarital education. For example, an Oklahoma statute provides that parties who complete a premarital education program pay a reduced fee for their marriage license. Also, in Lenawee County, Michigan, local civil servants and clergy have made a pact: they will not marry a couple unless that couple has attended marriage education classes. Other states that are considering policies to require or encourage premarital education include Arizona, Illinois, Iowa, Maryland, Minnesota, Mississippi, Missouri, Oregon, and Washington.

Proposed policies include not only mandating premarital education and lowering marriage license fees for those who attend courses but also imposing delays on issuing marriage licenses for those who refuse premarital education. However, "no state mandates premarital counseling as a prerequisite to obtaining a license" (Licata, 2002, p. 525).

Traditionally, most Protestant pastors and Catholic priests require premarital counseling before they will perform marriage ceremonies. Couples who do not want to participate in premarital education can simply get married in secular ceremonies (justice of the peace).

Advocates of mandatory premarital education emphasize that it will reduce marital discord. However, questions remain about who will offer what courses and whether couples will take seriously the content of such courses. Indeed, persons contemplating marriage are often narcotized with love and would doubtless not take any such instruction seriously. Love myths such as "divorce is something that happens to other people" and "our love will overcome any obstacles" work against the serious consideration of such courses.

Your Opinion?

1. To what degree do you believe premarital education should be required before the state issues a marriage license?
2. How effective do you feel such programs are for persons in a hurry to marry?
3. How receptive do you feel individuals in love are to marriage education?

Sources

Carroll, J. S. and W. J. Doherty. 2003. Evaluating the effectiveness of premarital prevention programs: A meta-analytic review of outcome research. *Family Relations* 52:105–08.

Licata, N. 2002. Should premarital counseling be mandatory as a requisite to obtaining a marriage license? *Family Court Review* 40:518–32.

be-spouse to protect the firm from intrusion by the spouse if the marriage does not work out.

Prenuptial contracts do have a value beyond the legal implications. Their greatest value may be that they facilitate the partners' discussing with each other their expectations of the relationship. In the absence of such an agreement, many couples may never discuss the issues they may later face.

There are disadvantages of signing a prenuptial agreement. They are often legally challenged ("My partner forced me to sign it or call off the wedding"), and not all issues can be foreseen ("Who gets the time-share vacation property?").

Prenuptial agreements also are not very romantic ("I love you, but sign here and see what you get if you don't please me") and may serve as a self-fulfilling prophecy ("We were already thinking about divorce."). Indeed, almost 20 percent (19.2%) of 1027 undergraduates agreed "I would not marry someone who required me to sign a prenuptial agreement" and a similar percentage (22.7%) feel that couples who have a prenuptial agreement are more likely to get divorced (Knox and Zusman, 2006). Prenuptial contracts are almost nonexistent in first marriages and are still rare in second marriages. Whether or not it is a good idea to sign a prenuptial agreement depends on the circumstances. Some individuals who do sign an agreement later regret it. Sherry, a never-married 22-year-old at the time, signed such an agreement:

Paul was adamant about my signing the contract. He said he loved me but would never consider marrying me unless I signed a prenuptial agreement stating that he would never be responsible for alimony in case of a divorce. I was so much in love, it didn't seem to matter. I didn't realize that basically he was and is a selfish person. Now, five years later after our divorce, I live in a mobile home and he lives in a big house overlooking the lake with his new wife.

The husband viewed it differently. He was glad that she had signed the agreement and that his economic liability to her was limited. He could afford the new house by the lake with his new wife because he was not sending money to Sherry. Billionaire Donald Trump attributed his economic survival of two divorces to prenuptial agreements with his ex-wives.

Couples who decide to develop a prenuptial agreement need separate attorneys to look out for their respective interests. The laws regulating marriage and divorce vary by state, and only attorneys in those states can help ensure that the document drawn up will be honored. Full disclosure of assets is also important. If one partner hides assets, the prenuptial can be thrown out of court. One husband recommended that the issue of the premarital agreement should be brought up and that it be signed a minimum of six months before the wedding. "This gives the issue time to settle rather than being an explosive emotional issue if it is brought up a few weeks before the wedding." Indeed, as noted above, if a prenuptial agreement is signed within two weeks of the wedding, that is grounds enough for the agreement to be thrown out of court, since it is assumed that the document was executed under pressure.

While individuals are deciding whether to have a prenuptial agreement, states are deciding whether to increase marriage license requirements, this chapter's social policy issue (see page 212).

Consider Calling Off the Wedding If . . .

"No matter how far you have gone on the wrong road, turn back" is a Turkish proverb. If your engagement is characterized by the factors identified below, consider prolonging your engagement and delaying the marriage at least until the most distressing issues have been resolved. Indeed, rather than defend a course of action that does not feel right to you, stop and reverse directions.

Age 18 or Younger
The strongest predictor of getting divorced is getting married as an adolescent. Individuals who marry in their teens have a greater risk of divorce than those who delay marriage into their mid-20s. Teenagers may be more at risk for marrying to escape an unhappy home and may be more likely to engage in impulsive decision-making and behavior. Early marriage is also associated with an end to one's education, social isolation from peer networks, early pregnancy/parenting, and locking one's self into a low income.

No man knows he is young while he is young.
Unknown

International Data
Worldwide, 51 million girls and young women between the ages of 15 and 19 are currently married (Child Marriages Projected to Increase, 2004).

Research by Meehan and Negy (2003) on being married while in college revealed higher marital distress among spouses who were also students. In addition, when married college students were compared with single college students, the married students reported more difficulty adjusting to the demands of higher education. The researchers conclude "these findings suggest that individuals opting to attend college while being married are at risk for compromising their marital happiness and may be jeopardizing their education" (p. 688).

Hence, waiting until one is older and through college not only may result in a less stressful marriage but also may be associated with less economic stress.

She bid me take love easy, as the leaves grow on the tree; But I, being young and foolish, with her would not agree.

W. B. Yeats

Known Partner Less Than Two Years

Almost a third (31.6%) of 1027 undergraduates agreed "If I were really in love, I would marry someone I had known for only a short time" (Knox and Zusman, 2006). Impulsive marriages in which the partners had known each other for less than a month are associated with a higher-than-average divorce rate. Indeed, partners who date each other for at least two years (25 months is best) before getting married report the highest level of marital satisfaction and are less likely to divorce (Huston et al., 2001). A short courtship does not allow the partners to observe and scrutinize each other's behavior in a variety of settings. Indeed, some individuals may be more prone to fall in love at first sight and to want to hurry the partner into a committed love relationship.

Suggestions include making a joint decision that getting to know each other over several years is important, taking a five-day "primitive" camping trip, taking a 15-mile hike together, wallpapering a small room together, or spending several days together when one partner has the flu. If the couple plan to have children, they may want to take care of a six-month-old together for a weekend. Time should also be spent with each other's friends.

Abusive Relationship

As we will discuss in Chapter 14, Violence and Abuse in Relationships, partners who emotionally and/or physically abuse their partners while dating and living together continue these behaviors in marriage. Abusive lovers become abusive spouses, with predictable negative outcomes. Though extricating oneself from an abusive relationship is difficult before the wedding, it becomes even more difficult after marriage. The price to oneself of being in an abusive relationship spreads to one's children.

One aspect of abuse to be aware of is a partner who attempts to isolate you from family or friends. Some partners attempt to systematically detach their intended spouse from all other relationships ("I don't want you spending time with your family and friends—you should be here with me"). This is a serious flag of impending relationship doom and should not be overlooked.

Numerous Significant Differences

Before you run in double harness, look well to the other horse.

Ovid

Relentless conflict often arises from numerous significant differences. Though all spouses are different from each other in some ways, those who have numerous differences in key areas such as race, religion, social class, education, values, and goals are less likely to report being happy and to have durable relationships.

Persons who report the greatest degree of satisfaction in durable relationships have a great deal in common (Wilson and Cousins, 2005). Conversely, Skowron (2000) found that the less couples had in common, the more their marital distress.

On-and-Off Relationship

A roller-coaster premarital relationship is predictive of a marital relationship that will follow the same pattern. Partners who break up and get back together several times have developed a pattern in which the dissatisfactions in the relationship become so frustrating that separation becomes the antidote for relief. In courtship, separations are of less social significance than marital separations. "Breaking up" in courtship is called "divorce" in marriage. Couples who routinely break up and get back together should examine the issues that continue to recur in their relationship and attempt to resolve them.

Dramatic Parental Disapproval

A parent recalled, "I knew when I met the guy it wouldn't work out. I told my daughter and pleaded that she not marry him. She did, and they divorced." Such parental predictions (whether positive or negative) often come true. If the predictions are negative, they sometimes contribute to stress and conflict once the couple marries.

Even though parents who reject the commitment choice of their offspring are often regarded as uninformed and unfair, their opinions should not be taken lightly. The parents' own experience in marriage and their intimate knowledge of their offspring combine to put them in a unique position to assess how their child might get along with a particular mate. If the parents of either partner disapprove of the marital choice, the partners should try to evaluate these concerns objectively. The insights might prove valuable. The value of parental approval is illustrated in a study of Chinese marriages. Pimentel (2000) found that higher marital quality was associated with parents' approving of the mate choice of their offspring.

Low Sexual Satisfaction

Sexual satisfaction is linked to relationship satisfaction, love, and commitment. Sprecher (2002) followed 101 dating couples across time and found that low sexual satisfaction (for both women and men) was related to reporting low relationship quality, less love, lower commitment, and breaking up. Hence, couples who are dissatisfied with their sexual relationship might explore ways of improving it (alone or through counseling) or consider the impact of such dissatisfaction on the future of their relationship.

Sex is a drive where there are too many reckless drivers.

Unknown

Basically it is time to end a relationship when the gain/advantages of staying together no longer outweigh the pain/disadvantages of staying . . . . . and the pain of leaving is less than the pain of staying. Of course, all relationships go through periods of time when the disadvantages outweigh the benefits, so one should not bail out without careful consideration.

Marrying for the Wrong Reason

Some reasons for getting married are more questionable than others. These reasons include the following.

1. *Rebound.* About half of undergraduates report getting involved with someone on the rebound. Over half (53.4%) of 1027 undergraduates agreed "I have become involved in a relationship when I was clearly on the rebound" (Knox and Zusman, 2006). A rebound marriage results when you marry someone immediately after another person has ended a relationship with you. It is a frantic attempt on your part to reestablish your desirability in your own eyes and in the eyes of the partner who dropped you. One man said, "After she told me she wouldn't marry me, I became desperate. I called up an old girlfriend to see if I could get the relationship going again. We were married within a month. I know it was foolish, but I was very hurt and couldn't stop myself." To marry on the rebound is questionable because the marriage is made in reference to the previous partner and not to the partner being married. In reality, you are using the person you intend to marry to establish yourself as the "winner" in the previous relationship.

To avoid the negative consequences of marrying on the rebound, wait until the negative memories of your past relationship have been replaced by positive aspects of your current relationship. In other words, marry when the satisfactions of being with your current partner outweigh any feelings of revenge. This normally takes between 12 and 18 months.

2. *Escape.* A person might marry to escape an unhappy home situation in which the parents are often seen as oppressive, overbearing, conflictual, or abusive. The parents' continued bickering might be highly aversive, causing the person to marry to flee the home. A family with an alcoholic parent might also create a context from which a person might want to escape. One woman said, "I couldn't wait to get away from home. Ever since my parents divorced, my mother has been drinking and watching me like a hawk. 'Be home early, don't drink, and watch out for those horrible men,' she would always say. I admit it. I married the first guy that would have me. Marriage was my ticket out of there." Marriage for escape is a bad idea. It is far better to continue the relationship with the partner until mutual love and respect, rather than the desire to escape an unhappy situation, becomes the dominant forces propelling you toward marriage. In this way you can evaluate the marital relationship in terms of its own potential and not solely as an alternative to an unhappy situation. Undergraduates do not seem vulnerable to getting married for reasons of escape. Only 5.5 percent of 1027 agreed "I am capable of marrying someone to escape an unhappy home life with my parents" (Knox and Zusman, 2006).

3. *Unanticipated pregnancy.* Getting married just because a partner becomes pregnant is usually a bad idea. Indeed, the decision of whether to marry should be kept separate from the fact that there is now a pregnancy. Adoption, abortion, single parenthood, and unmarried parenthood (the couple can remain together as an unmarried couple and have the baby) are all alternatives to simply deciding to marry if a partner becomes pregnant. Avoiding feelings of being trapped or "You married me just because of the pregnancy" is one of the reasons for not rushing into marriage because of pregnancy.

4. *Psychological blackmail.* Some individuals get married because their partner takes the position that "I can't live without you" or "I will commit suicide if you leave me." Because the person fears that the partner may commit suicide, he or she agrees to the wedding. The problem with such a marriage is that one partner has learned to manipulate the relationship to get what he or she wants.

Use of such power often creates resentment in the other partner, who feels trapped in the marriage. Escaping from the marriage becomes even more difficult. One way of coping with a psychological blackmail situation is to encourage the person to go with you to a counselor to "discuss the relationship." Once inside the therapy room, you can tell the counselor that you feel pressured to get married because of the suicide threat. Counselors are trained to respond to such a situation.

5. *Pity.* Some partners marry because they feel guilty about terminating a relationship with someone whom they pity. The fiancé of one woman got drunk one Halloween evening and began to light fireworks on the roof of his fraternity house. As he was running away from a Roman candle he had just ignited, he tripped and fell off the roof. He landed on his head and was in a coma for three weeks. A year after the accident his speech and muscle coordination were still adversely affected. The woman said she did not love him anymore but felt guilty about terminating the relationship now that he had become physically afflicted.

She was ambivalent. She felt it was her duty to marry her fiancé, but her feelings were no longer love feelings. Pity may also have a social basis. For example, a partner may fail to achieve a lifetime career goal (for example, he or she may flunk out of medical school). Regardless of the reason, if one partner loses a limb, becomes brain-damaged, or fails in the pursuit of a major goal, it is important to keep the issue of pity separate from the advisability of the marriage. The decision to marry should be based on factors other than pity for the partner.

Oh, what a tangled web we weave when first we practice to conceive.

Don Herold

6. *Filling a void.* A former student in the authors' classes noted that her father died of cancer. She acknowledged that his death created a vacuum, which she felt driven to fill immediately by getting married so that she would have a man in her life. Because she was focused on filling the void, she had paid little attention to the personality characteristics of or her relationship with the man who had asked to marry her.

She reported that she discovered on her wedding night that her new husband had several other girlfriends whom he had no intention of giving up. The marriage was annulled.

In deciding whether to continue or terminate a relationship, listen to what your senses tell you ("Does it feel right?"), listen to your heart ("Do you love this person or do you question whether you love this person?"), and evaluate your similarities ("Are we similar in terms of core values, goals, view of life?"). Also, be realistic. It would be unusual if none of the factors listed above applied to you. Indeed, most people have some negative and some positive indicators before they marry. For example, actor Russell Crowe had known his girlfriend, Danielle Spencer, for 13 years when they married (a positive predictor). However, their relationship had been described as an "on-again, off-again" one (a negative predictor) (Crawley, 2003).

Breaking up is never easy. Almost a third of 279 undergraduates reported that they "sometimes" remained in relationships they thought should end (31%) or that became "unhappy" (32.3%). This finding reflects the ambivalence students sometimes feel in ending an unsatisfactory relationship and their reluctance to do so (Knox, Zusman, McGinty, and Davis, 2002).

Ending an Unsatisfactory Relationship

Endings are as common as beginnings in relationships. Before pulling the plug, there are some considerations.

Cold Feet? The Jennifer Wilbanks Story and More

In April 2005 Jennifer Wilbanks, of Duluth, Georgia, was scheduled to marry at a wedding to which 600 guests had been invited. Instead, she fled on a bus to Las Vegas (while authorities searched for her, at a cost of $40,000 to $60,000). Initially she said she had been kidnapped, but she later confessed she had cold feet about the wedding. It had happened before: she had been involved in a previous engagement that she had broken.

Having cold feet about getting married is not unusual. Students in the authors' classes reported their experiences:

"I knew the day of the wedding that I did not want to marry. I told my dad, and he said, 'Be a man.' I went through with the marriage and regretted it ever since."

"I said 'Holy Jesus' just before I walked down the aisle with my dad. He said, 'What's the matter, honey?' I couldn't tell him, went through with the wedding and later divorced."

"I never really believed I was getting married till I saw my name in the paper that I was soon to be married. It scared me. I called it off after the announcements had been sent out and we had been to see the preacher. It was a real mess. She kept the ring."

Before Ending a Relationship

All relationships have difficulties, and all necessitate careful consideration of various issues before they are ended. These considerations include the following.

Why is it that we can't always recognize the moment that love begins, but we always know when it ends?

Steve Martin, actor, comedian, author

Jennifer Wilbanks ran away just days before her wedding, which 600 guests were to attend. Having anxiety about one's wedding is normal. Getting cold feet to the point of not showing up on one's wedding day is unusual.

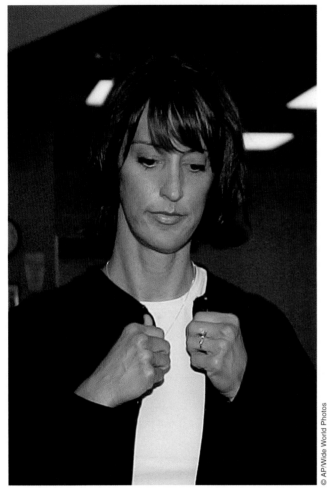

© AP/Wide World Photos

*It's not love's going hurts my days
But that it went in little ways.*

Edna St. Vincent Millay

1. *Is there any desire or hope to revive and improve the relationship?* In some cases, people end relationships and later regret having done so. Setting unrealistically high standards may eliminate an array of individuals who might be superb partners, companions, and mates. If the reason for ending a relationship is conflict over an issue or set of issues, an alternative to ending the relationship is to attempt to resolve the issues through negotiating differences, compromising, and giving the relationship more time. (We do not recommend giving an abusive relationship more time, as abuse, once started, tends to increase in frequency and intensity. Nor do we recommend staying in a relationship in which there are great differences in interests, styles, personalities, and levels of attachment. It is often more prudent to find someone with whom you are more compatible than to try to "remake" the person in your life now.)

2. *Acknowledge and accept that terminating a relationship may be painful for both partners.* There may be no way you can stop the hurt. One person said, "I can't live with him anymore, but I don't want to hurt him either." The two feelings are incompatible. To end a relationship with someone who loves you is usually hurtful to both partners. Persons who end a relationship usually conclude that the pain and suffering of staying in a relationship is more than they will experience from leaving.

3. *Blame yourself for the end.* One way to end a relationship is to blame yourself by giving a reason that is specific to you ("I need more freedom," "I want to go to graduate school in another state," "I'm not ready to settle down," etc.). If you blame your partner or give your partner a way to make things better, the relationship may continue because *you* may feel obligated to give your partner a second chance if he or she promises change.

4. *Cut off the relationship completely.* If you are the person ending the relationship, it will probably be easier for you to continue to see the other person without feeling too hurt. But the other person will probably have a more difficult time and will heal faster if you stay away completely. Alternatively, some people are skilled at ending love relationships and turning them into friendships. Though this is difficult and infrequent, it can be rewarding across time for the respective partners.

5. *Learn from the terminated relationship.* Included in the reasons a relationship may end are being too controlling, being oversensitive/jealous/too picky, cheating, fearing commitment, and being unable to compromise and negotiate conflict. Recognizing one's own contribution to the breakup and working on any characteristics that might be a source of problems are some of the benefits of terminating a relationship. Otherwise, one might repeat the process.

6. *Allow time to grieve over the end of the relationship.* Ending a love relationship is painful. It is okay to feel this pain, to hurt, to cry. Allowing yourself time to experience such grief will help you become healed for the next relationship. It takes 12 to 18 months to recover from a serious relationship.

Recovering from a Broken Heart

A sample of 410 freshmen and sophomores at a large southeastern university completed a confidential survey revealing their recovery from a previous love relationship. Some of the findings were as follows (Knox et al., 2000).

1. *Sex differences in relationship termination.* Women were significantly more likely (50%) than men (40%) to report that they initiated the breakup. Sociologists explain such a phenomenon (seeking a better partner) as investment in a potentially good father for offspring. One student recalled, "I got tired of his lack of ambition—I just thought I could do better. He's a nice guy but living in a trailer is not my idea of a life."

2. *Sex differences in relationship recovery.* Though recovery was not traumatic for either men or women, men reported more difficulty than women did in adjusting to a breakup. When respondents were asked to rate their level of difficulty from "no problem" (0) to "complete devastation" (10), women scored 4.35 and men scored 4.96. In explaining why men might have more difficulty adjusting to terminated relationships, some of the students said, "Men have such inflated egos, they can't believe that a woman would actually dump them." Others said, "Men are oblivious to what is happening in a relationship and may not have a clue that it is heading toward an abrupt end. When it does end, they are in shock."

3. *Time/new partner as factors in recovery.* The passage of time and involvement with a new partner were identified as the most helpful factors in getting over a love relationship that ended. Though the difference was not statistically significant, men more than women reported "a new partner" was more helpful in relationship recovery (34% vs. 29%). Similarly, women more than men reported that "time" was more helpful in relationship recovery (34% vs. 29%).

4. *Other findings.* Other factors associated with recovery for women and men were "moving to a new location" (13% vs. 10%) and recalling that "the previous partner lied to me" (7% vs. 5%). Men more than women were much more likely to use alcohol to help them get over a previous partner (9% vs. 2%). Neither men nor women reported using therapy with any frequency to help them get over a partner (1% vs. 2%). Shakespeare noted "true love never did run smooth." These data suggest that breaking up is not terribly difficult for most individuals (but more difficult for men than women) and that recovery is enabled by both time and a new partner.

Some partners seek revenge as a way of recovering from a relationship that was ended by the other partner. Breed et al (2007) detailed one example of a former girlfriend who downloaded Internet child pornography on her partner's home computer, called the cops and had him arrested for possession of child pornography. The penalty was 15 months in prison per image. She loaded 18 on his computer. He was found guilty.

> *Love is like war: easy to begin but hard to end.*
>
> Anonymous

SUMMARY

What are the cultural factors that influence mate selection?

Two types of cultural influences in mate selection are endogamy (to marry someone inside one's own social group—race, religion, social class) and exogamy (to marry someone outside one's own family).

What are the sociological factors that influence mate selection?

Sociological aspects of mate selection include homogamy (people prefer someone like themselves in race, age, and education) and spirituality/religion. Couples who have a lot in common are more likely to have a durable relationship.

What are the psychological factors operative in mate selection?

Psychological aspects of mate selection include complementary needs, exchange theory, and parental image. Complementary-needs theory suggests that people select others who have characteristics opposite to their own. For example, a highly disciplined, well-organized individual might select a free-and-easy, worry-about-nothing mate. Most researchers find little evidence for complementary-needs theory.

Exchange theory suggests that one individual selects another on the basis of rewards and costs. As long as an individual derives more profit from a relationship with one partner than with another, the relationship will continue. Exchange concepts influence who dates whom, the conditions of the dating relationship, and the decision to marry. Parental image theory suggests that individuals select a partner similar to the opposite-sex parent.

Personality characteristics of a potential mate desired by both men and women include being warm, kind, and open and having a sense of humor. Negative personality characteristics to avoid in a potential mate include disagreeableness/expressing few positives, poor impulse control, hypersensitive to criticism, inflated ego, neurotic (perfectionist)/insecure, and controlled by someone else (e.g., parents). Paranoid, schizoid, and borderline personalities are also to be avoided.

What are the sociobiological factors operative in mate selection?

The sociobiological view of mate selection suggests that men and women select each other on the basis of their biological capacity to produce and support healthy offspring. Men seek young women with healthy bodies, and women seek ambitious men who will provide economic support for their offspring. There is considerable controversy about the validity of this theory.

What factors should be considered when becoming engaged?

The engagement period is the time to ask specific questions about the partner's values, goals, and marital agenda, to visit each other's parents to assess parental models, and to consider involvement in premarital educational programs and/or counseling. Negative reasons for getting married include being on the rebound, escaping from an unhappy home life, psychological blackmail, and pity.

Some couples (particularly those with children from previous marriages) decide to write a prenuptial agreement to specify who gets what and the extent of spousal support in the event of a divorce. To be valid, the document should be developed by an attorney in accordance with the laws of the state in which the partners reside. Last-minute prenuptial agreements put enormous emotional strain on the couple and are often considered invalid by the courts. Discussing a prenuptial agreement six months in advance is recommended.

What factors suggest you might consider calling off the wedding?

Factors suggesting that a couple may not be ready for marriage include being in their teens, having known each other less than two years, and having a relationship characterized by significant differences and/or dramatic parental disapproval. Some research suggests that partners with the greatest number of similarities in values, goals, and common interests are most likely to have happy and durable marriages.

Ending an unsatisfactory relationship.

Getting cold feet and wanting to back out of a wedding sometimes occurs (e.g. Jennifer Wilbanks). More often, individuals end a relationship before planning a wedding. Considerations in ending a relationship include improving it, acknowledging that endings are often painful, and taking the blame. Recovery involves time (12 to 18 months) and a new partner. Some partners seek devastating revenge.

KEY TERMS

complementary-needs theory   exogamy
endogamy   homogamy
exchange theory

The Companion Website for *Choices in Relationships: An Introduction to Marriage and the Family*, Ninth Edition

http://thomson.edu/sociology/knox

Supplement your review of this chapter by going to the companion website to take one of the Tutorial Quizzes, use the flash cards to master key terms, and check out the many other study aids you'll find there. You'll also find special features such as the Marriage and Family Resource Center, Census 2000 information, and other data and resources at your fingertips to help you with that special project or to do some research on your own.

WEBLINKS

PAIR Project

www.utexas.edu/research/pair

RightMate

http://www.heartchoice.com/rightmate/

REFERENCES

Arrindell, W. A. and F. Luteijn. 2000. Similarity between intimate partners for personality traits as related to individual levels of satisfaction with life. *Personality and Individual Differences* 28:629–37.

Breed, R., D. Knox, and M. Zusman 2007 Hell hath no fury . . . Legal consequences of having Internet child pornography on one's computer. Poster, Southern Sociological Society, Atlanta, April.

Butler, M. H., J. A. Stout, and B. C. Gardner. 2002. Prayer as a conflict resolution ritual: Clinical implications of religious couple's report of relationship softening, healing perspective, and change responsibility. *American Journal of Family Therapy* 30:19–37.

Carroll, J. S. and W. J. Doherty. 2003. Evaluating the effectiveness of premarital prevention programs: A meta-analytic review of outcome research. *Family Relations* 52:105–18.

Cawley, J. 2003. Russell Crowe in command. *Biography Magazine* December, 50–54.

Child marriages projected to increase. 2004. *Popline: World Population News Service* July-August 26:1.

Crowell, J. A., D. Treboux, and E. Waters. 2002. Stability of attachment representations: The transition to marriage. *Developmental Psychology* 38:467–79.

DeCuzzi, A., D. Knox, and M. Zusman. 2006. Racial differences in perceptions of women and men. *College Student Journal* 40: 343–349

Dehle, C. and R. L. Weiss. 2002 Associations between anxiety and marital adjustment. *Journal of Psychology* 136:328–38.

Delinsky, S. S. 2005. Cosmetic surgery: A common and accepted form of self-improvement? *Journal of Applied Social Psychology* 35:2012–28.

Edwards, T. M. 2000. Flying solo. *Time* 28 August, 47–53.

Fu, X. 2006. Impact of socioeconomic status on inter-racial mate selection and divorce. *Social Science Journal* 43: 239–258

Gattis, K. S., S. Berns, L. E. Simpson, and A. Christensen. 2004. Birds of a feather or strange birds? Ties among personality dimensions, similarity, and marital quality. *Journal of Family Psychology* 18:564–78.

Haring, M., P. L. Hewitt, and G. L. Flett. 2003. Perfectionism, coping, and quality of relationships. *Journal of Marriage and the Family* 65:143–59.

Huston, T. L., J. P. Caughlin, R. M. Houts, S. E. Smith, and L. J. George. 2001. The connubial crucible: Newlywed years as predictors of marital delight, distress, and divorce. *Journal of Personality and Social Psychology* 80:237–52.

Ingoldsby, B., P. Schvaneveldt, and C. Uribe. 2003. Perceptions of acceptable mate attributes in Ecuador. *Journal of Comparative Family Studies* 34:171–86.

Jepsen, L. K. and C. A. Jepsen. 2002. An empirical analysis of the matching patterns of same-sex and opposite-sex couples. *Demography* 39:435–53.

Kalmijn, M. and H. Flap. 2001. Assortive meeting and mating: Unintended consequences of organized settings for partner choices. *Social Forces* 79:1289–312.

Knox, D. and K. McGinty. 2004. Elimidate II: An in class exercise for marriage and family classes. Poster, Annual Meeting of the National Council on Family Relations, Orlando, Florida, November.

Knox, D. and Zusman, M. E. 2006. Relationship and sexual behaviors of a sample of 1027 university students. Unpublished data collected for this text. Greenville, NC: Department of Sociology, East Carolina University.

Knox, D., M. E. Zusman, M. Kaluzny, and C. Cooper. 2000. College student recovery from a broken heart. *College Student Journal* 34:322–24.

Knox, D., M. E. Zusman, and V. W. Daniels. 2002. College student attitudes toward interreligious marriage. *College Student Journal* 31:445–48.

Knox, D., M. E. Zusman, K. McGinty, and B. Davis. 2002. College student attitudes and behaviors toward ending an unsatisfactory relationship. *College Student Journal* 36:630–34.

Knox, D., M. E. Zusman, and W. Nieves. 1997. College students' homogamous preferences for a date and mate. *College Student Journal* 31:445–48.

Lewis, S. K. and V. K. Oppenheimer. 2000. Educational assortative mating across marriage markets: Non-Hispanic whites in the United States. *Demography* 37:29–40.

Licata, N. 2002. Should premarital counseling be mandatory as a requisite to obtaining a marriage license? *Family Court Review* 40:518–32.

Luo, S.H. and E. C. Klohnen. 2005. Assortative mating and marital quality in newlyweds: A couple-centered approach. *Journal of Personality and Social Psychology* 88:304–26.

Marcus, D. K., and R. S. Miller. 2003. Sex differences in judgments of physical attractiveness: A social relations analysis. *Personality and Social Psychology Bulletin* 29:325–35.

Medora, N. P., J. H. Larson, N. Hortacsu, and P. Dave. 2002. Perceived attitudes towards romanticism: A cross-cultural study of American, Asian-Indian, and Turkish young adults. *Journal of Comparative Family Studies* 33:155–78.

Medora, N. P. 2003. Mate selection in contemporary India: Love marriages versus arranged marriages. In *Mate Selection Across Cultures,* edited by R. R. Hamon and B. B. Ingoldsby. Thousand Oaks, California: Sage Publications, 209–30.

Meehan, D. and C. Negy. 2003. Undergraduate students' adaptation to college: Does being married make a difference? *Journal of College Student Development* 44:670–90.

Meyer, J. P. and S. Pepper. 1977. Need compatibility and marital adjustment in young married couples. *Journal of Personality and Social Psychology* 35:331–42.

Pimentel, E. E. 2000. Just how do I love thee? Marital relations in urban China. *Journal of Marriage and the Family* 62:32–47.

Porter, E. and M. O'Donnell. 2006. The new gender divide: Facing middle age with no degree, and no wife. *The New York Times,* Retrieved August 7. http://www.nytimes.com/2006/08/06/us/06marry.html?ex=1155614400&en=035e314005c8ac44&ei=5070&emc=eta1

Regan, P. C. and A. Joshi.2003. Ideal partner preferences among adolescents. *Social Behavior and Personality* 31:13–20.

Saint, D. J. 1994. Complementarity in marital relationships. *Journal of Social Psychology* 134:701–4.

Skowron, E. A. 2000. The role of differentiation of self in marital adjustment. *Journal of Counseling Psychology* 47:229–37.

Snyder, D. K. and J. M. Regts. 1990. Personality correlates of marital dissatisfaction: A comparison of psychiatric, maritally distressed, and nonclinic samples. *Journal of Sex and Marital Therapy* 90:34–43.

Sprecher, S. and P. C. Regan. 2002. Liking some things (in some people) more than others: Partner preferences in romantic relationships and friendships. *Journal of Social and Personal Relationships* 19:463–81.

Sprecher, S. 2002. Sexual satisfaction in premarital relationships: Associations with satisfaction, love, commitment, and stability. *Journal of Sex Research* 39:190–96.

Statistical abstract of the United States: 2006. 125th Ed. Washington, D.C.: U.S. Bureau of the Census.

Strassberg, D. S. and S. Holty. 2003. An experimental study of women's Internet personal ads. *Archives of Sexual Behavior* 32:253–61.

Swenson, D., J. G. Pankhurst, and S. K. Houseknecht. 2005. Links between families and religion. In *Sourcebook of family theory & research,* edited by V. L. Bengtson, A. C. Acock, K. R. Allen, P. Dilworth-Anderson, and D. M. Klein. Thousand Oaks, California: Sage Publications, 530–33.

Toro-Morn, M. and S. Sprecher. 2003. A crosscultural comparison of mate preferences among university students: The United States versus the People's Republic of China (PRC). *Journal of Comparative Family Studies* 34:151–62.

Twenge, J. 2006. *Generation me.* New York: Free Press.

Waller, W. and R. Hill. 1951. *The family: A dynamic interpretation.* New York: Holt, Rinehart and Winston.

Wilson, G. D. and J. M. Cousins. 2005. Measurement of partner compatibility: Further validation and refinement of the CQ test. *Sexual and Relationship Therapy* 20:421–29

Winch, R. F. 1955. The theory of complementary needs in mate selection: Final results on the test of the general hypothesis. *American Sociological Review* 20:552–55.

Xie, Y., J. M. Raymo, K. Govette, and A. Thornton. 2003. Economic potential and entry into marriage and cohabitation. *Demography* 40:351–64.

Zaidi, A. U. and M. Shuraydi. 2002. Perceptions of arranged marriages by young Pakistani Muslim women living in a western society. *Journal of Comparative Family Studies* 33:495–514.

Zusman, M. E., J. Gescheidler, D. Knox, J. Gescheidler, and K. McGinty. 2003. Dating manners among college students. *Journal of Indiana Academy of Social Sciences.* 7:28–32.

chapter 8

Love is a sweet dream, and marriage is the alarm clock.

Jewish proverb

Marriage Relationships

Contents

Authors

True or False?

1. Honeymoons are no longer important, as fewer than half of more than 1000 undergraduates reported that it was important to take a honeymoon.

2. Economic security is the greatest expected benefit of marriage in the United States.

3. About a third of states now offer covenant marriages and a third of people getting married in these states elect the covenant alternative.

4. Males and whites tend to report greater marital happiness than females and blacks.

5. Marital satisfaction increases across time—the longer spouses are married, the happier they report themselves with each other.

Answers: **1.** F **2.** F **3.** F **4.** T **5.** F

That more than 95 percent of both men and women in the United States marry by age 75 (and most of those who divorce remarry) testifies to the value accorded marriage in our society. In a sample of 1027 undergraduates, almost all (97.7%) of the students agreed, "Someday, I want to marry" (Knox and Zusman, 2006). In this chapter we review why people are motivated to marry and the social functions of their doing so. We also look at some unique marriages as well as the characteristics of successful marriages.

Motivations for and Functions of Marriage

In this section we discuss both why people marry and the functions that getting married serves for society.

Individual Motivations for Marriage

We have defined marriage as a legal contract between two heterosexual adults that regulates their economic and sexual interaction. However, individuals in the United States tend to think of marriage in more personal than legal terms. The following are some of the reasons people give for getting married.

Love   Many couples view marriage as the ultimate expression of their love for each other—the desire to spend their lives together in a secure, legal, committed relationship. In U.S. society, love is expected to precede marriage—thus, only couples in love consider marriage. Those not in love would be ashamed to admit it.

Personal Fulfillment   We marry because we feel a sense of personal fulfillment in doing so. We were born into a family (family of origin) and want to create a family of our own (family of procreation). We remain optimistic that our marriage will be a good one. Even if our parents divorced or we have friends who have done so, we feel that our relationship will be different.

Companionship   Talk show host Oprah Winfrey once said that lots of people want to ride in her limo, but what she wants is someone who will take the bus when the limo breaks down. One of the motivations for marriage is to enter a

I love being married. It's so great to find that one special person you want to annoy for the rest of your life.

Rita Rudner, comedian

structured relationship with a genuine companion, a person who will take the bus with you when the limo breaks down.

Although marriage does not ensure it, companionship is the greatest expected benefit of marriage in the United States. Coontz (2000) noted that it has become "the legitimate goal of marriage" (p. 11). Eating meals together is one of the most frequent normative behaviors of spouses. Indeed, although spouses may eat lunch apart, dinner together becomes an expected behavior. **Commensality** is eating with others, and one of the issues spouses negotiate is "who eats with us" (Sobal et al., 2002).

Parenthood   Most people want to have children. In response to the statement, "Someday, I want to have children," 95.1% of 1027 undergraduates at a large southeastern university responded "yes" (Knox and Zusman, 2006). And the amount of time parents spend in rearing their children has increased. Sayer et al. (2004) documented that, contrary to conventional wisdom, both mothers and fathers report spending greater amounts of time in child-care activities in the late 1990s than in the "family-oriented" 1960s.

We never know the love of the parent until we become parents ourselves.

Henry Ward Beecher

Although some people are willing to have children outside marriage (in a cohabiting relationship or in no relationship at all), most Americans prefer to have children in a marital context. Previously, a strong norm existed in our society (particularly for whites) that individuals should be married before they have children. This norm has relaxed, with more individuals willing to have children without being married. An Australian survey revealed that women who elect to remain childfree are viewed more negatively than women who express a desire to have children (Rowland, 2006).

Economic Security   Married persons report higher household incomes than do unmarried persons. Indeed, national data from the Health and Retirement Survey revealed that individuals who were not continuously married had significantly lower wealth than those who remained married throughout the life course. Remarriage offsets the negative effect of marital dissolution (Wilmoth and Koso, 2002).

Although individuals may be drawn to marriage for the preceding reasons on a conscious level, unconscious motivations may also be operative. Individuals reared in a happy family of origin may seek to duplicate this perceived state of warmth, affection, and sharing. Alternatively, individuals reared in unhappy, abusive, drug-dependent families may inadvertently seek to re-create a similar family because that is what they are familiar with. In addition, individuals are motivated to marry because of the fear of being alone, to better themselves economically, to avoid an out-of-wedlock birth, and to prove that someone wants them.

Just as most individuals want to marry (regardless of the motivation), most parents want their children to marry. If their children do not marry too young and if they marry someone they approve of, parents feel some relief from the economic responsibility of parenting, anticipate that marriage will have a positive, settling effect on their offspring, and look forward to the possibility of grandchildren.

Societal Functions of Marriage

As we noted in Chapter 1 of this text, the primary function of marriage is to bind a male and female together who will reproduce, provide physical care for their dependent young, and socialize them to be productive members of society who will replace those who die (Murdock, 1949). Marriage helps protect children by giving the state legal leverage to force parents to be responsible to their offspring whether or not they stay married. If couples did not have children, the state would have no interest in regulating marriage.

Love is blind and marriage is the institution for the blind.

James Graham

Additional functions include regulating sexual behavior (spouses have less exposure to sexually transmitted infections [STIs] than singles) and stabilizing adult personalities by providing a companion and "in house" counselor. In the past, marriage and family have served protective, educational, recreational, economic, and religious functions.

But as these functions have gradually been taken over by the police/legal system, schools, entertainment industry, workplace, and church/synagogue, only the companionship-intimacy function has remained virtually unchanged.

The emotional support each spouse derives from the other in the marital relationship remains one of the strongest and most basic functions of marriage (Coontz, 2000). In today's social world, which consists mainly of impersonal, secondary relationships, living in a context of mutual emotional support may be particularly important. Indeed, the companionship and intimacy needs of contemporary U.S. marriage have become so strong that many couples consider divorce when they no longer feel "in love" with their partner. Sixty-seven percent of 1027 undergraduates reported that they would divorce someone they did not love (Knox and Zusman, 2006).

The very nature of the marriage relationship has also changed from being very traditional or male-dominated to being very modern or egalitarian. A summary of these differences is presented in Table 8.1. Keep in mind that these are stereotypical marriages and that only a small percentage of today's modern marriages have all the traditional/egalitarian characteristics that are listed.

The traditional role of marriage would not work in this relationship.

Ophra Winfrey (said of her relationship with Stedman Graham to whom she was engaged but broke the engagement)

Marriage as a Commitment

Marriage represents a multilevel **commitment**—person-to-person, family-to-family, and couple-to-state.

Person-to-Person Commitment

Individuals commit themselves to someone they love, with whom they feel a sense of equality, and whom they feel is the best of the persons available to them (Crawford et al., 2003). Beyond the idea of commitment as an intent to maintain

Table 8.1 Traditional versus Egalitarian Marriages

Traditional Marriage	Egalitarian Marriage
There is limited expectation of husband to meet emotional needs of wife and children.	Husband is expected to meet emotional needs of wife and to be involved with children.
Wife is not expected to earn income.	Wife is expected to earn income.
Emphasis is on ritual and roles.	Emphasis is on companionship.
Couples do not live together before marriage.	Couples may live together before marriage.
Wife takes husband's last name.	Wife may keep her maiden name.
Husband is dominant; wife is submissive.	Neither spouse is dominant.
Roles for husband and wife are rigid.	Roles for spouses are flexible.
Husband initiates sex; wife complies.	Sex is initiated by either spouse.
Wife takes care of children.	Parents share childrearing.
Education is important for husband, not for wife.	Education is important for both spouses.
Husband's career decides family residence.	Career of either spouse determines family residence.

a relationship, behavioral indexes of commitment (928 of them) were identified by 248 people who were committed to someone. These behaviors were then coded into ten major categories and included providing affection, providing support, maintaining integrity, sharing companionship, making an effort to communicate, showing respect, creating a relational future, creating a positive relational atmosphere, working on relationship problems together, and expressing commitment (Weigel and Ballard-Reisch, 2002). Glover et al. (2006) noted that being committed to another was associated with relationship satisfaction and feeling predisposed toward caregiving for the partner.

Family-to-Family Commitment

Whereas love is private, marriage is public. It is the second of three times that one's name can be expected to appear in the local newspaper. When individuals marry, the parents and extended kin also become enmeshed. In many societies (e.g., Kenya), it is the families who arrange for the marriage of their offspring and the groom is expected to pay for his new bride. How much is a bride worth? In some parts of rural Kenya, premarital negotiations include the determination of **bride wealth**—this is the amount of money a prospective groom will pay to the parents of his bride to be. Such a payment is not seen as "buying the woman" but compensating the parents for the loss of labor from their daughter. Forms of payment include livestock ("I am worth many cows" said one Kenyan woman), food, and/or money. The man who raises the bride wealth also demonstrates not only that he is ready to care for a wife and children but also that he has the resources to do so (Wilson et al., 2003).

Marriage also involves commitments by each of the marriage partners to the family members of the spouse. Married couples are often expected to divide their holiday visits between both sets of parents.

Couple-to-State Commitment

In addition to making person-to-person and family-to-family commitments, spouses become legally committed to each other according to the laws of the state in which they reside. This means they cannot arbitrarily decide to terminate their own marital agreement.

When you marry, you go into captivity.

Sadie Wright, in a postcard to her son on the day of his wedding

PERSONAL CHOICES

Should a Married Couple Have Their Parents Live with Them?

This question is more often asked by individualized Westernized couples who live in isolated nuclear units. Asian couples reared in extended-family contexts expect to take care of their parents and consider it an honor to do so. As the parents of American spouses get older, a decision must often be made by the spouses about whether to have the parents live with them. Usually the person involved is the mother of either spouse, since the father is more likely to die first. One wife said: "We didn't have a choice. His mother is 82 and has Alzheimer's disease. We couldn't afford to put her in a nursing home at $5,200 a month, and she couldn't stay by herself. So we took her in. It's been a real strain on our marriage, since I end up taking care of her all day. I can't even leave her alone to go to the grocery store." Wakabayashi and Donato (2005) noted that the hit women take in their own employment earnings to care for elderly parents is considerable.

Some elderly persons have resources for nursing home care, or their married children can afford such care. But even in these circumstances, some spouses decide to have their parents live with them. "I couldn't live with myself if I knew my mother was propped up in a wheelchair eating Cheerios when I could be taking care of her," said one daughter.

When spouses disagree about parents in the home, the result can be devastating. According to one wife, "I told my husband that Mother was going to live with us. He told me she wasn't and that he would leave if she did. She moved in, and he moved out (we were divorced). Five months later, my mother died."

Strengthening Marriage through Divorce Law Reform

Some family scholars and policymakers advocate strengthening marriage by reforming divorce laws to make divorce harder to obtain. Since California became the first state to implement "no-fault" divorce laws in 1969, every state has passed similar laws allowing couples to divorce without proving in court that one spouse was at fault for the marital breakup. The intent of no-fault divorce legislation was to minimize the acrimony and legal costs involved in divorce, making it easier for unhappy spouses to get out of their marriage. Under the system of no-fault divorce, a partner who wanted a divorce could get one, usually by citing irreconcilable differences, even if his or her spouse did not want a divorce.

Other states believe the no-fault system has gone too far and have taken measures designed to make breaking up harder to do by requiring proof of fault (such as infidelity, physical or mental abuse, drug or alcohol abuse, and desertion) or extending the waiting period required before a divorce is granted. In most divorce law reform proposals, no-fault divorces would still be available to couples who mutually agree to end their marriages.

Opponents argue that divorce law reform measures would increase acrimony between divorcing spouses (which harms the children as well as the adults involved), increase the legal costs of getting a divorce (which leaves less money to support any children), and delay court decisions on child support and custody and distribution of assets. In addition, critics point out that ending no-fault divorce would add countless court cases to the dockets of an already overloaded court system. Efforts in many state legislatures to repeal no-fault divorce laws have largely failed.

The Louisiana legislature became the first in the nation to pass a law (in 1997) creating a new kind of marriage contract that would permit divorce only in narrow circumstances. Under the Louisiana law, couples can voluntarily choose between two types of marriage contracts: the standard contract that allows a no-fault divorce or a **covenant marriage** that permits divorce only under conditions of fault (such as abuse, adultery, or imprisonment on a felony) or after a marital separation of more than two years. Couples who choose a covenant marriage are also required to get premarital counseling from a clergy member or another counselor. The goal of a state offering a covenant marriage is to show the positive public regard for marriage (Byrne and Carr, 2005) and the disregard for single-parent families (Lyman, 2005). Covenant marriage is a unique way to begin married life, but there has been no rush on the part of couples to seek these marriages. Fewer than 3 percent of couples who marry in Louisiana have chosen to take on the extra restrictions of marriage by covenant (Licata, 2002). However, both Arizona and Arkansas now offer covenant marriages, which are primarily promoted through evangelical churches (Gilgoff, 2005).

Your Opinion?

1. To what degree do you believe the government can legislate "successful" marriage relationships?
2. How has no-fault divorce gone too far?
3. Why do you feel that covenant marriage is an idea that has not caught on?

References

Byrne, A. and D. Carr. 2005. Commentaries on: Singles in society and in science. *Psychological Inquiry* 16:84–141.

Gilgoff, D. 2005. Tying a tight knot. *U.S. News & World Report* 138:64–67.

Licata, N. 2002. Should premarital counseling be mandatory as a requisite to obtaining a marriage license? *Family Court Review* 40:518–32.

Lyman, R. 2005. Trying to strengthen an "I do" with a more binding legal tie. *New York Times* 154:A1–16

Love is blind, but marriage restores its sight.

Samuel Lichtenberg

Just as the state says who can marry (not close relatives, the insane, or the mentally deficient) and when (usually at age 18 or older), legal procedures must be instituted if the spouses want to divorce. The state's interest is that a couple stay married, have children, and take care of them. Should they divorce, the state will dictate how the parenting is to continue, both physically and economically.

Social policies designed to strengthen marriage through divorce law reform reflect the value the state places on stable, committed relationships (see Social Policy).

Marriage as a Rite of Passage

A **rite of passage** is an event that marks the transition from one social status to another. Starting school, getting a driver's license, and graduating from high school or college are events that mark major transitions in status (to student, to driver, and to graduate). The wedding itself is another rite of passage that marks the transition from fiancé to spouse. Preceding the wedding is the traditional bachelor party for the soon–to-be groom. What is new on the cultural landscape is the bachelorette party (sometimes more wild than the bachelor party), which conveys the message of equality and that great changes are ahead (Montemurro, 2006).

Weddings

The wedding is a rite of passage that is both religious and civil. To the Catholic Church, marriage is a sacrament that implies that the union is both sacred and indissoluble. According to Jewish and most Protestant faiths, marriage is a special bond between the husband and wife sanctified by God, but divorce and remarriage are permitted. Wedding ceremonies still reflect traditional cultural definitions of women as property. For example, the bride is usually walked down the aisle by her father and "handed over to the new husband." In some cultures, the bride is not even present at the time of the actual marriage (see Diversity in Other Countries). The lastest wedding expense is to have the wedding webcast so that family and friends afar can actually see and hear the wedding vows being said in real time. One such website is http://www.webcastmywedding.net/, where the would-be bride and groom can arrange the details.

I hope you will be as happy as we all thought we would be.

Said to a bride by a guest at her wedding reception

That marriage is a public experience is emphasized by the wedding in which family and friends of both parties are invited to participate. The wedding is a time for the respective families to learn how to cooperate with each other for the benefit of the respective daughter and son. Conflicts over number of bridesmaids and ushers, number of guests to invite, and place of the wedding are not uncommon. Though some families harmoniously negotiate all differences, others become so adamant about their preferences that the prospective bride and groom elope to escape or avoid the conflict. However, most families recognize the importance of the event in the life of their daughter or son and try to be helpful and nonconflictual.

Some states require that in order to obtain a marriage license, the partners must have a blood test to certify that neither has an STI. The document is then taken to the county courthouse, where the couple applies for a marriage license. Two-thirds of the states require a waiting period between the issuance of the license and the wedding. Eighty percent of couples are married by a member of the clergy; 20 percent (primarily in remarriages) go to a justice of the peace, judge, or magistrate.

Some brides wear something old, new, borrowed, and blue. The "old" (e.g., a gold locket) is something that represents the durability of the impending marriage. The "new," perhaps in the form of new, unlaundered undergarments, emphasizes the new life to begin. The "borrowed" (e.g., a wedding veil) is something

A private moment of a couple on their wedding day.

Authors

that has already been worn by a currently happy bride. The "blue" (e.g., ribbons) represents fidelity (those dressed in blue have lovers true). The bride's throwing her floral bouquet signifies the end of girlhood; the rice thrown by the guests at the newly married couple signifies fertility.

It is no longer unusual for couples to have weddings that are neither religious nor traditional. In the exchange of vows, neither partner promises to obey the other, and the couple's relationship is spelled out by the partners rather than by tradition. Vows often include the couple's feelings about equality, individualism, humanism, and openness to change. In 2006, the average wedding cost between $17,640 and $23,520 and included 188 guests. A second wedding costs almost as much as the first (Kirn and Cole, 2000).

Honeymoons

Traditionally, another rite of passage follows immediately after the wedding—the honeymoon. Almost 85% (84.2%) of 1027 undergraduates at a large southeastern university reported that it was important for them to take a honeymoon (Knox and Zusman, 2006). The functions of the honeymoon are both personal and social. The personal function is to provide a period of recuperation from the usually exhausting demands of preparing for and being involved in a wedding ceremony and reception. The social function is to provide a time for the couple to be alone to solidify the change in their identity from that of an unmarried to a married couple. And, now that they are married, their sexual expression with each other achieves full social approval and legitimacy. Now the couple can have children with complete societal approval.

Changes after Marriage

After the wedding and honeymoon, the new spouses begin to experience changes in their legal, personal, and marital relationship.

Legal Changes

Unless the partners have signed a prenuptial agreement specifying that their earnings and property will remain separate, after the wedding, each spouse becomes part owner of what the other earns in income and accumulates in property. Although the laws on domestic relations differ from state to state, courts typically award to each spouse half of the assets accumulated during the marriage (even though one of the partners may have contributed a smaller proportion).

For example, if a couple buy a house together, even though one spouse invested more money in the initial purchase, the other will likely be awarded half of the value of the house if they divorce. (Having children complicates the distribution of assets, since the house is often awarded to the custodial parent.) In the case of death of the spouse, the remaining spouse is legally entitled to inherit between one-third and one-half of the partner's estate, unless a will specifies otherwise.

Personal Changes

New spouses experience an array of personal changes in their lives. One initial consequence of getting married may be an enhanced self-concept. Parents and close friends usually arrange their schedules to participate in your wedding and

Getting married, like getting hanged, is a great deal less dreadful than it has been made out.

H. L. Mencken

How College Students View Weddings

Once a couple make a commitment to marry and become engaged, they begin to discuss the wedding, which may occasion their first set of realized differences. A team of researchers documented the ways in which undergraduate women and men differ in their attitudes toward weddings.

Sample

The sample consisted of 196 undergraduates from a large southeastern university who responded to an anonymous 47-item questionnaire designed to assess wedding attitudes and perceptions. Seventy percent of the respondents were female; 30 percent were male. The median age was 20, and almost half (47.9%) were involved with someone at the time of the survey.

Findings and Discussion

There were several differences between women and men in their attitudes toward weddings.

1. Women prepared more. Women were significantly more likely than men (55.9% vs. 17.6%) to report that they were intent on preparing for their wedding by reading *Modern Bride Magazine,* which they felt would be helpful to them in planning their wedding. College women also prepared by watching programs on television such as "A Wedding Story."

2. The wedding was for the bride's family. In terms of who the wedding is for, 16.4 percent of the men, compared with only 11.6 percent of the women, saw the wedding as being for the bride's family and not for the bride and groom. Hence, although both the potential groom and bride viewed the wedding as being for them as a couple, male college students were likely to observe the influence of the bride's mother in suggesting certain arrangements and conclude that the wedding was more for the bride's family.

3. The bride wanted the wedding documented. Female respondents were significantly more likely than male respondents (84.8% vs. 70.6%) to report their desire to "hire a professional photographer to take photos and videotape the wedding."

4. The bride preferred a formal wedding. Female respondents were significantly more likely than male respondents (77.4% vs. 61.4%) to report that they did not want a civil wedding ceremony with a justice of the peace officiating. There were some unusual "dream" weddings identified by the respondents: "on a cruise," "on a cliff over the Pacific at sunset," and "in Charlottesville, Virginia, with the Dave Matthews Band." One respondent wrote, "one where everyone attends."

5. Both parents should be invited if they are still married. When asked if they would invite both parents to the wedding, 97.3 percent of those students with both parents still married to each other answered yes, versus only 77 percent of students whose parents were divorced. One wonders how much the divorce of one's parents influences the wedding preparation and pleasure of the soon-to-be spouses.

6. Racial background affected the perception of who should pay for the wedding. One's racial background was also associated with significant differences. White students were significantly (58.4% vs. 21.0%) more likely than nonwhite students to report that parents should be responsible for paying for the wedding expenses.

There are two implications from these data. One, individuals who decide to marry should not be surprised that there are gender differences in how the woman and the man view the wedding. The woman has had a long period of socialization to think of the wedding as "her day," whereas the man has had zero socialization to view it as a positive. Two, both should not be surprised that the wedding is also in reference to the woman's mother. Conflicts between the two women over various aspects of the wedding are to be anticipated.

Source

Adapted from *Weddings: Some Data on College Student Perceptions* by D. Knox, M. E. Zusman, K. McGinty, and D. A. Abowitz 2003. *College Student Journal* 37:197–200. Used by permission of *College Student Journal*.

---

give gifts to express their approval of you and your marriage. In addition, the strong evidence that your spouse approves of you and is willing to spend a lifetime with you also tells you that you are a desirable person.

The married person also begins adopting new values and behaviors consistent with the married role. Although new spouses often vow that "marriage won't change me," it does. For example, rather than stay out all night at a party, which is not uncommon for singles who may be looking for a partner, spouses (who are already paired off) tend to go home early. Their roles of spouse, employee, and parent result in their adopting more regular, alcohol/drug-free hours.

At a wedding, the happiest couple in the world are sometimes the bride and groom, but more often the bride's parents.

Evan Esar

Friendship Changes

Marriage also affects relationships with friends of the same and other sex. Less time will be spent with friends because of the new role demands as a spouse. More time will be spent with other married couples, who will become powerful influences on the new couple's relationship.

What spouses give up in friendships, they gain in developing an intimate relationship with each other. However, abandoning one's friends after marriage may be problematic, because one's spouse cannot be expected to satisfy all of one's social needs. And because many marriages end in divorce, friendships that have been maintained throughout the marriage can become a vital source of support for a person adjusting to a divorce.

PERSONAL CHOICES

Is "Partner's Night Out" a Good Idea?

Although spouses may want to spend time together, they may also want to spend time with their friends—shopping, having a drink, fishing, golfing, seeing a movie, or whatever. Some spouses have a flexible policy based on trust with each other. Other spouses are very suspicious of each other. One husband said, "I didn't want her going out to bars with her girlfriends after we were married. You never know what someone will do when they get three drinks in them." For partner's night out to have a positive impact on the couple's relationship, it is important that the partners maintain emotional and sexual fidelity to each other, that each partner have a night out, and that the partners spend some nights alone with each other. Friendships can enhance a marriage relationship by making the individual partners happier, but friendships cannot replace the marriage relationship. Spouses must spend time alone to nurture their relationship.

Marital Changes

A happily married couple of 45 years spoke to the authors' class and began their presentation with "Marriage is one of life's biggest disappointments." They spoke of the difference between all the hype and cultural ideal of what marriage is supposed to be . . . and the reality. One effect of getting married is **disenchantment**—the transition from a state of newness and high expectation to a state of mundaneness tempered by reality. It may not happen in the first few weeks or months of marriage, but it is almost inevitable. Whereas courtship is the anticipation of a life together, marriage is the day-to-day reality of that life together—and reality does not always fit the dream. "Moonlight and roses become daylight and dishes" is an old adage reflecting the realities of marriage. Disenchantment after marriage is also related to the partners' shifting their focus away from each other to work or children; each partner usually gives and gets less attention in marriage than in courtship. Most college students do not anticipate a nosedive toward disenchantment; fewer than 20% (18.8%) of 1027 respondents agreed that "most couples become disenchanted with marriage within five years." Only 1.2% "strongly agreed" (Knox and Zusman, 2006).

A couple will experience other changes when they marry, as detailed below.

1. Change in how money is spent. Entertainment expenses in courtship become allocated to living expenses and setting up a household together.

2. Discovering that one's mate is different from one's date. Courtship is a context of deception. Marriage is one of reality. Spouses sometimes say, "He (she) is not the person I married."

3. Experiencing loss of freedom. Although premarital norms permit relative freedom to move in and out of relationships, marriage involves a binding legal contract. The new sense of confinement to one person and a routine life may bring out the worst in partners who thrived on freedom.

4. Sexual changes. The sexual relationship of the couple also undergoes changes with marriage. First, because spouses are more sexually faithful to each other than are dating partners or cohabitants (Treas and Giesen, 2000), their number of sexual partners will decline dramatically. Second, the frequency with which they have sex with each other decreases. According to one wife:

> The urgency to have sex disappears after you're married. After a while you discover that your husband isn't going to vanish back to his apartment at midnight. He's going to be with you all night, every night. You don't have to have sex every minute because you know you've got plenty of time. Also, you've got work and children and other responsibilities, so sex takes a lower priority than before you were married.

After ecstasy, the laundry.
Zen saying

Although married couples may have intercourse less frequently than they did before marriage, marital sex is still the most satisfying of all sexual contexts. Eighty-eight percent of married people in a national sample reported that they experienced extreme physical pleasure, and 85 percent reported experiencing extreme emotional satisfaction, with their spouses. In contrast, 54 percent of individuals who were not married or not living with anyone said that they experienced extreme physical pleasure with their partners, and 30 percent said that they were extremely emotionally satisfied (Michael et al., 1994).

5. Power changes. The distribution of power changes after marriage and across time. The way wives and husbands perceive and interact with each other continues to change throughout the course of the marriage. Two researchers studied 238 spouses who had been married more than 30 years and observed that (across time) men changed from being patriarchal to collaborating with their wives and that women changed from deferring to their husbands' authority to challenging that authority (Huyck and Gutmann, 1992). In effect, men tend to lose power and women gain power. But such power changes may not always occur. In abusive relationships, the abusive partner may increase the display of power because he or she fears the partner will try to escape from being controlled.

Parents and In-Law Changes

Marriage affects relationships with parents. Time spent with parents and extended kin radically increases when the couple has children. Indeed, a major difference between couples with and without children is the amount of time they spend with relatives. Parents and kin rally to help with the newborn and are typically there for birthdays and family celebrations.

Emotional separation from one's parents is an important developmental task in building a successful marriage. When choices must be made between one's parents and one's spouse, more long-term positive consequences for the married couple are associated with choosing the spouse over the parents. But such choices become more complicated and difficult when one's parents are old, ill, or widowed.

Nuner (2004) interviewed 23 daughters-in-law married between 5 and 10 years (with no previous marriages and at least one child from the marriage) and found that most reported positive relationships with their mother-in-law. The evaluation by the daughter depended on the role of the mother-in-law within the family (e.g., the mother-in-law as grandmother was perceived more positively). Mother/son relationships were described by the daughters-in-law to be close, "mama's boy," polite, or distant. The next section details a daughter-in-law's relationship with her mother-in-law; she described the relationship her husband has with his mother as "polite."

A Wife's Experience with Her Mother-in-Law: A Mini-Story

> I am a recently married in-love-with-my-husband-woman and had always heard that when you marry your true love, you also wed their family. But I never realized how true this statement actually was until I said 'I do.' As the "girlfriend" in my dating

relationship, I suppose I did have some insight as to what my future relationship with my mother-in-law might be like. After all, the first time I met her, she only spoke to me when I specifically addressed a question to her, and even then, it was feigned interest in response. She even made a cutting remark that day about how "some people" might like the salad I had made to take to a Fourth of July picnic. Her tone in stating this was just sly enough to evade my husband's senses, but directed enough to let me know she didn't approve of me or my choice of covered dish. Although it seemed that she was deliberately creating a feeling of tension between me and her, I downplayed my own feelings about this initial meeting in her defense. After all, I was dating her only son. I hoped that she would warm to me as I proved to her how devoted I was to him. I feel that I am a genuinely nice person and have always won people over with little effort.

I have since realized, however, that with my in-law, I could have been a combination of Queen Elizabeth, Mother Teresa, and Jackie Kennedy all rolled into one, and I might not have been good enough for her son. At first, the criticisms that she had were shocking to me. On my wedding day, for example, she asked my sister, who was applying my makeup, if she had to go to clown school to learn how to do that. Was this her feeble attempt at some sense of humor to connect with my family or had she just called me a clown?

Later, I would find from her sometimes weekly unannounced visits to my house that she seemed to find some pleasure in pointing out my misgivings or those hidden flaws in myself and my home. No matter how much I had cleaned before she has arrived, she has pointed out such things as a cobweb on the ceiling, a spot on the floor, or a paint drip in the far corner of the room. This is not because she is a neat freak herself. She also expresses very specific ideas about how my pets shouldn't be allowed on the furniture, how my future children should be raised in a Stalin-like dictatorship with laughter allowed only at specified "fun" times, how every meal could be more perfect if I would only add this or leave out that, and how my husband and I should spend our free time. She has even gone so far as to sign us up to take an art class with her without asking us if we wanted to go or if we were free. To top it all off, she instructed us that we would pay for all meals and expenses of the day. I politely declined, but my husband dutifully attended.

My husband and I have experienced some tension in our relationship as a result of my mother-in-law's interference, particularly in that he will never say "no" to her unless I ask him to do so. Her spring cleaning has resulted in hideous furniture and revolting knick knacks finding their way into our home that we have neither needed nor wanted. She never asks, but assumes that we want everything she drops off or calls him to come get, including, most recently, her cat! Luckily, we had just gotten a second dog and were able to refuse this "gift," much to her resentment.

My biggest fear is that she will one day ask to live with us, and my husband will, again, not be able to say no or find an alternative. Had my mother-in-law been the warm, loving "second mom" that I had envisioned in my youthful dreams of what marriage would be like, I would have no problems taking her in. Instead, I cannot imagine living in the same household with the woman whom I feel would slowly pick my life apart and divide my relationship with my husband simultaneously.

While her comments still annoy me at times, I have come to expect them. I suspect it is a reflection of her insecurities of being replaced and not needed by her only child that pushes her to remind her son that his wife is not as perfect as his mother and never will be. I am disappointed that she will not accept that I would rather be her ally than her enemy. Thank goodness for long term care insurance.

Financial Changes

An old joke about money in marriage says that "two can live as cheaply as one as long as one doesn't eat." The reality behind the joke is that marriage involves the need for spouses to discuss and negotiate how they are going to get and spend

Don't let your in-laws become your outlaws.

Ed Hartz, sociologist

money in their relationship. Henry et al. (2005) analyzed data on 105 older married couples, who revealed that finances was the number three issue over which they disagreed and had difficulty ("leisure activities" and "intimacy" were one and two, respectively).

Some spouses bring considerable debt into the marriage. In a sample of 1027 undergraduates 11 percent reported that they owed "over a thousand dollars on one or more credit cards" (Knox and Zusman, 2006). Special Topic 2 on Money and Debt Management at the end of the text details the importance of budgeting—identifying what money is available, how it will be spent, and how much will be saved for what.

Diversity in Marriage

The tragedies of September 11, 2001 (9/11) emphasized the need to understand other cultures and an appreciation for military families who make personal sacrifices for the larger societal good. In this section we review Muslim American families and Military families. We also look at other examples of family diversity: interracial, interreligious, cross-national, and age-discrepant.

Muslim American Families

Although Islam (the religious foundation for Muslim families in 60 nations) is the third-largest religion in North America, 9/11 resulted in an increased awareness that Muslim families are part of American demographics. These families hardly represent the extremists responsible for terrorism, but more than 5.8 million adults in the United States and 1.3 billion worldwide self-identify with the Islamic religion (there are now more Muslims than Christians in the world). The three

To paint a monolithic picture of "the" Muslim family in North America would deny the important variations within and between types of families. North American Muslim families are similar but they are also different.

Sharon McIrvin Abu-Laban

This is a Muslim family living in the United States. In the center, holding the baby, is the grandfather, and to his left is the grandmother. Their two sons and daughters-in-law flank them on either side. The children on the grass and standing in the back are those of the younger married couples.

largest American Muslim groups in the United States are African Americans, Arabs, and South Asians (e.g., from Pakistan, Bangladesh, Afghanistan, India).

The five "pillars" of Islam provide the basis for the values and perspectives of its followers: faith (there is no god but Allah—One True God—and Muhammad is His Messenger); prayer (five times a day); alms (giving 2.5% annually of one's wealth to other Muslims in need, such as a Muslim orphanage); fasting for the month of Ramadan (the ninth of the month in the Islamic Lunar year); and going for Hajj (Pilgrimage) to Mecca (in Saudi Arabia, where Kaabah is located) once in one's life if physically and financially possible. Although Muslims vary in devoutness and practice, this discussion will focus on Muslim families in America who attempt to maintain their traditions based on the *Qur'an* (the holy book of Islam) and the *Hadith* (teachings of Prophet Mohammad).

Islamic tradition emphasizes close family ties with the nuclear and extended family, social activities with family members, and respect for the authority of the elderly and parents. Religion and family are strong sources of a Muslim's personal identity. Following one's religious and family codes results in a strong sense of emotional and social support. Breaking from one's religion and family comes at a great cost because alternatives are perceived as limited. Parents of Muslim children who are reared in America struggle to maintain traditional values while allowing their children (particularly sons) to pursue higher education and professional training.

One of the striking features of Muslim-American families is the strong influence parents have over the behavior of their children. Because the families control the property and economic resources and generally provide total financial support to the children, and because the offspring may not be able to find adequate work outside the family system, they generally acquiesce to their parent's wishes. Such acquiescence does not imply the nonexistence of genuine love and affection children may have for their parents, however. Table 8.2 details the various practices, values, and beliefs involved in Muslim family life.

Table 8.2 Core Values of Muslim American Families

Courtship is tightly controlled. Intimate mixing of sexes is against Islamic teaching. Therefore, mixed-sex gatherings of children, teens, and young adults do not occur unless in the presence of adults. At such gatherings there is "separate seating" next to one's same sex peers. "Dating" in the sense of being alone with a partner to explore romance and sex is prohibited and contrary to the Muslim idea that marriage is between two families.

Mate Choice. Offspring are taught early to consider marrying only a person who shares their religion/culture and to defer to their parents and kin, whose experience qualifies them to guide the choice of a mate. Offspring (particularly women) are expected to marry another Muslim (e.g., marriage to a cousin on either parent's side is permissible and not unusual). A man may marry outside of Islam only to a Christian or Jewish woman (who is not required to convert to Islam) as long as the partner agrees to rear children in the Islamic traditions. In the usual case, the parents of one Muslim family contact parents of another Muslim family for their respective offspring to meet. They may go off alone for a brief time to talk, but their time alone is limited and elders are not far away. Each has the right of refusal. In effect their only "choice" is to reject a preselected choice. Parents prefer to select a son-in-law whose education and financial resources are equal to or higher than their own and who has close family ties to the girl's family. The ideal daughter-in-law is one who is integrated into the parent's family (she knows her husband's parents) and who accepts the norm of patrilocal residence (she expects to live near her husband's parents). Independent dating/mate selection is prohibited. A Muslim daughter who sees men without supervision (or who continually rejects the marriage choices offered by her parents) can be ostracized by her family, and, in some cases, have her support cut off and inheritance withdrawn. Her chances for marriage to a Muslim may be lost because she may be regarded as morally depraved or promiscuous and will not be accepted by the Muslim family of a Muslim man. The threat of unmarriageability is a frightening prospect for Muslim young women and bears a strong force of social control over her behavior. Singlehood for both women and men is viewed as unnatural and abnormal.

Love. Because marriage is generally seen as a merging of two families and the individuals getting married are allowed to be alone together for only infrequent, brief periods, love is expected to follow, not precede marriage. Hence, romantic love in the American sense is nonexistent. For Muslims the selection and approval by parents/kin and satisfactory financial circumstances are the necessary ingredients of a successful marriage.

Table 8.2 Core Values of Muslim American Families—cont'd

Sexual behavior. Holding hands, kissing, and sexual intercourse are strictly forbidden before marriage. Indeed, the *Qur'an* implores individuals to "not go near fornication, as it is immoral and an evil way" (XVII:32). This translates into daughters avoiding provocative dress—no lipstick, makeup, tank tops, short dresses, high heels, or dancing (except for the pleasure of her husband). Sexual behavior after marriage allows only for penile vaginal penetration (which is procreative). Mutual manual stimulation and oral and anal sex are prohibited (these are non-procreative acts). Adultery by married adults is punishable by being publicly stoned to death. Adultery committed by unmarried adults is punishable by flogging in public.

Marriage. The ceremony involves two male witnesses for the bridegroom, a guardian (vakil) for the bride, and a payment by the husband of a dowry for the marriage to be valid. The dowry (also known as mehr) is an amount of money or property given by the husband to the wife at the time of the wedding (or that he promises to give her on demand). It symbolizes respect for the woman and becomes her property to spend as she wishes and remains hers even in the case of divorce. Indeed, it provides a sense of security should the marriage fail. The marital relationship is focused on "friendship accompanied by mutual benevolence" (Danespour, 1989). It is not focused on romantic sexual enjoyment, but "intellectual, spiritual, and moral compatibility" (Danespour, 1989).

Gender roles. Equality between husbands and wives is emphasized. Although fathers, husbands, and sons are in the role of protector/guardian of daughters, wives, and sisters, the teachings of the Prophet Mohammad specify protections and rights for women. They can keep their own name, inherit (although one-half as much as sons), own/buy/sell/inherit property, refuse a marriage proposal, initiate a divorce, and be awarded custody of the children in case of divorce. The husband is obligated to provide for his wife and children. The wife is expected to be responsible for childcare, maintain her chastity, and manage the household. She may be employed if she wants or in the event of financial necessity but not in jobs such as waitressing, which might involve short dresses or alcohol.

Rearing children. Children are highly valued, loved, and indulged. Parents tightly regulate/monitor their children's leisure activities (after-school activities, mixed parties, television, and playing cards are prohibited), particularly their daughters', and ensure that they spend as much time with other family members or other Muslims. Such restrictions are also designed to prohibit the development of a love relationship with a non-Muslim. This value is held so strongly that some families send their daughters to school in the homeland or move the whole family back home during the child's adolescence or young adulthood. *Communication* is the key word for parent-child relationships in the Muslim home. Parents always make themselves available to their children and keep a close watch on them. Parents also show Islamic values to their children by their behavior, not just by what they say. Girls are also kept busy performing domestic chores (washing dishes, cooking, cleaning the house, and watching younger siblings). Boys are usually spared such chores and are given more freedom. Sisters and brothers may feel that they live in separate worlds.

Elderly. Children are expected to respect and be kind and dutiful toward their parents. Sons are specifically obligated financially and otherwise to take care of their elderly parents in need (daughters are not as they receive only one-half the inheritance share of a son). The son will be looked down on by other Muslims if he fails in his responsibility to take care of his parents. Most will take their parents into their home or put them in an apartment nearby. A nursing home for the Muslim elderly is a rare exception in the United States. Many elderly Muslims have financial resources and require only frequent visits and help with some chores.

Alcohol. All Muslims (men, women, young, old) are prohibited from consuming alcohol or alcoholic products (e.g., some cough syrups). Also, it is a sin to produce, buy, sell, or give/receive alcohol. It is the worst nightmare for Muslim parents if their young adult children drink wine or any alcoholic beverage.

Birth Control is not generally accepted, but it is possible to limit the number of children in a family by *coitus interruptus*. Any birth control method that is known to interrupt pregnancy after fertilization is prohibited.

Abortion is allowed only to save the life of the mother.

Divorce. Although spouses are expected to stay together unless doing so becomes intolerable, either spouse may request divorce. A wife may divorce because of her husband's impotence, refusal to provide economic subsistence or clothing, change of religion, or infectious disease. A husband may divorce his wife for several reasons, such as infidelity, cultural incompatibility, persistent refusal to abide by Shariah (Islamic jurisprudence), or chronic complaints in spite of his sincere efforts to fulfill her legitimate needs. Custody and care of the children usually go to the husband, who is responsible for the expenses of child-rearing even if the wife is willing to do it.

This table was developed for this text with the assistance of Dr. Saeed Dar, a Muslim and Professor of Pharmacology and Toxicology, East Carolina University School of Medicine. Dr. Dar has lectured on Muslim Family Values as well as Islamic Sexuality.

Military Families

Newspaper headlines, nightly television news, and Presidential news conferences remind us that America is at war and individuals are coping with danger and the sacrifices of deployments in Iraq, Afghanistan, and other peacekeeping and humanitarian operations. Military personnel on active duty include 1.5 million in the military reserve, 1.1 million in the ready reserve, and 7 million in the National Guard (*Statistical Abstract of the United States: 2006,* Tables 505, 506, 507, 508). About 60 percent of military personnel are married and/or have children (NCFR Policy Brief, 2004). Sixty percent of Americans in a national sample report that they have a family member, close friend, or office worker in Iraq (Page, 2006).

Military families are unique in several ways:

1. Traditional Sex Roles. Although both men and women are members of the military service, there are considerably more men than women (Caforio, 2003). In the typical military family, it is the husband who is deployed (sent away) and the wife who is expected to "understand" his military obligations and to take care of the family in his absence. In the case of wives/mothers who are deployed, it is the rare husband who is able to switch roles and become Mr. Mom. One military career wife said of her husband, whom she left behind when she was deployed, "What a joke. He found out what taking care of kids and running a family was really like and he was awful. He fed the kids SpaghettiOs for the entire time I was deployed."

President Bush says he understands. Until he deploys his daughters to Iraq, he hasn't a clue.

Mother of son in Iraq

2. Loss of Control—Deployment. Military families have little control over their lives, as the specter of deployment is ever-present. Where one of the spouses will be next week and for how long, are beyond the control of the spouses and parents. Saleska (2004) interviewed wives of Air Force men (enlisted and officers) who emphasized the difficulty of being faced with the ever-present possibility that their husbands could leave immediately for an indeterminate period of time. And once gone, they may be relegated to a ten-minute phone call every two weeks. The needs of the military (referred to as a "Greedy Institution") come first, and military personnel are expected to be obedient and to do whatever is necessary to comply and get through the ordeal. "You can't believe what it's like to have an empty chair at the dinner table sprung on you and not know where he is or when he'll be back," one respondent said.

Compounding the loss of control is the fear of being captured, imprisoned, shot, killed by a suicide bomber, or beheaded. Not only may the deployed soldier have such fears, but his or her spouse, parents, and children may look at the evening news in stark terror and fear that their beloved will be the next to die. Sleeplessness, irritability, and depression may result in those who are left behind to carry on their jobs and parenting. Children may also become anxious and depressed over the absence of their deployed parent, who more often is the father (Cozza et al,, 2005).

3. Infidelity. Although most spouses are faithful to each other, the context of separation from each other for months (sometimes years) at a time increases the vulnerability of both spouses to infidelity. The double standard may also be operative, whereby "men are expected to have other women when they are away" and "women are expected to remain faithful and be understanding." Separated spouses try to bridge the time they are apart with e-mails and phone calls (when possible), but sometimes the loneliness becomes more difficult than anticipated. One enlisted husband said that he returned home after a year–and-a-half deployment to be confronted with the fact that his wife had become involved with someone else. "I absolutely couldn't believe it," he noted. "In retrospect, I think the separation was more difficult for her than it was for me."

4. Separation from Extended Family/Close Friends. In addition to infidelity being related to deployment, being separated from one's extended family is a major blow to military families. Parents no longer have doting grandparents available to help them rear their children. And although other military families become a community of support for each other, the consistency of such support may be lacking. "We moved seven states away from my parents to a town in North Dakota," said one wife. "It was dreadful."

Similar to being separated from parents and siblings is the separation from one's lifelong friends. Although new friendships and new supportive relationships develop within the military community to which the family moves, the relationships are sometimes tenuous and temporary as the new families move on. The result is the absence of a stable predictable social structure of support, which may result in a feeling of alienation and not belonging in either the military or the civilian community. And the more frequent the moves, the more difficult the transition and the more likely the alienation. "A higher divorce rate among military families is no surprise," notes Donald Wolfe (2006), who is a marriage/family counselor who specializes in military marriages. And among military marriages, whites are as likely to divorce as blacks (Lindquist, 2004).

5. Resilient Military Families. In spite of these difficulties, most military families are amazingly resilient. Not only do they anticipate and expect mobilization and deployment as part of their military obligation, they respond with pride. Indeed, some re-enlist eagerly and volunteer to return to military life even when retired. One military Captain stationed at Ft. Bragg, in Fayetteville, NC, noted, "It is part of being an American to defend your country. Somebody's got to do it and I've always been willing to do my part." He and his wife made a presentation in the authors' classes. She said, "I'm proud that he cares for our country and I support his decision to return to Iraq to help as needed. And most military wives that I know feel the same way."*

Interracial Marriages

Interracial marriages may involve many combinations, including American white, American black, Indian, Chinese, Japanese, Korean, Mexican, Malaysian, and Hindu mates. Almost half (48.9%) of a sample of 1027 undergraduates at a southeastern university reported a preference for marrying someone of their same race (Knox and Zusman, 2006). However, actual interracial marriages are rare in the United States—fewer than 5 percent of all marriages in the United States are interracial. Of these, fewer than 1 percent are of a black person and a white person (*Statistical Abstract of the United States: 2006,* Table 54). Examples of African-American men who are married to Caucasian women are Tiger Woods and Charles Barkley. Segregation in religion (the races worship in separate churches), housing (white and black neighborhoods), and education (white and black colleges), not to speak of parental and peer endogamous pressure to marry within one's own race, are factors that help to explain the low percentage of interracial black/white marriages.

The spouses in black–white couples are more likely to have been married before, to be age-discrepant, to live far away from their families of orientation, to have been reared in racially tolerant homes, and to have educations beyond high school. Some may also belong to religions that encourage interracial unions. The Baha'i religion, which has more than 6 million members worldwide and 84,000 in the United States, teaches that God is particularly pleased with interracial unions. Finally, interracial spouses may tend to seek contexts of diversity. "I have been reared in a military family, been everywhere and met people of different races and nationalities throughout my life. I seek diversity," noted one student.

*Appreciation is expressed to Samuel W. Lynch (stationed in Iraq) and Mark Fisch (stationed in Kosovo) for their insights into military marriages.

College students tend to be open to interracial dating.

Authors

Attitudes toward Interracial Dating Scale

Interracial dating or marrying is the dating or marrying of two people from different races. The purpose of this survey is to gain a better understanding of what people think and feel about interracial relationships. Please read each item carefully, and in each space, score your response using the following scale. There are no right or wrong answers to any of these statements.

1	2	3	4	5	6	7
Strongly Disagree						Strongly Agree

_____ 1. I believe that interracial couples date outside their race to get attention.

_____ 2. I feel that interracial couples have little in common.

_____ 3. When I see an interracial couple I find myself evaluating them negatively.

_____ 4. People date outside their own race because they feel inferior.

_____ 5. Dating interracially shows a lack of respect for one's own race.

_____ 6. I would be upset with a family member who dated outside his/her race.

_____ 7. I would be upset with a close friend who dated outside his/her race.

_____ 8. I feel uneasy around an interracial couple.

_____ 9. People of different races should associate only in nondating settings.

_____10. I am offended when I see an interracial couple.

_____11. Interracial couples are more likely to have low self-esteem.

_____12. Interracial dating interferes with my fundamental beliefs.

_____13. People should date only within their race.

_____14. I dislike seeing interracial couples together.

_____15. I would not pursue a relationship with someone of a different race, regardless of my feelings for him/her.

_____16. Interracial dating interferes with my concept of cultural identity.

_____17. I support dating between people with the same skin color, but not with a different skin color.

_____18. I can imagine myself in a long-term relationship with someone of another race.

_____19. As long as the people involved love each other, I do not have a problem with interracial dating.

_____20. I think interracial dating is a good thing.

Scoring

First, reverse the scores for items 18, 19, and 20 by switching them to the opposite side of the spectrum. For example, if you selected 7 for item 18, replace it with a 1; if you selected 3, replace it with a 5, etc. Next, add your scores and divide by 20. Possible final scores range from 1 to 7, with 1 representing the most positive attitudes toward interracial dating and 7 representing the most negative attitudes toward interracial dating.

Norms

The norming sample was based upon 113 male and 200 female students attending Valdosta State University. The participants completing the Attitudes toward Interracial Dating Scale (IRDS) received no compensation for their participation. All participants were United States citizens. The average age was 23.02 years (standard deviation [SD] = 5.09), and participants ranged in age from 18 to 50 years. The ethnic composition of the sample was 62.9 percent white, 32.6 percent black, 1 percent Asian, 0.6 percent Hispanic, and 2.2 percent other. The classification of the sample was 9.3 percent freshmen, 16.3 percent sophomores, 29.1 percent juniors, 37.1 percent seniors, and 2.9 percent graduate students. The average score on the IRDS was 2.88 (SD = 1.48), and scores ranged from 1.00 to 6.60, suggesting very positive views of interracial dating. Men scored an average of 2.97 (SD = 1.58), and women, 2.84 (SD = 1.42). There were no significant differences between the responses of women and men.

The Attitudes Toward Interracial Dating Scale, copyright 2004 by Mark Whatley, Ph.D., Department of Psychology, Valdosta State University, Valdosta, Georgia 31698. Information on validity and reliability may be obtained from Dr. Whatley. The scale is used by permission of Dr. Whatley (mwhatley@valdosta.edu). Other uses of this scale by written permission only.

And so this is Christmas
For black and for white,
For yellow and red,
Let's stop all the fight.

John Lennon

Kennedy (2003) identified three reactions to a black–white couple who cross racial lines to marry: approval (increases racial open-mindedness, decreases social segregation), indifference (interracial marriage is seen as a private choice), and disapproval (reflects racial disloyalty, impedes perpetuation of black culture). As Kennedy notes, "The argument that intermarriage is destructive of racial solidarity has been the principal basis of black opposition" (p. 115). There is also the concern for the biracial identity of offspring of mixed-race parents. Although most mixed-race parents identify their child as having minority race status, there is a trend toward identifying their child as multiracial or white (Brunsma, 2005).

Interracial partners sometimes experience negative reactions to their relationship. Blacks partnered with whites have their blackness and racial identity challenged by blacks. Whites partnered with blacks may lose their white status and have their awareness of whiteness heightened more than ever before. At the same time, one partner is not given full status as a member of the other partner's race (Hill and Thomas, 2000). Gaines and Leaver (2002) also note that the pairing of a black male and a white female is regarded as "less appropriate" than that of a white male and a black female. In the former, the black male "often is perceived as attaining higher social status (i.e., the white woman is viewed as the black man's 'prize,' stolen from the more deserving white man)" (p. 68). In the latter, when a white male pairs with a black female, "no fundamental change in power within the American social structure is perceived as taking place" (p. 68). Interracial marriages are also more likely to dissolve than same-race marriages (Fu, 2006).

Black-white interracial marriages are likely to increase—slowly. Not only has white prejudice against African-Americans in general declined, but segregation in school, at work, and in housing has decreased, permitting greater contact between the races. The Self-Assessment on page 240 allows you to assess your openness to involvement in an interracial relationship.

Interreligious Marriages

Although religion may be a central focus of some individuals and their marriage, Americans in general have become more secular, and as a result religion has become less influential as a criterion for selecting a partner. In a survey of 1027 undergraduates, only around 40 percent (40.7%) reported that it was important for them to marry someone of the same religion (Knox and Zusman, 2006).

Are people in interreligious marriages less satisfied with their marriages than those who marry someone of the same faith? The answer depends on a number of factors. First, people in marriages in which one or both spouses profess "no religion" tend to report lower levels of marital satisfaction than those in which at least one spouse has a religious tie. People with no religion are often more liberal and less bound by traditional societal norms and values; they feel less constrained to stay married for reasons of social propriety.

The impact of a mixed religious marriage may also depend more on the devoutness of the partners than on the fact that the partners are of different religions. If both spouses are devout in their religious beliefs, they may expect some problems in the relationship (although not necessarily). Less problematic is the relationship in which one spouse is devout but the partner is not. If neither spouse in an interfaith marriage is devout, problems regarding religious differences may be minimal or nonexistent. One interfaith couple who married (he Christian, she Jewish) said in their marriage vows that they viewed their different religions as an opportunity to strengthen their connections to their respective faiths and to each other. "Our marriage ceremony seeks to celebrate both the Jewish and Christian traditions, just as we plan to in our life together."

Cross-National Marriages

More than 60 percent (62.7%) of 1027 undergraduates reported that they would be willing to marry someone from another country (Knox and Zusman, 2006). The opportunity to meet someone from another country is increasing as more than half a million foreign students are studying at American colleges and universities.

National Data

Approximately 600,000 foreign students are enrolled at more than 2,500 colleges and universities in the United States. Most (60%) are from Asia (*Statistical Abstract of the United States: 2006*, Table 269).

The purpose of all the major religious traditions is not to construct big temples on the outside, but to create temples of goodness and compassion inside, in our hearts.

The Dalai Lama, *The Good Heart*

My parents don't know.

Pakistani student in love with an American

This wife is 20 years younger than her husband. They had 20 years together before her husband died.

At the age of 18, she became the fourth wife of 54-year-old Charlie Chaplin. The May–December alliance was expected to last the requisite six months, but they confounded skeptics by staying together, raising eight children, and remaining, in their words, blissfully happy.

Jane Scovell of Oona Chaplin, wife of Charles Chaplin, from the book *Oona: Living in the Shadows*

Because American students take classes with foreign students, there is the opportunity for dating and romance between the two groups, which may lead to marriage. Some persons from foreign countries marry an American citizen to gain citizenship in the United States, but immigration laws now require the marriage to last two years before citizenship is granted. If the marriage ends before the two years, the foreigner must prove good faith (he or she did not marry just to gain entry into the country) or will be asked to leave the country.

When the international student is male, more likely than not his cultural mores will prevail and will clash strongly with his American bride's expectations, especially if the couple should return to his country. One female American student described her experience of marriage to a Pakistani, who violated his parents' wishes by not marrying the bride they had chosen for him in childhood. The marriage produced two children before the four of them returned to Pakistan.

The woman felt that her in-laws did not accept her and were hostile toward her. The in-laws also imposed their religious beliefs on her children and took control of their upbringing. When this situation became intolerable, the woman wanted to return to the United States. Because the children were viewed as being "owned" by their father, she was not allowed to take them with her and was banned from even seeing them. Like many international students, the husband was from a wealthy, high-status family, and the woman was powerless to fight the family. The woman has not seen her children in six years.

Cultural differences do not necessarily cause stress in cross-national marriage; the degree of cultural difference is not necessarily related to degree of stress. Much of the stress is related to society's intolerance of cross-national marriages, as manifested in attitudes of friends and family. Japan and Korea place an extraordinarily high value on racial purity. At the other extreme is the racial tolerance evident in Hawaii, where a high level of out-group marriage is normative.

Age-Discrepant Relationships and Marriages

Among all sexually active women aged 15 to 44 in one study, 20 percent had a partner who was three to five years older, and 18 percent had a partner who was six or more years older (Darroch et al., 2000). Silverthorne and Quinsey (2000) asked 192 adults to express their age preferences for a preferred partner. Both heterosexual and homosexual men preferred younger partners (homosexual women preferred older partners).

When the partners are ten or more years apart in age, the union is regarded as an age-discrepant relationship. A study (Knox et al., 1997) of 77 female university faculty and their female students who were involved with men ten to 25 years older revealed five themes in these age-discrepant relationships.

1. Age-discrepant relationships are happy. Eighty percent reported that they were happy in their relationships. Forty percent agreed with the statement, "I am happy in my current relationship" and 40 percent reported "strong agreement." Only 4 percent disagreed with the statement. More than 60 percent said that they would become involved in another age-discrepant relationship if their current relationship ended.

2. Age-discrepant relationships lack social approval and support. Only a quarter of the respondents reported that their friends, mothers, and fathers provided clear support for their relationship. Fathers were least approving, with more than 40 percent not approving.

3. Age-discrepant relationships are not without problems. In addition to lack of support, the respondents in this study reported a range of problems they attributed to the age difference with their partners, including money, in-laws, and recreation. Having a child may also be an issue. Most older men have had children in previous relationships. Garrison Keillor, creator and host of *A Prairie Home Companion,* had a son almost age 30 when Keillor's new wife wanted a child. He said, "To have a child in your mid-50's is not for the timid. But I felt that if you are with a woman who wants very much to have a child, you must say yes or you must break up." (Engleman, 2005).

4. Women perceive benefits from involvement with older partners. Respondents noted financial security (58%), maturity (58%), and dependability (51%) as the primary advantages of involvement with an older man. Higher status was regarded as less important: only 28 percent of the respondents identified this.

5. Friends of the couple are joint friends. More than 70 percent (71%) of respondents reported that when they did something recreational, the friends were likely to be both of theirs. However, if the friends were friends of only one of them, it was more likely to be the man (22%) than the woman (5%).

Aside from the above study on age-discrepant relationships in a sample from academia, Brietman et al. (2004) looked at national data and found that where there is a large discrepancy between the ages (where the man is at least 16 years older or the woman at least 10 years older), there is a higher risk for intimate-partner homicide by either partner.

Some age-discrepant dating relationships become age-discrepant marriage relationships, also referred to as age-dissimilar marriages (ADMs). When the man is considerably older than the woman, such marriages are referred to as **May–December marriages.** Typically, she is in the spring of her youth (May), and he is in the later years of his life (December). There have been a number of age-discrepant celebrity marriages, including that of Celine Dion, who is 26 years younger than Rene Angelil (in 2007, aged 39 and 65). Larry King is also 26 years older than his seventh wife, Shawn (in 2007 he was 72). Michael Douglas is 25 years older than his wife, Catherine Zeta Jones, and Ellen DeGeneres is 15 years older than Portia de Rossi, her partner.

Though the situation is less common, some women are older than their partners. In 2005, Demi Moore married Ashton Kutcher—she is 16 years older (age 40, vs. his 27, at the time of the wedding). Levesque and Caron (2004) examined the dating experiences of heterosexual women between the ages of 35 and 50 years and compared them to a group of younger women between the ages of 20 and 25. The researchers found a tendency of older women to choose younger men and postulated that the definition of a "suitable" partner may change as a woman ages.

Valerie Gibson (2002) is the author of *Cougar: A Guide for Older Women Dating Younger Men.* She noted that the current use of the term *cougars* refers to "women, usually in their 30s and 40s, who are financially stable and mentally independent and looking for a younger man to have fun with" (ABC News, 2005). Financially independent women need not select a man in reference to his breadwinning capabilities. Instead, these "cougars" are looking for men, not to marry, but to enjoy. The downside to the woman being older comes if the man gets serious and wants to have children, which may spell the end of the relationship.

Gibson also referred to a 2003 AARP (American Association of Retired Persons) survey that revealed that one-third of women between the ages of 40 and 60 are dating younger men. In addition to Demi Moore's involvement with a younger man, comedienne Fran Drescher, 47, also dated a man for four years

A cougar is a single older woman who prefers to date younger men and is proud of the choice . . . I didn't hit my stride till 40.

Valerie Gibson, *Cougar: A Guide for Older Women Dating Younger Men*

This couple is celebrating their 50th wedding anniversary.

Authors

who was 16 years her junior. Although these celebrity relationships are visible, no current national figures are available on age-discrepant relationships in which the woman is older.

Success in Marriage

Marriage—as its veterans know well—is the continuous process of getting used to things you hadn't expected.

Tom Mullen, *A Very Good Marriage*

Successful marriage has been a focus of researchers in marriage and the family (DeOllos, 2005). But what is a successful marriage and what are the characteristics?

Characteristics of Successful Marriages

Marital success is measured in terms of marital stability and marital happiness. Stability refers to how long the spouses have been married and how permanent they view their relationship, whereas marital happiness refers to more subjective aspects of the relationship.

In describing marital success, researchers have used the terms *satisfaction, quality, adjustment, lack of distress,* and *integration.* Marital success is often measured by asking spouses how happy they are, how often they spend their free time together, how often they agree about various issues, how easily they resolve conflict, how sexually satisfied they are, and how often they have considered separation or divorce. The degree to which the spouses enjoy each other's companionship is another variable of marital success. Not all couples, even those recently married, achieve high-quality marriages. In a national sample comparing marrieds with the unmarried, Princeton Survey Research Associates International found that 43% of the spouses reported that they were "very happy," compared with 24% of unmarrieds (Stuckey and Gonzalez, 2006).

Corra et al. (2006) analyzed data collected over a 30-year period, from the 1972 to 2002 General Social Surveys, to discover the influence of sex (male or female) and race (white or black) on the level of reported marital happiness. Findings indicated greater levels of marital happiness among males and whites than among females and blacks. The researchers suggested that males make fewer accommodations in marriage and that whites are not burdened with racism and have less economic stress. Barrett (2005) also noted that wives feel less in control of their marriages.

Wallerstein and Blakeslee (1995) studied 50 financially secure couples in stable (from 10 to 40 years), happy marriages with at least one child. These couples defined marital happiness as feeling respected and cherished. They also regarded their marriage as a work in progress that needed continued attention lest it become stale. No couple said that they were happy all the time. Rather, a good marriage is a process. Frye and Karney (2002) noted that spouses generally perceive themselves to be better off than other couples in terms of having a good marriage and feel that their problems will remain stable (not get worse) over time.

Billingsley et al. (1995) interviewed 30 happily married couples who had been wed an average of 32 years and had an average of 2.5 children. Various themes of couples who stay together and who enjoy each other and their relationship include the following.

1. Commitment. Divorce was not considered an option. The spouses were committed to each other for personal reasons rather than societal pressure. In addition, the spouses were committed to maintain the marriage out of emotional rather than economic need (DeOllos, 2005).

2. Common interests. The spouses talked of sharing interests, values, goals, children, and the desire to be together. Wilson and Cousins (2005) confirmed that similarity of the spouses (in contrast to complementarity) is predictive of long-term relationship success.

3. Communication. Gottman and Carrere (2000) studied the communication patterns of couples over an 11-year period and emphasized that those spouses who stay together are five times more likely to lace their arguments with positives ("I'm sorry I hurt your feelings") and to consciously choose to say things to each other that nurture the relationship rather than destroy it. Successful spouses also feel comfortable telling each other what they want and not being defensive at feedback from the partner.

4. Religiosity. A strong religious orientation provided the couples with social, spiritual, and emotional support from church members and with moral guidance in working out problems (DeOllos, 2005).

5. Trust. Trust in the partner provided a stable floor of security for the respective partners and their relationship. Neither partner feared that the other partner would leave or become involved in another relationship.

6. Not materialistic. Being nonmaterialistic, being disciplined, and being flexible with each other's work schedules and commitments were characteristic of these happily married couples.

7. Role models. The couples spoke of positive role models in their parents. Good marriages beget good marriages—good marriages run in families. It is said that the best gift you can give your children is a good marriage. Greenfield and Marks (2006) analyzed data on the linkage of parents to their children's lives and found that parents who were happy in their own relationships tended to have children who mirrored similar well being. Conversely, parental problems and difficulties were reflected in those of their children.

8. Sexual desire. Regan (2000) studied 25 men and 25 women involved in dating relationships and found that sexual desire was related to a greater desire for the partner, relationship stability, being faithful, and not being attracted to others. Wilson and Cousins (2005) also confirmed that partners' similar rankings of sexual desire as important is predictive of long-term relationship success. Earlier, we noted the superiority of marital sex over sex in other relationship contexts in terms of both emotion and physical pleasure.

I am certain that I have one of America's better marriages; and yet the challenge of keeping it successful never dims, for Camille and I may be blinded by love, but we have Braille for each other's flaws.

Bill Cosby, Comedian

If there is something you need that you don't see, please let me know, and I'll show you how to do without it.

Richard Herman, sign on a wall in his one-room cabin in the woods

9. Equitable relationships. Donaghue and Fallon (2003) found that individuals in less traditional and more equitable relationships reported higher levels of relationship satisfaction. Spouses who feel that they are contributing more to the relationship that is not reciprocated by the partner may become resentful (DeOllos, 2005).

10. Absence of negative attributions. Spouses who do not attribute negative motives to their partner's behavior report higher levels of marital satisfaction than spouses who do ruminate about negative motives. Dowd et al. (2005) studied 127 husbands and 132 wives and found that the absence of negative attributions was associated with higher marital quality.

11. Sacrifice. A team of researchers (Stanley, et al., 2006) found that viewing sacrifice for the relationship as rewarding was predictive of less distress and higher marital adjustment. Hence, the antithesis of selfishness and individualism is a positive indicator of marital satisfaction over time.

Theoretical Views of Marital Happiness and Success

Interactionists, developmentalists, exchange theorists, and functionalists view marital happiness and success differently. Symbolic interactionists emphasize the subjective nature of marital happiness and point out that the definition of the situation is critical.

Only when spouses define the verbal and nonverbal behavior of their partner as positive, and only when they label themselves as being in love, may a happy marriage exist. Hence, marital happiness is not defined by the existence of eight or more specific criteria but is subjectively defined by the respective partners.

Family developmental theorists emphasize the developmental tasks that must be accomplished to enable a couple to have a happy marriage. Wallerstein and Blakeslee (1995) identified several of these tasks, including separating emotionally from one's parents, building a sense of "we-ness," establishing an imaginative and pleasurable sex life, and making the relationship safe for expressing differences.

Exchange theorists focus on the exchange of behavior of a kind and at a rate that is mutually satisfactory to both spouses. When spouses exchange positive behaviors at a high rate, they are more likely to feel marital happiness than when the exchange is characterized by high-frequency negative behavior (Turner, 2005).

Structural functionalists see marital happiness as contributing to marital stability, which is functional for society. When two parents are in love and happy, the likelihood that they will stay together to provide physical care and emotional nurturing for their offspring is increased. Furthermore, when spouses take care of their own children, society is not burdened with having to pay for the children's care through welfare payments, paying foster parents, or paying for institutional management (group homes) when all else fails. Happy marriages also involve limiting sex to each other. In their national sex survey, Michael and colleagues reported, "[H]appiness is clearly linked to having just one partner—which may not be too surprising since that is the situation that society smiles upon" (1994, p. 130). Fewer HIV cases also mean lower medical bills for society. Similarly, marriage is associated with improved health (Stack and Eshleman, 1998), because spouses monitor each other's health and encourage/facilitate medical treatment as indicated.

Marital Happiness across the Family Life Cycle

Although a successful marriage is one in which the partners are happy and in love across time, spouses report that some periods are happier than others. Figure 8.1 provides a retrospective of marriage by 52 white college-educated husbands and wives over a period of 35 years together. The couples reported the

To marry unequally is to suffer equally.

Henri F. Amiel

He who laughs, lasts.

Mary Poolegive

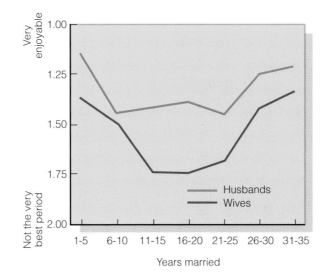

Figure 8.1

Source: Caroline O. Vaillant and George E. Vaillant. 1993. Is the U-curve of marital satisfaction an illusion? A 40-year study of marriage. *Journal of Marriage and the Family* 55:237 (Figure 6, copyrighted 1993 by the National Council on Family Relations, 3989 Central Avenue N. E., Suite 550, Minneapolis, MN 55421.)

After seven years of marriage, I'm sure of two things—first, never wallpaper together, and second, you'll need two bathrooms . . . both for her.

Dennis Miller, comedian

most enjoyment with their relationship in the beginning, followed by less enjoyment during the childbearing stages, and a return to feeling more satisfied after the children left home. Corra et al. (2006) found a similar curvilinear relationship in their analysis of survey data covering a period of 30 years.

The Value of Marriage Education Programs

A team of researchers (Macomber et al., 2005) evaluated the numerous programs available in our society designed to strengthen marriage. They found an incredible diversity of programs offered through a variety of settings (e.g., churches, mental health clinics), focused on a variety of populations (e.g., couples, parents), by personnel with a range of academic training (from none to PhD). The lack of systematic content, control groups, and follow-ups makes it impossible to evaluate the myriad programs, so their value remains an open question. DeMaria (2005) also confirmed the lack of data as to the effectiveness of marriage education programs.

Nevertheless, Congress passed legislation in 2005 (Deficit Reduction Act) including $750 million for a five-year program to fund marriage education programs including marriage skills training, high school education programs on the value of marriage, and programs encouraging responsible fatherhood. The funds may not be used to promote same-sex relationships/marriage (Healthy Marriage Initiative, 2006).

What can we learn from our knowledge about the characteristics of successful marriages? Commitment, common interests, communication skills, and a nonmaterialistic view of life all are factors in maintaining a successful marriage. Couples might strive to include these as part of their relationship.

As we have noted, durability is only one criterion for a successful marriage. Satisfaction is another. Researchers in the early nineties analyzed cross-sectional data and concluded that marital satisfaction drops across time, reaches a low point during the years the couple has teens in the house, and then returns to preteen satisfaction levels (Vaillant and Vaillant, 1993). More recent researchers have studied longitudinal data

Diversity in Other Countries

Halford and Simons (2005) reviewed the marriage education programs for couples in Australia. Although there is government support for such programs, there is little research to confirm the positive outcome of such programs.

and found that marital satisfaction consistently drops across time (with the steepest declines in the early and later years) (Vanlaningham et al., 2001).

Healthy Marriage Resource Center

The Healthy Marriage Resource Center has been set up in Washington, DC. It is the result of a $4.5-million grant from the U.S. Department of Health and Human Services Administration of Children and Families (ACF) to The National Council on Family Relations of Minneapolis, Minnesota. The goal of the Center is to collect and disseminate information on the skills and knowledge to build and sustain a healthy marriage. An annual Marriage Summit is also held to advise the center and discuss the state of marriage. The primary method of exposure to knowledge about healthy marriage is via the Web (http://www.healthymarriageinfo.org/). Individuals interested in national and regional events related to healthy marriages can use a point-and-click map that will direct them to healthy marriage opportunities in their area.

SUMMARY

What are individual motivations and social functions of marriage?

Individuals' motives for marriage include personal fulfillment, companionship, legitimacy of parenthood, and emotional and financial security. Social functions include continuing to provide society with socialized members, regulating sexual behavior, and stabilizing adult personalities.

What are three levels of commitment in marriage?

Marriage involves a commitment—person-to-person, family-to-family, and couple-to-state.

What are two rites of passage associated with marriage?

The wedding is a rite of passage signifying the change from the role of fiancé to the role of spouse. Women, more than men, are more invested in preparation for the wedding, the wedding is more for the bride's family, and women prefer a traditional wedding. The honeymoon is a time of personal recuperation and making the transition to the new role of spouse.

What changes might a person anticipate after marriage?

Changes after the wedding are legal (each becomes part owner of all income and property accumulated during the marriage), personal (enhanced self-concept), social (less time with friends), economic (money spent on entertainment in courtship is diverted to living expenses and setting up a household), sexual (less frequency), and parental (improved relationship with parents).

How do marriage patterns differ among racial and cultural groups?

Muslim-American families involve norms that involve no premarital sex, close monitoring of children, and intense nurturing of the parental-child bond. Military families cope with deployment, the double standard, and limited income. Mixed marriages include interracial, interreligious, and age-discrepant.

What are the characteristics associated with successful marriages?

Marital success is defined in terms of both quality and durability. Characteristics associated with marital success include commitment, common interests, communication, religiosity, trust, being nonmaterialistic, having positive role models, low stress levels, and sexual desire. Marriage education programs are designed to improve marriage. There are few empirical studies on marriage education programs.

KEY TERMS

bride wealth	commitment	disenchantment	May–December marriage
commensality	covenant marriage	marital success	rite of passage

The Companion Website for *Choices in Relationships: An Introduction to Marriage and the Family,* Ninth Edition
www.thomson.edu/sociology/knox

Supplement your review of this chapter by going to the companion website to take one of the Tutorial Quizzes, use the flash cards to master key terms, and check out the many other study aids you'll find there. You'll also find special features such as the Marriage and Family Resource Center, Census 2000 information, and other data and resources at your fingertips to help you with that special project or to do some research on your own.

WEBLINKS

American Marriage
http://www.heartchoice.com/marriage/

Bridal Registry
http://www.theknot.com

Brides and Grooms
http://www.bridesandgrooms.com/

Facts about Marriage
http://www.cdc.gov/nchs/fastats/marriage.htm

Healthy Marriage Resource Center
http://www.healthymarriageinfo.org/

Military Marriages
http://www.defenselink.mil/ The Department of Defense website

http://www.nmfa.org/ The National Military Family Association

http://www.mfrc-dodqol.org/ Military Family Resource Center

Smart Marriages
http://www.smartmarriages.com/

REFERENCES

ABC News. 2005. Are more older women with younger men? May 5 broadcast.
Barrett, A. E. 2005. Gendered experiences in midlife: Implications for age identity. *Journal of Aging Studies* 19:163–83.
Billingsley, S., M. Lim, and G. Jennings. 1995. Themes of long-term, satisfied marriages consummated between 1952–1967. *Family Perspective* 29:283–95.
Brietman, N., T. K. Shackelford, and C. R. Block. 2004. Couple age discrepancy and risk of intimate partner homicide. *Violence and Victims* 19:321–42.
Brunsma, D. L. 2005. Interracial families and the racial identification of mixed-race children: Evidence from the early childhood longitudinal study. *Social Forces* 84:1131–57.
Byrne, A. and D. Carr. 2005. Commentaries on: Singles in society and in science. *Psychological Inquiry* 16:84–141.
Caforio, G. 2003. *Handbook on the sociology of the military.* New York: Kluwer Academic.
Coontz, S. 2000. Marriage: Then and now. *Phi Kappa Phi Journal* 80:16–20.
Corra, M., J. S. Carter, and D. Knox. 2006. Marital happiness by sex and race: A second look. Paper, American Sociological Association, Annual Meeting, New York, August.
Cozza, S. J., R. S. Chun, and J. A. Polo. 2005. Military families and children during operation Iraqi freedom. *Psychiatric Quarterly* 76:371–78.
Crawford, D. W., D. Feng, J. L. Fischer, and L. K. Diana. 2003. The influence of love, equity, and alternatives on commitment in romantic relationships. *Family and Consumer Sciences Research Journal* 31:253–71.
Darroch, J. E., D. J. Landry, and S. Oslak. 2000. Age differences between sexual partners in the United States. *Family Planning Perspectives* 31:160–67.
DeMaria, R. M. 2005. Distressed couples and marriage education. *Family Relations* 54:242–53.
DeOllos, I. Y. 2005. Predicting marital success or failure: Burgess and beyond. In *Sourcebook of family theory & research,* edited by Vern L. Bengtson, Alan C. Acock, Katherine R. Allen, Peggye Dilworth-Anderson, and David M. Klein. Thousand Oaks, CA: Sage Publications, 134–36.
Donaghue, N., and B. J. Fallon. 2003. Gender-role self-stereotyping and the relationship between equity and satisfaction in close relationships. *Sex Roles: A Journal of Research* 48:217–31.

Dowd, D. A., M. J. Means, J. F. Pope, and J. H. Humphries. 2005. Attributions and marital satisfaction: The mediated effects of self-disclosure. *Journal of Family and Consumer Sciences* 97:22–27.

Engleman, P. 2005. An interview with Garrison Keillor: Keillor instinct. *AARP The Magazine,* 48:25–33.

Frye, N. E., and B. R. Karney. 2002. Being better or getting better? Social and temporal comparisons as coping mechanisms in close relationships. *Personality and Social Psychology Bulletin* 28:1287–99.

Fu, X. 2006. Impact of socioeconomic status on inter-racial mate selection and divorce. *Social Science Journal* 43: 239–258

Gaines, S. O., Jr., and J. Leaver. 2002. Interracial relationships. In *Inappropriate relationships: The unconventional, the disapproved, and the forbidden,* edited by R. Goodwin and D. Cramer. Mahwah, NJ: Lawrence Erlbaum, 65–78.

Gibson, V. 2002. *Cougar: A Guide for Older Women Dating Younger Men.* Boston, MA: Firefly Books.

Glover, K. R., M. Phelps, A. Sean Burleson, and A. Dessie. 2006. Friends or lovers: Understanding the cognitive underpinnings of the transition from affiliation to care-giving in romantic relationships. 4th Annual ECU Research and Creative Activities Symposium, April 21, East Carolina University, Greenville, NC.

Gottman, J., and S. Carrere. 2000. Welcome to the love lab. *Psychology Today.* September/October, 42 passim.

Greenfield, E. A. and N. F. Marks. (2006) Linked lives: Adult children's problems and their parents' psychological and relational well-being. *Journal of Marriage and Family* 68:442–454.

Guo, B. R. and J. Huang. 2005. Marital and sexual satisfaction in Chinese families: Exploring the moderating effects. *Journal of Sex & Marital Therapy* 31:21–29.

Halford, W. K. and M. Simons. 2005. Couple relationship education in Australia. *Family Process* 44:147–59.

Healthy Marriage Initiative, 2006. http://www.acf.hhs.gov/healthymarriage/about/mission.html Retrieved Feb 21, 2006.

Henry, R. G., R. B. Miller, and R. Giarrusso. 2005. Difficulties, disagreements, and disappointments in late-life marriages. *International Journal of Aging & Human Development* 61:243–65.

Hill, M. R., and V. Thomas. 2000. Strategies for racial identity development: Narratives of black and white women in interracial partner relationships. *Family Relations* 49:193–200.

Huyck, M. H., and D. L. Gutmann. 1992. Thirtysomething years of marriage: Understanding experiences of women and men in enduring family relationships. *Family Perspective* 26:249–65.

Kennedy, R. 2003. *Interracial intimacies.* New York: Pantheon.

Kirn, W., and W. Cole. 2000. Twice as nice. *Time* 19 June, 53.

Knox, D., T. Britton, and B. Crisp. 1997. Age discrepant relationships reported by university faculty and students. *College Student Journal* 31:290–93.

Knox, D. and Zusman, M. E. 2006. Relationship and sexual behaviors of a sample of 1027 university students. Unpublished data collected for this text. Department of Sociology, East Carolina University, Greenville, NC.

Levesque, L. M. and S. L. Caron. 2004. Dating preferences of women born between 1945 and 1960. *Journal of Family Issues* 25:833–46.

Licata, N. 2002. Should premarital counseling be mandatory as a requisite to obtaining a marriage license? *Family Court Review* 40:518–32.

Lindquist, J. H. 2004. When race makes no difference: Marriage and the military. *Social Forces* 83:731–57.

Macomber, J. E., J. Murray, and M. Stagner. 2005. Investigation of programs to strengthen and support healthy marriages. Feb 11. Available at http://www.urban.org/url.cfm?ID=411141.

Michael, R. T., J. H. Gagnon, E. O. Laumann, and G. Kolata. 1994. *Sex in America: A definitive survey.* Boston: Little, Brown.

Montemurro, B. 2006. *Something old, something bold.* New Brunswick, NJ: Rutgers University Press.

Murdock, G. P. 1949. *Social structure.* New York: Free Press.

Murray, C. I. and N. Kimura. 2003. Multiplicity of paths to couple formation in Japan. In *Mate selection across cultures* edited by R. R. Hamon and B. B. Ingoldsby. Thousand Oaks, CA: Sage Publications, 247–68.

NCFR Policy Brief. 2004. *Building strong communities for military families.* Minneapolis, MN: National Council on Family Relations.

Nuner, J. E. 2004. A qualitative study of mother-in-law/daughter-in-law relationships. *Dissertation Abstracts International, A: The Humanities and Social Sciences,* 65:August, 712A–13A.

Page, S. 2006. War has hurt USA. *USA Today* March 17, A1.

Pimentel, E. E. 2000. Just how do I love thee? Marital relations in urban China. *Journal of Marriage and the Family* 62:32–47.

Regan, P. C. 2000. The role of sexual desire and sexual activity in dating relationships. *Social Behavior and Personality* 28:51–60.

Rowland, I. 2006. Choosing to have children or choosing to be childfree: Australian students' attitudes towards the decisions of heterosexual and lesbian women. *Australian Psychologist* 41:55–59.

Saleska, S. 2004. Exploratory study of problems and stresses dependent military spouses experience. Paper, Second Annual East Carolina University Research and Scholarship Day, March 26, Greenville, NC.

Sayer, L. C., S. M. Bianchi, and J. P. Robinson. 2004. Are parents investing less in children? Trends in mothers' and father's time with children. *American Journal of Sociology* 110:1–43.

Scovell, J. 1998. *Living in the shadows: A biography of Oona O'Neill Chaplin.* New York: Warner Books.

Sherif-Trask, B. 2003. Love, courtship, and marriage from a cross-cultural perspective: The upper middle class Egyptian example. In *Mate Selection Across Cultures* edited by R. R. Hamon and B. B. Ingoldsby. Thousand Oaks, CA: Sage Publications, 121–36.

Silverthorne, Z. A., and V. L. Quinsey. 2000. Sexual partner age preferences of homosexual and heterosexual men and women. *Archives of Sexual Behavior* 29:67–76.

Sobal, J., C. F. Bove, and B. S. Rauschenbach. 2002. Commensal careers at entry into marriage: Establishing commensal units and managing commensal circles. *Sociological Review* 50:378–97.

Sousa, Lori A. 1995. Interfaith marriage and the individual and family life cycle. *Family Therapy* 22:97–104.

Stack, S., and J. R. Eshleman. 1998. Marital happiness: A 17-nation study. *Journal of Marriage and the Family* 60:527–36.

Stanley, S. M., S. W. Whitton, S. L. Sadberry,, R. L. Clements and H. J. Markman. 2006. Sacrifice as a predictor of marital outcomes. *Family Process* 45:289–303

Statistical Abstract of the United States: 2006. 125th ed. Washington, D.C.: U.S. Bureau of the Census.

Stuckey, D. and A. Gonzalez. 2006. Princeton Survey Research Associates Poll on marriage happiness. *USA Today* March 7, 1A.

Treas, J., and D. Giesen. 2000. Sexual infidelity among married and cohabiting Americans. *Journal of Marriage and the Family* 62:48–60.

Turner, A. J. 2005. Personal communication, Huntsville, Alabama, September. Used by permission.

Vaillant, C. O., and G. E. Vaillant. 1993. Is the U-curve of marital satisfaction an illusion? A 40-year study of marriage. *Journal of Marriage and the Family* 55:230–39.

Vanlaningham, J., D. R. Johnson, and P. Amato. 2001. Marital happiness, marital duration, and the U-shaped curve: Evidence from a five-year wave panel study. *Social Forces* 79:1313–41.

Wakabayashi, C. and K. Donato. 2005. The consequences of caregiving: Effects on women's employment and earnings. *Population Research and Policy Review* 24:467–88.

Wallerstein, J., and S. Blakeslee. 1995. *The good marriage.* Boston: Houghton-Mifflin.

Walters, L. H., P. Skeen, W. Warzywoda-Krusynska, and T. Kurko. 1997. Marital happiness in young families: Similarities and differences across countries. Paper, National Council on Family Relations conference, Crystal City, VA.

Weigel, D. J., and D. S. Ballard-Reisch. 2002. Investigating the behavioral indicators of relational commitment. *Journal of Social and Personal Relationships* 19:403–23.

Wilmoth, J., and G. Koso. 2002. Does marital history matter? Marital status and wealth outcomes among preretirement adults. *Journal of Marriage and the Family* 64:254–68.

Wilson, G. and J. Cousins. 2005. Measurement of partner compatibility; further validation and refinement of the CQ test. *Sexual and Relationship Therapy* 20:421–29.

Wilson, S. M., L. W. Ngige, and L. J. Trollinger. 2003. Kamba and Maasai paths to marriage in Kenya. In *Mate selection across cultures*, edited by R. R. Hamon and B. B. Ingoldsby. Thousand Oaks, CA: Sage Publications, 95–117.

Wolfe, D. 2006. Personal communication. Jacksonville, NC: Camp LeJune Military Base.

Kelly Lewis

chapter 9

I know you believe
You understand
What you think I said,
But I am not sure you realize
That what you heard
Is not what I meant.

Anonymous

Communication in Relationships

Contents

True or False?

1. It is possible to predict who is likely to be unfaithful in a relationship.

2. Sharing tasks with one's partner is associated with feeling emotionally close.

3. A parent's demand is often met with an adolescent's withdrawal.

4. Lying about the number of previous partners was the most frequently reported lie told by a sample of university students.

5. One's physiological makeup may enhance or impede one's potential to learn communication skills.

Answers: **1.** T **2.** T **3.** T **4.** T **5.** T

Communication is used as a barometer of a couple's relationship. Couples talk of communication as the factor that confirms the quality of their relationship ("We can talk all night about anything and everything") or condemns their relationship ("We have nothing to say to each other; we are getting a divorce"). In this chapter we examine various issues related to communication and identify some communication principles and skills. We begin by looking at the nature of interpersonal communication.

The Nature of Interpersonal Communication

Communication can be defined as the process of exchanging information and feelings between two people. Although most communication is focused on verbal content, much (estimated to be as high as 80%) interpersonal communication is nonverbal. Regardless of what a person says, crossed arms and lack of eye contact will convey a very different meaning than the same words accompanied with a gentle touch and eye-to-eye contact. We often attend to the nonverbal cues

One learns people through the heart, not the eyes or the intellect.

Mark Twain, humorist

The touch of your hand says you'll catch me whenever I fall.
You say it best when you say nothing at all.

Alison Krause, country western singer

Regardless of what this couple is saying, they are communicating negative nonverbal feelings by their behavior.

in interaction and assign them more importance (Preston, 2005). In effect, we like to hear sweet words but we feel more confident when we see behavior that supports the words. "Show me the money, honey" is a phrase that reflects a partner's focus on behavior rather than words. A team of researchers (Knobloch et al., 2006) studied 120 dating couples and found that the more they perceived a romantic interest in each other, the greater the amount of relationship talk.

Principles and Techniques of Effective Communication

The problem with communication is the illusion that it has occurred.

George Bernard Shaw

Persons who want effective communication in their relationship follow various principles and techniques, including the following.

1. Make communication a priority. Communicating effectively implies making communication an important priority in a couple's relationship. When communication is a priority, partners make time for it to occur in a setting without interruptions: they are alone; they do not answer the phone; they turn the television off. Making communication a priority results in the exchange of more information between the partners, which increases the knowledge each partner has about the other.

Negative relationship outcomes occur when partners do not prioritize communication with each other but are passionately and obsessively interacting with others via the Internet. Seguin-Levesque et al. (2003) found that it is not use of the Internet *per se* but the obsessive passion of involvement with the Internet that is destructive.

2. Establish and maintain eye contact. Shakespeare noted that a person's eyes are the "mirrors to the soul." Partners who look at each other when they are talking not only communicate an interest in each other but also are able to gain information about the partner's feelings and responses to what is being said. Not looking at your partner may be interpreted as lack of interest and prevents you from observing nonverbal cues.

3. Ask open-ended questions. When your goal is to find out your partner's thoughts and feelings about an issue, it is best to use open-ended questions. Such questions (e.g., "How do you feel about me?") encourage your partner to give an answer that contains a lot of information. Closed-ended questions (e.g., "Do you love me?"), which elicit a one-word answer such as yes or no, do not provide the opportunity for the partner to express a range of thoughts and feelings.

4. Use reflective listening. Effective communication requires being a good listener. One of the skills of a good listener is the ability to use the technique of **reflective listening,** which involves paraphrasing or restating what the person has said to you while being sensitive to what the partner is feeling. For example, suppose you ask your partner, "How was your day?" and your partner responds, "I felt exploited today at work because I went in early and stayed late and a memo from my new boss said that future bonuses would be eliminated because of a company takeover." Listening to what your partner is both saying and feeling, you might respond, "You feel frustrated because you really worked hard and felt unappreciated . . . and it's going to get worse."

My wife said I don't listen—at least I think that's what she said.

Laurence Peter, *Humorist*

Reflective listening serves the following functions: (1) it creates the feeling for the speaker that she or he is being listened to and is being understood and (2) it increases the accuracy of the listener's understanding of what the speaker is saying. If a reflective statement does not accurately reflect what the speaker thinks and feels, the speaker can correct the inaccuracy by restating her or his thoughts and feelings.

An important quality of reflective statements is that they are nonjudgmental. For example, suppose two lovers are arguing about spending time with their respective friends and one says, "I'd like to spend one night each week with my friends and not feel guilty about it." The partner may respond by making a statement that is judgmental (critical or evaluative), such as those exemplified in Table 9.1. Judgmental responses serve to punish or criticize someone for what he or she thinks, feels, or wants and often result in frustration and resentment.

Table 9.1 also provides several examples of nonjudgmental reflective statements.

5. Use "I" statements. **"I" statements** focus on the feelings and thoughts of the communicator without making a judgment on others. Because "I" statements are a clear and nonthreatening way of expressing what you want and how you feel, they are likely to result in a positive change in the listener's behavior.

In contrast, **"you" statements** blame or criticize the listener and often result in increasing negative feelings and behavior in the relationship. For example, suppose you are angry with your partner for being late. Rather than say, "You are always late and irresponsible" (which is a "you" statement), you might respond with, "I get upset when you are late and will feel better if you call me when you will be delayed." The latter focuses on your feelings and a desirable future behavior rather than blaming the partner for being late.

6. Avoid negative expressivity. Rayer and Volling (2005) studied the levels of emotional expressivity (both positive and negative) and found that negative expressivity had a strong impact on marital love and **conflict.** Because intimate partners are capable of hurting each other so intensely, be careful how you criticize or communicate disapproval to your partner.

7. Say positive things about your partner. A team of researchers found that emotional expressiveness was strongly related to marital adjustment, particularly when coupled with the suppression of negative statements (Ingoldsby et al., 2005).

People like to hear others say positive things about them. These positive statements may be in the form of compliments (e.g., "You look terrific!") or appreciation (e.g., "Thanks for putting gas in the car"). Gable et al. (2003) asked 58 heterosexual dating couples to monitor their interaction with each other. The respondents observed that they were overwhelmingly positive, at a five-to-one ratio.

8. Tell your partner what you want. Focus on what you want rather than on what you don't want. Rather than say, "You always leave the bathroom a wreck," an alternative might be "Please hang up your towel after you take a shower." Rather than say, "You never call me when you are going to be late," say "Please call me when you are going to be late."

9. Stay focused on the issue. **Branching** refers to going out on different limbs of an issue rather than staying focused on the issue. If you are discussing the

Give every man thine ear, but few thy voice; take each man's censure but reserve thy judgement.

William Shakespeare

He tells me I'm mad when I'm not. This makes me mad.

A woman, of her live-in partner

Table 9.1 Judgmental and Nonjudgmental Responses to Your Partner's Saying "I'd Like to Spend One Evening a Week with my Friends"

Nonjudgmental Reflective Statements	Judgmental Statements
It sounds like you really miss your friends.	You only think about what you want.
You think it is healthy for us to be with our friends some of the time.	Your friends are more important to you than I am.
You really enjoy your friends and want to spend some time with them.	You just want a night out so that you can meet someone new.
You think it is important that we not abandon our friends just because we are involved.	You just want to get away so you can drink.
You think that our being apart one night each week will make us even closer.	You are selfish.

overdrawn checkbook, stay focused on the checkbook. To remind your partner that he or she is equally irresponsible when it comes to getting things repaired or doing housework is to get off the issue of the checkbook. Stay focused.

10. Make specific resolutions to disagreements. To prevent the same issues or problems from recurring, it is important to agree on what each partner will do in similar circumstances in the future. For example, if going to a party together results in one partner's drinking too much and drifting off with someone else, what needs to be done in the future to ensure an enjoyable evening together? (For example, decide how many drinks the partner will have within a given time period.)

11. Give congruent messages. **Congruent messages** are those in which the verbal and nonverbal behaviors match. A person who says, "Okay, you're right" and smiles as he or she embraces the partner is communicating a congruent message. In contrast, the same words accompanied by leaving the room and slamming the door communicate a very different message. Walther et al. (2005) compared affect in on-line and face-to-face interaction and found few differences.

12. Share power. One of the greatest sources of dissatisfaction in a relationship is a power imbalance and conflict over power (Kurdek, 1994). **Power** is the ability to impose one's will on the partner and to avoid being influenced by the partner. Expressions of power are numerous and include the following.

Withdrawal (not speaking to the partner)

Guilt induction ("How could you ask me to do this?")

Being pleasant ("Kiss me and help me move the sofa.")

Negotiation ("We can go to the movie if we study for a couple of hours before we go.")

Deception (running up credit card debts that the partner is unaware of)

Blackmail ("I'll find someone else if you won't have sex with me.")

Physical abuse or verbal threats ("I'll kill you if you leave.")

Criticism ("I can't think of anything good about you.")

In general, the spouse with the more prestigious occupation, higher income, and more education exerts the greater influence on family decisions. Indeed, Dunbar and Burgoon (2005) noted that the greater the perception of one's own power, the more dominant one was in conversation with the partner.

But power may also take the form of love and sex. The person in the relationship who loves less and who needs sex less has enormous power over the partner who is very much in love and who is dependent on the partner for sex. This pattern reflects the principle of least interest we discussed earlier in the text.

13. Keep the process of communication going. Communication includes both content (verbal and nonverbal information) and process (interaction). It is important not to allow difficult content to shut down the communication process (Turner, 2005). To ensure that the process continues, the partners should focus on the fact that the sharing of information is essential and reinforce each other for keeping the process alive. For example, if your partner tells you something that you do that bothers him or her, it is important to thank him or her for telling you that rather than becoming defensive. In this way, your partner's feelings about you stay out in the open rather than hidden behind a wall of resentment. Otherwise, if you punish such disclosure because you don't like the content, subsequent disclosure will stop.

An underlying principle in all of these techniques is to engage in supportive communication. Supportive communication exists when the partners feel comfortable discussing a range of issues with each other. The Self-Assessment allows you to assess the degree to which your relationship is characterized by supportive communication.

Although effective communication skills can be learned, Robbins (2005) noted that there might be physiological capacities that may enhance or impede

Don't speak unless you can improve on the silence.

Spanish proverb

Chapter 9 Communication in Relationships

Supportive Communication Scale

This scale is designed to assess the degree to which partners experience supportive communication in their relationships. After reading each item, circle the number that best approximates your answer.

0 = strongly disagree (SD)
1 = disagree (D)
2 = undecided (UN)
3 = agree (A)
4 = strongly agree (SA)

	SD	D	UN	A	SA
1. My partner listens to me when I need someone to talk to.	0	1	2	3	4
2. My partner helps me clarify my thoughts.	0	1	2	3	4
3. I can state my feelings without his/her getting defensive.	0	1	2	3	4
4. When it comes to having a serious discussion, it seems we have little in common (reverse scored).	0	1	2	3	4
5. I feel put down in a serious conversation with my partner (reverse scored).	0	1	2	3	4
6. I feel it is useless to discuss some things with my partner (reverse scored).	0	1	2	3	4
7. My partner and I understand each other completely.	0	1	2	3	4
8. We have an endless number of things to talk about.	0	1	2	3	4

Scoring

Look at the numbers you circled. Reverse score the numbers for questions 4, 5, and 6. For example, if you circled a 0, give yourself a 4; if you circled a 3, give yourself a 1, etc. Add the numbers and divide by 8, the total number of items. The lowest possible score would be 0, reflecting the complete absence of supportive communication; the highest score would be 4, reflecting complete supportive communication. The average score of 94 male partners who took the scale was 3.01; the average score of 94 female partners was 3.07. Thirty-nine percent of the couples were married, 38 percent were single, and 23 percent were living together. The average age was just over 24.

Source

Sprecher, Susan, Sandra Metts, Brant Burelson, Elaine Hatfield, and Alicia Thompson. 1995. Domains of expressive interaction in intimate relationships: Associations with satisfaction and commitment. *Family Relations* 44:203–10. Copyright (1995) by the National Council on Family Relations, 3989 Central Ave. NE, Suite 550, Minneapolis, MN 55421.

the acquisition of these skills. She noted that persons with attention deficit hyperactivity disorder (ADHD) might have deficiencies in basic communication and social skills. Being able to communicate effectively is valuable. Rosof (2005) found that positive couple communication is related to feelings of individual fulfillment in an intimate relationship. Not only is a relationship enhanced by healthy communication, but also the individuals involved in the relationship benefit.

Self-Disclosure, Honesty, and Lying

Shakespeare noted in MacBeth that "the false face must hide what the false heart doth know," suggesting that withholding and dishonesty may affect the way one feels about one's self and one's relationship with others. All of us make choices, consciously or unconsciously, about the degree to which we disclose, are honest, and/or lie.

Lying is done with words and also with silence.

Adrienne Rich

Self-Disclosure in Intimate Relationships

One aspect of intimacy in relationships is self-disclosure, which involves revealing personal information and feelings about oneself to another person. McKenna et al. (2002) found that a positive function of meeting on-line is that persons were better able to express themselves and were more disclosing on the Internet than in person. Gibbs et al. (2006) also noted that persons seeking a partner on the Internet were more disclosing and more honest about their disclosures if they had the goal of a long-term relationship.

Relationships become more stable when individuals disclose themselves—their formative years, previous relationships (positive and negative), experiences of elation and sadness/depression, and goals (achieved and thwarted). We noted in the discussion of love in Chapter 3 that self-disclosure is a psychological condition necessary for the development of love. To the degree that you disclose yourself to another, you invest yourself in that person and feel closer to him or her. Persons who disclose nothing are investing nothing and remain aloof. One way to encourage disclosure in one's partner is to make disclosures about one's own life and then ask about his or her life. Patford (2000) found that the higher the level of disclosure, the more committed the spouses were to each other.

Honesty in Intimate Relationships

Lying is pervasive in our society. The government lies to its citizens ("Saddam has weapons of mass destruction") and citizens lie to the government (via cheating on taxes). Teachers lie to students ("The test will be easy") and students lie to teachers ("I studied all night"). Parents lie to their children ("It won't hurt") and children lie to their parents about where they have been, whom they were with, and what they did. Dating partners lie to each other ("I've had a couple of previous sex partners"), women lie to men ("I had an orgasm") and men lie to women ("I'll call"). But the price of lying is high—distrust and alienation. A student in the authors' class wrote:

> At this moment in my life I do not have any love relationship. I find college dating to be very hard. The guys here lie to you about anything and you wouldn't know the truth. I find it's mostly about sex here and having a good time before you really have to get serious. That is fine, but that is just not what I am all about.

In addition to lying to gain sexual access, is the behavior of cheating—having sex with another while involved in a relationship. McAlister et al. (2005) noted that extradyadic activity (defined as kissing or "sexual activity") among young adults who were dating could be predicted. Those young adults who had a high number of previous sexual partners, who were impulsive, who were not satisfied in their current relationship, and who had attractive alternatives were more vulnerable to being unfaithful.

Forms of Dishonesty and Deception

Dishonesty and deception take various forms. In addition to telling an outright lie, people may exaggerate the truth, pretend, conceal the truth, or withhold information. Regarding the latter, in virtually every relationship, there are things that partners have not shared with each other about themselves or their past. We often withhold information or keep secrets in our intimate relationships for what we believe are good reasons—we believe that we are protecting our partners from anxiety or hurt feelings, protecting ourselves from criticism and rejection, and protecting our relationships from conflict and disintegration. Finkenauer and Hazam (2000) found that happy relationships depend on withholding information. The researchers contend, "Nobody wants to be criticized (e.g., 'You're really too fat') or talk about topics that are known to be conflictive (e.g., 'You should not have spent that much money')."

Should One Partner Disclose Human Immunodeficiency Virus (HIV)/Sexually Transmitted Infection (STI) Status to Another?

Five percent of 1027 undergraduates at a large southeastern university reported that they had an STI (Knox and Zusman, 2006). Individuals often struggle over whether, or how, to tell a partner if they have an STI, including HIV infection. If a person in a committed relationship acquires an STI, then

that individual, or his or her partner, may have been unfaithful and have had sex with someone outside the relationship. Thus, disclosure about an STI may also mean confessing one's own infidelity or confronting one's partner about his or her possible infidelity. (However, the infection may have occurred prior to the current relationship but gone undetected.) Individuals who have an STI and who are beginning a new relationship face a different set of concerns. Will their new partners view them negatively? Will they want to continue the relationship? One Internet ad began "I have herpes—Now that that is out of the way . . . "

Although telling a partner about having an STI may be difficult and embarrassing, avoiding disclosure or lying about having an STI represents a serious ethical violation. The responsibility to inform a partner that one has an STI—before having sex with that partner—is a moral one. But there are also legal reasons for disclosing one's sexual health condition to a partner. If you have an STI and you do not tell your partner, you may be liable for damages if you transmit it to your partner. Ayres and Baker (2004) proposed a new crime, reckless sexual conduct, of which a person would be guilty if he or she did not use a condom the first time of intercourse with a person. The penalty for the perpetrator would be three months in prison.

Khalsa (2006) noted that reporting HIV infection and acquired immunodeficiency syndrome (AIDS) is mandatory in most states, although partner notification laws vary from state to state. New York has a strong partner notification law that requires health care providers to either notify any partners the infected person names or to forward the information about partners to the Department of Health, where public health officers notify the partners that they have been exposed to an STI and schedule an appointment for STI testing. The privacy of the infected individual is protected by not revealing his or her name to the partner being notified of potential infection. In cases where the infected person refuses to identify partners, standard partner notification laws require doctors to undertake notification without cooperation if they know who the sexual partner or spouse is.

Your Opinion?

1. What do you think the percentage chance is that one undergraduate (who knowingly has an STI) would have sex with another undergraduate and not tell the partner?
2. What do you think the penalty should be for deliberately exposing a person to an STI?
3. What partner notification law do you recommend?

Sources

Ayres, I. and K. Baker. 2004. A separate crime of reckless sex. *Yale Law School, Public Law Working Paper No. 80.*

Khalsa, A. M. 2006. Preventive counseling, screening, and therapy for the patient with newly diagnosed HIV infection. *American Family Physician* 73:271–80.

Lying among College Students

Lying occurs in college student relationships. Number of previous partners was the most frequent lie told to a partner (Knox et al., 1993). And, when lying is defined as cheating in a relationship, 36% of 1027 undergraduates agreed "I have cheated on a partner I was involved with" (Knox and Zusman, 2006).

One of the ways in which college students deceive their partners is by failure to disclose that they have a sexually transmitted infection (STI). It is estimated that 25 percent of college students will contract an STI while they are in college (Jeffries, 2006). Because the potential to harm an unsuspecting partner is considerable, should we have a national social policy regarding such disclosure?

Gender Differences in Communication

Women and men differ in their approach to and patterns of communication. Women are more communicative about relationship issues, view a situation emotionally, and initiate discussions about relationship problems. Deborah Tannen

While women speak a language of connection and intimacy, men speak a language of status and independence—in effect they speak different genderlects.

Deborah Tannen, *You Just Don't Understand*

How Much Do I Tell My Partner about My Past?
Because of the fear of HIV infection and other STIs, some partners want to know the details of each other's previous sex life, including how many partners they have had sex with and in what contexts. Those who are asked will need to make a decision about whether to disclose the requested information, which may include one's sexual orientation, present or past sexually transmitted diseases, and any sexual proclivities or preferences the partner might find bizarre (e.g., bondage and discipline). Ample evidence suggests that individuals are sometimes dishonest with regard to the sexual information they provide to their partners. We have noted that "number of previous sexual partners" is the most frequent lie undergraduates report telling each other.

In deciding whether or not to talk honestly about your past to your partner, you may want to consider the following questions: How important is it to your partner to know about your past? Do you want your partner to tell you (honestly) about her or his past?

A husband read an article to his wife about how many words women use in a day: 30,000 to men's 15,000. The wife replied, "The reason has to be because we have to repeat everything to men." The husband then turned to his wife and asked, "What?"

Internet humor

When a woman says, "Sure . . . go ahead," what she means is "I don't want you to." When a woman says, "I'm sorry," what she means is "You'll be sorry." When a woman says, "I'll be ready in a minute," what she means is "Kick off your shoes and start watching a football game on TV."

Internet humor

(1990, 2006) is a specialist in communication. She observed that to women, conversations are negotiations for closeness in which they try "to seek and give confirmations and support, and to reach consensus" (1990, p. 25). A woman's goal is to preserve intimacy and avoid isolation. To men, conversations are about winning and achieving the upper hand.

Women also tend to approach a situation emotionally. A husband might react to a seriously ill child by putting pressure on the wife to be mature (stop crying) about the situation and by encouraging stoicism (asking her not to feel sorry for herself). Wives, on the other hand, want their husbands to be more emotional (by asking them to cry to show that they really care that their child is ill).

The sexes also respond differently when confronted. Miller and Roloff (2005) asked women and men to estimate how they would respond when they were verbally attacked or teased by a partner in a romantic relationship. Women predicted that they would be more hurt and more willing to confront their romantic partner's attacks than did men. Similarly, men reported that they would be less likely to confront their partner.

Whether women and men differ in emotional expression is unclear. Ingoldsby et al. (2005) found no differences between spouses in their emotional expression. Finally, as noted earlier, women disclose more in their relationships than men do (Gallmeier et al., 1997). In this study of 360 undergraduates, women were more likely to disclose information about previous love relationships, previous sexual relationships, their love feelings for the partner, and what they wanted for the future of the relationship. They also wanted their partners to reciprocate their (the women's) disclosure, but such disclosure was not forthcoming. Punyanunt-Carter (2006) confirmed that female college students are more likely to disclose than are male college students.

Much has been written about communication differences between women and men. Behringer (2005) found that in spite of the fact that women and men may have different communication foci, they both value openness, honesty, respect, humor, and resolution as principal components of good communication. And they each endeavor to create a common reality. Hence, while spouses may be on different pages, they are reading the same book.

How Close Do You Want to Be?

Individuals differ in their capacity for and interest in an emotionally close/disclosing relationship.

These preferences may vary over time; the partners may want closeness at some times and distance at other times. Individuals frequently choose partners according to an "emotional fit"—agreement about the amount of closeness they desire in their relationship. Haas and Stafford (2005) noted that one of the ways both heterosexual and homosexual couples maintain emotional closeness is through task-sharing. Alexandrov (2005) found a link between couple attachment and marital quality.

In addition to emotional closeness is physical presence. Some partners prefer a pattern of complete togetherness (the current buzz word is co-dependency) in which all of their leisure and discretionary time is spent together. Others enjoy time alone and time with other friends and do not want to feel burdened by the demands of a partner with high companionship needs. Partners might consider their own choices and those of their partners in regard to emotional and spatial closeness.

Individuals vary in the degree to which they want to have an emotionally close relationship. This couple values emotional intensity.

Theories Applied to Relationship Communication

Symbolic interactionism and social exchange are theories that help to explain the communication process.

Symbolic Interactionism

Interactionists examine the process of communication between two actors in terms of the meanings each attaches to the actions of the other. Definition of the situation, the looking-glass self, and taking the role of the other (discussed in Chapter 1) are all relevant to understanding how partners communicate. With regard to resolving a conflict over how to spend the semester break (e.g., vacation alone or go to see parents), the respective partners must negotiate their definition of the situation (is it about their time together as a couple or their loyalty to their parents?). The looking-glass self involves looking at each other and

I was married for 15 years to a woman who spoke perfect English and I didn't understand a word she said.

Ivan Thompson, Cowboy Cupid, who subsequently married a Mexican woman seeing the reflected image of someone who is loved and cared for and someone with whom a productive resolution is sought. Taking the role of the other involves each partner's understanding the other's logic and feelings about how to spend the break.

Social Exchange

Exchange theorists suggest that the partners' communication can be described as a ratio of rewards to costs. Rewards are positive exchanges, such as compliments, compromises, and agreements. Costs refer to negative exchanges, such as critical remarks, complaints, and attacks. When the rewards are high and the costs are low, the outcome is likely to be positive for both partners (profit). When the costs are high and the rewards low, neither may be satisfied with the outcome (loss).

When discussing how to spend the semester break, the partners are continually in the process of exchange—not only in the words they use but also in the way they use them. If the communication is to continue, each partner needs to feel acknowledged for his or her point of view and to feel a sense of legitimacy and respect. Communication in abusive relationships is characterized by the parties criticizing and denigrating each other, which usually results in a shutdown of the communication process.

Conflicts in Relationships

Conflict can be defined as the process of interaction that results when the behavior of one person interferes with the behavior of another. A professor in a Marriage and Family class said, "If you haven't had a conflict with your partner, you haven't been going together long enough." This section explores the inevitability, desirability, sources, and styles of conflict in relationships.

The days are too short even for love; how can there be enough time for quarreling?

Alistair Grant

Inevitability of Conflict

If you are alone this Saturday evening from six o'clock until midnight, you are assured of six conflict-free hours. But if you plan to be with your partner, roommate, or spouse during that time, the potential for conflict exists. Whether you eat out, where you eat, where you go after dinner, and how long you stay must be negotiated. Although it may be relatively easy for you and your companion to agree on one evening's agenda, marriage involves the meshing of desires on an array of issues for potentially 60 years or more. Indeed, conflict is inevitable in any intimate relationship. DeMaria (2005) studied 129 married couples who signed up for a communication/marriage education workshop presumably designed for couples who were already functioning well and who were there for a "relationship tune-up." But analysis of the data revealed that the couples were highly distressed, conflicted, devitalized, and lacked communication skills.

Benefits of Conflict

Conflict can be healthy and productive for a couple's relationship. Ignoring an issue may result in the partners becoming increasingly resentful and dissatisfied with the relationship. Indeed, not talking about a concern can do more damage to a relationship than bringing up the issue and discussing it (Campbell, 2005). Couples in trouble are not those who disagree but those who never discuss their disagreements.

Daily communication provides the companionship most couples seek in marriage.

Sources of Conflict

Conflict has numerous sources, some of which are easily recognized, whereas others are hidden inside the web of marital interaction.

1. Behavior. Stanley et al. (2002) noted that money was the issue over which a national sample of couples reported that they argued the most. The behavioral expression of a money issue might include how the partner spends money (excessively), the lack of communication about spending (e.g., does not consult the partner), and the target (i.e., items considered unnecessary by the partner). But marital conflict is not limited to behavioral money issues. Stanley et al. (2002) found that remarried couples argued most about the children (e.g., rules for and discipline of). In a sample of 105 older married couples (average age, 69), the most often reported behavior problem was related to leisure activities (Henry et al., 2005). One 67-year-old wife noted, "My spouse watches too much football and after 48 years of hearing it, I get upset" (p. 249).

2. Cognitions and perceptions. Aside from your partner's actual behavior, your cognitions and perceptions of a behavior can be a source of satisfaction or dissatisfaction. One husband complained that his wife "had boxes of coupons everywhere and always kept the house a wreck." The wife made the husband aware that she saved $100 on their grocery bill every week and asked him to view the boxes and a mess as "saving money." He changed his view and the clutter ceased to be a problem.

3. Value differences. Because you and your partner have had different socialization experiences, you may also have different values—about religion (one feels religion is a central part of life; the other does not), money (one feels uncomfortable being in debt; the other has the buy-now-pay-later philosophy), in-laws (one feels responsible for parents when they are old; the other does not), and children (number, timing, discipline). The effect of value differences depends less on the degree of the difference than on the degree of rigidity with which each partner holds his or her values. Dogmatic and rigid thinkers, feeling threatened by value disagreement, may try to eliminate alternative views and thus produce more conflict. Partners who recognize the inevitability of difference may consider the positives of an alternative view and move toward acceptance. When both partners do this, the relationship takes priority and the value differences suddenly become less important.

A quarrel is quickly settled when deserted by one party; there is no battle unless there are two.

Seneca

4. Inconsistent rules. Partners in all relationships develop a set of rules to help them function smoothly. These unwritten but mutually understood rules include what time you are supposed to be home after work, whether you should call if you are going to be late, how often you can see friends alone, and when and how to make love. Conflict results when the partners disagree on the rules or when inconsistent rules develop in the relationship. For example, one wife expected her husband to take a second job so they could afford a new car, but she also expected him to spend more time at home with the family.

5. Leadership ambiguity. Unless a couple has an understanding about which partner will make decisions in which area (for example, the wife may make decisions about money management; the husband will make decisions about rearing the children), unnecessary conflict may result. Whereas some couples may want to discuss certain issues, others may want to develop a clear specification of roles.

Styles of Conflict

Spouses develop various styles of conflict. If you were watching a videotape of various spouses disagreeing over the same issue, you would notice at least six styles of conflict. These styles have been described by Greeff and De Bruyne (2000) as follows.

Competing Style   The partners are both assertive and uncooperative. Each tries to force his or her way on the other so that there is a winner and a loser. A couple arguing over whether to discipline a child with a spanking or time out would resolve the argument with the dominant partner's forcing a decision.

Collaborating Style   The respective partners are both assertive and cooperative. Each partner expresses his or her view and cooperates to find a solution. A spanking, time out, or just talking to the child might resolve the above issue, but both partners would be satisfied with the resolution.

Compromising Style   Here there would be an intermediate solution: both partners would find a middle ground they could live with—perhaps spanking the child for serious infractions such as playing with matches in the house and imposing time out for talking back.

Avoiding Style   The partners are neither assertive nor cooperative. They would avoid a confrontation and let either parent do what he or she wanted in disciplining the child. Thus the child might be both spanked and put in time out. Marchand and Hock (2000) noted that depressed spouses were particularly likely to use avoidance as a conflict-resolution strategy.

Accommodating Style   The respective partners are not assertive in their positions, but each accommodates to the other's point of view. Each attempts to soothe the other and to avoid conflict. Although the goal of this style is to rise above the conflict and keep harmony in the relationship, fundamental feelings about the "rightness" of one's own approach may be maintained.

Avoidance Style   Also referred to as parallel style, both partners deny, ignore, and retreat from addressing a problem issue. "Don't talk about it, and it will go away" is the theme of this conflict style. Gaps begin to develop in the relationship; neither partner feels free to talk, and both believe that they are misunderstood. They eventually become involved in separate activities rather than spending time together.

Greeff and De Bruyne (2000) studied 57 couples who had been married at least ten years and found that the collaborating style was associated with the

Marital Quality—Keep It High for Good Health

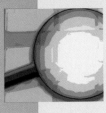

Spouses who do not communicate effectively to reduce the stress in their relationship end up in a poor-quality marriage. Being married has an important impact on one's health. Research has shown that "among the married, those in distressed marriages are in poorer health than those in nondistressed marriages . . . and individuals in low-quality marriages exhibit an even greater health risk than do divorced individuals" (Umberson et al., 2006, p. 1). To better understand the relationship between marriage and physical health, researchers conducted a study, described below, that attempts to answer the following questions: How do positive and negative aspects of marital quality affect physical health, and do these effects vary with age or gender?

Sample and Methods

Researchers used three waves of data from the Americans' Changing Lives (ACL) panel survey of U.S. adults (collected in 1986, 1989, and 1994) and found 1049 individuals who were continuously married across the 8-year period 1986–1994). Data from the ACL survey were obtained in 90-minute face-to-face interviews with respondents. In addition to looking at respondents' demographic variables, such as age, sex, education, race, and income, the researchers were interested in the degree to which respondents had positive and negative marital experiences and how respondents rated their physical health.

Positive marital experience was measured by respondents' answers to four questions: (1) How satisfied are you with your marriage? (2) How much does your husband/wife make you feel loved and cared for? (3) How much is he/she willing to listen when you need to talk about your worries or problems? and (4) Can you share your very private feelings and concerns with your spouse?

Negative marital experience was measured by asking respondents two questions: (1) How often do you feel bothered or upset by your marriage? and (2) How often would you say the two of you typically have unpleasant disagreements or conflicts? Self-rated physical health was measured by asking respondents the following question: Would you say your health in general is excellent, good, fair, or poor?

Selected Findings and Conclusions

Statistical procedures performed on the data that were collected across the 8-year period as described above revealed that marital quality tended to diminish over time: positive marital experiences generally decreased over the 8-year period, and negative marital experiences increased. Not surprisingly, self-rated health also diminished over time.

The study revealed some interesting findings in regard to the relationship between marital experiences and health. In examining both positive and negative marital experiences, the researchers found that only the negative experiences significantly affected self-rated health trajectories. The researchers explain that "this finding fits with previous research showing that negative aspects of relationships have stronger effects on psychological well-being than do positive aspects of relationships" (Umberson et al., 2006, p. 12). Further, negative marital experiences were more important to the health of older individuals than to younger ones. High levels of negative marital experiences in the oldest respondents were associated with a faster rate of decline in health than in younger respondents. The researchers explained that "the adverse effects of negative experiences may become apparent only at older ages either because they take a cumulative toll on health or because health status becomes more vulnerable to stress at older ages" (Umberson et al., 2006, p. 8). The findings also suggest that there may be greater benefits in the *absence* of marital strain at older ages. Regarding gender differences, this study found that the effects of marital quality on self-rated health are similar for men and women across the life course.

In conclusion, the authors state: "While self-rated health tends to decline over time for the sample as a whole, it appears that marital strain accelerates this decline in a representative sample of adults. Moreover, marital strain appears to matter more for health as individuals age" (Umberson et al., 2006, p. 11). Although this study provides evidence that marital quality affects subsequent assessments of health, the researchers note that self-rated health may also influence subsequent levels of marital quality. Statistical procedures performed on the data, however, support the interpretation that initial levels of negative marital experience affect health rather than vice versa.

What are the practical lessons of this study? The researchers suggest the following:

1. Unhappily married individuals have yet another reason to identify marital difficulties and seek to improve marital quality: their very health may depend on it.

2. Moreover, there is no reason for clinicians and policy-makers to think that marital quality is less important for older couples. In fact, the negative aspects of marriage appear to become more consequential for health as individuals age (Umberson et al., 2006, p. 13).

Source

Umberson, D., K. Williams, D. A. Powers, H. Liu, and B. Needham. 2006. You make me sick: Marital quality and health over the life course. *Journal of Health and Social Behavior* 47(March):1–16. Used by permission.

A high-quality marriage relationship is good for the spouses' health.

Authors

highest level of marital and spousal satisfaction. The competitive style, used by either partner, was associated with the lowest level of marital satisfaction. Regardless of the style of conflict, partners who say positive things to each other at a ratio of 5:1 (positives to negatives) seem to stay together (Gottman, 1994).

Fighting Fair: Seven Steps in Conflict Resolution

Whenever you're in conflict with someone, there is one factor that can make the difference between damaging your relationship and deepening it. That factor is attitude.

William James, American philosopher

When a disagreement ensues, it is important to establish rules for fighting that will leave the partners and their relationship undamaged after the disagreement. Such fair-fighting guidelines include not calling each other names, not bringing up past misdeeds, not attacking each other, and not beginning a heated discussion late at night. In some cases, a good night's sleep has a way of altering how a situation is viewed and may even result in the problem no longer being an issue.

Fighting fairly also involves keeping the interaction focused, respective, and moving toward a win-win outcome. If recurring issues are not discussed and resolved, conflict may create tension and distance in the relationship, with the result that the partners stop talking, stop spending time together, and stop being intimate. A conflictual, unsatisfactory marriage is similar to divorce in terms of its impact on the diminished psychological, social, and physical well-being of the partners (Hetherington, 2003). Developing and using fair fighting/conflict-resolution skills are critical for the maintenance of a good relationship.

Howard Markman is head of the Center for Marital and Family Studies at the University of Denver. He and his colleagues have been studying 150 couples at yearly intervals (beginning before marriage) to determine those factors most responsible for marital success. They have found that communication skills that reflect the ability to handle conflict, which they call "constructive arguing," are the single biggest predictor of marital success over time (Marano, 1992). According to Markman: "Many people believe that the causes of marital problems are the differences between people and problem areas such as money, sex, children. However, our findings indicate it is not the differences that are important, but how these differences and problems are handled, particularly early in marriage" (Marano, 1992, p. 53). The following sections identify fair fighting and steps for resolving interpersonal conflict.

Chapter 9 Communication in Relationships

Address Recurring, Disturbing Issues

It is important to address issues in the relationship. As noted earlier, couples who stack resentments rather than discuss conflictual issues do no service to their relationship. Indeed, the healthiest response to feeling upset about a partner's behavior is to engage the partner in a discussion about the behavior. Not to do so is to let the negative feelings fester, which will result in emotional and physical withdrawal from the relationship. For example, Pam is jealous that Mark spends more time with other people at parties than with her. "When we go someplace together," she blurts out, "he drops me to disappear with someone else for two hours." Her jealousy is spreading to other areas of their relationship. "When we are walking down the street and he turns his head to look at another woman, I get furious." If Pam and Mark don't discuss her feelings about Mark's behavior, their relationship may deteriorate as a result of a negative response cycle: He looks at another woman, she gets angry, he gets angry at her getting angry and finds that he is even more attracted to other women, she gets angrier because he escalates his looking at other women, and so on.

To bring the matter up, Pam might say, "I feel jealous when you spend more time with other women at parties than with me. I need some help in dealing with these feelings." By expressing her concern in this way, she has identified the problem from her perspective and asked her partner's cooperation in handling it.

When discussing difficult relationship issues, it is important to avoid attacking, blaming, or being negative. Such reactions reduce the motivation of the partner to talk about an issue and thus reduce the probability of a positive outcome.

It is also important to use good timing in discussing difficult issues with your partner. In general, it is best to discuss issues or conflicts when (1) you are alone with your partner in private rather than in public, (2) you and your partner have ample time to talk, and (3) you and your partner are rested and feeling generally good (avoid discussing conflict issues when one of you is tired, upset, or under unusual stress).

Identify New Desired Behaviors

Dealing with conflict is more likely to result in resolution if the partners focus on what they *want* rather than what they *don't want*. For example, rather than tell Mark she doesn't want him to spend so much time with other women at parties, Pam might tell him that she wants him to spend more time with her at parties.

Identify Perceptions to Change

Rather than change behavior, it may be easier and quicker to change one's perception of a behavior. Rather than expect one's partner to always be "on time," it may be easier to drop the expectation that one's partner be on time and to stop being mad about something that doesn't matter. Pam might also decide that it does not matter that Mark looks at and talks to other women. If she feels secure in his love for her, the behavior is inconsequential.

Summarize Your Partner's Perspective

We often assume that we know what our partner thinks and why he or she does things. Sometimes we are wrong. Rather than assume how our partner thinks and feels about a particular issue, we might ask our partner open-ended questions in an effort to get him or her to tell us thoughts and feelings about a particular situation.

Pam's words to Mark might be, "What is it like for you when we go to parties?" "How do you feel about my jealousy?" Once your partner has shared his or her thoughts about an issue with you, it is important for you to summarize your partner's perspective in a nonjudgmental way. After Mark has told Pam how he feels about their being at parties together, she can summarize his perspective by saying, "You feel that I cling to you more than I should, and you would like me to let you

It is better to debate a question without settling it than to settle it without debating it.

Joseph Joubert

wander around without feeling like you're making me angry." (She may not agree with his view, but she knows exactly what it is—and Mark knows that she knows.) In addition, Mark should summarize Pam's view—"You enjoy our being together and prefer when we go to parties that we hang relatively close to each other. You do not want me off in a corner talking to another girl or dancing."

Generate Alternative Win-Win Solutions

It is imperative to look for win-win solutions to conflicts. Solutions in which one person wins and the other person loses mean that one person is not getting his or her needs met. As a result, the person who loses may develop feelings of resentment, anger, hurt, and hostility toward the winner and may even look for ways to get even. In this way, the winner is also a loser. In intimate relationships, one winner really means two losers.

Generating win-win solutions to interpersonal conflict often requires **brainstorming.** The technique of brainstorming involves suggesting as many alternatives as possible without evaluating them. Brainstorming is crucial to conflict resolution because it shifts the partners' focus from criticizing each other's perspective to working together to develop alternative solutions.

The authors and their colleagues (Knox et al., 1995) studied the degree to which 200 college students who were involved in ongoing relationships were involved in win-win, win-lose, and lose-lose relationships. Descriptions of the various relationships follow.

Win-win relationships are those in which conflict is resolved so that each partner derives benefits from the resolution. For example, suppose a couple have a limited amount of money and disagree on whether to spend it on eating out or on seeing a current movie. One possible win-win solution might be for the couple to eat a relatively inexpensive dinner and rent a movie.

An example of a **win-lose solution** would be for one of the partners to get what he or she wanted (eat out or go to a movie), with the other partner getting nothing of what he or she wanted. Caughlin and Ramey (2005) studied demand-and-withdraw patterns in parent–adolescent dyads and found that the demand on the part of one of them was usually met by withdrawal on the part of the other partner. Such demand may reflect a win-lose interaction.

A **lose-lose solution** is one in which both partners get nothing that they want—in the scenario presented, the partners would neither go out to eat nor see a movie and would be mad at each other.

More than three-quarters (77.1%) of the students reported being involved in a win-win relationship, with men and women reporting similar percentages. Twenty percent of the respondents were involved in win-lose relationships. Only 2 percent reported that they were involved in lose-lose relationships. Eighty-five percent of the students in win-win relationships reported that they expected to continue their relationship, in contrast to only 15 percent of students in win-lose relationships. No student in a lose-lose relationship expected the relationship to last.

After a number of solutions are generated, each solution should be evaluated and the best one selected. In evaluating solutions to conflicts, it may be helpful to ask the following questions:

1. Does the solution satisfy both individuals? (Is it a win-win solution?)

2. Is the solution specific? Does it specify exactly who is to do what, how, and when?

3. Is the solution realistic? Can both parties realistically follow through with what they have agreed to do?

4. Does the solution prevent the problem from recurring?

5. Does the solution specify what is to happen if the problem recurs?

Kurdek (1995) emphasized that conflict-resolution styles that stress agreement, compromise, and humor are associated with marital satisfaction, whereas conflict engagement, withdrawal, and defensiveness styles are associated with

Soft words win hard hearts.
Proverb

Chapter 9 Communication in Relationships

lower marital satisfaction. In his own study of 155 married couples, the style in which the wife engaged the husband in conflict and the husband withdrew was particularly associated with low marital satisfaction for both spouses.

Forgive

Too little emphasis is placed on forgiveness as an emotional behavior that can move a couple from a deadlock to resolution. Forgiveness requires acknowledging to one's self and the partner that either can make a mistake and that the focus should be on moving beyond the transgression, mistake, accident, or whatever. . . . to "let it go." It takes more energy to hold on to a resentment than to move beyond it. One reason some people do not forgive a partner for a transgression is that one can use the fault to control the relationship. "I wasn't going to let him forget," said one woman of her husband's infidelity.

Gordon et al. (2005) emphasized that forgiveness is one of the important factors in a couple's recovery from infidelity on the part of one or both partners. Toussaint and Webb (2005) noted that women and men are equally forgiving. Day and Maltby (2005) found that individuals who are not capable of forgiveness tend to withdraw from social relationships and to become more lonely and/or socially isolated.

Be Alert to Defense Mechanisms

Effective conflict resolution is sometimes blocked by **defense mechanisms**— unconscious techniques that function to protect individuals from anxiety and to minimize emotional hurt. The following paragraphs discuss some common defense mechanisms.

Escapism is the simultaneous denial of and withdrawal from a problem. The usual form of escape is avoidance. The spouse becomes "busy" and "doesn't have time" to think about or deal with the problem, or the partner may escape into recreation, sleep, alcohol, marijuana, or work. Denying and withdrawing from problems in relationships offer no possibility for confronting and resolving the problems.

Rationalization is the cognitive justification for one's own behavior that unconsciously conceals one's true motives. For example, one wife complained that her husband spent too much time at the health club in the evenings. The underlying reason for the husband's going to the health club was to escape an unsatisfying home life. But the idea that he was in a dead marriage was too painful and difficult for the husband to face, so he rationalized to himself and his wife that he spent so much time at the health club because he made a lot of important business contacts there. Thus, the husband concealed his own true motives from himself (and his wife).

Projection occurs when one spouse unconsciously attributes his or her own feelings, attitudes, or desires to the partner. For example, the wife who desires to have an affair may accuse her husband of being unfaithful to her. Projection may be seen in such statements as "You spend too much money" (projection for "I spend too much money") and "You want to break up" (projection for "I want to break up"). Projection interferes with conflict resolution by creating a mood of hostility and defensiveness in both partners. The issues to be resolved in the relationship remain unchanged and become more difficult to discuss.

Displacement involves shifting your feelings, thoughts, or behaviors from the person who evokes them onto someone else. The wife who is turned down for a promotion and the husband who is driven to exhaustion by his boss may direct their hostilities (displace them) onto each other rather than toward their respective employers. Similarly, spouses who are angry at each other may displace this anger onto someone else, such as the children.

By knowing about defense mechanisms and their negative impact on resolving conflict, you can be alert to them in your own relationships. When a conflict continues without resolution, one or more defense mechanisms may be operating.

The most important thing in communication is to hear what isn't being said.

Peter F. Drucker

Should Parents Argue in front of the Children?

Parents may disagree about whether to argue in front of their children. One parent may feel that it is best to argue behind closed doors so as not to upset their children, but the other may feel it is best to expose children to the reality of relationships. This includes seeing parents argue and, hopefully, negotiating win-win solutions. Most therapists agree that being open is best. Children need to know that relationships involve conflict and how to resolve it. In the absence of such exposure, children may have an unrealistic view of relationships.

SUMMARY

What is the nature of interpersonal communication?

Communication is the exchange of information and feelings by two individuals. It involves both verbal and nonverbal messages. The nonverbal part of a message often carries more weight than the verbal part.

What are some principles and techniques of effective communication?

Some basic principles and techniques of effective communication include making communication a priority, maintaining eye contact, asking open-ended questions, using reflective listening, using "I" statements, complimenting each other, and sharing power. Partners must also be alert to keeping the dialogue (process) going even when they don't like what is being said (content).

How are relationships affected by self-disclosure, dishonesty, and lying?

Intimacy in relationships is influenced by the level of self-disclosure and honesty. High levels of self-disclosure are associated with increased intimacy. Most individuals value honesty in their relationships. Honest communication is associated with trust and intimacy.

Despite the importance of honesty in relationships, deception occurs frequently in interpersonal relationships. Partners sometimes lie to each other about previous sexual relationships, how they feel about each other, and how they experience each other sexually. Telling lies is not the only form of dishonesty. People exaggerate, minimize, tell partial truths, pretend, and engage in self-deception. Partners may withhold information or keep secrets in order to protect themselves and/or preserve the relationship. However, the more intimate the relationship, the greater our desire to share our most personal and private selves with our partner and the greater the emotional consequences of not sharing. In intimate relationships, keeping secrets can block opportunities for healing, resolution, self-acceptance, and a deeper intimacy with your partner.

What are gender differences in communication?

Men and women tend to focus on different content in their conversations. Men tend to focus on activities, information, logic, and negotiation and "to achieve and maintain the upper hand." To women, communication is emotion, relationships, interaction, and maintaining closeness. A woman's goal is to preserve intimacy and avoid isolation. Women are also more likely than men to initiate discussion of relationship problems, and women disclose more than men.

How are interactionist and exchange theories applied to relationship communication?

Symbolic interactionists examine the process of communication between two actors in terms of the meanings each attaches to the actions of the other. Definition of the situation, the looking-glass self, and taking the role of the other are all relevant to understanding how partners communicate.

Exchange theorists suggest that the partners' communication can be described as a ratio of rewards to costs. Rewards are positive exchanges, such as compliments, compromises, and agreements. Costs refer to negative exchanges, such as critical remarks, complaints, and attacks. When the rewards are high and the costs are low, the outcome is likely to be positive for both partners (profit). When the costs are high and the rewards low, neither may be satisfied with the outcome (loss).

What are various issues related to conflict in relationships?

Conflict is both inevitable and desirable. Unless individuals confront and resolve issues over which they disagree, one or both may become resentful and withdraw from the relationship. Conflict may result from one partner's doing something the other does not like, having different perceptions, or having different values. Sometimes it is easier for one partner to view a situation differently or alter a value than for the other partner to change the behavior causing the distress.

What are examples of fighting fair to resolve conflict?

The sequence of resolving conflict includes deciding to address recurring issues rather than suppressing them, asking the partner for help in resolving the issue, finding out the partner's point of view, summarizing in a nonjudgmental way the partner's perspective, and finding alternative win-win solutions. Defense mechanisms that interfere with conflict resolution include escapism, rationalization, projection, and displacement.

KEY TERMS

accommodating style of conflict	communication	displacement	projection
avoiding style of conflict	competing style of conflict	escapism	rationalization
brainstorming	compromising style of conflict	"I" statements	reflective listening
branching	conflict	lose-lose solution	win-lose solution
collaborating style of conflict	congruent message	parallel style of conflict	win-win relationships
	defense mechanisms	power	"you" statements

The Companion Website for Choices in Relationships: An Introduction to Marriage and the Family, Ninth Edition
www.thomson.edu/sociology/knox

Supplement your review of this chapter by going to the companion website to take one of the Tutorial Quizzes, use the flash cards to master key terms, and check out the many other study aids you'll find there. You'll also find special features such as the Marriage and Family Resource Center, Census 2000 information, and other data and resources at your fingertips to help you with that special project or to do some research on your own.

WEB LINKS

Department of Communication Resources
http://communication.ucsd.edu/resources/commlinks.html

Guidelines on Effective Communication, Healthy Relationships & Successful Living
http://www.drnadig.com/

Association for Couples in Marriage Enrichment
http://www.bettermarriages.org/

Episcopal Marriage Encounter
http://www.episcopalme.com/

REFERENCES

Alexandrov, E. Q. 2005. Couple attachment and the quality of marital relationships. *Attachment and Human Development* 7:123–52.

Bargh, J. A., K. Y. A. McKenna, and G. M. Fitzsimons. 2002. Can you see the real me? Activation and expression of the "true self" on the Internet. *Journal of Social Issues* 58:33–49.

Behringer, A. M. 2005. Bridging the gap between Mars and Venus: A study of communication meanings in marriage. *Dissertation Abstracts International, A: The Humanities and Social Sciences* 65:4007A–8A.

Campbell, S. 2005. *Seven keys to authentic communication and relationship satisfaction.* New York: New World Library.

Caughlin, J. P. and M. B. Ramey. 2005. The demand/withdraw pattern of communication in parent-adolescent dyads. *Personal Relationships* 12:337–55.

Day, L. and J. Maltby. 2005. Forgiveness. *Journal of Psychology* 139:553–55.

DeMaria, R. M. 2005. Distressed couples and marriage education. *Family Relations* 54:242–53.

Dunbar, N. E. and J. K. Burgoon. 2005. Perceptions of power and interactional dominance in interpersonal relationships. *Journal of Social and Personal Relationships* 22:207–33.

Finkenauer, C., and H. Hazam. 2000. Disclosure and secrecy in marriage: Do both contribute to marital satisfaction? *Journal of Social and Personal Relationships* 17:245–63.

Gable, S. L., H. T. Reis, and G. Downey. 2003. He said, she said: A Quasi-Signal detection analysis of daily interactions between close relationship partners. *Psychological Science* 14:100–05.

Gallmeier, C. P., M. E. Zusman, D. Knox, and L. Gibson. 1997. Can we talk? Gender differences in disclosure patterns and expectations. *Free Inquiry in Creative Sociology* 25:129–225.

Gibbs, J. L., N. B. Ellison and R. D. Heino. 2006. Self-presentation in online personals: The role of anticipated future interaction, self-disclosure, and perceived success in Internet dating. *Communication Research* 33:152–177.

Gordon, K. C., D. H. Baucom, and D. K. Snyder. 2005. Treating couples recovering from infidelity: An integrative approach. *Journal of Clinical Psychology* 61:1393–405.

Gottman, John. 1994. *Why marriages succeed or fail.* New York: Simon & Schuster.

Greeff, A. P., and T. De Bruyne. 2000. Conflict management style and marital satisfaction. *Journal of Sex and Marital Satisfaction* 26:321–34.

Haas, S. M. and L. Stafford. 2005. Maintenance behaviors in same-sex and marital relationships: A matched sample comparison. *Journal of Family Communication* 5:43–60.

Henry, R. G., R. B. Miller, and R. Giarrusso. 2005. Difficulties, disagreements, and disappointments in late-life marriages. *International Journal of Aging & Human Development* 61:243–65.

Hetherington, E. M. 2003. Intimate pathways: Changing patterns in close personal relationships across time. *Family Relations* 52:318–31.

Ingoldsby, B. B., G. T. Horlacher, P. L. Schvaneveldt, and M. Matthews. 2005. Emotional expressiveness and marital adjustment in Ecuador. *Marriage and Family Review* 38:25–44.

Jeffries, T. 2006. Sexually transmitted infections. Presentation to Courtship and Marriage class, Spring.

Knobloch, L. K., D. H. Solomon, and J. A. Theiss. 2006. The role of intimacy in the production and perception of relationship talk within courtship. *Communication Research* 33:211–241

Knox, D. and Zusman, M. E. 2006. Relationship and sexual behaviors of a sample of 1027 university students. Unpublished data collected for this text. Department of Sociology, East Carolina University, Greenville, NC.

Knox, D., C. Schacht, J. Holt, and J. Turner. 1993. Sexual lies among university students. *College Student Journal* 27:269–72.

Knox, D., C. Schacht, J. Turner, and P. Norris. 1995. College students' preference for win-win relationships. *College Student Journal* 29:44–46.

Kurdek, L. A. 1994. Areas of conflict for gay, lesbian, and heterosexual couples: What couples argue about influences relationship satisfaction. *Journal of Marriage and the Family* 56:923–34.

———. 1995. Predicting change in marital satisfaction from husbands' and wives' conflict resolution styles. *Journal of Marriage and the Family* 57:153–64.

Marano, H. E. 1992. The reinvention of marriage. *Psychology Today.* January/February, 49 passim.

Marchand, J. F. and E. Hock. 2000. Avoidance and attacking conflict-resolution strategies among married couples: Relations to depressive symptoms and marital satisfaction. *Family Relations* 49:201–06.

McAlister, A. R., N. Pachana, and C. J. Jackson. 2005. Predictors of young dating adults' inclination to engage in extradyadic sexual activities: A multi-perspective study. *British Journal of Psychology* 96:331–50.

McKenna, K. Y. A., A. S. Green, and M. E. J. Gleason. 2002. Relationship formation on the Internet: What's the big attraction? *Journal of Social Issues* 58:9–22.

Miller, C. W. and M. E. Roloff. 2005. Gender and willingness to confront hurtful messages from romantic partners. *Communication Quarterly* 53:323–37.

Patford, J. L. 2000. Partners and cross-sex friends: A preliminary study of the way marital and de facto partnerships affect verbal intimacy with cross-sex friends. *Journal of Family Studies* 6:106–19.

Preston, P. 2005. Nonverbal communication: Do you really say what your mean? *Journal of Healthcare Management* 50:83–87.

Punyanunt-Carter, N. N. (2006). An analysis of college students self-disclosure behaviors. *College Student Journal* 40:329–331.

Rayer, A. J. and B. L. Volling. 2005. The role of husbands' and wives' emotional expressivity in the marital relationship. *Sex Roles: A Journal of Research* 52:577–88.

Robbins, C. A. 2005. ADHD couple and family relationships: Enhancing communication and understanding through Imago Relationship Therapy. *Journal of Clinical Psychology* 61:565–78.

Rosof, F. 2005. An investigation of the therapeutic role of communication in couple relationships. Dissertation, Union Institute US. Dissertation Abstract International, Section B, Vol. 25 (8-B):4302.

Seguin-Levesque, C., M. L. N. Laliberte, L. G. Pelletier, C. Blanchard, and R. J. Vallerand. 2003. Harmonious and obsessive passion for the Internet: Their associations with the couple's relationship. *Journal of Applied Social Psychology* 33:197–221.

Stanley, S. M., H. J. Markman, and S. W. Whitton. 2002. Communication, conflict, and commitment: Insights on the foundations of relationship success from a national survey. *Interpersonal Relations* 41:659–66.

Tannen, D. 1990. *You just don't understand: Women and men in conversation.* London: Virago.

Tannen, D. (2006) *You're wearing that? Understanding mothers and daughters in conversation.* New York: Random House.

Toussaint, L. and J. R. Webb. 2005. Gender differences in the relationship between empathy and forgiveness. *Journal of Social Psychology* 145:673–85.

Turner, A. J. 2005. Communication basics. Personal communication.

Walther, J. B., T. Loh, and L. Granka 2005. Let me count the ways: The interchange of verbal and nonverbal cues in computer mediated and face to face affinity. *Journal of Language and Social Psychology* 34:36–65.

chapter 10

Contraceptives should be used on every conceivable occasion.

Spike Milligan, comedian

Planning Children and Contraception

Contents

True or False?

1. Most men of partners who had an abortion tended to regret the abortion.

2. Most Canadians and U.S. citizens are in favor of adoptees finding out about their parents.

3. Infertile women are more likely to blame their partners than themselves.

4. Children born using reproductive technology have a slightly higher risk of malformations.

5. Children of donor sperm want to find out more about their father because of economic motives.

Answers: **1.** F   **2.** T   **3.** F   **4.** T   **5.** F

Having children continues to be a major goal of most college students. Indeed, having a family is the top goal identified by more than 75% of 263,000 first-year students, representing a random sample from all U.S. colleges and universities (American Council on Education and University of California, 2005–2006). In a smaller, nonrandom sample of 1027 undergraduates at a large southeastern university, 95% percent agreed, "Someday I want to have children" (Knox and Zusman, 2006).

Planning children, or failing to do so, is a major societal issue. Planning when to become pregnant has benefits for both the mother and the child. Having several children at short intervals increases the chances of premature birth, infectious disease, and death of the mother or the baby. Would-be parents can minimize such risks by planning fewer children with longer intervals in between. Women who plan their pregnancies can also modify their behaviors and seek preconception care from a health-care practitioner to maximize their chances of having healthy pregnancies and babies. For example, women planning pregnancies can make sure they eat properly and avoid alcohol and other substances (such as cigarettes) that could harm developing fetuses. Partners who plan their children also benefit from family planning by pacing the financial demands of their offspring. Having children four years apart helps to avoid having more than one child in college at the same time.

Conscientious family planning will help to reduce the number of unwanted pregnancies. Bouchard (2005) compared individuals/couples with planned and unplanned pregnancies and found the latter more likely to be associated with individual depression/stress as well as low attachment and high stress for the couple. Unplanned children also have more negative outcomes. David et al. (2003) studied 220 children who were born to women who were twice denied abortion for the same pregnancy. The children were medically, psychologically, and socially evaluated at ages 9, 14–16, 21–23, 30, and 35 and found to have poorer mental health as adults than children born to mothers who wanted the pregnancy.

Making the decision to have a child is momentous. It is to decide forever to have your heart go walking around outside your body.

Elizabeth Stone, feminist

International Data

By 2050, the world population is expected to reach 8.9 billion—unplanned pregnancies will continue to increase throughout this period (Sitruk-Ware, R. 2006).

Your choices in regard to children and contraception have an important effect on your happiness, lifestyle, and resources. These choices, in large part, are influenced by social and cultural factors that may operate without your awareness. We now discuss these influences.

Do You Want to Have Children?

Beyond a biological drive to reproduce, societies socialize their members to have children. This section examines the social influences that motivate individuals to have children, the lifestyle changes that result from such a choice, and the costs of rearing children.

Diversity in Other Countries

Faced with declining fertility rates and the threat of national survival, print media in Britain have responded by pronatalist appeals to women. These have ranged from begging to lecturing to bribing in an effort for them to think of the national good. In effect, a society is dependent on socializing its members to reproduce; failure to do so is the end of the society (Brown and Ferree, 2005). In 2005 France increased the unpaid leave for a woman who leaves work to take care of a child at home, from $622 to between $850 and $1,215, with a guarantee that she can return to her job (Leichester, 2005).

Social Influences Motivating Individuals to Have Children

Our society tends to encourage childbearing, an attitude known as **pronatalism.** Our family, friends, religion, and government help to develop positive attitudes toward parenthood. Cultural observances also function to reinforce these attitudes.

Family Our experience of being reared in families encourages us to have families of our own. Our parents are our models. They married; we marry. They had children; we have children. Some parents exert a much more active influence. "I'm 73 and don't have much time. Will I ever see a grandchild?" asked the mother of an only child.

Friends Our friends who have children influence us to do likewise. After sharing an enjoyable weekend with friends who had a little girl, one husband wrote to the host and hostess, "Lucy and I are always affected by Karen—she is such a good child to have around. We haven't made up our minds yet, but our desire to have a child of our own always increases after we leave your home." This couple became parents 16 months later.

Diversity in Other Countries

Whereas the average number of children in North America is 1.9, in Europe it is 1.4; East Asia, 1.8; and sub-Saharan Africa, 6 (Townsend, 2003).

Religion Religion is a powerful influence on the decision to have children. Catholics are taught that having children is the basic purpose of marriage and gives meaning to the union. Mormonism and Judaism also have a strong family orientation.

Race Hispanics have the highest fertility rate, with 98 births per thousand women aged 15–44 years. Non-Hispanic blacks have the next highest rate, at 67 per thousand (Hamilton et al., 2005).

Diversity in Other Countries

In premodern societies (before industrialization and urbanization) and in some parts of China (despite China's one-child policy) and Korea today, spouses try to have as many children as possible because children are an economic asset. At an early age, children work as free labor for parents or for wages, which they give to their parents. Children are also expected to take care of their parents when the parents are elderly. In China and Korea, the eldest son is expected to take care of his aging parents by earning money for them and by marrying and bringing his wife into their home to physically care for his parents. The wife's parents need their own son and daughter-in-law to provide old-age insurance. Beyond its value in providing economic security and old-age insurance, having numerous children is regarded as a symbol of virility for the man, a source of prestige for the woman, and a sign of good fortune for the couple. In modern societies, having large numbers of children is less valued.

Government The tax structures imposed by our federal and state governments support parenthood. Married couples without children pay higher taxes than couples with children, although the reduction in taxes is not sufficient to offset the cost of rearing a child and is not large enough to be a primary inducement to have children.

Economy Times of affluence are associated with a high birth rate. Postwar expansion of the 1950s resulted in the oft-noted "baby boom" generation. Similarly, couples are less likely to decide to have a child during economically depressed times. In addition, the necessity of two wage earners in our postindustrial economy is associated with a reduction in the number of children.

Cultural Observances Our society reaffirms its approval of parents every year by identifying special days for Mom and Dad. Each year on Mother's Day and Father's Day (and now Grandparents' Day) parenthood is celebrated across the nation with cards, gifts, and embraces. There is no cultural counterpart (e.g., Childfree Day) for persons choosing not to have children. In addition to influencing individuals to have children, society and culture also influence feelings about the age parents should be when they have children. Recently, couples have been having children at later ages. Is this a good idea? The Social Policy on page 278 discusses this issue.

Individual Motivations for Having Children

Individual motivations, as well as social influences, play an important role in the decision to have children. Some of these are conscious, as in the desire for love and companionship with one's own offspring and in the desire to be personally fulfilled as an adult by having a child. Some also want to recapture their own childhood and youth by having a child.

The Self-Assessment on page 279 allows you to assess the degree to which having children is a value for you.

Unconscious motivations for parenthood may also be operative. Examples include wanting a child to avoid career tracking and to gain the acceptance and approval of one's parents and peers. Teenagers sometimes want to have a child

Aleta St. James was 57 when she had these twins. How old is too old to become a parent?

© AP/Wide World Photos

Companionship is a major motivation of parenthood. This father of two sons has shared experiences with his children since they were young. Here, the father and son are on a trip to New Zealand.

Authors

It is not flesh and blood but the heart which makes us fathers and sons.

Johann Schiller, German philosopher

How Old Is Too Old to Have a Child?

On November 10, 2004, Aleta St. James gave birth to twins 3 days before she turned 57 years old (Schienberg, 2004). She is thought to be the oldest woman in America to give birth to twins. Senator Strom Thurmond had a child in his 80s, and actor Tony Randall had a child at the age of 77 (both Thurmond and Randall are now deceased). Births to older parents are becoming more common, and questions are now being asked about the appropriateness of elderly individuals becoming parents. Should social policies on this issue be developed?

There are advantages and disadvantages of having a child as an elderly parent. The primary developmental advantage for the child of retirement-aged parents is the attention the parents can devote to their offspring. Not distracted by their careers, these parents have more time to nurture, play with, and teach their children. And, although they may have less energy, their experience and knowledge are doubtless better. Jokinen and Brown (2005) studied older parents of intellectually disabled children aged 40 and older. The parents reported a lifetime of care-giving and a high quality of family life.

The primary disadvantage of having a child in the later years is that the parents are likely to die before, or early in, the child's adult life. The daughter of James Dickey, the Southern writer, lamented the fact that her late father (to whom she was born when he was in his 50s) would not be present at her graduation or wedding. As noted above, the children of Strom Thurmond and Tony Randall knew their fathers for only a short time before these men died.

There are also medical concerns for both the mother and the baby during pregnancy in later life. They include an increased risk of morbidity (chronic illness and disease) and mortality (death) for the mother. These risks are typically a function of chronic disorders that go along with aging, such as diabetes, hypertension, and cardiac disease. Stillbirths, miscarriage, ectopic pregnancies, multiple births, and congenital malformations are also more frequent for women with advancing age. However, prenatal testing can identify some potential problems such as the risk of Down syndrome, and any chromosome abnormality and negative neonatal outcomes are not inevitable. Because an older woman can usually have a healthy baby, government regulations on the age at which a woman can become pregnant are not likely. Indeed, 255 births in 2000 were to women aged 50 to 54, a 50 percent increase over the year before (Regalado, 2002).

Age of the father may also be an issue in older parenting. Auger and Jouannet (2005) observed that a higher rate of miscarriages has been related to older fathers, and several studies have suggested that older fathers are at the origin of several diseases in the newborn. However, Romkens et al. (2005) reviewed both the medical and psychological literature of older men (age 50 years and up) having children and concluded that there are no medical or psychosocial data–based justifications to support an age limit for men having children. Given the lack of scientific support and the cultural norm that older men may fertilize younger women, governmental regulations on age limits for parenting are unlikely.

Your Opinion?

1. How old do you think is "too old" to begin being a parent?
2. Do you think the government should attempt to restrict having a biological child in one's fifties?
3. Who do you feel benefits most and least from having a child in later life?

Sources

Auger, J. and P. Jouannet. 2005. Age and male fertility: biological factors. *Review of Epidemiology et de Sante Publique* 53:2s25–35

Jokinen, N. S. and R. I. Brown. 2005. Family quality of life from the perspective of older parents. *Journal of Intellectual Disability Research* 49:789–93.

Regalado, A. 2002. Age is no barrier to motherhood: study gives latest proof older women can get pregnant with donor eggs. *Wall Street Journal* November 13:D3.

Romkens, M., B. Gordijn, C. M. Verhaak, E. J. H. Meuleman, and D. D. M. Braat. 2005. No arguments to support an age limit for men entering an in vitro fertilization or intracytoplasmic sperm injection programme. *Nederlands Tijdschrift Voor Geneeskunde* 149: 992–95.

Schienberg. J. 2004, November 9. Fifty-six-year-old gives birth to twins. CNN.com. Retrieved from http://www.cnn.com/2004/HEALTH/11/09/mom.56/.

to have someone to love them. On page 282 we detail teenage motherhood as a major social issue.

Lifestyle Changes and Economic Costs of Parenthood

Although there are numerous potential positive outcomes for becoming a parent, there are also drawbacks. Every parent knows that parenthood involves difficulties as well as joys. Some of the difficulties associated with parenthood are discussed next.

I don't know why they say "you have a baby." The baby has you.

Gallagher, comedian

Attitudes toward Parenthood Scale

The purpose of this survey is to assess your attitudes toward the role of parenthood. Please read each item carefully and consider what you believe and feel about parenthood. There are no right or wrong answers to any of these statements. After reading each statement, select the number that best reflects your answer, using the following scale:

1	2	3	4	5	6	7
Strongly Disagree						Strongly Agree

_____ 1. Parents should be involved in their children's school.

_____ 2. I think children add joy to a parent's life.

_____ 3. Parents attending functions such as sporting events and recitals of their children help build their child socially.

_____ 4. Parents are responsible for providing a healthy environment for their children.

_____ 5. When you become a parent, your children become your top priority.

_____ 6. Parenting is a job.

_____ 7. I feel that spending quality time with children is an important aspect of child rearing.

_____ 8. Mothers and fathers should share equal responsibilities in raising children.

_____ 9. The formal education of children should begin at as early an age as possible.

Scoring

After assigning a number to each item, add the numbers and divide by 9. The higher the number (7 is the highest possible), the more positive your view and the stronger your commitment to the role of parenthood. The lower the number (1 is the lowest possible), the more negative your view and the weaker your commitment to parenthood.

Norms

Norms for the scale are based upon 22 male and 72 female students attending Valdosta State University. The scores ranged from 3.89 to 7.00, and the average was 6.36 (standard deviation [SD] = 0.65; hence, the respondents had very positive views. There was no significant difference between male participants' attitudes toward parenthood (mean [M] = 6.15; SD = 0.70) and female participants' attitudes (M = 6.42; SD = 0.63). There were also no significant differences between ethnicities.

The average age of participants completing the Attitudes Toward Parenthood Scale was 22.09 years (SD = 3.85), and their ages ranged from 18 to 39. The ethnic composition of the sample was 73.4% white, 22.3% black, 2.1% Asian, 1.1% American Indian, and 1.1% other. The classification of the sample was 20.2% freshmen, 6.4% sophomores, 22.3% juniors, 47.9% seniors, and 3.2% graduate students.

Lifestyle Changes Becoming a parent often involves changes in lifestyle. Daily living routines become focused around the needs of the children. Living arrangements change to provide space for another person in the household. Some parents change their work schedule to allow them to be home more. Food shopping and menus change to accommodate the appetites of children. A major lifestyle change is the loss of freedom of activity and flexibility in one's personal schedule. Lifestyle changes are particularly dramatic for women. The time and effort required to be pregnant and rear children often compete with the time and energy needed to finish one's education. Building a career is also negatively impacted by the birth of children. Parents learn quickly that it is difficult to both be an involved, on-the-spot parent and climb the career ladder. The careers of women may suffer most.

Financial Costs Meeting the financial obligations of parenthood is difficult for many parents. The costs begin with prenatal care and continue at childbirth. For an uncomplicated vaginal delivery, with a 2-day hospital stay, the cost may total $10,000, whereas a cesarean section birth may cost $14,000 (Alexian Brothers Health System, 2005). The annual cost of a child less than 2 years old for middle-income parents ($41,700 to $70,200)—which includes housing, food, transportation, clothing, health care, and childcare—is $9,840. For a 15- to 17-year-

What I'm talking about is this constant, mindless yammering in the media, this neurotic fixation that suggests somehow everything—everything—has to revolve around the lives of children. It's completely out of balance.

George Carlin, comedian

No rational person would decide to have a child.

Ray Brannon, sociologist

old the cost is $10,900 (*Statistical Abstract of the United States: 2006*, Table 666). These costs do not include the wages lost when a parent drops out of the workforce to provide childcare.

How Many Children Do You Want?

Most people do not regard the lifestyle changes and economic costs of children as important when it comes to deciding to have children. While some opt for a child-free marriage, most (90%) want to have not only one child but several. When men and women are compared, women may want more children than men. Referred to as "the baby gap," twice as many women faculty members as men (38% compared to 18%) had fewer children than they wanted. This finding was based on a survey of 4,400 University of California faculty (Shellenbarger, 2006). Procreative liberty is the freedom to decide whether or not to have children.

Childfree Marriage?

Between 1980 and 2004, the percentage of childfree U.S. women aged 40–44 years nearly doubled, from 10.1% to 19.3% (Dye, 2005). Some evidence suggests that women are simply delaying having their children. From 2003 to 2004, the birth rate for women aged 40–44 years increased by 3 percent and the rate for women aged 45–49 years also increased (though at a slower rate) (Hamilton et al., 2005). Whether more women are *deciding* to delay having children or *deciding* against having them altogether is unknown. Hence, although fewer women are actually having children, we don't know why. Is it by choice or because of infertility? Aside from infertility, typical reasons individual couples give for not having children include the freedom to spend their time and money as they choose, to enjoy their partner without interference, to pursue their career, to avoid health problems associated with pregnancy, and to avoid passing on genetic disorders to a new generation (DeOllos and Kapinus, 2002). In regard to motives for not having children, Park (2005) noted that childfree men point to the sacrifices necessitated by children, whereas women point to never having had the "mother instinct."

Some people simply do not like children. Aspects of our society reflect **antinatalism** (against children). Indeed, there is a continuous fight to get corporations to implement or enforce any family policies (from family leaves to flex time to on-site day care). Profit and money—not children—are priorities. In addition, although people are generally tolerant of their own children, they often exhibit antinatalistic behavior in reference to the children of others. Notice the unwillingness of some individuals to sit next to a child on an airplane. What other examples illustrate an antinatalistic view?

One Child?

One in five women aged 40 to 44 has a single child (Roberts and Blanton, 2000). Twenty young adults (each an only child) were asked about the recommendations

This married couple delight in their childfree lifestyle. They own a nice home on a golf course, travel extensively, and have a network of close friends. They do have two dogs and a cat.

Authors

they would give to parents rearing single children today. These suggestions included facilitating connections with age mates, avoiding overindulgence, and encouraging independence (Roberts and Blanton, 2000).

Two Children?

The most preferred family size in the United States is the two-child family. Reasons for this preference include feeling that a family is "not complete" without two children, having a companion for the first child, having a child of each sex, and repeating the positive experience of parenthood they had with their first child. Some couples may not want to "put all their eggs in one basket." They may fear that if they have only one child and that child dies or turns out to be disappointing, they will not have another opportunity to enjoy parenting. Kramer and Ramsburg (2002) emphasized the importance of preparing existing children for the advent of new siblings. Such preparation requires the mother to deliberately set aside time to give to existing children as well as to encourage and reinforce them for prosocial behavior with the new infant.

Three or More Children?

Couples are more likely to have a third child, and to do so quickly, if they already have two girls rather than two boys. They are least likely to bear a third child if they already have a boy and a girl. Some individuals may want three children because they enjoy children and feel that "three is better than two." In some instances, a couple that have two children may simply want another child because they enjoy parenting and have the resources to do so.

Having a third child creates a "middle child." This child is sometimes neglected because parents of three children may focus more on the "baby" and the first-born than on the child in between. However, an advantage to being a middle child is the chance to experience both a younger and an older sibling. Each additional child also has a negative effect on the existing children by reducing the amount of parental time available to existing children. The economic resources for each child are also affected for each subsequent child.

Diversity in Other Countries

In India, the Indian Medical Association has recommended that the country implement a one-child policy in order to slow down population growth, which is estimated to overtake China's population before 2030 (Chatterjee, 2005).

Hispanics are more likely to want larger families than are whites or African-Americans. Larger families have complex interactional patterns and different values. The addition of each subsequent child dramatically increases the possible relationships in the family. For example, in the one-child family, four interpersonal relationships are possible: mother–father, mother–child, father–child, and father–mother–child. In a family of four, 11 relationships are possible; in a family of five, 26; and in a family of six, 57.

PERSONAL CHOICES

Is Genetic Testing for You?

Since each of us may have flawed genes that carry increased risk for diseases such as cancer and Alzheimer's, the question of whether to get a genetic test before becoming pregnant becomes relevant. The test involves giving a blood sample. The advantage is the knowledge of what defective genes you may have and what diseases you may pass to your children. The disadvantage is stress or anxiety (Dinc and Terzioglu, 2006) as well as feeling stigmatized (Sankar, 2006) for certain medical problems (e.g., breast cancer). And since no treatment may be available if the test results are positive, the knowledge that they might pass diseases to their children can be devastating to a couple. In addition, the results of genetic testing may become a part of one's medical history, which may be used to deny health insurance and employment. Finally, genetic testing is expensive, ranging from $200 to $2,400. The National Society of Genetic Counselors offers information about genetic testing. Its Web address is http://www.nsgc.org/.

We had a quicksand box in our back yard. I was an only child, eventually.

Steven Wright, comedian

No two children are ever born into the same family.

Seymour V. Reit, researcher/author

Teenage Motherhood

Almost seven percent (6.7%) of U.S. women between the ages of 15 and 19 have had a child (Dye, 2005). Reasons for these teenagers having a child include not being socialized as to the importance of contraception, having limited parental supervision, and perceiving few alternatives to parenthood. Indeed, motherhood may be one of the only remaining meaningful roles available to them. Geronimus (2003) emphasized that teen pregnancy was adaptive, particularly among African-Americans who "must contend with structural constraints that shorten healthy life expectancy" (p. 881). In addition, some teenagers feel lonely and unloved and have a baby to create a sense of being needed and wanted. In contrast, in Sweden, eligibility requirements for welfare payments make it almost necessary to complete an education and get a job before becoming a parent.

A teenager may not know how to raise children, but it is sure she knows how to raise parents.

Evan Esar

Problems Associated with Teenage Motherhood

Teenage parenthood is associated with various negative consequences, including the following.

1. *Stigmatized and marginalized.* Wilson and Huntington (2006) noted that because teen mothers resist the typical life trajectory of their middle-class peers, they are stigmatized and marginalized. In effect, they are a threat to societal goals of economic growth through higher education and increased female workforce participation. In spite of such stigmatization and marginalization, McDermott and Graham (2005) noted the resilient behaviors of teen mothers: they invest in the "good" mother identity, maintain kin relations, and prioritize the mother/child dyad in their life.

2. *Poverty among single teen mothers and their children.* Many teen mothers are unwed. Livermore and Powers (2006) studied a sample of 336 unwed mothers and found them plagued with financial stress; almost 20 percent had difficulty providing food for themselves and their children (18.5%), had their electricity cut off for nonpayment (19.7%), and had no medical care for their children (18.2%). Almost half (47%) reported experiencing "one or more financial stressor" (p. 6).

3. *Poor health habits.* Teenage unmarried mothers are less likely to seek prenatal care and more likely than older and married women to smoke, drink alcohol, and take other drugs. These factors have an adverse effect on the health of the baby. Indeed, babies born to unmarried teenage mothers are more likely to have low birth weights (less than 5 pounds, 5 ounces) and to be born prematurely. Children of teenage unmarried mothers are also more likely to be developmentally delayed. These outcomes are largely a result of the association between teenage unmarried childbearing and persistent poverty.

4. *Lower academic achievement.* Poor academic achievement is both a contributing factor and a potential outcome of teenage parenthood. Some studies note that between 30 and 70 percent of teen mothers drop out of high school before graduation (the schools may push them out and/or they may no longer feel motivated). Zachry (2005) interviewed 19 mothers and noted that although all dropped out of school, each evidenced a new appreciation for education as a way of providing a better future for their child. Wendy, one of the mothers, said, "I want to better my education for my kids, and myself . . . because I'm their role model. And they're only gonna learn from what they see from me" (p. 2566).

5. *Deficit nurturing skills?* Teen mothers may lack nurturing skills for their infants. One-hundred-nineteen first-time teen mothers (aged 18 or younger) were compared with 43 mature mothers (aged 25 or older) six weeks postpartum. Although teen mothers were more instrumental (e.g., changed diapers, adjusted clothes of their infant), they were less affectionate (e.g., stroked, held, and patted their infant) (Krpan et al., 2005).

6. *Anxious/depressed teen father?* Whereas considerable emphasis has been focused on teenage motherhood, Quinlivan and Condon (2005) compared teenage fathers with teenage nonfathers. The researchers found the former were more anxious and depressed, suggesting even greater fallout from teen childbirth than previously understood.

National Data

The birth rate for teen mothers has continued to drop. In 2004, the birth rate per 1000 women aged 15 to 19 was 41.2, a drop of 33% since 1991 (Hoyert et al., 2006).

As a result of increased public awareness of the need for pregnancy prevention via abstinence and responsible behavior, the birth rate for teenagers has been declining. However, the teenage birth rate in the United States, as measured by births per thousand teenagers, is nine times that of teens in the Netherlands (Fiejoo, 2006). Teens in countries such as France and Germany are as sexually active as U.S. teenagers, but the former grow up in a society that promotes responsible contraceptive use. In the Netherlands, for example, individuals are taught to use both the pill and the condom (an approach called "double Dutch") for prevention of pregnancy and sexually transmitted infections (STIs). A movement is also afoot to focus on teenage males as a way of decreasing teenage motherhood. Prevention programs have begun in high schools to target at-risk males (e.g., those who are failing a subject or repeating a grade; who have a history of STIs, drug use, cigarette smoking, or alcohol use; and who are from single-parent families) (Smith et al., 2003).

Baby Think It Over (see Research Application below) is an intervention program aimed at giving adolescent females a realistic view of parenting.

Baby Think It Over: Evaluation of an Infant Simulation Intervention for Adolescent Pregnancy Prevention

RESEARCH APPLICATION

Baby Think It Over (BTIO) is a pregnancy-prevention intervention in which adolescents take care of a realistic, life-sized computerized infant simulation doll to gain an understanding of the amount of time and effort involved in taking care of an infant and how an infant's needs might affect their daily lives and the lives of their family and significant others. Diane de Anda (2006) conducted research to evaluate the effects of participation in a BTIO intervention on adolescent participants' attitudes toward pregnancy and parenthood. Her research methods and findings are described below.

Sample and Methods: The sample consisted of 353 (140 male, 204 female, 9 no gender reported) predominantly ninth-grade and Latino students at a Los Angeles County high school. For two-and-a-half days, these students were responsible for taking care of a realistic, life-sized computerized infant simulator. The infant simulator is programmed to cry at random intervals, typically between eight and 12 times in 24 hours, with crying periods lasting typically between 10 and 15 minutes. The "baby"

stops crying only when the adolescent "attends" to the doll by inserting a key into a slot in the infant simulator's back until it stops crying. The key is attached to a bracelet, which is worn by the participant. The bracelets are designed so that an attempt to remove the bracelet is detectable. In certain situations, such as when a participant had a test in another class, another student with a key or the health class teacher who had extra keys was permitted to "babysit" a participant's infant simulator.

The infant simulator recorded data, including the amount of time the participant took to "attend" to the infant (insert the key) and any instances of "rough handling," such as dropping or hitting the doll. Students whose infant simulator recorded neglect and rough handling received a private counseling session with the health class teacher and were required to take a parenting class.

In addition to taking care of a computerized infant simulator, participants attended presentations and group discussions on such topics as the high incidence of adolescent pregnancy in the community, the factors that increase risk of adolescent pregnancy, and the costs of

Continued

Baby Think It Over: Evaluation of an Infant Simulation Intervention for Adolescent Pregnancy Prevention—cont'd

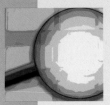

adolescent pregnancy and parenthood, with emphasis on the limitation of education and career opportunities and achievement. The health class teacher also offered a pregnancy-prevention education program in preparation for taking care of the infant simulator and a debriefing discussion period after everyone in the class had taken care of the infant simulator for two-and-a-half days.

The *Baby Think It Over* intervention in this study had seven major objectives, including to increase the degree to which the adolescent recognized that (1) caring for a baby affects an adolescent's academic and social life; (2) other family members are affected by having an adolescent with a baby in the family; (3) there are emotional risks for each parent in having a baby during adolescence; and (4) there are family and cultural values related to having a baby during adolescence. In addition, the intervention aimed to increase the number of adolescents planning to postpone parenthood until (5) a later age (for the majority, until graduation from high school); (6) education and career goals were met; or (7) marriage.

Data were collected through written surveys. The adolescents who participated in this research completed the surveys prior to the BTIO intervention. Survey responses obtained after the BTIO intervention provided post-test data.

Findings and Discussion: Statistical tests conducted on the data as well as the participants' self-report data revealed that the BTIO intervention was effective in changing perceptions of the time and effort involved in caring for an infant and in recognizing the significant effect having a baby has on all aspects of one's life. For example, significant gains from pretest to posttest were found on objectives 1, 2, and 3, suggesting that the BTIO intervention increases the degree to which adolescents recognize that caring for a baby affects their academic and social life; that other family members are affected by having an adolescent with a baby in the family; and that there are emotional risks for each parent in having a baby during adolescence. Self-report data indicated that more than half of participants agreed that BTIO changed their perceptions of what having a baby would be like. The most frequently cited reason was that taking care of a baby was much harder than they previously had thought. Nearly two-thirds reported that BTIO helped change their minds about using birth control.

Were the changes in the adolescents' perceptions and intentions a result of taking care of the BTIO infant simulator, or were these changes the result of the other educational components of the intervention? The present study does not answer this question. The researcher suggests that future research could ascertain the effects of the BTIO infant simulator alone by using a comparison group who receives all the educational components except the infant simulator.

Finally, the researcher notes that the results of her study indicate changes in perception and intention rather than longitudinal measures of actual behavior. Future research could assess long-term effects of the BTIO intervention by obtaining data on the pregnancy rates of the participants over the subsequent 3 years. De Anda (2006) further suggests that a comprehensive program that covers birth control methods (as well as abstinence) and that provides access to contraception "would provide adolescents with the knowledge and skills needed to actualize their intentions and the opportunity for choice in the means to accomplish this" (p. 33).

Source

Diane de Anda. 2006. "Baby Think It Over: evaluation of an infant simulation intervention for adolescent pregnancy prevention." *Health and Social Work* 31(1):26–35. Copyright © 2006 by the National Association of Social Workers, Inc. Reprinted by permission.

Computerized infant simulators, such as the one pictured here, are used in parenting education as well as pregnancy-prevention programs.

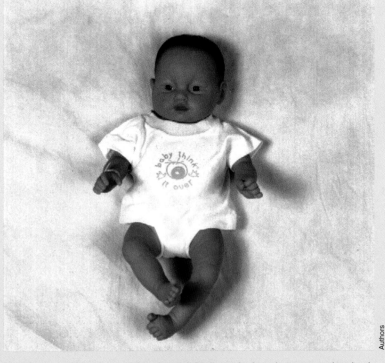

Authors

The *Baby Think It Over* (BTIO) infant simulator cited in the study, Standard Baby, is a discontinued model. Realityworks, Inc., the company that developed BTIO, currently produces more sophisticated and realistic infant simulators using wireless technology called *RealCare® Baby II and RealCare® Baby II-plus* (for more information, see www.realityworks.com).

Infertility

Infertility is defined as the inability to achieve a pregnancy after at least one year of regular sexual relations without birth control, or the inability to carry a pregnancy to a live birth. Different types of infertility include the following:

1. Primary infertility. The woman has never conceived even though she wants to and has had regular sexual relations for the past 12 months.

2. Secondary infertility. The woman has previously conceived but is currently unable to do so even though she wants to and has had regular sexual relations for the past 12 months.

3. Pregnancy wastage. The woman has been able to conceive but has been unable to produce a live birth.

National Data

Between 17% and 26% of all couples of reproductive age in industrialized societies are involuntarily childless. Half of these couples seek treatment for their infertility (Schmidt, 2006).

Causes of Infertility

Although popular usage does not differentiate between the terms *fertilization* and the *beginning of pregnancy,* **fertilization** or **conception** refers to the fusion of the egg and sperm, whereas **pregnancy** is not considered to begin until 5 to 7 days later, when the fertilized egg is implanted (typically in the uterine wall). Hence, not all fertilizations result in a pregnancy. An estimated 30–40 percent of conceptions are lost prior to or during implantation.

Forty percent of infertility problems are attributed to the woman, 40 percent to the man, and 20 percent to both of them. Some of the more common causes of infertility in men include low sperm production, poor semen motility, effects of STIs (such as chlamydia, gonorrhea, and syphilis), and interference with passage of sperm through the genital ducts due to an enlarged prostate. The causes of infertility in women include blocked fallopian tubes, endocrine imbalance that prevents ovulation, dysfunctional ovaries, chemically hostile cervical mucus that may kill sperm, and effects of STIs. Schmidt (2006) noted that women are more likely than men to blame themselves for the infertility.

The older a woman, the less likely she is to get pregnant or have a child. Increasingly, women are delaying having children. The 2004 birthrate for women aged 40 to 44 was nine births per 1,000 women—more than double what it was in 1981 (Hamilton et al., 2005). In 2004, births to women over age 40 topped 110,000 in a single year. First-birth rates for women aged 30 to 34 years, 35–39 years, and 40–44 years also increased in just 1 year, from 2003 to 2004 (Hamilton et al., 2005).

The psychological reaction to infertility is often depression, particularly if individuals link parenthood to happiness (Brothers and Maddux, 2003). When a couple view themselves as infertile, the woman is more likely than the man (21% vs. 9%) to view it as a stressful experience and to become depressed. When both partners experience the infertility with the same level of stress, there are higher levels of marital adjustment than when only one partner becomes stressed (Peterson et al., 2003). White and McQuillan (2006) also confirmed that closing the door on the possibility of having one's own child is, indeed, a very distressing event, particularly for educated women.

Assisted Reproductive Technology

A number of technological innovations are available to assist women and couples in becoming pregnant. These include hormonal therapy, artificial insemination, ovum transfer, in vitro fertilization, gamete intrafallopian transfer, and zygote intrafallopian transfer.

Families with babies and families without are so sorry for each other.

Ed Howe, country town philosopher

Hormone Therapy   Drug therapies are often used to treat hormonal imbalances, induce ovulation, and correct problems in the luteal phase of the menstrual cycle. Frequently used drugs include Clomid, Pergonal, and human chorionic gonadotropin (HCG), a hormone extracted from human placenta. These drugs stimulate the ovary to ripen and release an egg. Although they are fairly effective in stimulating ovulation, hyperstimulation can occur, which may result in permanent damage to the ovary.

Hormone therapy also increases the likelihood that multiple eggs will be released, resulting in multiple births. Such multiple births represent 3.3% of all births (Kochanek et al., 2005). The increase in triplets and higher-order multiple births over the past decade in the United States is largely attributed to the increased use of ovulation-inducing drugs for treating infertility. Infants of higher-order multiple births are at greater risk of having low birth weight and their mortality rates are higher. Mortality rates have improved for these babies, but these low-birth-weight survivors may need extensive neonatal medical and social services.

Artificial Insemination   When the sperm of the male partner are low in count or motility, sperm from several ejaculations may be pooled and placed directly into the cervix. This procedure is known as *artificial insemination by husband* (AIH). When sperm from someone other than the woman's partner are used to fertilize a woman, the technique is referred to as *artificial insemination by donor* (AID).

Lesbians who want to become pregnant may use sperm from a friend or from a sperm bank (some sperm banks cater exclusively to lesbians). Regardless of the source of the sperm, it should be screened for genetic abnormalities and STIs, quarantined for 180 days, and retested for human immunodeficiency virus (HIV); also, the donor should be younger than 50 to diminish hazards related to aging. These precautions are not routinely taken—let the buyer beware!

How do children from donor sperm feel about their fathers? A team of researchers (Scheib et al. 2005) studied 29 individuals (41% from lesbian couples, 38% from single women, and 21% from heterosexual couples) and found that most (75%) always knew about their origin and were comfortable with it. All but one reported a neutral to positive impact with the birth mother. Most (80%) indicated a moderate interest in learning more about the donor. No youths reported wanting money, and only 7% reported wanting a father-child relationship.

Artificial Insemination of a Surrogate Mother   In some instances, artificial insemination does not help a woman get pregnant. (Her fallopian tubes may be blocked, or her cervical mucus may be hostile to sperm.) The couple that still want a child and have decided against adoption may consider parenthood through a surrogate mother. There are two types of surrogate mothers. One is the contracted surrogate mother who supplies the egg, is impregnated with the male partner's sperm, carries the child to term, and gives the baby to the man and his partner. A second type is the surrogate mother who carries to term a baby to whom she is not genetically related (a fertilized egg is implanted in her uterus). As with AID, the motivation of the prospective parents is to have a child that is genetically related to at least one of them. For the surrogate mother, the primary motivation is to help childless couples achieve their aspirations of parenthood and to make money. (The surrogate mother is usually paid about $10,000.)

In Vitro Fertilization   About 2 million couples cannot have a baby because the woman's fallopian tubes are blocked or damaged, preventing the passage of eggs to the uterus. In some cases, blocked tubes can be opened via laser-beam surgery or by inflating a tiny balloon within the clogged passage. When these procedures are not successful (or when the woman decides to avoid invasive tests and exploratory surgery), *in vitro* (meaning "in glass") *fertilization* (IVF), also known as test-tube fertilization, is an alternative.

Using a laparoscope (a narrow, telescope-like instrument inserted through an incision just below the woman's naval to view tubes and ovaries), the physician is able to see a mature egg as it is released from the woman's ovary. The time of release can be predicted accurately within two hours. When the egg emerges, the physician uses an aspirator to remove the egg, placing it in a small tube containing stabilizing fluid. The egg is taken to the laboratory, put in a culture Petri dish, kept at a certain temperature-acidity level, and surrounded by sperm from the woman's partner (or donor). After one of these sperm fertilizes the egg, the egg divides and is implanted by the physician in the wall of the woman's uterus. Usually, several eggs are implanted in the hope one will survive. Early IVF attempts resulted in multiple births. These have been radically reduced.

Some couples want to ensure the sex of their baby. Called "family balancing" since the procedure is often used for couples that already have several children of one sex, the eggs of a woman are fertilized and the sex of the embryos 3 and 8 days old is identified. Only those of the desired sex are then implanted in the woman's uterus.

Alternatively, the Y chromosome of the male sperm can be identified and implanted. The procedure is accurate 75 percent of the time for producing a boy baby and 90 percent of the time for a girl baby. The Genetics and IVF Institute, in Fairfax, Virginia, specializes in the sperm sorting technique.

Occasionally, some fertilized eggs are frozen and implanted at a later time, if necessary. This procedure is known as *cryopreservation*. Separated or divorced couples may disagree over who owns the frozen embryos, and the legal system is still wrestling with the fate of their unused embryos, sperm, or ova where there has been a divorce or death.

Ovum Transfer   In conjunction with in vitro fertilization is ovum transfer, also referred to as embryo transfer. In this procedure an egg is donated, fertilized in vitro with the husband's sperm, and then transferred to his wife. Alternatively, the sperm of the male partner is placed by a physician in a surrogate woman. After about five days, her uterus is flushed out (endometrial lavage), and the contents are analyzed under a microscope to identify the presence of a fertilized ovum.

The fertilized ovum is then inserted into the uterus of the otherwise infertile partner. Although the embryo can also be frozen and implanted at another time, fresh embryos are more likely to result in successful implantation. Infertile couples that opt for ovum transfer do so because the baby will be biologically related to at least one of them (the father) and the partner will have the experience of pregnancy and childbirth. As noted earlier, the surrogate woman participates out of her desire to help an infertile couple or to make money.

Other Reproductive Technologies   A major problem with in vitro fertilization is that only about 15 to 20 percent of the fertilized eggs will implant on the uterine wall. To improve this implant percentage (to between 40 and 50 percent), physicians place the egg and the sperm directly into the fallopian tube, where they meet and fertilize. Then the fertilized egg travels down into the uterus and implants.

Since the term for sperm and egg together is *gamete*, this procedure is called *gamete intrafallopian transfer*, or GIFT. This procedure, as well as in vitro fertilization, is not without psychological costs to the couple.

Gestational surrogacy, another technique, involves fertilization in vitro of the wife's ovum and transfer to a surrogate. Trigametic IVF also involves the use of sperm in which the genetic material of another person has been inserted. This technique allows lesbian couples to have a child genetically related to both women. Infertile couples hoping to get pregnant through one of the more than 300 in vitro fertilization clinics should make informed choices by asking questions such as "What is the center's pregnancy rate for women with a similar diagnosis?"

What percentage of these women have a live birth? According to the Centers for Disease Control and Prevention, the typical success rate (live birth) for infertile couples who seek help in one of the 400 fertility clinics is 28% (Lee, 2006). In addition, although the absolute risks of birth defects may be small, babies conceived through assisted reproductive technology may have increased risk of malformations, low birth weight, and other perinatal complications (Shiota and Yamada, 2005).

Adoption

He changed everything, but in the most wonderful way. Everything that should matter, matters. He's absolutely the center of my life.

Angelina Jolie, on her adopted son, Maddox

Somehow destiny comes into play. These children end up with you and you end up with them. It's something quite magical.

Nicole Kidman, adoptive parent

Angelina Jolie and Nicole Kidman are celebrities who have given national visibility to adopting children. They are not alone in their desire to adopt children. The various routes to adoption are public (children from the child welfare system), private agency (children are placed with nonrelatives through agencies), independent adoption (children are placed directly by birth parents or through an intermediary such as a physician or attorney), kinship (children are placed in a family member's home), and stepparent (children are adopted by a spouse). Motives for adopting a child include wanting a child because of their inability to have a biological child (infertility), their desire to give an otherwise unwanted child a permanent loving home, or their desire to avoid contributing to overpopulation by having more biological children. Some couples may seek adoption for all of these motives.

Although parents can return a child they provisionally adopt, public opinion is against their doing so. Almost 60 percent (58%) of a national sample of U.S. respondents feel that parents who adopt a child should be required to keep that child (Hollingsworth, 2003).

National Data

Of U.S. children under the age of 18, two percent are adopted (Johnson D., 2005).

International Data

Of all adoptions in the United States, 15% are of children from other countries (Johnson D., 2005)

Demographic Characteristics of Persons Seeking to Adopt a Child

Whereas demographic characteristics of those who typically adopt are white, educated, and high-income, increasingly adoptees are being placed in nontraditional families including older, gay, and single individuals. Rosie O'Donnell represents the unmarried gay mother. She lives in Florida, where state law prohibits gay and lesbian individuals and couples from adopting children. Florida is not alone. Mississippi also bans adoption by gay couples (but not single gay individuals), and Utah has a similar law. Sixteen states have taken steps to ban adoption by gay couples on the grounds that since "marriage" is "heterosexual marriage," children do not belong in homosexual relationships (Stone, 2006).

Strong legal challenges have been made to prohibit the banning of gay adoption. Leung et al. (2005) compared children adopted or reared by gay/lesbian and

Diversity in Other Countries

Not all countries allow overseas adoptions. Romania emptied its orphanages and sent its children home to their parents or to foster care, which resulted in a large population of unwanted and abandoned children (Laffan, 2005).

heterosexual parents. They found no negative effects when the adoptive parents were gay/lesbian. Approval for adoption by same-sex couples is evident in the population, as half of the respondents in surveys in the United States report approval of same-sex adoptions (Maill and March, 2005).

Characteristics of Children Available for Adoption

Adoptees in the highest demand are infant white, healthy children. Those who are older, of a racial or ethnic group different from that of the adoptive parents, of a sibling group, or with physical or developmental disabilities have been difficult to place. Flower Kim (2003) noted that since the waiting period for a healthy white infant is from 5 to 10 years, couples are increasingly open to cross-racial adoptions. Of the 1.6 million adopted children younger than 18 living in U.S. households, the percentages adopted from other countries are as follows: Korea, 24 percent; China, 11 percent; Russia, 10 percent; and Mexico, 9 percent. International or cross-racial adoptions may complicate the adoptive child's identity. Children adopted after infancy may also experience developmental delays, attachment disturbances, and post–traumatic stress disorder (Nickman et al., 2005).

Transracial Adoption

Transracial adoption is defined as the practice of adopting children of a race different from that of the parents— for example, a white couple adopting a Korean or African-American child. In a study on transracial adoption attitudes of college students (using the Attitudes Toward Transracial Adoption Scale in this section), the scores of the 188 respondents reflected overwhelmingly positive attitudes toward transracial adoption. Significant differences included that women, persons willing to adopt a child at all, interracially experienced daters, and those open to interracial dating were more willing to adopt transracially than men, persons rejecting adoption as an optional route to parenthood, persons with no previous interracial dating experience, and persons closed to interracial dating (Ross et al., 2003). About one in five parents (21%) who adopt children from foster care adopt transracially (Urban Institute, 2003).

Transracial adoptions are controversial. Kennedy (2003) noted, "Whites who seek to adopt black children are widely regarded with suspicion. Are they ideologues, more interested in making a political point than in actually being parents?" (p. 447). Another controversy is whether it is beneficial for children to be adopted by parents of the same racial background (Hollingsworth, 2000). In regard to the adoption of African-American children by same-race parents, the National Association of Black Social Workers (NABSW) passed a resolution against transracial adoptions, citing that such adoptions prevented black children from developing a positive sense of themselves "that would be necessary to cope with racism and prejudice that would eventually occur" (Hollingsworth, 1997, p. 44). Madonna was criticized for her Malawain adoption attempt.

The counterargument is that healthy self-concepts, an appreciation for one's racial heritage, and coping with racism/prejudice can be learned in a variety of contexts. Legal restrictions on transracial adoptions have disappeared, and social approval for transracial adoptions is increasing. However, a substantial number of studies conclude that "same race placements are preferable and that special measures should be taken to facilitate such placements, even if it means delaying some adoptions" (Kennedy, 2003, p. 469).

One 26-year-old black female was asked how she felt about being reared by white parents and replied, "Again, they are my family and I love them, but I am black. I have to deal with my reality as a black woman" (Simon and Roorda, 2000, p. 41). A black man reared in a white home advised white parents considering a transracial adoption, "Make sure they have the influence of blacks in their lives; even if they have to go out and make friends with black families—it's a must"

I've found that academics—both black and white—are more likely to have problems with transracial adoption than the parents we've gotten to know through Cassie's school and extracurricular activities.

Susan Bordo, adoptive parent

Attitudes toward Transracial Adoption Scale

Transracial adoption is the adoption of children of a race other than that of the adoptive parents. Please read each item carefully and consider what you believe about each statement. There are no right or wrong answers to any of these statements, so please give your honest reaction and opinion. After reading each statement, select the number that best reflects your answer, using the following scale:

1	2	3	4	5	6	7
Strongly Disagree						Strongly Agree

_____ 1. Transracial adoption can interfere with a child's well-being.

_____ 2. Transracial adoption should not be allowed.

_____ 3. I would never adopt a child of another race.

_____ 4. I think that transracial adoption is unfair to the children.

_____ 5. I believe that adopting parents should adopt a child within their own race.

_____ 6. Only same-race couples should be allowed to adopt.

_____ 7. Biracial couples are not well prepared to raise children.

_____ 8. Transracially adopted children need to choose one culture over another.

_____ 9. Transracially adopted children feel as though they are not part of the family they live in.

_____ 10. Transracial adoption should occur only between certain races.

_____ 11. I am against transracial adoption.

_____ 12. A person has to be desperate to adopt a child of another race.

_____ 13. Children adopted by parents of a different race have more difficulty developing socially than children adopted by foster parents of the same race.

_____ 14. Members of multiracial families do not get along well.

_____ 15. Transracial adoption results in "cultural genocide."

Scoring

After assigning a number to each item, add the numbers and divide by 15. The lower the score (1 is the lowest possible), the more positive one's view of transracial adoptions. The higher the score (7 is the highest possible), the more negative one's view of transracial adoptions. The norming sample was based upon 34 male and 69 female students attending Valdosta State University. The average score was 2.27 (SD = 1.15), suggesting a generally positive view of transracial adoption by the respondents, and scores ranged from 1.00 to 6.60.

The average age of participants completing the scale was 22.22 years (SD = 4.23), and ages ranged from 18 to 48. The ethnic composition of the sample was 74.8% white, 20.4% black, 1.9% Asian, 1.0% Hispanic, 1.0% American Indian, and one person of nonindicated ethnicity. The classification of the sample was 15.5% freshmen, 6.8% sophomores, 32.0% juniors, 42.7% seniors, and 2.9% graduate students.

(p. 25). Indeed, Huh and Reid (2000) found that positive adjustment by adoptees was associated with participation in the cultural activities of the race of the parents who adopted them.

Open versus Closed Adoptions

Another controversy is whether adopted children should be allowed to obtain information about their biological parents. Surveys in both Canada and the United States reveal that about three-fourths of the respondents approve of some form of open adoption and of giving adult adoptees unlimited access to confidential information about their birth parents (Maill and March, 2005). In general, there are considerable benefits for having an open adoption—the biological parent has the opportunity to stay involved in the child's life. Adoptees learn early that they are adopted and who their biological parents are. Birth parents are more likely to avoid regret and to be able to stay in contact with their child. Adoptive parents have information about the genetic background of their adopted child.

Whether an adoption is open or closed, would-be adoptive parents who enroll in a preadoption course report being more emotionally ready to adopt, having more parenting knowledge, and knowing more about the adoption process. Such a course is offered through some private nonprofit agencies (Farber et al., 2003).

National Data
Private adoption usually costs from $15,000 to $35,000. Adopting a child from foster care costs nothing (Pertman, 2000).

Foster Parenting

Some individuals seek the role of parent via foster parenting. A **foster parent,** also known as a family caregiver, is neither a biological nor an adoptive parent but is a person who takes care of and fosters a child taken into custody. He or she has made a contract with the state for the service, has judicial status, and is reimbursed by the state. In general, the person is not required to have any degree or formal training (Isomaki, 2002).

About 600,000 children are in foster care (Urban Institute, 2003). Children placed in foster care have typically been removed from parents who are abusive, who are substance abusers, and/or who are mentally incompetent. Although foster parents are paid for taking care of children in their home, they are also motivated by "empathy, love, and generosity, a willingness to help" (p. 629). The goal of placing children in foster care is to remove them from a negative family context, improve that context, and return them, or find a more permanent home than foster care. In those cases where the return to the parents is successful, the parents were not substance abusers (Miller et al., 2006).

Contraception

Once individuals have decided on whether and when they want children, they need to make a choice about contraception. Pregnancy prevention, STI prevention, ease of use, and need to plan ahead are the primary criteria for selection among women. Men report that pregnancy prevention, STI prevention, and sexual pleasure are the primary criteria for their selection of a method (Grady et al., 2000). In a study of 1027 undergraduates at a southeastern university, 62.3% reported that they had used a form of birth control (other than withdrawal) during their most recent sexual intercourse (Knox and Zusman, 2006).

All contraceptive practices have one of two common purposes: to prevent the male sperm from fertilizing the female egg or to keep the fertilized egg from implanting itself in the uterus. About 5 to 7 days after fertilization, pregnancy begins. Although the fertilized egg will not develop into a human unless it implants on the uterine wall, pro-life supporters believe that conception has already occurred.

In this section, we look at the various methods of contraception. Whatever contraception a woman selects, change is likely; 40 percent of married women and 61 percent of unmarried women in one study reported switching their method of contraception over a 2-year period. The primary reason for switching was related to level of contraceptive effectiveness, health risks, and STI prevention (Grady et al., 2002).

The most effective birth control I know is spending the day with my kids.

Jill Bensley, financial feasibility consultant

(Appreciation is expressed to Beth Credle Burt, MAEd, CHES, a health education specialist with the Wake Area Health Education Center at WakeMed Hospital, Raleigh, North Carolina, who updated this section to provide state-of-the-art information.)

Hormonal Contraceptives

Hormonal methods of contraception currently available to women include "the pill," Jadelle®, Depo-Provera®, NuvaRing®, and Ortho Evra®.

Oral Contraceptive Agents (Birth Control Pills)   The birth control pill is the most commonly used method of all the nonsurgical forms of contraception. Although 8 percent of women who take the pill still become pregnant in the first year of use (Ranjit et al., 2001), it remains a desirable birth control option.

Oral contraceptives are available in basically two types: the combination pill, which contains varying levels of estrogen and progestin, and the minipill, which is progestin only. Combination pills work by raising the natural level of hormones in a woman's body, inhibiting ovulation, creating an environment where sperm cannot easily reach the egg, and hampering implantation of a fertilized egg.

The second type of birth control pill, the minipill, contains the same progesterone-like hormone found in the combination pill but does not contain estrogen. Progestin-only pills are taken every day, with no hormone-free interval. As with the combination pill, the progestin in the minipill provides a hostile environment for sperm and does not allow implantation of a fertilized egg in the uterus, but unlike the combination pill, the minipill does not always inhibit ovulation.

For this reason, the minipill is somewhat less effective than other types of birth control pills. The minipill has also been associated with a higher incidence of irregular bleeding. Neither the combination pill nor the minipill should be taken unless prescribed by a health-care provider who has detailed information about the woman's medical history. Contraindications—reasons for not prescribing birth control pills—include hypertension, impaired liver function, known or suspected tumors that are estrogen-dependent, undiagnosed abnormal genital bleeding, pregnancy at the time of the examination, and history of poor blood circulation or blood clotting. The major complications associated with taking oral contraceptives are blood clots and high blood pressure. Also, the risk of heart attack is increased for those who smoke or have other risk factors for heart disease.

If they smoke, women older than 35 should generally use other forms of contraception. Although the long-term negative consequences of taking birth control pills are still the subject of research, short-term negative effects are experienced by 25 percent of all women who use them. These side effects include increased susceptibility to vaginal infections, nausea, slight weight gain, vaginal bleeding between periods, breast tenderness, and mood changes (some women become depressed and experience a loss of sexual desire). A recent large population study confirmed that women who take oral contraceptives have an increased chance of suffering from both migraines and nonmigraine headaches (Aegidius, 2006). Women should also be aware of situations in which the pill is not effective, such as the first month of use, with certain prescription medications, and when pills are missed. On the positive side, pill use reduces the incidence of ectopic pregnancy and offers noncontraceptive benefits, such as reduced incidence of ovarian and endometrial cancers, pelvic inflammatory disease, anemia, and benign breast disease.

Finally, women should be aware that pill use is associated with an increased incidence of chlamydia and gonorrhea. One reason for the association of pill use and a higher incidence of STIs is that sexually active women who use the pill sometimes erroneously feel that because they are protected from becoming pregnant, they are also protected from contracting STIs. The pill provides no protection against STIs; the only methods that provide some protection against STIs are the male and female condoms.

Despite the widespread use of birth control pills, many women prefer a method that is longer acting and does not require daily action. Research con-

My best birth control now is to just leave the lights on.

Joan Rivers

tinues toward identifying safe, effective hormonal contraceptive delivery methods that are more convenient (Schwartz and Gabelnick, 2002). One such new hormonal contraceptive is the Food and Drug Administration (FDA)–approved Seasonale®, which reduces the number of periods a woman experiences from 13 to four per year. This hormonal contraceptive manages the menstrual cycles by skipping the hormone-free week and limiting women's menstrual periods to once every 3 months. However, unexpected menstrual bleeding occurs four times as often with this extended-cycle regimen, and women who use Seasonale are four times as likely to discontinue use in the first 26 weeks as a result (Wilson et al., 2005).

Jadelle® and Implanon®   **Jadelle** is a system of rod-shaped silicone implants that are inserted under the skin in the upper inner arm and provide time-release progestin into a woman's system for contraception. This method has replaced Norplant, the first hormonal implant system introduced on the market. The difference between Jadelle and Norplant is that Jadelle consists of only two thin, flexible silicone rods, whereas Norplant consisted of six capsules, and Jadelle is inserted with a needle rather than through a minor surgical procedure. Jadelle was originally approved for 3 years of constant contraceptive protection, but in July 2002, the FDA extended approval for use for 5 years of pregnancy protection.

Side effects are similar to those associated with Norplant; the most common is irregular menstrual bleeding (Population Council, 2003b). Population Council scientists are also developing a single-rod implant containing Nestrone® (a synthetic progestin similar to the natural hormone progesterone). It may be used by lactating women and may protect against pregnancy for 2 years.

Implanon®   The popularity of contraceptive implants has increased, with more than 11 million women around the world using hormonal implant systems. Research and development of these methods continue to produce improved, discreet, easier-to-use implants. Implanon, a single-rod system that provides pregnancy protection for 3 years after insertion, was developed in Indonesia in 1998 by the Dutch pharmaceutical company Organon and approved in November 2004 by the FDA for use in the United States. In clinical trials involving more than 2300 women, Implanon took an average of just 1.1 minutes to be inserted and 2.6 minutes to be removed (Organon International, Inc., 2004). Although both Jadelle and Implanon are approved for use by the FDA and are popular in Europe, these methods are not currently available to U.S. consumers, but they are expected to be launched in the near future, when a U.S distributor is found.

Depo-Provera®   Also known as "Depo" and "the shot," **Depo-Provera** is a synthetic compound similar to progesterone that is injected into the woman's arm or buttock and protects against pregnancy for 3 months by preventing ovulation. Depo-Provera has been used by 30 million women worldwide since it was introduced in the late 1960s (although it was not approved for use in the United States until 1992). A subcutaneous shot was approved for use in 2005, thus opening the door for a self-administered shot ("Injectable Update: Subcutaneous Form of Depo-Provera Is Approved," 2005). Side effects of Depo-Provera include menstrual spotting, irregular bleeding, and some heavy bleeding the first few months of use, although eight of ten women using Depo-Provera will eventually experience amenorrhea, or the absence of a menstrual period. Mood changes, headaches, dizziness, and fatigue have also been observed. Some women report a weight gain of 5 to 10 pounds.

Also, after the injections are stopped, it takes an average of 18 months before the woman will become pregnant at the same rate as women who have not used Depo-Provera. Slightly fewer than 3 percent of U.S. women report using the injectable contraceptive. The reasons they cite for not using "Depo" include lack

A baby crying is the best form of birth control.
Carole Tabron

of knowledge, fear of side effects or health hazards, and satisfaction with their current contraceptive method (Tanfer et al., 2000).

Recent research has shown that prolonged use of Depo-Provera is associated with significant loss of bone density, which may not be completely reversible after discontinuing use. In November 2004, the FDA announced a "black box" warning highlighting this information and recommending that Depo-Provera should be used only as a long-term contraceptive method (longer than 2 years) if other methods are inadequate (U. S. FDA, 2004). A new formulation of this drug, using the same active ingredient, has received FDA approval for managing pain associated with endometriosis (caused when endometrial tissue from the uterus moves to other pelvic organs, causing pain and infertility) (Pfizer, 2005).

Vaginal Rings NuvaRing®, which is a soft, flexible, transparent ring approximately 2 inches in diameter that is worn inside the vagina, provides month-long pregnancy protection. NuvaRing has two major advantages. First, because the hormones are delivered locally rather than systemically, very low levels are administered (the lowest of any of the hormonal contraceptives). Second, unlike oral contraceptives, with which the hormone levels rise and fall depending on when the pill is taken, the hormone level from the NuvaRing remains constant. The NuvaRing is a highly effective contraceptive when used according to the labeling. Out of a hundred women using NuvaRing for a year, one or two will become pregnant. This method is self-administered.

NuvaRing is inserted into the vagina and is designed to release hormones that are absorbed by the woman's body for 3 weeks. The ring is then removed for a week, at which time the menstrual cycle will occur; afterward the ring is replaced with a new ring. Side effects of NuvaRing are similar to those of the birth control pill, but a recent study showed NuvaRing to have a more favorable side effect profile and overall tolerability (van den Heuvel et al., 2005). In a 1-year study by the manufacturer, more than 2000 women tested NuvaRing; 85 percent of the women were satisfied, and 95 percent would recommend it to others (Roumen et al., 2001). In 2001 *Time* magazine recognized NuvaRing as one of the best health inventions of the year. NuvaRing became available in the United States in mid-2002.

Transdermal Applications Ortho Evra® is a contraceptive transdermal patch that delivers hormones to a woman's body through skin absorption. The contraceptive patch is worn on the buttocks, abdomen, upper torso (excluding the breasts), or on the outside of the upper arm for 3 weeks and is changed on a weekly basis. The fourth week is patch-free and the time when the menstrual cycle will occur (Mishell, 2002). Under no circumstances should a woman allow more than 7 days to lapse without wearing a patch (Madison and Zieman, 2002).

Ortho Evra provides pregnancy protection and has side effects similar to those of the pill. Ortho Evra simply offers a different delivery method of the hormones needed to prevent pregnancy from occurring. However, a major advantage of the patch over the pill is higher compliance rates (Mishell, 2002). In clinical trials, the contraceptive patch was found to keep its adhesiveness even through showers, workouts, and water activities, such as swimming. In November 2005, the FDA approved updating the labeling of Ortho Evra to warn users and healthcare providers that this product exposes women to about 60% more estrogen than most birth control pills. In general, increased estrogen exposure may increase the risk of blood clots (FDA, http://www.fda.gov/bbs/topics/news/2005/NEW01262.html). Another transdermal delivery system under development is a contraceptive gel that can be applied to a woman's abdomen. Current efforts in the development of this Nestorone-containing gel yielded a product that women apply to the abdomen for contraceptive or hormone therapy. Because marketing studies in a number of locales have demonstrated that gels are more acceptable in some cultures and patches in others, both Nestorone gel and patch formulations are currently being investigated and developed (Population Council, 2003a).

Male Hormonal Methods   Because of dissatisfaction with the few contraceptive options available to men, research and development are occurring in this area.

Numerous studies have found that administration of testosterone to men markedly reduces sperm count and is a very efficient and well tolerated method of contraception. Combinations with progestogens or with gonadotropin-releasing hormone antagonists are even more effective and suggest that hormonal contraception in men is feasible and may be as effective as the currently used methods. (Amory et al., 2006). Several promising male products under development rely on MENT™, a synthetic steroid that resembles testosterone. In contrast to testosterone, however, MENT does not have the effect of enlarging the prostate. A MENT implant and MENT transdermal gel and patch formulation are being developed for contraception. When male hormonal methods do become available, men will need access to clinical screening, prescriptions, and monitoring similar to the follow-up of women who are taking the pill (Alan Guttmacher Institute, 2002).

Male Condom

The condom is a thin sheath made of latex, polyurethane, or natural membranes. Latex condoms, which can be used only with water-based lubricants, historically have been more popular. They are also less likely to slip off and have a much lower chance of breakage (Walsh et al., 2003). However, the polyurethane condom, which is thinner but just as durable as latex, is growing in popularity.

Polyurethane condoms can be used with oil-based lubricants, are an option for some people who have latex-sensitive allergies, provide some protection against HIV and other STIs, and allow for greater sensitivity during intercourse. Condoms made of natural membranes (sheep intestinal lining) are not recommended because they are not effective in preventing transmission of HIV or other STIs. Individuals are more likely to use condoms with casual than with stable partners (Morrison et al., 2003).

The condom works by being rolled over and down the shaft of the erect penis before intercourse. When the man ejaculates, sperm are caught inside the condom. When used in combination with a spermicidal lubricant that is placed on the inside of the reservoir tip of the condom as well as a spermicidal (sperm-killing) agent that the woman inserts into her vagina, the condom is a highly effective contraceptive. Care should be taken to AVOID USING Non-oxynol-9 (N-9) as a contraceptive lubricant, because it has been shown to provide no protection against STIs or HIV. And N-9 products, such as condoms that have N-9 as a lubricant, should not be used rectally since doing so could INCREASE one's risk of getting HIV or other STIs.

Like any contraceptive, the condom is effective only when used properly. Practice putting the male condom on reduces the chance of the condom breaking or coming off during intercourse. A recent study found that the male condom breakage rate fell from 7% among first-time users to 2% among those who had used the method at least 15 times; the slippage rate dropped from 3% to 0.4% (Hollander, 2005).

The condom should be placed on the penis early enough to avoid any seminal leakage into the vagina. In addition, polyurethane or latex condoms with a reservoir tip are preferable, as they are less likely to break. Even when a condom has a reservoir tip, air should be squeezed out of the tip as it is being placed on the penis to reduce the chance of breaking during ejaculation. But such breakage does occur. Finally, the penis should be withdrawn from the vagina immediately after ejaculation, before it returns to its flaccid state. If the penis is not withdrawn and the erection subsides, semen may leak from the base of the condom

A birth control pill for men, that's fair. It makes more sense to take the bullets out of the gun than to wear a bulletproof vest.

Unknown

into the vaginal lips. Alternatively, when the erection subsides, the condom will come off when the man withdraws his penis if he does not hold onto the condom. Either way, the sperm will begin to travel up the vagina to the uterus and fertilization may occur.

In addition to furnishing extra protection, spermicides also provide lubrication, which permits easy entrance of the condom-covered penis into the vagina. If no spermicide is used and the condom is not of the prelubricated variety, a sterile lubricant (such as K-Y Jelly) may be needed. Vaseline or other kinds of petroleum jelly should not be used with condoms because vaginal infections and/or condom breakage may result. Though condoms should also be checked for visible damage and for the date of expiration, this is rarely done. Three-fourths of the respondents in Lane's (2003) study did not check for damage and 61 percent did not check the date of expiration.

Female Condom

The **female condom** resembles the male condom except that it fits in the woman's vagina to protect her from pregnancy, HIV infection, and other STIs. The female condom is a lubricated, polyurethane adaptation of the male version. It is about six inches long and has flexible rings at both ends. It is inserted like a diaphragm, with the inner ring fitting behind the pubic bone against the cervix; the outer ring remains outside the body and encircles the labial area. Like the male version, the female condom is not reusable. Female condoms have been approved by the FDA and are being marketed under the brand names Femidom and Reality. The one-size-fits-all device is available without a prescription.

The female condom is durable and may not tear as easily as latex male condoms. Some women may encounter some difficulty with first attempts to insert the female condom. A major advantage of the female condom is that, like the male counterpart, it helps protect against transmission of HIV and other STIs, giving women an option for protection if their partner refuses to wear a condom.

Placement may occur up to 8 hours before use, allowing greater spontaneity (Beksinska et al., 2001). A number of studies that have been conducted internationally have found that women assess the female condom as acceptable and have high rates of trial use with positive reactions (Artz et al., 2000). Women who have had an STI are more likely to use the female condom. Eighty-five percent of 895 women who attended an STI clinic reported that they had used the female condom during the previous 6 months (Macaluso et al., 2000). If women have instruction and training in the use of the female condom, including a chance to practice the method, this increases its use. Women who had a chance to practice skills on a pelvic model were more likely to rate the method favorably, use it, and use it correctly. Although not a method completely under the woman's control, the female condom gives women more control than does the male condom (Van Devanter et al., 2002).

There are problems with the female condom, however. London (2003) found in a study of 2232 female condom uses that slippage occurred in 10 percent of the cases and that women may have been exposed to semen in one in five uses. Problems also occur when there is a large disparity between the size of the woman's vagina and the size of her partner's penis and when intercourse is very active. The female condom is also very expensive in comparison with other nonreusable barrier methods. Researchers responding to

Female condom

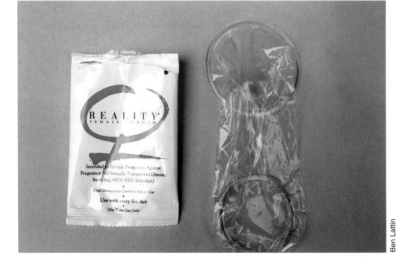

Ben Lattin

Chapter 10 Planning Children and Contraception

demands from studies involving women's focus groups with regard to better contraceptive barrier methods have several products under development that will offer improvements upon the existing female condom. The FC2 is based on the original female condom but is made of synthetic latex instead of polyurethane and could cost half as much. The VA feminine condom, also known as the Reddy female condom and as V-Amour, contains a soft sponge to hold it in place inside the vagina rather than a ring and has a V-shaped external rim. The PATH Woman's condom consists of a dissolving capsule intended to make insertion easier, a polyurethane condom pouch, and a soft outer ring, allowing for nearly universal fit. These methods are currently in clinical trials, and their application for U.S. FDA approval is expected (Upadhyay, 2005).

Vaginal Spermicides

A **spermicide** is a chemical that kills sperm. Vaginal spermicides come in several forms, including foam, cream, jelly, film, and suppository. In the United States, the active agent in most spermicides is N-9, which has previously been recommended for added STI protection. However, recent preliminary research in South African countries has shown that women who used N-9 became infected with HIV at approximately a 50 percent higher rate than women who used a placebo gel (Van Damme, 2000). Spermicidal creams or gels should be used with a diaphragm. Spermicidal foams, creams, gels, suppositories, and films may be used alone or with a condom.

Spermicides must be applied before the penis enters the vagina (appropriate applicators are included when the product is purchased), no more than 20 minutes before intercourse. Foam is effective immediately, but suppositories, creams, and jellies require a few minutes to allow the product to melt and spread inside the vagina (package instructions describe the exact time required). Each time intercourse is repeated, more spermicide must be applied. Spermicide must be left in place for at least 6 to 8 hours after intercourse; douching or rinsing the vagina should not be done during this period.

One advantage of using spermicides is that they are available without a prescription or medical examination. They also do not manipulate the woman's hormonal system and have few side effects. It was believed that a major noncontraceptive benefit of some spermicides is that they offer some protection against STIs. However, the 2002 guidelines for prevention and treatment of STIs from the Centers for Disease Control and Prevention (CDC) do not recommend spermicides for STI/HIV protection. Furthermore, the CDC emphasizes that condoms lubricated with spermicides offer no more protection from STIs than other lubricated condoms, and spermicidally lubricated condoms may have a shorter shelf life, may cost more, and may be associated with increased urinary tract infections in women (CDC, 2002).

Contraceptive Sponge

The Today contraceptive sponge is a disk-shaped polyurethane device containing the spermicide N-9. This small device, which has a 72 percent to 86 percent effectiveness rate, is dampened with water to activate the spermicide and then inserted into the vagina before intercourse begins. The sponge protects for repeated acts of intercourse for 24 hours without the need for supplying additional spermicide. It cannot be removed for at least 6 hours after intercourse, but it should not be left in place for more than 30 hours. Possible side effects that may occur with use include irritation, allergic reactions, or difficulty with removal; and the risk of toxic shock syndrome, a rare but serious infection, is greater when the device is kept in place longer than recommended. The sponge provides no protection from STIs. The Today sponge was taken off the market for 11 years due to manufacturing concerns. These concerns have been eliminated, and national distribution resumed in the summer of 2005.

Intrauterine Device (IUD)

Although not technically a barrier method, the **intrauterine device** (IUD) is a structural device that prevents implantation. It is inserted into the uterus by a physician to prevent the fertilized egg from implanting on the uterine wall or to dislodge the fertilized egg if it has already implanted. Two common IUDs sold in the United States are ParaGard Copper T 380A and the Progestasert Progesterone T. The Copper T is partly wrapped in copper and can remain in the uterus for 10 years. The Progesterone T contains a supply of progestin, which it continuously releases into the uterus in small amounts; after a year the supply runs out and a new IUD must be inserted. The Copper T has a lower failure rate than the Progesterone T. Recently, the FDA also approved an IUD called Mirena, widely used for years in Europe, for use in the United States. Mirena releases tiny amounts of the hormone levonorgestrel into the uterus and protects against pregnancy for 5 years.

Worldwide, the IUD is a popular method; 60 percent of women using IUDs continue their use after 2 years (Motamed, 2002). As a result of infertility and miscarriage associated with the Dalkon Shield IUD and subsequent lawsuits against its manufacturer by persons who were damaged by the device, use of all IUDs in the United States declined in the 1980s. However, other IUDs do not share the rates of pelvic inflammatory disease (PID) or resultant infertility associated with the Dalkon Shield. Nevertheless, other manufacturers voluntarily withdrew their IUDs from the U.S. market. In contrast to the 1 percent rate of use by women in the United States, the IUD accounts for 9 to 24 percent of all contraceptive use in five European countries (Italy, Spain, Poland, Germany, and Denmark).

The January 2001 reintroduction of the IUD in the United States and training of health-care providers in its proper screening, insertion, and follow-up care may prompt its revival in this country (Hubacher, 2002). The IUD is often an excellent choice for women who do not anticipate future pregnancies but do not wish to be sterilized or for women who are unable to use hormonal contraceptives. The IUD is not for women who have multiple sex partners.

Diaphragm

The **diaphragm** is a shallow rubber dome attached to a flexible, circular steel spring. Varying in diameter from 2 to 4 inches, the diaphragm covers the cervix and prevents sperm from moving beyond the vagina into the uterus. This device should always be used with a spermicidal jelly or cream.

To obtain a diaphragm, a woman must have an internal pelvic examination by a physician or nurse practitioner, who will select the appropriate size and instruct the woman on how to insert the diaphragm. The woman will be told to apply spermicidal cream or jelly on the inside of the diaphragm and around the rim before inserting it into the vagina (no more than 2 hours before intercourse).

The diaphragm must also be left in place for 6 to 8 hours after intercourse to permit any lingering sperm to be killed by the spermicidal agent. After the birth of a child, a miscarriage, abdominal surgery, or the gain or loss of 10 pounds, a woman who uses a diaphragm should consult her physician or health practitioner to ensure a continued good fit. In any case, the diaphragm should be checked every 2 years for fit.

A major advantage of the diaphragm is that it does not interfere with the woman's hormonal system and has few, if any, side effects. Also, for those couples who feel that menstruation diminishes their capacity to enjoy intercourse, the diaphragm may be used to catch the menstrual flow for a brief time. On the negative side, some women feel that use of the diaphragm with the spermicidal gel is messy and a nuisance, and it is possible that use of a spermicide may produce an allergic reaction. Furthermore, some partners feel that spermicides make oral genital contact less enjoyable. Finally, if the diaphragm does not fit properly or is left in place too long (more than 24 hours), pregnancy or toxic shock syndrome can result.

Cervical Cap

The **cervical cap** is a thimble-shaped contraceptive device made of rubber or polyethylene that fits tightly over the cervix and is held in place by suction. Like the diaphragm, the cervical cap, which is used in conjunction with spermicidal cream or jelly, prevents sperm from entering the uterus. Cervical caps have been widely available in Europe for some time and were approved for marketing in the United States in 1988. The cervical cap cannot be used during menstruation, since the suction cannot be maintained. The effectiveness, problems, risks, and advantages are similar to those of the diaphragm.

Natural Family Planning

Also referred to as **periodic abstinence,** rhythm method, and fertility awareness, **natural family planning** involves refraining from sexual intercourse during the 7 to 10 days each month when the woman is thought to be fertile. Women who use natural family planning must know their time of ovulation and avoid intercourse just before, during, and immediately after that time. Calculating the fertile period involves knowing when ovulation has occurred. This is usually the 14th day (plus or minus 2 days) before the onset of the next menstrual period. Numerous "home ovulation kits" are available without a prescription in a pharmacy; these allow the woman (by testing her urine) to identify 12 to 26 hours in advance when she will ovulate. Another method of identifying when ovulation has occurred is by observing an increase in the basal body temperature and a sticky cervical mucus. Calculating the time of ovulation may also be used as a method of becoming pregnant since it helps the couple to know when the woman is fertile.

Nonmethods: Withdrawal and Douching

Because withdrawal and douching are not effective in preventing pregnancy, we call them "nonmethods" of birth control. Also known as **coitus interruptus, withdrawal** is the practice whereby the man withdraws his penis from the vagina before he ejaculates. The advantages of coitus interruptus are that it requires no devices or chemicals, and it is always available. The disadvantages of withdrawal are that it does not provide protection from STIs, it may interrupt the sexual response cycle and diminish the pleasure for the couple, and it is very ineffective in preventing pregnancy.

Withdrawal is not a reliable form of contraception for two reasons. First, a man can unknowingly emit a small amount of pre-ejaculatory fluid, which may contain sperm. One drop can contain millions of sperm. In addition, the man may lack the self-control to withdraw his penis before ejaculation, or he may delay his withdrawal too long and inadvertently ejaculate some semen near the vaginal opening of his partner. Sperm deposited there can live in the moist vaginal lips and make their way up the vagina.

Though some women believe that douching is an effective form of contraception, it is not. Douching refers to rinsing or cleansing the vaginal canal. After intercourse, the woman fills a syringe with water, any of a variety of solutions that can be purchased over the counter, or a spermicidal agent and flushes (so she assumes) the sperm from her vagina. But in some cases, the fluid will actually force sperm up through the cervix. In other cases, a large number of sperm may already have passed through the cervix to the uterus, so the douche may do little good.

Sperm may be found in the cervical mucus within 90 seconds after ejaculation. In effect, douching does little to deter conception and may even encourage it. In addition, douching is associated with an increased risk for pelvic inflammatory disease and ectopic pregnancy.

Emergency Contraception

Also called post-coital contraception, **emergency contraception** (EC) refers to various types of morning-after pills that are used primarily in three circumstances: when a woman has unprotected intercourse, when a contraceptive method fails (such as condom breakage or slippage), and when a woman is raped. EC methods should be used in emergencies for those times when unprotected intercourse has occurred, and medication can be taken within 72 hours of exposure. Plan B, an emergency contraception, is now available over the counter for someone who is age 18 or above.

Fear that EC is being used as a routine method of contraception is not supported by data. Of 235 women who had received EC, 70 percent were using the pill and 73 percent were using the condom prior to their need for EC (Harvey et al., 2000). Also, in a recent study of 2117 women aged 15 to 24 years, women who were given access to the EC Plan B were no more likely to use EC than those with clinic access only. In addition, providing women with easy access to Plan B did not lead them to engage in more risky sexual behavior (Raine et al., 2005).

Combined Estrogen-Progesterone The most common morning-after pills are the combined estrogen-progesterone oral contraceptives routinely taken to prevent pregnancy. In higher doses, they serve to prevent ovulation, fertilization of the egg, or transportation of the egg to the uterus. They may also make the uterine lining inhospitable to implantation. Known as the "Yuzpe method" after the physician who proposed it, this method involves ingesting a certain number of tablets of combined estrogen-progesterone. *These pills must be taken within 72 hours of unprotected intercourse to be effective, and a pregnancy test is required.*

Common side effects of combined estrogen-progesterone EC pills (sold under the trade names Preven and Plan B) include nausea, vomiting, headaches, and breast tenderness, although some women also experience abdominal pain, headache, and dizziness. Side effects subside within a day or two after treatment is completed. The pregnancy rate is 1.2 percent if combined estrogen-progesterone is taken within 12 hours of unprotected intercourse, 2.3 percent if taken within 48 hours, and 4.9 percent if taken within 48 to 72 hours (Rosenfeld, 1997). Although the FDA has not yet approved the sale of this morning-after pill over the counter, the American Medical Association supports this action.

Postcoital IUD Insertion of a copper IUD within 5 to 7 days after ovulation in a cycle when unprotected intercourse has occurred is very effective for preventing pregnancy. This option, however, is used much less frequently than hormonal treatment because women who need EC often are not appropriate IUD candidates. Women who have chlamydia or gonorrhea when an IUD is inserted are at higher risk of developing pelvic inflammatory disease; therefore, testing is recommended prior to insertion (Johnson B., 2005).

Mifepristone (RU-486) **Mifepristone,** also known as **RU-486,** is a synthetic steroid that effectively inhibits implantation of a fertilized egg by making the endometrium unsuitable for implantation. The so-called abortion pill, approved by the FDA in the United States in 2000, is marketed under the name Mifeprex and can be given to induce abortion within 7 weeks of pregnancy. Side effects of RU-486 are usually very severe and may include cramping, nausea, vomiting, and breast tenderness. Potential serious adverse effects include the possibility of hemorrhage and infection. The pregnancy rate associated with RU-486 is 1.6 percent, which suggests that RU-486 is an effective means of EC (Rosenfeld, 1997). More than 90 percent of U.S. women who tried RU-486 would recommend it to others and choose it over surgery again. Its use remains controversial. To be clear,

Diversity in Other Countries

Today women in more than 20 European countries (including France, Belgium, Sweden, and Finland) can purchase emergency contraceptive pills over the counter (Wan and Lo, 2005).

an abortion is not a method of contraception since conception has already occurred. An abortion simply prevents a conception/fertilization/development of a zygote from progressing through a normal pregnancy.

Effectiveness of Various Contraceptives

In Table 10.1, we present data on the effectiveness of various contraceptive methods in preventing pregnancy and protecting against STIs. Table 10.1 also describes the benefits, disadvantages, and costs of various methods of contraception. Also included in the chart is the obvious and most effective form of birth control: abstinence. Its cost is the lowest, it is 100 percent effective for pregnancy prevention, and it also eliminates the risk of HIV and other STIs from intercourse. Abstinence can be practiced for a week, a month, several years, until marriage, or until someone finds the "right" sexual partner.

Sterilization

Unlike the temporary and reversible methods of contraception already discussed, **sterilization** is a permanent surgical procedure that prevents reproduction. It is the most prevalent method of contraception in the United States (Godecker et al., 2001). Sterilization may be a contraceptive method of choice when the woman should not have more children for health reasons or when individuals are certain about their desire to have no more children or to remain childfree. Most couples complete their intended childbearing in their late 20s or early 30s, leaving more than 15 years of continued risk of unwanted pregnancy. Because of the risk of pill use at older ages and the lower reliability of alternative birth control methods, sterilization has become the most popular method of contraception among married women who have completed their families.

Slightly more than half of all sterilizations are performed on women. Although male sterilization is easier and safer than female sterilization, women feel more certain they will not get pregnant if they are sterilized. "I'm the one that ends up being pregnant and having the baby," said one woman. "So I want to make sure that I never get pregnant again."

Female Sterilization Although a woman may be sterilized by removal of her ovaries (**oophorectomy**) or uterus (**hysterectomy**), these operations are not normally undertaken for the sole purpose of sterilization because the ovaries produce important hormones (as well as eggs) and because both procedures carry the risks of major surgery. But sometimes there is another medical problem requiring hysterectomy.

The usual procedures of female sterilization are the salpingectomy and a variant of it, the laparoscopy. **Salpingectomy,** also known as tubal ligation or tying the tubes (Figure 10.1), is often performed under a general anesthetic while the woman is in the hospital just after she has delivered a baby. An incision is made in the lower abdomen, just above the pubic line, and the fallopian tubes are brought into view one at a time. A part of each tube is cut out, and the ends are tied, clamped, or cauterized (burned). The operation takes about 30 minutes. About 700,000 such procedures are performed annually. The cost is around $2500.

A less expensive and quicker (about 15 minutes) form of salpingectomy, which is performed on an outpatient basis, is the **laparoscopy.** Often using local anesthesia, the surgeon inserts a small, lighted viewing instrument (laparoscope) through the woman's abdominal wall just below the navel, through which the uterus and the fallopian

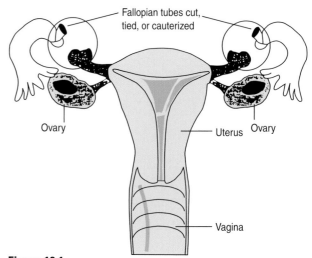

Figure 10.1
Female Sterilization: Tubal Sterilization

Table 10.1 Methods of Contraception and Sexually Transmitted Infection Protection

Method	Typical Use[1] Effectiveness Rates	STI Protection	Benefits	Disadvantages	Cost[2]
Oral contraceptive (The Pill)	92%	No	High effectiveness rate; 24-hour protection. Menstrual regulation.	Daily administration. Side effects possible. Medication interactions.	$10–42 per month
Jadelle® (2 rod, 3–5 year implant) and Implanon® (1 rod, 3-year implant)	99.95%	No	High effectiveness rate. Long-term protection.	Side effects possible. Menstrual changes.	$300–600 per insertion
Depo-Provera® (3-month injection)	97%	No	High effectiveness rate. Long-term protection.	Can seriously impact bone density. For most users, not recommended for use longer than 2 years. Side effects likely.	$45–75 per injection
Ortho Evra® (transdermal-patch)	92%	No	Same as oral contraceptives except use is weekly, not daily.	Patch changed weekly. Side effects possible; 60% more hormone exposure than pills.	$15–32 per month
NuvaRing® (vaginal ring)	92%	No	Lower hormone levels may mean fewer side effects than pills.	Must be comfortable with body for insertion.	$15–48 per month
Male condom	85%	Yes	Few or no side effects. Easy to purchase and use.	Can interrupt spontaneity.	$2–10 a box
Female condom	79%	Yes	Few or no side effects. Easy to purchase.	Decreased sensation. Insertion takes practice.	$4–10 a box
Spermicide	71%	No	Many forms to choose. Easy to purchase and use.	Can cause irritation. Can be messy.	$8–18 per box/tube/can
Today® Sponge[3]	68–84%	No	Few side effects. Effective for 24 hours after insertion.	Spermicide irritation possible.	$3–5 per sponge
Diaphragm and cervical cap[3]	68–84%	No	Few side effects. Can be inserted within 2 hours before intercourse.	Can be messy. Increased risk of vaginal/urinary tract infections.	$50–200 plus spermicide
Intrauterine device (IUD): Paraguard or Mirena	98.2–99%	No	Little maintenance. Longer-term protection.	Risk of pelvic inflammatory disease increased. Chance of expulsion.	$150–300
Withdrawal	73%	No	Requires little planning. Always available.	Pre-ejaculatory fluid can contain sperm.	$0
Periodic abstinence	75%	No	No side effects. Accepted in all religions/cultures.	Requires a lot of planning. Need ability to interpret fertility signs.	$0
Emergency contraception	75%	No	Provides an option after intercourse has occurred.	Must be taken within 72 hours. Side effects likely.	$10–32
Abstinence	100%	Yes	No risk of pregnancy or STDs.	Partners both have to agree to abstain.	$0

[1]Effectiveness rates are listed as percentages of women not experiencing an unintended pregnancy during the first year of typical use. Typical use refers to use under real-life conditions. Perfect-use effectiveness rates are higher.
[2]Costs may vary.
[3]Lower percentages apply to parous women (women who have given birth). Higher rates apply to nulliparous women (never given birth).
Source: Beth Credle Burt, MAEd, CHES, a health education specialist with the Wake Area Health Education Center at WakeMed Hospital, Raleigh, North Carolina.

tubes can be seen. The surgeon then makes another small incision in the lower abdomen and inserts a special pair of forceps that carry electricity to cauterize the tubes. The laparoscope and the forceps are then withdrawn, the small wounds are closed with a single stitch, and small bandages are placed over the closed incisions. (Laparoscopy is also known as "the Band-Aid operation.")

As an alternative to reaching the fallopian tubes through an opening below the navel, the surgeon may make a small incision in the back of the vaginal barrel (vaginal tubal ligation).

In late 2002, the FDA approved Essure, a permanent sterilization procedure that requires no cutting and only a local anesthetic in a half-hour procedure that blocks the fallopian tubes. Women typically may return home within 45 minutes (and to work the next day). Essure is already available in Australia, Europe, Singapore, and Canada.

These procedures for female sterilization are greater than 95 percent effective, but sometimes they have complications. In rare cases, a blood vessel in the abdomen is torn open during the sterilization and bleeds into the abdominal cavity. When this happens, another operation is necessary to find the bleeding vessel and tie it closed. Occasionally, injury occurs to the small or large intestine, which may cause nausea, vomiting, and loss of appetite. The fact that death may result, if only rarely, is a reminder that female sterilization is surgery and, like all surgery, involves some risks. In addition, although some female sterilizations may be reversed, a woman should become sterilized only if she does not want to have a biological child. Hillis et al. (2000) reported that women who were 30 or younger at the time they were sterilized were twice as likely to report feelings of regret as women who were older than 30.

Male Sterilization **Vasectomies** are the most frequent form of male sterilization. They are usually performed in the physician's office under a local anesthetic. In this operation the physician makes two small incisions, one on either side of the scrotum, so that a small portion of each vas deferens (the sperm-carrying ducts) can be cut out and tied closed. Sperm are still produced in the testicles, but since there is no tube to the penis, they remain in the epididymis and eventually dissolve.

The procedure takes about 15 minutes and costs about $800. The man can leave the physician's office within a short time. Since sperm do not disappear from the ejaculate immediately after a vasectomy (some remain in the vas deferens above the severed portion), a couple should use another method of contraception until the man has had about 20 ejaculations. In about 1 percent of the cases, the vas deferens grows back and the man becomes fertile again.

A vasectomy does not affect the man's desire for sex, ability to have an erection or an orgasm, amount of ejaculate (sperm account for only a minute portion of the seminal fluid), health, or chance of prostate cancer. Although in some instances a vasectomy may be reversed, a man should get a vasectomy only if he does not want to have a biological child.

Talking with a Partner about Contraception

Having a conversation about birth control is a good way to begin sharing responsibility for it; one can learn of the partner's interest in participating in the choice and use of a contraceptive method. Men can also share responsibility by purchasing and using condoms, paying for medical visits and the pharmacy bill, reminding his partner to use the method, assisting with insertion of barrier methods, checking contraceptive supplies, and having a vasectomy if that is an appropriate option (The Boston Women's Health Book Collective, 2005). However, in addition, women need to take steps to protect themselves from unwanted pregnancy and from exposure to STIs.

Katie Cushman

Abortion

An abortion may be either an **induced abortion,** which is the deliberate termination of a pregnancy through chemical or surgical means, or a **spontaneous abortion** (**miscarriage**), which is the unintended termination of a pregnancy. In this text, however, we will use the term *abortion* to refer to induced abortion. In general, abortion is legal in the United States. Abortion laws may be challenged, however, by new conservative pro-life appointments to the Supreme Court under the Bush administration. In 2006, South Dakota legislators banned abortion for all reasons (including rape and incest) except as necessary to save the woman's life. Abortion rights advocates view the ban as a symbol of more restrictive abortion legislation to come.

Incidence of Abortion

National Data

There are about 1.3 million abortions annually (*Statistical Abstract of the United States: 2006,* Table 94). Among unmarried women, 39 percent who get pregnant have an abortion; 7 percent of married women have an abortion (Ventura et al., 2003). However, these numbers do not reflect reality, because only about 35 to 50 percent of abortions are reported (Michael, 2001).

Table 10.2 reflects the abortion rates and ratios in the United States for selected years. **Abortion rate** refers to the number of abortions per thousand women aged 15 to 44; **abortion ratio** refers to the number of abortions per thousand live births. Compared with rates in 1990, the number of abortions per 1000 women aged 15 to 44 and the number of abortions per 1000 live births has dropped (*Statistical Abstract of the United States: 2006,* Table 94). Reasons for the decreasing number of abortions include a reduced rate of unintended pregnancies and more supportive attitudes toward women becoming single parents. In addition, part of the decline may be due to an increase in restrictive abortion policies, such as those requiring parental consent and mandatory waiting periods.

Chapter 10 Planning Children and Contraception

Table 10.2 Abortion Rates and Ratios for Selected Years

Year	Rate	Ratio
	Number of abortions per 1000 women aged 15–44	Number of abortions per 1000 live births
1980	25	359
1990	24	344
2002	16	246

Source: Strauss, L. T., J. Herndon, J. Chang, W. Y. Parker, S. V. Bowens, and C. J. Berg. 2005. Abortion Surveillance—United States, 2002. MMWR Morb Mortal Wkly Rep 54(SS07):1–31.

Who Gets an Abortion and Why

International Data

About a quarter of women in the world live in countries in which abortion is either completely prohibited or permitted only to save the woman's life (Adler et al., 2003).

A study in New York City identified women obtaining an abortion as unmarried, young (particularly those under age 16), having no religious affiliation, having had a previous abortion, and white (Michael, 2001; Hollander, 2001). Because there are proportionately more white women than minorities in the United States, about half of all abortions are by white women. Thirty-two states in the United States have parental notification laws that require explicit permission from one or both parents before an adolescent has an abortion (Adler et al., 2003).

A team of researchers (Finer et al., 2005) surveyed 1209 women who reported having had an abortion. The most frequently cited reasons were that having a child would interfere with a woman's education, work, or ability to care for dependents (74%); that she could not afford a baby now (73%); and that she did not want to be a single mother or was having relationship problems (48%). Nearly four in 10 women said they had completed their childbearing, and almost one-third were not ready to have a child. Fewer than 1% said their parents' or partner's desire for them to have an abortion was the most important reason.

Abortions performed to protect the life or health of the woman are called **therapeutic abortions.** However, there is disagreement over the definition. Garrett et al. (2001) noted, "Some physicians argue that an abortion is therapeutic if it prevents or alleviates a serious physical or mental illness, or even if it alleviates temporary emotional upsets. In short, the health of the pregnant woman is given such a broad definition that a very large number of abortions can be classified as therapeutic" (p. 218).

Some women with multifetal pregnancies (a common outcome of the use of fertility drugs) may have a procedure called *transabdominal first-trimester selective termination.* In this procedure, the lives of some fetuses are terminated to increase the chance of survival for the others or to minimize the health risks associated with multifetal pregnancy for the woman. For example, a woman carrying five fetuses may elect to abort three of them to minimize the health risks of the other two.

Republicans are against abortion until their daughter needs one; Democrats are for abortion until their daughter wants one.

Grace McGarvie

Pro-Life and Pro-Choice Abortion Positions

A dichotomy of attitudes toward abortion is reflected in two opposing groups of abortion activists. Individuals and groups who oppose abortion are commonly referred to as "pro-life" or "anti-abortion."

Pro-Life   Forty percent of 657 undergraduates in a random sample at a large southeastern university reported that they were pro-life in regard to their feelings about abortion (Bristol and Farmer, 2005). Pro-life groups favor abortion policies or a complete ban on abortion. They essentially believe the following:

The unborn fetus has a right to live and that right should be protected.

Abortion is a violent and immoral solution to unintended pregnancy.

The life of an unborn fetus is sacred and should be protected, even at the cost of individual difficulties for the pregnant woman.

Individuals who are over the age of 44, female, mothers of three or more children, married to white-collar workers, affiliated with a religion, and Catholic are most likely to be pro-life (Begue, 2001). Pro-life individuals emphasize the sanctity of human life and the moral obligation to protect it. The unborn fetus cannot protect itself so is literally dependent on others for life. Naomi Judd noted that if she had had an abortion she would have deprived the world of one of its greatest singers—Wynonna Judd.

Pro-Choice   Over half (55.2%) of 263,000 first-year students in a random sample of U.S. colleges and universities agreed that "abortion should be legal" (men, 55.8%; women, 54.8%) (American Council on Education and University of California, 2005–2006). In a large, nonrandom sample, 63.5% of 1027 undergraduates at a southeastern university reported that "abortion is acceptable under certain conditions" (Knox and Zusman, 2006). Pro-choice advocates support the legal availability of abortion for all women. They essentially believe the following:

Freedom of choice is a central value—the woman has a right to determine what happens to her own body.

Those who must personally bear the burden of their moral choices ought to have the right to make these choices.

Procreation choices must be free of governmental control.

People most likely to be pro-choice are female, are mothers of one or two children, have some college education, are employed, and have annual income of more than $50,000. Although many self-proclaimed feminists and women's organizations, such as the National Organization for Women (NOW), have been active in promoting abortion rights, not all feminists are pro-choice.

Diversity in Other Countries

There are wide variations in the range of cultural responses to the abortion issue. On one end of the continuum is the Kafir tribe in Central Asia, where an abortion is strictly the choice of the woman. In this preliterate society, there is no taboo or restriction with regard to abortion, and the woman is free to exercise her decision to terminate her pregnancy. One reason for the Kafirs' approval of abortion is that childbirth in the tribe is associated with high rates of maternal mortality. Since birthing children may threaten the life of significant numbers of adult women in the community, women may be encouraged to abort. Such encouragement is particularly strong in the case of women who are viewed as too young, too sick, too old, or too small to bear children.

Abortion may be encouraged by a tribe or society for a number of other reasons, including practicality, economics, lineage, and honor. Abortion is practical for women in migratory societies. Such women must control their pregnancies, since they are limited in the number of children they can nurse and transport. Economic motivations become apparent when resources (e.g., food) are scarce—the number of children born to a group must be controlled. Abortion for reasons of lineage or honor involves encouragement of an abortion in those cases in which a woman becomes impregnated in an adulterous relationship. To protect the lineage and honor of her family, the woman may have an abortion.

Physical Effects of Abortion

Part of the debate over abortion is related to the presumed effects of abortion. In regard to the physical effects, legal abortions, performed under safe medical conditions, are safer than continuing the pregnancy. The earlier in the pregnancy the abortion is performed, the safer it is.

Post-abortion complications include the possibility of incomplete abortion, which occurs when the initial procedure misses the fetus and the procedure must be repeated. Other possible complications include uterine infection; excessive bleeding; perforation or laceration of the uterus, bowel, or adjacent organs; and an adverse reaction to a medication or anesthetic. After having an abortion, women are advised to expect bleeding (usually not heavy) for up to 2 weeks and to return to their health-care provider 30 days after the abortion to check that all is well.

Abortion Attitude Scale

This is not a test. There are no wrong or right answers to any of the statements, so just answer as honestly as you can. The statements ask you to tell how you feel about legal abortion (the voluntary removal of a human fetus from the mother during the first 3 months of pregnancy by a qualified medical person). Tell how you feel about each statement by circling only one response. Use the following scale for your answers:

Strongly Disagree	Agree	Slightly Agree	Slightly Disagree	Disagree	Strongly Agree
5	4	3	2	1	0

____ 1. The Supreme Court should strike down legal abortions in the United States.

____ 2. Abortion is a good way of solving an unwanted pregnancy.

____ 3. A mother should feel obligated to bear a child she has conceived.

____ 4. Abortion is wrong no matter what the circumstances are.

____ 5. A fetus is not a person until it can live outside its mother's body.

____ 6. The decision to have an abortion should be the pregnant mother's.

____ 7. Every conceived child has the right to be born.

____ 8. A pregnant female not wanting to have a child should be encouraged to have an abortion.

____ 9. Abortion should be considered killing a person.

____10. People should not look down on those who choose to have abortions.

____11. Abortion should be an available alternative for unmarried pregnant teenagers.

____12. Persons should not have the power over the life or death of a fetus.

____13. Unwanted children should not be brought into the world.

____14. A fetus should be considered a person at the moment of conception.

Scoring and Interpretation

As its name indicates, this scale was developed to measure attitudes toward abortion. It was developed by Sloan (1983) for use with high school and college students. To compute your score, first reverse the point scale for items 1, 3, 4, 7, 9, 12, and 14. For example, if you selected a 5 for item one, this becomes a 0; if you selected a 1, this becomes a 4. After reverse-scoring the seven items specified, add the numbers you circled for all the items. Sloan provided the following categories for interpreting the results:

70–56	Strong pro-abortion
54–44	Moderate pro-abortion
43–27	Unsure
26–16	Moderate pro-life
15–0	Strong pro-life

Reliability and Validity

The Abortion Attitude Scale was administered to high school and college students, Right to Life group members, and abortion service personnel. Sloan (1983) reported a high total test estimate of reliability (0.92). Construct validity was supported in that the mean score for Right to Life members was 16.2; the mean score for abortion service personnel was 55.6; and other groups' scores fell between these values.

Source

Abortion Attitude Scale, by L. A Sloan. Reprinted with permission from the *Journal of Health Education* Vol. 14, No. 3, May/June 1983. The *Journal of Health Education* is a publication of the American Allegiance for Health, Physical Education, Recreation and Dance, 1900 Association Drive, Reston, VA 20191.

Vacuum aspiration, the method used for most U.S. abortions, does not increase the risks to future childbearing. However, late-term abortions do increase the risks of subsequent miscarriages, premature deliveries, and low-birth-weight babies (Baird, 2000).

Psychological Effects of Abortion

Of equal concern are the psychological effects of abortion. Fergusson et al. (2006) analyzed results from a 25-year longitudinal study of individuals in New Zealand. Of those young females who became pregnant by age 25 (41%), 15% had an abortion. Those having an abortion had elevated rates of subsequent mental health problems, including depression, anxiety, suicidal behaviors, and substance use disorders. Hollander (2001) studied 442 women 2 years after they had had an abortion and found that 16 percent regretted their decision. This number was up from 11 percent 1 month after the abortion. However, 69 percent reported that they would make the same decision to abort if faced with the same situation. The predominant emotion in reference to the abortion was relief, rather than either positive or negative emotions. Although 20 percent met the criteria for clinical depression, only 1 percent experienced post-traumatic stress disorder. The researcher concluded that "for most women, elective abortion . . . does not pose a risk to mental health" (p. 3).

The finding in the Hollander study is similar to the findings of a panel convened by the American Psychological Association, which reviewed the scientific literature on the mental health impact of abortion. For most women, a legal first-trimester abortion does not create psychological hazards, and symptoms of distress are within normal bounds. Bradshaw and Slade (2005) compared 98 women who had a first-trimester abortion with 51 never-pregnant women to assess the effect of an abortion on sexual attitudes. The researchers found that having an abortion was followed by more positive attitudes and feelings towards sexual matters (women who had had an abortion did not become sexual phobics).

Second-trimester abortions, which may reflect greater conflict about the pregnancy and involve more distressing medical procedures, are associated with a higher likelihood of negative response. Nevertheless, "well-designed studies of psychological responses following abortion have consistently shown that risk of psychological harm is low" (Adler et al., 2003, p. 211).

Post-Abortion Attitudes of Men

Researchers Kero and Lalos (2004) conducted interviews with men 4 and 12 months after their partners had had an abortion. Overwhelmingly, the men (at both time periods) were happy with the decision of their partners to have an abortion. More than half accompanied their partner to the abortion clinic (which they found less than welcoming); about a third were not using contraception a year later.

PERSONAL CHOICES

Should You Have an Abortion?

The decision to have an abortion continues to be a complex one. Women who are faced with the issue may benefit by considering the following guidelines:

1. **Consider all the alternatives available to you, realizing that no alternative may be all good or all bad.** As you consider each alternative, think about both the short-term and the long-term consequences of each course of action.

2. **Obtain information about each alternative course of action.** Inform yourself about the medical, financial, and legal aspects of abortion, childbearing, parenting, and placing the baby up for adoption.

3. **Talk with trusted family members, friends, or unbiased counselors.** Consider talking with the man who participated in the pregnancy. If possible, also talk with women who have had abortions as well as with women who have kept and reared their baby or placed their baby for adoption. If you feel that someone is pressuring you in your decision-making, look for help elsewhere.

4. **Consider your own personal and moral commitments in life.** Understand your own feelings, values, and beliefs concerning the fetus and weigh those against the circumstances surrounding your pregnancy.

SUMMARY

Why do people have children?

Having children continues to be a major goal of most college students. Social influences to have a child include family, friends, religion, government, favorable economic conditions, and cultural observances. The reasons people give for having children include love and companionship with one's own offspring, the desire to be personally fulfilled as an adult by having a child, and the desire to recapture one's youth. Having a child (particularly for women) reduces one's educational and career advancement. The cost for housing, food, transportation, clothing, health care, and child care for a child up to age 2 is almost $10,000 annually.

How many children do people have?

Between 1980 and 2004, the percentage of childfree U.S. women aged 40 to 44 nearly doubled, from 10.1% to 19.3%. Whether more women are deciding to delay having children or deciding against having them altogether is unknown. Reasons that spouses elect a childfree marriage include the freedom to spend their time and money as they choose, to enjoy their partner without interference, to pursue their career, to avoid health problems associated with pregnancy, and to avoid passing on genetic disorders to a new generation.

The most preferred type of family in the United States is the two-child family. Some of the factors in a couple's decision to have more than one child are the desire to repeat a good experience, the feeling that two children provide companionship for each other, and the desire to have a child of each sex.

Are teen mothers disadvantaged?

About seven percent of females aged 15 to 19 are mothers. They may view motherhood as one of the only viable roles open to them. Consequences associated with teenage motherhood include poverty, alcohol/drug abuse, lower-birth-weight babies, and lower academic achievement. As a result of increased public awareness of the need for pregnancy prevention via abstinence and responsible behavior, the birthrate for teenagers has been declining. However, the teenage birthrate in the United States, as measured by births per thousand teenagers, is nine times that of teens in the Netherlands.

What are the causes of infertility?

Infertility is defined as the inability to achieve a pregnancy after at least 1 year of regular sexual relations without birth control, or the inability to carry a pregnancy to a live birth. Forty percent of infertility problems are attributed to the woman, 40 percent to the man, and 20 percent to both of them. Some of the more common causes of infertility in men include low sperm production, poor semen motility, effects of STIs (such as chlamydia, gonorrhea, and syphilis), and interference with passage of sperm through the genital ducts due to an enlarged prostate.

The causes of infertility in women include blocked fallopian tubes, endocrine imbalance that prevents ovulation, dysfunctional ovaries, chemically hostile cervical mucus that may kill sperm, and effects of STIs. The psychological reaction to infertility is often depression over having to give up a lifetime goal.

A number of technological innovations are available to assist women and couples in becoming pregnant. These include hormonal therapy, artificial insemination, ovum transfer, in vitro fertilization, gamete intrafallopian transfer, and zygote intrafallopian transfer.

Why do people adopt?

Motives for adoption include a couple's inability to have a biological child (infertility), their desire to give an otherwise unwanted child a permanent loving home, or their desire to avoid contributing to overpopulation by having more biological children. There are approximately 100,000 adoptions annually in the United States.

Whereas demographic characteristics of those who typically adopt are white, educated, and high-income, increasingly adoptees are being placed in nontraditional families including older, gay, and single individuals; it is recognized that these individuals may also be white, educated, and high-income. Most college students are open to transracial adoption.

Some individuals seek the role of parent via foster parenting. A foster parent, also known as a family caregiver, is a person who at home, either alone or with a spouse, takes care of and fosters a child taken into custody. He or she has made a contract with the state for the service, has judicial status, and is reimbursed by the state.

What are various methods of contraception?

The primary methods of contraception include hormonal methods (the newest of which are Jadelle®, Depo-Provera®, NuvaRing®, and Ortho Evra®), which prevent ovulation; the IUD, which prevents implantation of the fertilized egg; condoms and diaphragms, which are barrier methods; and vaginal spermicides and the rhythm method. These and numerous new methods vary in effectiveness and safety.

Sterilization is a surgical procedure that prevents fertilization, usually by blocking the passage of eggs or sperm through the fallopian tubes or vas deferens, respectively. The procedure for female sterilization is called salpingectomy, or tubal ligation. Laparoscopy is another method of tubal ligation. The most frequent form of male sterilization is vasectomy.

What are the types of and motives for an abortion?

An abortion may be either an induced abortion, which is the deliberate termination of a pregnancy through chemical or surgical means, or a spontaneous abortion (miscarriage), which is the unintended termination of a pregnancy. In this text we use the term *abortion* to refer to induced abortion. In general, abortion is legal in the United States. The most frequently cited reasons were that having a child would interfere with a woman's education, work, or ability to care for dependents (74%); that she could not afford a baby now (73%); and that she did not want to be a single mother or was having relationship problems (48%). Nearly four in 10 women said they had completed their childbearing, and almost one-third were not ready to have a child. Fewer than 1% said their parents' or partner's desire for them to have an abortion was the most important reason.

KEY TERMS

abortion rate	female condom	Mifepristone	RU-486
abortion ratio	fertilization	miscarriage	salpingectomy
antinatalism	foster parent	natural family planning	spermicide
cervical cap	hysterectomy	NuvaRing®	spontaneous abortion
coitus interruptus	induced abortion	oophorectomy	sterilization
conception	infertility	Ortho Evra®	therapeutic abortion
Depo-Provera®	intrauterine device	periodic abstinence	transracial adoption
diaphragm	Jadelle®	pregnancy	vasectomy
emergency contraception	laparoscopy	pronatalism	withdrawal

The Companion Website for Choices in Relationships: An Introduction to Marriage and the Family, Ninth Edition
www.thomson.edu/sociology/knox

Supplement your review of this chapter by going to the companion website to take one of the Tutorial Quizzes, use the flash cards to master key terms, and check out the many other study aids you'll find there. You'll also find special features such as the Marriage and Family Resource Center, Census 2000 information, and other data and resources at your fingertips to help you with that special project or to do some research on your own.

WEB LINKS

Alan Guttmacher Institute
http://www.agi-usa.org

Childfree by Choice
http://www.childfreebychoice.com/

Engenderhealth
http://www.engenderhealth.org/

The Evan B. Donaldson Adoption Institute
http://www.adoptioninstitute.org/

Fetal Fotos (bonding with your fetus)
http://www.fetalfotosusa.com/

Georgia Reproduction Specialists (male infertility)
http://www.ivf.com/male.html

National Right to Life
http://www.nrlc.org/

Planned Parenthood Federation of America, Inc.
http://www.plannedparenthood.org

NARAL Pro-Choice America (reproductive freedom and choice)
http://www.naral.org/

REFERENCES

Adler, N. E., Em J. Ozer, and J. Tschann. 2003. Abortion among adolescents. *American Psychologist* 58:211–17.

Aegidius, K. 2006. Oral contraceptives and increased headache prevalence. *Neurology* 66:349–53.

Alan Guttmacher Institute. 2002. *In their own right: Addressing the sexual and reproductive health needs of American men.* New York: Alan Guttmacher Institute.

Alexian Brothers Health System. 2005. Social accountability report: Maternity assistance programs reduce financial burden of childbirth. Retrieved June 15, 2005, from http://www.alexianhealth-system.org/mission/accountreport/community.html.

American Council on Education and University of California. 2005-2006. *The American freshman: National norms for fall, 2005.* Los Angeles: Higher Education Research Institute. UCLA Graduate School of Education and Information Studies.

Amory, J., Page, S., and Bremner, W. 2006. Drug insight: recent advances in male hormonal contraception. *Nature Clinical Practice Endocrinology & Metabolism* 2:32–41.

Artz, L., Macaluso, M., Brill, I., Kelaghan, J., Austin, H., Fleenor, M. 2000. Effectiveness of an intervention promoting the female condom to patients at sexually transmitted disease clinics. *American Journal of Public Health* 90:237–44.

Assve, A. 2003. The impact of economic resources on premarital childbearing and subsequent marriage among young American women. *Demography* 40:105–26.

Baird, D. T. 2000. Therapeutic abortion. In *Family planning and reproductive healthcare,* edited by A. Glasier and A. Gebbie. London: Churchill Livingston, 249–62.

Begue, L. 2001. Social judgment of abortion: A black-sheep effect in a Catholic sheepfold. *Journal of Social Psychology* 141:640–50.

Beksinska, M. E., Rees, H. V., Dickson-Tetteh, K. E., Mqoqi, N., Kleinschmidt, I., and McIntyre, J. A. 2001. Structural integrity of the female condom after multiple uses, washing, drying, and relubrication. *Contraception* 63:33–36.

The Boston Women's Health Book Collective. (2005). *Our bodies, ourselves: A new edition for a new era.* New York: Touchstone.

Bouchard, G. 2005. Adult couples facing a planned or an unplanned pregnancy: Two realities. *Journal of Family Issues* 26:619–37.

Bradshaw, Z. and P. Slade. 2005. The relationships between induced abortion, attitudes toward sexuality and sexuality problems. *Sexual and Relationship Therapy* 20:391–406

Bristol, K. and B. Farmer. 2005. *Sexuality among southeastern university students: a survey.* Unpublished data, East Carolina University, Greenville, NC.

Brothers, Z., and J. E. Maddux. 2003. The goal of biological parenthood and emotional distress from infertility: Linking parenthood to happiness. *Journal of Applied Social Psychology* 33:248–262.

Centers for Disease Control and Prevention. 2002. Sexually transmitted diseases treatment guidelines 2002. *MMWR Morb Mortal Wkly Rep* 51(RR-6):1–84.

Chatterjee, P. 2005. Doctors' group proposes one-child policy for India. *Lancet* 365:1609.

David, H. P., Z. Dytrych, and Z. Matejcek 2003. Born unwanted: observations from the Prague study. *American Psychologist* 58:224–29.

Darroch, J. E., S. Singh, J. J. Frost, and The Study Team. 2001. Differences in teenage pregnancy rates among five developed countries: The roles of sexual activity and contraceptive use. *Family Planning Perspectives* 33:244–50.

DeOllos, I. Y. and C. A. Kapinus. 2002. Aging childless individuals and couples: Suggestions for new directions in research. *Sociological Inquiry* 72: 72-80

Dinc, L. and F. Terzioglu. 2006. The psychological impact of genetic testing on parents. *Journal of Clinical Nursing* 15:45–51.

Dye, J. L. 2005 (December). Fertility of American women: June 2004. Current Population Reports, P20-555. U.S. Census Bureau. Available at http:www.census.gov http://www.census.gov/prod/2005pubs/p20-555.pdf.

Farber, M. L. Z., E. Timberlake, H. P. Mudd, and L. Cullen. 2003. Preparing parents for adoption: An agency experience. *Child and Adolescent Social Work Journal* 20:175–96.

Fergusson, D. M., L. J. Horwood, and E. M. Ridder. 2006. Abortion in young women and subsequent mental health. *Journal of Child Psychology & Psychiatry & Allied Disciplines* 47:16–24.

Fiejoo, A. N. 2006. Adolescent sexual health in the U.S. and Europe: Why the difference? Updated article originally written by S. Alford and A. N. Fiejoo. Published by Youth Advocates, 2000.

Retrieved on Feb 22, 2006, from http://www.advocatesforyouth.org/PUBLICATIONS/ factsheet/fsest.pdf.

Finer, L. B., L. F. Frohwirth, L. a. Dauphinne, S. Singh, and A. M. Moore. 2005. Reasons U.S. women have abortions: quantitative and qualitative reasons. *Perspectives on Sexual and Reproductive Health* 37:110–18.

Flower Kim, K. M. 2003. We are family. Paper, 73rd Annual Meeting of the Eastern Sociological Society, Philadelphia, February 27.

Food and Drug Administration. 2005. FDA updates labeling for Ortho Evra contraceptive patch. FDA Publication No. P05-90. Washington, DC: FDA News. Retrieved from http://www.fda.gov/ bbs/topics/news/2005/NEW01262.html.

Garrett, T. M., Baillie, H. W., and Garrett, R. M. 2001. *Health care ethics.* 4th ed. Upper Saddle River, NJ: Prentice Hall.

Geronimus, A. T. 2003. Damned if you do: Culture, identity, privilege, and teenage childbearing in the United States. *Social Science and Medicine* 57:881–93.

Godecker, A. L., E. Thomson, and L. L. Bumpass. 2001. Union status, marital history and female contraceptive sterilization in the United States. *Family Planning Perspectives* 33:35–41.

Grady, W. R., J. O. G. Billy, and D. H. Klepinger. 2002. Contraception method switching in the United States. *Perspectives on Sexual and Reproductive Health* 34:135–45.

Grady, W. R., D. H. Klepinger, and A. Nelson-Wally. 2000. Contraceptive characteristics: The perceptions and priorities of men and women. *Family Planning Perspectives* 31:168–75.

Hamilton, B. E., J. A. Martin, S. J. Ventura, and P. D. Sutton. 2005 (December 29). Births: preliminary data for 2004. *National Vital Statistics Reports* 54(9).

Harvey, S. M., L. J. Beckman, C. Sherman, and D. Petitti. 2000. Women's experience and satisfaction with emergency contraception. *Family Planning Perspectives* 31:237–45.

Hillis, S. D. et al. 2000. Poststerilization regret: Findings from the United States collaborative review of sterilization. *Obstretics and Gynecology* 93:889–95.

Hollander, D. 2001. After abortion, mixed mental health. *Family Planning Perspectives* 33:1–3.

Hollander, D. 2005. Failure rates of male and female condoms fall with use. *International Family Planning Perspectives* 31(2, June).

Hollingsworth, L. D. 1997. Same race adoption among African Americans: a ten year empirical review. *African American Research Perspectives* 13:44–49.

———. 2000. Sociodemographic influences in the prediction of attitudes toward transracial adoption. *Families in Society* 81:92–100.

———. 2003. When an adoption disrupts: A study of public attitudes. *Family Relations* 52:161– 66.

Hoyert, D. L., T. J. Mathews, F. Menacker, D. M. Strobino, and B. Guyer. 2006. Annual summary of vital statistics. *Pediatrics* 117:168–83.

Hubacher, D. 2002. The checkered history and bright future of intrauterine contraception in the United States. *Perspectives on Sexual and Reproductive Health* 34:98–103.

Huh, N. S., and W. J. Reid. 2000. Intercountry, transracial adoption and ethnic identity: a Korean example. *International Social Work* 43:75–87.

Injectable update: subcutaneous form of Depo-Provera is approved. 2005. *Contraceptive Technology Update* May 1.

Isomaki, V. 2002 The fuzzy foster parenting—a theoretical approach. *Social Science Journal* 39:625–38.

Johnson, D. E. 2005. International adoption: What is fact, what is fiction, and what is the future? *Pediatric Clinics of North America* 52:1221.

Johnson, B. 2005. Insertion and removal of intrauterine devices. *American Family Physician,* 71(1):95–102.

Kennedy, R. 2003. *Interracial intimacies.* New York: Pantheon.

Kero, A. and A. Lalos. 2004. Reactions and reflections in men, 4 and 12 months post-abortion. *Journal of Psychosomatic Obstetrics and Gynecology* 25:135–43.

Knox, D. and Zusman, M. E. 2006. Relationship and sexual behaviors of a sample of 1027 university students. Unpublished data collected for this text. Department of Sociology, East Carolina University, Greenville, NC.

Kochanek, K. D., Strobino, D. M., Guyer, B., and MacDorman, M. F. (2005). Annual summary of vital statistics—2003. *Pediatrics* 115, 619–25.

Kramer, L., and D. Ramsburg. 2002. Advice given to parents on welcoming a second child: a critical review. *Family Relations* 51:2–14.

Krpan, K. M., R. Coombs, D. Zinga, W. Steiner, and A. S. Fleming. 2005. Experiential and hormonal correlates of maternal behavior in teen and adult mothers. *Hormones and Behavior* 47:112–22.

Lane, T. 2003. High proportion of college men using condoms report errors and problems. *Perspectives on Sexual and Reproductive Health* 35:50–52.

Laffan, G. 2005. Romania's policy of emptying its orphanages raises controversy. *BMJ* 331: 1360–63.

Lee, D. 2006 Device brings hope for fertility clinics. Indystar.com at http://www.indystar.com/ apps/pbcs.dll/article?AID=/20060221/BUSINESS/602210365/1003 Retrieved Feb 22, 2006.

Leichester, J. 2005. France boosts incentives for having children. Associated Press, September 21; retrieved October 19. http://news.findlaw.com/ap/i/629/09-21-2005/88040027109d1cde.html.

Leung, P., S. Erich and H. Kanenberg. 2005. A comparison of family functioning in gay/lesbian, heterosexual and special needs adoptions. *Children and Youth Services Review* 27:1031–44.

Livermore, M. M. and R. S. Powers. 2006. Unfulfilled plans and financial stress: unwed mothers and unemployment. *Journal of Human Behavior in the Social Environment* 13:1–17.

London, S. 2003. Method-related problems account for most failures of the female condom. *Perspectives on Sexual and Reproductive Health* 35:193–99.

Macaluso, M., M. Demand, L. Artz, M. Fleenor, L. Robey, J. Kelaghan, R. Cabral, and E. W. Hook. 2000. Female condom use among women at high risk of sexually transmitted disease. *Family Planning Perspectives* 32:138–44.

Madison, N.J. and M. Zieman 2002. Managing patients using the transdermal contraceptive system. *The Female Patient* August:26–32.

Maill, C. E. and K. March. 2005. Social support for changes in adoption practice: gay adoption, open adoption, birth reunions, and the release of confidential identifying information. *Families in Society* 86:83–92.

McDermott, E. and H. Graham. 2005. Resilient young mothering: social inequalities, late modernity and the problem if teenage motherhood. *Journal of Youth Studies* 8:59–79.

Michael, R. T. 2001. Abortion decisions in the United States. In *Sex, love, and health in America: Private choices and public policies,* edited by E. O. Laumann and R. T. Michael. Chicago: The University of Chicago Press, 377–438.

Miller, K. A., P. A. Fisher, B. Fetrow, and K. Jordan. 2006. Trouble on the journey home: reunification failures in foster care. *Children and Youth Services Review* 28:260–74.

Mishell, D. R. 2002.The transdermal contraceptive system. *The Female Patient* August:14–25.

Morrison, D. M., M. R. Gillmore, M. J. Hoppe, J. Gaylord, et al. 2003. Adolescent drinking and sex: findings from a daily diary study. *Perspectives on Sexual and Reproductive Health* 35:162–75.

Motamed, S. 2002. 100 million women can't be wrong: What most American women don't know about the IUD. Retrieved June 25, 2002, from www.plannedparenthood.org/articles/IUD/html.

Nickman, S. L., A. A. Rosenfeld, P. Fine, J. C. MacIntyre, D. J. Pilowsky, R. Howe, A. Dereyn, M. Gonzales, L. Forsythe and S. A. Sveda. 2005. Children in adoptive families: overview and update. *Journal of the American Academy of Child and Adolescent Psychiatry* 44:987–95.

Organon International Inc. (2004). Retrieved from http://www.who.int/reproductive-health/hrp/progress/61/news61.html, http://www.organon-usa.com/news/2004_11_04_implanon_approved_by_fda.asp, http://www.organon.com/products/contraception/Implanon.asp?.

Park, K. 2005. Choosing childlessness: Weber's typology of action and motives of the voluntarily childless. *Sociological Inquiry* 75:372–402.

Paulson, J. R., R. Boostanfar, P. Saadat, E. Mor, D. Tourgeman, C. C. Stater et al. 2002. Pregnancy in the sixth decade of life: obstetric outcomes in women in advanced reproductive age. *Journal of the American Medical Association* 288:2320–24.

Pertman, A. 2000. *Adoption nation: How the adoption revolution is transforming America.* New York: Basic Books.

Peterson, B. D., C. R. Newton, and K. H. Rosen. 2003. Examining congruence between partners' perceived infertility-related stress and its relationship to marital adjustment and depression in infertile couples. *Family Process* 42:59–70.

Pfizer. 2005. FDA approves Pfizer's Depo-Subq Provera 104™ for management of endometriosis pain. Retrieved June 9, 2005, from http://Pfizer.com/Pfizer/are/news_releases/2005pr/mn_2005_0329.jsp.

Pharmacy Access Partnership. 2006. Emergency contraception over-the-counter (OTC) status. Retrieved March 21, 2006 from http://www.pharmacyaccess.org/ECOTCStatus.htm.

Population Council. 2003a. Female contraceptive development. Retrieved February 10, 2003, from www.popcouncil.org/biomed/femalecontras.html.

———. 2003b. Jadele® Implants. Retrieved March 13, 2002, from www.popcouncil.org/faqs/adellefaq.html.

Raine, T., Harper, C., Rocca, C., Fischer, R., Padian, N., Klausner, J., and Darney, P. 2005. Direct access to emergency contraception through pharmacies and effect on unintended pregnancy and STIs. *JAMA* 293:54–62.

Quinlivan, J. A. and J. Condon. 2005. Anxiety and depression in fathers in teenage pregnancy. *Australian and New Zealand Journal of Psychiatry.* 39:915–20.

Ranjit, N., A. Bankole, J. E. Darroch, and S. Singh 2001. Contraceptive failure in the first two years of use: differences across socioeconomic subgroups. *Family Planning Perspectives* 33:19–27.

Regalado, A. 2002. Age is no barrier to motherhood: study gives latest proof older women can get pregnant with donor eggs. *Wall Street Journal* November 13: D3.

Roberts, L. C., and P. W. Blanton. 2000. Parenting education with one-child families. Poster, 62nd Annual Conference of the National Council on Family Relations, Minneapolis, November.

Rosenfeld, J. 1997. Postcoital contraception and abortion. In *Women's health in primary care,* edited by J. Rosenfeld. Baltimore: Williams & Wilkins, 315–29.

Ross, R., D. Knox, M. Whatley and J. N. Jahangardi. 2003. Transracial adoption: some college student data. Paper, 73rd Annual Meeting of the Eastern Sociological Society, Philadelphia.

Roumen, F., D. Apter, T. Mulders, and T. Dieben. 2001. Efficacy, tolerability and acceptability of a novel contraceptive vaginal ring releasing etonogestrel and ethinyl oestradiol. *Human Reproduction* 16:469–75.

Sankar, P. 2006. What is in a cause? Exploring the relationship between genetic cause and felt stigma. *Genetics in Medicine* 8:33–42.

Scheib, J. E., M. Riordan, and S. Rubin 2005. Adolescents with open-identity sperm donors: reports from 12-17 year olds. *Human Reproduction* 20:239–52.

Schmidt, L. 2006. Psychosocial burden of infertility and assisted reproduction. *The Lancet* 367:379–81.

Schwartz, J. L., and H. L. Gabelnick. 2002. Current contraceptive research. *Perspectives on Sexual and Reproductive Health* 34:310–16.

Shellenbarger, S. 2006. After lagging behind corporations, colleges bolster family-friendly benefits. *The Wall Street Journal* March 9:D1.

Shiota, K., and Yamada, S. (2005). Assisted reproductive technologies and birth defects. *Congenital Anomalies* 45:39–43.

Simon, R. J., and R. M. Roorda. 2000. *In their own voices: transracial adoptees tell their stories.* New York: Columbia University Press.

Sitruk-Ware, R. 2006. Contraception: an international perspective. *Contraception* 73:215–22.

Smith, P. B., R. S. Buzi, and M. L. Weinman. 2003. Targeting males for teenage pregnancy prevention in a school setting. *School of Social Work Journal* 27:23–36.

Statistical Abstract of the United States: 2006. 125th ed. Washington, D.C. U.S. Bureau of the Census.

Stone, A. 2006. Drives to ban gay adoption heat up. *USA Today* February 21:A1.

Tanfer, K., S. Wierzbicki, and B. Payn. 2000. Why are U.S. women not using long-acting contraceptives? *Family Planning Perspectives* 32:176–83.

Townsend, J. W. 2003. Reproductive behavior in the context of global population. *American Psychologist* 58:197–204.

Upadhyay, U.D. *New Contraceptive Choices. Population Reports,* Series M, No. 19. Baltimore, The Johns Hopkins Bloomberg School of Public Health, The INFO Project, April 2005.

Urban Institute. 2003. Who will adopt the foster care child left behind? *Caring for Children* Brief 2. June 1–2.

U.S. Food and Drug Administration. 2004. Depo-Provera black box warning. Retrieved from http://www.fda.gov/bbs/topics/ANSWERS/2004/ANS01325.html.

Van Damme, L. 2000. *Advances in topical microbicides.* Paper, 13th International AIDS Conference, July 9–14, Durban, South Africa.

van den Heuvel, M.W., van Bragt A.J., Alnabawy, A.K., and Kaptein, M.C. 2005. Comparison of ethinyl estradiol pharmacokinetics in three hormonal contraceptive formulations: the vaginal ring, the transdermal patch, and an oral contraceptive. *Contraception* 72:168–74.

Van Devanter, N., Gonzales, V., Merzel, C., Parikh, N. S., Celantano, D., and Greenberg, J. 2002. *American Journal of Public Health* 92:109–15.

Ventura, S. J., B. E. Hamilton, and P.D. Sutton. 2003. Revised birth and fertility rates for the United States, 2000 and 2001. *National Vital Statistics Reports* 51(4). Hyattsville, MD: National Center for Health Statistics.

Walsh, T. L., R. G. Frezieres, K. Peacock, A. L. Nelson et al. 2003. Evaluation of the efficacy of a nonlatex condom: results from a randomized, controlled clinical trial. *Perspectives on Sexual and Reproductive Health* 35:79–86.

Wan, R. S. F., and Lo, S. S. T. (2005). Are women ready for more liberal delivery of emergency contraceptive pills? *Contraception* 71:432–37.

White, L. and J. McQuillan 2006 No longer intending: The relationship between relinquished fertility intentions and distress. *Journal of Marriage and Family* 68: 478-490.

Wilson, S. A., and Kudis, H.A. 2005. Ethinyl estradiol/levinorgestrol (Seasonale) for oral contraception. *American Family Physician* 71(8):1581–86.

Wilson, H. and A. Huntington. 2006. Deviant mothers: the construction of teenage motherhood in contemporary discourse. *Journal of Social Policy* 35:59–76.

Zachry, E. M. 2005. Getting my education: teen mothers' experiences in school before and after motherhood. *Teachers College Record* 107:2566–98.

chapter 11

The child who is being raised by the book is probably the first edition.

Evan Esar, humorist

Parenting

Contents

True or False?

1. Mothers, more than fathers, are much more likely to overindulge their children.

2. In a national study, about a third of parents reported using a filtering or blocking system on the home computer.

3. An authoritative parenting style where children are held to clear expectations but in a context of warmth seems to have the best outcome for children.

4. Infants who sleep in their parents' bed are at significant risk of sudden infant death syndrome, in comparison with children who do not share a bed with their parents.

5. Parents report higher marital satisfaction than nonparents.

Answers: **1.** T **2.** T **3.** T **4.** F **5.** F

The experience described by John Wilmot, Earl of Rochester—"Before I got married, I had six theories about bringing up children; now I have six children and no theories"—is one with which all parents can connect. Although there are guidelines for effective parenting (and we will review them in this chapter), a number of factors (genetics, peers, health, economics, etc.) that affect how parenting turns out are beyond our control. Nevertheless, our focus in this chapter is on the issue of parenting choices, with the goal of facilitating happy, economically independent, socially contributing members to our society. We begin by looking at the various roles of parenting.

Roles Involved in Parenting

Although it is difficult to find one definition of **parenting,** there is general agreement about the roles parents play in the lives of their children. There are at least seven roles assumed by the new parent.

1. Caregiver. A major role of parents is the physical care of their children. From the moment of birth, when infants draw their first breath and are placed on the mother's stomach, parents stand ready to provide nourishment (milk), cleanliness (diapers), and temperature control (warm blanket). The need for such sustained care continues and becomes an accepted and anticipated role of parents. The parents who excuse themselves from a party early because they "need to check on the baby" are alerting the hostess of their commitment to the role of caregiver.

2. Emotional Resource. Beyond providing physical care, parents are sensitive to the emotional needs of children in terms of their need to belong, to be loved, and to develop positive self-concepts. Hugging, holding, and kissing their infant not only express their love for their infant but also reflect an awareness that such display of emotion is good for the child's sense of self-worth. The security that children feel when they are emotionally attached to their parents cuts across racial and ethnic identities. Arbona and Power (2003) found that securely attached adolescents in African-American, European-American, and Mexican-American families all reported positive self-esteem. In contrast, individuals who reported being emotionally neglected as children reported higher levels of psychological distress as adults (Wark et al., 2003).

Parents are regularly encouraged to spend "quality time" with their children, and it is implied that putting children in day care robs the child of this time. Booth et al. (2002) compared children in day care with those in home care in terms of time mother and child spent together per week. While the mothers of day-care children spent less time with their children than the mothers who cared for their children at home, the authors concluded that the "groups did not differ in the quality of mother-infant interaction" and that the difference in the "quality of the mother-infant interaction may be smaller than anticipated" (p. 16). Other research has suggested that social and emotional outcomes may not be compromised by day care (NICHD, 2003).

3. Teacher. All parents think they have a philosophy of life or set of principles they feel their children will benefit from. Parents later discover that their children may not be interested in their religion or philosophy—indeed, they may rebel against it. But this possibility does not deter them from their role as teacher. And children are forever learning from their parents, more often by observing their behavior.

4. Economic Resource. New parents are also acutely aware of the costs for medical care, food, and clothes for infants and seek ways to ensure that such resources are available to their children. Working longer hours, taking second jobs, and cutting back on leisure expenditures are attempts to ensure that money is available to meet the needs of the child. Parents may also take out additional life insurance or begin a college fund for their offspring.

In addition to providing philosophical and moral guidelines for children, parents also educate children about their culture. The War in Iraq is an ever-present reality in the media. A team of researchers interviewed 61 children (aged 3–18; median age of 9) and found that 65% reported that their parents had discussed the War in Iraq with them (Blankemeyer et al., 2004). Almost 10 percent (9.4%) of a sample of 1027 undergraduates reported that their parents were fearful that a terrorist attack would personally hit them (Knox and Zusman, 2006).

5. Protector. Parents also feel the need to protect their children from harm. This role may begin in pregnancy. Castrucci et al. (2006) interviewed 1451 women about their smoking behavior during pregnancy. Although 89% reduced their smoking during pregnancy, only about a quarter (24.9%) stopped smoking during pregnancy.

Other expressions of the protective role include insisting that children wear seat belts, protecting them from violence/nudity in the media, and protecting them from strangers. Diamond et al. (2006) studied 40 middle-class mothers of young children and identified 15 strategies they used to protect their children. Their three principal strategies were educate, control, and remove risk. The strategy used depended on the age and temperament of the child. For example, some parents feel it is important to protect their children from certain television content. This ranges from families that do not allow a television in their home to setting the V chip on their television to allowing only "G"-rated movies. Research confirms that parents do find media ratings helpful (Bushman and Cantor, 2003).

Learning how to fish begins with learning how to use the equipment. A hands-on lesson is the beginning of the skill.

Authors

Diversity in Other Countries

Six hundred fifty-four parents in Guyana identified the values they most wanted to teach their children as obedience, honesty, and mannerly conduct (Wilson et al., 2003).

Providing Housing for One's Adult Children?

After finishing high school or college, some young adults continue to live with their parents. Others have moved away but move back because of a return to school, loss of a job, or a divorce. In the case of the latter, they may bring young children with them back into the parental home. Saving money is the primary reason adult children reside with parents.

Parents vary widely in how they view, adapt to, and carry out such cohabiting. Some parents prefer that their children live with them, enjoy their company, and hope the arrangement continues indefinitely. They have no rules about their children living with them and expect nothing from them. The adult children can come and go as they please, pay for nothing, and have no responsibilities or chores. Other parents develop what is essentially a rental agreement, whereby their children are expected to pay rent, cook and clean the dishes, mow the lawn, and service the cars. No overnight guests are allowed, and a time limit is specified as to when the adult child is expected to move out.

Adult children also vary in terms of how they view the arrangement. Some enjoy living with their parents, volunteer to pay rent, take care of their own laundry, and participate in cooking/cleanup and housekeeping. Others are depressed that they are economically forced to live with their parents, embarrassed that they do so, pay nothing, and do nothing to contribute to the maintenance of the household. One effect of college students living with their parents instead of on campus is that the students drink less alcohol (Paschall et al., 2005).

Whether parents and adult children discuss the issues of their living together will depend on the respective parents and adult children. Although there is no best way, clarifying expectations might prevent some misunderstandings. For example, what is the norm about bringing new pets into the home? An example is a divorced son who moved in with his 6-year-old son *and* dog. His parents enjoyed being with them but were annoyed that the dog chewed on the furniture. Parental feelings eventually erupted that dismayed their adult child. He moved out, and the relationship with his parents became very strained.

Parenthood: That state of being better chaperoned than you were before marriage.

Marcelene Cox, writer

This mother is ensuring the safety of her child by putting her child on the school bus.

Authors

Some parents feel that protecting their children from harm implies appropriate discipline for inappropriate behavior. Galambos et al. (2003) noted "parents' firm behavioral control seemed to halt the upward trajectory in externalizing problems among adolescents with deviant peers." In other words, parents who intervened when they saw a negative context developing were able to help their children avoid negative peer influences.

6. Health promotion. The family is a major agent for health promotion (Novilla et al., 2006). Children learn from the family context about healthy food, the value of exercise, and the consequences of alcohol and other drugs. Protecting children from getting sunburned or overexposed to sunrays is another example of family health promotion that parents take seriously (Logan et al., 2005).

7. Ritual Bearer. In order to build a sense of family cohesiveness, parents often foster rituals to bind members together in emotion and in memory. Prayer at meals and before bedtime, birthday celebrations, and vacationing at the same place (beach, mountains, etc.) provide predictable times of togetherness and sharing.

The Choices Perspective of Parenting

Although both genetic and environmental factors are at work, it is the choices parents make that have a dramatic impact on their children. In this section we review the nature of parental choices and some of the basic choices parents make.

Nature of Parenting Choices

Parents might keep the following points in mind when they make choices about how to rear their children.

1. Not to make a parental decision is to make a decision. Parents are constantly making choices even when they think they are not doing so. When a child is impolite and the parent does not provide feedback and encourage polite behavior, the parent has chosen to teach his or her child that being impolite is acceptable. When a child makes a promise ("I'll call you when I get to my friend's house") and does not do as promised, the parent has chosen to allow the child to not take commitments seriously. Hence, parents cannot choose not to make choices in their parenting, since their inactivity is a choice that has as much impact as a deliberate decision to reinforce politeness and responsibility.

2. All parental choices involve trade-offs. Parents are also continually making trade-offs in the parenting choices they make. The decision to take on a second job or to work overtime to afford the larger house will come at the price of having less time to spend with one's children and being more exhausted when such time is available. The choice to enroll one's child in the highest-quality day care (which may also be the most expensive) will mean less money for family vacations. The choice to have an additional child will provide siblings for the existing children but will mean less time and fewer resources for those children. Parents should increase their awareness that no choice is without a tradeoff and should evaluate the costs and benefits in making such decisions.

3. Reframe "regretful" parental decisions. All parents regret a previous parental decision: they should have held their child back a year in school (or not done so); they should have intervened in a bad peer relationship; they should have handled their child's drug use differently, etc. Whatever the issue, parents chide themselves for their mistakes. Rather than berate themselves as parents, they might emphasize the positive outcome of their choices: not holding the child back made the child the "first" to experience some things among his or her peers; they made the best decision they could at the time, etc. Children might also be encouraged to view their own decisions positively.

Five Basic Parenting Choices

The five basic choices parents make include deciding whether to have a child, deciding the number of children, deciding the interval between children, deciding one's method of discipline and guidance, and deciding the degree to which one will be invested in the role of parent. Though all of these decisions are important, the relative importance one places on parenting as opposed to one's career will have implications for the parents, their children, and their children's children.

Whereas some parents focus their life around their children, others regard the children as only one aspect of their lives. One father of seven noted that while the first 5 years of parenting involved physical care-giving and intense emo-

The commonest fallacy among spouses is that simply having children makes them a parent— which is as absurd as believing that having a piano makes one a musician.

Sydney J. Harris, American journalist

tional bonding, the best parenting is to "create the context" and let the children flourish. In effect, he felt the seven children could best learn from one another with parents as the guardrails for safety.

Transition to Parenthood

Transition to parenthood refers to that period from the beginning of pregnancy through the first few months after the birth of a baby. The mother, father, and couple all undergo changes and adaptations during this period.

A woman's transition to the role of mother begins when she becomes pregnant.

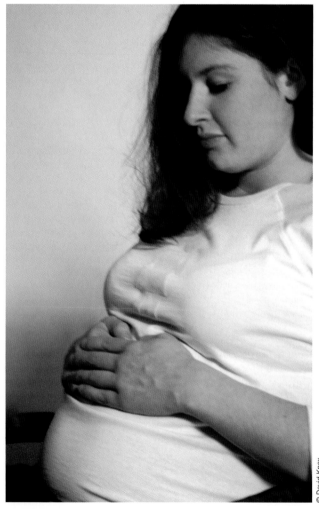

© David Knox

Transition to Motherhood

The Self-Assessment on page 321 examines one's view of traditional motherhood. Although childbirth is sometimes thought of as a painful ordeal, some women describe the experience as fantastic, joyful, and unsurpassed. A strong emotional bond between the mother and her baby usually develops early, and both the mother and infant resist separation. Sociobiologists suggest that there is a biological basis (survival) for the attachment between a mother and her offspring. The mother alone carries the fetus in her body for 9 months, lactates to provide milk, and produces **oxytocin**—a hormone from the pituitary gland—during the expulsive stage of labor that has been associated with the onset of maternal behavior in lower animals.

Not all mothers feel joyous after childbirth. Emotional bonding may be temporarily impeded by a mild depression, characterized by irritability, crying, loss of appetite, and difficulty in sleeping. The mother may also feel overwhelmed with the work of caring for an infant and feel a loss of a sense of mastery (Cassidy and Davies, 2003). Many new mothers experience **baby blues**—transitory symptoms of depression 24 to 48 hours after the baby is born. A few, about 10 percent, experience postpartum depression—a more severe reaction than baby blues.

Postpartum depression is believed to be a result of the numerous physiological and psychological changes occurring during pregnancy, labor, and delivery. Although the woman may become depressed during pregnancy or in the hospital, she more often experiences these feelings within the first month after returning home with her baby (sometimes the woman does not experience postpartum depression until a couple of years later). Most women recover within a short time; some (about 5 percent) seek therapy to speed their recovery.

Actress Brooke Shields (2005) recounts in *Down Came the Rain* her experience with postpartum depression, including this excerpt from the back cover:

I started to experience a sick sensation in my stomach; it was as if a vise were tightening around my chest. Instead of nervous anxiety that often accompanies panic, a feeling of devastation overcame me. I hardly moved. Sitting on my bed, I let

The Traditional Motherhood Scale

The purpose of this survey is to assess the degree to which students possess a traditional view of motherhood. Read each item carefully and consider what you believe. There are no right or wrong answers, so please give your honest reaction and opinion. After reading each statement, select the number that best reflects your level of agreement, using the following scale:

1	2	3	4	5	6	7
Strongly Disagree						Strongly Agree

_____ 1. The mother has a better relationship with her children than the father does.

_____ 2. A mother knows more about her child than the father, thereby being the better parent.

_____ 3. Motherhood is what brings women to their fullest potential.

_____ 4. A good mother should stay at home with her children for the first year.

_____ 5. Mothers should stay at home with the children.

_____ 6. Motherhood brings much joy and contentment to a woman.

_____ 7. A mother is needed in a child's life for nurturance and growth.

_____ 8. Motherhood is an essential part of a female's life.

_____ 9. I feel that all women should experience motherhood in some way.

_____10. Mothers are more nurturing than fathers.

_____11. Mothers have a stronger emotional bond with their children than do fathers.

_____12. Mothers are more sympathetic than fathers to children who have hurt themselves.

_____13. Mothers spend more time than fathers do with their children.

_____14. Mothers are more lenient than fathers toward their children.

_____15. Mothers are more affectionate than fathers toward their children.

_____16. The presence of the mother is vital to the child during the formative years.

_____17. Mothers play a larger role than fathers in raising children.

_____18. Women instinctively know what a baby needs.

Scoring

After assigning a number from 1 (strongly disagree) to 7 (strongly agree), add the numbers and divide by 18. The higher your score (7 is the highest possible score), the stronger the traditional view of motherhood. The lower your score (1 is the lowest possible score), the less traditional the view of motherhood.

Norms

The norming sample of this self-assessment was based upon 20 male and 86 female students attending Valdosta State University. The average age of participants completing the scale was 21.72 years (SD = 2.98), and ages ranged from 18 to 34. The ethnic composition of the sample was 80.2% white, 15.1% black, 1.9% Asian, 0.9% American Indian, and 1.9% other. The classification of the sample was 16.0% Freshmen, 15.1% Sophomores, 27.4% Juniors, 39.6% Seniors, and 1.9% graduate students.

Participants responded to each of the 18 items according to the seven-point scale. The most traditional score was 6.33; the score reflecting the least support for traditional motherhood was 1.78. The midpoint (average score) between the top and bottom score was 4.28 (SD = 1.04); thus, persons scoring above this number tended to have a more traditional view of motherhood and persons scoring below this number a less traditional view of motherhood.

There was a significant difference ($p < .05$) between female participants' scores (mean = 4.19; SD = 1.08) and male participants' scores (mean = 4.68; SD = 0.73), suggesting that males had more traditional views of motherhood than females.

Copyright

out a deep, slow, guttural wail. I wasn't simply emotional or weepy . . . this was something quite different. This was sadness of a shockingly different magnitude. It felt as if it would never go away.

To minimize baby blues and postpartum depression, antidepressants such as Paxil have been used. Brooke Shields benefited from Paxil, and the result was to give visibility to the issue of postpartum depression, its physiological basis, and the value of medication.

Postpartum psychosis, in which the woman harms her baby, is experienced by only one or two women per 1000 births (British Columbia Reproductive Mental Health Program, 2005). One must recognize that having misgivings about a new infant is normal. In addition, the woman who has negative feelings about her new role as mother should elicit help with the baby from her family or other support network so that she can continue to keep up her social contacts with friends and to spend time by herself and with her partner. Regardless of the cause, a team of researchers noted that maternal depression is associated with subsequent antisocial behavior in the child (Kim-Cohen et al., 2005).

Is transition to motherhood similar for lesbian and heterosexual mothers? Not according to Cornelius-Cozzi (2002), who interviewed lesbian mothers and found that the egalitarian norm of the lesbian relationship had been altered; for example, the biological mother became the primary caregiver, and the coparent, who often heard the biological mother refer to the child as "her child," suffered a lack of validation.

Transition to Fatherhood

Only when you become a father will you learn how to be a son.

Spanish proverb

The Self-Assessment on page 323 examines one's view of traditional fatherhood. While the likelihood that a male will live with his own children has been in decline (Eggebeen, 2002), the importance of the father in the lives of his children is enormous and goes beyond his economic contribution (Flouri and Buchanan, 2003; Aldous and Mulligan, 2002; Knox, 2000). Children from intact homes or those in which fathers maintained an active involvement in their lives after divorce tend to:

make good grades	have higher incomes as adults
be less involved in crime	have higher education levels
have good health/self-concept	form close friendships
have a strong work ethic	have stable jobs
have durable marriages	have fewer premarital births
have a strong moral conscience	have lower child sex abuse
have higher life satisfaction	exhibit fewer anorectic symptoms

A man can learn many things from his children until they grow old enough to know as little as he does.

Evan Esar

Strom et al. (2002) emphasized that the amount of time fathers spend with their children is important for being evaluated as a "successful father." How much time fathers actually spend with their children depends on whom you ask: fathers typically report spending more time with their children than mothers report they do (Coley and Morris, 2002). Gavin et al. (2002) noted that parental involvement was predicted most strongly by the quality of the parents' romantic relationship. If the father was emotionally and physically involved with the mother, he was more likely to take an active role in the child's life. Fathers whose wives worked more hours than the fathers worked also reported more involvement with their children (McBride et al., 2002).

Latino dads make up the largest number of fathers of any other ethnic group. Many have difficulty being "dad" since they were not born in this country and might not speak English. Nevertheless, "they place emphasis on cooperation, family unity, and child rearing—or in other words, familism" (Behnke, 2004).

Diversity in the United States

In the District of Columbia, more than 10 percent of African-American men between the ages of 18 and 35 are in prison; over half are under some form of correctional supervision; and 75 percent can expect to be incarcerated at some point in their lives. The result is less opportunity to function in the role of an active father for their children. African-American children are more vulnerable to growing up in homes without a father (Mauer and Chesney-Lind, 2003).

Transition from a Couple to a Family

Having a child is like throwing a hand grenade into a marriage.

Nora Ephron, writer/producer/director

Recent research suggests that parenthood decreases marital happiness. Bost et al. (2002) interviewed 137 couples before the birth of their first child and then at 3-, 12-, and 24-month periods. The spouses consistently reported depression and adjustment through 24 months postpartum. Twenge (2003) reviewed 148 samples representing 47,692 individuals in regard to the effect children have on marital satisfaction. They found that (1) parents (both women and men) reported lower marital satisfaction than nonparents; (2) mothers of infants reported the most significant drop in marital satisfaction; (3) the higher the num-

The Traditional Fatherhood Scale

The purpose of this survey is to assess the degree to which students have a traditional view of fatherhood. Read each item carefully and consider what you believe. There are no right or wrong answers, so please give your honest reaction and opinion. After reading each statement, select the number that best reflects your level of agreement, using the following scale:

1	2	3	4	5	6	7

Strongly   Strongly
Disagree   Agree

_____ 1. Fathers do not spend much time with their children.

_____ 2. Fathers should be the disciplinarians in the family.

_____ 3. Fathers should never stay at home with the children while the mother works.

_____ 4. The father's main contribution to his family is giving financially.

_____ 5. Fathers are less nurturing than mothers.

_____ 6. Fathers expect more from children than their mothers do.

_____ 7. Most men make horrible fathers.

_____ 8. Fathers punish children more than mothers do.

_____ 9. Fathers do not take a highly active role in their children's lives.

_____10. Fathers are very controlling.

Scoring

After assigning a number from 1 (strongly disagree) to 7 (strongly agree), add the numbers and divide by 10. The higher your score (7 is the highest possible score), the stronger the traditional view of fatherhood. The lower your score (1 is the lowest possible score), the less traditional the view of fatherhood.

Norms

The norming sample was based upon 24 male and 69 female students attending Valdosta State University. The average age of participants completing the Traditional Fatherhood Scale was 22.15 years (SD = 4.23), and ages ranged from 18 to 47. The ethnic composition of the sample was 77.4% white, 19.4% black, 1.1% Hispanic, and 2.2% other. The classification of the sample was 16.1% Freshmen, 11.8% Sophomores, 23.7% Juniors, 46.2% Seniors, and 2.2% graduate students.

Participants responded to each of the 10 items on the seven-point scale. The most traditional score was 5.50; the score representing the least support for traditional fatherhood was 1.00. The average score was 3.33 (SD = 1.03), suggesting a less than traditional view.

There was a significant difference ($p < .05$) between female participants' attitudes (mean = 3.20; SD = 1.01) and male participants' attitudes toward fatherhood (mean = 3.69; SD = 1.01), suggesting that males had more traditional views of fatherhood than females. There were no significant differences between ethnicities.

Copyright

ber of children, the lower the marital satisfaction; and (4) the factors that depressed marital satisfaction were conflict and loss of freedom.

For parents who experience a pattern of decreased happiness, it bottoms out during the teen years. Facer and Day (2004) found that adolescent problem behavior, particularly that of a daughter, is associated with increases in marital conflict. Of even greater impact was their perception of the child's emotional state. Parents who viewed their children as "happy" were less maritally affected by their adolescent's negative behavior.

Regardless of how children affect the feelings spouses have about their marriage, spouses report more commitment to their relationship once they have children (Stanley and Markman, 1992). Figure 11.1 illustrates that the more children a couple has, the more likely the couple will stay married. A primary reason for this increased commitment is the desire on the part of both parents to provide a stable family context for their children. In addition, parents of dependent children may keep their marriage together to maintain continued access to and a higher standard of living for their chil-

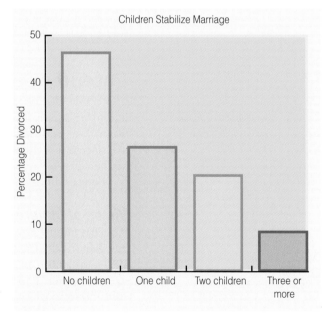

Figure 11.1
Percentage of Couples Getting Divorced by Number of Children

dren. Finally, people (especially mothers) with small children feel more pressure to stay married (if the partner provides sufficient economic resources) regardless of how unhappy they may be. Hence, though children may decrease happiness, they increase stability, since pressure exists to stay together.

PERSONAL CHOICES

Should Children Bed-Share with Parents?

Frequency of bed-sharing of parents and children ranges from never to every night. Children most likely to sleep in their parents' bed are young (1–12 weeks) and breast-fed. Indeed, most of the infants (60 to 90 percent) sharing a bed with their parents do so for ease of nighttime breast-feeding. Other reasons for regular bed-sharing include ideological factors, enjoyment, and lack of space. Children who bed-share with their parents on rare or specific occasions do so for reasons of illness, transient lack of space when traveling/visiting, or irritability (Ball, 2002).

In a longitudinal 10-year study (Jenni et al., 2005) of 493 Swiss children, ten percent slept with their parents their first year of life. This increased to 44 percent of children aged 2 to 7 who did so at least once a week. The authors concluded that bed-sharing with the parents was common in early childhood and that this practice should not be condemned by professionals who give advice to parents. Indeed, the child, the parents, and the context may suggest that bed-sharing should occur for the positive attachment needs of the child.

One concern about bed-sharing has been the risk of Sudden Infant Death Syndrome (SIDS). Hauck et al. (2003) studied 260 cases of SIDS and found that the usual bed-sharing in which one infant shares the bed with a parent is not associated with SIDS. However, when the parent slept on a sofa or when more than one child was in the bed, there was an increased risk of SIDS.

Sources

Ball, H. L. 2002. Reasons to bed-share: Why parents sleep with their infants. *Journal of Reproductive and Infant Psychology* 20:207–22.

Hauck, F. R., S. M. Herman, M. Donovan, C. M. Moore, S. Iyasu, E. Donoghue, R. H. Kirschner, and M. Willinger. 2003. Sleep environment and the risk of sudden death syndrome in an urban population: The Chicago Infant Mortality Study. *Pediatrics* 111:1207–15.

Jenni, O.G., H. Z. Fuhrer, I. Iglowstein, L. Molinari, and R. H. Largo. 2005. A longitudinal study of bed sharing and sleep problems among Swiss children in the first ten years of life. *Pediatrics* 115:233–40.

Parenthood: Some Facts

Parenting is only one stage in an individual's or couple's life (children typically live with an individual 30% of that person's life and with a couple 40% of their marriage). It involves responding to the varying needs of children as they grow up, and parents require help from family and friends in rearing their children.

Four additional facts of parenthood follow.

Views of Children Differ Historically

Whereas children of today are thought of as dependent, playful, and adventurous, historically they have been viewed quite differently (Mayall, 2002; Corsaro, 1997). Indeed, the concept of childhood, like gender, has been socially constructed rather than a fixed life stage. From the 13th through the 16th centuries, children were viewed as innocent, sweet, and a source of amusement for adults. From the 16th through the 18th centuries, they were viewed as in need of discipline and moral training. In the 19th century whippings were routine as a means of breaking children's spirit and bringing them to submission. Although remnants of both the innocent and moralistic views of children exist today, the lives of children are greatly improved. Child labor laws protect children from early forced labor, education laws ensure a basic education, and modern medicine has been able to increase the life span of children.

Each Child Is Unique

Children differ in their genetic makeup, physiological wiring, intelligence, tolerance for stress, capacity to learn, comfort in social situations, and interests. Parents soon become aware of the uniqueness of each child—of her or his difference from every other child they know. Parents of two or more children are often amazed at how children who have the same parents can be so different.

Bringing up children by the book is well enough, but remember, you need a different book for each child.
Unknown

National Data

In the United States, there are more than 6 million children with disabilities—almost half (47.4 percent) of these children have learning disabilities, 10 percent have mental retardation, and 8 percent have a serious emotional disturbance (*Statistical Abstract of the United States: 2006,* Table 249).

Children also differ in their mental and physical health. Mental and physical disabilities of children present emotional and financial challenges to their parents. Green (2003) discussed the potential stigma associated with disability and how parents cope with their children who are stigmatized.

Although parents often contend "we treat our children equally," Tucker et al. (2003) found that parents treat children differently, with firstborns usually receiving more privileges than children born later. However, the assignment of chores seems to be equal.

Parents Are Only One Influence in a Child's Development

Although parents often take the credit—and the blame—for the way their children turn out, they are only one among many influences on child development. Although parents are the first significant influence, peer influence becomes increasingly important during adolescence. Pinquart and Silbereisen (2002) studied 76 dyads of mothers and their 11- to 16-year-old adolescents and observed a decrease in connectedness between the children and their mothers and a movement toward their adolescent friends.

It's easy to make a buck. It's tougher to make a difference.
Tom Brokaw, former NBC news anchor

Siblings also have an important and sometimes lasting effect on each other's development. Siblings are social mirrors and models (depending on the age) for each other. They may also be sources of competition and can be jealous of each other.

Teachers are also a significant influence in the development of a child's values. Some parents send their children to religious schools to ensure that they will have teachers with conservative religious values. This may continue into the child's college and university education.

Media in the form of television, replete with MTV and "parental discretion advised" movies, are a major source of language, values, and lifestyles for children that may be different from those of the parents. Parents are also concerned about the violence (e.g., as seen in *The Sopranos*) that television exposes their children to.

Another influence of concern to parents is the Internet. Though parents may encourage their children to conduct research and write term papers using the Internet, they may fear the accessing of pornography and related sex sites. Parental supervision of teens on the Internet and the right of the teen for privacy remain potential conflict issues.

Government or Parental Control of Internet Content for Children?

Governmental censoring of content on the Internet is the focus of an ongoing public debate. At issue is what level of sexual content children should be exposed to—and should parents or the Government make the decision? This issue has already been decided in China, where the government ensures that all major search engines filter what is made available to the Internet user (Miller, 2006). In the United States, Congress passed the Communications Decency Act in 1996, which prohibited sending "indecent" messages over the Internet to people under age 18. But the Supreme Court struck down the law in 1997, holding that it was too broadly worded and violated free-speech rights by restricting too much material that adults might want access to. A similar ruling in 2002 has resulted in no government crackdown, even on virtual child pornography. In support of not limiting sexual content on the Internet, the American Civil Liberties Union emphasized that governmental restrictions threaten material protected by the First Amendment, including sexually explicit poetry and material educating disabled persons on how to experience sexual pleasure.

In 1998 Congress passed another law (the Child Online Protection Act), which makes it a crime to knowingly make available to people under age 17 any Web materials that, based on "contemporary community standards," are designed to pander to prurient interests. The law requires commercial operators to verify that a user is an adult through credit card information and adult access codes. Business owners who break the law are subject to a $50,000 fine and 6 months in jail. An inadvertent effect of the act was to require public libraries to install Internet filters to block access to objectionable sites. A U.S. appeals court and panel of federal judges have also struck down this law on the basis that it violates First Amendment rights.

In essence, many object to government control of sexual content on the Internet on the grounds of First Amendment rights. Many others believe government restrictions are necessary to protect children from inappropriate sexual content. The overriding question is, "Should parents or the government be in control of what children are exposed to?" Mitchell et al. (2005) found in a national survey of households that 33 percent of parents used a filtering or blocking system on their computer. The younger the child/children (10–15), the more likely such a system was used.

In the United States, parents, not the government, prefer to be responsible for regulating their children's use of the Internet. Beyond exposure to sexual content, children sometimes post information about themselves on the Internet which encourages pedophiles to email them. MySpace has more than 100 million profiles, with 230,000 new members signing up every day (Andrews, 2006). Some of these profiles may be more revealing than intended. Software products, such as Net Nanny, Surfwatch, CYBERsitter, CyberPatrol, and Time's Up, are being marketed to help parents control what their children view on the Internet. These software programs allow parents to block unapproved Web sites and categories (such as pornography), block transmission of personal data (such as address and telephone numbers), scan pages for sexual material before they are viewed, and track Internet usage. A more cumbersome solution is to require Internet users to provide passwords or identification numbers that would verify their ages before allowing access to certain Web sites.

Another alternative is for parents to use the Internet with their children both to monitor what their children are viewing and to teach their children values about what they believe is right and wrong on the Internet. Some parents believe that children must learn how to safely surf the Internet. Internet sites helpful in this regard are BlogSafety.com, forums for parents to discuss blogging and other aspects of social networking; NetSmartz.org service which teaches kids 5 to 17 how to be safe on the Internet, and WiredSafety.org which posts information on Internet safety

One parent reported that the Internet is like a busy street, and just as you must teach your children how to safely cross in traffic, you must teach them how to avoid giving information to strangers on the Internet.

Your Opinion?

1. To what degree do you believe the government should control Internet content?
2. To what degree do you believe that parents monitor what their children are exposed to on the Internet?
3. What can parents do to teach their children responsible use of the Internet?

Sources

Andrews, M. 2006. Decoding MySpace. *U.S. News and World Report* Sept 18, p. 46 et passim

Miller, M. J. 2006. Government versus the free Internet. *PC Magazine* 25:16–17.

Mitchell, K. J., D. Finkelhor, and J. Wolak. 2005. Protecting youth online: Family use of filtering and blocking software. *Child Abuse and Neglect* 29:753–65.

Parenting Styles Differ*

The study of parenting styles spans over half a century and has yielded fairly consistent dimensions centered on parental control and support (Barber et al., 2005). In this tradition, Diana Baumrind (1966) developed a typology of parenting styles that has become classic in the study of parenting. She noted that parenting behavior has two dimensions: responsiveness and demandingness. **Responsiveness** refers to the extent to which parents respond to and meet the needs of their children. In other words, how supportive are the parents? Warmth, reciprocity, person-centered communication, and attachment are all aspects of responsiveness. **Demandingness,** on the other hand, is the manner in which parents place demands on children in regard to expectations and discipline. How much control do they exert over their children? Monitoring and confrontation are also aspects of demandingness. Categorizing parents in terms of their responsiveness and their demandingness creates four categories of parenting styles: permissive (also known as indulgent), authoritarian, authoritative, and uninvolved.

1. Permissive parents are high on responsiveness and low on demandingness. They are very lenient and allow their children to largely regulate their own behavior.

2. Authoritarian parents are high on demandingness and low in responsiveness. They feel that children should obey their parents no matter what, and they provide a great deal of structure in the child's world.

3. Authoritative parents are both demanding and responsive. They impose appropriate limits on their children's behavior but emphasize reasoning and communication. This style offers a balance of warmth and control.

4. Uninvolved parents are low in responsiveness and demandingness. These parents are not invested in their children's lives.

Research suggests that authoritative parenting results in more positive child outcomes, including higher levels of social competence and lower levels of problem behavior (Darling, 1999). Similarly, there are negative effects from authoritarian parenting. Lagace-Seguin and Entremont (2006) assessed the effect of an authoritarian parenting style on 68 children slightly under 5 years of age. Not only were the children studied but also their mothers and teachers. Results showed that not only were the behaviors of the children negative (e.g., aggressive and asocial) but the mothers evidenced depression.

There is also some evidence to suggest that the outcomes of parenting styles may vary by ethnicity, with authoritarian parenting being more common among African-Americans (Hill and Bush, 2001; Smetana, 2000). However, other studies failed to demonstrate ethnic differences in parenting. In a study of youth nationwide, Amato and Fowler (2002) noted that a family context of support, monitoring, and avoidance of punitiveness—independent of race, ethnicity, family structure, education, income, and gender—is crucial to positive outcomes for children.

Barber et al. (2005) proposed a model of parenting behaviors and adolescent outcomes and tested the model with youth in 11 different countries, including the United States. They found that, across all cultures and subgroups studied, children benefit most from three key elements of parenting: connection, respect for individuality, and regulation. *Connection* is established within a relationship in which the child experiences support, closeness, belonging, and affection. The primary impact of connection is experienced in the child's ability to relate well to others (social competence), although connection also fosters psychological well-being and, to a lesser degree, conformity to parental and societal expectations. *Respect for individuality* is manifested by recognition, acceptance, and respect of the child's personal psychological boundaries; parents re-

While men take their countries to war, the way to end all wars is to get the mothers together.

Anonymous

In bringing up children, a good rule to follow is: Spend on them half as much money and twice as much time.

Unknown

*(Appreciation is expressed to Linda Robinson, family life specialist, East Carolina University for the development of this section.)

frain from intrusive psychological control such as attempts to change the child. Children who do not experience respect for their individuality tend to exhibit more internalizing problems (e.g., depression) or externalizing problems (e.g., antisocial behaviors) than their peers. *Regulation* refers to behavioral control in which parents are knowledgeable about and monitor their child's activities. Expectations are clear, consistent, and developmentally as well as contextually appropriate. Children whose parents provide adequate regulation are less likely to engage in antisocial behaviors.

In summary, parenting does not come as naturally to us humans as we might hope. After many decades of exploring how to provide an optimal environment for the development of children, social scientists are narrowing in on the essential elements, with more emphasis now on the specific ways in which elements of effective parenting are related to specific aspects of development for children and adolescents.

Principles of Effective Parenting

Numerous principles are involved in being an effective parent. Some of these follow.

I can soon learn how to do it if you'll show me how it's done. Fine counsel is confusing but example is always clear.

Edgar A. Guest

Give Time, Love, Praise, and Encouragement

Erma Bombeck once quipped that what children need is . . . your trust, your compassion, your binding love, and your car keys. But notice the order: children most need to feel that they are worth spending time with and that someone loves them. Since children depend first on their parents for the development of their sense of emotional security, it is critical that parents provide a warm emotional context in which the children can develop. Feeling loved as an infant also affects one's capacity to become involved in adult love relationships.

All children need love, especially those who don't deserve it.

Evan Esar

As the child matures, positive reinforcement for prosocial behavior also helps to encourage desirable behavior and a positive self-concept. Instead of focusing only on correcting or reprimanding bad behavior, parents should frequently comment on and reinforce good behavior. Comments like "I like the way you shared your toys," "You asked so politely," and "You did such a good job cleaning your room" help to reinforce positive social behavior and may enhance a child's self-concept. However, parents need to be careful not to over-praise their children, as too much praise may lead to the child's striving to please others rather than trying to please himself or herself.

PERSONAL CHOICES

Should Parents Reward Positive Behavior?

Most parents agree that some form of punishment is necessary to curb a child's inappropriate behavior, but there is disagreement over whether positive behavior (taking out the trash, cleaning one's room, making good grades) should be rewarded by praise, extra privileges, or money. Some parents feel that children should "do the right things anyway" and that to reward them is to bribe them. One parent said, "My children are going to do what I say because I say so, not because I am going to give them something for doing it." Other parents feel that both the child and the parent benefit when the parents reward the child for good behavior. Rewarding a child for a behavior will likely result in the child's engaging in that behavior more often and developing a set of positive behaviors. The parents, in turn, feel good about the child.

There is also concern among some professionals that parents praise their children too much. They feel that children develop an inflated ego and do not learn how to be realistic about themselves and to cope with failure.

Praise focuses on other people's judgments of a child's actions, whereas encouragement focuses more on the child's efforts. For example, telling a child who brings you his painting, "I love your picture; it is the best one that I have ever seen" is not as effective in building the child's confidence as saying, " You worked really hard on your painting. I notice that you used lots of different colors."

Avoid Overindulgence

Overindulgence is defined as giving children too much, too soon, too long—it is a form of child neglect in which children are not allowed to develop their own competences (www.overindulgence.info). Children who are overindulged think they are the center of the universe, don't work to earn what they want, and have an overblown sense of entitlement. Parents typically overindulge because they feel guilty or because they did not have certain material goods in their own youth. In a study designed to identify who overindulged, the researcher found mothers were four times more likely to overindulge as fathers (Clarke, 2004).

Monitor Child's Activities

Abundant research suggests that parents who monitor their children—know where their children are, who they are with, etc.—are less likely to report that their adolescents are involved in delinquent behavior and drinking alcohol (Laird et al., 2003; Longmore et al., 2001), poor academic performance (Jacobson and Crockett, 2000), and sexual activity (Hollander, 2003).

TV: The Third Parent
R. Buckminister Fuller

Set Limits and Discipline Children for Inappropriate Behavior

The goal of guidance is self-control. Parents want their children to be able to control their own behavior and to make good decisions without their parents.

Guidance may involve reinforcing desired behavior or providing limits to children's behavior. This sometimes involves disciplining children for negative behavior. Unless parents provide negative consequences for lying, stealing, and hitting, children can grow up to be dishonest, to steal, and to be inappropriately aggressive.

Time-out (a noncorporal form of punishment that involves removing the child from a context of reinforcement to a place of isolation for one minute for each year of the child's age) has been shown to be an effective consequence for inappropriate behavior. Withdrawal of privileges (watching television, playing with friends), pointing out the logical consequences of the misbehavior ("you were late; we won't go"), and positive language ("I know you meant well but . . .") are also effective methods of guiding children's behavior.

Physical punishment is less effective in reducing negative behavior; it teaches the child to be aggressive and encourages negative emotional feelings toward the parents. When using time out or the withdrawal of privileges, parents should make it clear to the child that they disapprove of the child's behavior, not the child. Some evidence suggests that consistent discipline has positive outcomes for children. Lengua et al. (2000) studied 231 mothers of 9- to 12-year-olds and found that inconsistent discipline was related to adjustment problems, particularly for children high in impulsivity.

Better a little chiding than a great deal of heartbreak.
Shakespeare, *The Merry Wives of Windsor*

Provide Security

Predictable responses from parents, a familiar bedroom or playroom, and an established routine help to encourage a feeling of security in children. Security provides children with the needed self-assurance to venture beyond the family. If the outside world becomes too frightening or difficult, a child can return to the safety of the family for support. Knowing it is always possible to return to an accepting environment enables a child to become more involved with the world beyond the family.

Should Parents Use Corporal Punishment?

Parents differ in the type of punishment they feel is appropriate for children. Some parents use corporal punishment as a means of disciplining their children, whereas others use "time out." The decision to choose a corporal or noncorporal method of punishment should be based on the consequences of use. In general, the use of time out and withholding of privileges seems to be more effective than corporal punishment in stopping undesirable behavior (Straus, 1994). Though beatings and whippings will temporarily decrease the negative verbal and nonverbal behaviors, they have major side effects. First, punishing children by inflicting violence teaches them that it is acceptable to physically hurt someone you love. Hence, parents may be inadvertently teaching their children to use violence in the family. Second, parents who beat their children should be aware that they are teaching their children to fear and avoid them. Third, children who grow up in homes in which corporal punishment is used are more likely to be aggressive and disobedient (Straus, 1994). In recognition of the negative consequences of corporal punishment, the law in Sweden forbids parents to spank their children.

So what kind of discipline is best? A team of researchers found that parents who reasoned with their children and then backed up their reasoning with punishment reported the most behavior change, as compared with parents who just used reasoning or punishment (Larzelere et al., 1998). Their data suggest that words backed up by some type of consequence have the most desirable outcome. We earlier noted that the authoritative parenting style (which emphasizes a balance of structure, control, and warmth) seems to have a good outcome for children and parents.

A review of some of the alternatives to corporal punishment include the following:

1. Be a positive role model. Children learn behaviors by observing their parents' actions, so parents must model the ways in which they want their children to behave. If a parent yells or hits, the child is likely to do the same.

2. Set rules and consequences. Make rules that are fair, realistic, and appropriate to a child's level of development. Explain the rules and the consequences of not following them. If children are old enough, they can be included in establishing the rules and consequences of breaking them.

3. Encourage and reward good behavior. When children are behaving appropriately, give them verbal praise and occasionally reward them with tangible objects, privileges, or increased responsibility.

4. Use charts. Charts to monitor and reward behavior can help children learn appropriate behavior. Charts should be simple and focus on one behavior at a time, for a certain length of time.

5. Use time out. "Time out" involves removing children from a situation following a negative behavior. This can help children calm down, end the inappropriate behavior, and re-enter the situation in a positive way. Explain what the inappropriate behavior is, why the time out is needed, when it will begin, and how long it will last. Set an appropriate length of time for the time out, based on age and level of development, usually 1 minute for each year of the child's age (see Self-Assessment on Spanking versus Time Out on page 331).

The old-fashioned parent believes that stern discipline means exactly where it is applied.

Unknown

Sources

Larzelere, R. E., P. R. Sather, W. N. Schneider, D. B. Larson, and P. L. Pike. 1998. Punishment enhances reasoning's effectiveness as a disciplinary response to toddlers. *Journal of Marriage and the Family* 60:388–403.

Straus, M. A. 1994. *Beating the devil out of them: Corporal punishment in American families.* San Francisco: Jossey-Bass.

This section on alternatives to spanking is based on the following source: National Mental Health Association, 2003. *Effective discipline techniques for parents: Alternatives to spanking.* Strengthening Families Fact Sheet. www.nmha.org.

Spanking versus Time-Out Scale

Parents discipline their children to help them develop self-control and correct misbehavior. Some parents spank their children; others use time-out. Spanking is a disciplinary technique whereby a mild slap (i.e., a "spank") is applied to the buttocks of a disobedient child. Time-out is a disciplinary technique whereby, when a child misbehaves, he/she is removed from the situation. The purpose of this survey is to assess the degree to which you prefer spanking versus time-out as a method of discipline. Please read each item carefully and select a number from 1 to 7 which represents your belief. There are no right or wrong answers; please give your honest opinion.

1	2	3	4	5	6	7
Strongly Disagree						Strongly Agree

_____ 1. Spanking is a better form of discipline than time-out.

_____ 2. Time-out does not have any effect on children.

_____ 3. When I have children, I will more likely spank them than use a time-out.

_____ 4. A threat of a time-out does not stop a child from misbehaving.

_____ 5. Lessons are learned better with spanking.

_____ 6. Time-out does not give the child an understanding of what he/she has done wrong.

_____ 7. Spanking teaches the child to respect authority.

_____ 8. Giving children time-outs is a waste of time.

_____ 9. Spanking has more of an impact on changing the behavior of children than time-out.

_____10. I do not believe "time-out" is a form of punishment.

_____11. Getting spanked as a child helps you become a responsible citizen.

_____12. Time-out is only used because parents are afraid to spank their kids.

_____13. Spanking can be an effective tool in disciplining a child.

_____14. Time-out is watered-down discipline.

Scoring

After you have selected a number from 1 to seven for each of the 14 items, if you want to know the degree to which you approve of spanking, reverse the number you selected for all odd numbered items (#1, #3, #5, #7, #9, #11, #13). For example, if you selected a 1 for item 1, change this number to a seven (1 = 7; 2 = 6; 3 = 5; 4 = 4; 5 = 3; 6 = 2; 7 = 1). Now add these seven numbers. The lower your score (7 is the lowest possible score) the lower

your approval of spanking and the higher your score (49 is the highest possible score), the greater your approval of spanking. A score of 21 places you at the midpoint between being very disapproving of or very accepting of spanking as a discipline strategy.

If you want to know the degree to which you approve of using time-out as a method of discipline, reverse the number you selected for all even numbered items (#2, #4, #6, #8, #10, #12, and #14. For example if you selected a 1 for item 2, change this number to a seven (i.e., 1 = 7; 2 = 6; 3 = 5; 4 = 4; 5 = 3; 6 = 2; 7 = 1). Now add these seven numbers. The lower your score (7 is the lowest possible score) the lower your approval of time-out and the higher your score (49 is the highest possible score), the greater your approval of time-out. A score of 21 places you at the midpoint between being very disapproving of or very accepting of time-out as a discipline strategy.

Scores of Other Students Who Completed the Scale

The scale was completed by 48 male and 168 female student volunteers at East Carolina University. Their ages ranged from 18 to 34 with a mean age of 19.65 ($SD = 2.06$). The ethnic background of the sample included 73.1% White, 17.1% African American, 2.8% Hispanic, 0.9% Asian, 3.7% from other ethnic backgrounds; 2.3% did not indicate ethnicity. The college classification level of the sample included 52.8% Freshman, 24.5% Sophomore, 13.9% Junior, and 8.8% Senior. The average score on the spanking dimension was 29.73 ($SD = 10.97$) and the time-out dimension was 22.93 ($SD = 8.86$) suggesting greater acceptance of spanking than time-out.

Time-out differences. In regard to sex of the participants, female participants were more positive about using time-out as a discipline strategy ($M = 33.72$, $SD = 8.76$) than were male participants ($M = 30.81$, $SD = 8.97$) ($p < .05$). In regard to ethnicity of the participants, White participants were more positive about using time-out as a discipline strategy ($M = 34.63$, $SD = 8.54$) than were Non-White participants ($M = 28.45$, $SD = 8.55$) ($p < .05$). In regard to year in school, Freshmen were more positive about using spanking as a discipline strategy ($M = 34.34$, $SD = 9.23$) than were Sophomores, Juniors, and Seniors ($M = 31.66$, $SD = 8.25$) ($p < .05$).

Spanking differences. In regard to ethnicity of the participants, Non-White participants were more positive about using spanking as a discipline strategy ($M = 35.09$, $SD = 10.02$) than were White participants ($M = 27.87$, $SD = 10.72$) ($p < .05$). In regard to year in school, Freshmen were less positive about using spanking as a discipline strategy ($M = 28.28$, $SD = 11.42$) than were Sophomores, Juniors, and Seniors ($M = 31.34$, $SD = 10.26$) ($p < .05$). There were no significant differences in regard to sex of the participants ($p > .05$) in the opinion of spanking.

Overall differences. There were no significant differences in overall attitudes to discipline in regards to sex of the participants, ethnicity, or year in school.

Copyright

The Spanking Versus Time-Out Scale, 2006, by Mark Whatley, PhD, Department of Psychology, Valdosta State University, Valdosta, Georgia 31698. Use of this scale is permitted only by prior written permission of Dr. Whatley (mwhatley@valdosta.edu).

Encourage Responsibility

Giving children increased responsibility encourages the autonomy and independence they need to be assertive and independent. Giving children more responsibility as they grow older can take the form of encouraging them to choose healthy snacks and letting them decide what to wear and when to return from playing with a friend (of course, the parents should praise appropriate choices).

Children who are not given any control and responsibility for their own lives remain dependent on others. Successful parents can be defined in terms of their ability to rear children who can function as independent adults. A dependent child is a vulnerable child.

Provide Sex Education

Parents are a powerful influence on the sexual behavior of their children. Although they are reluctant to discuss safe sex, their doing so often has positive consequences. In a sample of 237 Australian university students aged 16 to 19, researchers Troth and Peterson (2000) found that neither fathers nor mothers engaged in any substantial amount of education or communication with their offspring about the topic of safe sex. However, when parents do have such discussions, which are interpreted by their offspring as emphasizing that safe sex is important, the adolescents are more likely to follow through behaviorally by implementing frequent condom use. Remez (2003) found that mothers have a greater effect on daughters than sons and that mothers' disapproval of their daughters' having sex and mutual satisfaction with the mother-daughter relationship were associated with reduced risk of their daughters' becoming sexually active early in adolescence.

If parents have the goal of delaying the first intercourse experience of their children, they might consider taking the children to church. Karnehm (2000) studied the sexual debut of 815 children of relatively young mothers and found that children who reported at least monthly church attendance with parents at age 10 or 11 were more likely to delay their first sexual intercourse experience until at least age 16.

Express Confidence

"One of the greatest mistakes a parent can make," confided one mother, "is to be anxious all the time about your child, because the child interprets this as your lack of confidence in his or her ability to function independently." Rather, this mother noted that it is best to convey to the child that you know that he or she will be all right and that you are not going to worry about the child because you have confidence in him or her. "The effect on the child," said this mother, "is a heightened sense of self-confidence." Another way to conceptualize this parental principle is to think of the self-fulfilling prophecy as a mechanism that facilitates self-confidence. If the parents show the child that they have confidence in him or her, the child begins to accept these social definitions as real and becomes more self-confident.

Shellenbarger (2006) noted that some parents have become "helicopter parents" in that they are constantly hovering at school and in the workplace to ensure their child's "success." The workplace has become the new

Teenagers present a special challenge to parents; to begin with, teenagers sometimes put a low value on parents.

Melanie Coombs

field where parents negotiate the benefits and salaries of their children with employers they feel are out to take advantage of inexperienced workers. But employers may not appreciate the tampering, and the parents risk hampering their child's "ability to develop self-confidence" (p. D1).

Respond to the Teen Years Creatively

Parenting teenage children presents challenges that differ from those in parenting infants and young children. The teenage years have been characterized as a time when adolescents defy authority, act rebellious, and search for their own identity. Teenagers today are no longer viewed as innocent, naive children.

Conflicts between parents and teenagers often revolve around money and independence. The desires for a cell phone, DVD player, and high-definition TV can outstrip the budget of many parents. And teens increasingly want more freedom. But neither of these issues need to result in conflicts. And when they do, the effect on the parent-child relationship may be inconsequential. One parent tells his children, "I'm just being the parent, and you're just being who you are; it is OK for us to disagree—but you can't go." The following suggestions can help to keep conflicts with teenagers at a low level.

1. Catch them doing what you like rather than criticizing them for what you don't like. Adolescents are like everyone else—they don't like to be criticized but they do like to be noticed for what they do that is good.

2. Be direct when necessary. Though parents may want to ignore some behaviors of their children, addressing some issues directly may also be effective. Regarding the avoidance of STI/HIV infections, Dr. Louise Sammons (2003) tells her teenagers, "It is utterly imperative to require that any potential sex partner produce a certificate indicating no STIs or HIV infection and to require that a condom or dental dam be used before intercourse or oral sex."

3. Provide information rather than answers. When teens are confronted with a problem, try to avoid making a decision for them. Rather, it is helpful to provide information on which they may base a decision. What courses to take in high school and what college to apply for are decisions that might be made primarily by the adolescent. The role of the parent might best be that of providing information or helping the teenager obtain information.

4. Be tolerant of high activity levels. Some teenagers are constantly listening to loud music, going to each other's homes, and talking on cell phones for long periods of time. Parents often want to sit in their easy chairs and be quiet. Recognizing that it is not realistic to expect teenagers to be quiet and sedentary may be helpful in tolerating their disruptions.

5. Engage in some activity with your teenagers. Whether it is renting a video, eating a pizza, or taking a camping trip, it is important to structure some activities with your teenagers. Such activities permit a context in which to communicate with them.

Sometimes teenagers present challenges with which the parents feel unable to cope. Aside from monitoring their behavior closely, family therapy may be helpful. A major focus of such therapy is to increase the emotional bond between the parents and the teenagers and to encourage positive consequences (e.g., concert tickets) for desirable behavior (e.g., good grades) and negative consequences (loss of car privileges) for undesirable behavior (e.g., getting a speeding ticket).

Single-Parenting Issues

At least half of all children will spend one-fourth of their lives in a female-headed household (Webb, 2005). The stereotype of the single parent is the unmarried black single mother. In reality, 40 percent of single mothers are white and only 33 percent are black (Sugarman, 2003).

Sound travels slowly. Sometimes the things you say when your kids are teenagers don't reach them till they're in their forties.

Michael Hodgin

Obstinacy in children is like a kite; it is kept up just as long as we pull against it.

Marolene Cox

Diversity in Other Countries

There are wide variations in the percentage of children born to unmarried mothers. One half of Hawaiian mothers are not married when they give birth to their children. In contrast, only 8 percent of Chinese women and 10 percent of Japanese women are not married when they give birth to their children (*Statistical Abstract of the United States: 2006*, Table 80).

It is important to distinguish between a single-parent "family" and a single-parent "household." A single-parent family is one in which there is only one parent—the other parent is completely out of the child's life through death, sperm donation, or complete abandonment, and no contact is ever made with the other parent. In contrast, a single-parent household is one in which one parent typically has primary custody of the child or children but the parent living out of the house is still a part of the child's family. This is also referred to as a binuclear family. In most divorce cases where the mother has primary physical custody of the child, the child lives in a single-parent household, since he or she is still connected to the father, who remains part of the child's family. In cases in which one parent has died, the child or children live with the surviving parent in a single-parent family, since there is only one parent.

Single Mothers by Choice

Single parents enter their role though divorce/separation, widowhood, adoption, or deliberate choice to rear a child or children alone. Jodie Foster, Academy Award–winning actress, has elected to have children without a husband. She now has two children and smiles when asked, "Who's the father?" The implication is that she has a right to her private life and that choosing to have a single-parent family is a viable option. An organization for women who want children and who may or may not marry is Single Mothers by Choice.

Bock (2000) noted that single mothers by choice are, for the most part, in the middle to upper class, mature, well-employed, politically aware, and dedicated to motherhood. Interviews with 26 single mothers by choice revealed their struggle to avoid stigmatization and to seek legitimization for their choice. Most felt that their age (older), sense of responsibility, maturity, and fiscal capability justified their choice. Their self-concepts were those of competent, ethical, mainstream mothers.

Challenges Faced by Single Parents

The single-parent lifestyle involves numerous challenges, including some of the following issues.

Single parenting can be a lonely endeavor.

1. Responding to the demands of parenting with limited help. Perhaps the greatest challenge for single parents is taking care of the physical, emotional, and disciplinary needs of their children—alone. Many single parents resolve this problem by getting help from their parents or extended family.

2. Adult emotional needs. Single parents have emotional needs of their own that children are often incapable of satisfying. The unmet need to share an emotional relationship with an adult can weigh heavily on the single parent. One single mother said, "I'm working two jobs, taking care of my kids, and trying to go to school. Plus my mother has cancer. Who am I going to talk to about my life?" Many single women solve the dilemma with a network of friends.

3. Adult sexual needs. Some single parents regard their parental role as interfering with their sexual relationships. They may be concerned that their children will find out if they have a sexual encounter at home or be frustrated if they have to go away from home to enjoy a sexual relationship. Some choices with which they are confronted include "Do I wait until my children are asleep and then ask my lover to leave before morning?" "Do I openly acknowledge my lover's presence in my life to my children and ask them not to tell anybody?" "Suppose my kids get attached to my lover, who may not be a permanent part of our lives?"

4. Lack of money. Single-parent families, particularly those headed by women, report that money is always lacking.

> *I'm tired of love, I'm still more tired of rhyme, but money gives me pleasure all the time.*
>
> Hilaire Belloc

National Data

The median income of a single-woman householder is $26,550, much lower than that of a single-man householder ($38,032) or a married couple ($62,281) (*Statistical Abstract of the United States*: 2006, Table 682).

5. Guardianship. If the other parent is completely out of the child's life, the single parent needs to appoint a guardian to take care of her or his child in the event of her or his own death or disability.

6. Prenatal care. Single women who decide to have a child have poorer pregnancy outcomes than married women. The reason for such an association may be the lack of economic funds (no male partner with economic resources available) as well as the lack of social support for the pregnancy or the working conditions of the mothers, all of which result in less prenatal care for their babies.

7. Absence of a father. Another consequence for children of single-parent mothers is that they often do not have the opportunity to develop an emotionally supportive relationship with their father. Knox (2000) reported that children who grow up without fathers are more likely to drop out of school, be unemployed, abuse drugs, experience mental illness, and be a target of child sexual abuse. Conversely, those with fathers in their lives report higher life satisfaction, more stable marriages, and closer friendships. Ansel Adams, the late great photographer, attributed his personal and life success to his father, who steadfastly guided his development.

8. Negative life outcomes for the child in a single-parent family. Researcher Sarah McLanahan, herself a single mother, set out to prove that children reared by single parents were just as well off as those reared by two parents. McLanahan's data on 35,000 children of single parents led her to a different conclusion—children of only one parent were twice as likely as those reared by two married parents to drop out of high school, get pregnant before marriage, have drinking problems, and experience a host of other difficulties, including getting divorced themselves (McLanahan and Booth, 1989; McLanahan, 1991). Lack of supervision, fewer economic resources, and less extended family support were among the culprits. Other research suggests that negative outcomes are reduced or eliminated when income levels remain stable (Pong and Dong, 2000).

Though the risk of negative outcomes is higher for children in single-parent homes, most are happy and well-adjusted. Benefits to single parents themselves include a sense of pride and self-esteem that results from being independent.

Approaches to Childrearing

Parents are offered advice from professionals, their own parents, siblings, and friends. Early parenting education efforts were directed solely at women. Today, parenting information is targeted to fathers as well as mothers and is readily available to all parents. Sometimes, however, there is conflicting information that can be confusing to parents. Adding to the confusion is the fact that parenting advice may change from time to time. For example, parents in 1914 who wanted to know what to do about their child's thumb-sucking were told to try to control such a bad impulse by pinning the sleeves of the child to the bed if necessary. Today, parents are told that thumb-sucking meets an important psychological need for security and they should not try to prevent it. If the child's teeth become crooked as a result, an orthodontist should be consulted.

Table 11.1 Theories of Childrearing

Theory	Major Contributor	Basic Perspective	Focal Concerns	Criticisms
Developmental-Maturational	Arnold Gesell	Genetic basis for child passing through predictable stages	Motor behavior Adaptive behavior Language behavior Social behavior	Overemphasis on biological clock Inadequate sample to develop norms Demand schedule questionable Upper-middle-class bias
Behavioral	B. F. Skinner	Behavior is learned through operant and classical conditioning	Positive reinforcement Negative reinforcement Punishment Extinction Stimulus response	Deemphasis on cognitions of child Theory too complex to be accurately/appropriately applied by parent Too manipulative/controlling Difficult to know reinforcers and punishers in advance
Parent Effectiveness Training	Thomas Gordon	The child's worldview is the key to understanding the child	Change the environment before attempting to change the child's behavior Avoid hurting the child's self-esteem Avoid win-lose solutions	Parents must sometimes impose their will on the child's How to achieve win-win solutions is not specified
Socioteleological	Alfred Adler	Behavior is seen as attempt of child to secure a place in the family	Insecurity Compensation Power Revenge Social striving Natural consequences	Limited empirical support Child may be harmed taking "natural consequences"
Attachment	William Sears	Goal is to establish a firm emotional attachment with child	Connecting with baby Responding to cues Breast-feeding Wearing baby Sharing sleep	May result in spoiled, overly dependent child

There are several theoretical approaches to rearing children. A review of these approaches can be found in Table 11.1. In examining these approaches, it is important to keep in mind that no single approach is superior to another. What works for one child may not work for another. Any given approach may not even work with the same child at two different times. In addition, parents and the professional caregivers of children often differ in regard to childrearing approaches.

It is important for new parents to use a cafeteria approach when examining parenting advice from health care providers, family members, friends, parenting educators, and other well-meaning individuals. Take what makes sense, works, and feels right as a parent and leave the rest behind. Parents know their own child better than anyone else and should be encouraged to combine different approaches to find what works best for them and their unique child.

Developmental-Maturational Approach

For the past 60 years, Arnold Gesell and his colleagues at the Yale Clinic of Child Development have been known for their ages-and-stages approach to childrearing (Gesell et al., 1995). The **developmental-maturational approach** has been widely used in the United States. The basic perspective, some considerations for childrearing, and some criticisms of the approach follow.

Every child psychiatrist would like to bring up his or her children in the way (s)he advises other parents.

Unknown

Basic Perspective Gesell views what children do, think, and feel as being influenced by their genetic inheritance. Although genes dictate the gradual unfolding of a unique person, every individual passes through the same basic pattern of growth. This pattern includes four aspects of development: motor behavior (sitting, crawling, walking), adaptive behavior (picking up objects and walking around objects), language behavior (words and gestures), and personal-social behavior (cooperativeness and helpfulness). Through the observation of hundreds of normal infants and children, Gesell and his coworkers have identified norms of development. Although there may be large variations, these norms suggest the ages at which an average child displays various behaviors. For example, on the average, children begin to walk alone (although awkwardly) at age 13 months and use simple sentences between the ages of 2 and 3.

Considerations for Childrearing Gesell suggested that if parents are aware of their children's developmental clock, they will avoid unreasonable expectations. For example, a child cannot walk or talk until the neurological structures necessary for those behaviors have matured. Also, the hunger of a 4-week-old must be immediately appeased by food, but at 16 to 28 weeks, the child has some capacity to wait because the hunger pains are less intense. In view of this and other developmental patterns, Gesell suggested that infants need to be cared for on a demand schedule; instead of having to submit to a schedule imposed by parents, infants are fed, changed, put to bed, and allowed to play when they want. Children are likely to be resistant to a hard-and-fast schedule because they may be developmentally unable to cope with it.

In addition, Gesell emphasized that parents should be aware of the importance of the first years of a child's life. In Gesell's view, these early years assume the greatest significance because the child's first learning experiences occur during this period.

Criticisms of the Developmental-Maturational Approach Gesell's work has been criticized because of (1) its overemphasis on the idea of a biological clock, (2) the deficiencies of the sample he used to develop maturational norms, (3) his insistence on the merits of a demand schedule, and (4) the idea that environmental influences are weak.

Most of the children who were studied to establish the developmental norms were from the upper middle class. Children in other social classes are exposed to different environments, which influence their development. So norms established on upper-middle-class children may not adequately reflect the norms of children from other social classes.

Gesell's suggestion that parents do everything for the infant when the infant wants has also been criticized. Rearing an infant on the demand schedule can drastically interfere with the parents' personal and marital interests. In the United States, with its emphasis on individualism, many parents feed their infants on a demand schedule but put them to bed to accommodate the parents' schedule.

Behavioral Approach

The **behavioral approach** to childrearing, also known as the social learning approach, is based on the work of B. F. Skinner (Skinner et al., 1997). Behavioral approaches to child behavior have received the most empirical study. Public health officials and policymakers are encouraged to promote programs with empirically supported treatments.

We now review the basic perspective, considerations, and criticisms of this approach to childrearing.

Basic Perspective   Behavior is learned through classical and operant conditioning. Classical conditioning involves presenting a stimulus with a reward. For example, infants learn to associate the faces of their parents with food, warmth, and comfort. Although initially only the food and feeling warm will satisfy the infant, later just the approach of the parent will soothe the infant. This may be observed when a father hands his infant to a stranger. The infant may cry because the stranger is not associated with pleasant events. But when the stranger hands the infant back to the parent, the crying may subside because the parent represents positive events and the stimulus of the parent's face is associated with pleasurable feelings and emotional safety.

Other behaviors are learned through operant conditioning, which focuses on the consequences of behavior. Two principles of learning are basic to the operant explanation of behavior—reward and punishment. According to the reward principle, behaviors that are followed by a positive consequence will increase.

If the goal is to teach the child to say "please," doing something the child likes after he or she says "please" will increase the use of "please" by the child. Rewards may be in the form of attention, praise, desired activities, or privileges.

Whatever consequence increases the frequency of an occurrence is, by definition, a reward. If a particular consequence doesn't change the behavior in the desired way, a different reward needs to be tried.

The punishment principle is the opposite of the reward principle. A negative consequence following a behavior will decrease the frequency of that behavior; for example, the child could be isolated for 5 or 10 minutes following an undesirable behavior. The most effective way to change behavior is to use the reward and punishment principles together to influence a specific behavior. Praise children for what you want them to do and provide negative consequences for what they do that you do not like.

Considerations for Childrearing   Parents often ask, "Why does my child act this way, and what can I do to change it?" The behavioral approach to childrearing suggests the answer to both questions. The child's behavior has been learned through being rewarded for the behavior, and it can be changed by eliminating the reward for or punishing the undesirable behavior and rewarding the desirable behavior.

The child who cries when the parents are about to leave home to go to dinner or see a movie is often reinforced for crying by the parents' staying home longer. To teach the child not to cry when the parents leave, the parents should reward the child for not crying when they are gone for progressively longer periods of time. For example, they might initially tell the child they are going outside to walk around the house and they will give the child a treat when they get back if he or she plays until they return. The parents might then walk around the house and reward the child for not crying. If the child cries, they should be out of sight for only a few seconds and gradually increase the amount of time they are away. The essential point is that children learn to cry or not to cry depending on the consequences of crying. Because children learn what they are taught, parents might systematically structure learning experiences to achieve specific behavioral goals.

Criticisms of the Behavioral Approach Professionals and parents have attacked the behavioral approach to childrearing on the basis that it is deceptively simple and does not take cognitive issues into account. Although the behavioral approach is often presented as an easy-to-use set of procedures for child management, many parents do not have the background or skill to implement the procedures effectively. What constitutes an effective reward or punishment, how it should be presented, in what situation and with what child, to influence what behavior are all decisions that need to be made before attempting to increase or decrease the frequency of a behavior. Parents often do not know the questions to ask or lack the training to make appropriate decisions in the use of behavioral procedures. One parent locked her son in the closet for an hour to punish him for lying to her a week earlier—a gross misuse of learning principles.

Behavioral childrearing has also been said to be manipulative and controlling, thereby devaluing human dignity and individuality. Some professionals feel that humans should not be manipulated to behave in certain ways through the use of rewards and punishments.

Finally, the behavioral approach has been criticized because it de-emphasizes the influence of thought processes on behavior. Too much attention, say the critics, has been given to rewarding and punishing behavior and not enough attention has been given to how the child perceives a situation. For example, parents might think they are rewarding a child by giving her or him a bicycle for good behavior. But the child may prefer to upset the parents by rejecting the bicycle and may be more rewarded by their anger than by the gift.

Parent Effectiveness Training Approach

Thomas Gordon (2000) developed a model of childrearing based on **parent effectiveness training** (PET).

Basic Perspective Parent effectiveness training focuses on what children feel and experience in the here and now—how they see the world. The method of trying to understand what the child is experiencing is active listening, in which the parent reflects the child's feelings. For example, the parent who is told by the child, "I want to quit taking piano lessons because I don't like to practice" would reflect, "You're really bored with practicing the piano and would rather have fun doing something else." PET also focuses on the development of the child's positive self-concept.

To foster a positive self-concept in their child, parents should reflect positive images to the child—letting the child know he or she is loved, admired, and approved of.

Considerations for Childrearing To assist in the development of a child's positive self-concept and in the self-actualization of both children and parents, Gordon (2000) recommended managing the environment rather than the child, engaging in active listening, using "I" messages, and resolving conflicts through mutual negotiation.

An example of environmental management is putting breakables out of reach of young children. Rather than worry about how to teach children not to

How the twig is bent may be less important than the way it bends itself.

Joseph Wood Krutch

touch breakable knickknacks, it may be easier to simply move the items out of the children's reach.

The use of active listening becomes increasingly important as the child gets older. When Joanna is upset with her teacher, it is better for the parent to reflect the child's thoughts than to take sides with the child. Saying "You're angry that Mrs. Jones made the whole class miss play period because Becky was chewing gum" rather than saying "Mrs. Jones was unfair and should not have made the whole class miss play period" shows empathy with the child without blaming the teacher.

Gordon also suggested using "I" rather than "you" messages. Parents are encouraged to say "I get upset when you're late and don't call" rather than "You're an insensitive, irresponsible kid for not calling me when you said you would." The former avoids damaging the child's self-concept but still expresses the parent's feelings and encourages the desired behavior.

Gordon's fourth suggestion for parenting is the no-lose method of resolving conflicts. Gordon rejects the use of power by parent or child. In the authoritarian home, the parent dictates what the child is to do and the child is expected to obey. In such a system, the parent wins and the child loses. At the other extreme is the permissive home, in which the child wins and the parent loses. The alternative, recommended by Gordon, is for the parent and the child to seek a solution that is acceptable to both and to keep trying until they find one. In this way, neither parent nor child loses and both win.

Criticisms of the Parent Effectiveness Training Approach   Although much is commendable about PET, parents may have problems with two of Gordon's suggestions.

First, he recommends that because older children have a right to their own values, parents should not interfere with their dress, career plans, and sexual behavior. Some parents may feel they do have a right (and an obligation) to "interfere." Second, the no-lose method of resolving conflict is sometimes unrealistic.

Suppose a 16-year-old wants to spend the weekend at the beach with her boyfriend and her parents do not want her to. Gordon advises negotiating until a decision is reached that is acceptable to both. But what if neither the daughter nor the parents can suggest a compromise or shift their position? The specifics of how to resolve a particular situation are not always clear.

Socioteleological Approach

Alfred Adler, a physician and former student of Sigmund Freud, saw a parallel between psychological and physiological development. When a person loses her or his sight, the other senses (hearing, touch, taste) become more sensitive—they compensate for the loss. According to Adler, the same phenomenon occurs in the psychological realm. When individuals feel inferior in one area, they will strive to compensate and become superior in another. Rudolph Dreikurs, a student of Adler, developed an approach to childrearing that alerts parents as to how their children might be trying to compensate for feelings of inferiority (Soltz and Dreikurs, 1991). Dreikurs's **socioteleological approach** is based on Adler's theory.

Basic Perspective   According to Adler, it is understandable that most children feel they are inferior and weak. From the child's point of view, the world is filled with strong giants who tower above him or her. Because children feel powerless in the face of adult superiority, they try to compensate by gaining attention (making noise, becoming disruptive), exerting power (becoming aggressive, hostile), seeking revenge (becoming violent, hurting others), and acting inadequate (giving up, not trying). Adler suggested that such misbehavior is evidence that the child is discouraged or feels insecure about her or his place in the family. The term *socioteleological* refers to social striving or seeking a social goal. In the child's case, the goal is to find a secure place within the family—the first "society" the child experiences.

Considerations for Childrearing   When parents observe misbehavior in their children, they should recognize it as an attempt to find security. According to Dreikurs, parents should not fall into playing the child's game by, say, responding to a child's disruptiveness with anger but should encourage the child, hold regular family meetings, and let natural consequences occur. To encourage the child, the parents should be willing to let the child make mistakes. If John wants to help Dad carry logs to the fireplace, rather than saying, "You're too small to carry the logs," Dad should allow John to try and should encourage him to carry the size limb or stick that he can manage. Furthermore, Dad should praise John for his helpfulness.

As well as being constantly encouraged, the child should be included in a weekly family meeting. During this meeting, such family issues as bedtimes, the appropriateness of between-meal snacks, assignment of chores, and family fun are discussed. The meeting is democratic; each family member has a vote. Participation in family decision-making is designed to enhance the self-concept of each child. By allowing each child to vote on family decisions, parents respect the child as a person as well as the child's needs and feelings.

Resolutions to conflicts with the child might also be framed in terms of choices the child can make. "You can go outside and play only in the backyard, or you can play in the house" gives the child a choice. If the child strays from the backyard, he or she can be brought in and told, "You can go out again later." Such a framework teaches responsibility for and consequences of one's choices.

Finally, Driekurs suggested that the parents let natural consequences occur for their child's behavior. If a daughter misses the school bus, she walks or is charged taxi fare out of her allowance. If she won't wear a coat and boots in bad weather, she gets cold and wet. Of course, parents are to arrange logical consequences when natural consequences will not occur or would be dangerous if they did. For example, if a child leaves the television on overnight, access might be taken away for the next day or so.

Criticisms of the Socioteleological Approach   The socioteleological approach is sometimes regarded as impractical, since it teaches the importance of letting children take the natural consequences for their actions. Such a principle may be interpreted to let the child develop a sore throat if he or she wishes to go out in the rain without a raincoat. In reality, advocates of the method would not let the child make a dangerous decision. Rather, they would give the child a choice with a logical consequence such as "You can go outside wearing a raincoat, or you can stay inside—it is your choice."

Attachment Parenting

Dr. William Sears, along with his wife, Martha Sears, developed an approach to parenting called attachment parenting (Sears and Sears, 1993). This "common-sense parenting" approach focuses on parents connecting with their baby.

My mother loved children. She would have given anything if I had been one.

Groucho Marx

Basic Perspective   The emotional attachment process between mother and child is thought to begin prior to birth and continues to be established during the next 3 years. Sears identified three parenting goals: to know your child, to help your child feel right, and to enjoy parenting. He also suggested five concepts or tools (identified in the following section) that comprise attachment parenting that will help parents to achieve these goals. Overall, the ultimate goal is for parents to get connected with their baby. Once parents are connected, it is easy for parents to figure out what works for them and to develop a parenting style that fits them and their baby. Meeting a child's needs early in life will help him or her form a secure attachment with parents. This secure attachment will help the child to gain confidence and independence as he or she grows up. Attachment Parenting International is a nonprofit organization committed to educating society and parents about the critical emotional and psychological needs of infants and children.

Considerations for Childrearing The first attachment tool is for parents to connect with their baby early. The initial months of parenthood are a sensitive time for bonding with your baby and starting the process of attachment.

The second tool is to read and respond to the baby's cues. Parents should spend time getting to know their baby and learn to recognize his or her unique cues. Once a parent gets in tune with the baby's cues, it is easy to respond to the child's needs. Sears encourages parents to be open and responsive and to pick their baby up if he or she cries. Responding to a baby's cries helps the baby to develop trust and encourages good communication between child and parent. Eventually, babies who are responded to will internalize their security and will not be as demanding.

The third attachment tool is for mothers to breast-feed their babies and to do this on demand rather than trying to follow a schedule. He emphasizes the important role that fathers also play in successful breast-feeding by helping to create a supportive environment.

The fourth concept of attachment parenting is for parents to wear their baby by using a baby sling or carrier. The closeness is good for the baby and it makes life easier for the parent. Wearing your baby in a sling or carrier allows parents to engage in regular day-to-day activities and makes it easier to leave the house.

Finally, Sears advocates that parents let the child sleep in their bed with them since it allows parents to stay connected with their child throughout the night. However, some parents and babies often sleep better if the baby is in a separate crib, and Sears recognizes that wherever parents and their baby sleep best is the best policy. Wherever you choose to have your baby sleep, Sears is clear on one thing—it is never acceptable to let your baby cry when he or she is going to sleep! Parents need to parent their child to sleep rather than leaving him or her to cry. Requiring children to conform to rigid schedules of parents and using corporal punishment are considered abusive by most of today's child development research (Parker et al., 2004).

Criticisms of Attachment Parenting Some parents feel that responding to their baby's cries, carrying or wearing their baby, and sharing sleep with their baby will lead to a spoiled baby who is overly dependent. Some parents may feel more tied down using this parenting approach and may find it difficult to get their child on a schedule. Many women return to work after the baby is born and find some of the concepts difficult to follow. Some women choose not to breast-feed their children for a variety of reasons. Finally, the idea of sharing sleep has resulted in a lot of criticism. Some parents might be nervous that they might roll over on the child, that the child might disturb their sleep or intimacy with their partner, or that sharing a bed will mean that they will never get their child to sleep on his or her own. However, children who grow up in an emotionally secure environment and have strong attachment report less behavioral and substance abuse problems as adolescents (Elgar et al., 2003).

SUMMARY

What are the basic roles of parents?
Parenting includes providing physical care for children, loving them, being an economic resource, providing guidance as a teacher/model, and protecting them from harm.

What is a choices perspective of parenting?

Although both genetic and environmental factors are at work, it is the choices parents make that have a dramatic impact on their children. Parents who don't make a choice about parenting have already made one. The five basic choices parents make include deciding whether to have a child, deciding the number of children, deciding the interval between children, deciding one's method of discipline and guidance, and deciding the degree to which one will be invested in the role of parent.

What is the transition to parenthood like for women, men, and couples?

Transition to parenthood refers to that period of time from the beginning of pregnancy through the first few months after the birth of a baby. The mother, father, and couple all undergo changes and adaptations during this period. Most mothers relish their new role; some may experience the transitory feelings of baby blues; a few report postpartum depression.

The father's involvement with his children is sometimes predicted by the quality of the parents' romantic relationship. If the father is emotionally and physically involved with the mother, he is more likely to take an active role in the child's life. In recent years there has been a renewed cultural awareness of fatherhood.

A summary of almost 150 studies involving almost 50,000 respondents on the question of how children affect marital satisfaction revealed that parents (both women and men) reported lower marital satisfaction than nonparents. In addition, the higher the number of children, the lower the marital satisfaction; the factors that depressed marital satisfaction were conflict and loss of freedom.

What are several facts about parenthood?

Parenthood will involve about 40 percent of the time a couple live together, parents are only one influence on their children, each child is unique, and parenting styles differ. Research suggests that an authoritarian parenting style characterized by being both demanding and warm is associated with positive outcomes. In addition, being emotionally connected to the child, respecting his or her individuality, and monitoring the child's behavior to encourage positive contexts have positive outcomes.

What are the issues of single parenting?

About 40 percent of all children will spend one-fourth of their lives in a female-headed household. The challenges of single parenthood include taking care of the emotional and physical needs of a child alone, meeting one's own adult emotional and sexual needs, money, and rearing a child without a father (the influence of whom can be positive and beneficial).

What are some of the principles of effective parenting?

Giving time, love, praise, and encouragement; monitoring the activities of one's child; setting limits; encouraging responsibility; and providing sexuality education are aspects of effective parenting.

What are five theoretical approaches to childrearing?

There are several approaches to childrearing, including the developmental-maturational approach (children are influenced by their genetic inheritance), behavioral approach (consequences influence the behaviors children learn), socioteleological approach (children seek to gain attention from parents to overcome their feelings of powerlessness), and attachment parenting (the most important goal of parents is to become emotionally attached to their children).

baby blues

behavioral approach

demandingness

developmental-maturational
    training approach

oxytocin

parent effectiveness training

parenting

postpartum depression

responsiveness

socioteleological approach

time-out

transition to parenthood

The Companion Website for Choices in Relationships: An Introduction to Marriage and the
    Family, Ninth Edition
    www.thomson.edu/sociology/Knox

Supplement your review of this chapter by going to the companion website to take one of the
Tutorial Quizzes, use the flash cards to master key terms, and check out the many other
study aids you'll find there. You'll also find special features such as the Marriage and
Family Resource Center, Census 2000 information, and other data and resources at your
fingertips to help you with that special project or to do some research on your own.

Attachment Parenting International
    http://www.attachmentparenting.org/

Children, Youth and Family Consortium
    http://www.cyfc.umn.edu/

The Children's Partnership Online
    http://www.childrenspartnership.org

Mayberry USA
    http://www.mbusa.net/

Parenthood.com
    http://www.parenthoodweb.com/

REFERENCES

Aldous, J. and G. M. Mulligan. 2002. Father's childcare and children's behavior problems: A longi-
    tudinal study. *Journal of Family Issues* 23:624–47.

Amato, P. R. and F. Fowler. 2002. Parenting practices, child adjustment, and family diversity. *Journal
    of Marriage and the Family* 64:703–16.

Arbona, C., and T. G. Power. 2003. Parental attachment, self-esteem, and antisocial behaviors
    among African American, European American, and Mexican American adolescents. *Journal of
    Counseling Psychology* 50:40–51.

Ball, H. L. 2002. Reasons to bed-share: Why parents sleep with their infants. *Journal of Reproductive
    and Infant Psychology* 20:207–22.

Barber, B. K., Stolz, H. E., and Olsen, J. A. 2005. Parental support, psychological control, and be-
    havioral control: Assessing relevance across time, culture, and method. *Monographs of the Society
    for Research in Child Development,* 70(4):1–137.

Baumrind, D. 1966. Effects of authoritative parental control on child behavior. *Child Development*
    37:887–907.

Behnke, A. 2004. Latino dads: Structural inequalities and personal strengths. *NCFR Family Focus
    Report* 49(3):F6–F7.

Blankemeyer, M., C. O'Malley, and K. Walker. 2004. The War in Iraq: What are children learning?
    Poster, National Council on Family Relations Annual Meeting, Orlando, Florida, Nov. 17–22.

Bock, J. D. 2000. Doing the right thing? Single mothers by choice and the struggle for legitimacy.
    Gender and Society 14:62–86.

Booth, C. L., K. A. Clarke-Stewart, D. L. Vandell, K. McCartney, and M. T. Owen. 2002. Child-care
    usage and mother-infant "quality time." *Journal of Marriage and the Family* 64:16–26.

Bost, K. K., M. J. Cox, M. R. Burchinal, and C. Payne. 2002. Structural and supporting changes in
    couples' family and friendships networks across the transition to parenthood. *Journal of
    Marriage and the Family* 64:517–31.

British Columbia Reproductive Mental Health Program. 2005. Reproductive mental health:
    Psychosis. Retrieved June 15, 2005, from http://www.bcrmh.com/disorders/psychosis.htm.

Bushman, B. J., and J. Cantor. 2003. Media ratings for violence and sex. *American Psychologist*
    58:130–41.

Cassidy, G. L., and L. Davies. 2003. Explaining gender differences in mastery among married par-
    ents. *Social Psychology Quarterly* 66:48–61.

Castruccci, B. C., J. F. Culhane, E. K. Chung, I. Bennett, and K. F. McCollum 2006. Smoking in
    pregnancy: Patient and provider risk reduction behavior. *Journal of Public Health Management &
    Practice* 12:68–76.

Clarke, J. I. 2004. The overindulgence research literature: Implications for family life educators.
    Poster, National Council on Family Relations, Annual Meeting, Orlando, Florida, November.

Coley, R. L., and J. E. Morris. 2002. Comparing father and mother reports of father involvement
    among low-income minority families. *Journal of Marriage and the Family* 64:982–97.

Cornelius-Cozzi, T. 2002. Effects of parenthood on the relationships of lesbian couples. *PROGRESS:
    Family Systems Research and Therapy* 11:85–94.

Corsaro, W. A. 1997. *The sociology of childhood.* Thousand Oaks, CA: Sage.

Curtin, S.C., and J. A. Martin. 2000. Preliminary data for 1999. *National vital statistics reports.* Hyattsville, MD: National Center for Health Statistics, 48(14).

Darling, N. 1999. Parenting style and its correlates. *ERIC Digest* ED427896.

Diamond, A., J. Bowes, and G. Robertson. 2006. Mothers' safety intervention strategies with toddlers and their relationship to child characteristics. *Early Child Development and Care* 176:271–84

Eggebeen, D. J. 2002. The changing course of fatherhood: Men's experiences with children in demographic perspective. *Journal of Family Issues* 23:486–506.

Elgar, F. J., J. Knight, G. J. Worrall, and G. Sherman. 2003. Attachment characteristics and behavioral problems in rural and urban juvenile delinquents. *Child Psychiatry and Human Development* 34:35–48.

Facer, J. and R. Day. 2004. Explaining diminished marital satisfaction when parenting adolescents. Poster, Annual Meeting National Council on Family Relations, Orlando, Florida.

Flouri, E., and A. Buchanan. 2003. The role of father involvement and mother involvement in adolescents' psychological well-being. *British Journal of Social Work* 33:399–406.

Galambos, N. L., E. T. Barker, and D. M. Almeida. 2003. Parents do matter: Trajectories of change in externalizing and internalizing problems in early adolescent. *Child Development* 74:578–95.

Gavin, L. E., M. M. Black, S. Minor, Y. Abel, and M. E. Bentley. 2002. Young, disadvantaged fathers' involvement with their infants: An ecological perspective. *Journal of Adolescent Health* 31:266–76.

Gesell, A., F. L. Ilg, and L. B. Ames. 1995. *Infant and child in the culture of today.* Northvale, NJ: Jason Aronson.

Gordon, T. 2000. *Parent effectiveness training: The parents' program for raising responsible children.* New York: Random House.

Green, S. E. 2003. "What do you mean 'what's wrong with her?'" Stigma and the lives of families of children with disabilities. *Social Science & Medicine* 57:1361–74.

Hauck, F. R., S. M. Herman, M. Donovan, C. M. Moore, S. Iyasu, E. Donoghue, R. H. Kirschner, and M. Willinger. 2003. Sleep environment and the risk of sudden infant death syndrome in an urban population: The Chicago Infant Mortality study. *Pediatrics* 111:1207–15.

Hill, N. E., and Bush, K. R. 2001. Relationship between parenting environment and children's mental health among African American and European American mothers and children. *Journal of Marriage and the Family,* 63(4):954–66.

Hollander, D. 2003. Teenagers with the least adult supervision engage in the most sexual activity. *Perspectives on Sexual and Reproductive Health* 35:106–23.

Jacobson, K. C., and L. J. Crockett. 2000. Parental monitoring and adolescent adjustment: An ecological perspective. *Journal of Research on Adolescence* 10:65–97.

Jayson, S. 2004. It's time to grow up—later. *USA Today.* Sept 30, D1.

Jenni, O.G., H. Z. Fuhrer, I. Iglowstein, L. Molinari, and R. H. Largo. 2005. A longitudinal study of bed sharing and sleep problems among Swiss children in the first ten years of life. *Pediatrics* 115:233–40.

Karnehm, Amy L. 2000. *Adolescent sexual initiation: Are parents powerless to delay it?* Paper, Annual Conference of the National Council on Family Relations, Minneapolis, November.

Kim-Cohen, J., T. E. Moffitt, A. Taylor, S. J. Pawlby, and A. Caspi. 2005. Maternal depression and children's antisocial behavior: Nature and nurture effects. *Archives of General Psychiatry* 62:173–82.

Knox, D. (with Kermit Leggett). 2000. *The divorced dad's survival book: How to stay connected with your kids.* Reading, MA: Perseus Books.

Knox, D. and Zusman, M. E. 2006. Relationship and sexual behaviors of a sample of 1027 university students. Unpublished data collected for this text. Department of Sociology, East Carolina University, Greenville, NC.

Lagace-Seguin, D. and M.R. Entremont. 2006. Less than optimal parenting strategies predict maternal low level depression beyond that of child transgressions. *Early Child Development and Care* 176:343–55.

Laird, R. D., G. S. Pettit, J. E. Bates, and K. A. Dodge. 2003. Parents' monitoring, relevant knowledge and adolescents' delinquent behavior: Evidence of correlated developmental changes and reciprocal influences. *Child Development* 74:752–63.

Larzelere, R. E., P. R. Sather, W. N. Schneider, D. B. Larson, and P. L. Pike. 1998. Punishment enhances reasoning's effectiveness as a disciplinary response to toddlers. *Journal of Marriage and the Family* 60:388–403.

Lengua, L. J., S. A. Wolchik, I. N. Sandler, and S. G. West. 2000. The additive and interactive effects of parenting and temperament in predicting problems of children of divorce. *Journal of Clinical Child Psychology* 29:232–44.

Logan, C., C. Y. Lovato, B. Moffat, J. A. Shoveller and R. A. Young. 2005. Sun protection as a family health project in families with adolescents. *Journal of Health Psychology* 10:333–44.

Longmore, M. A., W. D. Manning, and P. C. Giordano. 2001. Preadolescent parenting strategies and teens' dating and sexual initiation: A longitudinal analysis. *Journal of Marriage and the Family* 63:322–55.

Mauer, M. and M. Chesney-Lind. 2003. *Invisible punishment: The collateral consequences of mass imprisonment.* New York: New Press.

Mayall, B. 2002. *Toward a sociology of childhood.* Philadelphia, PA: Open University Press.

McBride, B. A., S. J. Schoppe, and T. R. Rane. 2002. Child characteristics, parenting stress, and parental involvement: Fathers versus mothers. *Journal of Marriage and Family* 64:998–1011.

McLanahan,, S. S. 1991. The long term effects of family dissolution. In *When families fail: the social costs,* edited by Brice J. Christensen, New York: University Press of America for the Rockford Institute, 5-26.

McLanahan, S. S. and K. Booth. 1989. Mother-only families: Problems, prospects, and politics. *Journal of Marriage and the Family* 51: 557-580

NICHD Early Child Care Research Network. 2003. Does quality of child care affect child outcomes at age 41.2? *Developmental Psychology* 39:451–69.

Newsbytes. 2001. Germans seek to centralize Internet content control. 31 August, NWSBO1, 24300e.

Novilla, M., B. Lelinneth, M. D. Barnes, N. G. DeLa Cruz, P. N. Williams, and J. Rogers 2006. Public health perspectives on the family. *Family and Community Health* 29:28–42.

Parker, L., Curtner-Smith, M.E., and Bavolek, S. 2004. Attachment Parenting International support group attendance and mother's parenting attitudes, values, and knowledge. Poster, 66th Annual Conference, National Council on Family Relations, Orlando, Florida, November 17–20.

Paschall, M. J., M. Bersamin, and R. L. Flewelling. 2005. Racial/ethnic differences in the association between college attendance and heavy alcohol use: a national study. *Journal of Studies on Alcohol* 66:266–75.

Pinquart, M., and R. K. Silbereisen. 2002. Changes in adolescents' and mothers' autonomy and connectedness in conflict discussions: An observation study. *Journal of Adolescence* 25:509–22.

Pong, S. L. and B. Dong. 2000. The effects of change in family structure and income on dropping of out of middle and high school. *Journal of Family Issues* 21: 147-169.

Remez, L. 2003. Mothers exert more influence on timing of first intercourse among daughters than among sons. *Perspectives on Sexual and Reproductive Health* 35:55–56.

Sammons, L. 2003. Personal communication, Grand Junction, Colorado.

Sears, W. and M. Sears. 1993. *The Baby Book.* Boston: Little, Brown.

Shellenbarger, S. 2006. Helicopter parents go to work: Moms and dads are now hovering at the office. *The Wall Street Journal.* Thursday, March 16:D1.

Shields, B. 2005. *Down came the rain: My journey through postpartum depression.* New York: Hyperion.

Smetana, J. G. 2000. Middle-class African American adolescents' and parents' conceptions of parental authority and parenting practices: A longitudinal investigation. *Child Development,* 71(6):1672–86.

Skinner, B. F., P. B. Dews, C. B. Ferster, C. D. Cheney, and W. H. Morse. 1997. *Schedules of reinforcement.* New York: Paul and Co. Publications Consortium.

Soltz, V. and R. Dreikurs. 1991. *Children: The challenge.* New York: Penguin.

Stanley, S. M. and H. J. Markman. 1992. Assessing commitment in personal relationships. *Journal of Marriage and the Family* 54:595–608.

Statistical Abstract of the United States: 2006. 125th ed. Washington, DC: U.S. Bureau of the Census.

Straus, M. A. 1994. *Beating the devil out of them: Corporal punishment in American families.* San Francisco: Jossey-Bass.

Strom, R. D., T. E. Beckert, P. S. Strom, S. K. Strom, and D. L. Griswold. 2002. Evaluating the success of Causasian fathers in guiding adolescents. *Adolescence* 37:131–49.

Sugarman, S. D. 2003. Single-parent families. In *All our families, 2nd ed, New policies for a new century,* edited by M. A. Mason, A. Skolnick, and S. D. Sugarman. New York: Oxford University Press, 14–39.

Troth, A., and C. C. Peterson. 2000. Factors predicting safe-sex talk and condom use in early sexual relationships. *Health Communication* 12:195–218.

Tucker, C. J., S. M. McHale, and A. C. Crouter.2003. Dimensions of mothers' and fathers' differential treatment of siblings: Links with adolescents' sex-typed personal qualities. *Family Relations* 52:82–89.

Twenge, J. M., W. K. Campbell, and C. A. Foster. 2003. Parenthood and marital satisfaction: A meta-analytic review. *Journal of Marriage and Family* 65:574–83.

Wark, M.J., T. Kruczek, and A. Boley. 2003. Emotional neglect and family structure: Impact on student functioning. *Child Abuse and Neglect* 27:1033–43.

Webb, F. J. 2005. The new demographics of families. In *Sourcebook of family theory & research,* edited by Vern L. Bengtson, Alan C. Acock, Katherine R. Allen, Peggye Dilworth-Anderson, and David M. Klein. Thousand Oaks, California: Sage Publications, 101–02.

Wilson, L. C., C. M. Wilson, and L. Berkeley- Caines. 2003. Age, gender and socioeconomic differences in parental socialization preferences in Guyana. *Journal of Comparative Family Studies* 34:213–311.

chapter 12

If I returned home and had two "urgent" messages on my answering machine—one from CNN and one from my wife— guess which call I would return first?

Larry King, of his first six marriages

Balancing Work and Family Life

Contents

Authors

True or False?

1. Young attorneys who take time out to have and to rear their children are just as likely to eventually make partner in their firm as attorneys who do not take out time to have/rear children.

2. Most women with young children would rather be home with their children than be employed.

3. An employed unhappy wife is more likely to divorce than an employed happy wife.

4. When the job interferes with family life, both husbands and wives report decreased job satisfaction.

5. "Leisure activities" was the number one problem reported in a study on elderly couples.

Answers: **1.** F **2.** T **3.** T **4.** F **5.** T

"I slept, and dreamed that life was beauty; I awoke, and found that life was duty," wrote Ellen Sturgis Hooper. Her quote reflects the reality that although new couples enjoy the dream of love, soon they awake to the life of duty—work. This chapter is about that delicate balance between enjoying one's love relationship and working to earn money to sustain one's life/lifestyle.

This chapter is based on the premise that families are organized around work—where the couple lives is determined by where the spouses/parents can get jobs. What time they eat, which family members eat with whom, when they go to bed, and when, where, and for how long they vacation are all influenced by the job. We also emphasize that involvement in the workforce impacts the power and roles of family members. We begin this chapter by examining the meanings of money.

Meanings of Money

Trading your hours for a handful of dimes.

The Doors

Because spouses work for money to pay the bills, money becomes an important resource. Symbolic interactionists emphasize the social meanings associated with money, including security and avoiding poverty; self-esteem; power in relationships; love; and conflict.

Security/Economic Success

Success is a terrible attack on your sense of values. You get teed off because the heater of your swimming pool doesn't work.

Rod Serling, *The Twilight Zone*

College students want financial security. In a national survey of more than 263,000 first-year undergraduates at over 400 2- and 4-year universities in the United States, "being well off financially" was identified by almost three-quarters (74.5%) of the students as a top life goal (American Council on Education and University of California, 2005–2006). Table 12.1 reflects the reality in terms of the percentage of U.S. families at various income levels.

Poverty has traditionally been defined as the lack of resources necessary for material well-being. The lack of resources that leads to hunger and physical deprivation is known as **absolute poverty.** Poverty is devastating to families and particularly to children. Those living in poverty have poorer physical and mental health, report lower personal/marital satisfaction, and die sooner. The realities of poverty were examined by Barbara Ehrenreich (2001), a PhD journalist who ex-

Chapter 12 Balancing Work and Family Life

Table 12.1 Distribution of Income Level in U.S. Families

Income Level of Family	Percentage at This Level
Less than $15,000	9.6
$15,000–$24,999	11.1
$25,000–$34,999	11.4
$35,000–$49,999	15.0
$50,000–$74,999	20.1
$75,000–$99,999	13.3
$100,000 or more	19.4

Median family income = $52,680.

Source: *Statistical Abstract of the United States: 2006.* 125th ed. Washington, DC: U.S. Bureau of the Census, Table 686.

perienced minimum-wage life as a waitress cleaning tables, a maid scrubbing showers in motel rooms, and an employee restacking clothes in the women's wear department at Wal-Mart. She worked for $7 an hour at the latter and discovered that it takes two jobs to afford a decent place to live "and still have time to shower between them" (p. 39). These roles are held by the working poor, who "neglect their own children so the children of others will be cared for; they live in sub-standard housing so that other homes will be shiny and perfect. . . ." (p. 221).

What is the definition of poverty in terms of actual dollars? Table 12.2 reflects the Department of Health and Human Services' various poverty level guidelines by size of family and where the family lives. A significant proportion of families in the United States continue to be characterized by unemployment and low wages.

Poverty is no shame, but being ashamed of it is.

Benjamin Franklin

Self-Esteem

Money affects self-esteem because in our society human worth, particularly for men, is often equated with financial achievement. Persons who make $10,000 a year may think of themselves as worth less than those who make $100,000 a year. Persons inadvertently compare themselves with those who make more and less

Table 12.2 2006 HHS Poverty Guidelines

Size of Family Unit	48 Contiguous States, District of Columbia, and Outlying Jurisdictions	Alaska	Hawaii
1	$14,700	$18,375	$16,905
2	$19,800	$24,750	$22,770
3	$24,900	$31,125	$28,635
4	$30,000	$37,500	$34,500
5	$35,100	$43,875	$40,365
6	$40,200	$50,250	$46,230
7	$45,300	$56,625	$52,095
8	$50,400	$63,000	$57,960

For family units with more than 8 members, add the following amount for each additional family member: $5,100 for the 48 contiguous states, the District of Columbia, and outlying jurisdictions; $6,375 for Alaska; and $5,865 for Hawaii.

A person whose family's taxable income for the preceding year did not exceed 150 percent of the poverty level amount is considered low-income. The figures shown under family income represent amounts equal to 150 percent of the family income levels established by the Census Bureau for determining poverty status.

The poverty guidelines (effective February 2006 and until further notice) were published by the U.S. Department of Health and Human Services in the Federal Register, January 24, 2006, 71(15):3848–49.

money. Individuals also evaluate themselves on the degree to which they work. Alfred C. Kinsey, the famous sex researcher, is known for his work ethic and ignoring the advice of his physicians. "If I can't work, I would rather die," was his mantra. He died at age 62.

Power in Relationships

Money is a central issue in relationships because of its association with power, control, and dominance. Generally, the more money a partner makes, the more power that person has in the relationship. Males make considerably more money than females and generally have more power in relationships.

National Data

The average annual income of a male with some college who is working full-time is $46,332, compared with $31,655 for a female with the same education, also working full-time, year-round (*Statistical Abstract of the United States 2006*, Table 686).

Power is a tricky thing: first you use it, then you abuse it, and finally you lose it.

Unknown

When the wife earns an income, her power increases in the relationship. The authors know of a married couple in which the wife has recently begun to earn an income. Before doing so, her husband's fishing boat was in the protected carport. With her new job, and increased power in the relationship, her car is parked in the carport and her husband's fishing boat is sitting underneath the trees to the side of the house. Money also provides the employed woman the power to be independent and to leave an unhappy marriage. Indeed, the higher a wife's income, the more likely she is to leave an unhappy relationship (Schoen et al., 2002). Similarly, since adults are generally the only source of money in a family, they have considerable power over children, who have no money.

To some individuals, money also means love. While admiring the engagement ring of her friend, a woman said, "What a big diamond! He must really love you." The assumption is that a big diamond equals a high price equals deep love.

Similar assumptions are often made when gifts are given or received. People tend to spend more money on presents for the people they love, believing that the value of the gift symbolizes the depth of their love. People receiving gifts may make the same assumption. "She must love me more than I thought," mused one man. "I gave her a DVD for Christmas, but she gave me a DVD player. I felt embarrassed."

A large diamond ring is regarded as signifying "much love"—hence the association of love and money.

Chapter 12 Balancing Work and Family Life

Conflict

Money can also be a source of conflict in relationships. Couples argue about what to spend money on (new car? vacation? credit card debt?) and how much money to spend. One couple in marriage therapy reported that they argued over whether to buy a new air conditioner for their car. The husband thought it necessary; the wife thought they could roll down the windows. As noted earlier, conflicts over money in a relationship often signify conflict over power in the relationship.

Marital conflicts may also arise over religious differences—how much to give to the church and how money may be spent. One employed wife in the authors' classes said her husband, an evangelical Christian, told her she could not buy beer. They fought vigorously over this issue with the result that she ended up buying beer—"but it's a problem in our marriage," she said.

Dual-Earner Marriages: Considerations for Spouses

Coontz (1997) reviewed the woman in the workforce in the fifties. She noted that, in 1950, "only a quarter of all wives were in the paid labor force, and just 16 percent of all children had mothers who worked outside the home" (p. 51). In contrast to today, driven primarily by economic need, 61 percent of all U.S. wives are in the labor force. Most have children. Indeed, wives most likely to be in the labor force have children between the ages of 14 and 17 (79% of wives), when their children are relatively independent and the food and clothing expenses of teenagers are highest (*Statistical Abstract of the United States: 2006,* Table 587). The stereotypical family consisting of the husband who earns the income and the wife who stays at home with two or more children is no longer the norm.

Because women still take on more child-care and household responsibilities than men, women in dual-earner marriages are more likely than men to want to be employed part-time rather than full-time. If this is not possible, many women prefer to work only a portion of the year (the teaching profession allows employees to work about 10 months and to have 2 months in the summer free). Although many low-wage earners need two incomes to afford basic housing and a minimal standard of living, others have two incomes to afford expensive homes, cars, vacations, and educational opportunities for their children. Whether it makes economic sense is another issue.

Cost of the Wife Working Outside the Home

Some parents wonder if the money a wife earns by working outside the home is worth the sacrifices to earn it. Not only is the mother away from her children but also she must *pay* for strangers to care for her children. Sefton (1998) calculated that the value of the stay-at-home mother in terms of what a dual-income family spends to pay for all services that she provides (domestic cleaning, laundry, meal planning and preparation, shopping, providing transportation to child's activities, taking the children to the doctor, and running errands) is $36,000 per year. Adjusting for changes in the Consumer Price Index, this figure would be $44,291 in 2006.

This figure suggests working outside the home may not be as economically advantageous as one might think—that women may work outside the home for psychological (enhanced self-concept) and social (enlarged social network) benefits. And some may enjoy a lifestyle that is made possible by earning a significant income by outside employment. One wife noted that the only way she could afford the home she wanted was to help earn the money to pay for it.

There is no such thing as a non-working mother.

Hester Mundis, comedy writer for Joan Rivers

Feminists have challenged the term "working mother" with the slogan "every mother is a working mother," which affirms unpaid domestic labor as work.

M. Rivka Polatnick, Center for Working Families

The **mommy track** (stopping paid employment to spend time with young children) is another potential cost to those women who want to build a career. Taking time out to rear children in their formative years can derail a career. Noonan and Corcoran (2004) found that lawyers who took time out for child responsibilities were less likely to make partner and more likely to earn less money if they did make partner. Aware that executive women have found it difficult to re-enter the work force after being on the mommy track, the Harvard Business School created an executive training program to improve their technical skills and to help them return to the work force (Rosen, 2006).

Types of Dual-Career Marriages

A **dual-career marriage** is defined as one in which the spouses both pursue careers and maintain a life together that may or may not include dependents. A career is different from a job in that the former usually involves advanced education or training, full-time commitment, working nights and weekends "off the clock," and a willingness to relocate. Dual-career couples operate without a "wife"—a person who stays home to manage the home and care for dependents.

Nevertheless, three types of dual-career marriages are those in which the husband's career takes precedence, the wife's career takes precedence, or both careers are regarded equally. These career types are sometimes symbolized as HIS/her, HER/his, HIS/HER, and THEIR career.

When couples hold traditional gender role attitudes, the husband's career is likely to take precedence **(HIS/her career).** This situation translates into the wife's being willing to relocate and to disrupt her career for the advancement of her husband's career. In this arrangement, the wife may also have children early, which has an effect on the development of her career. Gordon and Whelan-Berry (2005) interviewed 36 professional women and found that in 22 percent of the marriages, the husband's career took precedence. The primary reasons for this arrangement were that the husband earned a higher salary, it was easier to go where the husband could earn the highest income since the wife could more easily find a job wherever he went (than vice versa), and ego needs (the husband needed to have the dominant career). This arrangement is sometimes at the expense of the wife's career. Mason and Goulden (2004) studied women in academia and found that those who had a child within 5 years of earning their PhD were less likely to achieve tenure.

For couples who do not have traditional gender role attitudes, the wife's career may take precedence **(HER/his career).** Almost 20 percent (19%) of the marriages in the Gordon and Whelan-Berry study (2005) mentioned above could be categorized as giving precedence to the wife's career. In such marriages, the husband is willing to relocate and to disrupt his career for his wife's. Such a pattern is also likely to occur when the wife earns considerably more money than her husband. In some cases the husband who is downsized or who prefers the role of full-time parent will become Mr. Mom. Over 100,000 husbands (and parents of at least one child under the age of six) are married to wives who work full time in the labor force (Tucker, 2005). "Mr. Moms" have increased almost 30 percent in the past 10 years ((*USA Today*, 2005); almost 80 percent of men in Europe report that they would be happy to stay at home with the kids (Pepper, 2006). A major advantage of men assuming this role is a more lasting emotional bond with their children (Tucker, 2005).

When the careers of both the wife and husband are given equal status in the relationship **(HIS/HER career),** they may have a commuter marriage in which they follow their respective careers wherever they lead. Alternatively, they hire domestic/child-care help so that neither spouse functions in the role of the wife. Almost sixty percent (58%) of the dual-career relationships studied by Gordon and Whelan-Berry (2005) gave equal precedence to the husband and wife's careers. In some cases a university may hire a husband/wife PhD pair (in the same

This man has a full-time career as an accountant; his wife is a PhD English Professor. They both have demanding careers.

Authors

or different departments) as a way of retaining both. The professors also benefit by having their vacation at the same time, not to speak of the advantage of scheduling their classes at different times so that one parent can be available to their children.

Finally, some couples have the same career (**THEIR career**) and may work together. Some news organizations hire both spouses to travel abroad to cover the same story. These careers are rare. In the following sections we look at the effects on women, men, their marriage, and their children when both spouses are employed outside the home.

Effects of the Wife's Employment on the Wife

A woman's work satisfaction is related to whether she has children, the age of her children, the nature of her work, and the support of her husband. In general, women without children or with older children who work in jobs they enjoy and are married to egalitarian husbands who support their employment are much happier than their counterparts. Employed women in dual-earner households who perceive inequality in the division of labor are less likely to feel happy in their marriage and are more likely to divorce.

Whether a wife is satisfied with her job is also related to the degree to which the job takes a toll on her family life. Grandey et al. (2005) found that when the wife's

A woman can have both a career and a home—if she knows how to put both of them first.

Evan Esar

employment interfered with her family life, the wife reported less job satisfaction. Such a relationship with husbands was not found—jobs that interfered with family life did not result in decreased job satisfaction for men. Consistent with this finding, Kiecolt (2003) noted that most women with young children much prefer to be at home and view home, not work, as their primary haven of satisfaction.

Indeed, family and work become spheres to manage, and sometimes there is **role overload**—not having the time or energy to meet the demands of one's role (wife/parent/worker) responsibilities. Because women have traditionally been responsible for most of the housework and childcare, employed women come home from work to what Hochschild (1989) calls the **second shift:** the housework and childcare that employed women do when they return home from their jobs.

> As a result, women tend to talk more intently about being overtired, sick, and "emotionally drained." Many women could not tear away from the topic of sleep. They talked about how much they could "get by on" . . . six and a half, seven, seven and a half, less, more. . . . Some apologized for how much sleep they needed. . . . They talked about how to avoid fully waking up when a child called them at night, and how to get back to sleep. These women talked about sleep the way a hungry person talks about food. (p. 9)

Another stressful aspect of employment for employed mothers in dual-earner marriages is **role conflict**—being confronted with incompatible role obligations. For example, the role of the employed mother is to stay late and get a report ready for tomorrow. But the role of the mother is to pick up the child from day care at 5 P.M. When these roles collide, there is role conflict. Although most women resolve their role conflict by giving preference to the mother role, some give priority to the career role and feel guilty about it. Mary Tyler Moore wrote in her biography that she spent more time with her TV son than her real son.

Role strain, the anxiety that results from being able to fulfill only a limited number of role obligations, occurs for both women and men in dual-earner marriages. There is no one at home to take care of housework and children while they are working, and they feel strained at not being able to do everything.

Effects of the Wife's Employment on Her Husband

Husbands also report benefits from their wives' employment. These include being relieved of the sole responsibility for the financial support of the family and having more freedom to quit jobs, change jobs, or go to school. Traditionally men had no options but to work full-time. Some may enjoy the role of Mr. Mom.

Men also benefit by having a spouse with whom to share the daily rewards and stresses of employment. And, to the degree that women find satisfaction in their work role, men benefit by having a happier partner. Finally, men benefit from a dual-earner marriage by increasing the potential to form a closer bond with their children through active childcare. Some prefer the role of househusband and stay-at-home dad. Patrick (2005) confirmed that stay-at-home dads have a stronger emotional bond with their children than traditional working dads. Though such dads are clearly the minority, their visibility is increasing.

However, not all husbands want their wives to be employed. Indeed, violence may erupt if the wife has a job, particularly if the husband is unemployed (Macmillan and Gartner, 2000). Some husbands of employed wives may also resent the fact that they are expected to provide more childcare (including taking children to various events/medical care), to vacuum/clean the house, and to prepare meals.

Effects on the Couple's Marriage of Having Two Earners

Are marriages in which both spouses earn an income more vulnerable to divorce? Not if the wife is happy; but if she is unhappy, her income will provide her a way to take care of herself when she leaves. Schoen et al. (2002) write, "Our results provide clear evidence that, at the individual level, women's employment does not destabilize happy marriages but increases the risk of disruption in unhappy marriages" (p. 643). Hence employment won't affect a happy marriage but it can do in an unhappy one. Further research by Schoen et al. (2006) revealed that full-time employment of the wife is actually associated with marital stability.

Couples may be particularly vulnerable when the wife earns more money than her husband. Thirty percent of working wives earn more than their husbands (Tyre and McGinn, 2003). This situation affects the couple's marital happiness in several ways. Marriages in which the spouses view the provider role as the man's responsibility, in which the husband cannot find employment, and in which the husband is jealous of his wife's employment are vulnerable to dissatisfaction. Cultural norms typically dictate that the man is supposed to earn more money than his partner and that something is wrong with him if he doesn't. Jalovaara (2003) found that the divorce rate was higher among couples where the wife's income exceeded her husband's. On the other hand, some men want an ambitious, economically independent woman who makes a high income. Over a third of men in a national *Newsweek Poll* said that if their wife earned more money, they'd consider quitting their job or reducing their hours (Tyre and McGinn, 2003).

Do the hours worked in terms of night or weekend make any difference? Yes. Strazdins et al. (2006) studied over 4000 families and compared families where parents worked standard weekday times with those where parents worked nonstandard schedules and found that couples working nonstandard schedules reported worse family functioning, more depressive symptoms, and less effective parenting. Their children were also more likely to have social and emotional difficulties.

Dual-Earner Marriages: Considerations for Children

Independent of the effect on the wife, husband, and marriage, what is the effect of the wife earning an income outside of the home? Individuals disagree on the effects of maternal employment on children. The Self-Assessment on page 356 provides a way to assess your beliefs in this regard.

Effects of the Wife's Employment on the Children

Mothers with young children are the least likely to be in the labor force. Table 12.3 reflects the percentage of employed mothers as related to the age of the child.

Table 12.1 Percentage of Employed Mothers, by Age of Child

Age of Child, Years	Percentage of Employed Mothers
≤1	54.7
2	61.3
3	63.6
4	64.1
5	66.9
6–13	75.1
14–17	81.2

Maternal Employment Scale

Directions

Using the following scale, please mark a number on the blank next to each statement to indicate how strongly you agree or disagree with it.

1	2	3	4	5	6
Disagree Very Strongly	Disagree Strongly	Disagree Slightly	Agree Slightly	Agree Strongly	Agree Very Strongly

_____ 1. Children are less likely to form a warm and secure relationship with a mother who is working full-time outside the home.

_____ 2. Children whose mothers work are more independent and able to do things for themselves.

_____ 3. Working mothers are more likely to have children with psychological problems than mothers who do not work outside the home.

_____ 4. Teenagers get into less trouble with the law if their mothers do not work full-time outside the home.

_____ 5. For young children, working mothers are good role models for leading busy and productive lives.

_____ 6. Boys whose mothers work are more likely to develop respect for women.

_____ 7. Young children learn more if their mothers stay at home with them.

_____ 8. Children whose mothers work learn valuable lessons about other people they can rely on.

_____ 9. Girls whose mothers work full-time outside the home develop stronger motivation to do well in school.

_____10. Daughters of working mothers are better prepared to combine work and motherhood if they choose to do both.

_____11. Children whose mothers work are more likely to be left alone and exposed to dangerous situations.

_____12. Children whose mothers work are more likely to pitch in and do tasks around the house.

_____13. Children do better in school if their mothers are not working full-time outside the home.

_____14. Children whose mothers work full-time outside the home develop more regard for women's intelligence and competence.

_____15. Children of working mothers are less well-nourished and don't eat the way they should.

_____16. Children whose mothers work are more likely to understand and appreciate the value of a dollar.

_____17. Children whose mothers work suffer because their mothers are not there when they need them.

_____18. Children of working mothers grow up to be less competent parents than other children because they have not had adequate parental role models.

_____19. Sons of working mothers are better prepared to cooperate with a wife who wants both to work and have children.

_____20. Children of mothers who work develop lower self-esteem because they think they are not worth devoting attention to.

_____21. Children whose mothers work are more likely to learn the importance of teamwork and cooperation among family members.

_____22. Children of working mothers are more likely than other children to experiment with alcohol, other drugs, and sex at an early age.

_____23. Children whose mothers work develop less stereotyped views about men's and women's roles.

_____24. Children whose mothers work full-time outside the home are more adaptable: they cope better with the unexpected and with changes in plans.

Scoring

Items 1, 3, 4, 7, 11, 13, 15, 17, 18, 20, and 22 refer to "costs" of maternal employment for children and yield a Costs Subscale score. High scores on the Costs Subscale reflect strong beliefs that maternal employment is costly to children. Items 2, 5, 6, 8, 9, 10, 12, 14, 16, 19, 21, 23, and 24 refer to "benefits" of maternal employment for children and yield a Benefits Subscale score. To obtain a Total Score, reverse score all items in the Benefits Subscale so that $1 = 6, 2 = 5, 3 = 4, 4 = 3, 5 = 2$, and $6 = 1$. The higher one's Total Score, the more one believes that maternal employment has negative consequences for children.

Source

E. Greenberger, W. A. Goldberg, T. J. Crawford, and J. Granger. Beliefs about the consequences of maternal employment for children. *Psychology of Women*, 1988, 12, 35–59. Used by permission of Blackwell Publishing.

Data from the National Survey of America's Families indicate that most children under the age of 5 whose parents are in the work force are in a nonparental-care arrangement. For whites, blacks, and Hispanic children, the proportions receiving such care are 87, 81, and 80 percent, respectively (Capizzano et al., 2006).

Dual-earner parents want to know how children are affected by maternal employment. An abundance of research has resulted in the finding of few negative effects (Perry-Jenkins et al., 2001). Hence children do not appear to suffer cognitively or emotionally as long as positive, consistent child-care alternatives are in place. However, less supervision of children by parents is an outcome of having two-earner parents. Leaving children to come home to an empty house is particularly problematic. In addition, in a study of 2246 adults who had lived with two biological parents until age 16, those whose mothers had been employed during most of their childhood reported less support and less discipline from both parents than those who had stay-at-home mothers (Nomaaguchi and Milkie, 2006).

Self-Care/Latchkey Children   Some dual-earner and single-parent families do not have the resources to afford childcare while the parent is working. Though relatives and friends may help, children whose mothers work over 30 hours a week are vulnerable to being unsupervised by an adult for some time after school (Lopoo, 2005). Although these self-care, or latchkey, children often fend for themselves very well, some are at risk. Over 230,000 are between the ages of 5 and 7 and are vulnerable to a lack of care in case of an accident or emergency. Children who must spend time alone at home should know the following:

1. How to reach their parents at work (the phone number, extension number, and name of the person to talk to if the parent is not there)

2. Their home address and phone number in case information must be given to the fire department or an ambulance service

3. How to call emergency services, such as the police and fire departments

4. The name and number of a relative or neighbor to call if the parent is unavailable

5. To keep the door locked and not let anyone in

6. Not to tell callers their parents are not at home; instead, they should tell them their parents are busy or can't come to the phone

7. Not to play with appliances, matches, or the fireplace

Parents should also consider the relationship of the children they leave alone. If the older one terrorizes the younger one, the children should not be left alone. Also, if the younger one is out of control, it is inadvisable to put the older one in the role of being responsible for the child. If something goes wrong (there is a serious accident), the older child may be unnecessarily burdened with guilt.

Day-Care Considerations

Parents going into the workforce or returning to the workforce is predicated on finding high-quality day care (Hand, 2005).

Quality of Day Care   More than half of U.S. children are in center-based childcare programs.

Parents are sometimes apprehensive about dropping off their child at day care. Essentially, they are leaving their children with paid strangers.

© Michael Newman/PhotoEdit

Most mothers prefer that the day care arrangement for their children is with relatives. Researchers Gordon and Hognas (2006) analyzed data from the National Institute of Child Health and Human Development Study of Early Child Care and found that nearly two thirds of the mothers reported such a preference for relatives, including 15 percent who preferred their spouse or partner, 28 percent who preferred another relative in their home, and 22 percent who preferred another relative in another home. An additional 16 percent preferred care by a nonrelative in their own home and 11 percent care by a nonrelative in another private home. Just 9 percent expressed a preference for center-based care. However, most children end up in center-based child care programs.

National Data

Forty-three percent of 3-year-olds and 73% of 5-year-old children are in center-based day-care programs (*Statistical Abstract of the United States 2006*, Table 568). Data from the National Survey of America's Families indicate that 44 percent of black, 32 percent of white, and 20 percent of Hispanic children whose parents are employed are in a center-based form of day care (Capizzano et al., 2006).

Employed parents are concerned that their children get good-quality care. Their concern is warranted. Warash (2005) reviewed the literature on day-care centers and found that the average quality of such centers is mediocre—"unsafe, unsanitary, non-educational, and inadequate in regard to the teacher-child ratio for a classroom." Care for infants was particularly lacking. Of 225 infant or toddler rooms observed, 40 percent were rated less than minimal with regard to hygiene and safety. Because of the low pay and stress of the occupation, the rate of turnover for family child-care providers is very high (estimated at between 33 and 50 percent) (Walker, 2000). Some parents want to *see* the care their children are getting. To meet this need, some day-care centers have installed video cameras throughout their center and beam video to parents via the Internet (Vanoverbeke, 2006).

Ahnert and Lamb (2003) emphasized that any potential negative outcomes in day care can be mitigated by attentive, sensitive, loving parents. "Home remains the center of children's lives even when children spend considerable amounts of time in child care. . . . [A]lthough it might be desirable to limit the amount of time children spend in child care, it is much more important for children to spend as much time as possible with supportive parents" (pp. 1047–48).

Cost of Day Care   Day-care costs are a factor in whether a low-income mother seeks employment, since the cost can absorb her paycheck (Baum, 2002). Even for dual-earner families, cost is a factor in choosing a day-care center. Day care costs vary widely—from nothing, where friends trade off taking care of their children, to very expensive institutionalized day care in large cities. The authors e-mailed a dual-earner metropolitan couple (who use day care for their two children) to inquire about the institutional costs of day care in their area (Baltimore).

The response from the father follows:

There is a sliding scale of costs based on quality of provider, age of child, full-time or part-time. Full-time infant care can be hard to find in this area. Many people put their names on a waiting list as soon as they know they are pregnant. We had our first child on a waiting list for about nine months before we got him into our first choice of providers.

Infant care runs $225 per week for high-quality day care. That works out to $11,700 per year—and you thought college was expensive.

The cost goes down when the child turns 2—to around $150–$180 per week. In Maryland, this is because the required ratio of teachers to children gets bigger at *age 2. This cost break doesn't last long. Preschool programs (more academic in structure than day care) begin at age 3 or 4. Our second child turns 4 this May and begins an academic preschool in September. The school year lasts nine and a half months, and it costs about $9,000. Summer camps or summer day-care costs must be added to this amount to get the true annual cost for the child. My wife and I have budgeted about $20,000 total for our two children to attend private school and summer camps this year.*

The news only gets worse when the children get older. A good private school for grades K–5 runs $10,000 per school year. Junior high (6–8) is about $11,000, and private high school here goes for $12,000 to $18,000 per nine-month school year.

Religious private schools run about half the costs above. However, they require that you be a member in good standing in their congregation and, of course, your child undergoes religious indoctrination.

There is always the argument of attending a good public school and thus not having to pay for private school. However, we have found that home prices in the "good school neighborhoods" were out of our price range. Some of the better-performing public schools also now have waiting lists—even if you move into that school's district.

In most urban and suburban areas, cost is secondary to admissions. Getting into any good private school is hard. In order to get into the good high schools, it's best to be in one of the private elementary or middle schools that serves as a "feeder" school. Of course, to get into the "right" elementary school, you must be in a good feeder kindergarten. And of course to get into the right kindergarten, you have to get into the right feeder preschool. Parents have A LOT of anxiety about getting into the right preschool— because this can put your child on the path to one of the better private high schools.

Getting into a good private preschool is not just about paying your money and filling out applications. Yes, there are entrance exams. Both of our children underwent the following process to get into preschool: First you must fill out an application. Second, your child's day-care records/transcripts are forwarded for review. If your child gets through this screening, you and your child are called in for a visit. This visit with the child lasts a few hours and your child goes through evaluation for physical, emotional, and academic development.

Then you wait several agonizing months to see if your child has been accepted. Though quality day care is expensive, parents delight in the satisfaction that they are doing what they feel is best for their child.

The quality of your life is the quality of your relationships.

Anthony Robbins, self/relationship enhancement guru

I think it is corrupting to care about success; and nothing could be more vulgar than to worry about prosperity.

Orson Welles, actor and director

Balancing Demands of Work and Family

For many people, leisure is not an option. This man has worked at this position as clerk of a hotel for 10 years.

One of the major concerns of employed parents and spouses is how to juggle the demands of work and family simultaneously and achieve a sense of accomplishment and satisfaction in each area. Halpern (2005) noted that, "despite changes in the workforce, the world of work is still largely organized for a family model that is increasingly rare—one with a stay-at-home caregiver" (p. 397).

Cinamon (2006) noted that women experience higher levels of work interfering with family. Women are more likely to resolve the conflict by giving precedence to family. Kiecolt (2003) examined national data and concluded that employed women with young children are "more likely to find home a haven, rather than finding work a haven or having high work-home satisfaction" (p. 33). As Faith Hill, the country-music superstar, has said, "Family comes first, always—what else is there?"

Hochschild (1997) had previously noted that there was a "cultural reversal" under way whereby work was becoming more satisfying than home for women. Kiecolt found "no evidence" for this reversal. Indeed, increasingly mothers are quitting their jobs and returning home to rear their children. In effect there has been "a switch in loyalties from paycheck to playpen" (Clark, 2002). And Abowitz (2005) found that the primary characteristic sought in a job by a random sample of college students was that it "be interesting." Chances for advancement, job security, and recognition/respect were much less important.

Nevertheless, the conflict between work and family is substantial, and various strategies are employed to cope with the stress of role overload and role conflict, including (1) the superperson strategy, (2) cognitive restructuring, (3) delegation of responsibility, (4) planning and time management, and (5) role compartmentalization (Stanfield, 1998).

Superperson Strategy

The superperson strategy involves working as hard and as efficiently as possible to meet the demands of work and family. The person who uses the superperson strategy often skips lunch and cuts back on sleep and leisure in order to have more time available for work. Women are particularly vulnerable because they feel that if they give too much attention to child-care concerns, they will be sidelined into lower-paying jobs with no opportunities.

Hochschild (1989) noted that the term **superwoman** or **supermom** is a cultural label that allows the woman to regard herself as very efficient, bright, and confident. However, Hochschild noted that this is a "cultural cover-up" for an overworked and frustrated woman. Not only does the woman have a job in the workplace (first shift), she comes home to another set of work demands in the form of house care and childcare (second shift). Finally, there is the "third shift" (Hochschild, 1997).

The **third shift** is the expense of emotional energy by a spouse or parent in dealing with various issues in family living. An example is the emotional energy needed for children who feel neglected by the absence of quality time. While young children need time and attention, responding to conflicts and problems with teenagers also involves a great deal of emotional energy—the third shift. Judkins and Presser (2005) noted that wives make excuses for why their husbands participate less in sustainability practices such as sorting recyclables by saying that they are "willing to help." In addition, husbands justify their behavior by saying that wives "have greater competence, skill, and concern" and will do a better job.

Cognitive Restructuring

Another strategy used by some women and men experiencing role overload and role conflict is cognitive restructuring, which involves viewing a situation in positive terms. Exhausted dual-career earners often justify their time away from their children by focusing on the benefits of their labor—their children live in a nice house in a safe neighborhood and attend the best of schools. Whether these outcomes offset the lack of "quality time" may be irrelevant—the beliefs serve simply to justify the two-earner lifestyle.

Delegation of Responsibility/Limiting Commitments

A third way couples manage the demands of work and family is to delegate responsibility to others for performing certain tasks. Because women tend to bear most of the responsibility for childcare and housework, they may choose to ask their partner to contribute more or to take responsibility for these tasks. Two researchers found that spouses who help each other perceive that their relationships are more equitable and that women who receive more help from their husbands are less angry (Van Willigen and Drentea, 2001).

Parents may also involve their children in household tasks, which not only relieves the parents but also benefits children by requiring them to learn domestic skills and the value of contributing to the family. This may sound like a good idea, but when 20 married self-employed mothers (Mary Kay consultants) were interviewed, they reported that their children contributed little to the household labor, and only a few of the mothers mentioned that they would like for their children to do more (Berke, 2000).

Another form of delegating responsibility involves the decision to reduce one's current responsibilities and not take on additional ones. For example, women and men may give up or limit agreeing to volunteer responsibilities or commitments. One woman noted that her life was being consumed by the responsibilities of her church; she had to change churches since the demands were relentless. In the realm of paid work, women and men can choose not to become involved in professional activities beyond those that are required.

Planning and Time Management

The use of planning and time management is another strategy for minimizing the conflicting demands of work and family (Grzywacz and Bass, 2003). This involves prioritizing and making lists of what needs to be done each day. Time planning also involves trying to anticipate stressful periods, planning ahead for them, and dividing responsibilities with the spouse. Such division of labor allows each spouse to focus on an activity that needs to get done (grocery shopping, picking up children at day care) and results in a smoothly functioning unit.

Having flexible jobs and/or careers is particularly beneficial for two-earner couples. Being self-employed, telecommuting, or working in academia permits flexibility of schedule so that individuals can cooperate on what needs to be done.

Alternatively, some dual-earner couples attempt to solve the problem of childcare by **shift work**—having one parent work during the day and the other parent work at night so that one parent can always be with the children. Shift workers often experience sleep deprivation and fatigue, which may make it difficult for them to fulfill domestic roles as a parent or spouse. Similarly, shift work may have a negative effect on a couple's relationship because of their limited time together.

Presser (2000) studied the work schedules of 3476 married couples and found that recent (married less than 5 years) husbands who had children and who worked at night were six times more likely to divorce than husbands/parents who worked days.

Role Compartmentalization

Some spouses use **role compartmentalization**—separating the roles of work and home so that they do not think about or dwell on the problems of one when they are at the physical place of the other. Spouses unable to compartmentalize their work and home feel role strain, role conflict, and role overload, with the result that their efficiency drops in both spheres.

Although the various mechanisms identified in this section are helpful in coping with job demands, regular exercise is also effective. We discuss exercise in more detail in the chapter on stress.

Work-Family Fit versus Balance of Work and Family

Family and consumer science specialists differentiate between the concepts of work-family fit and balance of work/family (Clark et al., 2004). Work-family fit refers to the way in which the family integrates the demands of work and family

When my husband asks what I have made for dinner, I reply, "What I make best—reservations."
Lisa Emmott, former financial analyst

When two-career couples today are asked what is the biggest challenge they face, they answer "too little time."
Mark Hunter, journalist

I can't give a hundred percent at home because of my work; I can't give a hundred percent at work because of my home; I live in fear of the moments, like these, when the two collide.
Jodi Picoult, in her novel *Perfect Match*

Government and Corporate Work-Family Policies and Programs

Under the Family and Medical Leave Act, all companies with 50 or more employees are required to provide each worker with up to 12 weeks of unpaid leave for reasons of family illness, birth, or adoption of a child. In a subsequent amendment, the Family Leave Act permits states to provide unemployment pay to workers who take unpaid time off to care for a newborn child or a sick relative. Nevertheless, the United States still lags behind other countries in providing paid time off for new parents. For instance, Germany provides 14 weeks off with 100 percent salary.

Aside from government-mandated work-family policies, corporations and employers have begun to initiate policies and programs that address the family concerns of their employees. Employer-provided assistance with childcare, assistance with elderly parent care, options in work schedules, and job relocation assistance are becoming more common.

Examples of the rare companies providing such benefits include those provided to employees at SAS Institute, a computer software company in Cary, North Carolina, where employees may eat meals with family members and guests at the company cafeteria. The older children at the company preschool can join their parents for lunch. Another example is Pepsi-Co, Inc., in Purchase, New York, which has an on-site concierge to do personal chores for its 800 employees.

As noted, family-friendly work policies include flexible working hours, on-site day care, and paid leave. Such policies offer a way to help employees balance work and family and to promote a more committed workforce (Blair-Loy and Warton, 2004). In the absence of such policies, employees not only report lower job satisfaction but are more likely to quit or seek more flexible employers on the premise that their children will benefit (Kinnoin, 2005). One researcher (Estes, 2004) examined the question of whether various flexibility options offered by employees actually benefit children. The answer is indirectly. The children benefit to the degree that flexible hours, reduced hours, and workplace social support enhance the psychological well-being of the mother.

Other researchers have examined parental leave policies. Tanaka (2005) found that job-protected paid leave has enormous positive effects on reducing post-neonatal mortality. Spencer-Dawe (2005) studied single mothers and found that in spite of paid leave, day care, and other benefits, they and their children were particularly challenged.

Not all "family-friendly" corporate policies are welcome, particularly among persons who do not have children. These workers often feel exploited in that they are transferred more often, asked more often to work weekends, and asked to work longer hours. Since they do not have families, they are assumed to be more available.

Your Opinion?

1. Argue for and against the fact that businesses benefit from having family-friendly policies.
2. Argue for and against the theory that childfree workers should work later and on holidays so that parents can be with their children.
3. Why do you think that the United States lags behind other industrialized nations in terms of paid leave for parents?

Sources

Blair-Loy, M. and A. S. Warton. 2004. Organizational commitment and constraints on work-family policy use: Corporate flexibility policies in a global firm. *Sociological Perspectives* 47:243–67.

Estes, S. B. 2004. How are family-responsive workplace arrangements family friendly? Employer accommodations, parenting, and children's socioemotional well-being. *The Sociological Quarterly* 45:637–61.

Kinnoin, C. M. 2005. An examination of the relationship between family-friendly policies and employee job satisfaction, intent to leave, and organizational commitment. Dissertation. Nova Southeastern University, December, DAI-A66/06:2290.

Spencer-Dawe, E. 2005. Lone mothers in employment: Seeking rational solutions to role strain. *The Journal of Social Welfare and Family Law* 27:251–67.

Tanaka, S. 2005. Effects of parental leave on child health and development. Dissertation. Columbia University, April, DAI-A65/10.

patterns. The focus is on the number of hours worked, the income generated from the work, and the household labor that is still accomplished. Balance of work/family is more focused on the frequency of family activities. A spouse/parent whose work schedule preludes time with the family reflects an unbalanced work/family outcome. The researchers found that job satisfaction and marital satisfaction predict both fit and balance; individuals having both job satisfaction and marital satisfaction tend to have achieved the fit with the work world that achieves the requisite number of hours to meet the demands of the job yet still have time to spend with the family. We noted earlier that when the wife's employment interfered with her family life, she reported that her job satisfaction decreased (Grandey et al., 2005).

Balancing Work and Leisure Time

The workplace has become the home place. Because of the technology of the workplace—laptops, cell phones, blackberries, etc.—some spouses work all the time, wherever they are. One couple noted that when they are at home they are working on their laptops and talking on their cell phones so that they rarely have any time when they are not working and communicating directly with each other. Although the family is supposed to be a place of relaxation and recovery, it has become another place where spouses work. Balance is needed. **Leisure** refers to the use of time to engage in freely chosen activity perceived as enjoyable and satisfying, including exercise.

Importance of Leisure

Leisure is becoming more important to people. In a nationwide poll conducted by Yankelovich Partners, almost half (48%) of the respondents reported that they would be willing to trade an increase in pay for more vacation time (Haralson and Mullins, 2000). More than one-quarter of the respondents (28%) reported that in the past 5 years they had voluntarily made changes in their lifestyle that resulted in making less money. Women, those with children, and those younger than 40 were more likely to do so. The positive value of leisure to a couple's marriage interaction and satisfaction is clear. Doumas et al. (2003) studied 49 dual-earner couples who reported more positive marital interaction on days they worked less. Claxton et al. (2004) also found that, at least for wives, the more leisure time they shared with their husbands, the more in love they were.

Functions of Leisure

Leisure fulfills important functions in our individual and interpersonal lives. Leisure activities may relieve work-related stress and pressure; facilitate social interaction and family togetherness; foster self-expression, personal growth, and skill development; and enhance overall social, physical, and emotional well-being.

These spouses work hard in their respective professions but balance off their careers with frequent relaxing vacations.

Though leisure represents a means of family togetherness and enjoyment, it may also represent an area of stress and conflict. Some couples function best when they are busy with work and childcare so that there is limited time to interact or for the relationship. Indeed, some prefer not to have a lot of time alone together.

Problems Related to Leisure

In a study of 102 older couples (average age of spouses was 69), "leisure activities" was the most frequently cited problem area (accounting for 23% of problems) (Henry et al., 2005). The ways in which this problem expressed itself included the following:

TV: he wanted to watch football and she wanted to watch *American Idol*.

Travel: she wanted to stay at home and he wanted to travel, or they differed in travel preferences.

Time: both were very busy with their lives and did not take time to spend together.

Preparing for Vacation

In planning a vacation: take along half as much baggage and twice as much money.

Unknown

Preparing for a vacation is sometimes problematic. A national sample identified some of the problems as taking care of one's work schedule (28%), arranging for pet care (23%), making travel plans (11%), and finding someone to look after one's home (11%) (Umminger and Parker, 2005). Some people may have trouble detaching from work; their cell phones provide access to job-related problems, and laptop computers are reminders of work to be done.

SUMMARY

What are the various meanings of money?

Money is a means of being secure and avoiding poverty. It also is a source of self-esteem, conveys power in relationships, is viewed as an expression of love, and can be a point of conflict over how it is spent.

What are the effects of the wife's employment on her, her husband, and the marriage?

In dual-earner relationships, benefits for the woman's employment include enhanced self-esteem, more power in the relationship, greater economic independence, and a wider set of social relationships. Negatives for the woman include exhaustion as a result of role overload and frustration or guilt caused by role conflict. Women all over the world do more housework than men, even if they are employed outside the home.

Men benefit from their wives' employment by being relieved of full responsibility for the financial support of the family, having the freedom to quit or to change jobs, and having the opportunity to be a househusband and/or full-time at-home parent. Men who report being dissatisfied with their wives' employment interpret such employment as a reflection of their own inadequacy to support the family. Some men are also torn between their work and family responsibilities.

Employed wives in unhappy marriages are more likely to leave the marriage than unemployed wives. Marriages in which the wife earns more money, the husband is unemployed, and his masculinity is threatened are also more vulnerable. Children reared in dual-earner families may have less time with the busy and frustrated parents and may experience less supervision of their behavior. Young children put in day-care centers for most of the day may be at risk.

What are the various strategies for balancing the demands of work and family?

Strategies used for balancing the demands of work and family include the superperson strategy, cognitive restructuring, delegation of responsibility, planning and time management, and role compartmentalization. Government and corporations have begun to respond to the family concerns of employees by implementing work-family policies and programs.

What is the importance of leisure and what are its functions?

Spouses and couples are increasingly beginning to value their leisure time. Leisure helps to relieve stress, facilitate social interaction and family togetherness, and foster personal growth and skill development. However, leisure time may also create conflict over how to use it.

KEY TERMS

absolute poverty	leisure	role overload	supermom
dual-career marriage	mommy track	role strain	superwoman
HER /his career	poverty	second shift	THEIR career
HIS/her career	role compartmentalization	shift work	third shift
HIS/HER career	role conflict		

The Companion Website for *Choices in Relationships: An Introduction to Marriage and the Family,* Ninth Edition
www.thomson.edu/sociology/knox

Supplement your review of this chapter by going to the companion website to take one of the Tutorial Quizzes, use the flash cards to master key terms, and check out the many other study aids you'll find there. You'll also find special features such as the Marriage and Family Resource Center, Census 2000 information, and other data and resources at your fingertips to help you with that special project or to do some research on your own.

WEB LINKS

Career.com (career planning)
http://www.career.com/

Slowlane.com (stay-at-home dads)
http://slowlane.com/

REFERENCES

Abowitz, D. A. 2005 Does money buy happiness? A look at gen Y college student beliefs. *Free Inquiry in Creative Sociology* 33: 119–130.

Ahnert, L. and M. E. Lamb. 2003. Shared care: Establishing a balance between home and child care settings. *Child Development* 74:1044–49.

American Council on Education and University of California. 2005—006. *The American freshman: National norms for fall, 2005.* Los Angeles: Higher Education Research Institute. UCLA Graduate School of Education and Information Studies.

Baum, C. L. 2002. A dynamic analysis of the effect of child care costs on the work decisions of low-income mothers with infants. *Demography* 39:139–64.

Berke, D. L. 2000. He does, she does, who does? Division of "family work" in couples where the wife is self-employed. Paper, Annual Meeting of the National Council on Family Relations, Minneapolis, November.

Capizzano, J., G. Adams, and J. Ost. 2006. The child care patterns of white, black and Hispanic children. Urban Institute (http://www.urban.org/url.cfm?ID=311285); retrieved April 2006.

Cinamon, R. G. 2006. Anticipated work-family conflict: effects of gender, self-efficacy, and family background. *Career Development Quarterly* 54:202–16.

Clark, K. 2002. Mommy's home: More parents choose to quit work to raise their kids. *U.S. News and World Report* 25 November 25:32 passim.

Clark, M. C., L. C. Koch, and E. J. Hill. 2004. The work-family interface: Differentiating balance and fit. *Family and Consumer Sciences Research Journal* 33:121–32.

Claxton, A., M. Perry-Jenkins, and J. Smith. 2004. Leisure activities and marital quality. Paper, Annual Conference of the National Council on Family Relations, Orlando, Florida, November.

Coontz, S. 1997. *The way we really are.* New York: Basic Books.

Doumas, D. M., G. Margolin, and R. S. John. 2003. The relationship between daily marital interaction, work, and health-promoting behaviors in dual-earner couples: An extension of the work-family spillover model. *Journal of Family Studies* 24:3–20.

Ehrenreich, B. 2001. *Nickel and dimed: On (not) getting by in America.* New York: Henry Holt.

Gordon, R. A. and R. S. Högnäs 2006 The best laid plans: Expectations, preferences, and stability of child-care arrangements. *Journal of Marriage and Family* 68:373–393.

Gordon, J. R. and K. S. Whelan-Berry. 2005. Contributions to family and household activities by the husbands of midlife professional women. *Journal of Family Studies* 26:899–923.

Grandey, A. A., B. L. Cordeiro and A. C. Crouter. 2005. A longitudinal and multi-source test of the work-family conflict and job satisfaction. *Journal of Occupational and Organizational Psychology* 78:305–23.

Grzywacz, J. G. and B. L. Bass. 2003. Work, family, and mental health: Testing different models of work-family fit. *Journal of Marriage and the Family* 65:248–62.

Halpern, D. 2005. Psychology at the intersection of work and family: Recommendations for employers, working families, and policymakers. *American Psychologist* 60:397–409.

Hand, K. 2005. Mother's views on using formal child care. *Family Matters* Autumn:10–17.

Haralson, D. and M. E. Mullins. 2000. Less pay, more play. *USA Today* 6 November, B1.

Henry, R. G., R. B. Miller, and R. Giarrusso. 2005. Difficulties, disagreements, and disappointments in late-life marriages. *International Journal of Aging & Human Development* 61:243–65.

Hochschild, A. R. 1989. *The second shift.* New York: Viking.

———. 1997. *The time bind.* New York: Metropolitan Books.

Jacobs, J. A. and J. C. Cognick. 2002. Hours of paid work in dual-earner couples: The United States in cross-national perspective. *Sociological Focus* 35:169–87.

Jalovaara, M. 2003. The joint effects of marriage partners' socioeconomic positions on the risk of divorce. *Demography* 40:67–81.

Judkins, B. and L. Presser 2005. Gender asymmetry in the division of eco-friendly domestic labor. Paper, Society of Social Problems, Philadelphia, Pennsylvania, August.

Kiecolt, K. J. 2003. Satisfaction with work and family life: No evidence of a cultural reversal. *Journal of Marriage and the Family* 65:23–35.

Lopoo, L. M. 2005. Maternal employment and latchkey adolescents. *Social Service Review* 79:602–23.

Macmillan, R. and Gartner. 2000. When she brings home the bacon: Labor-force participation and the risk of spousal violence against women. *Journal of Marriage and the Family* 61:947–58.

Mason, M.A. and M. Goulden. 2004. Do babies matter? The effect of family formation on the life-long careers of academic men and women. Annual Conference of the National Council on Family Relations, Orlando, Florida, November.

Mr. Mom reaches new peak. 2005. *USA Today* August 134(2723).

Noonan, M. C. and M. E. Corcoran. 2004. The mommy track and partnership: Temporary delay or dead end? *The Annals of the American Academy of Political and Social Science* 596:130–50.

Nomaaguchi, K. M. and M. A. Milkie. 2006. Maternal employment in childhood and adult's retrospective reports of parenting practices. *Journal of Marriage and the Family.* 68: 573–291.

Patrick, T. 2005. Stay-at-home dads. *Futurist* 39:12–13.

Pepper, T. 2006. Fatherhood: Trying to do it all. *Newsweek* (International ed.) Feb 27.

Perry-Jenkins, M., R. L. Repetti, and A. C. Crouter. 2001. Work and family in the 1990s. In *Understanding families into the new millennium: A decade in review,* edited by R. M. Milardo. Minneapolis: National Council on Family Relations, 200–17.

Presser, H. B. 2000. Nonstandard work schedules and marital instability. *Journal of Marriage and the Family* 62:93–110.

Rosen, E. 2006. Derailed on the mommy track? There's help to get going again. *New York Times* 155: Section 10:1–3.

Schoen, R., S. J. Rogers, and P. R. Amato. 2006. Wives' employment and spouses' marital happiness: Assessing the direction of influence using longitudinal couple data. *Journal of Family Issues* 27:506–28.

Schoen, R., N. M. Astone, K. Rothert, N. J. Standish, and Y. J. Kim. 2002. Women's employment, marital happiness, and divorce. *Social Forces* 81:643–62.

Sefton, B. W. 1998. The market value of the stay-at-home mother. *Mothering* 86:26–29.

Stanfield, J. B. 1998. Couples coping with dual careers: A description of flexible and rigid coping styles. *Social Science Journal* 35:53–62.

Strazdins, L., M. S. Clements, R. J. Korda, D. H. Broom and M. Rennie. 2006. Unsociable Work? Nonstandard Work Schedules, Family Relationships, and Children's Well-Being. *Journal of Marriage and Family* 68:394–410.

Statistical Abstract of the United States: 2006. 125th ed. Washington, DC: U.S. Bureau of the Census.

Travel Industry Association of America, 2006. Report "Fun Alone or with the Family" published in *USA Today,* July 7, 2006. P. 1.

Tucker, P. 2005. Stay-at-home dads. *The Futurist* 39:12–15.

Tyre, P. and D. McGinn. 2003. She works, he doesn't. *Newsweek* 12 May:45–52.

Umminger, A. and S. Parker. 2005. What hinders vacation? *USA Today* February 18:A1.

Vanoverbeke, C. 2006. Day care video streaming allows parents to tune in. *Tribune* (Mesa, AZ). March 21.

Van Willigen, M. and P. Drentea. 2001. Benefits of equitable relationships: The impact of fairness, household division of labor, and decision making power on social support. *Sex Roles* 44:571–97.

Walker, S. K. 2000. Making home work: Family factors related to stress in family child care providers. Poster, Annual Conference of the National Council on Family Relations, Minneapolis, November.

Warash, B. G., C. A. Markstrom, and B. Lucci. 2005. The early childhood environment rating scale-revised as a tool to improve child care centers. *Education* 126:240–50.

chapter 13

Don't let nothing get you plumb down.

Woody Guthrie, social activist singer

Stress and Crisis in Relationships

Contents

True or False?

1. Changing one's basic values is the strategy most helpful in coping with a crisis event.

2. College students who throw up, have hangovers, and have blackouts from drinking too much learn how much alcohol they can consume and adapt on subsequent drinking occasions.

3. Marriage/family therapists report that an extramarital affair is the most stressful event a married couple experience.

4. More than 60 percent of adults going through the middle years report having a "midlife crisis."

5. On-line marriage therapy (when both spouses are involved) has become as effective as face-to-face marital therapy.

Answers: **1.** T   **2.** F   **3.** F   **4.** F   **5.** F

Change is the word that best describes marriage. All spouses and couples experience stress and are confronted with crisis events with which they must cope. Indeed, none of us is immune to sudden, dramatic, sometimes shocking change. In this chapter we review some crisis events that are not uncommon to marriages and families. These include physical illness and disability, unemployment, drug abuse, extramarital affairs, and death.

Ann Landers was once asked what she would consider the single most useful bit of advice all people could profit from. She replied, "Expect trouble as an inevitable part of life, and when it comes, hold your head high, look it squarely in the eye and say, 'I will be bigger than you.'" Life indeed brings both triumphs and tragedies. Nearly half of all adults report experiencing at least one traumatic event at some point in their lives (Ozer et al., 2003). These are often turning points in one's life that may come at any time (Leonard and Burns, 2006). This chapter is about experiencing and coping with these events and the day-to-day stress we feel during the in-between time.

Sometimes life is full of laughs-
Sometimes it ain't funny.

Hoyt Axton, singer

Personal Stress and Crisis Events

In this section we review the definitions of crisis and stressful events, the characteristics of resilient families, and a framework for viewing a family's reaction to a crisis event.

Definitions of Stress and Crisis Events

Stress is a reaction of the body to substantial or unusual demands (physical, environmental, or interpersonal). Stress is often accompanied by irritability, high blood pressure, and depression. Stress is a process rather than a state. For example, a person will experience different levels of stress throughout a divorce—the stress involved in acknowledging that one's marriage is over, telling the children, leaving the home, getting the final decree, and seeing one's ex will result in varying levels of stress.

A **crisis** is a crucial situation that requires changes in normal patterns of behavior. A family crisis is a situation that upsets the normal functioning of the family and requires a new set of responses to the stressor. Sources of stress and crises

Hurricane Katrina, which hit New Orleans in 2005, was an external cause of a crisis event that devastated homes and displaced families to other cities and states. This photo was taken inside the living room of a house hit by Katrina.

can be external (e.g., hurricane, tornado, downsizing, military separation) or internal (e.g., alcoholism, extramarital affair, Alzheimer's disease of spouse or parents). Stressors or crises may also be categorized as expected or unexpected. Examples of expected family stressors include the need to care for aging parents and the death of one's parents. Unexpected stressors include contracting human immunodeficiency virus (HIV), a miscarriage, or the suicide of one's teenager.

Both stress and crises are a normal part of family life and sometimes reflect a developmental sequence. Pregnancy, childbirth, job changes or loss, children leaving home, retirement, and widowhood are all stressful and predictable for most couples and families. Crisis events may have a cumulative effect: the greater the number in rapid succession, the greater the stress.

Resilient Families

Just as the types of stress and crisis events vary, individuals and families vary in their abilities to respond successfully to crisis events. **Resiliency** refers to a family's strengths and ability to respond to a crisis in a positive way. Several characteristics associated with resilient families include having a joint cause or purpose, emotional/social support for each other, good problem-solving skills, the ability to delay gratification, flexibility, accessing residual resources, communication, and commitment (Raikes, 2005; Walsh, 2003; DeHaan, 2002). A family's ability to bounce back from a crisis (from loss of one's job to the death of a family member) reflects its level of resiliency. Resiliency may also be related to individuals' perception of the degree to which they are in control of their destiny. The Self-Assessment on page 370 measures this perception.

A Family Stress Model

Various theorists have explained how individuals and families experience and respond to stressors. Burr and Klein (1994) reviewed the ABC-X model of family stress, developed by Reuben Hill in the 1950s. The model can be explained as follows:

A = stressor event
B = family's management strategies, coping skills
C = family's perception, definition of the situation
X = family's adaptation to the event

When it is dark enough, you can see the stars.

Charles Beard, historian

Internality, Powerful Others, and Chance Scales

People have different feelings about their vulnerability to crisis events. The following scale addresses the degree to which you feel you have control, or feel others have control, or feel that chance has control of what happens to you.

Internality, Powerful Others, and Chance Scales

To assess the degree to which you believe that you have control over your own life (I = Internality), the degree to which you believe that other people control events in your life (P = Powerful Others), and the degree to which you believe that chance affects your experiences or outcomes (C = Chance), read each of the following statements and select a number from minus 3 to plus 3.

−3	−2	−1	+1	+2	+3
Strongly Disagree	Disagree	Slightly Disagree	Slightly Agree	Agree	Strongly Agree

Subscale

I   **1.** Whether or not I get to be a leader depends mostly on my ability.

C   **2.** To a great extent my life is controlled by accidental happenings.

P   **3.** I feel like what happens in my life is mostly determined by powerful people.

I   **4.** Whether or not I get into a car accident depends mostly on how good a driver I am.

I   **5.** When I make plans, I am almost certain to make them work.

C   **6.** Often there is no chance of protecting my personal interests from bad luck happenings.

C   **7.** When I get what I want, it's usually because I'm lucky.

P   **8.** Although I might have good ability, I will not be given leadership responsibility without appealing to those in positions of power.

I   **9.** How many friends I have depends on how nice a person I am.

C   **10.** I have often found that what is going to happen will happen.

P   **11.** My life is chiefly controlled by powerful others.

C   **12.** Whether or not I get into a car accident is mostly a matter of luck.

P   **13.** People like myself have very little chance of protecting our personal interests when they conflict with those of strong pressure groups.

C   **14.** It's not always wise for me to plan too far ahead because many things turn out to be a matter of good or bad fortune.

P   **15.** Getting what I want requires pleasing those people above me.

C   **16.** Whether or not I get to be a leader depends on whether I'm lucky enough to be in the right place at the right time.

P   **17.** If important people were to decide they didn't like me, I probably wouldn't make many friends.

I   **18.** I can pretty much determine what will happen in my life.

I   **19.** I am usually able to protect my personal interests.

P   **20.** Whether or not I get into a car accident depends mostly on the other driver.

I   **21.** When I get what I want, it's usually because I worked hard for it.

P   **22.** In order to have my plans work, I make sure that they fit in with the desires of other people who have power over me.

I   **23.** My life is determined by my own actions.

C   **24.** It's chiefly a matter of fate whether or not I have a few friends or many friends.

Scoring

Each of the subscales of Internality, Powerful Others, and Chance is scored on a six-point Likert format from minus 3 to plus 3. For example, the eight Internality items are 1, 4, 5, 9, 18, 19, 21, 23. A person who has strong agreement with all eight items would score a plus 24; strong disagreement, a minus 24. After adding and subtracting the item scores, add 24 to the total score to eliminate negative scores. Scores for Powerful Others and Chance are similarly derived.

Norms

For the Internality subscale, means range from the low 30s to the low 40s, with 35 being the modal mean (SD values approximating 7). The Powerful Others subscale has produced means ranging from 18 through 26, with 20 being characteristic of normal college student subjects (SD = 8.5). The Chance subscale produces means between 17 and 25, with 18 being a common mean among undergraduates (SD = 8).

Source

From *Research with the Locus of Control Construct*, by Hervert M. Lefcourt, 1981, pp. 57–59. Reprinted with permission from Elsevier.

A is the stressor event, which interacts with B, which is the family's coping ability, or crisis-meeting resources. Both A and B interact with C, which is the family's appraisal or perception of the stressor event. X is the family's adaptation to the crisis. Thus, a family that experiences a major stressor (e.g., spinal cord–injured spouse) but has great coping skills (e.g., religion/spirituality, love, communication, commitment) and perceives the event to be manageable will experience a moderate crisis. But a family that experiences a less critical stressor event (e.g., child makes Cs and Ds in school) but has minimal coping skills (e.g., everyone blames everyone else) and perceives the event to be catastrophic will experience an extreme crisis. Hence, how a family experiences and responds to stress depends not only on the event but also on the family's coping resources and perceptions of the event.

Positive Stress-Management Strategies

Researchers Burr and Klein (1994) administered an 80-item questionnaire to 78 adults to assess how families experiencing various stressors such as bankruptcy, infertility, disabled child, and a troubled teen used various coping strategies and how useful they evaluated these strategies. Below we detail some helpful stress-management strategies.

Changing Basic Values and Perspective

The strategy that the highest percentage of respondents reported as being helpful was changing basic values as a result of the crisis situation. Survivors of Hurricane Katrina, which devastated New Orleans, noted that focusing on the sparing of their lives and those close to them rather than the loss of their home or material possessions was an essential view in coping with the crisis. Sharpe and Curran (2006) confirmed that finding positive meaning in a crisis situation is associated with positive adjustment to the situation.

In responding to the crisis of bankruptcy, people may reevaluate the importance of money and conclude that relationships are more important. In coping with unemployment, people may decide that the amount of time they spend with family members is more valuable than the amount of time they spend making money. Individuals may also be realistic and accepting of crisis events. In one study of university students, over half (52%) of the respondents reported that they could "accept those things in life which they cannot change" (Lee et al., 2005). Buddhists have the saying "Pain is inevitable; suffering is not." This is another way of emphasizing that it is how one views a situation, not the situation itself, that determines its impact on you.

A team of researchers found that, among a sample of African-American college students, those with an optimistic attitude toward the future tended to report less perceived and global stress than their pessimistic counterparts (Baldwin et al., 2003). Hence, one's basic view of life has its benefits, independent of any particular event.

Since we cannot change reality, let us change the eyes which see reality.
Nikos Kazantzakis

Exercise

The Centers for Disease Control and Prevention (CDC) and the American College of Sports Medicine (ACSM) recommend that people aged 6 years and older engage regularly, preferably daily, in light to moderate physical activity for at least 30 minutes at a time. These recommendations are based on research that has shown the physical, emotional, and cognitive benefits of exercise (Szabo et al. 2005; Lee et al., 2005). Orr (2006) also found benefits from exercise in terms of reducing stress, depression, and maladaptive eating behaviors. In spite of such benefits, in a sample of university students, just slightly over a quarter (26.2%) of

Make an appointment with exercise and keep it.
Nolan Ryan, baseball player

the men and fewer than ten percent (9.3%) of the women reported intense vigorous exercise for at least 20 minutes a day (Lee et al., 2005).

When exercise is a part of leisure, there may be benefits for crisis coping. A team of researchers noted the positive effects of leisure: it improves one's ability to cope with problems, enhances performance, and facilitates interaction with intimates (Grafanaki et al., 2005).

Friends and Relatives

A network of relationships is associated with successful coping with various life transitions (Levitt et al., 2005). Women are more likely to feel connected to others and to feel that they can count on others in times of need (Weckwerth and Flynn, 2006). News media coving hurricanes, earthquakes, and tsunamis emphasized that as long as one's family was still together, individuals were less concerned about the loss of material possessions.

Love

A love relationship also helps an individual cope with stress. Being emotionally involved with another and sharing the experience with that person helps to insulate individuals from being devastated by a crisis event. Conversely, students just having gone through a breakup note how emotionally volatile they are; their capacity to cope with stress is reduced (Sbarra and Emery, 2005).

Religion and Spirituality

Religion may be helpful in adjusting to a crisis. Park (2006) found that religion is associated with providing meaning for experiencing a crisis and for adjusting to it. In addition, religion provides a framework for being less punitive. Rather than be in a rage and seek revenge against a perceived aggressor, the religious might "turn the other cheek" (Unnever et al., 2005).

Humor

A sense of humor is related to lower anxiety and a happier mood. Indeed, a team of researchers compared the effects of humor, aerobic exercise, and listening to music on the reduction of anxiety and found humor to be the most effective. Just sitting quietly seemed to have no effect on reducing one's anxiety (Szabo et al., 2005).

Sleep

Getting an adequate amount of sleep is also associated with lower stress levels (Mostaghimi et al, 2005). Persons who have had adequate sleep are also better at responding to novel situations and to multitasking (Nilsson et al., 2005). Even midday naps are associated with positive functioning, particularly memory performance (Schabus et al., 2005). Indeed, adequate sleep helps one to respond to daily stress and to crisis events.

Biofeedback

Biofeedback is a process in which information that is relayed back to the brain enables a person to change his or her biological functioning. Biofeedback treatment teaches a person to influence biological responses such as heart rate, nervous system arousal, muscle contrac-

Spending time fishing with one's beloved is one way to relieve stress.

Authors

tions, and even brain wave functioning. Biofeedback is used at about 1500 clinics and treatment centers worldwide. A typical session lasts about an hour and costs $60 to $150. There are several types of biofeedback:

1. Electromyographic (EMG) biofeedback, which measures electrical activity created by muscle contractions, is often used for relaxation training and for stress and pain management.

2. Thermal or temperature biofeedback uses a temperature sensor to detect changes in temperature of the fingertips or toes. Stress causes blood vessels in the fingers to constrict, reducing blood flow, leading to cooling. Thermal biofeedback, which trains people to quiet the nervous system arousal mechanisms that produce hand and/or foot cooling, is often used for stress, anxiety, and pain management.

3. Galvanic skin response (GSR) biofeedback utilizes a finger electrode to measure sweat gland activity. This measure is very useful for relaxation and stress management training and is also used in the treatment of attention deficit/hyperactivity disorder.

4. Neurofeedback, also called *neurobiofeedback* or *EEG (electroencephalogram) biofeedback,* may be particularly helpful for individuals coping with a crisis. It trains people to enhance their brain-wave functioning and has been found to be effective in treating a wide range of conditions, including anxiety, stress, depression, tension and migraine headaches, addictions, and high blood pressure.

Because neurofeedback is the fastest-growing field in biofeedback, we take a closer look at this treatment modality. Neurofeedback involves a series of sessions in which a client sits in a comfortable chair facing a specialized game computer. Small sensors are placed on the scalp to detect brain-wave activity and transmit this information to the computer. The neurofeedback therapist (in the same room) also sits in front of a computer that displays the client's brain-wave patterns in the form of an electroencephalogram (EEG). After a clinical assessment of the client's functioning, the therapist determines what kinds of brain-wave patterns are optimal for the client. During neurofeedback sessions, clients learn to produce desirable brain waves by controlling a computerized game or task, similar to playing a videogame, but instead of a joystick, it is the client's brain waves that control the game.

Neurofeedback is like an exercise of the brain, helping it to become more flexible and effective. But unlike body exercise, which when training is stopped

What it boils down to, really, is that I hate reality. And, you know, unfortunately it's the only place where you can get a good steak dinner.

Woody Allen, writer/director

will lose its benefits over time, brain-wave training generally does not. Once the brain is trained to function in its optimal state (which may take an average of 20 to 25 sessions), it generally remains in this more healthy state. Neurofeedback therapists liken the process to that of learning to ride a bicycle: once you learn to ride a bike, you can do so even if it has been years since you have ridden one. One neurofeedback therapist explains "clients speak often of their disorders—panic attacks, chronic pain, etc.—as if they were stuck in a certain pattern of response. Consistently in clinical practice, EEG biofeedback helps 'unstick' people from these unhealthy response patterns" (Carlson-Catalano, 2003).

Deep Muscle Relaxation

Tensing and relaxing one's muscles has been associated with an improved state of relaxation. Calling it abbreviated progressive muscle relaxation (APMR), Termini (2006) found that the cognitive benefits were particularly evident; persons who tensed and relaxed various muscle groups noticed a mental relaxation more than a physical relaxation.

Harmful Strategies

Some coping strategies not only are ineffective for resolving family problems but also add to the family's stress by making the problems worse. Respondents in the Burr and Klein (1994) research identified several strategies they regarded as harmful to overall family functioning. These included keeping feelings inside, taking out frustrations on or blaming others, and denying or avoiding the problem.

Burr and Klein's research also suggests that there are differences between women and men in their perceptions of the usefulness of various coping strategies. Women were more likely than men to view as helpful such strategies as sharing concerns with relatives and friends, becoming more involved in religion, and expressing emotions. Men were more likely than women to view potentially harmful strategies such as using alcohol, keeping feelings inside, keeping others from knowing how bad the situation was. Regarding alcohol, in a study of 79 students at a private coeducational Midwestern university, 75% of the males and 31% of the females reported three or more binge-drinking episodes in the prior 2 weeks (Biscaro et al., 2004).

Family Crisis Examples

Some of the more common crisis events faced by spouses and families include physical illness, mental challenges, an extramarital affair, unemployment, substance abuse, and death.

Physical Illness and Disability

Absence of physical illness is important for individuals to define themselves as being healthy overall. Although approximately 60% of those aged 85 and older define themselves as healthy (Ostbye et al., 2006), physical problems are not unusual. For men, prostate cancer is an example of a medical issue that can rock the foundation of their personal and marital well-being. The follow section details the experience of a husband coping with prostate cancer.

Reacting to Prostate Cancer: One Husband's Experience

The most common type of cancer for men is prostate cancer. A former student described his experience with prostate cancer as follows:

A serious illness is marriage's unspoken fear. The chances of a couple staying healthy together and dying at the same time are Las Vegas odds. Life is a dance you want to finish on the same beat. And I was the first to stumble.

Erma Bombeck, humorist

Since my father had prostate cancer, I was warned that it is genetic and to be alert as I reached age 50. At the age of 56, I noticed that I was getting up more frequently at night to urinate. My doc said it was probably just one of my usual prostate infections, and he prescribed the usual antibiotic. When the infection did not subside, a urologist did a transrectal ultrasound (TRUS) needle biopsy. The TRUS gives the urologist an image of the prostate while he takes about 10 tissue samples from the prostate with a thin, hollow needle.

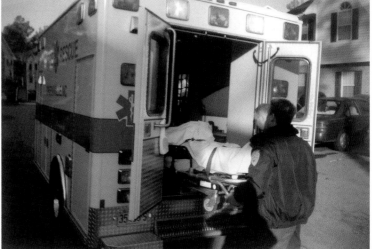

Authors

Treatment Options. *Two weeks later I learned the bad news (I had prostate cancer) and the good news (it had not spread). Since the prostate is very close to the spinal column, failure to act quickly can allow time for cancer cells to spread from the prostate to the bones. My urologist outlined a number of treatment options, including traditional surgery, laparoscopic surgery, radiation treatment, implantation of radioactive "seeds," cryotherapy (in which liquid nitrogen is used to freeze and kill prostate cancer cells), and hormone therapy, which blocks production of the male sex hormones that stimulate growth of prostate cancer cells. He recommended traditional surgery ("radical retropubic prostatectomy"), in which an incision is made between the belly button and the pubic bone to remove the prostate gland and nearby lymph nodes in the pelvis. This surgery is generally considered the "gold standard" when the disease is detected early. Within three weeks I had the surgery.*

Physical Effects Following Surgery. *Every patient awakes from radical prostate surgery with urinary incontinence and impotence—which can continue for a year, two years, or forever. Such patients also awake from surgery hoping that they are cancer-free. This is determined by laboratory analysis of tissue samples taken during surgery. The patient waits for a period of about two weeks, hoping to hear the medical term "negative margins" from his doctor. That finding means that the cancer cells were confined to the prostate and did not spread past the margins of the prostate. A finding of negative margins should be accompanied by a PSA [prostate specific antigen] score of zero, confirming that the body no longer detects the presence of cancer cells. I cannot describe the feeling of relief that accompanies such a report, and I am very fortunate to have heard those words used in my case.*

Psychological Effects Following Surgery. *The psychological effects have been devastating—more for me than for my partner. I have only been intimate with one woman—my wife—and having intercourse with her was one of the greatest pleasures in my life. For a year, I was left with no erection and an inability to have an orgasm. Afterwards I was able to have an erection (via caverject [$25], an injection) and an orgasm. Dealing with urinary incontinence (I refer to myself as Mr. Drippy) is a "wish it were otherwise" on my psyche.*

Evaluation. *When faced with the decision to live or die, the choice for most of us is clear. In my case, I am alive, cancer free, and enjoying the love of my life (now in our 40th year together).*

In addition to prostate cancer, chronic degenerative diseases (auto-immune dysfunction, rheumatoid arthritis, lupus, Crohn's disease, and chronic fatigue syndrome) can challenge a spouse's/mate's and couple's ability to cope. These particular illnesses are particularly invasive in that conventional medicine has little to offer besides pain medication (Selak and Overman, 2005). For example, spouses with

Attitudes toward Overweight Children Scale

Relatively little is known about college students' attitudes toward overweight children. The purpose of this survey is to gain a better understanding of what college students think and feel about children who are overweight. Please read each item carefully and consider how you feel about each statement. There are no right or wrong answers to any of these statements, so please give your honest reactions and opinions. Please read each statement carefully and respond by using the following scale:

1	2	3	4	5	6	7
Strongly Disagree						Strongly Agree

_____ 1. Overweight children are poorer athletes.

_____ 2. Overweight children are less happy than other children.

_____ 3. Overweight children eat as much as adults.

_____ 4. Overweight children lack self-control.

_____ 5. Overweight children are lazy.

_____ 6. Overweight children have few physically active interests.

_____ 7. Children who are overweight are unhealthy.

_____ 8. Overweight children lack motivation.

Scoring

Selecting a 1 reflects the least acceptance of negative beliefs about overweight children; selecting a 7 reflects the greatest acceptance of negative beliefs about overweight children. Once you have made your responses to the eight items, add the numbers. The lower your total score (8 is the lowest possible), the less accepting you are of negative beliefs about overweight children; the higher your total score (56 is the highest possible), the greater your acceptance of negative beliefs about overweight children. A score of 24 places you at the midpoint between being very disapproving of overweight children and very accepting of overweight children.

Scores of Other Students Who Completed the Scale

The scale was completed by 143 male and 145 female student volunteers at Valdosta State University. The average score on the scale was 32.11 (standard deviation [SD] = 8.25). Their ages ranged from 18 to 45, with a mean age of 21.64 (SD = 3.20). The ethnic background of the sample was as follows: 66.0% white, 27.8% African-American, 3.5% Hispanic, 1.0% Asian, 0.3% American Indian, and 1.4% other. The college classification level of the sample included 14.6% Freshmen, 17.7% Sophomores, 29.9% Juniors, 32.3% Seniors, 3.8% graduate students, and 1.7% post-baccalaureates. Male participants endorsed more negative beliefs about overweight children (mean = 33.28; SD = 7.50) than did female participants (mean = 30.96; SD = 8.80; $p < .05$). There were no significant differences in regard to race or college classification of the participants (p values $> .05$).

Source

Attitudes Toward Overweight Children Scale, 2006, by Mark Whatley, PhD, Department of Psychology, Valdosta State University, Valdosta, Georgia 31698-0100. Used by permission. Other uses of this scale by written permission of Dr. Whatley only. Information on the reliability and validity of this scale is available from Dr. Whatley (mwhatle@valdosta.edu).

chronic fatigue syndrome may experience financial consequences ("I could no longer meet the demands of my job so I quit"), gender role loss ("I couldn't cook for my family"/"I was no longer a provider"), and changed perceptions by their children ("They have seen me sick for so long they no longer ask me to do anything"). In addition, osteoporosis is a health threat for 44 million older women and results in a decline in their ability to perform routine activities (Roberto, 2005).

In those cases in which the illness is fatal, **palliative care** is helpful. This term describes the health care (focusing on relief of pain and suffering) for the individual who has a life-threatening illness and support for them and their loved ones. Such care may involve the person's physician or a palliative care specialist who works with the physician, nurse, social worker, and chaplain. Pharmacists or rehabilitation specialists may also be involved. The effects of such care are to approach the end of life with planning (how long should life be sustained on machines?) and forethought, to relieve pain, and to provide closure.

Another physical issue with which some parents cope is that of their children being overweight. Body mass index (BMI) is calculated as weight in kilograms divided by height in meters squared; overweight is indicated by a BMI of 25.0 to 29.9, and obese by 30.0 or higher (Pyle et al., 2006). Routh and Rao (2006) studied 225 9- to 10-year-olds in primary schools and found that 20 percent were overweight; 5 percent were obese. National figures reflect that about a third of youth (31.5%) are at risk for being overweight and that 16.5 percent could be classified as overweight. Minority females and individuals of lower socioeconomic status are more prone to being overweight or obese. Not only are there medical, life-threatening implications for this chronic physical health issue (Pyle et al. 2006) in both industrial and developing countries (Flynn et al. 2006), but also there is prejudice and discrimination toward overweight children. The Self-Assessment on page 376 allows you to assess your attitudes toward overweight children.

National Data

In a Gallup poll, 56 percent of Americans say they want to lose weight, including 18 percent who want to lose "a lot" of weight. Women are twice as likely to report that they want to lose a lot of weight (Moore, 2006).

Mental Illness

We have been discussing the management of physical illness. But not all illnesses are physical; some are mental. The stigma of mental illness is declining. In a survey conducted by the American Psychiatric Association, 90 percent of 1000 adults agreed that people with mental illness could have healthy lives. Seventy percent said that seeing a psychiatrist was a sign of strength. Only 20 percent said that they would not see a psychiatrist under any conditions (Frieden, 2005).

Indeed, it is recognized that everyone experiences problems in living, including relationship problems, criminal or domestic victimization, work-related problems, and low self-esteem. Based on household surveys of 9282 respondents, reported psychological problems and their percentages of lifetime incidence follow: anxiety (29%), impulse control disorder (25%), mood disorder (21%), and substance abuse (15%). Most of these mental maladies go untreated (Mahoney, 2005).

The toll of mental illness on a relationship can be immense. A major initial attraction of partners to each other includes intellectual and emotional qualities. When these are affected, the partner may feel that the mate has already died psychologically, since he or she is, literally, "not the same person I married." A depressed partner certainly changes the interaction patterns of a couple. Jeglic et al. (2005) found that spouses living with a partner with depressive symptoms had more symptoms of depression themselves. The researchers noted the need to treat not only the depression of the spouse but also that of the caregiver. Similar tension and difficulty was also noted by 16 wives caring for their veteran husbands, who were experiencing posttraumatic stress disorder (PTSD). Specifically, the wives reported tension resulting from being drawn into a fusion with their husbands and their struggle to maintain their independence from the illness (Jeglic et al., 2005).

Although we have focused this section on the physical and mental disability of spouses, the physical and mental disability of a child represents a similar crisis event for a couple and a family. Physical disabilities can begin at conception through the prenatal period and include cerebral palsy (10,000 babies per year), fetal alcohol syndrome (5,000 babies per year), and spina bifida (2 of every 10,000 births). Mental disabilities may involve autism (three to four times more common in boys), attention-deficit/hyperactivity disorder (4% of schoolchildren), and antisocial behavior (*Statistical Abstract of the United States: 2006*). These difficulties can stress spouses to the limit of their coping capacity, which may put an enormous strain on their marriage. Regardless of the physical and mental health malady, adequate health insurance can help insulate a family from a catastrophic medical bill that will wipe out savings, bankrupt them, or put them in debt for years.

Depression—it's like someone opened my skull and poured in a gallon of hopelessness.

E. Fred Johnson, Jr., retired

National Data

According to the Census Bureau's 2005 Current Population Survey (CPS), there were 45.8 million uninsured individuals in 2004, or 15.7 percent of the civilian non-institutionalized population. Persons most likely to be uninsured were single adults with no children (57%). Those least likely to be uninsured were parents (27%) (U.S. Department of Health and Human Services, 2005; retrieved August, 2006 from http://aspe.hhs.gov/health/reports/05/uninsured-cps/).

Middle Age Crazy (Mid-Life Crisis?)

The stereotypical explanation for the 45-year-old person who buys a convertible sports car, has an affair, marries a 20-year-old, or adopts a baby is that he or she is "having a midlife crisis." The label conveys that the person feels old, thinks that life is passing him or her by, and seizes one last great chance to do what he or she has always wanted. Indeed, one father (William Feather) noted, "Setting a good example for your children takes all the fun out of middle age."

However, a 10-year study of close to 8000 U.S. adults aged 25 to 74 by the MacArthur Foundation Research Network on Successful Midlife Development revealed that, for most respondents, the middle years were no crisis at all but a time of good health, productive activity, and community involvement. Fewer than a quarter (23%) reported a "crisis" in their lives. Those who did experience a crisis were going through a divorce. Two-thirds were accepting of their getting older; one-third did feel some personal turmoil related to the fact that they were aging (Goode, 1999).

Of those who initiated a divorce in midlife, 70 percent had no regrets and were confident that they did the right thing. This is the result of a study of 1147 respondents aged 40 to 79 who experienced a divorce in their 40s, 50s, or 60s. Indeed, midlife divorcers' levels of happiness/contentment were similar to those of singles their own age and those who remarried (Enright, 2004).

Some people embrace middle age. The Red Hat Society (http://www .redhatsociety.com/info/howitstarted.html) is a group of women who have decided to "greet middle age with verve, humor, and elan. We believe silliness is the comedy relief of life [and] share a bond of affection, forged by common life experiences and a genuine enthusiasm for wherever life takes us next." The society traces its beginning to Sue Ellen Cooper's impulsively buying a bright red fedora at a thrift shop in Tucson, enjoying the experience of wearing it out in public with a purple dress, and sharing the idea with friends. Soon these friends were sharing it with others, and the society had begun.

In the rest of this chapter, we examine how spouses cope with the crisis events of an extramarital affair, unemployment, drug abuse, and death. Each of these events can be viewed either as devastating and the end of meaning in one's life or as an opportunity and challenge to rise above.

Extramarital Affair

Extramarital affair refers to the emotional and sexual involvement of a spouse with someone other than the mate. Some research suggests that the impact of Internet infidelity that involves emotional betrayal is viewed as similar to that of sexual off-line infidelity (Whitty, 2005). Although prevalence rates vary, Adrian and Hartnett (2005) estimated that infidelity occurs in about 25 percent of committed relationships (and involves more men than women). Men report that the focus of extramarital encounters tends to be sex, but women report that it is emotion. In a *Psychology Today* survey, 26 percent of men reported having extramarital sex without being emotionally involved, versus 3 percent of the women (Marano, 2003). A self-assessment to measure your attitude toward infidelity is provided on page 379.

National Data

Of ever-married spouses in the United States, 23.2 percent of husbands and 13.18 of wives reported ever having had intercourse with someone to whom they were not married (Djamba et al., 2005).

Extradyadic involvement or extrarelational refers to emotional and sexual involvement of a pair-bonded

---

Middle age is the awkward period when Father Time catches up with Mother Nature.

Harold Coffin

Now there is somebody new; these dreams I've been dreaming have all fallen through.

Bonnie Raitt, *It's Too Soon to Tell*

Diversity in Other Countries

There is a double standard in regard to extramarital sex. In a study of 62 cultures, slightly over half allowed the husband to have extramarital sex but not the wife. In Jewish law, a wife is guilty of adultery if she has intercourse with a man not her husband; a man is guilty of adultery if he has sex with another man's wife. Up until 1979, a Brazilian man could kill his wife who had sex with another man (with no legal consequences—such "crimes of passion" were considered justified). In Morocco, a divorced woman brings shame on her family and becomes an outcast, pariah, and beggar (hence, a wife will not have sex with anyone other than her husband) (Scheinkman, 2005).

Attitudes toward Infidelity Scale

Infidelity can be defined as unfaithfulness in a committed monogamous relationship. Infidelity can affect anyone, regardless of race, color, or creed; it does not matter whether you are rich, whether you are attractive, where you live, or what your age is. The purpose of this survey is to gain a better understanding of what people think and feel about issues associated with infidelity. There are no right or wrong answers to any of these statements; we are interested in your honest reactions and opinions. Please read each statement carefully, and respond by using the following scale:

1	2	3	4	5	6	7
Strongly Disagree						Strongly Agree

_____ 1. Being unfaithful never hurt anyone.

_____ 2. Infidelity in a marital relationship is grounds for divorce.

_____ 3. Infidelity is acceptable for retaliation of infidelity.

_____ 4. It is natural for people to be unfaithful.

_____ 5. Online/Internet behavior (e.g., sex chatrooms, porn sites) is an act of infidelity.

_____ 6. Infidelity is morally wrong in all circumstances, regardless of the situation.

_____ 7. Being unfaithful in a relationship is one of the most dishonorable things a person can do.

_____ 8. Infidelity is unacceptable under any circumstances if the couple is married.

_____ 9. I would not mind if my significant other had an affair as long as I did not know about it.

_____10. It would be acceptable for me to have an affair, but not my significant other.

_____11. I would have an affair if I knew my significant other would never find out.

_____12. If I knew my significant other was guilty of infidelity, I would confront him/her.

Scoring

Selecting a 1 reflects the least acceptance of infidelity; selecting a 7 reflects the greatest acceptance of infidelity. Before adding the numbers you selected, reverse the scores for item numbers 2, 5, 6, 7, 8, and 12. For example, if you responded to item 2 with a "6," change this number to a "2"; if you responded with a "3," change this number to "5," and so on. After making these changes, add the numbers. The lower your total score (12 is the lowest possible), the less accepting you are of infidelity; the higher your total score (84 is the highest possible), the greater your acceptance of infidelity. A score of 48 places you at the midpoint between being very disapproving and very accepting of infidelity.

Scores of Other Students Who Completed the Scale

The scale was completed by 150 male and 136 female student volunteers at Valdosta State University. The average score on the scale was 27.85 (SD = 12.02). Their ages ranged from 18 to 49, with a mean age of 23.36 (SD = 5.13). The ethnic backgrounds of the sample included 60.8% white, 28.3% African-American, 2.4% Hispanic, 3.8% Asian, 0.3% American Indian, and 4.2% other. The college classification level of the sample included 11.5% Freshmen, 18.2% Sophomores, 20.6% Juniors, 37.8% Seniors, 7.7% graduate students, and 4.2% post-baccalaureates. Male participants reported more positive attitudes toward infidelity (mean = 31.53; SD = 11.86) than did female participants (mean = 23.78; SD = 10.86; $p < .05$). White participants had more negative attitudes toward infidelity (mean = 25.36; SD = 11.17) than did nonwhite participants (mean = 31.71; SD = 12.32; $p < .05$). There were no significant differences in regard to college classification.

Source

Attitudes Toward Infidelity Scale, 2006, by Mark Whatley, PhD, Department of Psychology, Valdosta State University, Valdosta, Georgia 31698-0100. Used by permission. Other uses of this scale by written permission of Dr. Whatley only. Information on the reliability and validity of this scale is available from Dr. Whatley (mwhatley@valdosta.edu).

individual with someone other than the partner. Extradyadic involvements are not uncommon. Thirty-six percent of 1027 undergraduates agreed, "I have cheated on a partner I was involved with." Over half (52.6%) agreed, "A partner I was involved with cheated on me" (Knox and Zusman, 2006).

An extramarital affair is ranked by marriage and family therapists as the second most stressful crisis event for a couple (physical abuse is number one) (Olson et al., 2002). Characteristics associated with spouses who are more likely to have extramarital sex include male gender, a strong interest in sex, permissive sexual values, low subjective satisfaction in the existing relationship, employment outside the home, low church attendance, greater sexual opportunities, and higher social status (power and money) (Smith, 2005; Olson et al., 2002; Treas and Giesen, 2000).

Extramarital affairs range from brief sexual encounters to full-blown romantic affairs. Becoming more common today is the computer affair. Although legally an extramarital affair does not exist unless two persons (one being married) have intercourse, an on-line computer affair can be just as disruptive to a marriage or a couple's relationship. Computer friendships may move to feelings of intimacy, involve secrecy (one's partner does not know the level of involvement), include sexual tension (even though there is no overt sex), and take time, attention, energy, and affection away from one's partner. Schneider (2000) studied 91 women who had experienced serious adverse consequences from their partner's cybersex involvement, including loss of interest in relational sex; feeling hurt, betrayed, rejected, abandoned, lonely, and jealous; and anger over being constantly lied to. These women noted that the cyber affair was as emotionally painful as an off-line affair and that the cybersex addiction of their partners was a major reason for their separation or divorce.

Effects of a cybersex affair also had negative effects on the children, involving the objectification of women, hearing conflicts between the parents, absence of attention (since the father was looking at porn rather than being with children), and break-up of the marriage (Schneider, 2003). Many men are unaware of the consequences of cybersex/affairs in general until it is too late. Married 68-year-old Boeing CEO Harry Stonecipher lost his job when it was discovered and publicized that he was having an affair with a female company executive (Acohido and O'Donnell, 2005).

Reasons for Extramarital Involvements   Spouses report a number of reasons why they become involved in an emotional or sexual encounter outside their marriage. Some of these reasons are discussed in the following subsections.

1. Variety, novelty, and excitement. Extradyadic sexual involvement may be motivated by the desire for variety, novelty, and excitement. One of the characteristics of sex in long-term committed relationships is the tendency for it to become routine. Early in a relationship, the partners cannot seem to have sex often enough. But with constant availability, the partners may achieve a level of satiation, and the attractiveness and excitement of sex with the primary partner seem to wane.

The **Coolidge Effect** is a term used to describe this waning of sexual excitement and the effect of novelty and variety on sexual arousal:

> *One day President and Mrs. Coolidge were visiting a government farm. Soon after their arrival, they were taken off on separate tours. When Mrs. Coolidge passed the chicken pens, she paused to ask the man in charge if the rooster copulated more than once each day. "Dozens of times," was the reply. "Please tell that to the President," Mrs. Coolidge requested. When the President passed the pens and was told about the rooster, he asked, "Same hen every time?" "Oh no, Mr. President, a different one each time." The President nodded slowly and then said, "Tell that to Mrs. Coolidge."(Bermant, 1976, p. 76–77)*

Whether or not individuals are biologically wired for monogamy continues to be debated. Monogamy among mammals is rare (from 3 to 10 percent) and monogamy tends to be the exception more often than the rule (Morell, 1998). Equally debated is whether, even if such biological wiring for plurality of partners does exist, such wiring justifies nonmonogamous behavior—that individuals are responsible for their decisions.

2. Workplace friendships. A common place for extramarital involvements to develop is the workplace. Coworkers share the same world 8 to 10 hours a day and over a period of time may develop good feelings for each other that eventually lead to a sexual relationship. Earlier we made reference to married Harry Stonecipher and the woman he became involved with at work. Tabloid reports regularly reflect that romances develop between married actors making a movie together (e.g., Brad Pitt and Angelina Jolie, who eventually married).

3. Relationship dissatisfaction. It is commonly believed that people who have affairs are not happy in their marriage. Spouses who feel misunderstood,

I'm sleepwalking into a nightmare.
Gabe (married), of his impending affair with a 21-year-old in Woody Allen's *Husbands and Wives*

unloved, and ignored sometimes turn to another who offers understanding, love, and attention. Djamba et al. (2005) analyzed national Social Survey data and found that unhappiness in one's marriage was the primary reason for becoming involved in an extramarital affair.

One source of relationship dissatisfaction is an unfulfilling sexual relationship. Some spouses engage in extramarital sex because their partner is not interested in sex. Others may go outside the relationship because their partners will not engage in the sexual behaviors they want and enjoy. The unwillingness of the spouse to engage in oral sex, anal intercourse, or a variety of sexual positions sometimes results in the other spouse's looking elsewhere for a more cooperative and willing sexual partner.

4. Revenge. Some extramarital sexual involvements are acts of revenge against one's spouse for having an affair. When partners find out that their mate has had or is having an affair, they are often hurt and angry. One response to this hurt and anger is to have an affair to get even with the unfaithful partner.

5. Homosexual relationship. Some individuals marry as a front for their homosexuality. Cole Porter, known for "I've Got You Under My Skin," "Night and Day," and "Easy to Love," was a homosexual who feared no one would buy or publish his music if his sexual orientation were known. He married Linda Lee Porter (alleged to be a lesbian), and they had an enduring marriage for 30 years.

Other gay individuals marry as a way of denying their homosexuality. These individuals are likely to feel unfulfilled in their marriage and may seek involvement in an extramarital homosexual relationship. Other individuals may marry and then discover later in life that they desire a homosexual relationship. Such individuals may feel that (1) they have been homosexual or bisexual all along, (2) their sexual orientation has changed from heterosexual to homosexual or bisexual, (3) they are unsure of their sexual orientation and want to explore a homosexual relationship, or (4) they are predominately heterosexual but wish to experience a homosexual relationship for variety. In Chapter 9 we identified the term *down low* as referring to African-American married men who have sex with men (Browder, 2005).

6. Aging. A frequent motive for intercourse outside marriage is the desire to return to the feeling of youth. Ageism, which is discrimination against the elderly, promotes the idea that it is good to be young and bad to be old. Sexual attractiveness is equated with youth, and having an affair may confirm to an older partner that he or she is still sexually desirable. Also, people may try to recapture the love, excitement, adventure, and romance associated with youth by having an affair.

7. Absence from partner. One factor that may predispose a spouse to an affair is prolonged separation from the partner. Some wives whose husbands are away for military service report that the loneliness can become unbearable. Some husbands who are away say it is difficult to be faithful. Partners in commuter relationships may also be vulnerable to extradyadic sexual relationships.

Reactions to the knowledge that one's spouse has been unfaithful vary. For most, the revelation is difficult. Below is an example of a wife's reaction to her husband's affairs.

> *My husband began to have affairs within six months of our being married. Some of the feelings I experienced were disbelief, doubt, humiliation and outright heart wrenching pain! When I confronted my husband he denied any such affair and said that I was suspicious, jealous and had no faith in him. In effect, I had the problem. He said that I should not listen to what others said because they did not want to see us happy but only wanted to cause trouble in our marriage. I was deeply in love with my husband and knew in my heart that he was guilty as sin; I lived in denial so I could continue our marriage.*

Sleeping with married men is always trouble; I have forsworn it.
Erica Jong

Of course, my husband continued to have affairs. Some of the effects on me included:

1. I lost the inability to trust my husband and, after my divorce, other men.

2. I developed a negative self concept—the reason he was having affairs is that something was wrong with me.

3. He robbed me of the innocence and my "VIRGINITY"—clearly he did not value the opportunity to be the only man to have experienced intimacy with me.

4. I developed an intense hatred for my husband.

It took years for me to recover from this crisis. I feel that through faith and religion I have emerged "whole" again. Years after the divorce my husband made a point of apologizing and letting me know that there was nothing wrong with me, that he was just young and stupid and not ready to be serious and committed to the marriage.

Linquist and Negy (2005) suggested that therapists might consider being less judgmental about clients who have affairs and consider the benefits: validation, enhancement of one's self-esteem, experiencing affection. They also suggested that marriages need not end over an affair. Indeed, Linquist, a therapist, noted treating couples "who have reported that they had actually benefited by one of the partners having an extrarelational affair" (by improving communication between the two primary partners).

No matter what everybody says, ultimately these things can harm us only by the way we react to them.

Jean Baptiste Poquelin

PERSONAL CHOICES

Should You Seek a Divorce If Your Partner Has an Affair?

About 20 percent of spouses face the decision of whether to stay with a mate who has had an extramarital affair. One alternative is to end the relationship immediately on the premise that trust has been broken and can never be mended. Persons who take this position regard fidelity as a core element of the marriage that if violated necessitates a divorce. College students disapprove of an extramarital affair. In a sample of 1019 undergraduates, 73 percent said that they would "divorce a spouse who had an affair" (Knox and Zusman, 2006). For some, emotional betrayal is equal to sexual betrayal (Sabini and Silver, 2005).

Other couples build into their relationship the fact that each will have external relationships. In Chapter 3 we discussed polyamory, in which partners are open and encouraging of multiple relationships at the same time. The term infidelity does not exist for polyamorous couples.

Even for traditional couples, infidelity need not be the end of a couple's marriage but the beginning of a new, enhanced, and more understanding relationship. Healing takes time, commitment on the part of the straying partner not to repeat the behavior, forgiveness by the partner (and not bringing it up again), and a new focus on improving the relationship. Couples who have been through the crisis of infidelity admonish, "Don't make any quick decisions" (Olson et al., 2002, p. 433). In spite of the difficulty of adjusting to an affair, most spouses are reluctant to end a marriage. Not one of 50 successful couples said that they would automatically end their marriage over adultery (Wallerstein and Blakeslee, 1995).

The spouse who chooses to have an affair is often judged as being unfaithful to the vows of the marriage, as being deceitful to the partner, and as inflicting enormous pain on the partner (and children). What is often not considered is that when an affair is defined in terms of giving emotional energy, time, and economic resources to something or someone outside the primary relationship, other types of "affairs" may be equally as devastating to a relationship. Spouses who choose to devote their lives to their children, careers, parents, friends, or recreational interests may deprive the partner of significant amounts of emotional energy, time, and money and create a context in which the partner may choose to become involved with a person who provides more attention and interest.

When traditional spouses do stay together after an affair, the price is high. In the Charney and Parnass (1995) study, 43 percent of the 66 percent who stayed together were judged to have difficulty coping: "The majority of the betrayed husbands and wives suffered significant damage to their self-image, personal confidence, or sexual confidence, feelings of abandonment, attacks on their sense of belonging, betrayals of trust, enraged feelings, and/or a surge of justification to leave their spouses" (p. 100).

Among the respondents, the researchers noted that only 14.5 percent of those who remained together after an affair described their relationship as characterized by improvement and growth. Of these, almost half were relationships in which the affair was a one-night stand, not a long-term romantic and sexual involvement. (Interestingly, 89 percent of the betrayed spouses "really knew" that their partner had had an affair before being told.)

Another issue in deciding whether to take the spouse back following extramarital sex is the concern over HIV. One spouse noted that though he was willing to forgive and try to forget his partner's indiscretion, he required that she be tested for HIV and that they use a condom for 6 months. She tested negative, but their use of a condom was a reminder, he said, that sex outside one's bonded relationship in today's world has a life-or-death meaning. Related to this issue is that wives are more likely to develop cervical cancer if their husbands have other sexual partners.

There is no single way to respond to a partner who has an extramarital relationship. Most partners are hurt and think of ways to work through the crisis. Some succeed.

Choose to forgive. Muster all your strength to let go of an offense so you can move on.

Kathleen Lawler, psychologist

Sources

Charney, I. W. and S. Parnass. 1995. The impact of extramarital relationships on the continuation of marriages. *Journal of Sex and Marital Therapy* 21:100–15.

Knox, D. and Zusman, M. E. 2006. Relationship and sexual behaviors of a sample of 1027 university students. Unpublished data collected for this text. Department of Sociology, East Carolina University, Greenville, NC.

Olson, M. M., C. S. Russell, M. Higgins-Kessler, and R. B. Miller. 2002. Emotional processes following disclosure of an extramarital affair. *Journal of Marital and Family Therapy* 28:423–34.

Wallerstein, J. S. and S. Blakeslee. 1995. *The good marriage.* Boston: Houghton Mifflin.

Unemployment

In the wake of hurricanes Katrina and Rita, there was massive unemployment in the Gulf States. The jobs of thousands were literally washed away. Coupled with these job losses, corporate America continues to downsize and outsource jobs to India and Mexico. The result is massive layoffs and insecurity in the lives of American workers. Forced unemployment or the threatened loss of one's job is a major stressor for individuals, couples, and families (Legerski et al., 2006). Also, when spouses or parents lose their jobs as a result of physical illness or disability, the family experiences a double blow—loss of income combined with high medical bills. Unless the unemployed spouse is covered by the partner's medical insurance, unemployment can also result in loss of health insurance for the family. Insurance for both health care and disability is very important to help protect the family from an economic disaster. Prolonged unemployment may result in bankruptcy (Miller, 2005).

Diversity in Other Countries

A man of the Masai tribe (200,000 of whom live in Arusha, Tanzania, East Africa), coming back from a long journey in the forest herding animals, can put a spear outside the house of another man's wife to alert the husband that he is inside having sex with his wife. Seeing the spear, the husband does not bother the visitor, since to do so would be impolite.

The United States and France also view extramarital involvements differently. In the Unites States, the Clinton/Lewinsky scandal was big news and threatened the presidency. Knowledge of the adulterous relationship of France's president François Mitterand in which he fathered a child was viewed as hardly worth mentioning in the media and certainly not cause for dismissal.

The effects of unemployment may be more severe for men than for women. Our society expects men to be the primary breadwinners in their families and equates masculine self-worth and identity with job and income. Stress, depression, suicide, alcohol abuse, and lowered self-esteem, as well as increased cigarette smoking (Falba et al., 2005) and seizures (Strine et al., 2005), are all associated with unemployment. Macmillan and Gartner (2000) also observed that men who are unemployed are more likely to be violent toward their working wives.

Women tend to adjust more easily to unemployment than men. Women are not burdened with the cultural expectation of the provider role, and their identity is less tied to their work role. Hence, women may view unemployment as an opportunity to spend more time with their families; many enjoy doing so.

Although unemployment can be stressful, increasing numbers of workers are experiencing job stress. With layoffs and downsizing, workers are given the work of two and told "we'll hire someone soon." Meanwhile, the employer learns it can pay once and get the work of two. The result is lowered morale, exhaustion, and stress that can escalate into violence. Workers note that taking frequent breaks or a day off is their best way of coping with job stress (Armour, 2003).

Substance Abuse

We haven't got a great deal to offer in the way of managing crack cocaine addiction.

Member of a drug therapy team

Spouses, parents, and children who abuse drugs contribute to the stress and conflict experienced in their respective marriages and families. Although some individuals abuse drugs to escape from unhappy relationships or the stress of family problems, substance abuse inevitably adds to the individual's marital/family problems when it results in health and medical problems, legal problems, loss of employment, financial ruin, school failure, divorce, and even death (due to accidents or poor health). The following reflects the experience of the wife of a crack addict.

> Marriage needs a foundation of trust and open communication, but when you are married to an addict, you won't find either. When I first met "Nate," I thought he was everything I wanted in a partner—good looks, a great sense of humor, intelligence, and ambition. It didn't take me long to realize that I was extremely attracted to this man. Since we knew some of the same people, we ended up at several parties together. We both drank and occasionally used cocaine but it wasn't a problem. I loved being with him and whenever I wasn't with him, I was thinking about him.

> After only four months of dating, we realized we were falling madly in love with each other and decided to get married. Neither of us had ever been married before but I had a five-year-old son from a previous relationship. Nate really took to my son and it seemed like we had the perfect marriage—that changed drastically and quickly. I didn't know it but my husband had been using crack cocaine the entire time we dated. I wasn't extremely familiar with this new drug and I knew even less about its addictive power. Of course, I soon found out more about it than I cared to know.

> Less than three months after we married, Nate went on his first binge. After leaving work one Friday night, he never came home. I was really worried and was calling all of his friends trying to find out where he was and what was wrong. I talked to the girlfriend of one of his best friends and found out everything. She told me that Nate had been smoking crack off and on for a long time and that he was probably out using again. I didn't know what to do or where to turn. I just stayed home all weekend waiting to hear from my husband.

> Finally, on Sunday afternoon, Nate came home. He looked like hell and I was mad as hell. I sent my son to play with his friends next door and as soon as he was out of hearing range, I lost it. I began screaming and crying, asking why he never came home over the weekend. He just hung his head in absolute shame. I found out later

that he had spent his entire paycheck, pawned his wedding ring, and had written checks off of a closed checking account. He went through over $1000 of "our" money in less than three days.

I was devastated and in shock. After I calmed down and my anger subsided, we discussed his addiction. He told me he loved me with all his heart and that he was so sorry for what he had done. He also swore he would never do it again. Well, this is when I became an enabler and I continued to enable my husband for three years. He would stay clean for a while and things would be great between us, then he would go on another binge. It was the roller-coaster ride from hell. Nate was on a downward spiral and he was dragging my son and me down along with him.

By the end of our third year together, we were more like roommates. Our once wonderful sex life was virtually nonexistent and what love I still had for him was quickly fading away. I knew I had to leave before my love turned to hate. It was obvious that my son had been pulling away from Nathaniel emotionally so it was a good time to end the nightmare. I packed our belongings and moved in with my sister.

Less than three months after I left, Nate went into a rehabilitation program. I was happy for him but I knew I could never go back; it was too little, too late. That was nine years ago and the last I heard, he was still struggling with his addiction, living a life of misery. I have no regrets about leaving but I am saddened by what crack cocaine had done to my once wonderful husband—it turned him into a thief and a liar and ended our marriage.

Family crises involving alcohol and/or drugs are not unusual. Alcohol is also a major problem on campus (over half of 772 college students who reported drinking alcohol reported having blacked out [White et al., 2002]). The context of the most intense drinking is at a fraternity social. In a study of 306 university students, the average number of drinks at a fraternity social was 5.91 compared to 4.04 at campus parties (Miley and Frank, 2006).

Some students have also grown up in homes where one or both parents abused alcohol. Haugland (2005) studied alcohol abuse by the father and found that the drunk father simply was not around the children during drinking episodes. University attempts (mostly unsuccessful) have been made to curb excessive drinking (see Social Policy on page 386). Miley and Frank (2006) suggested that fraternity socials might become a focus since alcohol consumption is the highest.

As indicated in Table 13.1, drug use is most prevalent among 18- to 25-year-olds. Drug use among teenagers under age 18 is also high. Because teenage drug use is common, it may compound the challenge parents may have with their teenagers.

Drug Abuse Support Groups   Although treatments for alcohol abuse are varied, a combination of medications (naltrexone and acamprosate) and behavioral interventions (e.g., control for social context) is a favored approach (Mattson and Litten, 2005). Alcoholics Anonymous (AA), a support group (national head-

People who drink to drown their sorrow should be aware that sorrow knows how to swim.

Ann Landers

Table 13.1   Drug Use (Ever), by Type of Drug and Age Group

Type of Drug Used	Age 12 to 17	Age 18 to 25	Age 26 to 34
Marijuana and hashish	20%	54%	51%
Cocaine	3%	15%	18%
Alcohol	43%	87%	no data
Cigarettes	31%	70%	no data

Source: Adapted from *Statistical Abstract of the United States: 2006.* 125th ed. Washington, DC: U.S. Bureau of the Census, Table 194.

Alcohol Abuse on Campus

Alcohol is the drug most frequently used by college students. About 40 percent of university students report heavy episodic drinking (Mitchel et al., 2005), with between 50 percent and 60 percent reporting having engaged in a "drinking game" (e.g., taking a drink whenever a certain word is mentioned in a song or on television) (Borsari and Bergen-Cico, 2003). College students do not learn from negative consequences to their drinking. Mallett et al. (2006) found that students who threw up, made unwise sexual decisions, or experienced a hangover or blackout underestimated the amount of alcohol they could drink before they experienced negative consequences. Continuing to drink was a given.

College students who drink alcohol heavily are more likely to miss class, make lower grades, have auto accidents, have low self-esteem, and be depressed (Williams et al., 2002). Nevertheless, unless a college student drinks every day, he or she is unlikely to define himself or herself as having a drinking problem (Lederman et al., 2003). Alcohol is officially sanctioned at some colleges. In a study of campus policies in Minnesota and Wisconsin, 29% of the colleges/universities reported having bars on campus and accepted gifts from the alcohol industry (Mitchel et al., 2005).

Campus policies throughout the United States include alcohol-free dorms, alcohol bans, enforcement and sanctions, peer support, and education. A minority (around 30%) of colleges and universities ban alcohol or its possession (Mitchel et al., 2005). Administrators fear that students will attend other colleges where they are allowed to drink. And some alumni want to drink at football games and view such university banning as intrusive. Some attorneys think colleges and universities can be held liable for not stopping dangerous drinking patterns, but others argue that college is a place for students to learn how to behave responsibly. Should sanctions (expel a student, close down a fraternity) be used? If police are too restrictive, drinking will go underground, where it will be more difficult to detect.

For the purposes of peer education, students at the University of Buffalo in New York created a video illustrating responsible drinking and discussions by students of their own alcohol poisoning. While such educational interventions may increase knowledge of alcohol use, subsequent reduction in alcohol is unknown (Williams et al., 2002). Rutgers regularly conducts campus focus groups and recommends replacement of the term "binge drinking" with "dangerous drinking," since the latter emphasizes outcomes (Lederman et al., 2003).

Some universities have hired full-time alcohol-education coordinators. Providing nonalcoholic ways to meet others and to socialize has also been suggested (Borsari and Bergen-Cico, 2003). Policies will continue to shift in reference to parental pressure to address the issue. The latest approach has been to inform parents of their son's or daughter's alcohol or drug abuse on campus. An alcohol or drug infraction is reported to the parents, who may intervene early in curbing abuse. Kuo et al. (2002) emphasized that parents may play a potentially important role in prevention efforts.

Your Opinion?

1. To what degree do you believe drinking on campus should be a concern of the administration?
2. What university policies do you recommend to reduce binge drinking?
3. Under what conditions should a student who abuses alcohol be expelled?

Sources

Borsari, B. and D. Bergen-Cico. 2003. Self-reported drinking-game participation of incoming college students. *Journal of American College Health* 51:149–54.

Kuo, M., E. M. Adlaf, H. Lee, L. Gliksman, A. Demers, and H. Wechsler. 2002. More Canadian students drink but American students drink more: Comparing college alcohol use in two countries. *Addiction* 97:1583–92.

Lederman, L. C., L. P. Stewart, F. W. Goodhart, and L. Laitman. 2003. A case against "binge" as a term of choice: Convincing college students to personalize messages about dangerous drinking. *Journal of Health Communication* 8:79–91.

Mallett, K. A., C. M. Lee, C. Neighbors, M. E. Larimer, and R. Turrisi. 2006. Do we learn from our mistakes? An examination of the impact of negative alcohol-related consequences on college students' drinking patterns and perceptions. *Journal of Studies on Alcohol* 67:269–76.

Mitchel, R. J., T. L. Toomey, and D. Erickson. 2005. Alcohol policies on college campuses. *Journal of American College Health* 53:149–57.

Williams, D. J., A. Thomas, W. C. Buboltz, Jr., and M. McKinney. 2002. Changing the attitudes that predict underage drinking in college students: A program evaluation. *Journal of College Counseling* 5:39–49.

quarters: www.alcoholics-anonymous.org), has also been helpful. There are over 15,000 AA chapters nationwide; the one in your community can be found through the Yellow Pages. The only requirement for membership is the desire to stop drinking.

Former abusers of drugs (other than alcohol) meet regularly in local chapters of Narcotics Anonymous (NA), patterned after Alcoholics Anonymous, to help each other continue to be drug-free. As with AA, the premise of NA is that the best person to help someone stop abusing drugs is someone who once abused drugs. NA members of all ages, social classes, and educational levels provide a sense of support for each other to remain drug-free.

Al-Anon is an organization that provides support for family members and friends of alcohol abusers. Spouses and parents of substance abusers learn how to live with and react to living with a substance abuser.

Parents who abuse drugs also may benefit from the Strengthening Families Program, which provides specific social skills training for both parents and children. After families attend a 5-hour retreat, parents and children are involved in face-to-face skills training over a 4-month period. A 12-month follow-up revealed that parenting skills remained improved and that reported heroin and cocaine use had declined.

Death

Even more devastating than drug abuse are family crises involving death—of one's child, parent, or loved one (we discuss the death of one's spouse in Chapter 17, Aging in Marriage and Family Relationships). The crisis is particularly acute when the death is a suicide.

Death of One's Child   A parent's worst fear is the death of a child. Most people expect the death of their parents but not the death of their children. Jiong Li et al. (2003) found that the loss is particularly devastating to mothers, who experience a higher mortality risk after their child's death.

Grief fills up the room with my absent child.

Shakespeare

Many parents experience the loss of their child even before it is born. A miscarriage is a major crisis for a couple (Abbound and Liamputtong, 2005). Although some would-be parents may be relieved by a miscarriage if the pregnancy was unwanted, many feel sadness, frustration, disappointment, and anger. Some women blame themselves for the miscarriage and believe that they are being punished for something they have done in the past.

Among infants less than 1 year of age, the leading causes of death in the United States are congenital anomalies, sudden infant death syndrome (SIDS), disorders relating to short gestation and low birth weight, and respiratory distress syndrome. Between the ages of 1 and 24 years, accidents are the primary cause of death. Types of accidental deaths include drowning and poisoning from ingesting household products or medication. But the most common cause of accidental death among youth is motor vehicle accidents (Anderson and Smith, 2003).

Mothers and fathers sometimes respond to the death of their child in different ways. When they do, the respective partners may interpret these differences in negative ways, leading to relationship conflict and unhappiness. For example, after the death of their 17-year-old son, one wife accused her husband of not sharing in her grief. The husband explained that while he was deeply grieved he poured his grief into working more as a way of distraction. To deal with these differences, spouses might need to be patient and practice tolerance in allowing each other to grieve in his or her own way.

Death comes to us all and is often a crisis for those we leave behind.

Death of One's Parent   Terminally ill parents may be taken care of by their children. Such care over a period of years can be emotionally stressful, financially draining, and exhausting. Hence, by the time the parent dies, a crisis has already occurred.

Reactions to the death of one's parents include depression, loss of concentration, and anger (Ellis and Granger, 2002). Michael and Snyder (2005) studied college students who had experienced the death of a loved one (most often a parent) and found that their constant ruminations about the deceased correlated with a lower sense of psychological well-being.

Whether the death is that of a child or a parent, Burke et al. (1999) noted that grief is not a one-time experience that people adjust to and move on. Rather, for some, there is "chronic sorrow" where grief-related feelings occur periodically throughout the lives of those left behind. Paul Newman, the aging Hollywood celebrity, was asked how he got over the death of his son who overdosed. He replied, "You never get over it." Grief feelings may be particularly acute on the anniversary of the death or when the bereaved individual thinks of what might have been had the person lived. Burke et al. (1999) noted that 97 percent of the individuals in one study who had experienced the death of a loved one 2 to 20 years earlier met the criteria for chronic sorrow. Field et al. (2003) also observed bereavement-related distress 5 years after the death of a spouse.

Surviving the Suicide of a Loved One   Worldwide, about 800,000 individuals elect to commit suicide annually (Mercy et al., 2003). In the United States, there are over 30,000 suicides annually (Minino et al., 2006). Families of suicide victims are often left with questions about why their loved ones killed themselves and what could have been done to prevent the suicide. Although such questions are generally unanswerable, they often linger for years, prolonging the grieving process. In addition, survivors of suicide victims struggle with how others will view them and their family.

Because some religions consider suicide a mortal sin, surviving family members may also struggle with what the fate of their deceased loved one might be in the hereafter. Suicide among young people is particularly devastating, not only because a young life is ended but also because of the impact on the victim's family members and friends. A team of researchers identified 15,555 suicides among 15- to 24-year-olds in 34 countries in a 1-year period and found an association between divorce rates and suicide rates (Johnson et al., 2000).

Family members who are left behind after the suicide of a loved one often feel that there is something they could have done to prevent the death. Singer Judy Collins lost her son to suicide and began to attend a support group for persons who had a loved one who had committed suicide. One of the members in the group answered the question of whether there was something she could have done with a resounding no.

> *I was sitting on his bed saying, "I love you, Jim. Don't do this. How can you do this?" I had my hand on his hand, my cheek on his cheek. He said excuse me, reached his other hand around, took the gun from under the pillow, and blew his head off. My face was inches from his. If somebody wants to kill himself or herself, there is nothing you can do to stop them (Collins, 1998, p. 210).*

Marriage and Family Therapy

Couples might consider marriage and family therapy rather than continuing to drag through dissatisfaction related to a crisis event. Signs to look for in your own relationship that suggest you might consider seeing a therapist include feeling distant and not wanting or being unable to communicate with your partner, avoiding each other, drinking heavily or taking other drugs, privately contemplating separation, being involved in an affair, and feeling depressed.

If you are experiencing one or more of these problems, it may be wise not to wait until the relationship reaches a stage beyond which repair is impossible.

Relationships are like boats. A small leak will not sink them. But if left unattended, the small leak may grow larger or new leaks may break through. Marriage therapy sometimes serves to mend relationship problems early by helping the partners to sort out values, make decisions, and begin new behaviors so that they can start feeling better about each other.

Availability of Marriage and Family Therapists

There are around 50,000 marriage and family therapists in the United States. About 40 percent are clinical members of the American Association for Marriage and Family Therapy (AAMFT). Currently there are 57 masters, 19 doctoral, and 16 postgraduate programs accredited by the AAMFT. All but six states (Delaware, Montana, New York, North Dakota, Ohio, and West Virginia) and the District of Columbia regulate (e.g., require a license) for a person to practice marriage and family therapy (Northey, 2002).

Therapists holding membership in AAMFT have had graduate training in marriage and family therapy and a thousand hours of direct client contact (with 200 hours of supervision). The phone number of AAMFT is 1-202-452-0109, and the Web address is www.aamft.org. Clients are customers and should feel comfortable with their therapists and the progress they are making. If they don't, they should switch therapists.

Managed care has resulted in private therapists' lowering their fees so as to compete with what insurance companies will pay (about $80 per session). Some mental health centers offer marital and family therapy on a sliding-fee basis so that spouses, parents, and their children can be seen for as little as $5. Marriage therapists generally see both the husband and the wife together (called conjoint therapy); about 60 percent will involve the children as necessary but often not as active participants (Lund et al., 2002).

To what degree do spouses, parents, and children benefit from marriage and family therapy? Shadish and Baldwin (2003) synthesized 20 intervention studies on marriage and family therapy and concluded that such interventions were "clearly efficacious compared to no treatment" (p. 566).

Whether a couple in therapy remain together will depend on their motivation to do so, how long they have been in conflict, the severity of the problem, and whether one or both partners are involved in an extramarital affair. Two moderately motivated partners with numerous conflicts over several years are less likely to work out their problems than a highly motivated couple with minor conflicts of short duration. Severe depression or alcoholism on the part of either spouse is a factor that will limit positive marital and family gains. In general, these issues must be resolved individually before the spouses can profit from marital therapy.

Some couples come to therapy with the goal of separating amicably. The therapist then discusses the couple's feelings about the impending separation, the definition of the separation (temporary or permanent), the "rules" for their interaction during the period of separation (e.g., see each other, date others), and whether to begin discussions with a divorce mediator or attorneys.

One alternative for enhancing one's relationship is the Association for Couples in Marriage Enrichment (ACME). This organization provides conferences for couples to enrich their marriage. The web address for ACME is www.bettermarriages.org.

Outcome-Based Therapeutic Perspective

There are more than 20 different treatment approaches used by members of AAMFT (Northey, 2002). The largest percentage (27%) report that they use a "cognitive-behavioral" approach, which means they focus on the cognitions or assumptions that underlie a

The average total cost of marital therapy is about a thousand dollars for eleven sessions. Even twice this amount seems certainly less than the cost of a divorce and pales in comparison to the cost of many medical procedures.

James Bray and Ernest Bray, marriage therapists

Diversity in Other Countries

Lundblad and Hansson (2006) assessed the effectiveness of couples therapy with an initial sample of 300 Swedish couples and found that gains of improved functioning were evident at a 2-year follow-up.

marriage or family, with the goal of ensuring that these are accurate and functional. The therapist also examines what behaviors family members want increased or decreased, initiated or terminated, and tries to negotiate ways to accomplish these behavioral goals (hence the cognitive-behavioral label).

Other therapists (Imago relationship therapists) may conceptualize relationship and communication difficulties as rooted in family dynamics in which one was reared, specify how these patterns are operative in one's present relationship, and look for ways of understanding and improvement (Gelbin, 2003).

In the past, marriage and family therapy models were not constrained by the economics of managed care. As the bottom line has become the driving force determining what services insurance companies will pay for, increasingly marriage and family therapy has become more time-bound, brief, and solution-focused. Increasingly managed care companies have insisted on "outcome based" perspectives to ensure that specific observable positive changes occur as a result of the therapeutic dollar (Nelson and Smock, 2005).

Some Caveats about Marriage and Family Therapy

In spite of the potential benefits of marriage therapy, some valid reasons exist for not becoming involved in such therapy. Not all spouses who become involved in marriage therapy regard the experience positively. Some feel that their marriage is worse as a result. Saying things the spouse can't forget, feeling hopeless at not being able to resolve a problem "even with a counselor," and feeling resentment over new demands made by the spouse in therapy are reasons some spouses cite for negative outcomes.

Therapists also may give clients an unrealistic picture of loving, cooperative, and growing relationships in which partners always treat each other with respect and understanding, share intimacy, and help each other become whoever each wants to be. In creating this idealistic image of the perfect relationship, therapists may inadvertently encourage clients to focus on the shortcomings in their relationship and to expect more of their marriage than is realistic. Dr. Robert Sammons (2004) calls this his first law of therapy: "That spouses always focus on what is missing rather than what they have . . indeed the only thing that is important to couples in therapy is that which is missing."

Couples and families in therapy must also guard against assuming that therapy will be a quick and easy fix. Changing one's way of viewing a situation (cognitions) and behavior requires a deliberate, consistent, relentless commitment to make things better. Without it, couples are wasting their time and money.

Couples who become involved in marriage counseling may also miss work, have to pay for child care, and be "exposed" at work if they use their employer's insurance policy to cover the cost of the therapy. Though these are not reasons to decide against seeing a counselor, they are issues that concern some couples.

On-Line Marriage Counseling

The Internet now features over 200 "on-line therapy" sites promising access to over 350 on-line counselors (*not* marriage counselors). Though these may be helpful for getting e-mail answers to e-mail questions, ongoing on-line marital therapy is virtually unknown. Since effective marriage counseling requires the participation and involvement of both spouses, on-line marital therapy is made difficult since both partners would need to be on-line at the same time. In addition, nonverbal interaction behaviors cannot be observed by the therapist. If these concerns are not enough to dissuade spouses from trying therapy on-line, they should be aware that e-mail communications are not always safe and the therapist might not be available in times of a crisis. One client noted that his counselor simply disappeared for 3 weeks, only to reemerge with the explanation that "my e-mail was jammed and messages were not getting through" (Segall, 2000, p. 43).

Chocolate is cheaper than therapy and you don't need an appointment.

Unknown

What is stress and what is a crisis event?

Stress is a reaction of the body to substantial or unusual demands (physical, environmental, or interpersonal). Stress is a process rather than a state. A crisis is a situation that requires changes in normal patterns of behavior. A family crisis is a situation that upsets the normal functioning of the family and requires a new set of responses to the stressor. Sources of stress and crises can be external (e.g., hurricane, tornado, downsizing, military separation) or internal (e.g., alcoholism, extramarital affair, Alzheimer's disease).

Resiliency refers to a family's strengths and ability to respond to a crisis in a positive way. Characteristics associated with resilient families include having a joint cause or purpose, emotional support for each other, good problem-solving skills, ability to delay gratification, flexibility, accessing residual resources, communication, and commitment.

What are positive stress management strategies?

Changing one's basic values and perspective is the most helpful strategy in reacting to a crisis. Viewing ill health as a challenge, bankruptcy as an opportunity to spend time with one's family, and infidelity as an opportunity to improve communication are examples. Other positive coping strategies are exercise, adequate sleep, love, religion, friends/relatives, multiple roles, and humor. Still other strategies include intervening early in a crisis, not blaming each other, keeping destructive impulses in check, and seeking opportunities for fun.

What are harmful strategies for reacting to a crisis?

Some harmful strategies include keeping feelings inside, taking out frustrations on others, and denying or avoiding the problem.

What are five of the major family crisis events?

Some of the more common crisis events faced by spouses and families include physical illness, an extramarital affair, unemployment, alcohol/drug abuse, and the death of one's spouse or children. The occurrence of a "mid-life crisis" is reported by less than a quarter of adults in the middle years. Those who did experience a crisis were going through a divorce.

What help is available from marriage and family therapists?

There are around 50,000 marriage and family therapists in the United States. About 40 percent are clinical members of the American Association for Marriage and Family Therapy (AAMFT). Whether a couple in therapy remain together will depend on their motivation to do so, how long they have been in conflict, the severity of the problem, and whether one or both partners are involved in an extramarital affair. Two moderately motivated partners with numerous conflicts over several years are less likely to work out their problems than a highly motivated couple with minor conflicts of short duration.

KEY TERMS

Al-Anon	crisis	extramarital affair	resiliency
biofeedback	extradyadic involvement	palliative care	stress
Coolidge Effect			

The Companion Website for *Choices in Relationships: An Introduction to Marriage and the Family,* Ninth Edition
www.thomson.edu/sociology/knox

Supplement your review of this chapter by going to the companion website to take one of the Tutorial Quizzes, use the flash cards to master key terms, and check out the many other study aids you'll find there. You'll also find special features such as the Marriage and Family Resource Center, Census 2000 information, and other data and resources at your fingertips to help you with that special project or to do some research on your own.

WEB LINKS

Association for Applied and Therapeutic Humor
http://www.aath.org

Association for Couples in Marriage Enrichment
http://www.bettermarriages.org/

All State Investigations, Inc. (help investigating infidelity)
http://www.infidelity.com/

American Association for Marriage and Family Therapy
http://www.aamft.org/

Association for Applied Psychophysiology and Biofeedback
http://www.aapb.org

Dear Peggy.com (coping with an affair)
http://www.vaughan-vaughan.com/

Red Hat Society
http://www.redhatsociety.com/info/howitstarted.html

U.S. Department of Health
http://www.health.gov/

REFERENCES

Abbound, L. and P. Liamputtong. 2005. When pregnancy fails: Coping strategies, support networks and experiences with health care of ethnic women and their partners. *Journal of Reproductive and Infant Psychology* 23:3–18.

Acohido, B. and J. O'Donnell. 2005. Extramarital affair topples Boeing CEO. *USA Today.* Tuesday March 8, Section B, p. 1.

Adrian, J. B. and K. Hartnett. 2005 Infidelity in committed relationships II: A substantive review. *Journal of Marital and Family Therapy.* 31:217–34.

Anderson, R.N., and B. L. Smith 2003. Leading causes of death for 2001. *National Vital Statistics Reports* 52 (1). Hyattsville, MD: National Center for Health Statistics.

Armour, S. 2003. Rising job stress could affect bottom line. *USA Today* 29 July,1B.

Baldwin, D. R., L. N. Chambliss, and K. Towler. 2003. Optimism and stress: An African- American college student perspective. *College Student Journal* 37:276–83.

Bermant, G. 1976. Sexual behavior: Hard times with the Coolidge Effect. In *Psychological research: The inside story,* edited by M. H. Siegel and H. P. Zeigler. New York: Harper and Row.

Biscaro, M, K. Broer, and N. Taylor. 2004. Self efficacy, alcohol expectancy and problem-solving appraisal as predictors of alcohol use in college students. *College Student Journal* 38:541–51.

Bosari, B., and D. Bergen-Cico. 2003. Self-reported drinking-game participation of incoming college students. *Journal of American College Health* 51:149–54.

Browder, B. S. 2005 *African-American woman who wrote On the Up and Up: A Survival Guide for Women Living with Men on the Down Low* New York: Kensington Publishers Corp.

Burke, M. L., G. G. Eakes, and M. A. Hainsworth. 1999. Milestones of chronic sorrow: Perspectives of chronically ill and bereaved persons and family caregivers. *Journal of Family Nursing* 5:387–84.

Burr, W. R., and S. R. Klein and associates.1994. *Reexamining family stress: New theory and research.* Thousand Oaks, CA: Sage.

Carlson-Catalano, J. 2003. Director, Clinical Biofeedback Services. *Health innovations.* Greenville, NC 27878. Personal communication, June 9.

Charney, I. W., and S. Parnass. 1995. The impact of extramarital relationships on the continuation of marriages. *Journal of Sex and Marital Therapy* 21:100–15.

Collins, J. 1998. *Singing lessons: A memoir of love, loss, hope, and healing.* New York: Pocket Books.

DeHaan, L. 2002. Book review of *Resilient marriages* by K. J. Shirley. *Family Relations* 51:185–86.

Djamba, Y. K., M. J. Crump, and A. G. Jackson. 2005. Levels and determinants of extramarital sex. Paper, Southern Sociological Society, Charlotte, NC. March.

Ellis, R. T., and J. M. Granger. 2002. African- American adults' perceptions of the effects of parental loss during adolescence. *Child and Adolescent Social Work Journal* 19:271–86.

Enright, E. 2004. A house divided. *AARP The Magazine* July/August. *60 et passim.*

Falba, T. M., J. L. Sindelar, and W. T. Gallo. 2005. The effect of involuntary job loss on smoking intensity and relapse. *Addiction.* 100:1330–39.

Field, N. P., Gal-Oz, E., and G. A. Bananno. 2003. Continuing bonds and adjustment at 5 years after the death of a spouse. *Journal of Consulting and Clinical Psychology* 71:110–17.

Flynn, M. A. T., D. A. McNeil, B. Maloff, D. Matasingwa, M. Wu, C. Ford and S. C. Tough. 2006. Reducing obesity and related chronic disease risk in children and youth: a synthesis of evidence with 'best practice' recommendations. *Obesity Reviews* 7:7–66.

Frieden, J. 2005. Declining mental illness stigma. *Clinical Psychiatry News* 33:72–73.

Gelbin, R. Personal communication. 2003. (Ms. Gelbin is a Certified Professional Counselor in Arizona who specializes in Imago Therapy; see Imagocouples.com).

Goode, E. 1999. New study finds middle age is prime of life. *New York Times.* 17 July 17, D6.

Grafanaki, S., D. Pearson, F. Cini, D. Godula, B. Mckenzie, S. Nason, and M. Anderegg. 2005. Sources of renewal: A qualitative study of the experience and role of leisure in the life of counselors and psychologists. *Counselling Psychology Quarterly* 18:31–41.

Haugland, B. S. M. 2005. Recurrent disruptions of rituals and routines in families with paternal alcohol abuse. *Family Relations* 54:225–41.

Jeglic, E. L., J. W. Griffith, C. M. Pepper, and A. B. Miller. 2005. A caregiving model of coping with a partner's depression. *Family Relations* 54:37–46.

Jiong Li, P., H. Dorthe, P. B. Mortensen, and J. Olsen. 2003. Mortality in parents after death of a child in Denmark: A nationwide follow-up study. *Lancet* 361:363–68.

Johnson, G. R., E. G. Krug, and L. B. Potter. 2000. Suicide among adolescents and young adults: A cross-national comparison of 34 countries. *Suicide and Life-Threatening Behavior* 30:74–82.

Knox, D. and Zusman, M. E. 2006. Relationship and sexual behaviors of a sample of 1027 university students. Unpublished data collected for this text. Department of Sociology, East Carolina University, Greenville, NC.

Kuo, M., E. Adlaf, H. Lee, L. Gliksman, A. Demers, and H. Wechsler. 2002. More Canadian students drink but American students drink more: Comparing college alcohol use in two countries. *Addiction* 97:1583–92.

Lederman, L. C., L. P. Stewart, F. W. Goodhart, and L. Laitman. 2003. A case against "binge" as a term of choice: Convincing college students to personalize messages about dangerous drinking. *Journal of Health Communication* 8:79–91.

Lee, R. L. T., A. J. Loke and T. Yuen 2005. Health-promoting behaviors and psychosocial well-being of university students in Hong Kong. *Public Health Nursing* 22:209–20.

Legerski, E. M., M. Cornwall, and B. O'Neil. 2006 Changing locus of control: Steelworkers adjusting to forced unemployment. *Social Forces* 84:1521–1537.

Leonard, R. and A. Burns. 2006 Turning points in the lives of midlife and older women: Five-year follow-up. *Australian Psychologist* 41:28–36.

Levitt, M. J., J. Levitt, G. L. Bustos, N. A. Crooks, J. D. Santos, P. Telan, J. Hodgetts and A. Milevsky. 2005. Patterns of social support in middle childhood to early adolescent transition: Implications for adjustment. *Social Development* 14:398–420.

Linquist, L. and C. Negy. 2005. Maximizing the experiences of an extrarelational affair: An unconventional approach to a common social convention. *Journal of Clinical Psychology/In Session* 61:1421–28.

Lund, L. K., T. S. Zimmerman, and S. A. Haddock. 2002. The theory, structure, and techniques for the inclusion of children in family therapy: A literature review. *Journal of Marital and Family Therapy* 28:445–54.

Lundblad, A. and K. Hansson. 2006. Couples therapy: effectiveness of treatment and long-term follow up. *Journal of Family Therapy* 28:136–52.

Macmillan, R., and Gartner. 2000. When she brings home the bacon: Labor-force participation and the risk of spousal violence against women. *Journal of Marriage and the Family* 61:947–58.

Mahoney, D. 2005. Mental illness prevalence high, despite advances. *Clinical Psychiatry News.* 33:1–2.

Marano, H. S. 2003. The new sex scorecard. *Psychology Today* 36:38 passim.

Mattson, M.E. and R. Z. Litten. 2005. Combining treatments for alcoholism: Why and how? *Journal of Studies on Alcohol.* July:8–16.

Mercy, J. A., E. G. Krug, L. L. Dahlberg, and A. B. Zwi. 2003. Violence and health: The United States in global perspective. *American Journal of Public Health* 93:256–61.

Michael, S. T. and C. R. Snyder. 2005. Getting unstuck: The roles of hope, finding meaning, and rumination in the adjustment to bereavement among college students. *Death Studies* 29:435–459.

Minino, A. M., R. N. Anderson, L. A. Fingerhut, M. A. Boudreault and M. Warner. 2006. Deaths: Injuries, 2002. National Vital Statistics Reports, Vol 54., No. 10. Hyattsville, Maryland: National Center for Health Statistics.

Miley, W. M. and W. Frank. 2006. Binge and non-binge college students' perceptions of other students' drinking habits. *College Student Journal* 40: 259-262

Miller, M. 2005. Where's the outrage? *Social Policy* 35:5–8.

Mitchel, R. J., T. L. Toomey, and D. Erickson. 2005. Alcohol policies on college campuses. *Journal of American College Health* 53:149–157.

Moore, D. W. 2006. Close to 6 in 10 Americans want to lose weight. Gallup poll. March 10. Retrieved April 20 from http://poll.gallup.com/content/default.aspx?ci=21859&pg=1.

Morell, V. 1998. A new look at monogamy. *Science* 281:1982 passim.

Mostaghimi, L., W. H. Obermeyer, B. Ballamudi, D. M. Gonzalez and R. M. Benca. 2005. Effects of sleep deprivation on wound healing. *Journal of Sleep Research* 12:213–19.

Nelson, T. S. and S. A. Smock. 2005. Challenges of an outcome-based perspective for marriage and family therapy education. *Family Process* 44:355–65.

Nilsson, J. P., M. Soderstrom, A. U. Karlsson, M. Lekander, T. Akerstedt, N. E. Lindroth, and J. Axelsson. 2005. Influence of midday naps on declarative memory performance and motivation. *Journal of Sleep Research* 14:1–6.

Northey, W. F. J. 2002. Characteristics and clinical practices of marriage and family therapists: A national survey. *Journal of Marriage and Family Therapy* 28:487–94.

Olson, M. M., C. S. Russell, M. Higgins-Kessler, and R. B. Miller. 2002. Emotional processes following disclosure of an extramarital affair. *Journal of Marital and Family Therapy* 28:423–34.

Orr, C. Racial differences in dieting and exercise behaviors, stress, and depression in college students. 4th Annual ECU Research and Creative Activities Symposium, April 21, East Carolina University, Greenville, NC.

Ostbye, T., K. M. Krause, M. C. Norton, J. Tschanz, L. Sanders, K. Hayden, C. Pieper, and K. A. Welsh-Bohmer. 2006. Ten dimensions of health and their relationships with overall self-reported health and survival in a predominately religiously active elderly population: The Cache County Memory Study. *Journal of the American Geriatrics Society* 54:199–209.

Ozer, E. J., S. R. Best, T. L. Lipsey, and D. S. Weiss. 2003. Predictors of posttrauamatic stress disorder and symptoms in adults: A meta-analysis. *Psychological Bulletin* 129:52–73.

Park, C. 2006. Exploring relations among religiousness, meaning, and adjustment to lifetime and current stressful encounters in later life. *Anxiety Stress & Coping* 19:33–45.

Pyle, S. A., J. Sharkey, G. Yetter, E. Felix, M. J. Furlog, W. S. Poston. 2006. Fighting an epidemic: The role of schools in reducing childhood obesity. *Psychology in the Schools* 43:361–76.

Raikes, H. A. 2005 Efficacy and social support as predictors of parenting stress among families in poverty. *Infant Mental Health Journal* 26 177–90.

Roberto, K. A. 2005. Families and policy: Health issues of older women. In *Sourcebook of family theory & research*, edited by Vern L. Bengtson, Alan C. Acock, Katherine R. Allen, Peggye Dilworth-Anderson, and David M. Klein. Thousand Oaks, CA: Sage Publications, 547–48.

Routh, K. and J. N. Rao. 2006. A simple, and potentially low-cost method for measuring the presence of childhood obesity. *Child: Care, Health & Development* 32:239–45.

Sabini, J. and M. Silver. 2005. Gender and jealousy. *Cognition and Emotion* 19:713–27.

Sammons, R. A., Jr. 2004. First law of therapy. Personal communication. Grand Junction, Colorado (Dr. Sammons is a psychiatrist in private practice).

Sbarra, D. A. and R. E. Emery. 2005. The emotional sequelae of nonmarital relationship dissolution: Analysis of change and intraindividual variability over time. *Personal Relationships* 12:213–32.

Schabus, M., K. Hodlmoser, T. Pecherstorfer, G. Klosch. 2005. Influence of midday nap on declarative memory performance and motivation. *Somnologie* 9:148–53.

Scheinkman, M. 2005. Beyond the trauma of betrayal: Reconsidering affairs in couples therapy. *Family Process* 44:227–44.

Schneider, J. P. 2000. Effects of cybersex addiction on the family: Results of a survey. *Sexual Addiction and Compulsivity* 7:31–58.

Schneider, J. P. 2003. The impact of compulsive cybersex behaviors on the family. *Sexual and Relationship Therapy* 18:329–55.

Segall, R. 2000. Online shrinks: The inside story. *Psychology Today* May/June, 38–43.

Selak, J. H. and S. S. Overman. 2005. *You don't LOOK sick!* New York: The Haworth Medical Press.

Shadish, W. R., and S. A. Baldwin. 2003. Metaanalysis of MTF interventions. *Journal of Marriage and the Family* 29:547–70.

Sharpe, L. and L. Curran. 2006. Understanding the process of adjustment to illness. *Social Science and Medicine* 62: 1153–66.

Smith, L. R. 2005. Infidelity and emotionally focused therapy: A program design. Dissertation Abstracts. International, Section B, The Sciences and Engineering, 65(10-B):5423.

Statistical Abstract of the United States: 2006. 125th ed. Washington, DC: U.S. Bureau of the Census.

Strine, T. W., R. Kobau, D. P. Chapman, D. J. Thurman, P. Price, and L. S. Balluz. 2005. Psychological distress, comorbidities, and health behaviors among U.S. adults with seizures: Results from the 2002 National Health interview survey. *Epilepsia* 46:1133–39.

Szabo, A., S. E. Ainsworth, and P. K. Danks. 2005. Experimental comparison of the psychological benefits of aerobic exercise, humor, and music. *International Journal of Humor Research* 18:235–46.

Termini, K. A. 2006. Reducing the negative psychological and physiological effects of chronic stress. 4th Annual ECU Research and Creative. Activities Symposium, April 21, East Carolina University, Greenville, NC.

Treas, J. and D. Giesen. 2000. Sexual infidelity among married and cohabiting Americans. *Journal of Marriage and the Family* 62: 48-60.

Unnever, J. D., F. T. Cullen, B. K. Applegate. 2005. Turning the other cheek: Reassessing the impact of religion on punitive ideology. *Justice Quarterly* 22:304–39.

U.S. Department of Health and Human Services (2005) An analysis of the 2005 Current Population Survey; retrieved August, 2006 from http://aspe.hhs.gov/health/reports/05/uninsured-cps.

Wallerstein, J. S., and S. Blakeslee. 1995. *The good marriage.* Boston: Houghton Mifflin.

Walsh, F. 2003. Clinical views of family normality, health, and dysfunction. In *Normal Family Processes: Growing Diversity and Complexity*, edited by F. Walsh. New York/London: Guilford Press, 27–57.

Weckwerth, A. C. and D. M. Flynn. 2006 Effect of sex on perceived support and burnout in university students. *College Student Journal* 40: 237-249

White, A. M., D. W. Jamieson-Drake, and H. S. Swartzwelder. 2002. Prevalence and correlates of alcohol-induced blackouts among college students: Results of an e-mail survey. *Journal of American College Health* 51:117–32.

Whitty, M. T. 2005. The realness of cybercheating: Men's and women's representations of unfaithful Internet relationships. *Social Science Computer Review* 23:57–67.

Williams, D. J., A. Thomas, W. C. Buboltz, Jr., and M. McKinney. 2002. Changing the attitudes that predict underage drinking in college students: A program evaluation. *Journal of College Counseling* 5:39–49.

chapter 14

. . . . there are some things that can't be the truth even if it did happen.

Ken Kesey, American author

Violence and Abuse in Relationships

Contents

True or False?

1. Women who dress seductively, even wives, are viewed by both undergraduate men and undergraduate women as partly responsible for being raped by their husbands.

2. As of 2006, a Pentagon survey of the Army, Navy, and Air Force military academies reported virtual elimination of sexual harassment of women.

3. Women are as likely to stalk a former lover as a man.

4. Women who took self-defense classes reported that men reacted to their attempt to resist rape with overwhelming force.

5. Women in abusive relationships report more psychological symptoms (depression/anxiety) than men in abusive relationships.

Answers: **1.** T   **2.** F   **3.** F   **4.** F   **5.** T

I think about her all the time. Not as [a celebrity], but as my daughter, my little girl.

Sharon Rocha, mother of Laci Peterson, who was murdered by Scott Peterson, her husband.

Television programs such as *American Justice, Cold Case Files,* and *Investigative Reports* (not to speak of the nightly news) are replete with stories of domestic violence, abuse, and murder. What these television/news stories have in common is that they reveal a frightening reality—the person most likely to be violent and abusive toward you (even kill you) is the person you are involved with in a romantic or sexual relationship. We expect our intimate relationships to be havens where we will be loved, protected, and cared for. The reality is sometimes different. In this chapter we look at intimate partners who hurt each other. We begin by defining terms.

Types and Incidence of Abuse

There are several types of abuse in relationships.

Lacy Peterson was the victim of uxoricide at the hands of Scott Peterson, who was convicted and sentenced to death.

Violence

Also referred to as **physical abuse, violence** may be defined as the intentional infliction of physical harm by either partner on the other. Examples of physical violence include pushing, throwing something at the partner, slapping, hitting, and forcing sex on the partner. **Intimate-partner violence** is an all-inclusive term that refers to crimes committed against current or former spouses, boyfriends, or girlfriends. **Battered-woman syndrome** refers to the general pattern of battering that a woman is subjected to and is defined in terms of the frequency, severity, and injury she experiences. Battering is severe if the person's injuries require medical treatment or the perpetrator could be prosecuted. Battering may lead to murder. **Uxoricide** is the murder of a woman by a romantic partner. The murder of Laci Peterson by her convicted husband Scott Peterson is an example of uxoricide.

As a prelude to murder, the man may hold his wife hostage (holding one or more

© AP/Wide World Photos

persons against their will with the actual or implied use of force). A typical example is an estranged spouse who re-enters the house of the former spouse and holds hostage his wife and children. These situations are potentially dangerous because the perpetrator often has a weapon and may use it on the victims, himself, or both. A negotiator is usually called in to resolve the situation.

National Data

In the U.S. adult population, one in five couples report experiencing violence in their relationships (Field and Caetano, 2005). A great deal more abuse occurs than is reported. In a national sample of women who had been physically assaulted, almost three-fourths (73%) did not report the incidence to the police. The primary reason for no report was the belief that the police could not help (Dugan et al., 2003).

You are going to burn in hell for this.

Dennis Rocha, father of Scott Peterson's pregnant wife.

Interpersonal violence is not unique to heterosexual couples. Physical violence is reported to occur in 11 to 12 percent of same-sex couples. Differences between violence in other-sex and same-sex couples include that the latter is more mild, the threat of "outing" is present, and there is greater isolation of the violence since it often occurs in a context of "being in the closet." Finally, victims of same-sex relationship abuse lack legal protections and services that are available to abuse victims in heterosexual couples. For example, most battered women's shelters do not serve lesbian women (or gay men), and in nine states, legal protections for victims of domestic violence are granted only when the relationship is between a man and a woman or between spouses or former spouses (Rohrbaugh, 2006).

Emotional Abuse

In addition to being physically violent, partners may also engage in **emotional abuse** (also known as **psychological abuse, verbal abuse,** or **symbolic aggression**). Whereas more than 10 percent of undergraduates (10.4%; 1019) reported being involved in a physically abusive relationship, almost 30 percent (29.5%) reported that they "had been involved in an *emotionally* abusive relationship with a partner" (Knox and Zusman, 2006). Although emotional abuse does not involve physical harm, it is designed to make the partner feel bad—to denigrate the partner, reduce the partner's status, and make the partner vulnerable to being controlled by the abuser. Although there is debate about what constitutes psychological abuse, examples include the following:

There is a big difference between criticism and abuse, but many people don't know it.

Evan Esar, humorist

- calling the partner obese, stupid, crazy, ugly, and repulsive
- controlling the partner's time with friends, siblings, and parents
- telling the partner she or he is pitiful or pathetic and lucky to have anyone
- controlling how the partner dresses and/or accusing the partner of dressing like a slut
- refusing to talk to or to touch the partner
- accusing the partner of infidelity
- threatening to leave the partner to enjoy being with someone else
- controlling the money of the partner to ensure dependence
- threatening to harm one's self if the partner leaves the relationship
- threatening to harm one's children or take them away
- demeaning or insulting the partner in front of others
- restricting the partner's mobility—for example, use of the car
- criticizing the partner's child care, food preparation, or job performance
- threatening to harm the partner, the partner's relatives, or the partner's pets
- telling the partner that he or she is a terrible lover
- threatening to have the partner committed to a mental institution
- demanding the partner does as he or she is told

Although women may be the abuser (see next section), it is more often the man. Of 227,941 restraining orders issued to adults in California, most were for domestic violence and most (72.2%) of the abusers were men, African-Americans, and 25- to 34-year-olds (Shen and Sorenson, 2005). Similarly, according to National Incident-Based Reporting System (NIBRS) data on intimate-partner violence, the victims tended to be young, female, and minority members (Vazquez et al., 2005).

Female Abuse of Partner: "I Regularly Abuse My Boyfriend"

Women also abuse their partners. Loy et al. (2005) noted that a history of trauma, limited support systems, and substance abuse were associated with women who abused their partners. The following was written by an abusive female.

> When people think of physical abuse in relationships, they get the picture of a man hitting or pushing the woman. Very few people, including myself, think about the "poor man" who is abused both physically and emotionally by his partner—but this is what I do to my boyfriend. We have been together for over a year now and things have gone from really good to terrible.
>
> The problem is that he lets me get away with taking out my anger on him. No matter what happens during the day, it's almost always his fault. It started out as kind of a joke. He would laugh and say "oh, how did I know that this was going to somehow be my fault." But then it turned into me screaming at him, and not letting him go out and be with his friends as his "punishment."
>
> Other times, I'll be talking to him or trying to understand his feelings about something that we're arguing about and he won't talk. He just shuts off and refuses to say anything but "alright." That makes me furious, so I start verbally and physically attacking him just to get a response.
>
> Or, an argument can start from just the smallest thing, like what we're going to watch on TV, and before you know it, I'm pushing him off the bed and stepping on his stomach as hard as I can and throwing the remote into the toilet. That really gets to him.

Stalking

Abuse may take the form of stalking. **Stalking** is defined as unwanted following or harassment that induces fear in the target person. Stalking is a crime. The stalker is one who has been rejected by a previous lover or is obsessed with a stranger or acquaintance who fails to return the stalker's romantic overtures. In about 80 percent of the cases, the stalker is a heterosexual male who follows his previous lover. Women who stalk are more likely to target a married male. Stalking is a pathological emotional/motivational state (Meloy and Fisher, 2005). Almost a third (32%) of 1027 university students reported that they had been stalked (followed and harassed) (Knox and Zusman, 2006). Such stalking is usually designed either to seek revenge or to win the partner back.

National Data

Between 8 and 15 percent of women and between 2 and 4 percent of men in the United States will be stalked in their lifetime (Meloy and Fisher, 2005).

Explanations for Violence and Abuse in Relationships

Research suggests that numerous factors contribute to violence and abuse in intimate relationships. These factors include those that occur at the cultural, community, and individual and family levels.

Cultural Factors

In many ways, American culture tolerates and even promotes violence. Violence in the family stems from the acceptance of violence in our society as a legitimate means of enforcing compliance and solving conflicts at interpersonal, familial, national, and international levels. Violence and abuse in the family may be linked to cultural factors, such as violence in the media, acceptance of corporal punishment, gender inequality, and the view of women and children as property. The context of stress is also conducive to violence.

Violence in the Media   One need only watch the evening news to see the violence in countries such as Iraq. In addition, feature films and TV movies regularly reflect themes of violence. As a society, we are inundated with violence in the media.

Corporal Punishment of Children   **Corporal punishment** is defined as the use of physical force with the intention of causing a child to experience pain, but not injury, for the purpose of correction or control of the child's behavior (Straus, 2000). In the United States it is legal in all 50 states for a parent to spank, hit, belt, paddle, whip, or otherwise inflict punitive pain on a child so long as the corporal punishment does not meet the individual state's definition of child abuse. Violence has become a part of our cultural heritage through the corporal punishment of children. In a review of the literature, 94 percent of parents of toddlers reported using corporal punishment (Straus, 2000). Spankers are more likely to be young parents and mothers and to have been hit when they were a child (Walsh, 2002). Children who are victims of corporal punishment display more antisocial behavior, are more violent, and have an increased incidence of depression as adults. Straus (2000) recommended an end to corporal punishment to reduce the risk of physical abuse and other harm to children.

> **Diversity in Other Countries**
>
> Sweden passed a law in 1979 that effectively abolished corporal punishment as a legitimate childrearing practice. Fifteen other countries, including Italy, Germany, and Ukraine, have banned all corporal punishment in all settings, including the home (Global Initiative to End All Corporal Punishment of Children, 2005).

Gender Inequality   Domestic violence and abuse may also stem from traditional gender roles. Nayak et al. (2003) found that individuals in countries espousing very restrictive roles for women (e.g., Kuwait) tend to be more accepting of violence against women. Traditionally, men have also been taught that they are superior to women and that they may use their aggression toward women, believing that women need to be "put in their place." The greater the inequality and dependence of the woman on the man, the more likely the abuse.

Traditional female gender roles have also taught women to be submissive to their male partners' control. DeMaris et al. (2003) found that when a nontraditional woman is paired with a traditional male, violence in their relationship is more likely.

Some occupations lend themselves to contexts of gender inequality. Military contexts are notorious for men devaluing, denigrating, and sexually harassing women. In spite of the rhetoric about gender equality in the military, half of the women in the Army, Navy, and Air Force academies in a 2004 Pentagon survey reported being sexually harassed (Komarow, 2005). These male perpetrators may not separate their work roles from their domestic roles. One student in the authors' classes noted that she was the wife of a Navy Seal and that "he knew how to torment someone and I was his victim."

Being a police officer may also be associated with abuse. Johnson et al. (2005) studied 413 officers and found that burnout, authoritarian style, alcohol use, and department withdrawal were associated with domestic violence. A team of researchers (Erwin et al., 2005) compared 106 police officers who had been charged with intimate-partner violence with 105 police officers without such an offense and found that minority status, being on the force for over 7 years, and being assigned to a high crime district were associated with domestic violence. The combined data from these two studies suggest that the stress of being a police officer seems to make one vulnerable to domestic abuse.

It is not surprising that marital violence is found to occur at a higher rate among the less educated (Verma, 2003), which is another context for inequality. Similarly, women who have higher incomes than their husbands report a higher frequency of beatings by their husbands (Verma, 2003).

In cultures where a man's "honor" is threatened if his wife is unfaithful, there is tolerance toward the husband's violence toward her. In some cases there are formal, legal traditions defending a man's right to beat or even to kill his wife in response to her infidelity (Vandello and Cohen, 2003). Unmarried women in Jordan who have intercourse are viewed as bringing shame on their parents and siblings and may be killed; this is referred to as an *honor crime* or *honor killing*. The legal consequence is minimal to nonexistent.

View of Women and Children as Property   Prior to the late nineteenth century, a married woman was considered the property of her husband. A husband had a legal right and marital obligation to discipline and control his wife through the use of physical force.

Stress   Our culture is also a context of stress. The stress associated with getting and holding a job, rearing children, staying out of debt, and paying bills may predispose one to lash out at others. Haskett et al. (2003) found that parenting stress was particularly predictive of child abuse. Persons who have learned to handle stress by being abusive toward others are vulnerable to becoming abusive partners.

Community Factors

Community factors that contribute to violence and abuse in the family include social isolation, poverty, and inaccessible or unaffordable health-care, day-care, elder-care, and respite-care services and facilities.

Social Isolation   Living in social isolation from extended family and community members increases the risk of being abused. A team of researchers (Grossman et al. 2005) also noted that the service needs of victims of abuse are greater for those living in rural areas. Spouses whose parents live nearby are least vulnerable.

Poverty   Abuse in adult relationships occurs among all socioeconomic groups. However, poverty and low socioeconomic development are associated with crime and higher incidences of violence. This violence may spill over into interpersonal relationships as well as the frustration of living in poverty (Tolan et al., 2003).

Inaccessible or Unaffordable Community Services   Failure to provide medical care to children and elderly family members sometimes results from the lack of accessible or affordable health-care services in the community. Failure to provide supervision for children and adults may result from inaccessible day-care and elder-care services. Without elder-care and respite-care facilities, families living in social isolation may not have any help with the stresses of caring for elderly family members and children.

Individual Factors

Individual factors associated with domestic violence and abuse include psychopathology, personality characteristics, and alcohol or substance abuse.

Personality Factors   A number of personality characteristics have been associated with persons who are abusive in their intimate relationships. Some of these characteristics follow:

1. Dependency. Therapists who work with batterers have observed that they are extremely dependent on their partners. Because the thought of being left by their partners induces panic and abandonment anxiety, batterers use physical aggression and threats of suicide to keep their partners with them.

2. Jealousy. Along with dependence, batterers exhibit jealousy, possessiveness, and suspicion. An abusive husband may express his possessiveness by isolating his wife from others; he may insist she stay at home, not work, and not socialize with others. His extreme, irrational jealousy may lead him to accuse his wife of infidelity and to beat her for her presumed affair. Indeed, jealousy is sometimes viewed as a sign of love (Puente and Cohen, 2003).

3. Need to control. Abusive partners have an excessive need to exercise power over their partners and to control them. The abusers do not let their partners make independent decisions, and they want to know where they are, whom they are with, and what they are doing. They like to be in charge of all aspects of family life, including finances and recreation.

4. Unhappiness and dissatisfaction. Abusive partners often report being unhappy and dissatisfied with their lives, both at home and at work. Many abusers have low self-esteem and high levels of anxiety, depression, and hostility. They may expect their partner to make them happy.

5. Anger and aggressiveness. Abusers tend to have a history of interpersonal aggressive behavior. They have poor impulse control and can become instantly enraged and lash out at the partner. Battered women report that episodes of violence are often triggered by minor events, such as a late meal or a shirt that has not been ironed.

6. Quick involvement. Because of feelings of insecurity, the potential batterer will move his partner quickly into a committed relationship. If the woman tries to break off the relationship, the man will often try to make her feel guilty for not giving him and the relationship a chance.

7. Blaming others for problems. The abuser takes little responsibility for his problems and blames everyone else. For example, when he makes a mistake, he will blame the woman for upsetting him and keeping him from concentrating on his work. He becomes upset because of what she said, hits her because she smirked at him, and kicks her in the stomach because she poured him too much alcohol. Abusive men typically report that their partners gave them "good reason for the abuse" (Henning et al., 2005).

8. Jekyll-and-Hyde personality. The abuser has sudden mood changes so that his partner is continually confused. One minute he is nice, and the next minute angry and accusatory. Explosiveness and moodiness are typical.

9. Isolation. The abusive person will try to cut off the person from all family, friends, and activities. Ties with anyone are prohibited. Isolation may reach the point at which the abuser tries to stop the victim from going to school, church, or work.

10. Alcohol and Other Drug Use. Whether alcohol reduces one's inhibitions to display violence, acts to allow one to avoid responsibility for being violent, or increases one's aggression, alcohol/substance abuse is associated with violence/abuse (DeMaris et al., 2003).

11. Emotional deficit. Some abusing spouses and parents may have been reared in contexts that did not provide them with the capacity to love, nurture, or be emotionally engaged. Rosenbaum and Leisring (2003) found that men who abuse women report having had limited love from their mothers, more punishment from their mothers, and less attention from their fathers than men who do not abuse their partners.

12. Other factors. Having traditional role relationship ideology and not being able to express feelings easily (Umberson et al., 2003) are other factors associ-

Well . . . you might as well know it all. The others beat me up too. Men think it's their great power over us. God help you, you'll find out soon enough, my darling daughter.

Betty Davis to her daughter, B.D. Hyman

He was calm, cool, nonchalant, polite, arrogant—thinks he is smarter than everybody.

Jim Brazelton, detective, describing Scott Peterson

ated with the potential to be abusive. Schaeffer et al. (2005) also noted that spouses who were depressed and stressed were more likely to be abusive.

Family Factors

Family factors associated with domestic violence and abuse include being abused as a child, having parents who abused each other, and not having a father in the home.

Child Abuse in Family of Origin   Individuals who were abused as children were more likely to be abusive toward their partners as adults.

Family Conflict   Schaeffer et al. (2005) found that high conflict between spouses, parents, and children was predictive of child abuse. Fathers who were not affectionate were also more vulnerable to being abusive.

Parents Who Abused Each Other   Men who witnessed their fathers abusing their mothers are more likely to become abusive partners themselves (Heyman and Slep, 2002). Similarly, women who witnessed their mothers being violent toward their fathers are more likely to be abusive toward their intimates and others (Babcock et al., 2003). However, a majority of those who witnessed abuse do not continue the pattern. A family history of violence is only one factor out of many that may be associated with a greater probability of adult violence.

Abuse in Dating Relationships

Violence in dating relationships begins in grade school and continues into college. Ten percent of the 7824 twelfth-grade females in Howard and Wang's (2003) study reported at least one aspect of dating violence. In the sample of 1027 university students, 10 percent reported physical violence and 30 percent reported emotional abuse (Knox and Zusman, 2006). In a study of 50 heterosexual dating couples, the researchers (Perry and Fromuth, 2005) found that 28 percent agreed that there was abuse in their relationship. When agreement was not required in the study, 60 percent of the relationships had at least one partner who reported abuse. Prevalence rates of dating violence are similar for heterosexual, gay, lesbian, and bisexual couples (Freedner et al., 2002). Although most abuse cases reported to the police are real, there are occasions in which no abuse has occurred (Watkins, 2005).

The Self-Assessment: Abusive Behavior Inventory on page 403 allows you to assess the degree of abuse in your current or most recent relationship.

Acquaintance and Date Rape

The word *rape* often evokes images of a stranger jumping out of the bushes or a dark alley to attack an unsuspecting victim. However, most rapes are perpetrated not by strangers but by persons who have a relationship with the victim. About 85 percent of rapes are perpetrated by someone the woman knows. This type of rape is known as **acquaintance rape,** which is defined as nonconsensual sex between adults (of same or other sex) who know each other. The behaviors of sexual coercion occur on a continuum from verbal pressure and threats to the use of physical force to obtain sexual acts, such as kissing, petting, or intercourse. A double standard is operative in the perception of the gender of the person engaging in sexual coercion. Whereas men are viewed as aggressive, women are viewed as promiscuous (Oswald and Russell, 2006).

One type of acquaintance rape is **date rape,** which refers to nonconsensual sex between people who are dating or on a date. Women who dress seductively,

Abusive Behavior Inventory

Circle the number that best represents your closest estimate of how often each of the behaviors happened in your relationship with your partner or former partner during the previous six months.

1 Never
2 Rarely
3 Occasionally
4 Frequently
5 Very frequently

1. Called you a name and/or criticized you.   1 2 3 4 5

2. Tried to keep you from doing something you wanted to do (e.g., going out with friends, going to meetings).   1 2 3 4 5

3. Gave you angry stares or looks.   1 2 3 4 5

4. Prevented you from having money for your own use.   1 2 3 4 5

5. Ended a discussion with you and made the decision himself/herself.   1 2 3 4 5

6. Threatened to hit or throw something at you.   1 2 3 4 5

7. Pushed, grabbed, or shoved you.   1 2 3 4 5

8. Put down your family and friends.   1 2 3 4 5

9. Accused you of paying too much attention to someone or something else.   1 2 3 4 5

10. Put you on an allowance.   1 2 3 4 5

11. Used your children to threaten you (e.g., told you that you would lose custody, said he/she would leave town with the children).   1 2 3 4 5

12. Became very upset with you because dinner, housework, or laundry was not done when he/she wanted it or done the way he/she thought it should be.   1 2 3 4 5

13. Said things to scare you (e.g., told you something "bad" would happen, threatened to commit suicide).   1 2 3 4 5

14. Slapped, hit, or punched you.   1 2 3 4 5

15. Made you do something humiliating or degrading (e.g., begging for forgiveness, having to ask his/her permission to use the car or to do something).   1 2 3 4 5

16. Checked up on you (e.g., listened to your phone calls, checked the mileage on your car, called you repeatedly at work).   1 2 3 4 5

17. Drove recklessly when you were in the car.   1 2 3 4 5

18. Pressured you to have sex in a way you didn't like or want.   1 2 3 4 5

19. Refused to do housework or child care.   1 2 3 4 5

20. Threatened you with a knife, gun, or other weapon.   1 2 3 4 5

21. Spanked you.   1 2 3 4 5

22. Told you that you were a bad parent.   1 2 3 4 5

23. Stopped you or tried to stop you from going to work or school.   1 2 3 4 5

24. Threw, hit, kicked, or smashed something.   1 2 3 4 5

25. Kicked you.   1 2 3 4 5

26. Physically forced you to have sex.   1 2 3 4 5

27. Threw you around.   1 2 3 4 5

28. Physically attacked the sexual parts of your body.   1 2 3 4 5

29. Choked or strangled you.   1 2 3 4 5

30. Used a knife, gun, or other weapon against you.   1 2 3 4 5

Scoring

Add the numbers you circled and divide the total by 30 to find your score. The higher your score (five = highest score), the more abusive your relationship.

The inventory was given to 100 men and 78 women equally divided into groups of abusers/abused and nonabusers/nonabused. The men were members of a chemical dependency treatment program in a veterans' hospital and the women were partners of these men. Abusing or abused men earned an average score of 1.8; abusing or abused women earned an average score of 2.3. Nonabusing/abused men and women earned scores of 1.3 and 1.6, respectively.

Source

Melanie F. Shepard and James A. Campbell. The abusive behavior inventory: A measure of psychological and physical abuse. *Journal of Interpersonal Violence*, September 1992, 7, no. 3, 291–305. Inventory is on pages 303–304. Used by permission of Sage Publications, 2455 Teller Road, Newbury Park, CA 91320.

A woman is most likely to be raped by a person she knows. . . . often by a person she has been seeing on a regular basis.

© Jeff Greenberg/PhotoEdit

even wives, are viewed by both undergraduate men and women as partly responsible for being raped by their partners (Whatley, 2005).

National Data

One in four women have been raped by a current or former partner, compared with one in 13 men (Mears and Visher, 2005). Abbey (2005) suggested that even higher rates might be revealed if individuals are allowed to give their incidence to a computer.

Both women and men may pressure a partner to have sex. Buddie and Testa (2005) studied sexual aggressiveness in women (both in and out of college) and found that women who engaged in heavy episodic drinking and who had a high number of sexual partners were most likely to be sexually aggressive and rape or attempt to rape their partners. Sexual arousal is one method a woman uses to be sexually aggressive with a male (see quote below).

I locked the room door that we were in. I kissed and touched him. I removed his shirt and unzipped his pants. He asked me to stop. I didn't. Then I sat on top of him. He had had two beers but wasn't drunk. (Struckman-Johnson et al., 2003, p. 84).

Although acts of sexual aggression of women on men do occur, they are less frequent than women being raped. An example follows.

I Was Raped by My Boyfriend

I was 13, a freshman in high school. He was 16, a junior. We were both in band, that's where we met. We started dating and I fell head over heels in love. He was my first real boyfriend. I thought I was so cool because I was a freshman dating a junior. Everything was great for about 6 maybe 8 months and that's when it all started. He began to verbally and emotionally abuse me. He would tell me things like I was fat and ugly and stupid. It hurt me but I was young and was trying so hard to fit in that I didn't really do anything about it.

When he began to drink and smoke marijuana excessively everything got worse. I would try to talk to him about it and he would only yell at me and tell me I was stupid. He began to get physical with the abuse about this time too. He would push me around, grab me really hard, and bruise me up, things like that.

I tried to break up with him and he would tell me that if I broke up with him I would be alone because I was so ugly and fat that no one else would want me. He would say that I was lucky to have him. He began to pressure me for sex. I wasn't ready for sex I was only 14 at that time. He told everyone that we were having sex and I never said anything different. I was scared to say anything different. It got to the point that I was trying my best to avoid being around him alone. I was afraid of him; I never knew what he would do to me.

Our school band had a competition in Myrtle Beach, South Carolina. When we were there we stayed overnight in a hotel. At the hotel he asked me to come by his room so we could go to dinner together. When I went by his room he pulled me into the room. He locked the door behind me. He pushed me into the bathroom and shut and locked the door behind us. He had his hand over my mouth and told me not to make a sound or he would hurt me. He said that it was time for him to get what everyone already thought he had.

He pushed me up against the counter where the sink was and pulled my clothes off. He put a condom on and began to rape me. I closed my eyes, tears streamed down my face. I had no idea what to do. When he was finished he cleaned himself up and turned on the shower. He pushed me into the shower and told me to take a shower and get ready for dinner. Before he left the room he told me that if I had given in to him and had sex earlier he wouldn't have had to force me. He also told me not to tell anyone or he'd hurt me plus no one would believe me because everyone already thought we had been having sex.

I sat in the shower and cried for a while. I thought it was my fault. I didn't tell anyone about it for two years. I decided not to press charges; I didn't want to go through it again. I regret that decision still to this day. I was an emotional wreck after I was raped. I didn't know how to act. He acted normal, like nothing had happened. A few months later he broke up with me for another girl. It was a relief to me and it made me angry all at the same time. It angered me that he would rape me and then break up with me.

I didn't know what to do with myself. I started working out all the time, several hours a day everyday. After high school I didn't have the time to work out so I developed an eating disorder. This went on for almost 6 years. I am just now dealing with moving past the rape and rebuilding my self-esteem, ability to trust men, and ability to enjoy sex in a new relationship. I encourage anyone who has been raped to speak up about it. Do not let the person get away with it. I also encourage anyone who has been raped to get immediate professional help, don't try to get past it on your own.

Rophypnol—The Date Rape Drug

Rophypnol—also known as the date rape drug, rope, roofies, Mexican valium, or the "forget (me) pill"—causes profound, prolonged sedation and short-term memory loss. Similar to Valium but 10 times as strong, it is a prescription drug in Europe and used as a potent sedative. It is sold in the United States for about $5, dropped in a woman's drink (where it is tasteless and odorless), and causes her to lose memory for 8 to 10 hours. During this time she may be raped yet be unaware until she notices signs of it. A former student in the authors' classes reported being drugged ("he put something in my drink") by a "family friend" when she was 16. She noticed blood in her panties the next morning but had no memory of the previous evening. The "friend" is currently being prosecuted.

The Drug-Induced Rape Prevention and Punishment Act of 1996 makes it a crime to give a controlled substance to anyone, without their knowledge, with the intent of committing a violent crime (such as rape). Violation of this law is punishable by up to 20 years in prison and a fine of $250,000.

The effect of rape, whether drug-induced or not, is negative to devastating. In addition to the loss of self-esteem, loss of trust, and ability to be sexual,

Campbell and Wasco (2005) emphasized that rape also affects the family, friends, and significant others of the victim. In regard to preventing rape, Brecklin and Ullman (2005) analyzed data on 1623 women and noted that those who had had self-defense or assertiveness training reported that their resistance stopped the offender or made him less aggressive than victims without such training. The women with the training also noted that they were less scared during the attack.

Abuse in Marriage Relationships

The chance of abuse in a relationship increases with marriage. Although most reported abuse cases are real, there are exceptions.

General Abuse in Marriage

Abuse in marriage differs from unmarried abuse in that the husband may feel "ownership" of the wife and feel the need to "control her." The following reveals the experience of horrific marital abuse by a woman who survived it:

Twenty-Three Years in an Abusive Marriage

My name is Jane and I am a survivor (though it took me forever to get out) of a physically and mentally abusive marriage. I met my husband-to-be in high school when I was 15. He was the most charming person that I had ever met. He would see me in the halls and always made a point of coming over to speak. I remember thinking how cool it was that an older guy would be interested in a little freshman like me. He asked me out but my parents would not let me go out with him until I turned 16 so we met at the skating rink every weekend until I was old enough to go out on a date with him. Needless to say, I was in love and we didn't date other people.

When we married I remember the wedding day very well as I was on top of the world because I was marrying the man of my dreams whom I loved very much. My family (they did not like him) tried for the longest time to talk me out of marrying him. My dad always told me that something was not right with him but I did not listen. I wish to God that I had listened to my dad and walked out of the church before I said "I do."

I remember the first slap which came six months after we had been married. I had burned the toast for his dinner. I was shocked when he slapped me, cried, and when he saw the blood coming from my lip he started crying and telling me he was sorry and that he would never do that again. But the beatings never stopped for 23 years. His most famous way of torturing me was to put a gun to my head while I was sleeping and wake me up and pull the trigger and say next time you might not be so lucky.

I left this man a total of 17 times but I always went back after I got well from my beatings. At one point during this time every bone in my body had been broken with the exception of my neck and back. Some bones more than once and sometimes more than one at a time. I was pushed down a flight of stairs and still suffer from the effects of it to this day. I was stabbed several times and I had to be hospitalized a total of six times.

After our children were born I thought the beatings would stop but they didn't. When our children got older he started on them so I took many beatings for my children just so he would leave them alone. Three years ago he beat me so badly that I almost died. I remember being in the hospital wanting to die because if I did I would not hurt anymore and I would finally be safe. At that time I was thinking that death would be a blessing since I would not have to go back to live with this monster.

While I was in the hospital a therapist came to see me and told me that I had two choices—go home and be killed or fight for my life and live. Instead of going back to the monster, I moved my children in with my mother. It took me three years to get completely clear of him, and even today he harasses me but I am free. I had to learn how to think for myself and how to love again. I am now remarried to a wonderful man.

If you are in an abusive relationship there is help out there and please don't be afraid to ask for it. I can tell you this much—the first step (deciding to leave) is the hardest to take. If and when you do leave I promise your life will be better—it may take awhile but you will get there.

Rape in Marriage

Marital rape, now recognized in all states as a crime, is forcible rape by one's spouse (Rousseve, 2005). The forced sex may take the form of sexual intercourse, fellatio, or anal intercourse. Sexual violence against women in an intimate relationship is often repeated. Rand (2003) analyzed data from the National Violence Against Women survey and noted that about half of the women raped by an intimate partner and two-thirds of the women physically assaulted by an intimate partner had been victimized multiple times.

PERSONAL CHOICES

Should You End an Abusive Marital Relationship?

Students in the authors' classes were asked whether they would end a marriage if the spouse hit or kicked them.

Most said, "Don't overreact and try to work it out."

Most felt that marriage was too strong a commitment to end if the abuse could be stopped.

I would not divorce my spouse if she hit or kicked me. I'm sure that there's always room for improvement in my behavior, although I don't think it's necessary to assault me. I recognize that under certain circumstances, it's the quickest way to draw my attention to the problems at hand. I would try to work through our difficulties with my spouse.

***

The physical contact would lead to a separation. During that time, I would expect him to feel sorry for what he had done and to seek counseling. My anger would be so great, it's quite hard to know exactly what I would do.

***

I wouldn't leave him right off. I would try to get him to a therapist. If we could not work through the problem, I would leave him. If there was no way we could live together, I guess divorce would be the answer.

***

I would not divorce my husband, because I don't believe in breaking the sacred vows of marriage. But I would separate from him and let him suffer!

***

I would tell her I was leaving but that she could keep me if she would agree for us to see a counselor to ensure that the abuse never happened again.

Some said, "Seek a divorce."

Those opting for divorce (a minority) felt they couldn't live with someone who had abused or might abuse them again. "I abhor violence of any kind, and since a marriage should be based on love, kicking is certainly unacceptable. I would lose all respect for my husband and I could never trust him again. It would be over."

Some therapists emphasize that a pattern of abuse develops and continues if such behavior is not addressed immediately. The first time abuse occurs, the couple should seek therapy. The second time it occurs, they should separate. The third time, they might consider a divorce. Later in the chapter we discussed disengaging from an abusive relationship.

Effects of Abuse

Abuse affects the physical and psychological well-being of the victim. Abuse between parents also affects the children.

Effects of Partner Abuse on Victims

The most obvious effect of physical abuse by an intimate partner is physical injury. Other, less obvious effects of abuse on a partner include fear, feelings of helplessness, confusion, isolation, humiliation, anxiety, symptoms of post-traumatic stress disorder (PTSD), and attempts to commit suicide (Lloyd and Emery, 2000). Depression and substance abuse have also been associated with partner violence (Anderson, 2002). Gender may also be a factor in how abuse is experienced. A team of researchers studied 280 college students in both abusive and nonabusive relationships and found that women in abusive relationships reported more psychological symptoms (depression/anxiety) than men in abusive relationships (Clements et al., 2005).

When the abuse by a partner is sexual, it may be more devastating than sexual abuse by a stranger. The primary effect on a woman is to destroy her ability to trust men in intimate interpersonal relationships. In addition, the woman raped by her husband lives with her rapist and may be subjected to repeated assaults.

Effects of Partner Abuse on Children

Abuse between adult partners affects children. The most dramatic effect is that some women are abused during their pregnancy, resulting in a high rate of miscarriage and birth defects. Negative effects may also accrue to children who just witness domestic abuse. Kitzmann et al. (2003) analyzed 118 studies to identify outcomes for children who were and were not exposed to interparental violence. The researchers found more negative outcomes (e.g., arguing, withdrawing, avoidance, overt hostility) among children who had witnessed such behavior than those who had not.

Howard et al. (2002) found that adolescents who had witnessed violence reported "intrusive thoughts, distraction, and feeling a lack of belonging" (p. 455). Children who witness high levels of parental violence are also more likely to blame themselves for the violence (Grych et al., 2003), to be violent toward their parents (particularly their mother) (Ulman and Straus, 2003), and to engage in aggressive delinquent behavior (Kernic et al., 2003).

It is not unusual for children to observe and to become involved in adult domestic violence. One-fourth of 114 battered mothers noted that their children yelled, called for help, or intervened when they (the mothers) were being physically abused by their adult partners (Edleson, 2003).

The Cycle of Abuse

The following reflects the cycle of abuse.

"I Got Flowers Today"

I got flowers today. It wasn't my birthday or any other special day. We had our first argument last night, and he said a lot of cruel things that really hurt me. I know he is sorry and didn't mean the things he said, because he sent me flowers today.

I got flowers today. It wasn't our anniversary or any other special day. Last night, he threw me into a wall and started to choke me. It seemed like a nightmare. I couldn't believe it was real. I woke up this morning sore and bruised all over. I know he must be sorry, because he sent me flowers today.

Last night, he beat me up again. And it was much worse than all the other times. If I leave him, what will I do? How will I take care of my kids? What about money? I'm afraid of him and scared to leave. But I know he must be sorry, because he sent me flowers today.

I got flowers today. Today was a very special day. It was the day of my funeral. Last night, he finally killed me. He beat me to death.

If only I had gathered enough courage and strength to leave him, I would not have gotten flowers today.

Author unknown

The cycle of abuse begins when a person is abused and the perpetrator feels regret, asks for forgiveness, and starts acting nice (e.g., gives flowers). The victim, who perceives few options and feels guilty terminating the relationship with the partner who asks for forgiveness, feels hope for the relationship at the contriteness of the abuser and does not call the police or file charges. Forgiving the partner and taking him back usually occurs seven times before the partner leaves for good. Shakespeare (in *As You Like It*) said of such forgiveness, "Thou prun'st a rotten tree."

After the forgiveness, there is usually a making-up or honeymoon period, during which the victim feels good again about his or her partner and is hopeful for a nonabusive future. But stress, anxiety, and tension mount again in the relationship, which is relieved by violence toward the victim. Such violence is followed by the familiar sense of regret and pleadings for forgiveness, accompanied by being nice (a new bouquet of flowers, etc.).

As the cycle of abuse reveals, some wives do not prosecute their partners who abuse them. To deal with this problem, Los Angeles has adopted a "zero tolerance" policy toward domestic violence. Under the law, an arrested person is required to stand trial and his victim required to testify against the perpetrator. The fine in Los Angeles County for partner abuse is up to 6 months in jail and a fine of $1,000.

Figure 14.1 illustrates this cycle, which occurs in clockwise fashion. In the rest of this section we discuss reasons why people stay in an abusive relationship and how to get out of such a relationship.

Why People Stay in Abusive Relationships

One of the most frequently asked questions of people who remain in abusive relationships is "Why do you stay?" Few and Rosen (2005) interviewed 25 women who had been involved in abusive dating relationships from 3 months to 9 years (average = 2.4 years) to find out why they stayed. The researchers conceptualized the women as entrapped—stuck and unable to extricate themselves from the abusive

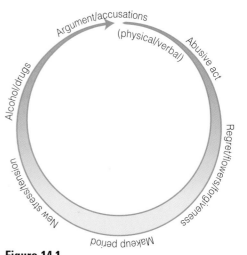

Figure 14.1
The Cycle of Abuse

partner. Indeed, these women escalated their commitment to stay in hopes that their doing so would eventually pay off. Among the factors of their perceived investment was the time they had already spent in the relationship, the sharing of self they had done with their partner, and the relationships they were connected to because of the partner. In effect, they had invested time with a partner they were in love with and wanted to turn the relationship around into a safe, nonabusive one. Here are some of the factors explaining how abused women become entrapped:

- fear of loneliness ("I'd rather be with someone who abuses me than alone")
- love ("I love him")
- emotional dependency ("I need him")
- commitment to the relationship ("I took a vow 'for better or for worse'")
- hope ("He will stop")
- a view of violence as legitimate ("All relationships include some abuse")
- guilt ("I can't leave a sick man")
- fear ("He'll kill me if I leave him")
- economic dependence ("I have no place to go")
- isolation ("I don't know anyone who can help me")

Battered women also stay in abusive relationships because they rarely have escape routes related to educational or employment opportunities, their relatives are critical of plans to leave the partner, they do not want to disrupt the lives of their children, and they may be so emotionally devastated by the abuse (anxious/depressed/low self-esteem) that they feel incapable of planning and executing their departure.

It is easier to stay out than to get out.

Mark Twain

Diversity in Other Countries

Rivers (2005) interviewed Navajo women who had been abused by their husbands and discovered the term *hozho*, which helps to explain cultural motivations for their staying with an abuser. Hozho means beauty and harmony, and when the husband is abusive to his wife there is disharmony. . . . but men and women belong together, to contribute to this ultimate beauty, so leaving the husband is not an easy consideration. Also, there is the belief that the husband is "out of balance," and the Navajo woman is less likely to leave a man who is "sick."

How One Leaves an Abusive Relationship

Leaving an abusive partner begins with the decision to do so. Such a choice often follows the belief that one will die or one's children will be harmed by staying. A plan (e.g., move in with a sister, mother, or friend or go to a homeless shelter) comes into being and the person acts. If the alternative is better than being in the abusive context, the person will stay away. Otherwise, the person may go back and start the cycle all over. As noted above, this leaving and returning typically happens seven times.

Sometimes the woman does not just disappear while the abuser is away but calls the police and has the man arrested for violence/abuse. While he is in jail, she may move out and leave town. In either case, disengagement from the abusive relationship takes a great deal of courage. Calling 800-799-7233, the national domestic violence 24-hour hotline, is a point of beginning.

Some women who withdraw go to abuse shelters. A team of researchers (Ham-Rowbottom et al., 2005) conducted a follow-up of 81 abuse victims who had escaped their situations and graduated from the abuse shelter. Although all of the women were now living independently, 43% and 75% reported clinical levels of depression and trauma symptoms, respectively. The researchers noted that earlier child sexual abuse accounted for much of the continued psychological depression and trauma.

Strategies to Prevent Domestic Abuse

Family violence and abuse prevention strategies are focused at three levels: the general population, specific groups thought to be at high risk for abuse, and families who have already experienced abuse. Public education and media campaigns aimed at the general population convey the criminal nature of domestic assault, suggest ways to prevent abuse (seek therapy for anger, jealousy, or dependency), and identify where abuse victims and perpetrators can get help.

Preventing or reducing family violence through education necessarily involves altering aspects of American culture that contribute to such violence. For example, violence in the media must be curbed or eliminated (not easy, with nightly video clips of bombing assaults in other countries), and traditional gender roles and views of women and children as property must be replaced with egalitarian gender roles and respect for women and children.

Another important cultural change is to reduce violence-provoking stress by reducing poverty and unemployment and by providing adequate housing, nutrition, medical care, and educational opportunities for everyone. Integrating families into networks of community and kin would also enhance family well-being and provide support for families under stress.

Treatment of Partner Abusers

Silvergleid and Mankowski (2006) identified learning a new way of masculinity and making a personal commitment to change as crucial to the rehabilitation of the male batterer. These new behaviors can be learned in individual or group therapy. Since alcohol or drug abuse and violence toward a partner are often related, addressing one's alcohol or substance abuse problem is often a prerequisite for treating partner abuse (Stuart, 2005).

When a girl tells me to stop, I stop.
Holden Caulfield, *The Catcher in the Rye*

In addition, some men stop abusing their partners only when their partners no longer put up with it. One abusive male said that his wife had to leave him before he learned not to be abusive toward women. "I've never touched my second wife," he said.

This concludes our discussion of abuse in adult relationships. In the pages to follow we discuss other forms of abuse, including child abuse and parent, sibling, and elder abuse.

General Child Abuse

National Data
The number of children investigated each year by protective services who are alleged to have been abused is around 3 million (*Statistical Abstract of the United States: 2006*, Table 333). More than one thousand children die annually as a result of child abuse (Trokel et al., 2006).

Child abuse may take many forms—physical abuse, neglect, and sexual abuse.

Physical Abuse and Neglect

A 25-pound 8-year-old daughter was found emaciated and near death in a locked closet in January 2002. Her mother, 30-year-old Barbara Atkins of Dallas, Texas, pleaded guilty to the offense and was sentenced to prison for life; she will be eligible for parole in 30 years. She said that her daughter was being punished and that she did not love her daughter as much as her other five children. The stepfather was also charged with sexual assault and serious bodily injury to a child (*Mother Gets Life for Abuse of Girl*, 2002).

Child abuse can be defined as any interaction or lack of interaction between a child and his or her parents or caregiver that results in nonaccidental harm to the child's physical or psychological well-being. Child abuse includes physical abuse, such as beating and burning; verbal abuse, such as insulting or demeaning the child; and neglect, such as failing to provide adequate food, hygiene, medical care, or adult supervision for the child. A child can also experience emotional neglect by his or her parents. Infants and children more likely to be abused are the result of unintended pregnancies and have poor health and developmental problems (Sidebotham et al., 2003).

The most prevalent form of child abuse in America today is neglect. People are working 18 hours a day, and our children are being neglected.
Laura Schlessinger, radio personality

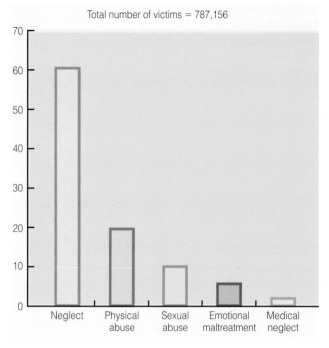

Total number of victims = 787,156

Figure 14.2
Child Abuse and Neglect Cases: 2003
Source: *Statistical Abstract of the United States, 2006.* 125th ed.
Washington, DC: U.S. Bureau of the Census, Table 332.

The percentages of various types of child abuse in substantiated victim cases are illustrated in Figure 14.2. Notice that "neglect" is the largest category of abuse. The children most likely to be neglected are young—between infancy and age 5 (Connell-Carrick, 2003). De Paul and Arruabarrena (2003) observed that these families are the most difficult to treat and have the highest dropout rates.

Although our discussion will focus on physical abuse, it is important to keep in mind that children often are not fed, are not given medical treatment, and are left to fend for themselves.

Factors Contributing to General Child Abuse

A variety of factors contribute to child abuse.

1. Parental psychopathology. Symptoms of parental psychopathology that may predispose a parent to abuse or neglect children include low frustration tolerance, inappropriate expression of anger, and alcohol or substance abuse.

Munchausen syndrome by proxy (MSP) is a rare form of child abuse whereby a parent (usually the mother) takes on the sick role indirectly (hence, by proxy) by inducing illness or sickness in her child. The parent may suffocate the child to the point of unconsciousness or scrub the child's skin with sandpaper to induce a rash. The goal of the caretaker (often the mother) is to find a fulfilling role, to get attention from friends and family for being heroic, or to get money from insurance companies. Sheridan (2003) analyzed 451 cases from 154 journals and found that the victims may be either males or females and are usually under 4 years of age. One-fourth of the victims' known siblings were *dead* and 61 percent of the siblings had similar illnesses. Not much is understood about MSP, and accuracy is important "to protect children from abuse and caretakers from false allegations" (p. 444). One way to assess whether MSP is operative is to separate the child from the mother. The children who return to health and show no symptoms of illness in the absence of the mother are victims of their mother's abuse (Wright, 2004).

2. Unrealistic expectations. Abusing parents often have unrealistic expectations of their children's behavior. For example, a parent might view the crying of a baby as a deliberate attempt on the part of the child to irritate the parent. **Shaken baby syndrome**—whereby the caretaker, most often the father, shakes the baby to the point of causing the child to experience brain or retinal hemorrhage—most often occurs in response to a baby who won't stop crying (Ricci et al., 2003). Most victims are younger than 6 months (Smith, 2003). **Abusive head trauma** (AHT) refers to nonaccidental head injury in infants and toddlers. It is estimated that there are over one thousand cases of AHT each year (Ricci et al., 2003, p. 279). Head injury is the leading cause of death among abused children (Rubin et al., 2003). Managing parents' unrealistic expectations and teaching them to cope with frustration is an important part of an overall plan to reduce physical child abuse.

3. History of abuse. Individuals who were abused as children are more likely to report abusing their own children (Dixon et al., 2005). The researchers observed that the absence of positive parenting carries over into the next generation. Although the majority do not, some parents who were themselves physically or verbally abused or neglected may duplicate these patterns in their own families. Although parents who were abused as children are somewhat more likely to repeat that behavior than parents who were not abused, we would like to emphasize that the majority of parents who were abused do *not* abuse their own children. Indeed, many parents who were abused as children are dedicated to ensuring nonviolent parenting of their own children precisely because of their own experience of abuse.

4. Displacement of aggression. One cartoon shows several panels consisting of a boss yelling at his employee, the employee yelling at his wife, the wife yelling at their child, and the child kicking the dog, who chases the cat up a tree. Indeed, frustration may spill over from the adults to the children, the latter being less able to defend themselves.

5. Social isolation. The saying "it takes a village to raise a child" is relevant to child abuse. Unlike inhabitants of most societies of the world, many Americans rear their children in closed and isolated nuclear units. In extended kinship societies, other relatives are always present to help with the task of childrearing. Isolation means no relief from the parenting role as well as no supervision by others who might interrupt observed child abuse.

6. Fatherless homes. Living in a home where the father is absent increases a child's risk for being abused. Some men may romance a single mother with three young daughters to get access to the daughters whom he may molest. This is largely because stepfathers and mothers' boyfriends are not constrained by the cultural incest taboo that prohibits biological fathers from having sex with their children.

7. Other factors. In addition to the factors just mentioned, the following factors are associated with child abuse and neglect:

   a. The pregnancy is premarital or unplanned, and the father or mother does not want the child.
   b. Mother-infant attachment is lacking.
   c. The child suffers from developmental disabilities or mental retardation.
   d. Childrearing techniques are harsh, with little positive reinforcement.
   e. The parents are unemployed.
   f. Abuse between the husband and wife is present.
   g. The children are adopted or are foster children.

Effects of General Child Abuse

How does being abused in general affect the victim as a child and later as an adult? In general, the effects are negative and vary according to the intensity and frequency of the abuse. Researchers have found that children who have been abused are more likely to display the following (Jonzon and Lindblad, 2005; Muller and Lemieux, 2000):

   1. Few close social relationships and an inability to love or trust
   2. Communication problems and learning disabilities
   3. Aggression, low self-esteem, depression, and low academic achievement
   4. Physical injuries that may result in disfigurement, physical disability, or death
   5. Increased risk of alcohol or substance abuse and suicidal tendencies as adults
   6. Posttraumatic stress disorder (PTSD)

What is the relative impact of parental emotional abuse, child sexual abuse, and parental substance abuse on a child's development as an adult? Melchert (2000) studied the psychological distress of 255 college students and discovered that the largest amount of distress was explained by previous emotional abuse and neglect.

When there is evidence of child abuse, social workers traditionally choose between leaving the child in the home or removing the child from the home and putting him or her in a foster care home. A new alternative is being tried in Minnesota, whereby an abusive parent is allowed to keep the child but is carefully monitored by friends and family. If the parent is drug-addicted, he or she must agree to treatment and to being drug-tested by friends or family (Coates, 2006).

Child Sexual Abuse

Often found in combination with other forms of child abuse is child sexual abuse (Dong et al., 2003). Two types are extrafamilial and intrafamilial.

Extrafamilial Child Sexual Abuse—The Catholic Example

In extrafamilial child sexual abuse the perpetrator is someone outside the family who is not related to the child. Extrafamilial child sexual abuse received national attention with the 2002 accusation that an estimated 2000 priests (McGeary et al., 2002) in the Catholic Church had had sex with young children; more than 3000 children have claimed abuse (Hewitt et al., 2002). The diocese of Orange County, south of Los Angeles, agreed to pay $100 million to some 90 alleged victims of priests and other church people. There are still claims of another 544 alleged victims in this area alone (*The Orange Approach*, 2005).

Intrafamilial Child Sexual Abuse

A more frequent type of child sexual abuse is intrafamilial (formerly referred to in professional literature as incest). This refers to exploitive sexual contact or attempted sexual contact between relatives before the victim is 18. Sexual contact or attempted sexual contact includes intercourse, fondling of the breasts and genitals, and oral sex. Relatives include biologically related individuals but may also include stepparents and stepsiblings.

Female children are more likely than male children to be sexually abused. Prevalence rates suggest that one out of four girls and one out of ten boys experience sex abuse (Fieldman and Crespi, 2002). However, Goodman et al. (2003) suggested that nearly 40 percent of abused adults fail to report their own documented child sexual abuse. About 60 percent of sexually abused children disclose their abuse to someone within 2 weeks after it happens—40 percent within 48 hours. Children who are younger, who did not feel responsible for the abuse, and who were abused by someone external to the family are more likely to disclose the abuse sooner than older children, those who felt partly responsible, and those who were abused by someone inside the family (Goodman-Brown et al., 2003).

Intrafamilial child sexual abuse, particularly when the perpetrator is a parent, involves an abuse of power and authority. The following describes the experience of one woman who was forced to have sexual relations with her father over a period of years:

> I was around 6 years old when I was sexually abused by my father. He was not drinking at that time; therefore, he had a clear mind as to what he was doing. On looking back, it seemed so well planned. For some reason, my father wanted me to go with him to the woods behind our house to help him saw wood. Once we got there, he looked around for a place to sit and wanted me to sit down with him. He said, "Susan, I want you to do something for Daddy. I want you to lie down, and we are going to play Mama and Daddy." Being a child, I said, "Okay," thinking it was going to be fun. I don't know what happened next and I can't remember if there was pain or whatever. I was threatened not to tell, and remembering how he beat my mother, I didn't want the same treatment. It happened a few more times. I remember not liking this at all.
>
> But what could I do? Until age 18, I was constantly on the run, hiding from him when I had to stay home alone with him, staying out of his way so he wouldn't

I was only 9 years old when I was raped by my 19-year-old cousin. He was the first of three family members to sexually molest me.

Oprah Winfrey, TV talk show host, publisher, and actress

touch me by hiding in the cornfields all day long, under the house, in the barns, and so on until my mother got back home, then getting punished by her for not doing the chores she had assigned to me that day. It was a miserable life, growing up in that environment.

When an ex-spouse accuses her husband of child sex abuse, more often than not, the sexual abuse did occur. In a study of 9000 families embroiled in contested divorce proceedings, 169 cases involved an allegation of child sexual abuse. Of these, only 14 percent were deliberately false allegations. Most were proven to be legitimate (Goldstein and Tyler, 1998).

Conviction of a child sex abuser is rare. In one study of 323 court cases, only 15 went to trial, and in only 6 cases was the offender convicted (Faller and Henry, 2000).

Effects of Child Sexual Abuse

Child sexual abuse may have serious negative long-term consequences. Not only has the child been violated physically, but he or she has lost an important social support. Adult-child sexual contact is, in most cases, a child's first introduction to adult sexuality. The sexual script acquired during such relationships forms the basis on which other sexual experiences are assimilated. The negative effects include the following (Koss et al., 2003; Walrath et al., 2003):

1. Early forced sex is associated with being withdrawn, anxious, and depressed and with substance abuse (Walrath et al., 2003). Women who have experienced childhood sexual abuse are 7.75 times more likely to attempt suicide (Anderson et al., 2003). PTSD is also frequently associated with childhood sexual abuse (Koss et al., 2003).

2. Daughters of mothers who have been sexually abused are 3.6 times more likely to be sexually victimized than are daughters whose mothers have not been abused (McCloskey and Bailey, 2000). It is possible that women who have been sexually abused develop an "internal working model" of sexual relationships that encompasses exploitative, coercive, and domineering behavior by men. If such a relationship "template" results from early exposure to sexual abuse, then these women might be more tolerant of men either in their households or in their social spheres who are potential abusers of their daughters (p. 1032).

3. Spouses who were physically and sexually abused as children report lower marital satisfaction, higher individual stress, and lower family cohesion than do couples with no abuse history (Nelson and Wampler, 2000).

4. Adult males who were sexually abused as children are more likely to become child molesters themselves (Lussier et al., 2005).

Successful movement beyond negative consequences of being sexually molested as a child involves disclosing the abuse to an accepting adult partner (Jonzon and Lindblad, 2005). In addition, the adult who was abused as a child must accept that the adult molester (not the child) was responsible for the abuse (some children feel that they "led the adult on" and are responsible for the abuse).

Strategies to Reduce Child Sexual Abuse

Strategies to reduce child sexual abuse include regendering cultural roles, providing specific information and training to children to be alert to inappropriate touching, improving the safety of neighborhoods, providing healthy sexuality information for both teachers and children in the public schools at regular intervals, and promoting public awareness campaigns.

From a larger societal preventive perspective, Bolen (2001) recommended that child sexual abuse should be viewed as a gendered problem. Indeed, "child sexual abuse is endemic within society and may be a result of the unequal power of males over females" (p. 249). Evidence for this claim includes the facts that fe-

The rape and murder of 7-year-old Megan Kanka resulted in Megan's Law.

© AP/Wide World Photos

Megan's Law And Beyond

SOCIAL POLICY

In 1994, Jesse Timmendequas lured 7-year-old Megan Kanka into his Hamilton Township house in New Jersey to see a puppy. He then raped and strangled her and left her body in a nearby park. Prior to his rape of Megan, Timmendequas had two prior convictions for sexually assaulting girls. Megan's mother, Maureen Kanka, argued that she would have kept her daughter away from her neighbor if she had known about his past sex offenses. She campaigned for a law, known as **Megan's Law,** requiring that communities be notified of a neighbor's previous sex convictions. New Jersey and 45 other states have enacted similar laws. A Federal law has been suggested but is being legally challenged.

The law requires that convicted sexual offenders register with local police in the communities in which they live. It also requires the police to go out and notify residents and certain institutions (such as schools) that a previously convicted sex offender has moved into the area. It is this provision of the law that has been challenged on the belief that individuals should not be punished forever for past deeds. Critics of the law argue that convicted child molesters who have been in prison have paid for their crime. To stigmatize them in communities as sex offenders may further alienate them from mainstream society and increase their vulnerability for repeat offenses.

In many states, Megan's Law is not operative because it is on appeal. Although parents ask, "Would you want a convicted sex offender, even one who has completed his prison sentence, living next door to your 8-year-old daughter?" the reality is that little notification is afforded parents in most states. Rather, the issue is tied up in court and will likely remain so until the Supreme Court decides it. A group of concerned parents (Parents for Megan's Law) are trying to implement Megan's Law nationwide. See the Web Links at the end of this chapter for the group's Web site address.

Parents for Megan's Law has also sought to enact legislation for the "civil commitment for a specific group of the highest-risk sexual predators who freely roam our streets, unwilling or unable to obtain proper treatment. This kind of predator commitment follows a criminal sentence and generally targets repeat sex offenders who then remain in a sexual predator treatment facility until it is safe to release them to a less restrictive environment or into the community." Indeed, this group not only wants parents to be notified of criminal sex offenders but also wants the offenders to be housed in treatment centers on release from prison.

Your Opinion?

1. To what degree do you believe the Supreme Court should uphold Megan's Law?
2. To what degree do you believe that convicted child molesters who have served their prison sentence should be free to live wherever they like without neighbors being aware of their past?
3. Independent of Megan's Law, how can parents protect their children from sex abuse?

males are at greater risk of abuse than males, males are more likely to offend, and child sexual abuse is most frequently heterosexual. The implication is that one way to discourage child sexual abuse is to change traditional notions of gender role relationships, masculinity, and male sexuality so that men respect the sexuality of women and children and take responsibility for their sexual behavior.

The public schools also help children acquire specific knowledge and skills to protect themselves from sexual abuse. A survey of 400 school districts in the United States revealed that 85 percent had offered such a prevention program in the past year (Davis and Gidyez, 2000). Through various presentations in the elementary schools, children are taught how to differentiate between appropriate and inappropriate touching by adults or siblings, to understand that it is okay to feel uncomfortable if they do not like the way someone else is touching them, to say no in potentially exploitative situations, and to tell other adults if the offending behavior occurs. Boyle and Lutzker (2005) also demonstrated that even young children aged 5 and 6 years can learn to discriminate appropriate and inappropriate touching.

Finally, helping to ensure that children live in neighborhoods safe from convicted child sex offenders may reduce child sexual abuse. Protecting children from former convicted child molesters is the basis of Megan's Law (see Social Policy on page 416).

Parent, Sibling, and Elder Abuse

As we have seen, intimate partners and children may be victims of relationship violence and abuse. Parents, siblings, and the elderly may also be abused by family members.

Parent Abuse

Some people assume that because parents are typically physically and socially more powerful than their children, they are immune from being abused by their children. But parents are often targets of their children's anger, hostility, and frustration. It is not uncommon for teenage and even younger children to physically and verbally lash out at their parents. Children have been known to push parents down stairs, set the house on fire while their parents are in it, and use weapons such as guns or knives to inflict serious injuries or even to kill a parent. Background characteristics of children who abuse their parents include observing parents' violence toward each other, having parents who used corporal punishment on their children as a method of discipline (Ulman and Straus, 2003) and having parents who were violent toward their children (Ulman, 2003).

Sibling Abuse

Observe a family with two or more children and you will likely observe some amount of sibling abuse. Even in "well-adjusted" families, some degree of fighting among the children is expected. Most incidents of sibling violence consist of slaps, pushes, kicks, bites, and punches. What passes for "normal," "acceptable," or "typical" behavior between siblings would often be regarded as violent and abusive behavior outside the family context. In this regard, sibling abuse, particularly when compared to parent-child abuse, is underreported and generally of limited concern to child-protective services (Caffaro and Conn-Caffaro, 2005).

Sibling abuse is said to be the most prevalent form of abuse. Ninety-eight percent of the females and 89 percent of the males in one study reported having received at least one type of emotionally aggressive behavior from a sibling; 88 percent of the females and 71 percent of the males reported having received at least one type of physically aggressive behavior from a sibling (Simonelli et al., 2002).

Sibling abuse may include sexual exploitation, whereby an older brother will coerce younger female siblings into nudity or sex. Though some sex between siblings is consensual, often it is not. "He did me *and* all my sisters before he was done," reported one woman. Another woman reported that (as a child and young adolescent) she performed oral sex on her brother three times a week for years since he told her that ingesting a man's semen was the only way a woman would be able to have babies as an adult. Even milder forms of sibling abuse seem to feed the cycle of abuse, as persons who were abused by their siblings report abusing others in their adulthood (Simonelli et al., 2002).

In addition to physical and sexual sibling abuse, there is **sibling relationship aggression.** This is behavior of one sibling toward another sibling that is intended to induce social harm or psychic pain in the sibling. Examples include social alienation/exclusion (e.g., not asking the sibling to go to a movie when a group is going), telling secrets/spreading rumors (e.g., revealing a sibling's sexual/drug past), and withholding support/acceptance (e.g., not acknowledging a sibling's achievements in school and sports). A team of researchers studied 185 sibling pairs—both younger siblings (mean age, 13.47 years) and older siblings (mean age, 15.95 years)—and their respective parents (married an average of 19.5 years). They found that relational sibling aggression was more likely to occur when intimate feelings between siblings were low and their negativity toward each other was high. Family contexts in which parents are warm and involved with their children are conducive to lower levels of relational aggression (Updegraff et al., 2005).

If you suspect elder abuse or are concerned about the well-being of an older person, call your state abuse hotline immediately.

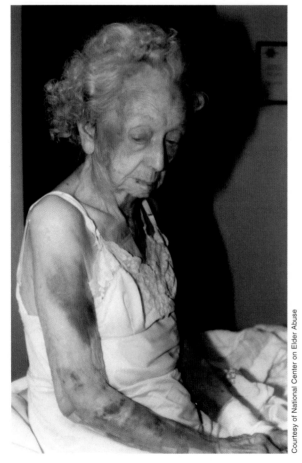

Courtesy of National Center on Elder Abuse

Elder Abuse

As increasing numbers of the elderly end up in the care of their children, abuse, though infrequent, is likely to increase. Currently it is estimated that about half a million elderly in the United States are abused (Bergeron and Gray, 2003). Cross-culturally, there are 600 million individuals over the age of 60, a figure that will double by 2025, and there is little to no visibility given to the vulnerability of these elderly to being abused (Patterson and Malley-Morrison, 2006). Various examples of abuse include:

1. Neglect: failing to buy or give the elderly needed medicine, failing to take them to receive necessary medical care, or failing to provide adequate food, clean clothes, and a clean bed. Neglect is the most frequent type of domestic elder abuse; more than 70 percent of all adult protective services reports are for this reason (Fulmer et al., 2005).

2. Physical abuse: inflicting injury or physical pain or sexual assault.

3. Psychological abuse: verbal abuse, deprivation of mental health services, harassment, and deception.

4. Social abuse: unreasonable confinement and isolation, lack of supervision, abandonment.

5. Legal abuse: improper or illegal use of the elder's resources.

Another type of elder abuse that has received recent media attention is **granny dumping.** Adult children or grandchildren who feel burdened with the care of their elderly parent or grandparent drive the elder to the entrance of a hospital and leave him or her there with no identification. If the hospital cannot identify responsible relatives, it is required by state law to take care of the abandoned elder or transfer the person to a nursing-home facility, which is paid for by state funds. Relatives of the dumped granny, hiding from financial responsibility, never visit or see granny again.

Adult children who are most likely to dump or abuse their parents tend to be under a great deal of stress and to use alcohol or other drugs. In some cases, parent abusers are getting back at their parents for mistreating them as children. In other cases, the children are frustrated with the burden of having to care for their elderly parents. Such frustration is likely to increase. As baby boomers age, they will drain already limited resources for the elderly, and their children will be forced to care for them with little governmental support. Prevention of elder abuse involves reducing the stress for caregivers by linking caregivers to community services (Bergeron and Gray, 2003).

SUMMARY

How are violence and abuse defined?

Violence/physical abuse may be defined as the intentional infliction of physical harm by either partner on the other. *Intimate partner violence* is an all-inclusive term that refers to crimes committed against current or former spouses, boyfriends, or girlfriends. Battered-woman syndrome refers to the general pattern of battering that a woman is subjected to and is defined in terms of the frequency, severity, and injury she experiences. Uxoricide is the murder of a woman by a romantic partner.

Emotional abuse (also known as psychological abuse, verbal abuse, or symbolic aggression) is designed to denigrate the partner, reduce the partner's status, and make the partner vulnerable, thereby giving the abuser more control.

Stalking is unwanted following or harassment that induces fear in the target person. The stalker is most often a heterosexual male who has been rejected by someone who fails to return his advances. Stalking is typically designed either to seek revenge or to win the partner back.

What are explanations for violence in relationships?

Cultural explanations for violence include violence in the media, corporal punishment in childhood, gender inequality, and stress. Community explanations involve social isolation of individuals and spouses from extended family, poverty, inaccessible community services, and lack of violence prevention programs. Individual factors include psychopathology of the person (antisocial), personality (dependent/jealous), and alcohol abuse. Family factors include child abuse by one's parents and observing parents who abuse each other.

How does abuse in dating relationships manifest itself?

Violence in dating relationships begins as early as grade school, is mutual, and escalates with emotional involvement. Violence occurs more often among couples in which the partners disagree about each other's level of emotional commitment and when the perpetrator has been drinking alcohol or using drugs. Acquaintance rape is defined as nonconsensual sex between adults who know each other.

One type of acquaintance rape is date rape, which refers to nonconsensual sex between people who are dating or on a date. Rophypnol, known as the date rape drug, causes profound sedation so that the person may not remember being raped. Most women do not report being raped by an acquaintance or date.

How does abuse in marriage relationships manifest itself?

Abuse in marriage is born out of the need to control the partner and may include repeated rape. About half of the women raped by an intimate partner and two-thirds of the women physically assaulted by an intimate partner have been victimized multiple times.

What are the effects of abuse?

The most obvious effect of physical abuse by an intimate partner is physical injury. As many as 35 percent of women who seek hospital emergency room services are suffering from injuries incurred by battering. Other, less obvious effects of abuse by one's intimate partner include fear; feelings of helplessness, confusion, isolation, and humiliation; anxiety; stress-induced illness; symptoms of posttraumatic stress disorder; and suicide attempts.

When the abuse by a partner is sexual, it may be more devastating than sexual abuse by a stranger. The primary effect on a woman is to destroy her ability to trust men in intimate interpersonal relationships. In addition, the woman raped by her husband lives with her rapist and may be subjected to repeated assaults.

Abuse between adult partners affects children. Some women are abused during their pregnancy, resulting in a high rate of miscarriage and birth defects. The March of Dimes has concluded that the physical abuse of pregnant women causes more birth defects than all the diseases put together for which children are usually immunized.

What is the cycle of abuse and why do people stay in an abusive relationship?

The cycle of abuse begins when a person is abused and the perpetrator feels regret, asks for forgiveness, and starts acting nice (e.g., gives flowers). The victim, who perceives few options and feels guilty terminating the relationship with the partner who asks for forgiveness, feels hope for the relationship at the contriteness of the abuser and does not call the police or file charges. There is usually a making-up or honeymoon period, during which the person feels good again about his or her partner. But tensions mount again and are released in the form of violence. Such violence is followed by the familiar sense of regret and pleadings for forgiveness, accompanied by being nice (a new bouquet of flowers, etc.).

The reasons people stay in abusive relationships include love, emotional dependency, commitment to the relationship, hope, view of violence as legitimate, guilt, fear, economic dependency, and isolation. The catalyst for breaking free combines the sustained aversiveness of staying, the perception that they and their children will be harmed by doing so, and the awareness of an alternative path or of help in seeking one.

What is child abuse and what factors contribute to it?

Child abuse can be defined as any interaction or lack of interaction between a child and his or her parents or caregiver that results in nonaccidental harm to the child's physical or psychological well-being. Child abuse includes physical abuse, such as beating and burning; verbal abuse, such as insulting or demeaning the child; and neglect, such as failing to provide adequate food, hygiene, medical care, or adult supervision for the child. A child can also experience emotional neglect by his or her parents.

Some of the factors that contribute to child abuse include parental psychopathology, a history of abuse, displacement of aggression, and social isolation.

What are the effects of general child abuse?

The negative effects of child abuse include impaired social relationships, difficulty in trusting others, aggression, low self-esteem, depression, low academic achievement, and posttraumatic stress disorder (PTSD). Physical injuries may result in disfigurement, physical disability, and even death.

One of the most devastating types of child abuse is child sexual abuse, which has serious negative long-term consequences. The effects include being withdrawn, anxious, and depressed; delinquency; suicide attempts; and substance abuse. PTSD, which means recurrent experiencing of the event, is common, as is heavy alcohol or substance abuse. Strategies to reduce child sexual abuse include regendering cultural roles, providing specific information to children on sex abuse, improving the safety of neighborhoods, providing healthy sexuality information for both teachers and children in the public schools at regular intervals, and promoting public awareness campaigns.

What is the nature of parent, sibling, and elder abuse?

Parent abuse is the deliberate harm (physical or verbal) of parents by their children. Ten percent of parents report that they have been hit, bitten, or kicked at least once by their children. About 300 parents are killed by their children annually. The Menendez brothers of California brutally murdered both parents.

Sibling abuse is the most severe prevalent form of abuse. What passes for "normal," "acceptable," or "typical" behavior between siblings would often be regarded as violent and abusive behavior outside the family context.

Elder abuse is another form of abuse in relationships. Granny dumping is a new form of abuse in which children or grandchildren who feel burdened with the care of their elderly parents or grandparents drive them to the emergency entrance of a hospital and dump them. If the relatives of the elderly patient cannot be identified, the hospital will put the patient in a nursing home at state expense.

KEY TERMS

abusive head trauma	emotional abuse	Munchausen syndrome by	stalking
acquaintance rape	granny dumping	proxy	uxoricide
battered-woman syndrome	intimate partner violence	physical abuse	sibling relationship aggression
battering rape	intrafamilial child sexual abuse	psychological abuse	symbolic aggression
child sexual abuse	marital rape	Rophypnol	verbal abuse
corporal punishment	Megan's Law	shaken baby syndrome	violence
date rape			

The Companion Website for *Choices in Relationships: An Introduction to Marriage and the Family,* Ninth Edition

www.thomson.edu/sociology/knox

Supplement your review of this chapter by going to the companion website to take one of the Tutorial Quizzes, use the flash cards to master key terms, and check out the many other study aids you'll find there. You'll also find special features such as the Marriage and Family Resource Center, Census 2000 information, and other data and resources at your fingertips to help you with that special project or to do some research on your own.

WEB LINKS

Childabuse.com
http://www.childabuse.com/

Male Survivor
http://www.malesurvivor.org/

Minnesota Center Against Violence and Abuse
http://www.mincava.umn.edu

National Sex Offender Data Base
http://www.nationalalertregistry.com

Parents for Megan's Law
http://www.parentsformeganslaw.com

Rape, Abuse & Incest National Network (RAINN)
http://www.rainn.org/

Stop It Now! The Campaign to Prevent Child Sexual Abuse
http://www.stopitnow.com/

V-Day
http://www.vday.org/

REFERENCES

Abbey, A. 2005. Lessons learned and unanswered questions about sexual assault perpetration. *Journal of Interpersonal Violence* 20:39–42.

Anderson, K. L. 2002. Perpetrator or victim? Relationships between intimate partner violence and well-being. *Journal of Marriage and Family* 64:851–63.

Anderson, P. L., J. A. Tiro, A. W. Price, M. A. Bender, and N. J. Kaslow. 2003. Additive impact of childhood emotional, physical, and sexual abuse on suicide attempts among low-income African American woman. *Suicide and Life-Threatening Behavior* 32:131–38.

Babcock, J. C., S. A. Miller, and C. Siard. 2003. Toward a typology of abusive women: Differences between partner-only and generally violent women in the use of violence. *Psychology of Women Quarterly* 27:153–61.

Bergeron, L. R., and B. Gray. 2003. Ethical dilemmas of reporting suspected elder abuse. *Social Work* 48:96–106.

Bolen, R. M. 2001. *Child sexual abuse: Its scope and our failure.* New York: Kluwer Academic/Plenum Publishers.

Boyle, C. L. and J. R. Lutzker. 2005. Teaching young children to discriminate abusive from nonabusive situations using multiple exemplars in a modified discrete trial teaching format. *Journal of Family Violence* 20:55–70.

Brecklin, L. R. and S. E. Ullman. 2005. Self-defense or assertiveness training and women's responses to sexual attacks. *Journal of Interpersonal Violence* 20:738–62.

Buddie, A. M. and M. Testa. 2005. Rates and predictors of sexual aggression among students and nonstudents. *Journal of Interpersonal Violence* 20:713–24.

Caffaro, J. V. and A. Conn-Caffaro. 2005. Treating sibling abuse families. *Aggression and Violent Behavior* 10:604–23.

Campbell, R. and S. M. Wasco. 2005. Understanding rape and sexual assault: 20 years of progress and future directions. *Journal of Interpersonal Violence* 20:127–31.

Clements, C., R. Ogle, and C. Saboruin. 2005. Perceived control and emotional status in abusive college student relationships: An exploration of gender differences. *Journal of Interpersonal Violence* 20:1058–1077.

Coates, T. P. 2006. When parents are the threat. *Time,* May 8, Special issue.

Connell-Carrick, K. 2003. A critical review of the empirical literature: Identifying correlates of child neglect *Child and Adolescent Social Work Journal* 20:389–925.

Davis, M. K., and C. A. Gidyez. 2000. Child sexual abuse prevention programs: A meta-analysis. *Journal of Clinical Child Psychology* 29:257–66.

DeMaris, A., M. L. Benson, G. L. Fox, T. Hill, and J. V. Wyk. 2003. Distal and proximal factors in domestic violence: A test of an integrated model. *Journal of Marriage and Family* 65:652–67.

De Paul, J., and I. Arruabarrena. 2003. Evaluation of a treatment program for abusive and high-risk families in Spain. *Child Welfare* 82:413–42.

Dixon, L., C. Hamilton-Giachritsis, and K. Browne. 2005. Attributions and behaviors of parents abused as children: a mediational analysis of the intergenerational continuity of child maltreatment (Part II). *Journal of Child Psychology and Psychiatry and Allied Disciplines* 46:58–73.

Dong, M., R. F. Anda, S. R. Dube, W. H. Giles, and V. J. Felitti. 2003. The relationship of exposure to childhood sexual abuse to other forms of abuse, neglect, and household dysfunction during childhood. *Child Abuse and Neglect* 27:625–39.

Dugan, L., D. S. Nagin, and R. Rosenfeld 2003. Exposure reduction or retaliation? The effects of domestic violence resources on intimate-partner homicide. *Law & Society Review* 37:169–98.

Edleson, J. L., L. F. Mbilinyi, S. K. Beeman, and A. K. Hagemeister. 2003. How children are involved in adult domestic violence. *Journal of Interpersonal Violence* 18:18–32.

Erwin, M. J., R. R. M. Gershon, M. Tiburzi, and S. Lin. 2005. Reports of intimate partner violence made against police officers. *Journal of Family Violence* 20:13–20.

Faller, K. C., and J. Henry. 2000. Child sexual abuse: A case study in community collaboration. *Child Abuse and Neglect* 24:1215–25.

Few, A. L. and K. H. Rosen 2005. Victims of chronic dating violence: How women's vulnerabilities link to their decisions to stay *Family Relations* 54:265–279.

Field, C. A. and R. Caetano. 2005. Intimate partner violence in the U.S. general population: Progress and future generations *Journal of Interpersonal Violence* 20:463–69.

Fieldman, J. P., and T. D. Crespi. 2002. Child sexual abuse: Offenders, disclosure, and school-based initiatives. *Adolescence* 37:151–60.

Freedner, N., L. H. Freed, W. Yang, and S. B. Austin. 2002 Dating violence among gay, lesbian, and bisexual adolescents: Results from a community survey. *Journal of Adolescent Health* 31:469–74.

Fulmer, T., G. Paveza, C. VandeWeerd, L. Guadagno, S. Fairchild, R. Norman, V. Abraham, and M. Bolton-Blatt. 2005. Neglect assessment in urban emergency departments and confirmation by an expert clinical team. *Journals of Gerontology Series A–Biological Sciences and Medical Sciences* 60:1002–06.

Global Initiative to End All Corporal Punishment of Children. 2005. March. Legality of Corporal Punishment Worldwide. Available at http://www.endcorporalpunishment.org.

Goldstein, S. L., and R. P. Tyler. 1998. Frustrations of inquiry: Child sexual abuse allegations in divorce and custody cases. *FBI Law Enforcement Bulletin* 67:1–6.

Goodman, G. S., Simona Ghetti, J. A. Quas, R. S. Edelstein, K. W. Alexander, A. D. Redlich, I. M. Cordon, and D. P. H. Jones. 2003. A prospective study of memory for child sexual abuse: New findings relevant to the repressed-memory controversy. *Psychological Science* 14:113–18.

Goodman-Brown, T. B., R. S. Edelstein, G. S. Goodman, D. P. H. Jones, and D. S. Gordon. 2003. Why children tell: A model of children's disclosure of sexual abuse. *Child Abuse and Neglect* 27:525–40.

Grossman, S. F., S. Hinkley, A. Kawalski and C. Margrave. 2005. Rural versus urban victims of violence: Interplay of race and region. *Journal of Family Violence* 20:71–82.

Grych, J. H., G. T. Harold, and C. J. Miles. 2003. A prospective investigation of appraisals as mediators of the link between interparental conflict and child adjustment. *Child Development* 74:1176–96.

Ham-Rowbottom, K. A., E. E. Gordon, K. L. Jarvis, and R. W. Novaco. 2005. Life constraints and psychological well-being of domestic violence shelter graduates. *Journal of Family Violence* 20:109–22.

Haskett, M. E., S. S. Scott, R. Grant, C. S. Ward, and Canby Robinson. 2003. Child-related cognitions and affective functioning of physically abusive and comparison parents. *Child Abuse and Neglect* 27:663–86.

Henning, K., A. Jones and R. Holdford. 2005. "I didn't do it, but if I did I had a good reason": Minimization, denial, and attributions of blame among male and female domestic violence offenders F. *Journal of Family Violence* 20:131–40.

Hewitt, B., K. Klise, L. Comander, M Schorr, A. Hardy, T. Duffy et al. 2002. Breaking the silence. *People* 57:56 passim.

Heyman, R. E., and A. M. S. Slep. 2002 Do child abuse and interpersonal violence lead to adult family violence? *Journal of Marriage and Family* 64:864–70.

Howard, D. E., S. Feigelman, X. Li, S. Gross, and L. Rachuba. 2002. The relationship among violence victimization, witnessing violence, and youth distress. *Journal of Adolescent Health* 31:455–62.

Howard, D.E. and M. Q. Wang. 2003. Risk profiles of adolescent girls who were victims of dating violence. *Adolescence* 38:1–19.

Johnson, L., M. Todd, and G. Subramanian. 2005. Violence in police families: Work-family spillover. *Journal of Family Violence* 20:3–13.

Jonzon, E. and F. Lindblad. 2005. Adult female victims of child sexual abuse: Multitype maltreatment and disclosure characteristics related to subjective health. *Journal of Interpersonal Violence* 20:651–66.

Kernic, M. A., M. E. Wolfe, V. L. Holt, B. McKnight, C. E. Huebner, and F. P. Rivara. 2003. Behavioral problems among children whose mothers are abused by an intimate partner. *Child Abuse & Neglect* 27:1231–46.

Kim, J., and C. Emery. 2003. Marital power: Conflict, norm consensus, and marital violence in a nationally representative sample of Korean couples. *Journal of Interpersonal Violence* 18:197–219.

Kitzmann, K. M., N. K. Gaylord, A. R. Holt, and E. D. Kenny. 2003. Child witnesses to domestic violence: A meta-analytic review. *Journal of Clinical and Consulting Psychology* 71:339–52.

Knox, D. and Zusman, M. E. 2006. Relationship and sexual behaviors of a sample of 1027 university students. Unpublished data collected for this text. Department of Sociology, East Carolina University, Greenville, NC.

Komarow, S. 2005. Report: Military women devalued. *USA Today* August 26.

Koss, M. P., J. A. Bailey, N. P. Yaun, V. M. Herrera, and E. L. Lichter. 2003. Depression and PTSD in survivors of male violence: Research and training initiatives to facilitate recovery. *Psychology of Women Quarterly* 27:130–42.

Lloyd, S. A., and B. C. Emery. 2000. *The dark side of courtship: Physical and sexual aggression.* Thousand Oaks, CA: Sage.

Loy, E., L. Machen, M. Beaulieu, and G. L. Greif. 2005. Common themes in clinical work with women who are domestically violent. *The American Journal of Family Therapy* 33:33–42.

Lussier, P., E. Beauregard, J. Proulx, and N. Alexandre. 2005. Developmental factors related to deviant sexual preferences in child molesters. *Journal of Interpersonal Violence* 20:999–1017.

Mears, D. P. and C. A. Visher. 2005. Trends in understanding and addressing domestic violence. *Journal of Interpersonal Violence* 20:204–11.

Meloy, J. R. and H. Fisher. 2005. Some thoughts on the neurobiology of stalking. *Journal of Forensic Science* 50:1472–80.

McCloskey, L. A., and J. A. Bailey. 2000. The intergenerational transmission of risk for child sexual abuse. *Journal of Interpersonal Violence* 15:1019–35.

McGeary, J., R. Winters, S. Morrissey, S. Scully, M. Sieger, S. Crittle. et al. 2002. Can the church be saved? *Time Atlantic* 159:52–62.

Melchert, T. P. 2000. Clarifying the effects of parental abuse, child sexual abuse, and parental caregiving on adult adjustment. *Professional Psychology: Research and Practice* 31:64–69.

Muller, R. T., and K. E. Lemieux. 2000. Social support, attachment, and psychopathology in high risk formerly maltreated adults. *Child Abuse and Neglect* 24:883–900.

Nayak, M. B., C. A. Byrne, M. K. Martin, and A. G. Abraham. 2003. Attitudes toward violence against women: A cross-nation study. *Sex Roles: A Journal of Research* 49:333–43.

Nelson, B. S., and K. S. Wampler. 2000. Systemic effects of trauma in clinic couples: An exploratory study of secondary trauma resulting from childhood abuse. *Journal of Marriage and Family Counseling* 26:171–84.

Oswald, D. L. and B. L. Russell. 2006. Perceptions of sexual coercion in heterosexual dating relationships: the role of aggressor gender and tactics. *The Journal of Sex Research* 43:87–98.

Patterson, M. and K. Malley-Morrison. 2006. A cognitive-ecological approach to elder abuse in five cultures: Human rights and education. *Educational Gerontology* 32:73–82.

Perry, A. R. and M. E. Fromuth. 2005. Courtship violence using couple data characteristics and perceptions. *Journal of Interpersonal Violence* 20:1078–95.

Puente, S., and D. Cohen. 2003. Jealousy and the meaning (or nonmeaning) of violence. *Personality and Social Psychology Bulletin* 29:449–60.

Rand, M. R. 2003. The nature and extent of recurring intimate partner violence against women in the United States. *Journal of Comparative Family Studies* 34:137–46.

Ricci, L., A. Giantris, P. Merriam, S. Hodge, and T. Doyle. 2003. Abusive head trauma in Maine infants: Medical, child protective, and law enforcement analysis. *Child Abuse and Neglect* 27:271–83.

Rivers, M. J. 2005. Navajo women and abuse: The context for their troubled relationships. *Journal of Family Violence* 20:83–90.

Rohrbaugh, J.B. 2006. Domestic violence in same-gender relationships. *Family Court Review* 44:287–99.

Rosenbaum, A., and P. A. Leisring. 2003. Beyond power and control: Towards an understanding of partner abusive men. *Journal of Comparative Family Studies* 34:7–21.

Rousseve, A. 2005. Domestic violence in the United States. *Georgetown Journal of Gender & the Law* 6:431–58.

Rubin, D. M., C. W. Christian, L. T. Bilaniuk, K. A. Zaxyczny, and D. R. Durbin. 2003. Occult head injury in high-risk abused children. *Pediatrics* 111:1382–86.

Schaeffer, C. M., P. C. Alexander, K. Bethke, and L. S. Kretz. 2005. Predictors of child abuse potential among military parents: Comparing mothers and fathers. *Journal of Family Violence* 20:123–130.

Shen, H. and S. B. Sorenson. 2005. Restraining orders in California: A look at statewide data. *Violence Against Women* 11:912–33.

Sheridan, M. S. 2003. The deceit continues: An updated literature review of Munchausen syndrome by proxy. *Child Abuse & Neglect* 27:431–51.

Sidebotham, P., J. Heron, and the ALSPAC study team. 2003. Child maltreatment in the "children of the nineties:" The role of the child. *Child Abuse & Neglect* 27:337–52.

Silvergleid, C. and E. S. Mankowski. 2006. How batterer intervention programs work: Participant and facilitator accounts of processes of change. *Journal of Interpersonal Violence* 21:139–59.

Simonelli, C. J., T. Mullis, A. N. Elliott, and T. W. Pierce. 2002. Abuse by siblings and subsequent experiences of violence within the dating relationship. *Journal of Interpersonal Violence* 17:103–21.

Smith, J. 2003. Shaken baby syndrome. *Orthopaedic Nursing* 22:196–205.

Statistical Abstract of the United States: 2006. 125th ed. Washington, DC: U.S. Bureau of the Census.

Struckman-Johnson, C., D. Struckman-Johnson, and P. B. Anderson. 2003. Tactics of sexual coercion: When men and women won't take no for an answer. *Journal of Sex Research* 40:76–86.

Straus, M. A. 2000. Corporal punishment and primary prevention of physical abuse. *Child Abuse and Neglect* 24:1109–14.

Stuart, G. L. 2005. Improving violence intervention outcomes by integrating alcohol treatment. *Journal of Interpersonal Violence* 20:388–93.

The Orange Approach 2005. *The Economist* 374:31–33.

Tolan, P. H., D. Gorman-Smith, and D. B. Henry. 2003. The developmental ecology of urban males' youth violence. *Developmental Psychology* 39:274–91.

Trokel, M., W. Anthony, J. Griffith, and R. Sege. 2006. Variation in the diagnosis of child abuse in severely injured infants. *Pediatrics* 722–29.

Ulman, A. 2003. Violence by children against mothers in relation to violence between parents and corporal punishment by parents. *Journal of Comparative Family Studies* 34:41–56.

Ulman, A., and M. A. Straus. 2003. Violence by children against mothers in relation to violence between parents and corporal punishment by parents. *Journal of Comparative Family Studies* 34:41–56.

Umberson, D. K., L. Anderson, K. Williams, and M. D. Chen. 2003. Relationship dynamics, emotional state, and domestic violence: A stress and masculine perspective. *Journal of Marriage and the Family* 65:233–47.

Updegraff, K. A., S. M. Thayer, S. D. Whiteman, D. J. Denning and S. M. McHale. 2005. Relational aggression in adolescents' sibling relationships: Links to sibling and parent-adolescent relationship quality. *Family Relations* 54:373–86.

Vandello, J. A., and D. Cohen. 2003. Male honor and female fidelity: Implicit cultural scripts that perpetuate domestic violence. *Journal of Personality and Social Psychology* 84:997–1010.

Vazquez, S., M. K. Stohr, K. Skow, and M. Purkiss. 2005. Why is a woman still not safe when she's home? Seven years of NIBRS data on victims and offenders of intimate partner violence. *Criminal Justice Studies: A Critical Journal of Crime, Law and Society* 18:125–46.

Verma, R. K. 2003. Wife beating and the link with poor sexual health and risk behavior among men in urban slums in India. *Journal of Comparative Family Studies* 34:1–61.

Walrath, C., M. Ybarra, E. W. Holden, Q. Liao, R. Santiago, and P. Leaf. 2003. Children with reported histories of sexual abuse. *Child Abuse and Neglect* 27:509–24.

Walsh, W. 2002. Spankers and nonspankers: Where they get information on spanking. *Family Relations* 51:81–88.

Watkins, P. 2005. Police perspective: Discovering hidden truths in domestic violence intervention. *Journal of Family Violence* 20:47–55.

Whatley, M. 2005. The effect of participant sex, victim dress, and traditional attitudes on casual judgments for marital rape victims. *Journal of Family Violence* 20:191–201.

Wright, J. W. Jr. 2004. Personal communication, Monroe, La. Dr. Wright is an attorney who has been involved in litigation of Munchausen syndrome by proxy lawsuits.

Zoellner, L. A., N. C. Feeny, J. Alvarez, C. Watlington, M. L. O'Neill, R. Zager, and E. B. Foa. 2000. Factors associated with completion of the restraining order process in female victims of partner violence. *Journal of Interpersonal Violence* 15:1081–99.

chapter 15

> Getting a divorce is like pulling a decayed tooth—it hurts, and you miss your spouse, but you will feel better.
>
> *Count Basie, jazz musician*

Divorce

Contents

© Joel Gordon

True or False?

1. The divorce rate has continued to increase so that by 2006, six in ten new marriages were predicted to end in divorce.

2. Over half the states now provide some form of covenant marriage that seeks to slow couples down (on their way to divorce) and to encourage them to work through their problems.

3. After 15 years of marriage, about 43 percent of couples will have divorced.

4. Spouses in loveless, conflicted, unhappy marriages are less healthy/happy than those who divorced/left these type relationships.

5. Whether a spouse initiates the divorce (the "dumper") or reacts to a spouse who does (the "dumpee") makes no difference in terms of adjustment to the divorce.

Answers: **1.** F   **2.** F   **3.** T   **4.** T   **5.** F

Divorce remains an event most individuals are sad about. No one is "happy" for one's own divorce or for the divorce of a close friend. Divorce signals trauma, unhappiness, and a period of adjustment. With over forty percent (42 percent, to be specific) of marriages begun today predicted to end in divorce, divorce is a stable part of the American relationship landscape. In this chapter we examine the reasons (societal and individual) for divorce and the effects on spouses and children. First, let's look at how divorce rates are measured.

Ways of Measuring Divorce Prevalence

Divorce is the legal ending of a valid marriage contract. There are three ways of measuring the degree to which divorce occurs in our society.

Crude Divorce Rate

The **crude divorce rate** is a statement of how many divorces have occurred for every 1,000 people in the population. In 2003, there were 3.8 divorces per 1,000 population. This rate reflects a drop in the divorce rate (as it was 4.0 the two previous years) (*Statistical Abstract of the United States, 2006*, Table 72).

Refined Divorce Rate

More revealing than the crude divorce rate is the **refined divorce rate**, which is an expression of the number of divorces and annulments in a given year divided by the number of married women in the population times 1,000. For example, there are approximately 1.1 divorces/annulments per year and 58 million married women, which translates into .0189 times 1,000, or 18.9—the refined divorce rate.

Diversity in Other Countries

Compared with other countries, the United States has one of the highest divorce rates in the world. Only Belarus of Eastern Europe comes close when crude divorce rates are compared. However, one of the reasons for a high U.S. divorce rate is a very high marriage rate (Blossfeld & Muller, 2002).

Percentage of Marriages Ending in Divorce

A final way of estimating divorce is to identify the percentage of people who are married who eventually get divorced. The problem with this statistic is the period of time considered in identifying those who divorced. The percentage of those who divorced within five years would be lower than the percentage of those who divorced within a ten-year period. Current estimates suggest that about 40 percent of those who married in the past couple of decades will divorce (Hawkins et al., 2002). Goodwin (2003) noted that 20 percent of first marriages will end in divorce within five years, 33 percent within 10 years, and 43 percent within fifteen years of marriage.

Regardless of how one measures the rate of divorce in the United States, "divorce rates have been stable or dropping for two decades" (Coltrane & Adams, 2003, p. 363). The principal factor for such decline is that persons are delaying marriage so that they are older at the time of marriage. According to Heaton (2002), "age at marriage plays the greatest role in accounting for trends in marital dissolution. . . . [W]omen who marry at older ages have more stable marriages."

National Data
There are about 1 million divorces each year (*Statistical Abstract of the United States: 2006*, Table 72). A CNN/USA Today Gallup poll finds that 30 percent of adult Americans say that they have been divorced at some point (Carroll, 2006).

University students in the United States are no strangers to divorce. About a quarter have parents who are divorced (American Council on Education and the University of California, 2006.). A major question is "why?" There are both macro and micro factors that help to explain why spouses divorce.

It's never too late—in fiction or in life—to revise.

Nancy Thayer

Macro Factors Contributing to Divorce

Sociologists emphasize that social context creates outcomes. This is best illustrated in the statistic that there was an average of only one divorce per year in Massachusetts among the Puritans from 1639 to 1760 (Morgan, 1944). The social context of that era involved strong pro family values and strict divorce laws with the result that divorce was almost nonexistent. In contrast, divorce occurs more frequently today as a result of various structural and cultural factors, also known as macro factors (Lowenstein, 2005).

Everywhere, marriage is becoming more fragile.

Stephanie Coontz, family historian

Increased Economic Independence of Women
In the past, the unemployed wife was dependent on her husband for food and shelter. No matter how unhappy her marriage was, she stayed married because she was economically dependent on her husband. Her husband literally represented her lifeline. Finding gainful employment outside the home made it possible for the wife to afford to leave her husband if she wanted to. Now that about three fourths of wives are employed, fewer and fewer wives are economically trapped in an unhappy marriage relationship. As we noted earlier, the wife's employment does not increase the risk of divorce in a happy marriage. But it does provide an avenue of escape for women in unhappy or abusive marriages (Kesselring and Bremmer, 2006).

Employed wives are also more likely to require an egalitarian relationship; while some husbands prefer this role relationship, others are unsettled by it. Another effect of the wife's employment is that she may meet someone new in the workplace so that she

Mistakes are part of the dues one pays for a full life.

Sophia Loren

becomes aware of an alternative to her current partner. Finally, unhappy husbands may be more likely to divorce if their wives are employed and able to be financially independent (less alimony and child support).

Changing Family Functions and Structure

Many of the protective, religious, educational, and recreational functions of the family have been largely taken over by outside agencies. Family members may now look to the police for protection, the church or synagogue for meaning, the school for education, and commercial recreational facilities for fun rather than to each other within the family for fulfilling these needs. The result is that although meeting emotional needs remains a primary function of the family, fewer reasons exist to keep the family together.

In addition to the changing functions of the family brought on by the Industrial Revolution, the family structure has changed from that of larger extended families in rural communities to smaller nuclear families in urban communities. In the former, individuals could turn to a lot of people in times of stress; in the latter, more stress necessarily fell on fewer shoulders. Cohen and Savaya (2003) documented an increased divorce rate among Muslim Palestinian citizens of Israel due to an increased acceptance of the modern views of marriage—increased emphasis on happiness and compatibility.

Liberalized Divorce Laws

No-fault divorce was first made available in California in 1970. All states now recognize some form of no-fault divorce. Although the legal grounds for it are irreconcilable differences and incompatibility, the reality is that spouses can get a divorce if they want to without having to prove that one of the partners is at fault (e.g., through adultery). The effect of no-fault divorce laws (neither party is assigned blame for the divorce) has been "to make divorce less acrimonious and restrictive, rendering the legal environment neutral and noncoercive" (Vlosky and Monroe, 2002, p. 317). And, no-fault divorce is associated with more divorce (Allen, Pendakur, and Suen, 2006). However, a backlash has occurred in response to the fact that divorce has become too easy to obtain, and a movement is afoot to make divorce harder to get. One divorced spouse said, "I should have stayed married when the hard times hit—it was just too easy to walk out" (author's files).

Fewer Moral and Religious Sanctions

Many priests and clergy recognize that divorce may be the best alternative in a particular marital relationship and attempt to minimize the guilt that members of their congregation may feel at the failure of their marriage. Increasingly, marriage is viewed in secular rather than religious terms. Hence, divorce has become more acceptable.

Starter Marriages

In her book *The Starter Marriage and the Future of Matrimony,* Paul (2002) noted that marriage is sometimes perceived as a growth experience whereby people marry, mature, and divorce before children arrive. Such a view of marriage reflects our sometimes cultural disregard for marriage as a permanent life-sustaining relationship. Indeed, the label of "starter" suggests something temporary, like a starter house from which the owners will eventually move. Comedian Rita Rudner quips in one of her stand up routines, "I select a man to marry on the basis of whether I think he will be a good father every other weekend."

More Divorce Models

As the number of divorced individuals in our society increases, the probability increases that a person's friends, parents, siblings, or children will be divorced. The more divorced people a person knows, the more normal divorce will seem to that person. The less deviant the person perceives divorce to be, the greater the probability the person will divorce if that person's own marriage becomes strained. Divorce has become so common that there are numerous websites exclusively for the divorced (e.g., www.heartchoice.com/divorce)

Mobility and Anonymity

That individuals are highly mobile results in fewer roots in a community and greater anonymity. Spouses who move away from their respective family and friends often discover that they are surrounded by strangers who don't care if they stay married or not. Divorce thrives when pro-marriage social expectations are not operative. In addition, the factors of mobility and anonymity also result in the removal of a consistent support system to help spouses deal with the difficulties they may encounter in marriage.

Individualistic Cultural Goal of Happiness

Unlike familistic values in Asian cultures, individualistic values in American culture emphasize the goal of personal happiness in marriage. When spouses stop having fun (when individualistic goals are no longer met), they sometimes feel that there is no reason to stay married. Only 5 percent of 1027 undergraduates at a large southeastern university agreed with the statement "I would not divorce my spouse for any reason." (Knox and Zusman, 2006). Reflecting an individualistic philosophy, Geraldo Rivera asked of his divorces, "Who cares if I've been married five times?"

Getting divorced just because you don't love a man is almost as silly as getting married just because you do.

Zsa Zsa Gabor, actress

Micro Factors Contributing to Divorce

Although macro factors may make divorce a viable cultural alternative to marital unhappiness, they are not sufficient to "cause" a divorce. One spouse must make a choice to divorce and initiate proceedings. Such a view is micro in that it focuses on the individual decisions and interactions within specific family units. The following subsections discuss some of the micro factors that may be operative in influencing a couple toward divorce.

American couples are socialized to marry for love and to stay married as long as love is alive. When love dies, spouses are vulnerable to divorce.

Falling Out of Love

"Falling out of love" was a top reason men reported as their reason for divorce (Enright, 2004) Previti and Amato (2003) noted that the absence of love was also associated with both men and women being more likely to divorce. Indeed, love, respect, friendship,

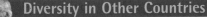
and good communication were mentioned by 60 percent of the respondents as reasons for staying married.

Negative Behavior

Physical or emotional abuse and alcohol or drug abuse were among the top behavioral reasons identified by women who divorced in mid life (Enright, 2004). Ostermann et al. (2005) confirmed that a discrepancy in the amount of alcohol consumed by spouses is associated with divorce. When the presence of negative behavior is coupled with the absence of positive behavior, the combination can be deadly. Shumway and Wampler (2002) emphasized that the absence of positive behaviors such as "small talk," "reminiscing about shared times together," and "encouragement" increases a couple's marital dissatisfaction.

People marry because they anticipate greater rewards from being married than from being single. During courtship, each partner engages in a high frequency of positive verbal (compliments) and nonverbal (eye contact, physical affection) behavior toward the other. The good feelings the partners experience as a result of these positive behaviors encourage them to marry to "lock in" these feelings across time. Just as love feelings are based on positive behavior from the partner, negative feelings are created when the partner engages in a high frequency of negative behavior and thoughts of divorce (to escape the negative behavior) occur.

Affair

Never underestimate the power of passion.

Eve Sayer

An extramarital affair has been associated with subsequent divorce (O'Leary, 2005). The spouse having the affair may feel unloved at home where there is little to no sex. Involvement in an affair may bring both love and sex and speed the spouse toward divorce. Alternatively, the at home spouse may become indignant and demand that they partner leave. Although most spouses do not leave their mates for a lover, the existence of an extramarital relationship may weaken the emotional tie between the spouses so that they are less inclined to stay married. Combined with the partner having the affair is the spouse who may feel betrayed and terminate the marriage. Seventy-three percent of 1027 undergraduates agreed with the statement "I would divorce a spouse who had an affair" and 74 percent reported having "previously ended a relationship with a partner who cheated on" him/her (Knox and Zusman, 2006).

Lack of Conflict Resolution Skills

Managing differences and conflict in a relationship helps to reduce the negative feelings that develop in a relationship. Some partners respond to conflict by withdrawing emotionally from their relationship; others respond by attacking, blaming, and failing to listen to their partner's point of view. Ways to negotiate differences and reduce conflict were discussed in the chapter on Communication in Relationships.

Value Changes

Both spouses change throughout the marriage. "He's not the same person I married" is a frequent observation of persons contemplating divorce. People may undergo radical value changes after marriage. One minister married and decided seven years later that he did not like the confines of the marriage role. He left the ministry, earned a Ph.D., and began to drink and have affairs. His wife, who had married him when he was a minister, now found herself married to a clinical psychologist who spent his evenings at bars with other women. The couple divorced.

Because people change throughout their lives, the person that one selects at one point in life may not be the same partner one would select at another.

Margaret Mead, the famous anthropologist, noted that her first marriage was a student marriage; her second, a professional partnership; and her third, an intellectual marriage to her soul mate, with whom she had her only child. At each of several stages in her life, she experienced a different set of needs and selected a mate who fulfilled those needs.

Satiation

Satiation, also referred to as habituation, refers to the state in which a stimulus loses its value with repeated exposure. Spouses may tire of each other. Their stories are no longer new, their sex is repetitive, and their presence no longer stimulates excitement as it did in courtship. Some persons, feeling trapped by the boredom of constancy, divorce and seek what they believe to be more excitement by a return to singlehood and, potentially, new partners. A developmental task of marriage is for couples to be creative to maintain excitement in their marriage.

Going new places, doing new things, and making time for intimacy in the face of rearing children and sustaining careers become increasingly important across time for spouses. Alternatively, spouses need to be realistic and not expect every evening to be a New Year's Eve. Comedian George Carlin says in one of his routines, "If all of your needs are not being met, drop some of your needs." If spouses did not expect so much of marriage, maybe they would not be disappointed.

Perception That One Would Be Happier If Divorced

Brinig and Allen (2000) noted that two-thirds of the applications for divorce are filed by women. Their behavior may be based on the perception that they will achieve greater power over their own life, money (in the form of child support and/or alimony) without having the liability of dealing with an unsupportive husband on a daily basis, and greater control over their children, since in 80 percent of the cases custody is awarded to the woman. The researchers argue that "who gets the children" is by far the greatest predictor of who files for divorce, and they contend that *if* the law presumes that joint custody will follow divorce, there will be fewer women filing for divorce since women will have less to gain.

Top Twenty Factors Associated with Divorce

Researchers have identified the characteristics of those most likely to divorce (Lowenstein, 2005; Goodwin, 2003; Orbuch et al., 2002; Wills, 2002; Teachman, 2002). Some of the more significant associations include the following:

1. Courtship of less than two years (partners know less about each other)
2. Having little in common (similar interests serve as a bond between people)
3. Marrying in teens (associated with low education and income and lack of maturity)
4. Differences in race, education, religion, social class, age, values, and libido (widens gap between spouses)
5. Not being religiously devout (less bound by traditional values)
6. A cohabitation history (established history of breaking social norms)
7. Previous marriage (less fearful of divorce)
8. No children (less reason to stay married)
9. Limited education (associated with lower income, more stress, less happiness)
10. Urban residence (more anonymity, less social control in urban environment)
11. Infidelity (trust broken, emotional reason to leave relationship)
12. Divorced parents (models for ending rather than repairing relationship; may have inherited traits such as alcoholism that are detrimental to staying married) or parents who never married and never lived together (being unmarried is normative)

He didn't stop loving her and she didn't stop loving him, but something turned off like an old-fashioned watch piece that one forgets to wind.

Biographer Howard Teichmann of Henry Fonda and his wife Susan

Marriage is not Christmas morning. It's Christmas afternoon.

Tim Britton, sociologist

13. Poor communication skills (issues go unresolved and accumulate)

14. Unemployment of husband (his self-esteem lost, loss of respect by wife who expects breadwinner)

15. Employment of wife (wife more independent and can afford to leave an unhappy marriage; husband feels threatened)

16. Depression, alcoholism, physical illness of spouse or child (partners undergo grave change from earlier in relationship) as well as imprisonment (extended separation erodes relationship between the spouses (Lopoo and Western, 2005)

17. Having seriously ill child (impacts stress, finances, couple time)

18. Low self-esteem of spouses (associated with higher jealousy, less ability to love and accept love)

19. Race (African-Americans live under more oppressive conditions, which increases stress in the marital relationship; 32 percent of European-American marriages, in contrast to 47 percent of African-American marriages, will end within ten years) (Goodwin, 2003).

20. Retirement (children gone, more couple time together, discovery that job was functional in keeping the spouses busy and away from each other; nothing to talk about).

The more of these factors that exist in a marriage, the more vulnerable a couple is to divorce. Regardless of the various factors associated with divorce, there is debate about the character of people who divorce. Are they selfish, amoral people who are incapable of making good on a commitment to each other and who wreck the lives of their children? Or are they individuals who care a great deal about relationships and won't settle for a bad marriage? Indeed, they may divorce precisely because they value marriage and want to rescue their children from being reared in an unhappy home.

Gender Differences in Filing for Divorce

I don't believe man is woman's natural enemy. Perhaps his lawyer is.

Shana Alexander

Enright (2004) noted that two-thirds of their 1,147 respondents who filed for divorce were women. They felt that their life would be better without their husbands (few worried they would be separated from their children). Indeed, the primary reason for delay was financial and once they could see a way to survive financially, they were on their way out of the marriage. Feeling a sense of renewed "self identity" was the top reward women reported on the other side of the divorce.

Men are less likely to seek a divorce since they view the cost as separation from their children. Eighty percent of divorces end with the woman getting primary custody of the children. Men who are not bonded with their children and/or those who are involved in a new relationship are more likely to seek a divorce.

Regardless of who files for divorce, there is regret in doing so. Over two thirds (67%) of both women and men who had been separated an average of 7 and 10 months, respectively, reported that their "top" advice for others was to "work it out/see a counselor together." This advice comes from 58 separated/divorcing individuals who completed an Internet survey on their experience with separation/divorce (Corte and Knox, 2007).

Consequences for Spouses Who Remain Unhappily Married

Hawkins and Booth (2005) analyzed longitudinal data of spouses in unhappy marriages over a 12-year period and found that those who stayed unhappily married had lower life satisfaction, self-esteem, and overall health compared to those who divorced whether or not they remarried. The authors suggest that there is no evidence that spouses in unhappy marriages are better off in any aspect of overall well-being than those who divorce. Similarly, Gardner and Oswald (2006) found that the psychological functioning/happiness of spouses going through divorce improved after the divorce—hence divorce was good for them. However, as we discuss in the following sections, divorce is not an easy process for spouses or for children.

Consequences for Spouses Who Divorce

Divorce is often an emotional and financial disaster (see Table 15.1, which identifies stages and issues of the divorce process) for both women and men. In addition to the death of a spouse, separation and divorce are among the most difficult of life's crisis events.

Psychological Consequences of Divorce

While spouses in thriving marriages are healthier and happier than the divorced, those in loveless, adversarial, abusive, unhappy marriages are less healthy/happy than the divorced (Gardner and Oswald, 2006). Although we tend to think of divorce as an intrinsically stressful event, continuing an unhappy, abusive, conflictual marriage may be more stressful than the divorce. Indeed, divorce may actually reduce rather than increase one's stress. The words of a t-shirt telegraph this theme: "How do you spell DIVORCE? R-E-L-I-E-F. Sakraida (2005) interviewed women who both initiated and were the recipient of a terminated relationship and found that the person being dropped (the "dumpee") was more vulnerable to depression and ruminated more about the divorce. Thuen and Rise (2006) found that perceived control was associated with positive psychological adjustment to divorce (the "dumpee" typically feels less control).

How do women and men differ in their emotional and psychological adjustment to divorce? While Hilton and Kopera-Frye (2004) disagree that there is a gender difference, researchers (Siegler and Costa, 2000) noted that women fare better emotionally after separation/divorce than do men. They note that women are more likely than men to not only have a stronger network of supportive relationships but also to profit from divorce by developing a new sense of self-esteem and confidence, since they are thrust into a more independent role. On the other hand, men are more likely to have been dependent on their wives for domestic and emotional support and to have a weaker external emotional support system. As a result, divorced men are more likely than divorced women to date more partners sooner and to remarry more quickly. Hence, there are gender differences in adjustment to divorce but these balance out over time.

We have been discussing the personal emotional consequences of divorce for the respective spouses, but the extended family and friends also feel the impact of a couple's divorce. Whereas some parents are happy and relieved that their offspring are divorcing, others grieve. Either way, the relationship with their grandchildren may be jeopardized. Friends often feel torn and divided in their loyalties. The courts divide the property and grant custody of the children. But who gets the friends?

Financial Consequences

Both women and men experience a drop in income following divorce, but women may suffer more (Hilton and Kopera-Freye, 2004). Since men usually have greater financial resources, they may take all they can with them when they leave. The only money they may continue to give to the ex-wife is court-ordered child support or spousal support (alimony). The latter is rare. Some states, such

Divorce is like cutting off a gangrened arm to save the rest of the body.

Jack Wright, sociologist

When I can no longer bear to think of the victims of broken homes, I begin to think of the victims of intact ones.

Peter de Vries, novelist

Table 15.1 Stages and Issues of Divorce Adjustment

Stage 1: Pre-separation

Issues

Personal: Consider seeing a marriage counselor with your spouse to improve your marriage. If you must seek a divorce, get a formal/legal separation agreement drawn up and signed before you and your spouse begin to live in separate residences.

Spouse: If divorce is inevitable, adopt the perspective that you will nurture as positive a relationship as possible with your soon-to-be-former spouse. You, your former spouse, and your children benefit from such a relationship.

Children: Tell your children nothing until you make a definite decision to get divorced and have begun to develop a separation agreement.

Relatives: Same as for your children.

Finances: Anticipate a drop in income. Look for alternative housing. If unemployed, get a job.

Legal: Contact a divorce mediator if your relationship with your spouse is civil to develop the terms of your separation agreement. If mediation is not an option, each spouse needs to hire the best attorney he/she can afford.

Stage 2: Separation

Issues

Personal: Consider seeing a therapist alone to help you through the emotional devastation of divorce. Don't separate (move out) until you have a formal/legal signed agreement.

Spouse: Endeavor to cooperate/be civil with your former partner.

Children: Tell your children of your decision to divorce. Make clear to them that they are not to blame and that the divorce will not change your love and care for them.

Relatives: Tell your parents/friends of your decision to divorce and tell them you will need their support during this period of transition. Reach out to the parents of your soon-to-be-former partner and tell them that in spite of the divorce you would like to maintain a positive relationship with them—which your children will benefit from.

Finances: Your divorce mediator or attorney will instruct you to develop an inventory of what you own. Open up a separate savings and checking account.

Legal: Complete divorce mediation with a mediator. If this is not possible, ask your attorney to develop a legal separation agreement you can live with. Be reasonable. The more you and your spouse can agree on, the more time and money you save in legal fees. Mediated divorces cost about $1,500 and take two to three months. A litigated divorce costs $18,000 and takes about three years.

Stage 3: Divorce

Issues

Personal: Nurture your relationships with friends and family who will provide support during the process of your divorce.

Spouse: Continue to nurture as civil a relationship as possible with your soon-to-be-former partner.

Children: Ensure that they have frequent and regular contact with each parent. Nurture their relationship with the other parent and their grandparents on both sides.

Relatives: Continue to have regular contact with friends/family who provide emotional support.

Finances: Be frugal.

Legal: Do what your attorney tells you.

Stage 4: Postdivorce

Issues

Personal: Seek other relationships but go slow in terms of commitment to a new partner. Give yourself at least 18 months before making a new commitment to a new partner.

Former Spouse: Continue as civil a relationship as possible. Encourage his/her involvement in another relationship. Be positive about a new partner in his/her life. You will want the same from your former partner some day.

Children: Spend individual time with each child. Do not require your children to like/enjoy your new partner.

Relatives: Provide both sets of grandparents access to your children.

Finances: Continue to be frugal. Move toward getting out of debt.

Legal: Make a payment plan with your attorney—begin to reduce this debt.

Adapted from The Divorce Room at Heartchoice.com. Used by permission.

as Texas, do not provide alimony at all. However, most states do provide for an equitable distribution of property, whereby property is divided according to what seems fairest to each party, on the basis of a number of factors (like ability to earn a living, fault in breaking up the marriage, etc.).

Although 56 percent of custodial mothers are awarded child support, the amount is usually inadequate, infrequent, and not dependable, and the woman is forced to work (sometimes at more than one job) to take financial care of her children. A comparison of divorced mothers, single mothers by choice, and married mothers revealed that economic stability and involvement of the father were more important than family structure in determining the quality of life for the mothers (Segal-Engelchin and Wozner, 2005).

How the money is divided depends on whether there was a prenuptial agreement or a **postnuptial agreement.** Such agreements are most likely to be upheld if the attorney insists on four conditions—full disclosure by both parties; independent representation by separate counsel; absence of coercion or duress; and terms that are fair and equitable (Abut, 2005).

Community property—they give her the property and throw you out of the community.

A divorced father

Fathers' Separation from Children

". . . divorce transforms family power from intact patriarchy to post-divorce matriarchy" (Finley, 2004, p. F9) where women are typically given custody of the children and child support. The result is that about 5 million divorced dads wake up every morning in an apartment or home while their children are waking up with their separated/divorced mother. These are noncustodial fathers who may be allowed to see their children only at specified times (e.g., two weekends a month).

Ahrons and Tanner (2003) found that it is not divorce but low father involvement, which has a negative impact on the father child relationship. Fathers who stay involved in the lives of their children (in spite of being a non custodial parent, having an adversarial former spouse, and a remarriage) emotionally, physically, and economically mitigate any negative effects on the relationship with their children. Indeed, some relationships with the father may improve since the father may spend more one-on-one time with his children.

Shared Parenting Dysfunction

Shared parenting dysfunction refers to the set of behaviors on the part of each parent (embroiled in a divorce) that are focused on hurting the other parent and are counterproductive for the well-being of the children. Examples identified by Turkat (2002) include:

- *a parent that forced the children to sleep in a car to prove the other parent had bankrupted them*

- *a non-custodial parent who burned down the house of the primary residential parent after losing a court battle over custody of the children*

- *a parent who bought a cat for the children because the other parent was highly allergic to cats.*

Finally, some of the most destructive displays of Shared Parenting Dysfunction may include kidnapping, physical abuse, and murder (p. 390).

Parental Alienation Syndrome

Shared parenting dysfunction may lead to parental alienation syndrome. **Parental alienation syndrome** is a disturbance in which children are obsessively preoccupied with deprecation and/or criticism of a parent, denigration that is unjustified and/or exaggerated (Gardner, 1998).

Although the "alienators" are fairly evenly balanced between fathers and mothers (Gardner, 1998), the custodial parent (more often the mother) has more opportunity and control to alienate the child from the other parent. A couple need not necessarily be going through divorce for parental alienation syndrome to occur—this phenomenon may also occur in intact families (Baker, 2006).

Schacht (2000) identified several types of behavior that either parent may engage in to alienate the child from the other parent:

1. Minimizing the importance of contact and the relationship with the other parent

2. Excessively rigid boundaries: rudeness, refusal to speak to or inability to tolerate the presence of the other parent, even at events important to the child; refusal to allow the other parent near the home for drop-off or pick-up visitations

3. No concern about missed visits with the other parent

4. No positive interest in the child's activities or experiences during visits with the other parent

5. Granting autonomy to the point of apparent indifference ("It's up to you if you want to see your dad, I don't care.")

6. Overt expressions of dislike of visitation ("OK, visit, but you know how I feel about it. . .")

7. Refusal to discuss anything about the other parent ("I don't want to hear about . . . ") or selective willingness to discuss only negative matters

8. Innuendo and accusations against the other parent, including statements that are false

9. Portraying the child as an actual or potential victim of the other parent's behavior

10. Demanding that the child keep secrets from the other parent

11. Destruction of gifts or memorabilia of the other parent

12. Promoting loyalty conflicts (such as by offering an opportunity for a desired activity that conflicts with scheduled visitation)

The most telling sign of a child who has been alienated from a parent is the irrational behavior of the child, who for no properly explained reason says that he or she wants nothing further to do with one of the parents.

Children who are alienated from the other parent are sometimes unable to see through the alienation process and regard their negative feelings as natural. Such children are similar to those who have been brainwashed by cult leaders to view outsiders negatively. Baker (2005) interviewed 38 adults who had experienced parental alienation as a child and observed several areas of impact, including low self-esteem, depression, drug/alcohol abuse, lack of trust, alienation from own children and divorce.

People who fight fire with fire end up with ashes.

Abigail Van Buren

PERSONAL CHOICES

Choosing to Maintain a Civil Relationship with Your Former Spouse

Spouses who divorce must choose the kind of relationship they will have with each other. This choice is crucial in that it affects not only their own but also their children's lives. Constance Ahrons (1995) identified four types of ex-spouse relationships, including "fiery foes," which 25 percent of her divorcing spouses exemplified. Another 25 percent were categorized as "angry associates." Hence, half of her respondents had adversarial relationships with their former partners. Other patterns included "perfect pals" (12%) and cooperative colleagues (38%).

Everyone loses when the "fiery foes" and "angry associates" pattern develops and continues. The parents continue to harbor negative feelings for each other, and the children are caught in the crossfire. They aren't free to develop or express love for either parent out of fear of disapproval from the other. Ex-spouses might consider the costs to their children of continuing their hostility and do whatever is necessary to maintain a civil relationship with their former partner. We emphasize the benefits of coparenting after divorce in the chapter on remarriage.

Sometimes the parent intent on alienating their child from the other parent may discover that the child resents the custodial parent for such deprivation. In addition, the child may feel deceived if they are told negative things about the other parent and later learn that these were designed to foster a negative relationship with that parent. The result is often a strained and distanced relationship with the custodial parent when the child grows up—an unintended consequence. Alternatively, the negative socialization toward the other parent may create a lifelong bias against that parent

Effects of Divorce on Children

One million children annually experience the divorce of their parents (Wallace & Koerner, 2003). The Children's Beliefs about Parental Divorce Scale (see page 438) provides a way to measure the perceived effects of divorce on children.

Wallerstein (2000) conducted a longitudinal study of ninety-three children from divorced homes and found negative effects that surfaced in the child's 20s and 30s in the form of difficulty in relationship formation and fears of loss. Not only were the children of divorce more likely to have difficulty deciding whom they wanted to marry, but they also expressed concerns over whether to have children.

Wallerstein's study has been criticized since she did NOT compare her subjects with those of similar age from intact homes. Hence, one cannot assume that children from intact homes do not also have concerns about whom to marry and whether to have children. In addition, some research contradicts Wallerstein's predictions. Gordon (2005) noted that divorce may actually benefit children. When parental conflict is very high prior to divorce, children benefit by no longer being subjected to such relentless anger and emotional abuse they observe between their parents. Indeed, when children whose conflicted parents divorced were compared with children whose parents were still together, the children were very similar. DeCuzzi, Knox, and Zusman (2004) also reported that while a quarter (26%) of the undergraduates in their study reported that the divorce of their parents had a negative effect, almost a third (32.9%) reported a *positive* effect. Finally, when children of divorced parents become involved in their own marriage to a supportive, well-adjusted partner, the negative effects of parental divorce are mitigated (Hetherington, 2003).

Kelly and Emery (2003) reviewed the literature on the effect of divorce on children and concluded:

> . . . . *it is important to emphasize that approximately 75-80% of children and young adults do not suffer from major psychological problems, including depression; have achieved their education and career goals; and retain close ties to their families. They enjoy intimate relationships, have not divorced, and do not appear to be scarred with immutable negative effects from divorce (p. 357–358)*

Indeed, Coltrane and Adams (2003) emphasized that claiming that divorce seriously damages children is a "symbolic tool used to defend a specific moral vision for families and gender roles within them" (p. 369). They go on to state that "Understanding this allows us to see divorce not as the universal moral evil depicted by divorce reformers, but as a highly individualized process that engenders different experiences and reactions among various family members . . ." (p. 370). Nevertheless, there is negative fallout from divorce for some children. Kilmann et al.(2006) compared 147 college females with biological intact parents with 157 college females whose parents had divorced. Compared to those with intact parents, females whose parents had divorced had lower self-esteem and rated both their biological fathers and mothers more negatively.

Children's Beliefs about Parental Divorce Scale

The following are some statements about children and their separated parents. Some of the statements are true about how you think and feel, so you will want to check YES. Some are NOT TRUE about how you think or feel, so you will want to check NO. There are no right or wrong answers. Your answers will just tell us some of the things you are thinking now about your parents' separation.

1. It would upset me if other kids asked a lot of questions about about my parents. ___ Yes ___ No

2. It was usually my father's fault when my parents had a fight. ___ Yes ___ No

3. I sometimes worry that both my parents will want to live without me. ___ Yes ___ No

4. When my family was unhappy it was usually because of my mother. ___ Yes ___ No

5. My parents will always live apart. ___ Yes ___ No

6. My parents often argue with each other after I misbehave. ___ Yes ___ No

7. I like talking to my friends as much now as I used to. ___ Yes ___ No

8. My father is usually a nice person. ___ Yes ___ No

9. It's possible that both my parents will never want to see me again. ___ Yes ___ No

10. My mother is usually a nice person. ___ Yes ___ No

11. If I behave better I might be able to bring my family back together. ___ Yes ___ No

12. My parents would probably be happier if I were never born. ___ Yes ___ No

13. I like playing with my friends as much now as I used to. ___ Yes ___ No

14. When my family was unhappy it was usually because of something my father said or did. ___ Yes ___ No

15. I sometimes worry that I'll be left all alone. ___ Yes ___ No

16. Often I have a bad time when I'm with my mother. ___ Yes ___ No

17. My family will probably do things together just like before. ___ Yes ___ No

18. My parents probably argue more when I'm with them than when I'm gone. ___ Yes ___ No

19. I'd rather be alone than play with other kids. ___ Yes ___ No

20. My father caused most of the trouble in my family. ___ Yes ___ No

21. I feel that my parents still love me. ___ Yes ___ No

22. My mother caused most of the trouble in my family. ___ Yes ___ No

23. My parents will probably see that they have made a mistake and get back together again. ___ Yes ___ No

24. My parents are happier when I'm with them than when I'm not. ___ Yes ___ No

25. My friends and I do many things together. ___ Yes ___ No

26. There are a lot of things about my father I like. ___ Yes ___ No

27. I sometimes think that one day I may have to go live with a friend or relative. ___ Yes ___ No

28. My mother is more good than bad. ___ Yes ___ No

29. I sometimes think that my parents will one day live together again. ___ Yes ___ No

30. I can make my parents unhappy with each other by what I say or do. ___ Yes ___ No

31. My friends understand how I feel about my parents. ___ Yes ___ No

32. My father is more good than bad. ___ Yes ___ No

33. I feel my parents still like me. ___ Yes ___ No

34. There are a lot of things about my mother I like. ___ Yes ___ No

35. I sometimes think that once my parents realize how much I want them to they'll live together again. ___ Yes ___ No

36. My parents would probably still be living together if it weren't for me. ___ Yes ___ No

Scoring

The CBAPS identifies problematic responding. A "yes" response on items 1, 2, 3, 4, 6, 9, 11, 12, 14–20, 22, 23, 27, 29, 30, 35, 36 and a "no" response on items 5, 7, 8, 10, 13, 21, 24–26, 28, 31–34 indicate a problematic reaction to one's parents divorcing. A total score is derived by summing the number of problematic beliefs across all items, with a total score of 36. The higher the score, the more problematic the beliefs about parental divorce.

Norms: A total of 170 schoolchildren, 84 boys and 86 girls, with a mean age of 11 whose parents were divorced, completed the scale. The mean for the total score was 8.20, with a standard deviation of 4.98.

Source

Kurdek, L. A., and B. Berg. 1987. Children's beliefs about parental divorce scale: Psychometric characteristics and concurrent validity. *Journal of Consulting and Clinical Psychology* 55: 712–18. Copyright ©, Professor Larry Kurdek, Department of Psychology, State University, Dayton, OH 45435-0001. Used by permission of Dr. Kurdek.

While the primary factor that determines the effect of divorce on children is the degree to which the divorcing parents are civil or adversarial (see personal choices section), legal and physical custody are important issues. The following section details how judges go about making this decision.

Who Gets the Children?*

Judges who are assigned to hear initial child custody cases must make a judicial determination regarding whether one or both parents will have decisional authority on major issues affecting the child (called "**legal custody**"), and the distribution of parenting time (called "visitation" or "**physical custody**"). Toward this end, judges in all states are guided by the statutory dictum called "best interests of the child." In some states (Florida, Michigan, California, New Jersey, and others) specific statutory custody factors have been enacted to guide the judge in making a "best interest" determination.

In a highly contested custody case, the judge will often appoint a mental health professional to conduct a custody evaluation in order to assist the judge in determining what will be the best future arrangement for the child. Of course, each custody case is different because the circumstances of the children are different, but some of the frequently employed custody factors are:

1. The child's age, maturity, sex, activities, including culture and religion. Since the focus of custody is on the child's best interests, this factor considers all relevant information about the child's life.

2. The wishes expressed by the child, particularly the older child. Judges will often interview in chambers children six years old or older.

3. Each parent's capacity to care for and provide for the emotional, intellectual, financial, religious, and other needs of the child. This factor will often consider the work schedules of the parents.

4. The parents' ability to agree, communicate and cooperate in matters relating to the child.

5. The nature of the child's relationship with the parents. This factor also takes into consideration the child's relationship with other significant people, such as members of the child's extended family.

6. The need to protect the child from physical and psychological harm caused by abuse or ill treatment. This factor focuses on the issue of domestic violence.

7. The past and present parental attitudes and behavior. This factor deals with issues of parenting skills and personalities.

8. The proposed plans for caring for the child. Custody is not ownership of the child, so the judge will want to know how each parent proposes to raise the child, including proposed parenting times for the other parent.

These and other custody factors, whether presented by the custody evaluator or by testimony, will become the basis for the judge's custody determination.

How Does Parental Divorce Affect the Romantic Relationships of College Students?

One of the greatest hopes of individuals whose parents have a happy and enduring marriage, is that they too, will end up in such an endearing relationship. Similarly, one of the greatest fears of individuals whose parents are divorced or who have an unhappy and dysfunctional relationship, is that they too, will have such an unhappy/unfulfilling marriage. These hopes and fears are common

Diversity in Other Countries

A team of Norwegian researchers (Storksen et al. 2006) examined 8,984 adolescents ages 13 to 19 comparing their levels of anxiety and depression in homes where their parents were not distressed and together and where their parents were distressed and divorcing. Children from the former homes were less likely to report anxiety/depression than in the divorcing homes (14% and 30%, respectively).

In some divorces, the wife wants custody of the children; in others, the custody of the money.

Evan Esar

---

* This section was written by Dr. Ken Lewis, a custody evaluator, and author of *Five Stages of Child Custody* (CCES Press, PO Box 202, Glenside, PA 19038).

emotions of offspring who contemplate their futures and wonder of the degree to which their parent's divorce will affect their own.

There are about 22 million divorced individuals in the United States (*Statistical Abstract of the United States: 2006*, Table 50) most (60%) of whom have children. Indeed, about a quarter of first year college students report that their parents are separated or divorced (American Council on Education and University of California. 2006). Extrapolating to the 16 million college students in the United States, over four million college students have parents who are separated or divorced (*Statistical Abstract of the United States: 2006*).

Decuzzi, Knox, and Zusman (2004) analyzed data from 333 undergraduates, 30% of whom had divorced and/or remarried parents. They compared students whose parents were divorced with students whose parents were still married. Several significant findings were revealed:

1. Student was more unhappy if divorced parents had remarried. Eighteen percent of those with divorced parents and both parents remarried reported feeling less happy about life in contrast to 4% whose parents were still married. Hence, it seems that the happiest students were those whose parents were still together, next were those where one parent had remarried, and lastly, those where both parents had remarried. It is commonly thought that many children whose parents divorce hold out hope that maybe their parents will get back together and they will be a "family" again. The remarriage of both parents dooms forever that possibility, which may be reflected in a lower level of happiness.

2. Student was less close to divorced biological father if he (the father) had remarried. Both divorce and remarriage weaken the relationship with one's biological father. Students from homes in which their parents divorced (in contrast to homes where the parents were still married) were less likely to agree that "I feel really close to my biological father." Sixty-three percent of those with divorced parents reported a close relationship with their dad in contrast to 86% of students whose parents were still together.

3. Student less close to biological mother if she (the mother) divorced. Students from homes in which their parents divorced (in contrast to homes were the parents were still married) were less likely to agree "I feel really close to my biological mother." Eighty-one percent of those with divorced parents reported a close relationship with their mother in contrast to 95% of students whose parents were still together.

4. Students of divorced parents preferred long-term relationships. Students whose parents were divorced seemed to get in and stay in relationships longer than those whose parents were married. Seventy-six percent of the students from divorced families, in contrast to 57 percent of students with married parents reported that their longest relationship was more than a year. It is possible that experiencing the divorce of one's parents was so traumatic that it sensitized the students to maintain their own romantic relationships.

Minimizing Negative Effects of Divorce on Children

Researchers have identified the conditions under which a divorce has the fewest negative consequences for children (Wallerstein, 2003). Some of these follow.

1. Healthy parental psychological functioning. To the degree that parents remain psychologically fit and positive, and socialize their children to view the divorce as a "challenge to learn from," children benefit. Parents who nurture self-pity, abuse alcohol or drugs, and socialize their children to view the divorce as a tragedy from which they will never recover create negative outcomes for their children. Some divorcing parents can benefit from therapy as a method for coping with their anger/depression and making choices in the best interest of their children.

Some parents also enroll their children in the "new beginnings program"— "an empirically driven prevention program designed to promote child resilience during the post divorce period" (Hipke et al., 2002, 121). The program focuses

How life catches up to us and teaches us to forgive each other.

Judy Collins

on improving the quality of the primary residential mother-child relationship, ensuring continued discipline, reducing exposure to parental conflict, and providing access to the nonresidential father. Outcome data reveal that not all children benefit, particularly those with "poor regulatory skills" and "demoralized" mothers (p. 127).

2. A cooperative relationship between the parents. The most important variable in a child's positive adjustment to divorce is that the child's parents continue to maintain a cooperative relationship throughout the separation, divorce, and post-divorce period. In contrast, bitter parental conflict places the children in the middle. One daughter of divorced parents said: "My father told me, 'If you love me, you would come visit me,' but my mom told me, 'If you love me, you won't visit him.'" Baum (2003) confirmed that the longer and more conflictual the legal proceedings, the worse the coparental relationship. Bream and Buchanan (2003) noted that children of conflictual divorcing parents are "children in need." Luecken, Rodriguez, and Applehans (2005) noted that the blood pressure of these children may be elevated. Numerous states mandate parenting classes as part of the divorce process. The court approved "Positive Parenting Through Divorce" program may also be taken on line http://www.positiveparentingthroughdivorce.com/.

3. Parents' attention to the children and allowing them to grieve. Children benefit when both the custodial and the noncustodial parent continue to spend time with them and to communicate to them that they love them and are interested in them. Parents also need to be aware that their children do not want the divorce and to allow them to grieve over the loss of their family as they knew it. Indeed, children do not want their parents to separate. Some children are devastated to the point of suicide. A team of researchers identified 15,555 suicides among 15- to 24-year olds in thirty-four countries in a one-year period and found an association between divorce rates and suicide rates (Johnson, Krug, and Potter, 2000).

4. Encouragement to see noncustodial parent. Children who usually live with custodial mothers following divorce are encouraged by the mother to maintain a regular and stable visitation schedule with their father.

5. Attention from the noncustodial parent. Noncustodial parents, usually the fathers, establish frequent and consistent times to be with the children. Noncustodial parents who do not show up at regular intervals exacerbate their children's emotional insecurity by teaching them, once again, that parents cannot be depended on. Parents who show up often and consistently teach their children to feel loved and secure. Sometimes joint custody solves the problem of children's access to their parents. Hsu et al. (2002) noted the devastating effect on children who grow up without a father.

6. Assertion of parental authority. Both parents continue to assert their parental authority with their children and continue to support the discipline practices of each other to their children.

7. Regular and consistent child support payments. Support payments (usually from the father to the mother) are associated with economic stability for the child.

8. Stability. The parents don't move the children to a new location. Moving them causes them to be cut off from their friends, neighbors, and teachers. It is important to keep their life as stable as possible during a divorce.

9. No new children in a new marriage. Manning and Smock (2000) found that divorced noncustodial fathers who remarried and who had children in the new marriages were more likely to shift their emotional and economic resources to the new family unit than were fathers who did not have new biological children. Fathers might be alert to this potential and consider each child, regardless of when he or she was born and with which wife in which marriage, as worthy of a father's continued love, time, and support.

Learn the wisdom of compromise, for it is better to bend a little than to break.

Jane Wells

Telling children about divorce is never easy but telling them together gives children confirmation that both parents want this to happen and "neither parent is the bad parent." This context also allows the spouses to emphasize their love for and continued care for their children.

© Howard Grey/Getty Images

10. Age and reflection on the part of children of divorce. Sometimes children whose parents are divorced benefit from growing older and reflecting on their parents divorce as an adult rather than a child. Nielsen (2004) emphasized that daughters who feel distant from their fathers can benefit from examining the divorce from the viewpoint of the father (was he alienated by the mother?), the cultural bias against fathers (they are "deadbeat dads" who "abandon their families for a younger woman"), and the facts about divorced dads (they are more likely to be depressed and suicidal following divorce than mothers).

11. Divorce Education Program for Children. Gilman, Schneider, and Shulak (2005) found that Kid's Turn, a San Francisco area divorce education program for children, was effective in reducing the conflict between parents and children of a group of 60 seven-to-nine year old children. However, the children also had more reconciliation fantasies.

What Divorcing Parents Might Tell Their Children: an Example

As emphasized above, a cooperative relationship between the divorcing spouses is crucial to enhancing the adjustment of children. Below are the words of the parents (the mother did the talking) of two children (8 and 12) telling them of their divorce. It assumes that both parents take some responsibility for the divorce and are willing to provide a united front to the children. The script should be adapted for one's own unique situation.

Daddy and I want to talk to you about a big decision that we have made. Awhile back we told you that we were having a really hard time getting along, and that we were having meetings with someone called a therapist who has been helping us talk about our feelings, and deciding what to do about them.

We also told you that the trouble we are having is not about either of you. Our trouble getting along is about our grown up relationship with each other. That is still true. We both love you very much, and love being your parents. We want to be the best parents we can be.

Daddy and I have realized that we don't get along so much, and disagree about so many things all the time, that we want to live separately, and not be married to each other anymore. This is called getting divorced. Daddy and I care about each other but we don't love each other in the way that happily married people do. We are sad about that. We want to be happy, and want each other to be happy. So to be happy we have to be true to our feelings.

It is not your fault that we are going to get divorced. And it's not our fault. We tried for a very long time to get along living together but it just got too hard for both of us.

We are a family and will always be your family. Many things in your life will stay the same. Mommy will stay living at our house here, and Daddy will move to an apartment close by. You both will continue to live with mommy and daddy but in two different places. You will keep your same rooms here, and will have a room at Daddy's apartment. You will be with one of us every day, and sometimes we will all be together, like to celebrate somebody's birthday, special events at school, or scouts.

You will still go to your same school, have the same friends, go to soccer, baseball, and so on. You will still be part of the same family and will see your aunts, uncles, and cousins.

The most important things we want you both to know are that we love you, and we will always be your mom and dad . . . nothing will change that. It's hard to understand sometimes why some people stop getting along and decide not to be friends anymore, or if they are married decide to get divorced. You will probably have lots of different feelings about this. While you can't do anything to change the decision that daddy and I have made; we both care very much about your feelings. Your feelings may change a lot. Sometimes you might feel happy and relieved that you don't have to see and feel daddy and I not getting along. Then sometimes you might feel sad, scared, or angry. Whatever you are feeling at any time is ok. Daddy and I hope you will tell us about your feelings, and its OK to ask us about ours. This is going to take some time to get used to. You will have lots of questions in the days to come. You may have some right now. Please ask any question at any time.

Daddy and I are here for you. Today, tomorrow, and always.

We love you with our heart and soul.

PERSONAL CHOICES

Is Joint Custody a Good Idea?

Traditionally, sole custody to the mother was the only option considered by the courts for divorcing parents. The presumption was made that the "best interests of the child" were served if they were with their mother (the presumed more involved, more caring parent). As men have become more involved in the nurturing of their children, the courts no longer assume that "parent" means "mother." Indeed, a new **family relations doctrine** is emerging that suggests that even nonbiological parents may be awarded custody or visitation rights if they have been economically and emotionally involved in the life of the child (Holtzman 2002). A stepparent is an example.

Given that fathers are no longer routinely excluded from custody considerations, over half of the states have enacted legislation authorizing joint custody. About 16 percent of separated and divorced couples actually have a joint custody arrangement. In a typical joint physical custody arrangement, the parents continue to live in close proximity to each other. The children may spend part of each week with each parent or may spend alternating weeks with each parent.

New terminology is being introduced in the lives of divorcing spouses and in the courts. The term joint custody, which implies ownership, is being replaced with shared parenting, which implies cooperation in taking care of children.

There are several advantages of joint custody-shared parenting. Ex-spouses may fight less if they have joint custody because there is no inequity in terms of their involvement in their children's lives. Children will benefit from the resultant decrease in hostility between parents who have both "won" them. Unlike sole-parent custody, in which one parent (usually the mother) wins and the other parent loses, joint custody allows children to continue to benefit from the love and attention of both parents. Children in homes where joint custody has been awarded might also have greater financial resources available to them than children in sole-custody homes.

Joint physical custody may also be advantageous in that the stress of parenting does not fall on one parent but rather is shared. One mother who has a joint custody arrangement with her ex-husband said, "When my kids are with their Dad, I get a break from the parenting role, and I have a chance to do things for myself. I love my kids, but I also love having time away from them." Another joint-parenting father said, "When you live with your kids every day, you can get very frustrated and are not always happy to be with them. But after you haven't seen them for three days, it feels good to see them again."

A disadvantage of joint custody is that it tends to put hostile ex-spouses in more frequent contact with each other, and the marital war continues. When sole custody is given to one parent, there is minimal to no contact between the spouses. One parent said that she and her ex hadn't spoken in three years . . . that they just met at McDonald's . . . one parent would walk in with the kids, see the other parent and leave. Joint custody usually means multiple contacts over a lot of issues every week.

Depending on the level of hostility between the ex-partners, their motivations for seeking sole or joint custody, and their relationship with their children, any arrangement could have positive or negative consequences for the ex-spouses as well as for the children. In those cases in which the spouses exhibit minimal hostility toward each other, have strong emotional attachments to their children, and want to remain an active influence in their children's lives, joint custody may be the best of all possible choices.

Joyal, Queniart, Gijseghem (2005) interviewed both parents and children where a joint custody arrangement was in place. They found positive outcomes for the children, who were even able to adapt to half siblings.

Conditions of a "Successful" Divorce

There is no system ever devised by humankind that is guaranteed to rip husband and wife or father, mother and child apart so bitterly as our present Family Court System.

Judge B. Lindsay, Retired Supreme Court Judge, New York

While acknowledging that divorce is usually an emotional and economic disaster, it is possible to have a "successful" divorce. Indeed, most people are resilient and "are able to adapt constructively to their new life situation within 2 to 3 years following divorce, a minority being defeated by the marital breakup, and a substantial group of women being enhanced" (Hetherington, 2003, p. 318). The following are some of the behaviors spouses can engage in to achieve this.

1. Mediate rather than litigate the divorce. Divorce mediators encourage a civil, cooperative, compromising relationship while moving the couple toward an agreement on the division of property, custody, and child support. By contrast, attorneys make their money by encouraging hostility so that spouses will prolong the conflict, thus running up higher legal bills. In addition, money spent on divorce attorneys (average is more than $15,000) cannot be divided by the couple. Ricci (2000) noted that "the best advice is still to do as much as you can to settle the issues about children outside the court, where parents can maintain control of the process and keep their discussions private, and where they, not a judge, decide what is best for their child" (p. F12). Because the greatest damage to children from a divorce is a continuing hostile and bitter relationship between their parents, some states require **divorce mediation** as a mechanism to encourage civility in working out differences and to clear the court calendar from protracted court battles. The Social Policy on page 445 focuses on divorce mediation

2. Coparent with your ex-spouse. Setting negative feelings about your ex-spouse aside so as to cooperatively coparent not only facilitates parental adjustment but also takes children out of the line of fire. Such coparenting translates into being cooperative when the other parent needs to change the child care schedule, sitting together during a performance by the children, and showing appreciation for the other parent's skill in responding to a crisis with the children.

© Michael Newman/PhotoEdit

Should Divorce Mediation Be Required before Litigation?*

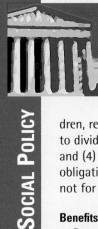

Divorce mediation is a process in which spouses who have decided to separate or divorce meet with a neutral third party (mediator) to negotiate the issues of (1) How they will parent their children, which is referred to as child custody and visitation, (2) How they are going to financially support their children, referred to as child support, (3) How they are going to divide their property, known as property settlement, and (4) How each one is going to meet their financial obligations referred to as spousal support. Mediation is not for everyone.

Benefits of Mediation

1. Better relationship. Spouses who choose to mediate their divorce have a better chance for a more civil relationship because they cooperate in specifying the conditions of their separation or divorce. Mediation emphasizes negotiation and cooperation between the divorcing partners. Such cooperation is particularly important if the couple has children in that it provides a positive basis for discussing issues in reference to the children and how they will be parented across time.

2. Economic benefits. Mediation is less expensive than litigation. The cost of hiring an attorney and going to court over issues of child custody and division of property is around $18,000. A mediated divorce costs about an average of $1500 (six 2-hour sessions at $250 per session) (Haswell, 2006). What the couple spend in legal fees they cannot keep as assets to later divide.

3. Less time-consuming process. A mediated divorce takes two to three months versus two to three years if the case is litigated. "But the process can take place in one session from 8:00 AM until both parties are satisfied with the terms. This process is called a mediated settlement conference" (Haswell, 2006)

4. Avoidance of public exposure. Some spouses do not want to discuss their private lives and finances in open court. Mediation occurs in a private and confidential setting.

5. Greater overall satisfaction. Mediation results in an agreement developed by the spouses, not one imposed by a judge or the court system. A comparison of couples who chose mediation with couples who chose litigation found that those who mediated their own settlement were much more satisfied with the conditions of their agreement. In addition, children of mediated divorces are exposed to less marital conflict, which may facilitate their long-term adjustment to divorce.

Basic Mediation Guidelines

1. Children. What is best for a couple's children should be the major concern of the parents since they know their children far better than a judge or the mediator. Children of divorced parents adjust best under three conditions: (1)

that both parents have regular and frequent access to the children, (2) the children see the parents relating in a polite and positive way, and (3) each parent talks positively about the other parent and neither parent talks negatively about the other to the children. Sometimes children are included in the mediation. They may be interviewed without the parents present to provide information to the mediator about their perceptions/preferences (Schoffer, 2005)

2. Fairness. It is important that the agreement be fair, with neither party being exploited or punished. It is fair for both parents to contribute financially to the children. It is fair for both parents to have regular access to their children.

3. Open disclosure. The spouses will be asked to disclose all facts, records, and documents to ensure an informed and fair agreement regarding property, assets, and debts.

4. Other professionals. During mediation the spouses may be asked to consult an accountant regarding tax laws. In addition, spouses are encouraged to consult an attorney throughout the mediation and to have the attorney review the written agreements that result from the mediation. However, during the mediation sessions, all forms of legal action by the spouses against each other should be stopped.

Another term for involving a range of professionals in a divorce is **collaborative practice,** a process that brings a team of professionals (lawyer, psychologist, mediator, social worker, financial counselor) together to help a couple separate and divorce in a humane and cost-effective way.

5. Confidentiality. The mediator will not divulge anything the spouses say during the mediation sessions without their permission. The spouses are asked to sign a document stating that should they not complete mediation, they agree not to empower any attorney to subpoena the mediator or any records resulting from the mediation for use in any legal action.

Such an agreement is necessary for the spouses to feel free to talk about all aspects of their relationship without fear of legal action against them for such disclosures.

Divorce mediation is not for every couple. It does not work where there is a history of spouse abuse, where the parties do not disclose their financial information, where one party is controlled by someone else (e.g., a parent), where there is the desire for revenge, or where the mediator is biased. Mediation should be differentiated from **negotiation** (the spouses discuss and resolve the issues themselves), **arbitration** (a third party listens to both spouses and makes a decision about custody, division of property, etc), and **litigation** (a judge hears arguments from lawyers representing the respective spouses and decides issues of custody, child support, division of property, etc.).

Haswell (2006) noted a continuum of consequences from negotiation to litigation.

Negotiation	Mediation	Arbitration	Litigation
Cooperative..			Competitive
Low Cost...			High Cost
Private			Public
Protects Relationships..			Damages Relationships
Future Focused			Focus on the Past
Parties in Control...			Parties Lose Control

Your Opinion?:

1. To what degree do you believe the government should be involved in mandating divorce mediation?
2. How can divorce mediation go wrong? Why should a couple not want to mediate their divorce?
3. What are the advantages for children for their parents mediating their divorce?

*Appreciation is expressed to Mike Haswell for contributing to this section. See www.haswellmeditation.com.

Think wrongly if you please, but in all cases think for yourself.

Doris Lessing

3. Take some responsibility for the divorce. Since marriage is an interaction between the spouses, one person is seldom totally to blame for a divorce. Rather, both spouses share reasons for the demise of the relationship. Take some responsibility for what went wrong.

4. Learn from the divorce. View the divorce as an opportunity to improve yourself for future relationships. What did you do that you might consider doing differently in the next relationship?

5. Create positive thoughts. Divorced people are susceptible to feeling as though they are failures. They see themselves as Divorced persons with a capital D, a situation sometimes referred to as "hardening of the categories" disease. Improving their self-esteem is important for divorced persons. They can do this by systematically thinking positive thoughts about themselves. One technique is to write down twenty-one positive statements about your self ("I am honest," "I have strong family values," "I am a good parent," etc.) and transfer them to seven three-by-five cards, each containing three statements. Take one of the cards with you each day and read the thoughts at three regularly spaced intervals (e.g., seven in the morning, one in the afternoon, seven at night). This ensures that you are thinking good things about yourself and are not allowing yourself to drift into a negative set of thoughts ("I am a failure"; "no one wants a divorced person").

6. Avoid alcohol and other drugs. The stress and despair that some people feel following a divorce make them particularly vulnerable to the use of alcohol or other drugs. These should be avoided because they produce an endless negative cycle. For example, stress is relieved by alcohol; alcohol produces a hangover and negative feelings; the negative feelings are relieved by more alcohol, producing more negative feelings, etc.

7. Relax without drugs. Deep muscle relaxation can be achieved by systematically tensing and relaxing each of the major muscle groups in the body. Alternatively, yoga, transcendental meditation, and getting a massage can induce a state of relaxation in some people. Whatever the form, it is important to schedule a time each day to get relaxed.

8. Engage in aerobic exercise. Exercise helps one not only to counteract stress but also to avoid it. Jogging, swimming, riding an exercise bike, or other similar exercise for thirty minutes every day increases the oxygen to the brain and helps facilitate clear thinking. In addition, aerobic exercise produces endorphins in the brain, which create a sense of euphoria ("runner's high").

9. Engage in fun activities. Some divorced people sit at home and brood over their "failed" relationship. This only compounds their depression. Doing what they have previously found enjoyable—swimming, horseback riding, sking, sporting events with friends—provides an alternative to sitting on the couch alone.

10. Continue interpersonal connections. Adjustment to divorce is facilitated when relationships with friends and family are continued. These individuals provide emotional support and help buffer the feeling of isolation/aloneness.

11. Let go of the anger for your ex-partner. Former spouses who stay negatively attached to their ex by harboring resentment and trying to get back at the ex prolong their adjustment to divorce. The old adage that you can't get ahead by getting even is relevant to divorce adjustment.

12. Allow time to heal. Since self-esteem usually drops after divorce, a person is often vulnerable to making commitments before working through feelings about the divorce. The time period most people need to adjust to divorce is between twelve and eighteen months. Although being available to others may help to repair one's self-esteem, getting remarried during this time should be considered cautiously. Two years between marriages is recommended.

Alternatives to Divorce

Divorce is not the only means of terminating a marriage. Others include annulment, separation (legal or informal), and desertion.

Annulment

An **annulment** returns the spouses to their premarital status. In effect, an annulment means that the marriage never existed in the first place. The concept of annulment is both religious and civil. As a religious matter, annulment is a technical mechanism that allows Catholics to remarry. The Roman Catholic Church views marriage as a sacrament that is indissoluble except by death. Hence, while the Church does not recognize divorce, it does recognize annulment. A Catholic who wants to get out of a marriage and remain a good Catholic in the eyes of the church (be able to participate in Communion) cannot divorce but can petition for an annulment. This involves providing the right "reason" for an annulment and paying a fee. Almost all annulments are granted, even those to spouses who have been married for decades and have numerous children.

The basis used by the Catholic Church for granting an annulment is usually "lack of due discretion," which means that one or both parties lacked the ability needed to consent to the "essential obligations of matrimony." Personality or psychiatric disorders, premarital pregnancy, and problems in one's family of origin that one is trying to escape are examples of factors that are said by the Catholic Church to interfere with a person's ability to fulfill the essential matrimonial obligations, such as a permanent partnership, faithfulness, and sharing.

Religious annulments are not recognized by the state in which the couple resides. Even though a Catholic couple's marriage is annulled by the Church, the couple must still get a civil annulment (or a divorce) for the action to be legal and for the couple to legally remarry. An annulment granted by a civil court specifies that no valid marriage ever existed and returns both parties to their premarital status. Any property the couple exchanged as part of the marriage arrangement is returned to the original owner. Neither party is obligated to support the other economically.

Common reasons for granting civil annulments are fraud, bigamy, being under legal age, erectile failure, and insanity. As an example of fraud, a university professor became involved in a relationship with one of his colleagues. During courtship, he promised her that they would rear a "houseful of babies." But after the marriage, the woman discovered that the man had had a vasectomy several years earlier and had no intention of having more children. The marriage was annulled on the basis of fraud—the man misrepresented himself to the woman. Most annulments are for fraud.

Divorce was always an option—not murder.

Sharon Rocha (mother of Laci Peterson who was murdered by her husband Scott Peterson)

Bigamy is another basis for annulment. In our society, a person is allowed to be married to only one spouse at a time. If another marriage is contracted at the time a person is already married, the new spouse can have the marriage annulled.

Most states have age requirements for marriage. When individuals younger than the minimum age marry without parental consent, the marriage may be annulled if either set of parents does not approve of the union. However, if neither set of parents or guardians disapproves of the marriage, the marriage may be regarded as legal; it is not automatically annulled.

Intercourse is a legal right of marriage. In some states, if a spouse is impotent, refuses to have intercourse, or is unable to do so for physical or psychological reasons, the other spouse can seek and may be granted an annulment.

Insanity and a lack of understanding of the marriage agreement are also reasons for annulment. Someone who is mentally deficient and incapable of understanding the meaning of a marriage ceremony can have a marriage annulled. Annulments are usually not granted if one or both of the parties is drunk at the time of the wedding ceremony or get married as a lark. However, Britney Spears (pop singer) and Jason Allen Alexander married in early 2004 and sought an annulment (on grounds of that she "lacked understanding of her actions") less than twelve hours after their marriage (Chen, 2004).

Separation

There are two types of separation—formal and informal. Typical items in a **formal separation** agreement include the following: (1) the husband and wife live separately; (2) their right to sexual intercourse with each other is ended; (3) the economic responsibilities of the spouses to each other are limited to those in the separation agreement; and (4) custody of the children is specified in the agreement, with visitation privileges granted to the noncustodial parent. The spouses may have emotional and sexual relationships with others, but neither party has the right to remarry. Although some couples live under this agreement until the death of one spouse, others draw up a separation agreement as a prelude to divorce. In some states, being legally separated for one year is a ground for divorce.

An **informal separation** (which is much more common) is similar to a legal separation except that no agreement is filed in the courthouse. The husband and wife settle the issues of custody, visitation, alimony, and child support between themselves. Because no legal papers are drawn up, the couple is still married from the state's point of view.

Attorneys advise against an informal separation (unless it is temporary) to avoid subsequent legal problems. For example, after three years of an informal separation, a mother decided that she wanted custody of her son. Although the father would have been willing earlier to sign a separation agreement that would have given her legal custody of her son, he was now unwilling to do so. Each spouse hired a lawyer, and a bitter and expensive court fight ensued.

Desertion

Desertion differs from informal separation in that the deserter walks out and breaks off all contact. Although either spouse may desert, it is usually the husband who does so. A major reason for deserting is to escape the increasing financial demands of a family. Desertion usually results in nonsupport, which is a crime.

The sudden desertion by a husband sometimes has more severe negative consequences for the wife than divorce. Unlike the divorced woman, the deserted woman is usually not free to remarry for several years. In addition, she receives no child support or alimony payments, and the children are deprived of a father.

Desertion is not unique to husbands. Although rare, wives and mothers also leave their husbands and children. Their primary reason for doing so is to escape an intolerable marriage and the sense of being trapped in the role of mother. "I'm tired of having to think about my children and my husband all the time—I want a life for myself," said one woman who deserted her family. "I want to live, too." But such desertion is not without its consequences as social norms dictate parental role responsibility, particularly for mothers. Most mothers who desert their children feel extremely guilty.

Divorce Prevention

Divorce remains stigmatized in our society, as evidenced by the term **divorcism**—the belief that divorce is a disaster. In view of this cultural attitude, a number of attempts have been made to reduce it. Marriage education workshops provide an opportunity for couples to meet with other couples and a leader who provides instruction in communication, conflict resolution, and parenting skills. Stanley et al. (2005) studied couples in marriage education classes provided for the U.S. Army and found positive outcomes in marital functioning.

Another attempt at divorce prevention is **covenant marriage** (now available in Louisiana, Arizona, and Arkansas), which emphasizes the importance of staying married (Byrne and Carr, 2005). In these states a couple agrees to the following when they marry: (1) marriage preparation (meeting with a counselor who discusses marriage and their relationship), (2) full disclosure of all information that could reasonably affect the partner's decision to marry (e.g., previous marriages, children, STIs, one's homosexuality), (3) an oath that their marriage is a lifelong commitment, (4) an agreement to consider divorce only for "serious" reasons such as abuse, adultery, and imprisonment for a felony or separation of more than two years, (5) an agreement to see a marriage counselor if problems threaten the marriage, and (6) not to divorce until after a two-year "cooling off" period ("Yes, I really do" 2005).

While most of a sample of 1,324 adults in a telephone survey in Louisiana, Arizona, and Minnesota were positive about covenant marriage (Hawkins et al., 2002), fewer than 3 percent of marrying couples elected covenant marriages when given the opportunity to do so (Licata, 2002). And, though already married couples can convert their standard marriages to covenant marriages, there are no data on how many have done so (Hawkins et al., 2002).

We have to find ways to make marriage work, if for no other reason than marriage is extraordinarily important to children.

David Popenoe, sociologist

SUMMARY

What is the nature of divorce in the United States?

Divorce is the legal ending of a valid marriage contract. While the divorce rate has dropped somewhat, about 43 percent of couples who marry will divorce after 15 years. A quarter of college students experience the divorce of their parents.

What are macro factors contributing to divorce?

Macro factors contributing to divorce include increased economic independence of women, liberal divorce laws, fewer religious sanctions, more divorce models, and the individualistic cultural goal of happiness.

What are micro factors contributing to divorce?

Micro factors include falling out of love, negative behavior, lack of conflict resolution skills, satiation, and extramarital relationships. Having a courtship of less than two years, having little in common, and having divorced parents are all associated with subsequent divorce.

What are gender differences in filing for divorce?

Women are more likely to file for divorce since they see that by getting divorced, they get the husband's money (via division of property/child support), the children, and the husband out of the house. Husbands are less likely to seek divorce since they more often end up without the house, half their money, and separated from their children.

What are the consequences for spouses who remain unhappily married?

Spouses who remain unhappily married are less happy and healthy than spouses who divorce.

What are the consequences of divorce for spouses?

The psychological consequences for divorcing spouses depend on how unhappy the marriage was. Spouses who were miserable while in a loveless conflictual marriage often regard the divorce as a relief. Spouses who were left (e.g., spouse leaves for another partner) may be devastated and suicidal. Women tend to fare better emotionally after separation and divorce than do men. Women are more likely than men not only to have a stronger network of supportive relationships but also to profit from divorce by developing a new sense of self-esteem and confidence, since they are thrust into a more independent role.

Factors associated with a quicker adjustment on the part of both spouses include mediating rather than litigating the divorce, coparenting their children, avoiding alcohol or other drugs, reducing stress through exercise, engaging in enjoyable activities with friends, and delaying a new marriage for two years.

What are the effects of divorce on children?

Although researchers agree that a civil, cooperative, coparenting relationship between ex-spouses is the greatest predictor of a positive outcome for children, researchers disagree on the long-term negative effects of divorce on children. Divorce mediation encourages civility between divorcing spouses who negotiate the issues of division of property, custody, visitation, child support, and spousal support.

What are alternatives to divorce?

Alternatives to divorce include annulment, separation, and desertion. Annulment returns the parties to their premarital state and is both a religious and civil concept. The Catholic Church does not recognize divorce but does recognize annulment, which in effect says that the parties were never married. However, the parties are still legally married and must have their marriage legally annulled or they must get a divorce. Reasons for legal annulment include bigamy and being underage.

What are strategies to prevent divorce?

In response to the alarming incidence of divorce and the devastation to spouses and children that follows, divorce prevention strategies include a divorce tax whereby all property settlements in divorces would be subject to an additional tax. In addition, three states (Louisiana, Arizona, and Arkansas) offer covenant marriages, in which spouses agree to divorce only for serious reasons such as imprisonment on a felony or separation of more than two years. They also agree to see a marriage counselor if problems threaten the marriage. When given the option to choose a covenant marriage, few couples do so.

KEY TERMS

arbitration	desertion	informal separation	physical custody
annulment	divorce	litigation	postnuptial agreement
bigamy	divorce mediation	legal custody	refined divorce rate
collaborative practice	divorcism	negotiation	satiation
covenant marriage	family relations doctrine	no-fault divorce	shared parenting dysfunction
crude divorce rate	formal separation	parental alienation syndrome	

The Companion Website for Choices in Relationships: An Introduction to Marriage and the Family, Ninth Edition
www.thomson.edu/sociology/knox

Supplement your review of this chapter by going to the companion website to take one of the Tutorial Quizzes, use the flash cards to master key terms, and check out the many other study aids you'll find there. You'll also find special features such as the Marriage and Family Resource Center, Census 2000 information, and other data and resources at your fingertips to help you with that special project or to do some research on your own.

WEB LINKS

Association for Conflict Resolution
http://www.acrnet.org

Divorce at Heartchoice.com
http://heartchoice.com/divorce/index.php

Divorcesource.com
http://www.divorcesource.com/

Divorcesupport.com
http://www.divorcesupport.com/index.html

Divorceinfo.com (divorce Weblinks)
http://www.divorceinfo.com/

Family Mediation
http://familymediators.org

Parental Alienation Syndrome
http://www.coeffic.demon.co.uk/pas.htm

Positive Parenting Through Divorce On-Line Course
http://www.positiveparentingthroughdivorce.com/

National Center for Health Statistics (marriage and divorce data)
http://www.cdc.gov/nchs/

REFERENCES

Abut, C. C. 2005. Ten common questions about postnuptial agreements. *New Jersey Law Journal.* August 15.

Ahrons, C. R., and J. L. Tanner. 2003. Adult children and their fathers: Relationship changes 20 years after parental divorce. *Family Relations* 52:340–51.

Ahrons, C. R. 1995. *The good divorce: Keeping your family together when your marriage comes apart.* New York: HarperCollins.

Allen, D. W., K. Pendakur, and W. Suen. 2006. No-fault divorce and the compression of marriage ages. *Economic Inquiry* 44: 547–559.

American Council on Education and University of California. 2006. *The American freshman: National norms for fall, 2005.* Los Angeles: Los Angeles Higher Education Research Institute.

Baker, A. L. 2005. The long term effectrs of parental aienation on adult children: A qualitative research study. *American Journal of Family Therapy* 33: 289–302.

Baker, A. J. L. 2006. Patterns of Parental Alienation Syndrome: A qualitative study of adults who were alienated from a parent as a child. *American Journal of Family Therapy,* 34: 63–78.

Baum, N. 2003. Divorce process variables and the co-parental relationship and parental role fulfillment of divorced parents. *Family Process* 42: 117–131.

Blossfeld, H.P., and R. Muller. 2002. Union disruption in comparative perspective: The role of assertive partner choice and careers of couples. *International Journal of Sociology* 32:3–35.

Bream, V., and A. Buchanan. 2003. Distress among children whose separated or divorced parents cannot agree on arrangements for them. *British Journal of Social Work* 33: 227–38.

Brinig, M. F. and D. W. Allen. 2000. "These boots are made for walking": Why most divorce filers are women. *American Law and Economic Association* 2: 126–169.

Byrne, A. and D. Carr. 2005. Commentaries on: Singles in society and science. *Psychological Inquiry* 16: 84–141.

Carroll, J. 2006. Women more likely than men to say they've been divorced. Cnn/USA Today Gallup poll. Poll taken April 5. Retrieved April 20. http://poll.gallup.com/content/default.aspx?ci=22264&pg=1.

Chen, J. 2004. Britney-Jason union wouldn't last. *USA Today.* 5 January, D1.

Cohen, O., and R. Savaya. 2003. Lifestyle differences in traditionalism and modernity and reasons for divorce among Muslim Palestinian citizens of Israel. *Journal of Comparative Family Studies* 34:283–94.

Coltrane, S., and M. Adams. 2003. The social construction of the divorce "problem": Morality, child victims, and the politics of gender. *Family Relations* 52:363–72.

Corte, U. and Knox, D. 2007. "Work it out: Advice from separated spouses" Unpublished data. Department of Sociology, East Carolina University.

DeGraff, P. M. and M. Kalmijn. 2006. Divorce motives in a period of rising divorce- Evidence from a Dutch life-history survey. *Journal of Family Issues* 27: 485–505.

Decuzzi, A., D. Knox, and M. Zusman. 2004. The effect of parental divorce on relationships with parents and romantic partners of college students. Roundtable, Southern Sociological Society, Atlanta, April 17.

Enright, E. 2004. A house divided. *AARP The Magazine July/August 60- et passim.*

Finley, G. E. 2004. Divorce inequities. *NCFR Family Focus Report.* Vol 49:3, F7.

Gardner, J. and A. J. Oswald. 2006. Do divorcing couples become happier by breaking up? *Journal of the Royal Statistical Society: Series A (Statistics and Society).* 169: 319–336.

Gardner, R. A. 1998. *The parental alienation syndrome.* 2d ed. Cresskill, N.J.: Creative Therapeutics.

Gilman, J., D. Schneider, and R. Shulak. 2005. Children's ability to cope post-divorce: The effects of Kids' Turn Intervention Program on 7 to 9 year olds. *Journal of Divorce and Remarriage.* 42: 109–126.

Goodwin, P. Y. 2003. African American and European American women's health marital well-being. *Journal of Marriage and Family* 65:550–60.

Gordon, R. M. 2005. The doom and gloom of divorce research- Comment on Wallerstein and Lewis (2004) *Psychoanalytic Psychology* 22: 450–451.

Haswell, W. M. 2006. Mediation: An alternative to litigation.Presentation, Department of Sociology, East Carolina University, April 5. www.HaswellMediation.com

Hawkins, A. J., S. L. Nock, J. C. Wilson, L. Sanchez ,and J. D. Wright. 2002. Attitudes about covenant marriage and divorce: Policy implications from a three state comparison. *Family Relations* 51:166–75.

Hawkins, D. N. and A. Booth. 2005. Unhappily ever after: Effects of long-term, low-quality marriages on well-being. *Social Forces* 84: 445–465.

Heaton, T. B. 2002. Factors contributing to increasing marital stability in the United States. *Journal of Family Issues* 23: 392–409.

Hetherington, E. M. 2003. Intimate pathways: Changing patterns in close personal relationships across time. *Family Relations* 52:318–31.

Hilton, J. M. and Kopera-Frye, K. 2004. Patterns of psychological adjustment among divorced custodial parents. Poster, 66th Annual National Council on Family Relations, Orlando, Florida

Hipke, K. N., S. A. Wolchik, I. N. Sandler, and S. L. Braver. 2002. Predictors of children's intervention-induced resilience in a parenting program for divorced mothers. *Family Relations* 51:121–29.

Holtzman, M. 2002. The "family relations" doctrine: Extending Supreme Court precedent to custody disputes between biological and nonbiological parents. *Family Relations* 51:335–343.

Hsu, M., D. L. Kahn, and C. Huang. 2002. No more the same: The lives of adolescents in Taiwan who have lost fathers. *Family Community Health* 25: 43–56.

Johnson, G. R., E. G. Krug, and L. B. Potter. 2000. Suicide among adolescents and young adults. A cross-national comparison of 34 countries. *Suicide and Life-Threatening Behavior* 30:74–82.

Joyal, R., A. Queniart, and H. Gijseghem. 2005. Children in joint custody: Some questions and answers. *Intervention* 122: 51–59.

Kelly, J. B., and R. E. Emery. 2003. Children's adjustment following divorce: Risk and resilience perspectives. *Family Relations* 52:352–62.

Kesselring, R. G. and D. Bremmer. 2006. Female income and the divorce decision: Evidence from micro data. *Applied Economics* 38: 1605–1617.

Kilmann, P. R., L. V. Carranza and J. M. C. Vendemia. 2006. Recollections of parent characteristics and attachment patterns for college women of intact vs. non-intact families. *Journal of Adolescence* 29: 89–102.

Knox, D. and Zusman, M. E. 2006. Relationship and sexual behaviors of a sample of 1027 university students. Unpublished data collected for this text. Department of Sociology, East Carolina University, Greenville, NC.

Licata, N. 2002. Should premarital counseling be mandatory as a requisite to obtaining a marriage license? *Family Court Review* 40: 518–32.

Lopoo, L. M. and B. Western. 2005. Incarceration and the formation and stability of marital unions. *Journal of Marriage and the Family* 67: 721–735.

Luecken, L. J., A. P. Rodriguez, and B. M. Applehans. 2005. Cardiovascular stress responses in young adulthood associated with family of origin in relationship experiences. *Psychosomatic Medicine* 67: 514–521.

Manning, W. D., and P. J. Smock. 2000. "Swapping" families: Serial parenting and economic support for children. *Journal of Marriage and the Family* 62:111–122.

Morgan, E. S. 1944. *The Puritan family.* Boston: Public Library.

Nielsen, L. 2004. *Embracing your father: How to build the relationship you always wanted with your dad.* New York: McGraw-Hill.

Lowenstein, L. F. 2005. Causes and associated factors of divorce as seen by recent research. *Journal of Divorce and Remarriage.* 42: 153–171.

O'Leary, K. D. 2005. Commentary on intrapersonal, interpersonal, and contextual factors in extramarital involvement. *Clinical Psychology: Science and Practice.* 12: 131–133.

Orbuch, T. L., J. Veroff, H. Hassan, and J. Horrocks. 2002. Who will divorce: A 14-year longitudinal study of black couples and white couples. *Journal of Social and Personal Relationships* 19:179–202.

Ostermann, J., F. A. Sloan, and D. H. Taylor. 2005. Heavy alcohol use and marital dissolution in the USA. *Social Science and Medicine* 61: 2304–2320.

Paul, P. 2002 *The starter marriage and the future of matrimony.* New York: Random House, 2002.

Previti, D., and P. R. Amato. 2003. Why stay married? Rewards, barriers, and marital stability. *Journal of Marriage and Family* 65:561–73.

Ricci, Isolina. 2000. The courts, child custody and visitation. *Family Focus.* September, F12–F13.

Sakraida, T. 2005. Divorce transition differences of midlife women. *Issues in Mental Health Nursing.* 26: 225–249.

Schacht, T. E. 2000. Protection strategies to protect professionals and families involved in high-conflict divorce. *UALR Law Review* 22 (3):565–592.

Chapter 15 Divorce

Schoffer, M. J. 2005. Bringing children to the mediation table: Defining a child's best interest in divorce mediation. *Family Court Review* 43: 323–338

Segal-Engelchin, D. and Y. Wozner. 2005. Quality of life to single mothers in Israel: A comparison to single mothers and divorced mothers. *Marriage and Family Review* 37: 7–28.

Shumway, S T., and R. S. Wampler. 2002. A behaviorally focused measure for relationships: The couple behavior report (CBR). *The American Journal of Family Therapy* 30:311–21.

Siegler, I., and P. Costa. 2000. Divorce in midlife. Paper presented at the Annual Meeting of the American Psychological Association, Boston.

Stanley, S. M., E. S. Allen, H. J. Markman, C. Saiz, G. Bloomstrom, R. Thomas and W. R. Schumm. 2005. Dissemination and evaluation of marriage education in the Army. *Family Process* 44: 187–201.

Statistical Abstract of the United States: 2006. 125th ed. Washington, D.C.: U.S. Bureau of the Census.

Storksen, I, E. Roysamb, T. L. Holmen, and K. Tambs. 2006. Adolescent adjustment and well-being: Effects of parental divorce and distress. *Scandinavian Journal of Psychology* 47: 75–84.

Teachman, J. D. 2002. Childhood living arrangements and the intergenerational transmission of divorce. *Journal of Marriage and the Family* 64:717–29.

Thuen, F. and J. Rise. 2006. Psychological adaptation after marital disruption: The effects of optimism and perceived control. *Scandinavian Journal of Psychology* 47: 121–128.

Turkat, I. D. 2002. Shared parenting dysfunction. *The American Journal of Family Therapy* 30: 385–393.

Vlosky, D. A., and P. A. Monroe. 2002. The effective dates of no-fault divorce laws in the 50 states. *Family Relations* 51:317–26.

Wallace, S. R., and S. S. Koerner. 2003. Influence of child and family factors on judicial decisions in contested custody cases. *Family Relations* 52:160–88.

Wallerstein, J. S. 2000. *The unexpected legacy of divorce: A 25-year landmark study.* New York: Hyperion.

———. 2003. Children of divorce: A society in search of a policy. In *All our families: New Policies for a new century,* 2d ed., edited by M. A. Mason, A. Skolnick, and S. D. Sugarman. New York: Oxford University Press, 66–95.

Wills, J. 2002. Research updates: Transmitting divorce across generations. *Stepfamilies* 21: 1–2

"Yes, I really do" (no author). 2005. *Economist* 374: 31–32.

© Royalty-Free/Corbis

chapter 16

Divorce reorganizes a family; it does not destroy it.

Constance Ahrons, We're Still Family

Remarriage and Stepfamilies

Contents

True or False?

1. About 2 years is recommended between the end of a marriage and a remarriage.

2. From the viewpoint of the child, the stepmother is more difficult to adjust to than the stepfather.

3. Stepmothers react to the "wicked stepmother" stereotype by viewing the biological mother as mentally unstable and themselves as good for the child.

4. The divorced who live with a new partner before remarriage report a happier remarriage than those who do not live together before a remarriage.

5. Stepparents are encouraged to discipline their own children.

Answers: **1.** T **2.** T **3.** T **4.** F **5.** T

Today, one in four marriages is a remarriage. Most often the new spouses bring children from previous relationships into the marriage; thus, stepfamilies are very evident in our society. Indeed, former presidential candidate John Kerry and his wife Teresa Heinz-Kerry both were previously married (and both have children from these marriages: he has two daughters and she has three sons) and now live in a stepfamily. Over a quarter (25.1%) of 1027 undergraduates at a large university reported that they have lived in a stepfamily (Knox and Zusman, 2006). We begin with a discussion of remarriage.

Remarriage

Although divorced spouses may continue to be in contact with each other 10 years after their divorce (Fischer et al., 2005), they usually waste little time getting involved in a new relationship. Indeed, one-fourth are dating someone new before the divorce is final. And, within 2 years, 75 percent of the women and 80 percent of the men are in a serious, exclusive relationship (Enright, 2004). Persons who left their previous marriage, in comparison with persons who were left by their spouse, are more likely to remarry first (Sweeney, 2002). Men who were domestically dependent on their partners are also more likely to repartner (DeGraaf and Kalmijn, 2003). When the divorced who have remarried and the divorced who have not remarried are compared, the remarried report greater personal and relationship happiness.

I guess I'm an optimist, because I believe that love has the same kind of selective amnesia that childbirth does (you completely forget how much it hurt the last time around until you're in the delivery room again).

Jody Picoult, novelist

Remarriage for the Divorced

Ninety percent of remarriages are of persons who are divorced rather than widowed. The principle of homogamy is illustrated in the selection of a new spouse (never-marrieds tend to marry never-marrieds, the divorced tend to marry the divorced, and the widowed tend to marry the widowed). The majority of divorced persons remarry for many of the same reasons as for a first marriage—love, companionship, emotional security, and a regular sex partner. Other reasons are unique to remarriage and include financial security (particularly for the wife with children), help in rearing one's children, the desire to provide a "social" father or mother for one's children, escape from the stigma associated with the label "divorced person," and legal threats regarding the custody of one's children. With regard to the latter, the courts view a parent seeking custody of a child more favorably if he or she is married.

Regardless of the reason for remarriage, it is best to proceed slowly into a remarriage. Grieving over the loss of a first spouse through divorce or death and developing a relationship with a new partner takes time. At least 2 years is the recommended interval between the end of one's marriage and a remarriage (Marano, 2000). Older divorced women (over 40) are less likely than younger divorced women to remarry. Not only are there fewer available men, but also the mating gradient whereby men tend to marry women younger than themselves is operative. In addition, older women are more likely to be economically independent, to enjoy living alone, to value the freedom of singlehood, and to want to avoid the restrictions of marriage.

Divorced persons getting remarried are usually about 10 years older than those marrying for the first time. Persons in their mid-30s who are considering remarriage are more likely than first-marrieds to have finished school and to be established in a job or career.

Courtship for the previously married is usually short and takes into account the individuals' respective work schedules and career commitments. Because each partner may have children, much of the couple's time together includes their children. The practice of going out alone on an expensive dinner date during courtship before the first marriage is often replaced with eating pizza at home and renting a DVD to watch with the kids.

Preparation for Remarriage

Although there are exceptions (Swallow, 2004), most couples going into a second marriage do so with little preparation. Timmer and Veroff (2000) emphasized the importance of getting to know the partner's family (particularly wives getting to know their husbands' families) and nurturing such relationships; they pointed out that doing so is related to increases in reported marital happiness of remarried partners. Some live with the new partner before the remarriage. But as with never-married cohabitants, living together is associated with subsequent instability in the new marriage. Xu et al. (2006) analyzed national data on 3480 remarried respondents and found that those who had lived together prior to remarriage reported lower marital happiness than those who had not done so.

Persons getting remarried are more likely than never-marrieds to develop a prenuptial agreement. These soon-to-be remarried spouses may have assets to protect as well as wanting to ensure that their respective children get whatever portion of their estate they desire. See the Appendix for an example of a prenuptial agreement developed by a couple getting remarried.

Issues of Remarriage

People who remarry are challenged by several issues (Swallow, 2004; Ganong and Coleman, 1994; Goetting, 1982).

Boundary Maintenance   Movement from divorce to remarriage is not a static event that happens in a brief ceremony and is over. Rather, ghosts of the first marriage in terms of the ex-spouse and, possibly, the children must be dealt with. A parent must decide how to relate to his or her ex-spouse in order to maintain a good parenting relationship for the biological children while keeping an emotional distance to prevent problems from developing with the new partner. Some spouses continue to be emotionally attached to the ex-spouse and have difficulty breaking away. These former spouses have what Masheter (1999) terms a "negative commitment" in that they:

> have decided to remain [emotionally] in this relationship and to invest considerable amounts of time, money, and effort in it . . . [T]hese individuals do not take respon-

Love is not a business deal.

Christina Onassis, daughter of Aristotle Onassis

sibility for their own feelings and actions, and often remain "stuck," unable to move forward in their lives (p. 297).

One former spouse recalled a conversation with his ex from whom he had been divorced for more than 17 years. "Her anger coming through the phone seemed like she was still feeling the divorce as though it had happened this morning," he said.

Emotional Remarriage Remarriage involves beginning to trust and love another person in a new relationship. Such feelings may come slowly as a result of negative experiences in the first marriage.

Psychic Remarriage Divorced individuals considering remarriage may find it difficult to give up the freedom and autonomy of being single again and to develop a mental set conducive to pairing. This transition may be particularly difficult for people who sought a divorce as a means to personal growth and autonomy. These individuals may fear that getting remarried will put unwanted constraints on them.

Community Remarriage This stage involves a change in focus from single friends to a new mate and other couples with whom the new pair will interact. The bonds of friendship established during the divorce period may be particularly valuable because they have lent support at a time of personal crisis. Care should be taken not to drop these friendships.

Parental Remarriage Because most remarriages involve children, people must work out the nuances of living with someone else's children. Since mothers are usually awarded primary physical custody, this translates into the new stepfather's adjusting to the mother's children and vice versa. If a person has children from a previous marriage who do not live primarily with him or her, the new spouse must adjust to these children on weekends, holidays, and vacations or at other visitation times.

This never-married woman with no children is marrying a man with six children. Of her experience she said, "It hasn't always been easy . . . but it really is worth it."

Economic and Legal Remarriage The second marriage may begin with economic responsibilities to the first marriage. Alimony and child support often threaten the harmony and sometimes even the economic survival of second marriages. Although legally the income of a new wife is not used to decide the amount her new husband is required to pay in child support for his children of a former marriage, his ex-wife may petition the court for more child support. The premise of her doing so is that his living expenses are reduced with the new wife and therefore he can afford to pay more child support. Although the ex-wife is not likely to win, she can force the new wife to court and a disclosure of her income (all with considerable investment of time and legal fees for the newly remarried couple).

Economic issues in a remarriage may become evident in another way. A remarried woman who is receiving inadequate child support from her ex-spouse and needs money for her child's braces, for instance, might wrestle with how much money to ask her new husband for.

Consequences of a Woman Marrying a Divorced Man with Children

Women considering marriage to a divorced man with children should be cautious. In a study of 274 questionnaires of second wives—women who married men who had been married before—39 percent reported having thought about divorcing their husbands, and one in four reported that, with what she now knew, she would not have married him (Knox and Zusman, 2001). Implications of the study included the following:

1. **Acknowledge that a second marriage is vulnerable.** Be realistic about the degree to which his children will accept you and/or your children. And how do you really feel about his children?

2. **Question whether living together is beneficial to future marital success.** Eighty-three percent of second wives reported that they had lived with their husbands before they married. Abundant evidence suggests that living together is not associated with positive relationship outcomes.

3. **Delay marriage to a person who has been married before.** Over a quarter (27.4%) of the wives reported marrying their current husband less than a year after his divorce became final. We have previously noted the importance of knowing a person at least 2 years before getting married.

4. **Consider a fresh start in a new home.** Eighty-two percent of the second wives reported that their new husbands had moved into their homes—homes the former husband of the new wife had vacated in the prior divorce. Not one of the second wives reported that she currently lived in a newly bought or rented place with her new husband. Only 16 percent reported that they (the second wives) had moved into their husband's home or condo.

Remarriage for the Widowed

Only 10 percent of remarriages are of widows or widowers. Nevertheless, remarriage for the widowed is usually very different from remarriage for the divorced. Unlike the divorced, they are usually much older and their children are grown. A widow or widower may marry someone of similar age or someone who is considerably older or younger. Marriages in which one spouse is considerably older than the other are referred to as May-December marriages (discussed in Chapter 8, Marriage Relationships). Here we will discuss only **December marriages,** in which both spouses are elderly.

A study of 24 elderly couples found that the need to escape loneliness or the need for companionship was the primary motivation for remarriage (Vinick, 1978). The men reported a greater need to remarry than the women.

Most of the spouses (75 percent) met through a mutual friend or relative and married less than a year after their partner's death (63 percent) Increasingly the elderly are meeting online. Some sites cater to older individuals seeking a partner, including seniorfriendfinder.com and thirdage.com.

The children of the couples in Vinick's study had mixed reactions to their parent's remarriage. Most of the children were happy that their parent was happy and felt relieved that the companionship needs of their elderly parent would now be met by someone on a more regular basis. But some children also disapproved of the marriage out of concern for their inheritance rights. "If that woman marries Dad," said a woman with two children, "she'll get everything when he dies. I love

Increasingly, the widowed are using the Internet to meet new partners.

© Peter Dazeley/zefa/Corbis

him and hope he lives forever, but when he's gone, I want the house I grew up in." Though children may be less than approving of the remarriage of their widowed parent, adult friends of the couple, including the kin of the deceased spouses, are usually very approving (Ganong and Coleman, 1994).

Stages of Involvement with a New Partner

After a legal separation or divorce (or being widowed), the parent who becomes involved in a new relationship passes through various transitions (Table 16.1) (Anderson and Greene, 2005). These not only affect the individuals and their relationship but any children and/or extended family.

Stability of Remarriages

National data reflect that remarriages are more likely than first marriages to end in divorce in the early years of remarriage (Clarke and Wilson, 1994). This vulnerability of second marriages to divorce is because once-divorced individuals are less fearful of divorce than never-divorced individuals. So, rather than stay in an unhappy second marriage, they leave.

Though remarried persons are more vulnerable to divorce in the early years of their subsequent marriage, after 15 years of staying in the second marriage, they are less likely to divorce than those in first marriages (Clarke and Wilson, 1994). Hence, these spouses are likely to remain married because they want to, not because they fear divorce.

Stepfamilies

Stepfamilies, also known as blended, binuclear, step, remarried, or reconstituted families, represent the fastest-growing type of family in the United States. Indeed, more than half of Americans are members of a blended family (Christian, 2005). A **blended family** is one in which the spouses in a new marriage relationship are blended with the children of at least one of the spouses from a previous marriage. (Another popular term is **binuclear,** which refers to a family that spans two households; when a married couple with children divorce, their family unit spreads into two households.)

Families don't blend. Not like ice cream or fruit drinks, anyway. But more like patchwork quilts. Take the pieces and sew them tenderly together. Don't deny that you're two families; rejoice in those differences.

Van Chapman, reseacher

Table 16.1   Stages of Parental Repartnering

Relationship Transition	Definition
Dating initiation	The parent begins to date.
Child introduction	The child and new dating partner meet.
Serious involvement	The parent begins to present the relationship as "serious" to the child.
Sleepover	The parent and the partner begin to spend nights together when the child is present.
Cohabitation	The parent and the partner combine households.
Break up of a serious relationship.	The relationship experiences a temporary or permanent disruption.
Pregnancy in the new relationship.	A planned or unexpected pregnancy occurs.
Engagement.	The parent announces plans to remarry.
Remarriage	The parent and partner create a legal or civil union.

Anderson, E. R. and S. M. Greene. 2005. Transitions in parental repartnering after divorce. *Journal of Divorce & Remarriage* 43:49 (http://www.haworthpress.com/web/jdr/).

There is a movement away from the use of the term *blended* since stepfamilies really do not blend. The term **stepfamily** (sometimes referred to as **step relationships**) is the term currently in vogue. Leon and Angst (2005) reviewed U.S. films over a 13-year period and found that stepfamilies and remarriages were depicted in negative or mixed ways. This section examines how stepfamilies differ from nuclear families; how they are experienced from the viewpoints of women, men, and children; and the developmental tasks that must be accomplished to make a successful stepfamily.

Definition and Types of Stepfamilies

The bond that links your true family is not one of blood, but of respect and joy in each other's life.

Richard Bach, American writer

Although there are various types of stepfamilies, the most common is a family in which the partners bring children from previous relationships into the home, where they may also have a child of their own. The couple may be married or living together, heterosexual or homosexual. Although a stepfamily can be created when a never-married or a widowed parent with children marries a person with or without children, most stepfamilies today are composed of spouses who are divorced and who bring children into the new marriage. This is different from stepfamilies characteristic of the early twentieth century, which more often were composed of spouses who had been widowed.

As noted, stepfamilies may be both heterosexual and homosexual. Lesbian stepfamilies model gender flexibility in that the biological mother and the stepmother tend to share parenting (in contrast to the traditional family, in which the mother may take primary responsibility for parenting and the father is less involved). This allows the biological mother some freedom from motherhood as well as support in it. In gay male stepfamilies, the gay men may also share equally in the work of parenting.

Myths of Stepfamilies

Various myths regarding stepfamilies include that there will be instant emotional bonding between the new family members, that children in stepfamilies are damaged and do not recover, that stepfamilies are not "real" families, and that stepmothers are "wicked" and "home-wreckers." The Stepfamily Association of America has identified other myths, such that part-time (weekend) stepfamilies are easier than full-time stepfamilies, that it is easier/better if the biological parents withdraw, and that stepfamilies formed after the death of a parent rather than the divorce of a parent are "easier" (http://www.saafamilies.org/faqs/myths.htm, 2005).

Unique Aspects of Stepfamilies

Stepfamilies differ from nuclear families in a number of ways (see Table 16.2 on page 463). To begin with, the children in a nuclear family are biologically related to both parents, whereas the children in a stepfamily are biologically related to only one parent. Also, in a nuclear family, both biological parents live with their children, whereas only one biological parent in a stepfamily lives with the children. In some cases, the children alternate living with each parent.

Though nuclear families are not immune to loss, everyone in a stepfamily has experienced the loss of a love partner, which results in grief. About 70 percent of children are living without their biological father (who some children desperately hope will reappear and reunite the family). The respective spouses may also have experienced emotional disengagement and physical separation from a once-loved partner. Stepfamily members may also experience losses because of having moved away from the house in which they lived, their familiar neighborhood, and their circle of friends.

Stepfamily members also are connected psychologically to others outside their unit. Bray and Kelly (1998) referred to these relationships as the "Ghosts at the table":

Children are bound to absent parents, adults to past lives and past marriages. These invisible psychological bonds are the Ghosts at the table and because they play on the most elemental emotions—emotions like love and loyalty and guilt and fear—they have the power to tear a marriage and stepfamily apart (p. 4).

Children in nuclear families have also been exposed to a relatively consistent set of beliefs, values, and behavior patterns. When children enter a stepfamily, they "inherit" a new parent, who may bring a new set of values and beliefs and a new way of living into the family unit. Likewise, the stepparent now lives with children who may have been reared differently from the way in which biological parents reared the child. "His kids had never been to church," reported one frustrated stepparent.

Another unique aspect of stepfamilies is that the relationship between the biological parent and the children has existed longer than the relationship between the adults in the remarriage. Jane and her twin children have a 9-year relationship and are emotionally bonded to each other. But Jane has known her new partner only a year, and although her children like their new stepfather, they hardly know him.

In addition, the relationship between the biological parent and his or her children is of longer duration than that of the stepparent and stepchildren. The short history of the relationship between the child and the stepparent is one factor that may contribute to increased conflict between these two during the child's adolescence. Children may also become confused and wonder whether they are disloyal to their biological parent if they become friends with the stepparent.

Another unique feature of stepfamilies is that unlike children in the nuclear family, who have one home they regard as theirs, children in stepfamilies have two homes they regard as theirs. In some cases of joint custody, children spend part of each week with one parent and part with the other; they live with two sets of adult parents in two separate homes.

Money, or lack of it, from the ex-spouse (usually the husband) may be a source of conflict. In some stepfamilies, the ex-spouse (usually the father) is expected to send child support payments to the parent who has custody of the children. Fewer than one-half of these fathers send any money; those who do may be irregular in their payments. Fathers who pay regular child support tend to have higher incomes, to have remarried, to live close to their children, and to visit them regularly. They are also more likely to have legal shared or joint custody, which helps to ensure that they will have access to their children.

Fathers who do not voluntarily pay child support and are delinquent by more than one month may have their wages garnished by the state. Some fathers change jobs frequently and move around to make it difficult for the government to keep up with them. Such dodging of the law is frustrating to custodial mothers who need the child support money. Added to the frustration is the fact that fathers are legally entitled to see their children even when they do not pay court-ordered child support. This angers the mother, who must give up her child on weekends and holidays to a man who is not supporting his child financially. Such distress on the part of the mother is probably conveyed to the child.

New relationships in stepfamilies experience almost constant flux. Each member of a new stepfamily has many adjustments to make. Issues that must be dealt with include how the mate feels about the partner's children from a former marriage, how the children feel about the new stepparent, and how the newly married spouse feels about the spouse's sending alimony and child support payments to an ex-spouse. In general, families in the Bray and Kelly (1998) study did not begin to think and act like a family until the end of the second or third year.

These early years are the most vulnerable, with a quarter of the stepfamilies ending during that period.

Stepfamilies are also stigmatized. **Stepism** is the assumption that stepfamilies are inferior to biological families. Stepism, like racism, heterosexism, sexism, and ageism, involves prejudice and discrimination. Social changes need to be made to give support to stepfamilies. For example, "If there's a banquet, is there an opportunity for a child to invite all four of his or her parents?"(Everett, 2000). More often, children are forced to choose, which usually results in selecting the biological parent and ignoring the stepparents. The more adults (e.g., both biological and stepparents) children have who love and support them, the better, and our society should support this.

Stepparents also have no childfree period. Unlike the newly married couple in the nuclear family, who typically have their first child about two and one-half years after their wedding, the remarried couple begins their marriage with children in the house.

Profound legal differences exist between nuclear and blended families. Whereas biological parents in nuclear families are required in all states to support their children, only five states require stepparents to provide financial support for their stepchildren. However, when there is a divorce, this and other discretionary types of economic support usually stop. Other legal matters with regard to nuclear families versus stepfamilies involve inheritance rights and child custody. Stepchildren do not automatically inherit from their stepparents, and courts have been reluctant to give stepparents legal access to stepchildren in the event of a divorce. In general, U.S. law does not consistently recognize stepparents' roles, rights, and obligations regarding their stepchildren (Malia, 2005). Without legal support to ensure such access, these relationships tend to become more distant and nonfunctional.

Finally, extended family networks in nuclear families are smooth and comfortable, whereas those in stepfamilies often become complex and strained. Table 16.2 summarizes the differences between nuclear families and stepfamilies.

Stepfamilies in Theoretical Perspective

Structural functionalists, conflict theorists, and symbolic interactionists view stepfamilies from different points of view.

1. Structural-Functional Perspective. To the structural functionalist, integration or stability of the system is highly valued. The very structure of the stepfamily system can be a threat to the integration and stability of the family system. The social structure of stepfamilies consists of a stepparent, a biological parent, biological children, and stepchildren. Functionalists view the stepfamily system as vulnerable to an alliance between the biological parent and the biological children who have a history together.

In 75 percent of the cases, the mother and children create the alliance. The stepfather, as an outsider, may view this alliance between the mother and her children as the mother's giving the children too much status or power in the family. Whereas the mother relates to her children as equals, the stepfather relates to the children as unequals whom he attempts to discipline. The result is a fragmented parental subsystem whereby he accuses her of being too soft and she accuses him of being too harsh.

Structural family therapists suggest that parents should have more power than children and that they should align themselves with each other. Not to do so is to give children family power, which they may use to splinter the parents off from each other and create another divorce.

2. Conflict Perspective. Conflict theorists view conflict as normal, natural, and inevitable as well as functional in that it leads to change. Conflict in the stepfamily system is seen as desirable in that it leads to equality and individual autonomy.

Conflict is a normal part of stepfamily living. The spouses, parents, children, and stepchildren are constantly in conflict for the limited resources of space,

Table 16.2 Differences between Nuclear Families and Stepfamilies

Nuclear Families	Stepfamilies
1. Children are (usually) biologically related to both parents.	1. Children are biologically related to only one parent.
2. Both biological parents live together with children.	2. As a result of divorce or death, one biological parent does not live with the children. In the case of joint physical custody, the children may live with both parents, alternating between them.
3. Beliefs and values of members tend to be similar.	3. Beliefs and values of members are more likely to be different because of different backgrounds.
4. Relationship between adults has existed longer than relationship between children and parents.	4. Relationship between children and parents has existed longer than relationship between adults.
5. Children have one home they regard as theirs.	5. Children may have two homes they regard as theirs.
6. The family's economic resources come from within the family unit.	6. Some economic resources may come from ex-spouse.
7. All money generated stays in the family.	7. Some money generated may leave the family in the form of alimony or child support.
8. Relationships are relatively stable.	8. Relationships are in flux: new adults adjusting to each other; children adjusting to stepparent; stepparent adjusting to stepchildren; stepchildren adjusting to each other.
9. No stigma is attached to nuclear family.	9. Stepfamilies are stigmatized.
10. Spouses had childfree period.	10. Spouses had no childfree period.
11. Inheritance rights are automatic.	11. Stepchildren do not automatically inherit from stepparents.
12. Rights to custody of children are assumed if divorce occurs.	12. Rights to custody of stepchildren are usually not considered.
13. Extended family networks are smooth and comfortable.	13. Extended family networks become complex and strained.
14. Nuclear family may not have experienced loss.	14. Stepfamily has experienced loss.

time, and money. Space refers to territory (rooms) or property (television, CD player, PlayStation®) in the house that the stepchildren may fight over. Time refers to the amount of time that the parents will spend with each other, with their biological children, and with their stepchildren. Money must be allocated in a reasonably equitable way so that each member of the family has a sense of being treated fairly.

Problems arise when space, time, and money are limited. Two new spouses who each bring a child from a former marriage into the house have a situation fraught with potential conflict. Who sleeps in which room? Who gets to watch which channel on television?

To further complicate the situation, suppose the couple have a baby. Where does the baby sleep? And since both parents may have full-time jobs, the time they have for the three children is scarce, not to speak of the fact that the baby will require a major portion of their available time. As for money, the cost of the baby's needs, such as formula and disposable diapers, will compete with the economic needs of the older children. Meanwhile, the spouses may need to spend time alone and may want to spend money as they wish. All these conflicts are functional, since they increase the chance that a greater range of needs will be met within the stepfamily.

3. Interactionist Perspective. Symbolic interactionists emphasize the meanings and interpretations that members of the stepfamily develop for events and interactions in the family. Children may blame themselves for their parents' divorce and feel that they and their stepfamily are stigmatized; parents may view stepchildren as spoiled.

Stepfamily members also nurture certain myths. Stepchildren sometimes hope that their parents will reconcile and that their nightmare of divorce and stepfamily living will end. This is the myth of reconciliation. Another is the myth of instant love, usually held by each biological parent, who hopes that his or her children will instantly love the new partner. Although this does happen, particularly if the child is young and has no negative influences from the other parent, it is unlikely.

Stages in Becoming a Stepfamily

Just as a person must pass through various developmental stages in becoming an adult, a stepfamily goes through a number of stages as it overcomes various obstacles. Researchers such as Bray and Kelly (1998) and Papernow (1988) have identified various stages of development in stepfamilies. These stages include the following.

Stage 1: Fantasy Both spouses and children bring rich fantasies into the new marriage. Spouses fantasize that their new marriage will be better than the previous one. If the person they are marrying has adult children, they assume that these children will be open to a rewarding relationship with them. Young children have their own fantasy—they hope that their biological parents will somehow get back together and that the stepfamily is temporary.

Stage 2: Reality Instead of realizing their fantasies, new spouses may find that stepchildren ignore or are rude to them. Young children may wish that the stepparent would disappear and that the biological parents would get back together. Indeed, stepparents may feel that they are outsiders in an already-functioning unit (the biological parent and child).

Stage 3: Being Assertive Initially the stepparent assumes a passive role and accepts the frustrations and tensions of stepfamily life. Eventually, however, the resentment reaches a level where the stepparent is driven to make changes. The stepparent makes the partner aware of the frustrations and suggests that the marital relationship should have priority some of the time. The stepparent may also make specific requests, such as reducing the number of conversations the partner has with the ex-spouse, not allowing the dog on the furniture, or requiring the stepchildren to use better table manners. This stage is successful to the degree that the partner supports the recommendations for change. A crisis may ensue.

Stage 4: Strengthening Pair Ties During this stage the remarried couple solidify their relationship by making it a priority. At the same time, the biological parent must back away somewhat from the parent-child relationship so that the new partner can have the opportunity to establish a relationship with the stepchildren.

This relationship is the product of small units of interaction and develops slowly across time. Many day-to-day activities, such as watching television together, eating meals together, and riding in the car together, provide opportunities for the stepparent-stepchild relationship to develop. It is important that the stepparent not attempt to replace the relationship that the stepchildren have with their biological parents.

Stage 5: Recurring Change A hallmark of all families is change, but this is even more true of stepfamilies. Bray and Kelly (1998) note that even though a stepfamily may be functioning well when the children are preadolescent, when they become teenagers, a new era has begun as the children may begin to question how the family is organized and run. But such questioning by adolescents is not unique to stepchildren.

Michaels (2000) noted that spouses who become aware of the stages through which stepfamilies pass report that they feel less isolated and unique. Involvement in stepfamily discussion groups such as the Stepfamily Enrichment Program provides enormous benefits.

Strengths of Stepfamilies

Stepfamilies have both strengths and weaknesses. Strengths include children's exposure to a variety of behavior patterns, their observation of a happy remarriage, adaptation to stepsibling relationships inside the family unit, and greater objectivity on the part of the stepparent.

Exposure to a Variety of Behavior Patterns

Children in stepfamilies experience a variety of behaviors, values, and lifestyles. They have the advantage of living on the inside of two families. And although this may be confusing and challenging for a child or adolescent, they are learning early how different family patterns can be. For example, one 12-year-old had never seen a couple pray at the dinner table until his mom remarried a man who was accustomed to a "blessing" before the meal.

Happier Parents

While children may want their parents to get back together, they often come to observe that their parents are happier alone than married. And if the parent remarries, the child may see the parent in a new and happier relationship. Such happiness may spill over into the context of family living so that there is less tension and conflict.

Opportunity for New Relationship with Stepsiblings

Though some children reject their new stepsiblings, others are enriched by the opportunity to live with a new person to whom they are now "related." One 14-year-old remarked, "I have never had an older brother to do things with. We both like to do the same things and I couldn't be happier about the new situation." Some stepsibling relationships are maintained throughout adulthood.

More Objective Stepparents

Because of the emotional tie between a parent and a child, some parents have difficulty discussing certain issues or topics. A stepparent often has the advantage of being less emotionally involved and can relate to the child at a different level. One 13-year-old said of the relationship with his new stepmom, "She went through her own parents' divorce and knew what it was like for me to be going through my dad's divorce. She was the only one I could really talk to about this issue. Both my dad and mom were too angry about the subject to be able to talk about it."

The New Stepmother

Almost every little girl grows up hoping to be a mother someday. But has there ever been one who dreamed of becoming a stepmother?

Evelyn Bassoff, psychologist

Christian (2005) studied 69 posts by stepmothers on the Stepfamily Association of America Web site forum (www.saafamilies.org) and noted that they felt they were negatively stereotyped as the "wicked stepmother" and reacted to this stereotype by characterizing the biological mother as mentally unstable. Of one biological mother, a stepmother noted:

Diversity in Other Countries

Not all countries view stepmothers negatively. The French have stopped using the pejorative word for stepmother (*marâtre*) and replaced it with a new word—*belle-mère*, which literally means "beautiful mother."

> She has chased us down and followed us to the store and places, then proceeded to cuss me out. She tried to fist fight with me at Alex's school, right in front of Alex.

The role of stepmom played by Julia Roberts provided insight into this very difficult role.

© Columbia TriStar Pictures/Photofest

Or the biological mothers were seen as incompetent:

The birth mom is an alcoholic who had hardly nothing to do with these children for several years. . .

The stepmothers also viewed themselves as a martyr for helping to save the child:

Whether BioMom likes it or not, I am more of a mother to her daughter than she is.

Specific sources of frustration for the woman in the role of stepmother include accepting and being accepted by her new partner's children, adjusting to alimony and child support payments made by her husband to an ex-wife, having the new partner accept her children and having her children accept him, and having another child in the new marriage.

Accepting Partner's Children

"She'd better think a long time before she marries a guy with kids," said one 29-year-old woman who had done so. This stepmother went on to explain:

It's really difficult to love someone else's children. Particularly if the kid isn't very likable. A year after we were married, my husband's 9-year-old daughter visited us for a summer. It was a nightmare. She didn't like anything I cooked, was always dragging around making us late when we had to go somewhere, kept her room a mess, and acted like a gum-chewing smart aleck. I hated her, but felt guilty because I wanted to have feelings of love and tenderness. Instead, I was jealous of the relationship she had with her father, and I wanted to get rid of her. I began counting how many days until she would be gone.

You can hide your dislike for a while, but eventually you must tell your partner how you feel. I was lucky. My husband also thought his daughter was horrible to live with and wasn't turned off by my feelings. He told her if she couldn't act more civil, she couldn't come back. The message to every woman about to marry a guy with kids is to be aware that your man is a package deal and that the kid is in the package.

Partner's Children Accepting Stepmother

That's why we are called STEPMOTHERS. I have footprints all over my body.

A stepmother

Children may blame the new stepmother for destroying the family as they knew it. Particularly if their biological mother encourages this perspective, it will be difficult for the stepmother to change this notion until the children grow up, leave home, and evaluate their stepmother independently. There is also the historical cultural view of the stepmother. This negative view of stepmothers is reflected in the prefix *step*, which in Old English referred to a family relationship caused by death. A stepchild, in essence, was an orphan child, and a stepmother was one who took care of an orphan. This view was enhanced by the structure of remarriages in early European households.

The legacy of stepmother folklore may have contributed to the fact that children have more difficulty accepting a new stepmother than a new stepfather. (It may also be that she is the more active parent, which provides greater opportunity for conflict.) In addition, a stepchild may feel the need to keep an emotional distance in the relationship with the stepmother so as not to incur the anger of the biological mother.

Lou Everett (2000), a specialist who works with stepfamilies, recommends that children remember their stepmother on Mother's Day. "The card doesn't have to say 'Mom' on it. Just something to express your appreciation . . . how great the cookies were that she baked or how happy she makes your dad. Those things mean so much to the stepmother."

A team of researchers (Henry et al., 2005) studied adolescent girls in remarried families and found that they were more likely to be concerned for the welfare of others when they themselves felt supported by the residential parent. Such support must be genuine and not controlling or manipulative.

Resenting Alimony and Child Support

In addition to the potential problems of not liking the partner's children, there may be problems of alimony and child support. As noted earlier, it is not unusual for a wife to become upset when her husband mails one-quarter or one-third of his income to his former wife. This amount of money is often equal to the current wife's earnings. Some wives in this position see themselves as working for their husband's ex-wife—a perception that may create negative feelings.

Her New Partner and How Her Children Accept Him

For some remarried wives, two main concerns are how her new husband accepts her children and how her children accept him. The ages of the children are important in these adjustments. If the children are young (age 3 or younger), they will usually accept any new adult in the natural parent's life. On the other hand, if the children are in adolescence, they are most likely struggling for independence from their natural parents and may not want any new authority figures in their lives.

In regard to whether the new spouse will accept and invest in her children, Hofferth and Anderson (2003) found that the new husband was more likely to do so if he did not already have and was not already supporting nonresidential children from a prior relationship. The authors hypothesized that, from a biological perspective, such a father did not invest as heavily in the new family because it did not facilitate reproduction of new offspring.

Having Another Child

A national study of women in second marriages revealed that about two-thirds have a baby within 6 years after remarriage (Wineberg, 1992). When this baby is a first biological baby, stepmothers report greater dissatisfaction with their stepchildren. Their time with their stepchildren is also diminished with the birth of a new baby (hence, stepchildren do not seem to benefit from a half-sibling) (Stewart, 2005). This same phenomenon is also true for first-time fathers and may be a function of having to choose between one's biological children and stepchildren in making certain decisions. "Evolutionary views of parental psychology suggest that stepparents have a genetic propensity to express greater solicitude toward their biological children than toward stepchildren" (MacDonald and DeMaris, 1996, p. 23).

In spite of the decreased satisfaction with one's stepchildren, having a child in a second marriage is associated with increasing the stability of the relationship and reducing the probability of divorce (Wineberg, 1992). The researcher suggested that "couples with a mutual birth may tolerate more marital stress before considering divorce than couples with no mutual children" (p. 885). Remarried spouses who have another child soon discover how biased our society is against remarriages and the children born into these marriages. The Social Policy on page 468 discusses this issue.

Although we have been discussing the stepmother and her adjustment to her stepchild, the biological mother has her own unique feelings and adjustments.

Child Support and Second Families

The court system is biased against men who have children in second marriages. A man who has been married before and who has children from that marriage is required to pay child support for those children at the expense of children he has in a second marriage. For example, a divorced father of two children earning $40,000 per year would normally be required to pay 25 percent of his gross income, or $10,000, annually in child support (and his ex-wife may still get the tax deduction of the children as dependents). Should he remarry and have two additional children with his new wife, the needs of those children in the second marriage are irrelevant—the man must still pay $10,000 in child support for his first two children.

Moreover, if the man's income increases or the needs of the children in his first marriage increase (excessive medical bills), the man's ex-wife can petition the court and have his child support payments increased. Again, the fact that the man has other children in his second marriage who have clothing, food, and medical needs is not recognized by the courts. Indeed, "second family children are considered hardships and their needs are given consideration only after the needs of the children in the first family are satisfied," notes Dianna Thompson. She is the founder of Second Wives Crusade, an organization for women married to men who have children from previous marriages. She discovered, to her dismay, that her husband's child support obligations to the children in his first marriage were calculated without consideration for their own children. This is the law in all 50 states. Ms. Thompson is lobbying legislators to recognize the bias against children in second marriages and encouraging them to adopt new laws to protect all children—not just those who happen to be born of first marriages.

Your Opinion?

1. To what degree do you believe a father should be financially obligated to his first children before providing money for his second set of children?
2. What laws should be passed in regard to how money is to be distributed?
3. Should the second wife's income be considered in deciding how much a father should pay for child support to his first wife?

Source

Dianna Thompson, founder of Second Wives Crusade

One such mother lamented that she now had to share her daughter with the woman her husband left her for and that this was a very difficult adjustment. "I had a particularly hard time when she took my daughter shopping with her on their vacation during Christmas. . . . I wanted to share that with her. And his new wife's name was 'Star' and almost my daughter's age, to top it off" (author's files).

The New Stepfather

My early reaction to my new stepfamily was typical: "If I can just survive until the children leave home."

James Eckler, stepfather

Men in stepfamilies may or may not have children from a previous relationship. Three possible stepfamily combinations include a man with children married to a woman without children, a man with children married to a woman with children, and a man without children married to a woman with children. Men in stepfamilies are, in some ways, like men in biological families. They tend to be less involved in child care and spend less time with children than the stepmother or the biological mother (Kyungok, 1994). Men in stepfamilies are also different from those in nuclear families in that stepchildren often want them to respect the relationship they have with their biological dad rather than the new stepfather be their new dad (Bray and Kelly, 1998).

Parental status refers to the degree to which the stepparent is considered to be a parent to a stepchild. The Self-Assessment on page 470 assesses the degree to which a stepchild views his or her father as a parent. (Although the scale was developed for stepfathers and would not be valid for other uses, you might give it a try for other family relationships such as stepmothers, stepsiblings, and stepgrandparents.)

This stepfather has been the "real" father in this bride's life. He is giving his stepdaughter away on her wedding day.

Man with Children Who Marries a Woman without Children

Men with biological children enter stepfamilies with an appreciation for the role of parent, some skills (one would hope) in fathering, and a bond with a child or children who usually live with the biological mother/ex-wife. The latter may mean a grieving father who not only misses his children but also experiences a number of push-pull factors that influence how involved he is in the lives of his stepchildren.

Men with children typically want their new wife to bond with their children and actively participate in their care. Such an expectation places the new wife in the role of the "instant mother," which may backfire as a result of his children rejecting her. Since the new wife in the role of the new mother has typically spent limited time with her partner's children, it is important that he allow sufficient time (2 to 7 years) for a relationship to develop. Although this is rare, some women may want nothing to do with their partner's children and may be jealous of the time and money he spends on them. One husband reported, "She was jealous of everything I did with my kids; our marriage ended over it."

Another concern of the father who marries a woman without children is whether she will want children of her own. Often she does (and maybe more than one), but the man who already has children may be less interested. Nevertheless, it is not unusual for men to have additional children with a new wife.

Man with Children Who Marries a Woman with Children

A man living with children (from a previous relationship) is most likely to marry a woman with children (Goldscheider and Sassler, 2006). His doing so not only results in multiple sets of relationships but increases his involvement with his new wife's issues. This was the conclusion of a team of researchers who studied the family context of 60 stepfathers:

> [S]tepfathers may be drawn closer to their stepchildren, and they may have fewer negative attitudes toward them because of the strategies they adopt in striving to treat both sets of children in an equitable manner. The presence of their biological children in the household may, in effect, force them to parent to a greater extent than if they had merely been absorbed into a pre-existing family. It becomes incumbent upon them to constitute a viable living pattern and take a more active role with regard to all children in the household. This may lead them to minimize negative thoughts and feelings about their stepchildren and also to exaggerate positive attitudes about stepchildren in the interests of fairness (Palisi et al., 1991, p. 102).

Parental Status Inventory (PSI)*

The Parental Status Inventory (PSI) is a 14-item inventory that measures on an 11-point scale from 0 to 100% the degree to which respondents consider their stepfather to be a parent. Read each of the following statements and circle the percentage indicating the degree to which you regard the statement as true.

1. I think of him as my father. (0%, 10, 20, 30, 40, 50%, 60, 70, 80, 90, 100%)

2. I am comfortable when someone else refers to him as my father/dad. (0%, 10, 20, 30, 40, 50%, 60, 70, 80, 90, 100%)

3. I think of myself as his daughter/son. (0%, 10, 20, 30, 40, 50%, 60, 70, 80, 90, 100%)

4. I refer to him as my father/dad. (0%, 10, 20, 30, 40, 50%, 60, 70, 80, 90, 100%)

5. He introduces me as his son/daughter. (0%, 10, 20, 30, 40, 50%, 60, 70, 80, 90, 100%)

6. I introduce my mother and him as my parents. (0%, 10, 20, 30, 40, 50%, 60, 70, 80, 90, 100%)

7. He and I are just like father and son/daughter. (0%, 10, 20, 30, 40, 50%, 60, 70, 80, 90, 100%)

8. I introduce him as "my father" or "my dad." (0%, 10, 20, 30, 40, 50%, 60, 70, 80, 90, 100%)

9. I would feel comfortable if he and I were to attend a father-daughter/father-son function alone together, such as a banquet, baseball game, -or cookout. (0%, 10, 20, 30, 40, 50%, 60, 70, 80, 90, 100%)

10. I introduce him as "my mother's husband" or "my mother's partner." (0%, 10, 20, 30, 40, 50%, 60, 70, 80, 90, 100%)

11. When I think of my mother's house, I consider him and my mother to be parents to the same degree. (0%, 10, 20, 30, 40, 50%, 60, 70, 80, 90, 100%)

12. I consider him to be a father to me. (0%, 10, 20, 30, 40, 50%, 60, 70, 80, 90, 100%)

13. I address him by his first name. (0%, 10, 20, 30, 40, 50%, 60, 70, 80, 90, 100%)

14. If I were choosing a greeting card for him, the inclusion of the word 'father' or 'dad' in the inscription would prevent me from choosing the card. (0%, 10, 20, 30, 40, 50%, 60, 70, 80, 90, 100%)

Scoring

First, reverse the scores for items 10, 13, and 14. For example, if you circled a 90, change the number to 10, if you circled a 60, change the number to 40, and so on. Now add the percentages and divide by 14. A 0% reflects that you do not regard your stepdad as your parent at all. A 100% reflects that you totally regard your stepdad as your parent. The percentages between 0% and 100% show the gradations from no regard to total regard of your stepdad as parent.

Norms

The scale was completed by respondents in two studies. The numbers of respondents in the studies were 159 and 156, respectively, and the average score in the respective studies was 45.66. Between 40 and 50 percent of both Canadians and Americans viewed their stepfather as their parent.

*Developed by Dr. Susan Gamache (2000), Hycroft Medical Centre, #217, 3195 Granville Street, Vancouver, B.C., Canada, V6H 3K2 (gamache @interchange.ubc.ca). Details on construction of the scale, including validity and reliability, are available from Dr. Gamache. The PSI Scale is used in this text with the permission of Dr. Gamache and may not be used otherwise (except as an in-class student exercise) without written permission.

MacDonald and DeMaris (2002) also found that the less involved the biological father, the more involved the stepfather. Moreover, while stepfathers tend to expect their stepchildren to be "obedient" (Bernhardt et al., 2002), stepfathers also report increased satisfaction in their role as stepfathers when they frequently engage in authoritative parenting behaviors. Hence, the more engaged the stepfather in disciplining his stepchildren, the greater the level of reported satisfaction (Fine et al., 1997).

Man without Children Who Marries a Woman with Children

A man without children is less likely to marry a woman with children than to marry a woman without children (Goldscheider and Sassler, 2006). The adjustments of the never-married, divorced, or widowed man without children who marries a woman with children are probably the most difficult and are related to her children, their acceptance of him, and his awareness that his wife is emotionally bonded to her children. Unlike childfree marital partners, who are bonded only to each other, the husband entering a relationship with a woman who has children must accept her attachment with her children from the outset. New partners may view such bonding differently. One man said that such concern for her own children was a sign of a caring and nurturing person. "I wouldn't want to live with anyone who didn't care about her kids." But another said, "I felt like an outsider. . . . I couldn't stand the noise and disruption that children involve. I was used to a very quiet, ordered life—that is the exact opposite of what children in the house are like."

Lou Everett (1998) interviewed six stepfathers and identified two primary factors that contribute to a positive stepfather-stepchild relationship:

1. Active involvement in teaching the stepchild something mutually valued. Just spending time with the stepchild (eating meals, watching television) had no positive effect on the relationship. Teaching the stepchild (how to skate, fish, fly a kite) made the stepfather feel as though he was contributing to the development of the stepchild and endeared the stepchild to the stepfather.

2. An intense love relationship between the stepfather and the biological mother of the stepchild. "Men who love their wives are more tolerant of their stepchildren," observed Dr. Everett. Marsiglio (1992) analyzed data from 195 men in stepfamilies and observed that the men, like the stepfathers in the Everett study, reported more positive relationships with stepchildren to the degree that they were happy with their partner. In addition, 55 percent reported that it was "somewhat true" or "definitely true" that "having stepchildren is just as satisfying as having your own children" (p. 204). Men who were living with their partners had perceptions of their stepfathering role similar to those of married men.

Here are some questions a man without children (who is considering marriage) might ask a woman who has children:

1. How do you expect me to relate to your children? Am I supposed to be their friend, daddy, or something else? Men who develop a relationship first with their stepchild before they start disciplining the child report more positive outcomes (Bray and Kelly, 1998).

2. How do you feel about having another child? How many additional children are you interested in having?

3. How much money do you get in alimony and child support from your ex-husband? When do these payments stop?

4. What expenses of the children do you expect me to pay for? Who is going to pay for their college expenses? In essence, the mother receiving child support from the biological father may create the illusion for the potential stepfather that the former husband will take care of the children's expenses. In reality, child-support payments cover only a fraction of what is actually spent on a child; consequently, the new stepfather may feel burdened with more financial responsibility for his stepchildren than he bargained for. This may engender negative feelings toward the wife in the new marriage relationship.

Children in Stepfamilies

National Data

About 10 percent of all children live in stepfamilies (Furukawa, 1994). Of these children in stepfamilies, 86 percent live with their biological mother and stepfather; 14 percent live with their biological father and stepmother (Rutter, 1994).

Research confirms that of the children growing up in stepfamilies (see National Data) "80 percent . . . are doing well. Children in stable stepfamilies look very much like those reared in stable first families" (Pasley, 2000, p. 6). Mackay (2005) compared children from intact and nonintact homes and found that the latter were "worse off." However the differences between the groups were not large and were moderated by loss of income following the divorce, conflict between the parents, declines in mental health of custodial mothers, and compromised parenting (reduced attentiveness of parents due to their coping with their own adult issues). Similarly, Ruschena et al. (2005) compared longitudinal data on adolescents whose parents divorced/remarried versus those whose parents stayed married and found no significant group differences with regard to behavioral and emotional adjustment concurrently or across time, nor on academic outcomes and social competence. The researchers commented on the amazing resiliency of children as they transition across different family contexts. And, as noted above, the psychological well-being of the parents who have gone through divorce/remarriage is predictive of a positive outcome for the adjustment of the children in the new stepfamily structure (Willetts and Maroules, 2005). A specific benefit for children, particularly nonminority children, from the remarriage or cohabitation of their parents is economic (Manning and Brown, 2006).

Stepchildren have viewpoints and must make adjustments of their own when their parents remarry. They have experienced the transition from living in a family with both biological parents to either living alone with one parent (usually the mother) or alternating between the homes of both parents. When their parents remarry, they must adjust again to living with two new sets of stepparents/stepsiblings.

"The biggest source of problems for kids in stepfamilies is parental conflict left over from the first marriage" (Rutter, 1994, p. 32). "Study after study shows that divorce and remarriage do not harm children—parental conflict does" (p. 33). Children caught in the conflict between their parents find it impossible to be loyal to both parents. The child temporarily resolves the conflict by siding with one parent at the expense of the relationship with the other parent. No one wins—the child feels bad for abandoning a loving parent, the parent who has been tossed aside feels deprived of the opportunity for a close parent-child relationship, and the custodial parent runs the risk that the children, as adults, may resent being prevented from developing or continuing a relationship with the other parent.

Divided Loyalties, Discipline, Stepsiblings

Other problems experienced by children in stepfamilies often revolve around feeling abandoned, having divided loyalties, discipline, and stepsiblings. Some stepchildren feel that they have been abandoned twice—once when their parents got divorced and again when the parents turned their attention to their new marital partners. One adolescent explained:

> It hurt me when my parents got divorced and my dad moved out. I really missed him and felt he really didn't care about me. But my sister and me adjusted with just my mom, and when everything was going right again, she got involved with this new guy and we were left with baby-sitters all the time. My dad also got involved with a new woman. I feel my sister and I have lost both parents in two years.

Coping with feelings of abandonment is not easy. It is best if the parents assure the children that the divorce was not their fault and that they are loved a great deal by both parents. In addition, the parents should be careful to find a balance between spending time with their new partner and spending time with their children. This translates into spending some alone time with their children.

Some children experience abandonment yet again if their parents' second marriage ends in divorce. They may have established a close relationship with the new stepparent only to find the relationship disrupted. Relationships with the stepgrandparents may also become strained. Stepgrandparents may be an enormous source of emotional support for the child, but a divorce can interrupt this support.

I've discovered that stepfamilies are kind of like salad dressing. You can shake us up and we blend really well for moments, days, months even. After a while, though, we have a tendency to settle back into original family lines and loyalties.

Natalie Nichols Gillespie, author of *Stepfamily Survival*

Divided loyalties represent another issue children must deal with in stepfamilies. Sometimes children develop an attachment to a stepparent that is more positive than the relationship with the natural parent of the same sex. When these feelings develop, the children may feel they are in a bind. One adolescent boy explained:

My real dad left my mother when I was 6, and my mom remarried. My stepdad has always been good to me, and I really prefer to be with him. When my dad comes to pick me up on weekends, I have to avoid talking about my stepdad because my dad doesn't like him. I guess I love my dad, but I have a better relationship with my stepdad.

For some adolescents, the more they care for the stepparent, the more guilty they feel, so they may try to hide their attachment. The stepparent may be aware of both positive and negative feelings coming from the child. Ideally, both the biological parent and the stepparent should encourage the child to have a close relationship with the other parent.

Discipline is another issue for stepchildren. "Adjusting to living with a new set of rules from your stepparent," "accepting discipline from a stepparent," and "dealing with the expectations of your stepparent" are situations that 80 percent of more than a hundred adolescents in stepfamilies said they had experienced (Lutz, 1983). At least two studies have identified the stepmother as the more difficult stepparent to adjust to from the child's point of view (Fine and Kurdek, 1992). This is probably because the woman is more often in the role of the active parent, which increases the potential for conflict with the child. There are also data that stepfather families monitor children less than stepmother families, so the opportunity for conflict with stepfathers may be less (Fisher et al., 2003). Therapists who work with stepfamilies encourage stepparents to discipline *their own* children only.

Siblings can also be a problem for stepchildren as they compete for parental approval, space, and TV access. The children who are already in the house may feel imposed upon and threatened. The entering children may feel out of place and that they do not belong.

Moving between Households

Other issues unique to stepchildren include moving back and forth between households. Though sometimes this is a structured transition, it may also be a mechanism used by the child to manipulate the parents. In effect they communicate to each parent that if they don't get their way, they will move in with the other parent. Parents in conflict with each other are easy prey for this destructive ploy. They should be alert to manipulation and avoid it.

Ambiguity of the Extended Family

A final issue for children in stepfamilies is their ambiguous place in the extended family system of the new stepparent. While some parents and siblings of the new stepparent welcome the new stepchild into the extended family system, others may ignore the child since they are busy enough with their own grandchildren and children, respectively. In other cases, the biological parent may socialize the children to not accept the new stepgrandparents and stepaunts/uncles: "My mom told us not to be fooled by dad's new wife's parents and siblings— that they really didn't care about us. I now know that she was just trying to get back at my dad. She knew that our not being open to a new family would make him sad. In effect my mom deprived us of getting to know some really great people." Stewart (2005) noted that boundary ambiguity, who is regarded as in or out of the family, is much more prevalent in stepfamilies.

Stepfamily living is often difficult for everyone involved: remarried spouses, children, even ex-spouses, grandparents, and in-laws. Though some of the problems begin to level out after a few years, others may take longer. Many couples become impatient with the unanticipated problems that are slow to abate, and they divorce.

Developmental Tasks for Stepfamilies

A **developmental task** is a skill that, if mastered, allows the family to grow as a cohesive unit. Developmental tasks that are not mastered will edge the family closer to the point of disintegration. Some of the more important developmental tasks for stepfamilies are discussed in this section.

Acknowledge Losses and Changes

As noted earlier, each stepfamily member has experienced the loss of a spouse or a biological parent in the home.

These are losses of an attachment figure and are significant (Marano, 2000). These losses are sometimes compounded by home, school, neighborhood, and job changes. Feelings about these losses and changes should be acknowledged as important and consequential. In addition, children should not be required to love their new stepparent or stepsiblings (and vice versa). Such feelings will develop only as a consequence of positive interaction over an extended period of time.

Preserving original relationships is helpful in reducing the child's grief over loss. It is sometimes helpful for the biological parent and child to take time to nurture their relationship apart from stepfamily activities. This will reduce the child's sense of loss and any feelings of jealousy toward new stepsiblings.

Nurture the New Marriage Relationship

It is critical to the healthy functioning of a new stepfamily that the new spouses nurture each other and form a strong unit. Indeed, the adult dyad is vulnerable since the couple had no childfree time upon which to build a common base (Pacey, 2005). Once a couple develop a core relationship, they can communicate, cooperate, and compromise with regard to the various issues in their new blended family. Too often spouses become child-focused and neglect the relationship on which the rest of the family depends. Such nurturing translates into spending time alone with each other, sharing each other's lives, and having fun with each other. One remarried couple goes out to dinner at least once a week without the children. "If you don't spend time alone with your spouse, you won't have one," says one stepparent. Two researchers studied 115 stepfamily couples and found that the husbands tended to rank "spouse" as their number one role (over parent or employee), whereas wives tended to rank "parent" as their number one role (Degarmo and Forgatch, 2002).

Integrate the Stepfather into the Child's Life

Stepfathers who become interested in what their stepchild does and who spend time alone with the stepchild report greater integration into the life of the stepchild and the stepfamily. The benefits to both the father and the stepchild in terms of emotional bonding are enormous. In addition, the mother of the child feels closer to her new husband if he has bonded with her offspring.

Allow Time for Relationship between Partner and Children to Develop

In an effort to escape single parenthood and to live with one's beloved, some individuals rush into remarriage without getting to know each other. Not only do they have limited information about each other, but their respective children may have spent little or no time with their future stepparent. One stepdaughter

The principal challenge of stepfamily life is building an emotionally satisfying marriage.

James Bray and John Kelly,
Stepfamilies

remarked, "I came home one afternoon to find a bunch of plastic bags in the living room with my soon-to-be-stepdad's clothes in them. I had no idea he was moving in. It hasn't been easy." Both adults and children should have had meals together and spent some time in the same house before becoming bonded by marriage as a family.

As noted earlier, stepparents are encouraged to discipline their own children. The rationale for this is that there has usually been insufficient time for stepparent and stepchild to experience a relationship that allows for discipline from the stepparent.

Have Realistic Expectations

Because of the complexity of meshing the numerous relationships involved in a stepfamily, it is important to be realistic. Dreams of one big happy family often set up stepparents for disappointment, bitterness, jealousy, and guilt. As noted earlier, stepfamily members do not begin to feel comfortable with each other until the third year (Bray and Kelly, 1998). Just as nuclear and single-parent families do not always run smoothly, neither do stepfamilies.

Accept Your Stepchildren

Rather than wishing your stepchildren were different, it is more productive to accept them. All children have positive qualities; find them and make them the focus of your thinking. Stepparents may communicate acceptance of their stepchildren through verbal praise and positive or affectionate statements and gestures. In addition, stepparents may communicate acceptance by engaging in pleasurable activities with their stepchildren and participating in daily activities such as homework, bedtime preparation, and transportation to after-school activities.

Funder (1991) studied 313 parents who had been separated for 5 to 8 years and who had become involved with new partners. In general, the new partners were very willing to be involved in the parenting of their new spouses' children. Such involvement was highest when the children lived in the household.

Establish Your Own Family Rituals

One of the bonding elements of nuclear families is its rituals. Stepfamilies may integrate the various family members by establishing common rituals, such as summer vacations, visits to and from extended kin, and religious celebrations. These rituals are most effective if they are new and unique, not mirrors of rituals in the previous marriages and families.

Decide about Money

Money is an issue of potential conflict in stepfamilies because it is a scarce resource and several people want to use it for their respective needs. The father wants a new computer; the mother wants a new car; the mother's children want bunk beds, dance lessons, and a satellite dish; the father's children want a larger room, clothes, and a phone. How do the newly married couple and their children decide how money should be spent?

Some stepfamilies put all their resources into one bank and draw out money as necessary without regard for whose money it is or for whose child the money is being spent. Others keep their money separate; the parents have separate incomes and spend them on their respective biological children. Although no one pattern is superior to the other, it is important for remarried spouses to agree on whatever financial arrangements they live by.

In addition to deciding how to allocate resources fairly in a stepfamily, remarried couples may face decisions regarding sending the children and stepchildren to college. Remarried couples may also be making a will that is fair to all family members.

Give Parental Authority to Your Spouse

How much authority the stepparent will exercise over the children should be discussed by the adults before they get married. Some couples divide the authority—each spouse disciplining his or her own children. This is the strategy we recommend. The downside is that stepchildren may test the stepparent in such an arrangement when the biological parent is not around. One stepmother said, "Joe's kids are wild when he isn't here because I'm not supposed to discipline them."

Support Child's Relationship with Absent Parent

A continued relationship with both biological parents is critical to the emotional well-being of the child. Ex-spouses and stepparents should encourage children to have a positive relationship with both biological parents. Respect should also be shown for the biological parent's values. Bray and Kelly (1998) note that this is "particularly difficult" to exercise. "But asking a child about an absent parent's policy on movies or curfews shows the child that, despite their differences, his/her mother and father still respect each other" (p. 92).

Cooperate with the Child's Biological Parent and Coparent

A cooperative, supportive, and amicable co-parenting relationship between the biological and stepparents is a win-win for the children and parents. Otherwise, children are continually caught between the crossfire of the conflicted parental sets.

Support Child's Relationship with Grandparents

It is important to support children's continued relationships with their natural grandparents on both sides of the family. This is one of the more stable relationships in the child's changing world of adult relationships. Regardless of how ex-spouses feel about their former in-laws, they should encourage their children to have positive feelings for their grandparents. One mother said, "Although I am uncomfortable around my ex-in-laws, I know they are good to my children, so I encourage and support my children spending time with them."

Anticipate Great Diversity

Stepfamilies are as diverse as nuclear families. It is important to let each family develop its own uniqueness. One remarried spouse said, "The kids were grown when the divorces and remarriages occurred, and none of the kids seem particularly interested in getting involved with the others" (Rutter, 1994, p. 68). Ganong et al. (1998) also emphasized the importance of resisting the notion that stepfamilies are not "real" families and that adoption makes them a real family. Indeed, social policies need to be developed that allow for "the establishment of some legal ties between the stepparent and stepchild without relinquishing the biological parent's legal ties" (p. 69).

SUMMARY

What is the nature of remarriage in the United States?

About 40 percent of marriages today are remarriages, and most of these involve children or the creation of stepfamilies. When the divorced who have remarried and the divorced who have not remarried are compared, the remarried report greater personal and relationship happiness. Ninety percent of remarriages are of persons who are divorced rather than widowed.

It is not uncommon for persons who are divorced to live together with a new partner before remarriage. Aside from living together, they (like most couples in courtship, whether for a first or second marriage) do little else to prepare for their new marriage. National data reflect that remarriages are more likely than first marriages to end in divorce in the early years of remarriage. But after 15 years, second marriages are more stable than first marriages.

What is the nature of stepfamilies in the United States?

Stepfamilies represent the fastest-growing type of family in the United States. A blended family is one in which the spouses in a new marriage relationship are blended with the children of at least one of the spouses from a previous marriage.

There is a movement away from the use of the term *blended,* since stepfamilies really do not blend. Although a stepfamily can be created when a never-married or a widowed parent with children marries a person with or without children, most stepfamilies today are composed of spouses who were once divorced.

Stepfamilies differ from nuclear families: the children in a nuclear family are biologically related to both parents, whereas the children in a stepfamily are biologically related to only one parent. Also, in a nuclear family, both biological parents live with their children, whereas only one biological parent in a stepfamily lives with the children. In some cases, the children alternate living with each parent.

Stepism is the assumption that stepfamilies are inferior to biological families. Stepism, like racism, heterosexism, sexism, and ageism, involves prejudice and discrimination.

What are the strengths of stepfamilies?

The strengths of stepfamilies include exposure to a variety of behavior patterns, a happier parent, and greater objectivity on the part of the stepparent.

What can women do to ease their transition into stepfamily living?

Learning to get along with the husband's children, not being resentful of his relationship with his children, and adapting to the fact that one-third to one-half of his net income may be sent to his ex-wife as alimony or child support are skills the new wife must develop. The new wife may also want children with her new partner or may bring her own children into the marriage. In the latter case, she is anxious that her new husband will accept her children.

What can men do to ease their transition into stepfamily living?

Getting along with their wife's children, paying for many of the expenses of their stepchildren, having their new partner accept their own children, and dealing with the issue of having more children are among the issues to be confronted by the new stepfather.

What are the challenges for children in stepfamilies?

Children must cope with feeling abandoned and with problems of divided loyalties, discipline, and stepsiblings. They may also move between households, and their role in the extended stepfamily may be ambiguous.

What are the developmental tasks of stepfamilies?

Developmental tasks for stepfamilies include nurturing the new marriage relationship, allowing time for partners and children to get to know each other, deciding whose money will be spent on whose children, deciding who will discipline the children and how, and supporting the child's relationship with both parents and natural grandparents. Both sets of parents and stepparents should form a parenting coalition in which they cooperate and actively participate in childrearing.

The Companion Website for Choices in Relationships: An Introduction to Marriage and the Family, ninth edition
www.thomson.edu/sociology/knox

Supplement your review of this chapter by going to the companion website to take one of the Tutorial Quizzes, use the flash cards to master key terms, and check out the many other study aids you'll find there. You'll also find special features such as the Marriage and Family Resource Center, Census 2000 information, and other data and resources at your fingertips to help you with that special project or to do some research on your own.

KEY TERMS

binuclear	December marriage	parental status	stepism
blended family	developmental task	stepfamily	step relationships

WEB LINKS

Second Wives Club
http://www.secondwivesclub.com/

Stepfamily Association of America
http://www.stepfam.org

Stepfamily Network
http://www.stepfamily.net/

Online Stepfamily Magazine
http://www.saafamilies.org/yourstepfamily/index.php

Stepmothers.org/Stepmothers International
http://www.saafamilies.org/smi/old/resources.htm

REFERENCES

Ahrons, C. 2004. *We're still family: What grown children have to say about their parents' divorce.* New York: Harper Collins.

Anderson, E. R. and S. M. Greene. 2005. Transitions in parental repartnering after divorce. *Journal of Divorce & Remarriage* 43:47–62.

Bernhardt, E. M., F. K. Goldscheider, M. L. Rogers, and H. Koball. 2002. Qualities men prefer for children in the U.S. and Sweden: Differences among biological, step and informal fathers. *Journal of Comparative Family Studies* 33:235–47.

Bray, J. H. and J. Kelly. 1998. *Stepfamilies: Love, marriage and parenting in the first decade.* New York: Broadway Books.

Christian, A. 2005. Contesting the myth of the 'Wicked stepmother': Narrative analysis of an only stepfamily support group. *Western Journal of Communication* 69:27–48.

Clarke, S. C. and B. F. Wilson. 1994. The relative stability of remarriages: A cohort approach using vital statistics. *Family Relations* 43:305–10.

Degarmo, D. S. and M. S. Forgatch. 2002. Identity salience as a moderator of psychological and marital distress in stepfather families. *Social Psychology Quarterly* 65:266–84.

DeGraaf, P. M. and M. Kalmijn. 2003. Alternative routes in the remarriage market: Competing-risk analysis of union formation after divorce. *Social Forces* 81:1459–98.

Enright, E. 2004. A house divided. *AARP The Magazine* July/August:60 et passim.

Everett, Lou. 1998. Factors that contribute to satisfaction or dissatisfaction in stepfather-stepchild relationships. *Perspectives in Psychiatric Care* 34(2):25–35.

———. 2000. Personal communication, East Carolina University, October 12.

Fischer, T. F. C., P. M. DeGraaf, and M. Kalmijn. 2005. Friendly and antagonistic contact between former spouses after divorce: Patterns and determinants. *Journal of Family Issues* 26(8, Nov):1131–63.

Fine, M. A., L. H. Ganong, and M. Coleman. 1997. The relation between role constructions and adjustment among stepfathers. *Journal of Family Issues* 18:503–25.

Fine, M. A., and L. A. Kurdek. 1992. The adjustment of adolescents in stepfather and stepmother families. *Journal of Marriage and the Family* 54:725–36.

Fisher, P. A., L. D. Leve, C. C. O'Leary, and C. Leve. 2003. Parental monitoring of children's behavior: Variation across stepmother, stepfather, and two-parent biological families. *Family Relations* 52:45–52.

Funder, K. 1991. New partners as co-parents. *Family Matters* April:44–46.

Furukawa, S. 1994. The diverse living arrangements of children: Summer 1991. *Current Population Reports*, Series P70, no. 38. Washington, DC: U.S. Bureau of the Census.

Ganong, L. H., and M. Coleman. 1994. *Remarried family relationships.* Thousand Oaks, CA: Sage.

Ganong, L. H., M. Coleman, M. Fine, and A. K. McDaniel. 1998. Issues considered in contemplating stepchild adoption. *Family Relations* 47:63–71.

Goetting, A. 1982. The six stations of remarriage: The developmental tasks of remarriage after divorce. *The Family Coordinator* 31:213–22.

Goldscheider, F. and S. Sassler. 2006. Creating stepfamilies: Integrating children into the study of union formation. *Journal of Marriage and the Family* 68:275–91.

Henry, C. S, J. P. Nichols, L. C. Robinson, and R. A. Neal. 2005. Parent and stepparent support and psychological control in remarried families and adolescent empathic concern. *Journal of Divorce and Remarriage* 43:29–46.

Hofferth, S. L. and K. G. Anderson. 2002. All dads are equal? Biology versus marriage as a basis for paternal investment. *Journal of Marriage and the Family* 65:213–32.

Knox, D. and M. E. Zusman. 2001. Marrying a man with "baggage": Implications for second wives. *Journal of Divorce and Remarriage* 35:67–80.

Knox, D. and Zusman, M. E. 2006. Relationship and sexual behaviors of a sample of 1027 university students. Unpublished data collected for this text. Department of Sociology, East Carolina University, Greenville, NC.

Kyungok, Huh. 1994. Father's child care time across family types. In *Families and justice: From neighborhoods to nations.* Proceedings, Annual Conference of the National Council on Family Relations, 4:22.

Leon, K. and E. Angst. 2005. Portrayals of stepfamilies in film: Using media images in remarriage education. *Family Relations* 54:3–23.

Lutz, P. 1983. The stepfamily: An adolescent perspective. *Family Relations* 32:367–75.

MacDonald, W., and A. DeMaris. 1996. Parenting stepchildren and biological children: The effects of stepparent's gender and new biological children. *Journal of Family Issues* 17:5–25.

———. 2002. Stepfather-stepchild relationship quality. *Journal of Family Issues* 23:121–37.

Mackay, R. 2005. The impact of family structure and family change on child outcomes: a personal reading of the research literature. *Social Policy Journal of New Zealand.* March 111–34.

Malia, S. E. C. Balancing family members' interests regarding stepparent rights and obligations: a social policy challenge. *Family Relations* 54:298–320.

Manning, W. D. and S. Brown. 2006. Children's economic well-being in married and cohabiting parent families. *Journal of Marriage and the Family* 68:345–62.

Marano, H. E. 2000. Divorced? Don't even think of remarrying until you read this. *Psychology Today* March/April, 56–64.

Marsiglio, W. 1992. Stepfathers with minor children living at home. *Journal of Family Issues* 13:195–214.

Masheter, C. 1999. Examples of commitment in postdivorce relationships between spouses. In *Handbook of interpersonal commitment and relationship stability,* edited by J. M. Adams and W. H. Jones. New York: Academic/Plenum Publishers, 293–306.

Michaels, M. L. 2000. The stepfamily enrichment program: A preliminary evaluation using focus groups. *American Journal of Family Therapy* 28:61–73.

Pacey, S. 2005. Step change: the interplay of sexual and parenting problems when couples form stepfamilies. *Sexual & Relationship Therapy* 20:359–369.

Palisi, B. J., M. Orleans, D. Caddell, and B. Korn. 1991. Adjustment to stepfatherhood: The effects of marital history and relations with children. *Journal of Divorce and Remarriage* 14:89–106.

Papernow, P. L. 1988. Stepparent role development: From outsider to intimate. In *Relative strangers,* edited by William R. Beer. Lanham, MD: Rowman and Littlefield, 54–82.

Pasley, K. 2000. Stepfamilies doing well despite challenges. *National Council on Family Relations Report* 45:6–7.

Ruschena, E., M. Prior, A. Sanson, and D. Smart. 2005. A longitudinal study of adolescent adjustment following family transitions. *Journal of Child Psychology & Psychiatry & Allied Disciplines* 46:353–63.

Rutter, Virginia. 1994. Lessons from stepfamilies. *Psychology Today* May/June 27:30 passim.

Stepfamily Myths. Retrieved September 5, 2005 (http://www.saafamilies.org/faqs/myths.htm).

Stewart, S. D. 2005. How the birth of a child affects involvement with stepchildren. *Journal of Marriage and the Family* 67:461–73.

Stewart, S. D. 2005. Boundary ambiguity in stepfamilies. *Journal of Family Issues* 26:1002–29.

Swallow, W. 2004. *The Triumph of Love over Experience: A Memoir of Remarriage.* New York: Hyperion/Theia.

Sweeney, M. M. 2002. Remarriage and the nature of divorce: Does it matter which spouse chose to leave? *Journal of Family Issues* 23:410–40.

Timmer, S. G., and J. Veroff. 2000. Family ties and the discontinuity of divorce in black and white newlywed couples. *Journal of Marriage and the Family* 62:349–61.

Vinick, B. 1978. Remarriage in old age. *The Family Coordinator* 27:359–63.

Willetts, M. C. and N. G. Maroules. 2005. Parental reports of adolescent well-being: Does marital status matter? *Journal of Divorce and Remarriage* 43:129–49.

Wineberg, H. 1992. Childbearing and dissolution of the second marriage. *Journal of Marriage and the Family* 54:879–87.

Xu, X.H., C. D. Hudspeth, and J. P. Bartkowski. 2006. The role of cohabitation in remarriage. *Journal of Marriage and the Family* 68:261–74.

chapter 17

Age is a question of mind over matter. If you don't mind, it doesn't matter.

Satchel Paige, baseball player

Aging in Marriage and Family Relationships

Contents

E. Fred Johnson, Jr.

True or False?

1. Elderly men tend to report higher marital happiness than elderly women.

2. The process of death, rather than death itself, is the primary fear of the elderly.

3. Good physical health is the primary determinant of an elderly person's reported happiness.

4. "Mastery/competence" is the way elderly grandparents describe the experience of taking care of grandchildren.

5. For the elderly, attitudes about sex are a more significant influence on sexual desire than biomedical factors.

Answers: **1.** T **2.** T **3.** T **4.** F **5.** T

A s a person gets older he or she is subject to jokes and jabs about being "over the hill," "an old coot," or "an old biddy." Adult birthday cards mock the elderly's mobility, memory, and sex drive. But the reality of aging is no joke. Contemporary concerns such as the cost of even a brief stay in the hospital, the cost of prescription medication, and care for the elderly in a retirement or long-term-care facility are a major concern for the elderly in U.S. society. Although most couples getting married today rarely give a thought to their own aging, the care of their aging parents is an issue that looms ahead in their future. In this chapter, we focus on the factors that confront couples as they age and the dilemma of how to care for their aging parents. We begin by looking at the concept of age.

Turned off by shuffleboard and managed care communities in the sun, this generation longs to discover new passions and new careers, new adventures and new friends.

Ken Dychtwald and Daniel J. Kadlec,
The Power Years

Age and Ageism

All societies have a way to categorize their members by age. And all societies provide social definitions for particular ages.

The Concept of Age

A person's **age** may be defined chronologically, physiologically, psychologically, sociologically, and culturally. Chronologically, an "old" person is defined as one who has lived a certain number of years. How many years it takes to be regarded as old varies with one's own age. A child of 12 may regard a sibling of 18 as old—and his or her parents as "ancient." But the teenagers and the parents may regard themselves as "young" and reserve the label "old" for their grandparents' generation.

Chronological age has obvious practical significance in everyday life. Bureaucratic organizations and social programs identify chronological age as a criterion of certain social rights and responsibilities. One's age determines the right to drive, vote, buy alcohol or cigarettes, and receive Social Security and Medicare benefits.

This couple is in their 80s, and they have been married over 60 years (they met on a train); they meet the cultural definition of being old.

Lucy Johnson

Age also determines retirement. Federal law requires airline pilots to retire at age 65, the bureaucratic chronological definition of "old." In ancient Greece and Rome, where the average life expectancy was 20 years, one was old at 18; similarly, one was old at 30 in medieval Europe and at age 40 in the United States in 1850. In the United States today, however, people are usually not considered old until they reach age 65. Current life expectancy is shown in Table 17.1.

Our society is moving toward new chronological definitions of "old." Three groups of the elderly are the "young-old," the "middle-old," and the "old-old." The young-old are typically between the ages of 65 and 74; the middle-old, 75 to 84, and the old-old, 85 and beyond. Individuals aged 18 to 35 identify 50 as the age when the average man or woman becomes "old." However, those between the ages of 65 and 74 define "old" as 80 (Cutler, 2002). Tanner (2005) noted that views of the aging are changing from needy and dependent to active and resourceful. Kelley-Moore et al. (2006) noted that the elderly begin to see themselves as disabled when their driver's licenses are taken away and when home health care workers begin to come into their homes to help take care of them. This finding comes from a study of a large sample of older adults 72 and older.

Research is underway to extend life (see websites under Life Extension at the end of the chapter). Dr. Aubrey de Grey (Department of Genetics, University of Cambridge) predicts that continued research will make possible the ability to add HUNDREDS of years to one's life. Mercer (2006) notes that the most optimistic predictions are that we are two or three decades away from significant breakthroughs (Ray Kurzweil says we are 49 years away) but notes the paths are known via genetic engineering (we must learn how to switch off the genes responsible for aging), tissue/organ replacement (knees and kidneys are already being replaced) and the merging of computer technology with human biology (hearing and seeing are now being improved). Should the human lifespan be extended hundreds of years, imagine the impact on marriage—"till death do us part"?

Ballard and Morris (2003) found that interest in family life education topics varies by age, with those 85 and beyond least likely to be interested in family and relationship issues. Where an interest does exist (widowhood was of particular interest to this age group), print material in the form of newsletters, brochures, and self-help books were most preferred (Morris and Ballard, 2003).

Some individuals seem to have been successful in delaying the aging process. Frank Lloyd Wright, the famous architect, enjoyed the most productive period of his life between 80 and 92. George Burns and Bob Hope were active into their 90s. Boulton-Lewis et al. (2006) found that continuing to learn was associated with the elderly continuing to feel healthy.

Physiologically, a person is old when his or her auditory, visual, respiratory, and cognitive capabilities decline significantly. Individuals also decline in strength, with older women being weaker than older men (Musselman and Brouwer, 2005). Persons who need full-time nursing care for eating, bathing, and taking medication properly and who are placed in nursing homes are thought of as being old. Failing health is the criterion used by the elderly to define them-

The young know the rules, but the old know the exceptions.

Oliver Wendell Holmes, American jurist

Table 17.1	Life Expectancy			
Year	Caucasoid Males	African-American Males	Caucasoid Females	African-American Females
2000	74.9	68.3	80.1	75.2
2005	75.4	69.9	81.1	76.8
2010	76.1	70.9	81.8	77.8

Source: *Statistical Abstract of the United States: 2006.* 125th ed. Washington, DC: U.S. Bureau of the Census, 2006, Table 96.

selves as old (O'Reilly, 1997), and successful aging is typically defined as maintaining one's health, independence, and cognitive ability. Jorm et al. (1998) observed that the prevalence of successful aging declines steeply from age 70 to 74 to age 80.

Persons who have certain diseases are also regarded as old. Although younger individuals may suffer from Alzheimer's, arthritis, and heart problems, these ailments are more often associated with aging. As medical science conquers more diseases, the physiological definition of aging changes so that it takes longer for people to be defined as "old."

Psychologically, a person's self-concept is important in defining how old that person is. As individuals begin to fulfill the roles associated with the elderly—retiree, grandparent, nursing home resident—they begin to see themselves as aging. Sociologically, once they occupy these roles, others begin to see them as "old." Barrett (2005) analyzed national data of 2681 mid-life respondents to assess how women and men differed in their age identity. Since women were typically pair-bonded with older partners, they typically had younger age identities than men.

Culturally, the society in which an individual lives defines when and if a person becomes old and what being old means. In U.S. society, the period from age 18 through 64 is generally subdivided into young adulthood, adulthood, and middle age. Cultures also differ in terms of how they view and take care of their elderly. Spain is particularly noteworthy in terms of care for the elderly, with eight of 10 elderly persons receiving care from family members and other relatives. The elderly in Spain report very high levels of satisfaction in the relationships with their children, grandchildren, and friends (Fernandez-Ballesteros, 2003).

A person is not old until regrets take the place of dreams.

John Barrymore, actor

Perceptions of the Elderly by College Students

A team of researchers (Knox et al., 2005) examined how 441 undergraduates viewed the elderly and found gender differences in their perceptions. Significant differences included the age at which a person becomes "old" (men selected a younger age than women [58 vs. 62]), strength (men saw less decline in strength), reaction time (women saw less decline in reaction time), and the perception of the elderly as "dangerous" drivers (women saw the elderly as less dangerous). Analyzing the same data set, Kimuna et al. (2005) reported that the older the student, the more likely the student was to select a higher age for that at which a person becomes "old." For example, a first-year student (age 18) might think old is 55, whereas a senior (age 21) might think old begins at 60. In addition, Kimuna et al. (2005) found that the more exposure a student had to the elderly, the more positive the perception.

Ageism

Every society has some form of **ageism**—the systematic persecution and degradation of people because they are old. Ageism is similar to sexism, racism, and heterosexism. The elderly are shunned, discriminated against in employment, and sometimes victims of abuse. Media portrayals contribute to the negative image of the elderly. They are portrayed as difficult, complaining, and burdensome and are often underrepresented in commercials and comic strips.

Negative stereotypes and media images of the elderly engender **gerontophobia**—a shared fear or dread of the elderly, which may create a self-fulfilling prophecy. For example, an elderly person forgets something and attributes his or her behavior to age. A younger person, however, engaging in the same behavior, is unlikely to attribute forgetfulness to age, given cultural definitions surrounding the age of the onset of senility.

The negative meanings associated with aging underlie the obsession of many Americans to conceal their age by altering their appearance. With the hope of

The trick is to grow up without growing old.

Frank Lloyd Wright, architect

holding on to youth a little bit longer, aging Americans spend millions of dollars each year on exercise equipment, hair products, facial creams, and plastic surgery.

The latest attempt to reset the aging clock is to have regular injections of human growth hormone (HGH), which promises to lower blood pressure, build muscles without extra exercise, increase the skin's elasticity, thicken hair,

Table 17.2 Theories of Aging

Name of Theory	Level of Theory	Theorists	Basic Assumptions	Criticisms
Disengagement	Macro	Elaine Cumming William Henry	The gradual and mutual withdrawal of the elderly and society from each other is a natural process. It is also necessary and functional for society that the elderly disengage so that new people can be phased in to replace them in an orderly transition.	Not all people want to disengage; some want to stay active and involved. Disengagement does not specify what happens when the elderly stay involved.
Activity	Macro	Robert Havighurst	People continue the level of activity they had in middle age into their later years. Though high levels of activity are unrelated to living longer, they are related to reporting high levels of life satisfaction.	Ill health may force people to curtail their level of activity. The older a person, the more likely the person is to curtail activity.
Conflict	Macro	Karl Marx Max Weber	The elderly compete with youth for jobs and social resources such as government programs (Medicare).	The elderly are presented as disadvantaged. Their power to organize and mobilize political resources such as the American Association of Retired Persons is underestimated.
Age stratification	Macro	M. W. Riley	The elderly represent a powerful cohort of individuals passing through the social system that both affect and are affected by social change.	Too much emphasis is put on age, and little recognition is given to other variables within a cohort such as gender, race, and socioeconomic differences.
Modernization	Macro	Donald Cowgill	The status of the elderly is in reference to the evolution of the society toward modernization. The elderly in premodern societies have more status because what they have to offer in the form of cultural wisdom is more valued. The elderly in modern technologically advanced societies have low status since they have little to offer.	Cultural values for the elderly, not level of modernization, dictate the status of the elderly. Japan has high respect for the elderly and yet is highly technological and modernized.
Symbolic interaction	Micro	Arlie Hochschild	The elderly socially construct meaning in their interactions with others and society. Developing social bonds with other elderly can ward off being isolated and abandoned. Meaning is in the interpretation, not in the event.	The power of the larger social system and larger social structures to affect the lives of the elderly is minimized.
Continuity	Micro	Bernice Neugarten	The earlier habit patterns, values, attitudes of the individual are carried forward as the person ages. The only personality change that occurs with aging is the tendency to turn one's attention and interest on the self.	Other factors than one's personality affect aging outcomes. The social structure influences the life of the elderly rather than vice versa.

heighten sexual potency, etc. It is part of the regimen of clinics such as Lifespan (Beverly Hills), which costs $1,000 a month after an initial workup of $5,000. In the absence of long-term data, many physicians remain skeptical—there is no fountain of youth.

Theories of Aging

Gerontology is the study of aging. Table 17.2 on page 484 identifies several theories, the level (macro or micro) of the theory, the theorists typically associated with the theory, assumptions, and criticisms. As noted, there are diverse ways of conceptualizing the elderly. Currently in vogue are the age-stratification and life-course perspectives (Mitchell, 2003).

Caregiving for the Frail Elderly— The "Sandwich Generation"

An elderly person is defined as **frail** if he or she has difficulty with at least one personal care activity or other activity related to independent living; the severely disabled are unable to complete three or more personal care activities. These personal care activities include bathing, dressing, getting in and out of bed, shopping for groceries, and taking medications. About 6.1 percent of the U.S. adult population over the age of 65, or 2.0 million people, are defined as being severely disabled and are not living in nursing homes. Only 6.8 percent of these frail elderly have long-term health-care insurance (Johnson and Wiener, 2006). Hence, the bulk of their care falls to the children of these elderly. And the term "children" typically means female children. Indeed, women account for about two-thirds of all unpaid caregivers (Johnson and Wiener, 2006). The adults who provide **family caregiving** to these elderly parents (and their own children simultaneously) are known as the "sandwich generation" since they are in the middle of taking care of the needs of both parents and children.

The typical caregiver is a middle-aged married woman who works outside the home. High levels of stress and fatigue may accompany caring for one's elders. Martire and Stephens (2003) noted even higher levels of fatigue and competing demands among women who were both employed and caring for an aging parent. Fifty-five percent of a sample of women at midlife (most of whom had children) reported that they were providing care to their mothers; 34 percent were caring for their fathers (Peterson, 2002). The number of individuals in the sandwich generation will increase for the following reasons:

1. Longevity. The over-85 age group, the segment of the population most in need of care, is the fastest-growing segment of our population.

2. Chronic disease. In the past, diseases took the elderly quickly. Today, diseases such as arthritis and Alzheimer's are associated with not an immediate death sentence but a lifetime of managing the illness and being cared for by others. Family caregivers of parents with Alzheimer's note the difficulty of the role. "He's not the man I married," lamented one wife.

3. Fewer siblings to help. The current generation of elderly had fewer children than the elderly in previous generations. Hence, the number of adult siblings to help look after parents is more limited. Only children are more likely to feel the weight of caring for elderly parents alone.

4. Commitment to parental care. Contrary to the myth that adult children in the United States abrogate responsibility for taking care of their elderly parents, most children institutionalize their parents only as a last resort. Furthermore, Wells (2000) identified some benefits of caregiving, including a closer relationship to the dependent person and a feeling of enhanced self-esteem. Most of Peterson's (2002) sample of women caring for their parents did not view doing

Because this woman's mother and brother were in a nursing home, she felt comfortable moving to one. In this photo, the daughter-in-law (who lives out of state) is visiting her.

Diversity in Other Countries

Whereas female children in the United States have the most frequent contact and are more involved in the caregiving of their elderly parents than male children, it is the daughter-in-law in Japan who offers the most help to elderly individuals. The female child who is married gives her attention to the parents of her husband (Ikegami, 1998). Eastern cultures emphasize filial piety, which is love and respect toward their parents. **Filial piety** involves respecting parents, bringing no dishonor to parents, and taking good care of parents (Jang and Detzner, 1998). Western cultures are characterized by **filial responsibility** emphasizing duty, protection, care, and financial support.

so as a burden. Asian children, specifically Chinese children, are socialized to expect to take care of their elderly. Zhan and Montgomery (2003) observed, "Children were raised for the security of old age" (p. 209).

5. Lack of support for the caregiver. Caring for a dependent, aging parent requires a great deal of effort, sacrifice, and decision-making on the part of more than 14 million adults in the United States who are challenged with this situation. The emotional toll on the caregiver may be heavy. Guilt (over not doing enough), resentment (over feeling burdened), and exhaustion (over the relentless care demands) are common feelings that are sometimes mixed. One caregiver adult child said, "I must be an awful person to begrudge taking my mother supper, but I feel that my life is consumed by the demands she makes on me, and I have no time for myself, my children, or my husband." Marks et al. (2002) noted an increase in symptoms of depression among a national sample of caregivers (of a child, parent, or spouse). Older caregivers also risk their own health in caring for their aging parents (Wallsten, 2000). Caregiving can also be expensive and can devastate the family budget.

Some reduce the strain of caring for an elderly parent by arranging for home health care. This involves having a nurse go to the home of the parent and provide such services as bathing the parent and giving medication. Other services may include taking meals to the elderly (e.g., through Meals on Wheels). The National Family Caregiver Support Program, enacted in 2000, provides support services for individuals (including grandparents) who provide family caregiving services. Such services might include elder-care resource and referral services, caregiver support groups, and classes on how to care for an aging parent. In addition, increasingly, states are providing family caregivers a tax credit or deduction.

Offspring who have no help may become overwhelmed and frustrated. Elder abuse, an expression of such frustration, is not unheard of (we discussed this in the chapter on abuse). Many wrestle with the decision to put their parents in a nursing home or other long-term care facility. We discuss this issue in the Personal Choices section.

PERSONAL CHOICES

Should I Put My Parents in a Long-Term-Care Facility?
Over 1.5 million individuals (almost three times as many women as men) over the age of 65 are in a nursing home (Statistical *Abstract of the United States: 2006*, Table 68). Factors relevant in deciding whether to care for an elderly parent at home, arrange for nursing home care, or provide another form of long-term care include the following.

1. Level of care needed. As parents age, the level of care that they need increases.
An elderly parent who cannot bathe, dress, prepare meals, or be depended on to take medication responsibly needs either full-time in-home care or a skilled nursing facility that provides 24-hour nursing supervision by registered or licensed vocational

nurses. Commonly referred to as "nursing homes" or "convalescent hospitals," these facilities provide medical, nursing, dietary, pharmacy, and activity services.

An intermediate-care facility provides eight hours of nursing supervision per day. Intermediate care is less extensive and expensive and generally serves patients who are ambulatory and who do not need care throughout the night.

A skilled nursing facility for special disabilities provides a "protective" or "security" environment to persons with mental disabilities. Many of these facilities have locked areas where patients reside for their own protection.

An assisted living facility is for individuals who are no longer able to live independently but who do not need the level of care that a nursing home provides. Although nurses and other health care providers are available, assistance is more typically in the form of meals and housekeeping.

Retirement communities involve a range of options, from apartments where residents live independently to skilled nursing care. These communities allow older adults to remain in one place and still receive the care they need as they age.

2. Temperament of parent. Some elderly parents have become paranoid, accusatory, and angry with their caregivers. Family members no longer capable of coping with the abuse may arrange for their parents to be taken care of in a nursing home or other facility.

3. Philosophy of adult child. Most children feel a sense of filial responsibility—a sense of personal obligation for the well-being of aging parents. Theoretical explanations for such responsibility include the norm of reciprocity (adult children reciprocate the care they received from their parents), attachment theory (caring results from positive emotions for one's parents), and a moral imperative (caring for one's elderly parents is the right thing to do).

One only child promised his dying father that he would take care of the father's spouse (the child's mother) and not put her in a nursing home. When the mother became 82 and unable to care for herself, the son bought a bed and made the living room of his home the place for his mother to spend the last days of her life. "No nursing home for my mother," he said. This man also had the same philosophy for his mother-in-law and moved her into his home when she was 94.

Whereas some adult children have the philosophy that they must care for their elderly parents themselves, others prefer to hire the help needed. Their philosophy, in combination with the amount of care needed, the amount of help from other family members, and the commitment to other responsibilities (e.g., work), influences their decision.

A crisis in care for the elderly may be looming. In the past, women have taken care of their elderly parents. But these were women who lived in traditional families where one paycheck took care of a family's economic needs. Women today work out of economic necessity, and quitting work to take care of an elderly parent is becoming less of an option. As more women enter the labor force, less free labor is available to take care of the elderly. Government programs are not in place to take care of the legions of elderly Americans. Who will care for them when both spouses are working full-time?

4. Length of time for providing care. Offspring must also consider how long they will be in the role of caring for an aging parent. The duration of caregiving can last from less than a year to more than 40 years. About 40% of caregivers provide assistance for 5 or more years; nearly a fifth for 10 or more years (Family Caregiver Alliance, 2006).

5. Privacy needs of caregivers. Some spouses take care of their elderly at home but note the effect on their own marital privacy. A wife who took care of her husband's mother for 12 years in her home said that it was a relief for his mother to die and for them to get their privacy back. Chadiha, Rafferty, & Pickard (2003) found that lower levels of caretaking burden were associated with higher levels of marital functioning. The impact on one's marriage needs to be considered.

6. Cost. For private full-time nursing home care, including room, board, medical care, etc., count on spending $1,000 to $1,500 a week. **Medicare,** a federal health in-

continued

The best we can do—and it is a lot—is to accept the inevitability of aging and try to adapt to it to be in the best health we can at any age.

Andrew Weil, aging guru

surance program for persons 65 and older, was developed for short-term acute hospital care. Medicare generally does not pay for long-term nursing care. In practice, adult children who arrange for their aging parent to be cared for in a nursing home end up paying for it out of the elder's own funds. After all of these economic resources are depleted, **Medicaid,** a state welfare program for low-income individuals, will pay for the cost of care. A federal law prohibits offspring from shifting the assets of an elderly parent so as to become eligible for Medicaid.

7. Sexual orientation. Homosexual elders may be resistant to going to a nursing home because they fear prejudice and discrimination from workers and patients at the facility. Some feel they have to "go back in the closet" (Gallanis, 2002).

The elderly should be included in the decision to be cared for in a long-term-care facility. Wielink et al. (1997) found that the more frail the elderly person, the more willing the person was to go to a nursing home. Once a decision is made for nursing home care, it is important to assess several facilities. Taking a tour of the facility, eating a meal at the facility, and meeting staff are also helpful in making a decision and a smooth transition. Since there is an acute nursing shortage, it is important to find out the ratio of registered nurses per patient. Some facilities hide their number of registered nurses by lumping them with those without training in a category called "nursing staff" (Jacoby, 2003).

Some elderly adapt well to living in a residential facility, such as a nursing home. They enjoy the community of others of similar age, enjoy visiting others in the nursing home, and reach out to others to make new friends. In essence, they find positive meaning in the nursing home experience. Ritblatt and Drager (2000) found that the elderly who were in age-segregated living arrangements reported higher levels of satisfaction and social support.

8. Other issues. Whether or not the decision is made to put one's parent (or spouse) in a nursing home, a document detailing the conditions under which life support measures should be used (do you want your parent to be sustained on a respirator?), called an **advance directive** (also known as a **living will**), should be completed by the elderly person or those with power of attorney. These decisions, made ahead of time, spare the adult children the responsibility of making them in crisis contexts and give clear directives to the medical staff in charge of the elderly person. For example, the elderly person can direct that should he or she be unable to feed himself or herself that a feeding tube should not be used. Hence, by making this decision, the children are spared the decision regarding a feeding tube. A **durable power of attorney,** which gives the adult children complete authority to act on behalf of the elderly, is also advised. These documents also help to save countless legal hours, time, and money for those responsible for the elderly. Copies of the living will and durable power of attorney are in Appendixes D and E.

Finally, adult children may consider buying long-term-care insurance (LTCI) to cover what Medicare and many private health care plans do not—"nonmedical" day-to-day care such as bathing or eating for an Alzheimer's parent, as well as nursing home costs. Cost of LTCI begin at about $1000 a year but vary a great deal depending on the age and health of the insured.

Sources

Gallanis, T. P. 2002. Aging and the nontraditional family. The University of Memphis Law Center, 32:607–42.

Jacoby, S. 2003. The nursing squeeze. *AARP Bulletin.* May:6 passim.

Ritblatt, S. N., and L. M. Drager. 2000. The relationship between living arrangement and perception of social support, depression, and life satisfaction among the elderly. Poster, 62nd Annual Conference of the National Council on Family Relations, Minneapolis, November 12.

Statistical Abstract of the United States: 2006. 125th ed. Washington, DC: U.S. Bureau of the Census.

Wielink, G., Huijsman R., and McDonnell J. 1997. A study of the elders living independently in the Netherlands. *Research on Aging* 19:174–198.

Issues Confronting the Elderly

Numerous issues become concerns as a person ages. In middle age, the issues are early retirement (sometimes forced), job layoffs (recession-related cutbacks), **age discrimination** (older persons are not hired and younger workers are hired to take their place), separation or divorce from spouse, and adjustment to children's leaving home. For some in middle age, grandparenting is an issue if they become the primary caregiver for their grandchildren.

Though these may be difficult issues during middle age, there are positive counterbalances such as relief from worrying about pregnancy, voluntary grandparenting, opportunities to try new things, and freedom now that the children have left home (Glazer et al., 2002). As the couple moves from the middle to the later years, the issues become more focused on income, housing, health, retirement, and sexuality.

Income

For most individuals, the end of life is characterized by reduced income. Social Security and pension benefits, when they exist, are rarely equal to the income a retired person formerly earned.

National Data

The median income of men aged 65 and older is $20,363; women, $11,845 (*Statistical Abstract of the United States: 2006,* Table 685).

Financial planning to provide end-of-life income is important. Kemp et al. (2005) interviewed 51 mid- and later-life individuals about their financial planning to identify the conditions under which such planning is initiated. They found catalysts for planning were employer programs and the offering of retirement seminars. Constraints on such planning were losing one's job or unforeseen expenses (in both cases, there were no resources to plan around). Other life events that could be either a catalyst or a constraint were health changes, death of a spouse, divorce, or remarriage. The Special Topics section at the end of the text reviews basics of financial planning in terms of budget management, investing, life insurance, and uses of credit.

To be caught at the end of life without adequate resources is not unusual. Women are particularly disadvantaged since their work history has often been discontinuous, part-time, and low-paying. Social Security and private pension plans favor those with continuous full-time work histories.

Housing

Almost 80 percent (78.8%) of those over the age of 75 live in and own their own home (*Statistical Abstract of the United States: 2006*, Table 953) (hence, most do not live in a nursing home). The home typically is the one the individual has lived in for years or in the same neighborhood. Even those who do not live in their own home are likely to live in a family setting.

For the most part, the physical housing of the elderly is adequate. Indeed, only 6 percent of the housing units inhabited by the elderly are inadequate. Where deficiencies exist, the most common are inadequate plumbing and heating. However, as individuals age, they find themselves living in homes that are not "elder-friendly." Such homes would feature bathroom doors wide enough for a wheelchair, grab bars in the bathrooms, and the absence of stairs (Nicholson, 2003).

Though most elderly own and live in their own homes, as they age, their need for care increases. The

Diversity in Other Countries

Loss of income is minimized among the elderly Chinese, who receive money from their children. Logan and Bian (2003) noted that money from "children accounts for nearly a third of parents' incomes" (p. 85).

most recent trend in housing for the elderly is home health care (mentioned earlier), as an alternative to nursing home care. In this situation the elderly person lives in a single-family dwelling, and other people are hired to come in regularly to help with various needs.

Other elderly individuals live in group living or shared housing arrangements. Referred to as **cohousing,** residents essentially plan their own communities where they own their own units and a common area where they meet for meals several times a week (they typically rotate responsibilities of cooking these meals). The plus side of cohousing is companionship with others. The downside is the need to make decisions by consensus and what to do when members can no longer take care of themselves. Silver Sage Village (www.silversagevillage.com) and ElderSpirit Community (www.elderspirit.net) are two such cohousing arrangements.

Some elderly singles choose to maintain their own home even though they are in a partnered relationship. Karlsson and Borell (2005) interviewed 116 elderly men and women living in an increasingly popular arrangement: living apart together (LAT), in which partners retain their own homes although they are involved in a long-term intimate relationship. The authors focused on the motives of the women who sought to use their homes as a way of establishing boundaries in order to influence their interaction with partners, friends, and kin. All the women studied seem to prioritize the possibility of keeping their various social relations separate from one another.

Physical Health

Our current cultural value for health is in great contrast to previous times. The biographer of 40's screen legend Clark Gable noted that he drank and smoked heavily and that he was going to live until he died. When journalists asked him about a rumor that his doctor ordered him off cigarettes and liquor, he quipped, "Yeah, I heard about it. I was having a highball at the time." (Tornabene, 1976, p. 372) [Gable died of a coronary at age 59.]

Good physical health is the single most important determinant of an elderly person's reported happiness (Smith et. al., 2002). Most elderly individuals, even those of advanced years, continue to define themselves as being in good health. Ostbye et al. (2006) studied an elderly population in Cache County, Utah, on 10 dimensions of health (independent living, vision, hearing, activities of daily living, instrumental activities of daily living, absence of physical illness, cognition, healthy mood, social support and participation, and religious participation and spirituality) and found that 80 percent to 90 percent of those aged 65 to 75 were healthy on all 10 dimensions. Prevalence of excellent and good self-reported health decreased with age, to approximately 60 percent among those aged 85 and older. Most (over 90%) of the elderly do not exercise.

Even individuals who have a chronic, debilitating illness maintain a perception of good health as long as they are able to function relatively well. However, when elders' vision, hearing, physical mobility, and strength are markedly diminished, there is a significant impact on their sense of well-being (Smith et al., 2002). For many, the ability to experience the positive side of life seems to become compromised after age 80 (Smith et al., 2002). Driving accidents also increase with aging. However, unlike teenagers, who also have a high percentage of automobile accidents, the elderly are less likely to die in an accident since they are usually driving at a much slower speed (Pope, 2003).

Some elderly become so physically debilitated that questions about the quality of life sometimes lead to consideration of physician-assisted suicide. Some debilitated elderly ask their physicians to end their lives. Spouses or adult children may also be asked by the debilitated elderly for help in ending their life. Physician-assisted suicide is addressed in the Social Policy on page 491.

The first wealth is health.

Ralph Waldo Emerson, philosopher/clergyman

If I had known I was going to live this long, I would have taken better care of myself.

Mickey Mantle, baseball player

I know a man who gave up smoking, drinking, sex, and food. He was healthy right up to the time he killed himself.

Johnny Carson (former host of the *Tonight Show*)

Chapter 17 Aging in Marriage and Family Relationships

Physician-Assisted Suicide for Terminally Ill Family Members

One's parents may experience a significant drop in quality of life and a total loss of independence due to illness, accident, or, more typically, aging. Adult children or spouses are often asked their recommendations about withdrawing life support (food, water, or mechanical ventilation), starting medications to end life (intravenous vasopressors), or withholding certain procedures that would prolong life (cardiopulmonary resuscitation). Sometimes the patient asks for death. The top reasons cited by patients for wanting to end their lives are losing autonomy (84%), decreasing ability to participate in activities they enjoyed (84%), and losing control of bodily functions (47%) (Chan, 2003). The term **quality of life** is often used in making decisions regarding whether to end one's life. Degenholt et al. (2006) noted that quality of life has been negatively associated with physical function, visual acuity, continence, mobility, and mental/emotional health.

Physician-assisted suicide is legal in the Netherlands. In a national study of 1379 citizens, 15 percent felt that a "suicide pill" should be made available for the elderly who want to end their life in a dignified way. Of 87 relatives whose loved one had died via euthanasia or physician-assisted suicide, 36 percent thought such a pill should be available (Rurup et al., 2005).

Fourteen percent of a sample of 1255 physicians in New Zealand (where euthanasia is illegal) reported that they had taken actions to hasten the death of a patient. Those doing so were typically older and less likely to report a religious affiliation (Mitchell and Owens, 2003). In a study of 1117 physicians in Tennessee, almost half (47%) did not favor euthanasia or physician-assisted suicide (PAS) and would oppose the legalization of such procedures (Essinger, 2003). Where the practice did exist, the physicians were in favor of restrictions and safeguards to protect vulnerable patients. In the United States, 60 percent of adults in a Gallup Poll reported that they approved of physician-assisted suicide (Carroll, 2006).

All 50 states now have living-will statutes permitting individuals to decide (or family members to decide on their behalf) to withhold artificial nutrition (food) and hydration (water) from a patient who is wasting away. In practice, this means not putting in a feeding tube. For elderly, frail patients who may have a stroke or heart attack, do-not-resuscitate (DNR) orders may also be put in place (Cardozo, 2006).

Euthanasia is from the Greek words meaning "good death," or dying without suffering. Euthanasia may be passive, where medical treatment is withdrawn and nothing is done to prolong the life of the patient, or active, which involves deliberate actions to end a person's life. Active euthanasia received nationwide attention through the actions of Dr. Jack Kevorkian (convicted and now in prison), who was involved in over 130 physician-assisted suicides at the patients' request.

The Supreme Court has ruled that state law will apply in regard to physician-assisted suicide. In January 2006, The Supreme Court ruled that Oregon has a right to physician-assisted suicide. Its Death with Dignity Act requires that two physicians must agree that the patient is terminally ill and is expected to die within 6 months, the patient must ask three times for death both orally and in writing, and the patient must swallow the barbiturates themselves rather than be injected with a drug by the physician. The number of physician-assisted suicides increased from 21 in 2001 to 38 in 2002 (an 81% increase). Most patients had cancer and were more likely to be white, male, and well educated (Chan, 2003).

The official position of the American Medical Association (AMA) is that physicians must respect the patient's decision to forgo life-sustaining treatment but that they should not participate in patient-assisted suicide: "PAS is fundamentally incompatible with the physician's role as healer." Thus, the AMA disagrees with Dr. Kevorkian. Arguments against PAS emphasize that the practice can be abused by people who want to end the life of those they feel burdened by (or worse, to get the person's money) and that since physicians make mistakes, what is diagnosed as "terminal" may not in fact be terminal.

Physician-assisted suicide has been legal in Holland for 15 years. A concern has been the potential to misuse the law. But a study of 25 years of euthanasia and physician-assisted suicide in Dutch general practice in which there were about 5000 requests per year concluded, "Some people feared that the lives of increasing numbers of patients would end through medical intervention, without their consent and before all palliative options were exhausted. Our results, albeit based on requests only, suggest that this fear is not justified" (Marquet et al., 2003, p. 202).

Your Opinion?

1. Suppose your father has Alzheimer's and is in a nursing home. He is 88 and no longer recognizes you. He has stopped eating. Would you have a feeding tube inserted to keep him alive?
2. To what degree do you agree with the Death with Dignity policy operative in Oregon?
3. What do you think the position of the government should be in regard to physician-assisted suicide?

Sources

Carroll, J. 2006. Public continues to support right-to-die for terminally ill patients. Gallop Poll. Dated June 19; retrieved June 21 from www.galluppoll.com/content/CI=23356

Chan, S. 2003. Rates of assisted suicides rise sharply in Oregon. *Student BMJ* 11:137–38.

Essinger, D. 2003. Attitudes of Tennessee physicians toward euthanasia and assisted death. *Southern Medical Journal* 96:427–35.

Marquet, R., A. Bartelds, G. J. Visser, P. Spreeuwenberg, and I. Peters. 2003. Twenty-five years of requests for euthanasia and physician assisted suicide in Dutch practice: Trend analysis. *British Medical Journal* 327:201–02.

Mitchell, K. and R. G. Owens. 2003. National survey of medical decisions at end of life made by New Zealand practitioners. *British Medical Journal* 327:202–03.

Rurup, M. L., B. D. Onwuteaka-Philipsen, G. van der Wal, A. van der Heide, and P. van der Maas 2005. A "suicide pill" for older people: Attitudes of physicians, the general population, and relatives of patients who died after euthanasia or physician-assisted suicide in the Netherlands. *Death Studies* 29:519–35.

Mental Health

Mental processes are also affected by aging. Elderly persons (particularly those 85 and older) more often have a reduced capacity for processing information quickly, for cognitive attention to a specific task, for retention, and for motivation to focus on a task. However, judgment may not be affected, and experience and perspective are benefits to decision-making. A team of researchers (Vance et al. 2005) noted that being socially active, which often leads to physical activity, is associated with keeping one's cognitive functioning.

Mental health may worsen for some elderly. Mood disorders, with depression being the most frequent, are more common among the elderly. Such depression is usually related to the fact that the older a person gets, the more likely it is that he or she will experience chronic health problems and related disabilities. Specifically, Schnittker (2005) found that the presence of various illnesses (e.g., cancer, stroke, heart condition, chronic obstructed pulmonary disease, diabetes, high blood pressure, and arthritis) or disability (in activities of daily living, mobility, or strength) was associated with the individual being depressed. And the younger the person at the onset of these illnesses/disabilities, the more the depressive symptoms.

Regrets may also be related to depression. Elderly women who have not had children and who have not accepted their childlessness also report more mental distress than those who had children or those who have accepted their childfree status (Wu and Hart, 2002). The mental health of elderly men seems unaffected by their parental status.

Regardless of the source of depression, it has a negative impact on the elderly couple's relationship. Sandberg et al. (2002) studied depression in 26 elderly couples and concluded, "The most striking finding was the frequent mention of marital conflict and confrontation among the depressed couples and the almost complete absence of it among the nondepressed couples" (p. 261).

National Data

Depression among the elderly may be linked to suicide. Suicide rates are among the highest for white males aged 85 or older. Eighteen percent of all suicides in 2000 were among individuals aged 65 and over (National Institute of Mental Health, 2003).

Dementia, which includes Alzheimer's disease, is the mental disorder most associated with aging. Its presence is assessed in numerous ways, including clinician rating methods, questionnaire-based methods, and performance-based methods (Clare et al., 2005). In spite of the association, only 3 percent of the aged population experience severe cognitive impairment—the most common symptom is loss of memory. It can be devastating to the individual and the partner. An 87-year-old woman, who was caring for her 97-year-old demented husband, said, "After 56 years of marriage, I am waiting for him to die, so I can follow him. At this point, I feel like he'd be better off dead. I can't go before him and abandon him" (Johnson and Barer, 1997, p. 47).

The Self-Assessment on page 493 reflects some of the misconceptions about Alzheimer's disease.

Retirement

Retirement represents a rite of passage through which most elderly pass. In 1983 Congress increased the retirement age at which individuals can receive full Social Security benefits, from 65 (for those born before 1938) to 67 (for those born after 1960). It is possible to take early retirement at age 62 with reduced benefits. Retirement affects the individual's status, income, privileges, power, and prestige.

Alzheimer's Quiz

	True	False	Don't Know
1. Alzheimer's disease can be contagious.	___	___	___
2. People will almost certainly get Alzheimer's if they just live long enough.	___	___	___
3. Alzheimer's disease is a form of insanity.	___	___	___
4. Alzheimer's disease is a normal part of getting older, like gray hair or wrinkles.	___	___	___
5. There is no cure for Alzheimer's disease at present.	___	___	___
6. A person who has Alzheimer's disease will experience both mental and physical decline.	___	___	___
7. The primary symptom of Alzheimer's disease is memory loss.	___	___	___
8. Among persons older than age 75, forgetfulness most likely indicates the beginning of Alzheimer's disease.	___	___	___
9. When the husband or wife of an older person dies, the surviving spouse may suffer from a kind of depression that looks like Alzheimer's disease.	___	___	___
10. Stuttering is an inevitable part of Alzheimer's disease.	___	___	___
11. An older man is more likely to develop Alzheimer's disease than an older woman.	___	___	___
12. Alzheimer's disease is usually fatal.	___	___	___
13. The vast majority of persons suffering from Alzheimer's disease live in nursing homes.	___	___	___
14. Aluminum has been identified as a significant cause of Alzheimer's disease.	___	___	___
15. Alzheimer's disease can be diagnosed by a blood test.	___	___	___
16. Nursing-home expenses for Alzheimer's disease patients are covered by Medicare.	___	___	___
17. Medicine taken for high blood pressure can cause symptoms that look like Alzheimer's disease.	___	___	___

Answers: 1–4, 8, 10, 11, 13–16 False; remaining items True.

Source

Copyright by Neal E. Cutler, Boettner/Gregg Professor of Financial Gerontology, Widener University. Originally published in Psychology Today, 20th Anniversary Issue, "Life Flow: A Special Report—The Alzheimer's Quiz" 1987, Vol. 21, No. 5, pp. 89, 93. Used by permission of Neal E. Cutler. The scale was completed by sixty-nine undergraduates at East Carolina University in 1998. Forty percent of the respondents identified less than 50 percent of the items correctly.

Persons least likely to retire are unmarried, widowed, single-parent women who need to continue working because they have no pension or even Social Security benefits—if they don't work or continue to work, they will have no income, so retirement is not an option. Some workers experience what is called **blurred retirement** rather than a clear-cut one. A blurred retirement means the individual works part-time before completely retiring or takes a bridge job that provides a transition between a lifelong career and full retirement. About half of all people between ages 55 and 64 who quit their primary careers take some kind of bridge job (Dychtwald and Kadlec, 2005).

Individuals who have a positive attitude toward retirement are those who have a pension waiting for them, are married (and thus have social support for

This man enjoys being retired and loves his birds. When he and his wife go on vacation, they have someone come into their home and talk to their parrots.

Authors

the transition), have planned for retirement, are in good health, and have high self-esteem. Those who regard retirement negatively have no pension waiting for them, have no spouse, gave no thought to retirement, have bad health, and have negative self-esteem (Mutran et al., 1997). In general, most retired individuals enjoy it. Few regard continuing to work as a privilege. Although they still prefer to be active, they want to pick their own activity and pace, as they could not do when they were working. Collins and Smyer (2005) found that the loss of one's job via retirement was not a major issue and most were very resilient in their feelings about life and self.

Retirement may have positive consequences for the marriage. Szinovacz and Schaffer (2000) analyzed national data and concluded that husbands viewed the retirement of their wives as associated with a reduction in heated arguments. And if both spouses were retired, there was a perceived reduction of disagreements. Webber et al. (2000) found that spouses had an easier time adjusting to retirement if they had similar retirement goals or expectations in the areas of leisure, family relationships, friendships, and finances.

Some individuals experience disenchantment during retirement; it is not enjoyable and does not live up to their expectations. Lee Iacocca, former president of Chrysler Corporation, reported that he was bored with retirement, missed "the action," and warned others "never to retire" but to stay busy and involved. Some heed his advice and either "die in harness" or go back to work. Dychtwald and Kadlec (2005) recommend that the retired benefit from volunteering—giving back their time and money to attack poverty, illiteracy, oppression, crime, etc. Examples of volunteer organizations are the Service Core of Retired Executives, known as SCORE (www.score.org), Experience Works (www.experienceworks.org), and Generations United (www.gu.org).

Sexuality

Levitra, Cialis, and **Viagra** (prescription drugs that help a man obtain and maintain an erection) are advertised regularly on television and have given cultural visibility to the issue of sexuality among the elderly. Though there are physiological changes for the elderly (both men and women) that impact sexuality, older adults often continue their interest in sexual activity. Indeed, there is a new era of opportunity to continue sexual functioning into one's later years, which has replaced the earlier script of decline and end to sexual functioning as a natural part of aging (Potts et al., 2006).

Table 17.3 on page 495 describes the physiological changes that elderly men experience during the sexual response cycle. Table 17.4 describes the physical changes elderly women experience during the sexual response cycle.

DeLamater and Sill (2005) studied a sample of 1384 and focused on those older than 60. They found that the level of sexual desire for women was related to age and the presence of a sexual partner. For men the level of sexual desire was related to age and the value they attributed to sex. The researchers also found in this older sample that attitudes about sex were more significant influences on sexual desire than biomedical factors. Beckman et al. (2006) also reported in their sample of over 500 persons in their seventies that for 95 percent, sexual interest continued. Alford-Cooper (2006) studied the sexuality of couples married for

Table 17.3 Physiological Sexual Changes in Elderly Men

Phases of Sexual Response	Changes in Men
Excitement phase	As men age, it takes them longer to get an erection. While the young man may get an erection within 10 seconds, elderly men may take several minutes (10 to 30). During this time, they usually need intense stimulation (manual or oral). Unaware that the greater delay in getting erect is a normal consequence of aging, men who experience this for the first time may panic and have erectile dysfunction.
Plateau phase	The erection may be less rigid than when the man was younger, and there is usually a longer delay before ejaculation. This latter change is usually regarded as an advantage by both the man and his partner.
Orgasm phase	Orgasm in the elderly male is usually less intense, with fewer contractions and less fluid. However, orgasm remains an enjoyable experience, as over 70 percent of older men in one study reported that having a climax was very important when having a sexual experience.
Resolution phase	The elderly man loses his erection rather quickly after ejaculation. In some cases, the erection will be lost while the penis is still in the woman's vagina and she is thrusting to cause her orgasm. The refractory period is also increased. Whereas the young male needs only a short time after ejaculation to get an erection, the elderly man may need considerably longer.

Source: Adapted from W. Boskin, G. Graf, and V. Kreisworth. *Health dynamics: Attitudes and behaviors.* St. Paul: West, 1990, p. 209. Used by permission.

Table 17.4 Physiological Sexual Changes in Elderly Women

Phases of Sexual Response	Changes in Women
Excitement phase	Vaginal lubrication takes several minutes or longer, as opposed to 10 to 30 seconds when younger. Both the length and the width of the vagina decrease. Considerable decreased lubrication and vaginal size are associated with pain during intercourse. Some women report decreased sexual desire and unusual sensitivity of the clitoris.
Plateau phase	Little change occurs as the woman ages. During this phase, the vaginal orgasmic platform is formed and the uterus elevates.
Orgasm phase	Elderly women continue to experience and enjoy orgasm. Of women aged 60 to 91, almost 70 percent reported that having an orgasm made for a good sexual experience. With regard to their frequency of orgasm now as opposed to when they were younger, 65 percent said "unchanged," 20 percent "increased," and 14 percent "decreased."
Resolution phase	Defined as a return to the preexcitement state, the resolution phase of the sexual response phase happens more quickly in elderly than in younger women. Clitoral retraction and orgasmic platform disappear quickly after orgasm. This is most likely a result of less pelvic vasocongestion to begin with during the arousal phase.

Source: Adapted from W. Boskin, G. Graf, and V. Kreisworth. *Health dynamics: Attitudes and behaviors.* St. Paul: West, 1990, p. 210. Used by permission.

more than 50 years and found that sexual interest/activity/capacity declined with age but that marital satisfaction did not decrease with these changes.

There may be ethnic differences in sexual interest in midlife. In a study of women aged 42 to 52, in regard to the importance of sex, African-American women were the most likely to report that sex was "very important"; Japanese and Chinese women were least likely to report this. Caucasian women were between the extremes (Cain et al., 2003).

Although sex may occur less often, it remains satisfying for most elderly. Winterich (2003) found that in spite of vaginal, libido, and orgasm changes past menopause, women of both sexual orientations reported that they continued to enjoy active, enjoyable sex lives due to open communication with their partners.

The debate continues about whether menopausal women should become involved in estrogen and estrogen-progestin replacement therapy (ERT) (Friel, 2005). Schairer et al. (2000) studied 2082 cases of breast cancer and concluded that the estrogen-progestin regimen increased the risk of breast cancer beyond that associated with estrogen alone. Beginning in 2003, women were no longer routinely encouraged to take ERT, and the effects of their not doing so on their physical and emotional health were minimal. A major study (Hays et al., 2003) of more than 16,608 postmenopausal women aged 50 to 79 found no significant benefits from ERT in terms of quality-of-life outcomes. Those with severe symptoms (e.g., hot flashes, sleep disturbances, irritability) do seem to benefit without negative outcomes. Indeed, one woman reported, "I'd rather be on estrogen so my husband can stand me (and I can stand myself)." New data suggest that women who have had a hysterectomy can benefit from estrogen-alone therapy without raising their breast cancer risk.

Successful Aging

Researchers who worked on the Landmark Harvard Study of Adult Development (Valliant, 2002) followed 824 men and women from their teens into their 80s and identified those factors associated with successful aging. These include not smoking (or quitting early), developing a positive view of life and life's crises, avoiding alcohol and substance abuse, maintaining healthy weight, exercising daily, continuing to educate oneself, and having a happy marriage. Indeed, those who were identified as "happy and well" were six times more likely to be in a good marriage than those who were identified as "sad and sick." Conversely, spouses with higher levels of negative marital behavior had more chronic health problems, physical disability, and poorer perceived health (Bookwala, 2005). Those (particularly from lower socioeconomic backgrounds) with negative marital interaction were more likely to die of a heart attack than spouses where no such marital interaction existed (Krause, 2005).

Crosnoe and Elder (2002) identified success in one's career as a factor associated with successful aging. Indeed, one of the regrets a person is most likely to have is not pursuing a specific career. Almost two-thirds of a sample of individuals reported that they regretted not pursuing a particular degree or profession (Wrosch and Heckhausen, 2002).

Not smoking is "probably the single most significant factor in terms of health" according to Valliant (2002), of the Landmark Harvard Study on Adult Development. Smokers who quit before age 50 were as healthy at 70 as those who had never smoked.

Exercise is one of the most beneficial activities the elderly can engage in to help them maintain good health. Yet there are significant decreases in exercise as a person ages. A team of researchers (Lees et al., 2005) interviewed 57 adults over the age of 65 about their exercise behavior. Those who did not exercise noted fear of falling and inertia for avoiding exercise. Those who did exercise identified inertia, time constraints, and physical ailments as being the most significant barriers to exercise. Jack LaLanne, the health fitness guru, remarked at age 90, "I hate to exercise; I love the results." At his 90th birthday party, he challenged his well-wishers to be at his 115th birthday party.

Relationships and the Elderly

Relationships continue into old age. Here we examine relationships with one's spouse, siblings, and children.

Relationship with Spouse at Age 85 and Beyond

Marriages that survive into late life are characterized by little conflict, considerable companionship, and mutual supportiveness. All but one of the 31 spouses over age 85 in the Johnson and Barer (1997) study reported "high expressive rewards" from their mate.

There are no shortcuts. No magic. The truth is you have to make smart food choices and exercise, not once in a while, but every day.

Jack LaLanne, fitness guru, age 93

Those who love deeply never grow old; they may die of old age, but they die young.

Sir Arthur Wing Pinero, English dramatist

Difficulties, Disagreements, and Disappointments in Late-Life Marriages

Over the last century, life expectancy in the United States has increased dramatically, from about 47 years in 1900 to nearly 80 years in 2010 (*Statistical Abstract of the United States: 2006*. Table 96). Because people are living longer, the elderly population has grown considerably and so has the number of married elderly couples. The research study presented here attempts to explore the difficulties, disagreements, and disappointments experienced by husbands and wives in "late-life" marriages (Henry et al., 2005).

Sample and Methods

This study used data from the University of Southern California Longitudinal Study of Generations and included 105 older couples. The participants' ages ranged from 52 to 86, with the average age being 69. Total household income of participants averaged between 50,000 and 59,000. Most of the individuals in the sample (more than 86%) reported being in either excellent or good health, and about one-fourth (27%) were in the workforce. The sample consisted of about 78% Caucasians and 22% from other groups.

The research participants were asked to write their answer to the following question: "In the last few years, what are some of the things on which you have differed, disagreed, or been disappointed about (even if not openly discussed) with your spouse?" Researchers read a total of 329 responses to this question, identified the main points or topics for each response, and condensed the main topics into 10 themes.

In addition, a 10-item scale that assessed positive interaction was used to measure marital quality, and the health of the participants was measured by the question, "Compared to people your own age, how would you rate your overall physical health at the present time?" Respondents could mark *excellent, good, fair,* or *poor* to describe their health. Individuals who marked *excellent* or *good* were considered to be healthy, and those marking *fair* or *poor* were considered unhealthy.

Selected Findings and Conclusions

The average number of problems reported by each respondent was 1.40. The most common theme identified as a problem in the marriage was leisure activities; nearly one-fourth (23%) of respondents identified this as a problem. The second most common problem involved the theme of intimacy (13%), followed by finances (11%), no problems (11%), personality (9%), intergenerational relations (9%), household concerns (9%), personal habits (7%), health issues (6%), and work/retirement (2%). Among the 14% of participants who were rated as unhealthy, the most common issues were financial issues, health issues, and intergenerational relations. The following table describes each theme and presents examples of comments made by husbands and wives that reflect each theme.

Theme and % Reporting	Main Topics of Theme	Examples of Comments by Spouses
Leisure (23%)	Time spent with spouse, politics, religion, hobbies/ interests, traveling	"My spouse watches too much football and after 48 years of hearing it, I get upset." "We are both so busy with our own schedules that we neglect doing enough things together."
Intimacy (13%)	Emotional intimacy, physical intimacy	"He doesn't talk enough to me." "Sex life is nil. We are affectionate but don't 'screw.'"
Financial matters (11%)	Spending and investing	"Most disagreements are about money. She wants to eat out a lot, go to movies, plays and then can't understand why I don't have any more money."
No problems (11%)	No problems in marriage	"I really can't think of anything or remember any time when we have not been able to agree on decisions . . ." "I guess we are the exception; I am very happy with my best friend."
Personality (9%)	Attitude and temperament	"He does things that bother me a great deal and his attitude is 'learn to live with it.'" "[She has an] ability to experience extreme anger over what I consider relatively minor provocations."
Intergenerational relations (9%)	Discipline of grandchildren, support of children and grandchildren, relationships	"The only thing we disagree on is how to handle grandchildren . . . like how much money to give them"

Household concerns (9%)	Where to live, home repairs	"I would like to move from [a city], but my husband likes it here." "We disagree on decorating our house or changing the landscape."
Personal habits (7%)	Grooming, driving habits, and substance use	"I wish he would take a shower every night instead of every other night." "When traveling together in the car, we sometimes disagree about directions or driving habits. . . ." "He smokes behind my back." "We disagree on what constitutes a problem with alcohol."
Health problems (6%)	General health problems, hearing and memory loss, doctor appointments, care-giving issues	"It's just a bit hard at times because [strokes] have made it harder for him to talk . . . It can be a little lonely." "I don't think he takes care of himself. He has a poor diet and no exercise. He will try to tough it out rather than make a doctor appointment." "I need help caring for my husband now, and get virtually no assistance from his children. . . ."
Work/retirement (2%)	Current employment, retirement	"I would like my spouse to retire but he wants to keep working."

Although wives and husbands reported almost identical numbers of issues, wives were more likely than husbands to complain about personal habits (9% vs. 5%) and health issues (9% vs. 4%), whereas husbands were more likely than wives to report financial issues (13% vs. 8%). Husbands were also more likely than wives to report that they had no problems in the marriage (15% vs. 7%). Gender had no effect on how often the other themes were reported. Not surprisingly, couples in happy marriages reported fewer problems than those in unhappy marriages, with happy couples reporting no problems 12% of the time and unhappy couples reporting no problems only 4% of the time.

Source

Henry, R. G., R. B. Miller, and R. Giarrusso. 2005. Difficulties, disagreements, and disappointments in late-life marriages. *The International Journal of Aging & Human Development* 61(3):243–64. Used by permission.

Field and Weishaus (1992) reported interview data on 17 couples who had been married an average of 59 years and found that the husbands and wives viewed their marriages very differently. Men tended to report more marital satisfaction, more pleasure in the way their relationships had been across time, more pleasure in shared activities, and closer affectional ties. Whereas club activities, including church attendance, were related to marital satisfaction, financial stability, amount of education, health, and intelligence were not. Sex was also more important to the husbands. Every man in the study reported that sex was always an important part of the relationship with his wife, but only four of the 17 wives had the same report.

The wives, according to the researchers, presented a much more realistic view of their long-term marriage. For this generation of women, "it was never as important for them to put the best face on things" (Field and Weishaus, 1992, p. 273). Hence, many of these wives were not unhappy; they were just more willing to report disagreements and changes in their marriages across time.

Only a small percentage (8%) of individuals older than 100 are married. Most married centenarians are men in their second or third marriage. Many have outlived some of their children. Marital satisfaction in these elderly marriages is related to a high frequency of expressing love feelings to one's partner. Though it is assumed that spouses who have been married for a long time should know how their partners feel, this is often not the case. Telling each other "I love you" is very important to these elderly spouses.

Renewing an Old Love Relationship

Some widowed or divorced elderly try to find and renew an earlier love relationship. Researcher Nancy Kalish (1997) surveyed 1001 individuals who reported that they had renewed an old love relationship. Two-thirds of those who had contacted a previous love (mostly by phone or letter) were female; one-third were male. Almost two-thirds (62%) reported that the person they made contact with was their first love, and 72 percent reported that they were still in love with and together with the person with whom they had renewed their relationship. An example of one reunited couple follows:

> He was my first boyfriend. It was all very innocent. Mostly we were good friends. We walked home from school together, and went out a couple of times. My parents didn't like him for some reason. Neither of us remembers why we broke up.
>
> We went on to long marriages with other people. Sixty-three years passed, and we were both widowed when we went to our high school reunion. I knew as soon as we saw each other. We got to talking and nothing else mattered. It seemed like something that was destined to happen.
>
> We were married on my eightieth birthday. He's so loving and kind and caring and understanding and peaceful. He doesn't argue. He's very respectful, a perfect gentlemen. I just love everything about him. He's all I ever wanted. (p. 11)

Weintraub (2006) suggested some cautions in renewing an old love relationship. These include discussing the original breakup/what went wrong; if one partner was hurt badly, the other must take responsibility/make amends; and go slow in reviving the relationship/take your time. She also identified some successful couples who reunited after a long interval. Harry Kullijian contacted Carol Channing 70 years after they left high school. Carol's mother had broken up the relationship. Kullijian and Channing married in 2003.

Relationship with Siblings at Age 85 and Beyond

Relationships the elderly have with their siblings are primarily emotional (enjoying time together) rather than functional (the sibling provides money or services). Earlier in the text we noted that the sibling relationship, particularly that between sisters is the most enduring of all relationships.

Relationship with One's Own Children at Age 85 and Beyond

In regard to relationships of the elderly with their children, emotional and expressive rewards are high. Actual caregiving is rare. Only 12 percent of the Johnson and Barer (1997) sample of adults older than 85 lived with their children. Most preferred to be independent and to live in their own residence. "This independent stance is carried over to social supports; many prefer to hire help rather than bother their children. When hired help is used, children function more as mediators than regular helpers, but most are very attentive in filling the gaps in the service network" (Johnson and Barer, 1997, p. 86).

Relationships among multiple generations will increase. Whereas three-generation families have been the norm, increasingly four- and five-generation families will become the norm. These changes have already become visible.

Grandparenthood

Another significant role for the elderly is grandparenting. Among adults aged 40 and older who had children, close to 95 percent are grandparents and most have, on average, five or six grandchildren. Grandparents may be in the full-time active role of taking care of their grandchildren, as grandparents in a multigen-

Grandparents are people who play with children whether they are busy or not.

Lanie Carter, "America's first professional grandmother"

This grandmother delights in her new grandbaby.

© David Knox

erational family where they provide supplemental help, as part-time helpers on an occasional basis, or as occasional visitors to see their grandchildren. Grandparents see themselves as caretakers, emotional/economic resources, teachers, and historical connections on the family tree.

The average age of becoming a grandparent is 48; the age at which grandparents become caregivers of their grandchildren is from 53 to 59 (Landry-Meyer, 2000). Fifteen percent of grandparents provide childcare services for their grandchildren while the parents work (Davies, 2002). Most grandparents see their grandchildren two to four times a month (Davies, 2002).

Grandparents in Full-Time Role of Raising Grandchildren

About 6 million grandparents are housing and rearing their own grandchildren. This represents about 6 percent of all children being reared full time by their grandparents (Hayslip and Kaminski, 2005). Although both grandparents may be involved, grandparenthood is primarily a "woman's issue" (Mills et al., 2005), and their psychological distress is in reference to economic stress and social support. Musil and Standing (2005) reported on the diaries of grandmothers who revealed stress in their full-time role of grandmother as they coped with their grandchildren's daily activities.

Grandfathers may also be involved in caring for grandchildren. Bullock (2005) studied 21 grandfathers over the age of 65 who were involved in the active care of at least one grandchild. "Powerless" was the term used by these grandfathers to describe their experience. In effect, they felt overwhelmed. Hayslip and Kaminski (2005) spoke to the need for help for both grandparents and grandchildren.

Styles of Grandparenting

Grandparents also have different styles of relating to grandchildren. Whereas some grandparents are formal and rigid, others are informal and playful, and authority lines are irrelevant. Still others are surrogate parents providing considerable care for working mothers and/or single parents. Some grandparents have regular contact with their grandchildren; others are distant and show up only for special events like birthdays. E-mail is helping grandparents to stay connected to their grandchildren. Davies (2002) reported that 35 percent of grandparents use e-mail to communicate with their grandchildren.

Age seems to be a factor in determining how grandparents relate to their grandchildren. Grandparents over the age of 65 are less likely to be playful and fun-seeking than those under 65. This may be because the older grandparents are less physically able to engage in playful activities with their grandchildren. Indeed, according to Johnson and Barer (1997),

> . . . members of the oldest generation in our study place more emphasis on their relationship with their own children over their grandchildren. This situation could stem from the fact that as grandchildren reach adulthood, they become more independent from their own parent. That parent then is freed up to strengthen the relationship with their oldest old parent, at the same time they, as the middle generation, maintain a lineage bridge linking their parent to their child (p. 89).

Finally, the quality of the grandparent-grandchild relationship is affected by the parents' relationship to their own parents. If a child's parents are estranged from their parents, it is unlikely that the child will have an opportunity to develop a relationship with the grandparents.

Effect of Divorce on Grandparent-Child Relationship

The degree of involvement of grandparents in the life of a grandchild is sometimes related to whether the grandparents are divorced (King, 2003). "Divorced grandparents have less contact with grandchildren and participate in fewer shared activities with them" (p. 180).

In addition, grandparental involvement with grandchildren is also related to whether their own children are divorced and whether the grandchild is on the mother's or father's side. Since mothers end up with custody in about 85 percent of the cases, the maternal grandparents may escalate their involvement with their grandchildren. However, since divorcing fathers may have less access to their children, the paternal grandparents may find that their time with their grandchildren is radically reduced. Sometimes a custodial parent will purposefully try to sever the relationship in an attempt to seek revenge or vent hostility against his or her former spouse.

Indeed, when their children divorce, some grandparents are not allowed to see their grandchildren. Although all 50 states have laws granting grandmothers and grandfathers the right to petition for visitation with their grandchildren over the protests of the parent or parents (Hill, 2000), the role of the grandparent has limited legal and political support. By a vote of six to three, the Supreme Court in 2000 (*Troxel v. Granville*) sided with the parents and virtually denied 60 million grandparents the right to see their grandchildren. The court viewed parents as having a fundamental right to make decisions about who their children could spend time with. However, some grandparents have petitioned the court and won the right to see their grandchildren (Henderson, 2005). Grandparents should not give up. Stepgrandparents have no legal rights to their stepgrandchildren.

Benefits to Grandchildren

Over 80 percent (81.3 percent) of a sample of 1027 university students agreed that "I have a loving relationship with my grandmother" (71.8 percent had a loving relationship with their grandfather) (Knox and Zusman, 2006). Grandchildren report enormous benefits from having a close relationship with grandparents, including development of a sense of family ideals, moral beliefs, and a work ethic. Kennedy (1997) focused on the memories grandchildren had of their grandparents and found that "love and companionship" was identified most frequently by the grandchildren. In addition, a theme running through a fourth of the memories was that the grandchild felt that he or she was regarded as "special" by the grandparent, either because of being the first or last grandchild, having personality characteristics similar to the grandparent's, or being the child of a favorite son or daughter of the grandparent. Taylor et. al. (2005) surveyed 70 international college students in the Unites States regarding per-

This grandson benefits from the happiness he sees in and adoration he feels from his granddad.

Kelly Lewis

Some elderly enjoy finding bargains at tag sales.

ceived grandparental influence and relationship satisfaction with their "closest" grandparent. The students reported that, despite distance and infrequent contact, the grandchildren reported significant satisfaction and influence from this relationship.

The End of One's Life

The end of one's life sometimes involves the death of one's spouse.

Death of One's Spouse

The death of one's spouse is the most stressful life event a person ever experiences (Hobson et al., 1998). Because women tend to live longer than men, and because women are often younger than their husbands, women are more likely than men to experience the death of their marital partner. Antonucci et al.

(2002) noted that women in several countries—e.g., Germany, Japan, France—as well as in the United States are all more likely than men to experience widowhood, illness, and financial strain.

Although individual reactions and coping mechanisms for dealing with the death of a loved one vary, several reactions to death are common. These include shock, disbelief and denial, confusion and disorientation, grief and sadness, anger, numbness, physiological symptoms such as insomnia or lack of appetite, withdrawal from activities, immersion in activities, depression, and guilt. Eventually, surviving the death of a loved one involves the recognition that life must go on, the need to make sense out of the loss, and the establishment of a new identity. However, research is not consistent on the degree to which individuals are best served by continuing or relinquishing the emotional bonds with the deceased (Stroebe and Schut, 2005).

Women and men tend to have different ways of reacting to and coping with the death of a loved one. Women are more likely than men to express and share feelings with family and friends and are also more likely to seek and accept help, such as attending support groups of other grievers. Initial responses of men are often cognitive rather than emotional. From early childhood, males are taught to be in control, to be strong and courageous under adversity, and to be able to take charge and fix things. Showing emotions is labeled weak.

Men sometimes respond to the death of their spouse in behavioral rather than emotional ways. Sometimes they immerse themselves in work or become involved in physical action in response to the loss. For example, a widower immersed himself in repairing a beach cottage he and his wife had recently bought. Later, he described this activity as crucial to getting him through those first two months. Another coping mechanism for men is the increased use of alcohol and other drugs.

Women's response to the death of their husbands may necessarily involve practical considerations. Johnson and Barer (1997) identified two major problems of widows—the economic effects of losing a spouse and the practical problems of maintaining a home alone. The latter involves such practical issues as cleaning the gutters, painting the house, and changing the filters in the furnace.

Whether a spouse dies suddenly or after a prolonged illness has an impact on the reaction of the remaining spouse. The sudden rather than prolonged death of one's spouse is associated with being less at peace with death and being more angry. Widows of spouses who die a "painful" death report higher anxiety and intrusive thoughts (Carr, 2003).

The age at which one experiences the death of a spouse is also a factor in one's adjustment. Persons in their 80s may be so consumed with their own health and disability concerns that they have little emotional energy left to grieve. On the other hand, a team of researchers (Hansson et al., 1999) noted that death does not end the relationship with the deceased. Some widows and widowers report a year after the death of their beloved a feeling that their spouses are with them at times and are watching out for them. They may also dream of them, talk to their photographs, and remain interested in carrying out their wishes. Such continuation of the relationship may be adaptive by providing meaning and purpose for the living or maladaptive in that it may prevent one from establishing new relationships.

Involvement with New Partners at Age 80 and Beyond

Most women who live to age 80 have lost their husbands. At age 80 there are only 53 men for every 100 women. Patterns women use to adjust to this lopsided man-woman ratio include dating younger men, romance without marriage, and "share-a-man" relationships. Most individuals in their 80s who have lost a partner do not remarry (Stevens, 2002). Those who do report as the primary reasons a need for companionship and to provide meaning in life. Stevens (2002) noted "evidence of

You can't turn back the clock. But you can wind it up again.

Bonnie Prudden, advocate of physical health/founder of Trigger Point Therapy continuing loyalty to the deceased" in all the late-life partnerships she studied, suggesting that new partners do not simply "replace" former partners.

Over 3 million women are married to men who are at least 10 years younger than they. Although traditionally women were socialized to seek older, financially established men, the sheer shortage of men has encouraged many women to seek younger partners. These age-discrepant relationships were discussed in the chapter on Marriage Relationships.

Faced with a shortage of men but reluctant to marry those who are available, some elderly women are willing to share a man. In elderly retirement communities such as Palm Beach, Florida, women count themselves lucky to have a man who will come for lunch, take them to a movie, or be an escort to a dance. They accept the fact that the man may also have lunch, go to a movie, or go dancing with other women.

But finding a new spouse is usually not a goal. Women in their later years have also moved away from the idea that they must remarry and have become more accepting of the idea that they can enjoy the romance of a relationship without the obligations of a marriage. Many enjoy their economic independence, their control over their life space, and their freedom not to be a nurse to an aging partner.

To avoid marriage, some elderly couples live together. Parenthood is no longer a goal, and many do not want to entangle their assets. For some, marriage would mean the end of their Social Security benefits or other pension moneys.

Preparing for One's Own Death

What is it like for those near the end of life to think about death? To what degree do they go about actually "preparing" for death? Johnson and Barer (1997) interviewed 48 individuals with an average age of 93 to find out their perspective on death. Most were women (77%) who lived alone (56%), but most had some sort of support in terms of children or one or more social support services (73%). The findings to follow are specific to those who died within a year after the interview.

Thoughts in the Last Year of Life   Most had thought about death and saw their life as one that would soon end. Most did so without remorse or anxiety. With their spouses and friends dead and their health failing, they accepted death as the next stage in life. Some comments follow (Johnson and Barer, 1997, p. 205).

If I die tomorrow, it would be all right. I've had a beautiful life, but I'm ready to go.

My husband is gone, my children are gone, and my friends are gone.

That's what is so wonderful about living to be so old. You know death is near and you don't even care.

I've just been diagnosed with cancer, but it's no big deal. At my age, I have to die of something.

The major fear expressed by these respondents was not the fear of death but the dying process. Dying in a nursing home after a long illness is a dreaded fear. Sadly, almost 60 percent of the respondents died after a long, progressive illness. They had become frail, fatigued, and burdened by living. They identified dying in their sleep as the ideal way to die. Some hasten their death by no longer taking their medications or wish they could terminate their own life. "I'm feeling kind of useless. I don't enjoy anything anymore. . . . What the heck am I living for? I'm ready to go anytime—straight to hell. I'd take lots of sleeping pills if I could get them" (Johnson and Barer, 1998, p. 204). Two researchers (Nakashima and Canda, 2005) studied 16 hospice patients and found that having a positive death was associated with coming to terms with one's own mortality, finding meaning in death and dying, and spiritual well-being.

It is nothing to die, but it is frightful not to live.

Jean Valjean, *Les Miserables*

It's not dark yet, but it's getting there.

Bob Dylan, *Not Dark Yet*

Chapter 17 Aging in Marriage and Family Relationships

Behaviors in the Last Year of Life   Aware that they are going to die, most simplify their life, disengage from social relationships, and leave final instructions. In simplifying their life, they sell their home and belongings and move to smaller quarters. One 81-year-old woman sold her home, gave her car away to a friend, and moved into a nursing home. The extent of her belongings became a chair, lamp, and TV.

Disengaging from social relationships is selective. Some maintain close relationships with children and friends but others "let go." Christmas cards are no longer sent out, letters stop, and phone calls become the source of social connections. Some leave final instructions in the form of a will or handwritten note expressing wishes of where to be buried, handling costs associated with disposal of the body, and what to do about pets. One of Johnson and Barer's (1997) respondents left $30,000 to specific caregivers to take care of each of several pets (p. 204).

The elderly who had counted on children to take care of them may discover that there are too few children, who may have scattered because of job changes, divorce, or both (Kutza, 2005). The result is that the elderly may have to fend for themselves.

Well, he started retreating more and more and more and more and more. By retreating, I mean not communicating as much, not being interested in things, not doing things.

Kelly, about her father, actor Jimmy Stewart, at age 88 (he died at age 89)

At the time of death, the best parting gift is peace of mind.

The Dalai Lama, *Imagine All the People*

SUMMARY

What is meant by the terms age and ageism?

Age is defined chronologically (time), physiologically (capacity to see, hear, etc.), psychologically (self-concept), sociologically (social roles), and culturally (value placed on elderly). Ageism is the denigration of the elderly, and gerontophobia is the dreaded fear of being elderly. Theories of aging range from disengagement (individuals and societies mutually disengage from each other) to continuity (the habit patterns of youth are continued in old age). Age stratification and life-course theoretical perspectives are currently in vogue.

What is the "sandwich generation"?

Elder care combined with child care is becoming common among the sandwich generation, adult children responsible for the needs of both their parents and their children. Exhaustion over the relentless demands, guilt over not doing enough, and resentment over feeling burdened are among the feelings reported by members of the sandwich generation.

Deciding whether to arrange for an elderly parent's care in a nursing home requires attention to a number of factors, including the level of care needed by the parent, the philosophy and time availability of the adult child, and the resources of the adult children and other siblings. Full-time nursing care (not including medication) is over $1000 a week.

Elderly parents who are dying from terminal illnesses incur enormous medical bills. Some want to die and ask for help. Our society continues to wrestle with physician-assisted suicide and euthanasia. Only Oregon allows for physician-assisted suicide.

What issues confront the elderly?

Issues of concern to the elderly include housing, health, retirement, and sexuality. Most elderly live in their own homes, which they have paid for. Most housing of the elderly is adequate, although repair becomes a problem with the age of the person. Health concerns are paramount for the elderly. Good health is the single most important factor associated with an elderly person's perceived life satisfaction.

Hearing and visual impairments, arthritis, heart conditions, and high blood pressure are all common to the elderly. Mental problems may also occur with mood disorders; depression is the most common.

Though the elderly are thought to be wealthy and living in luxury, most are not. The median household income of persons over the age of 65 is less than half of what the couple earned in the prime of their lives. The most impoverished elderly are those who have lived the longest, who are widowed, and who live alone. Women are also particularly economically disadvantaged since their work history has often been discontinuous, part-time, and low-paying.

Sexuality among the elderly involves, for most women and men, lower reported interest, activity, and capacity. Fear of the inability to have an erection is the sexual problem most frequently reported by elderly men (Viagra, Levitra, and Cialis have helped allay this fear). The absence of a sexual partner is the most frequently reported sexual problem among elderly women.

What factors are associated with successful aging?

Factors associated with successful aging include not smoking (or quitting early), developing a positive view of life and life's crises, avoiding alcohol and substance abuse, maintaining healthy weight, exercising daily, continuing to educate oneself, and having a happy marriage. Indeed, those who were identified as "happy and well" were six times more likely to be in a good marriage than those who were identified as "sad and sick." Success in one's career is also associated with successful aging.

What are relationships like for the elderly?

Marriages that survive into old age (beyond age 85) tend to have limited conflict, considerable companionship, and mutual supportiveness. Relationships with siblings are primarily emotional rather than functional. In regard to relationships of the elderly with their children, emotional and expressive rewards are high. Caregiving help is available but rare. Only 12 percent of one sample of adults older than 85 lived with their children.

What is grandparenthood like?

Among adults aged 40 and older who had children, close to 95 percent are grandparents. There is considerable variation in role definition and involvement.

Whereas some delight in seeing their lineage carried forward in their grandchildren and provide emotional and economic support, others are focused on their own lives or on their own children and relate formally and at a distance to their grandchildren. When grandparents are involved in their lives, grandchildren benefit in terms of positive psychological and economic benefits.

How do the elderly face the end of life?

The end of life involves adjusting to the death of one's spouse and to the gradual decline of one's health. Most elderly are satisfied with their life, relationships, and health. Declines begin when persons are in their 80s. Most fear not death but the process of dying.

KEY TERMS

advance directive	cohousing	filial responsibility	Medicaid
age	dementia	frail	Medicare
age discrimination	durable power of attorney	gerontology	quality of life
ageism	euthanasia	gerontophobia	sandwich generation
blurred retirement	family caregiving	Levitra	Viagra
Cialis	filial piety	living will	

The Companion Website for *Choices in Relationships: An Introduction to Marriage and the Family,* Ninth Edition

www.thomson.edu/sociology/knox

Supplement your review of this chapter by going to the companion website to take one of the Tutorial Quizzes, use the flash cards to master key terms, and check out the many other study aids you'll find there. You'll also find special features such as the Marriage and Family Resource Center, Census 2000 information, and other data and resources at your fingertips to help you with that special project or to do some research on your own.

WEB LINKS

AARP (widowed persons services)
http://www.aarp.org

Calculate Your Life Expectancy
www.livingto100.com

ElderSpirit Community
www.elderspirit.net

ElderWeb
http://www.elderweb.com

Family Caregiver Alliance
http://caregiver.org/caregiver/jsp/home.jsp

Foundation for Grandparenting
http://www.grandparenting.org/

Generations United
http://www.gu.org

GROWW (Grief Recovery Online)
http://www.groww.com

Life Extension
//www.sens.or:/
//www.kurzweilai.net/index.html?flash:1

Nolo: Law for All (wills and legal issues)
http://www.nolo.com

Silver Sage Village
http://www.silversagevillage.com

REFERENCES

Alford-Cooper, F. 2006. Where has all the sex gone? Sexual activity in lifetime marriage. Paper, Southern Sociological Society, New Orleans, March 23–26.

Antonucci, T. C, J. E. Lansford, H. Akiyama, J. Smith, M. M. Baltes, K. Takahaski, R. Fuhrer, and J. Francois Dartigues. 2002. Differences between men and women in social relations, resource deficits, and depressive symptomatology during later life in four nations. *Journal of Social Issues* 58:767–84.

Ballard, S. M. and M. L. Morris. 2003. The family life education needs of midlife and older adults. *Family Relations* 52:129–36.

Barrett, A. E. 2005. Gendered experiences in midlife: Implications for age identity. *Journal of Aging Studies* 19:163–183.

Beckman, N. M., M. Waern, I. Skoog, and The Sahlgrenska Academy at Goteborg University, Sweden. 2006. Determinants of sexuality in 70 year olds. *The Journal of Sex Research* 43:2–3.

Bookwala, J. 2005. The role of marital quality in physical health during the mature years. *Journal of Aging and Health* 17:85–97.

Boulton-Lewis, G. M., L. K. Buys, and J. Kitchin. 2006. Learning and active aging. *Educational Gerontology* 32 271–82.

Bullock, K. 2005. Grandfathers and the impact of raising grandchildren. *Journal of Sociology and Social Welfare* 32:43–59.

Cain, V. S., C. B. Johannes, N. E. Avis, B. Mohr, M. Schocken, J. Skurnick, and M. Ory. 2003. Sexual functioning and practices in a multiethnic study of midlife women: Baseline results from SWAN. *Journal of Sex Research* 40:266–76.

Cardozo, M. 2006. What is a good death? Issues to examine in critical care. *British Journal of Nursing* 14:1056–60.

Carr, D. 2003. A "good death" for whom? Quality of spouses' death and psychological distress among older widowed persons. *Journal of Health and Social Behavior* 44:215–25.

Chadiha, L., J. Rafferty, and J. Pickard. 2003. The influence of caregiving stressors, social support, and caregiving appraisal on marital function among African American wife caregivers. *Journal of Marital and Family Therapy* 29:479–90.

Chan, S. 2003. Rates of assisted suicides rise sharply in Oregon. *Student MBJ* 11:137–38.

Clare, L., I. Markova, F. Verhey, and G. Kenny. 2005. Awareness in dementia: A review of assessment methods and measures. *Aging and Mental Health* 9:394–404.

Collins, A. L. and M. A. Smyer. 2005. The resilience of self-esteem in late adulthood. *Journal of Aging and Health* 17:471–90.

Crosnoe, R. and G. H. Elder, Jr. 2002. Successful adaptation in the later years: A life course approach to aging. *Social Psychology Quarterly* 65:309–28.

Cutler, N. E. 2002. *Advising mature clients.* New York: Wiley.

Davies, C. 2002. The grandparent study 2002 report. Washington, DC: AARP.

DeLamater, J. D. and M. Sill 2005. Sexual desire in later life. *Journal of Sex Research* 42:138–49.

Degenholt, H. B., R. A. Kane, R. L. Kane, B. Bershadsky, K C. Kling 2006. Predicting nursing facility residents' quality of life using external indicators. *Health Services Research* 41:335–57.

Dychtwald, K. and D. J. Kadlec. 2005. *The power years: A user's guide to the rest of your life.* New York: Wiley.

Essinger, D. 2003. Attitudes of Tennessee physicians toward euthanasia and assisted death. *Southern Medical Journal* 96:427–35.

Family Caregiver Alliance. Retrieved April 10, 2006 (http://caregiver.org/caregiver/jsp/content_node.jsp?nodeid=439).

Fernandez-Ballesteros, R. 2003. Social support and quality of life among older people in Spain. *Journal of Social Issues* 58:645–60.

Field, D., and S. Weishaus. 1992. Marriage over half a century: A longitudinal study. In *Changing lives,* edited by M. Bloom. Columbia, SC: University of South Carolina Press, 269–73.

Friel, P. 2005. Estrogen replacement therapy: how much is too much? Townsend letter for doctors and patients. *The Townsend Letter Group.* January(258):105–07.

Gallanis, T. P. 2002. Aging and the nontraditional family. *The University of Memphis Law Center* 32:607–42.

Glazer, G., R. Zeller, L. Delumbia, C. Kalinyak, S. Hobfoll, and P. Hartman. 2002. The Ohio midlife women's study. *Health Care for Women International* 23:612–30.

Hansson, R. O., J. O. Berry, and M. E. Berry. 1999. The bereavement experience: Continuing commitment after the loss of a loved one. In *Handbook of interpersonal commitment and relationship stability,* edited by J. M. Adams and W. H. Jones. New York: Academic/Plenum Publishers, 281–91.

Hayslip, B. and P. L. Kaminski 2005. Grandparents raising their grandchildren. *Marriage and Family Review* 37:171–90.

Hays, J., J. K. Ockene, R. L. Brunner, J. M. Kotchen, J. E. Manson, R. E. Patterson, A. K. Aragki, M. S., S. A. Shumaker, R. G. Bryzyski, et al. 2003. Effects of estrogen plus progestin on health-related quality of life. *The New England Journal of Medicine* 348:1839–54.

Henderson, T. L. 2005. Grandparent visitation rights: Successful acquisition of court-ordered visitation. *Journal of Family Issues* 26:107–16.

Henry, R. G., R. B. Miller, and R. Giarrusso. 2005. Difficulties, disagreements, and disappointments in late-life marriages. *International Journal of Aging & Human Development* 61:243–265.

Hill, Twyla J. 2000. Legally extending the family: An event history analysis of grandparent visitation rights laws. *Journal of Family Issues* 21:246–61.

Hobson, C. J., J. Kamen, J. Szostek, C. M. Nethercut, J. W. Tiedmann, and S. Wojnarowiez. 1998. Stressful life events: A revision and update of the social readjustment rating scale. *International Journal of Stress Management* 5:1–23.

Ikegami, N. 1998. Growing old in Japan. *Age and Ageing* 27:277–78.

Jacoby, S. 2003. The nursing squeeze. *AARP Bulletin* May:6 passim.

Jang, S., and D. F. Detzner. 1998. Filial responsibility in cross-cultural context. Poster, Annual Conference of the National Council on Family Relations, Milwaukee, Wisconsin.

Johnson, C. L., and B. M. Barer. 1997. *Life beyond 85 years: The aura of survivorship.* New York: Springer Publishing.

Johnson, R. W. and J. M. Wiener. 2006. A profile of frail older Americans and their caregivers. Urban Institute Report. Posted March 1 (http://www.urban.org/url.cfm?ID=311284).

Jorm, A. F., H. Christensen, A. S. Henderson, P. A. Jacomb, A. E. Korten, and A. Mackinnon. 1998. Factors associated with successful ageing. *Australian Journal of Ageing* 17:33–37.

Kalish, N. 1997. *Lost & found lovers: Facts and fantasies of rekindled romances.* New York: William Morrow and Company.

Karlsson, S. G. and K. Borell 2005. A home of their own. Women's boundary work in LAT-relationships. *Journal of Aging Studies* 19:73–84.

Kelley-Moore, J. A., J. G. Schumacher, E. Kahana, and B. Kahana. 2006. When do older adults become 'disabled'? Social and health antecedents of perceived disability in a panel study of the oldest old. *Journal of Health and Social Behavior* 47: 126–42.

Kemp, C. L., C. J. Rosenthal, and M. Denton. 2005. Financial planning for later life: Subjective understandings of catalysts and constraints. *Journal of Aging Studies* 19:273–90.

Kennedy, G. E. 1997. Grandchildren's memories: A window into relationship meaning. Paper, Annual Conference of the National Council on Family Relations, Crystal City, Virginia.

Kimuna, S., D. Knox, and M. Zusman. 2005. College students perceptions about older people and aging. *Educational Gerontology* 31:563–72.

King, V. 2003. The legacy of a grandparent's divorce: Consequences for ties between grandparents and grandchildren. *Journal of Marriage and the Family* 65:170–83.

Knox, D. and Zusman, M. E. 2006. Relationship and sexual behaviors of a sample of 1027 university students. Unpublished data collected for this text. Department of Sociology, East Carolina University, Greenville, NC.

Knox, D., S. Kimuna, and M. Zusman. 2005. College student views of the elderly: Some gender differences. *College Student Journal* 39:14–16.

Krause, N. 2005. Negative interaction and heart disease in late life: Exploring variations by socio-economic status. *Journal of Aging and Health* 17:28–35.

Kutza, E. A. 2005. The intersection of economics and family status in later life: Implications for the future. *Marriage and Family Review* 37:3–8.

Landry-Meyer, L. 2000. Grandparents as parents: What they need to be successful. *National Council on Family Relations* 45:89.

Lees, F.D., P. G. Clark, C. R. Nigg, and P. Newman. 2005. Barriers to exercise behavior among older adults: A focus-group study. *Journal of Aging and Physical Activity* 13:23–34.

Leichtentritt, R. D., and K. D. Rettig. 2000. Conflicting value considerations for end-of-life decisions. Poster, 62nd Annual Conference of the National Council on Family Relations, Minneapolis, November 12.

Logan, J. R. and F. Bian. 2003. Parents' needs, family structure, and regular international financial exchange in Chinese cities. *Sociological Forum* 18:85–101.

Marks, N. F., J. D. Lambert, and H. Choi. 2002. Transitions to caregiving, gender, and psychological well-being: A prospective U.S. national study. *Journal of Marriage and Family* 64:657–67.

Marquet, R., A. Bartelds, G. J. Visser, P. Spreeuwenberg, and I. Peters. 2003. Twenty-five years of requests for euthanasia and physician assisted suicide in Dutch practice: Trend analysis. *British Medical Journal* 327:201–02.

Martire, L. M., and M. A. P. Stephens. 2003. Juggling parent care and employment responsibilities: The dilemmas of adult daughter caregivers in the workforce. *Sex Roles: A Journal of Research* 48:167–74.

Mercer, C. 2006 Practical immortality. Submitted to *The Christian Century* for publication,

Mills, T. L., Z. Gomez-Smith, and J. M. DeLeon. 2005. Skipped generation families: Sources of psychological distress among grandmothers of grandchildren who live in homes where neither parent is present. *Marriage and Family Review* 37:191–212.

Mitchell, J. 2003. Personal communication. Center for Aging, East Carolina University, Greenville, North Carolina.

Mitchell, K. and R. G. Owens. 2003. National Survey of medical decisions at end of life made by New Zealand practitioners. *British Medical Journal* 327:202–03.

Morris, M. L. and S. M. Ballard. 2003. Instructional techniques and environmental considerations in family life education programming for midlife and older adults. *Family Relations* 52:167–73.

Musil, C. M. and T. Standing. 2005. Grandmothers' diaries: A glimpse at daily lives *International Journal of Aging and Human Development* 60:317–29.

Musselman, K. and B. Brouwer. 2005. Gender-related differences in physical performance among seniors. *Journal of Aging & Physical Activity* 13:239–54.

Mutran, E. J., D. Reitzes, and M. E. Fernandez. 1997. Factors that influence attitudes toward retirement. *Research on Aging* 19:251–73.

Nakashima, M. and E. R. Canda. 2005. Positive dying and resiliency in later life: A qualitative study. *Journal of Aging Studies* 19:109–22.

National Institute of Mental Health. 2003. Older adults: Depression and suicide facts. HIH Publication No. 03–4593. Retrieved on July 23, 2003 from www.nimh.nih.gov/publicat/elderlydep suicide.cfm.

Nicholson, T. 2003. Homeowners fail to prepare for aging. *AARP Bulletin* 44:7.

O'Reilly, E. M. 1997. *Decoding the cultural stereotypes about aging: New perspectives on aging talk and aging issues*. New York: Garland.

Ostbye, T., K. M. Krause, M. C. Norton, J Tschanz, L. Sanders, K. Hayden, C. Pieper, and K. A. Welsh-Bohmer. 2006. Ten dimensions of health and their relationships with overall self-reported health and survival in a predominately religiously active elderly population: *The Cache County Memory Study. Journal of the American Geriatrics Society* 54:199–209.

Peterson, B. E. 2002. Longitudinal analysis of midlife generativity, intergenerational roles, and caregiving. *Psychology and Aging* 17:161–68.

Pope, E. 2003. MIT study: Older drivers know when to slow down. *AARP Bulletin* 44:11–12.

Potts, A., V. M. Grace, T. Vares and N. Gavey. 2006. 'Sex for life'? Men's counter-stories on 'erectile dysfunction', male sexuality and ageing. *Sociology and Health and Illness* 28:306–29.

Ritblatt, S. N., and L. M. Drager. 2000. The relationship between living arrangement and perception of social support, depression, and life satisfaction among the elderly. Poster session at the 62nd Annual Conference of the National Council on Family Relations, Minneapolis, November 12.

Sandberg, J. G., R. B. Miller, and J. M. Harper. 2002. A qualitative study of marital process and depression in older couples. *Family Relations* 51:256–64.

Schairer, C., J. Lubin, R. Troisi, S. Sturgeon, L. Brinton, and R. Hoover. 2000. Menopausal estrogen and estrogen-progestin replacement therapy and breast cancer risk. *Journal of the American Medical Association* 283:485–91.

Schnittker, J. 2005. Chronic illness and depressive symptoms in late life. *Social Science and Medicine* 60:13–24.

Smith, J., M. Borchelt, H. Maier, and D. Jopp. 2002. Health and well-being in the young old and oldest old. *Journal of Social Issues* 58:715–33.

Statistical Abstract of the United States: 2006. 125th ed. Washington, DC: U.S. Bureau of the Census.

Stevens, N. 2002. Re-engaging: New partnerships in late-life widowhood. *Ageing International* 27:27–42.

Szinovacz, M. E. and A. M. Schaffer. 2000. Effects of retirement on marital tactics. *Journal of Family Issues* 21:367–89.

Stroebe, M. and H. Schut. 2005. To continue or relinquish bonds: A review of consequences for the bereaved. *Death Studies* 29:477–95.

Tanner, D. 2005. Promoting the well-being of older people: Messages for social workers. *Practice* 17:191–205.

Taylor, A. C., M. Robila, and L. Hae Seung. 2005. Distance, contact, and intergenerational relationships: Grandparents and adult grandchildren from an international perspective. *Journal of Adult Development* 12:33–41.

Tornabene, L. 1976 *Long live the king: A biography of Clark Gable*. New York: Pocket Books.

Valliant, G. E. 2002. *Aging well: Surprising guideposts to a happier life from the Landmark Harvard study on adult development.* New York: Little, Brown.

Vance, D. E., V. G. Wadley, K. K. Ball, D. L. Roenker, and M. Rizzo. 2005. The effects of physical activity and sedentary behavior on cognitive health in older adults. *Journal of Aging & Physical Activity* 13:294–314.

Wallsten, S. S. 2000. Effects of care giving, gender, and race on the health, mutuality, and social supports of older couples. *Journal of Aging and Health* 12:90–111.

Weintraub, P. (2006) Guess who's back? *Psychology Today* 39 (4):79–84.

Webber, S., J. P. Scott, R. Wampler. 2000. Perceived congruency of goals as a predictor of marital satisfaction and adjustment in retirement. Poster, 62nd Annual Conference of the National Council on Family Relations, Minneapolis, November 12.

Wells, Y. D. 2000. Intentions to care for spouse: Gender differences in anticipated willingness to care and expected burden. *Journal of Family Studies* 5:220–34.

Wielink, G., R. Huijsman, and J. McDonnell. 1997. A study of the elders living independently in the Netherlands. *Research on Aging* 19:174–98.

Winterich, J. A. 2003. Sex, menopause, and culture: Sexual orientation and the meaning of menopause for women's sex lives. *Gender and Society* 17:627–42.

Wrosch, C. and J. Heckhausen. 2002. Perceived control of life regrets: Good for young and bad for old adults. *Psychology and Aging* 17:340–50.

Wu, Z., and R. Hart. 2002 The mental health of the childless elderly. *Sociological Inquiry* 72:21–42.

Zhan, H. J., and R. J. V. Montgomery. 2003. Gender and elder care in China: The influence of filial piety and structural constraints. *Gender and Society* 17:209–29.

The Future of Marriage and the Family

What can we predict about marriage and the family as we move toward 2010? We have no crystal ball but think it reasonable to suggest the following:

Marriage: Marriage will continue to be the lifestyle of choice for most (over 95%) U.S. adults. Though individuals are putting off getting married (to complete their educations, launch their careers, and/or become economically independent) until their mid- to late 20s (McGinn, 2006), there is no evidence that they intend to avoid marriage completely (*Statistical Abstract of the United States: 2006,* Table 50). The almost 5 million who marry annually and for the over 125 million already married, they will be among the happiest, healthiest, and most sexually fulfilled in our society.

Children: Children will become less of a requirement for personal/marital fulfillment. Coontz (2005) noted, "Most women have always loved their children. But women were also more aware of the sacrifices involved in child-rearing and much more interested than their husbands in limiting the number of children they bore" (p. 300). Similarly, while men have become more involved in rearing children, they have become more aware of the restrictions on their freedom children impose. What has changed is that the cultural imperative to have children is disappearing and the child-free option is being met with less disapproval.

While about 65 percent of children will continue to be born into married, two-parent homes, there will be increased acceptability for conceiving and rearing children outside legal marriage into single-parent families (both heterosexual and homosexual). Children born to married couples will not be immune to being reared in single-parent families since around 40 – 45 percent of their parents will divorce.

Singlehood: Singlehood will (in the cultural spirit of diversity) lose its stigma, slightly more will choose this option, and most of those who do will find satisfaction in it.

Gay marriage: Gay marriage in the United States will continue to be a controversial issue with equally powerful opposing forces. Massachusetts remains the only state recognizing same-sex marriage. But there is no Federal recognition of gay marriage, meaning that no Social Security benefits are available to survivors of deceased partners. The federal government and over half the states have enacted laws barring the recognition of gay marriages. About sixty percent (57.9%) of first-year college students throughout the U.S. agree that "Same-sex couples should have the right to legal marital status" (American Council on Education and University of California, 2005–2006).

Living together: Cohabitation has become an accepted, and for many, a predictable stage of courtship. The link between cohabitation and subsequent divorce will dissolve as more individuals elect to cohabit before marriage. Previously, only risk-takers and persons willing to abandon traditional norms lived together before marriage. In the future, mainstream individuals will increasingly cohabit.

Dual-earner relationships: The one-income family will continue to be in the minority. The prices of goods and services are such that it often takes two incomes to pay for housing, food, cars, and so on. Allowing time with the family will continue to take precedence over work time, and job flexibility will therefore remain an important factor in selecting a career and a job. Women will continue to give greater priority to family life than men and pay a greater cost in terms of decreased wages and career advancement.

Day care: As more children will spend more time in day care, increasing pressure will be put on the industry to provide "quality" care. But the sheer demand will not be followed by the funds to pay for trained staff in smaller classes, with the result that more children will spend time in substandard day-care facilities. Concerned parents will put pressure on Congress to address this national crisis.

Violence and abuse: Abuse in relationships will continue to occur behind closed doors. But cultural visibility of abusive relationships, support for leaving abusive partners, and the availability of hotlines and shelters will enable increasing numbers of individuals to bravely leave these relationships.

Divorce: Divorce will continue to end more marriages than the death of a spouse. Between 40 and 45 percent of persons beginning their lives together as spouses will end up as ex-spouses haggling over custody, child support, and visitation. The college-educated woman is less likely to divorce than the woman with no college education. While the former may be more able to leave, she has the skills to negotiate a more satisfying/egalitarian relationship with her husband (Coontz, 2006, p. 49). Divorce mediation will become a valuable alternative to some litigated endings. The stigma of divorce will remain. One is surrounded by friends and family at a wedding; most divorcing spouses go to court alone.

Elderly: Elderly individuals, particularly those in their 80s, will continue to find the end of life difficult. Health problems (and the attendant lack of health care), lower incomes, and the death of close loved ones reflect the reality for many elderly in the United States today. As our population continues to age, attention to problems of the elderly will continue and give hope to improvement of the final days.

"Nothing endures but change" is the summary statement for the future of marriage and the family. We embrace the future.

American Council on Education and University of California. 2005–2006. *The American freshman: National norms for fall, 2005.* Los Angeles: Higher Education Research Institute. U.C.L.A. Graduate School of Education and Information Studies.

Coontz, S. 2005. *Marriage, A History: How love conquered marriage.* New York: Penguin Books.

Coontz, S. 2006. Three "rules" that don't apply. *Newsweek* June 5, p. 49.

McGinn, D. 2006 Marriage by the numbers. *Newsweek* June 5, 40–48.

The best career advice is to find a career you like and get someone to pay you to do it.

Katherine Whithorn

Careers in Marriage and the Family

Authors

Contents

Students who take courses in marriage and the family sometimes express an interest in working with people and ask what careers are available if they major in marriage and family studies. In this Special Topics section we review some of these career alternatives, including family life education, marriage and family therapy, child and family services, and family mediation. These careers often overlap, so you might engage in more than one of these at the same time. For example, you may work in family services but participate in family life education as part of your job responsibilities.

For all the careers discussed in this section, it is helpful to have a bachelor's degree in a family-related field such as family science, sociology, or social work. Family science programs are the only academic programs that have a focus specifically on families and approach working with people from a family systems perspective. These programs have many different names, including child and family studies, human development and family studies, child development and family relations, and family and consumer sciences. Marriage and family programs are offered through sociology departments; family service programs are typically offered through departments of social work as well as through family science departments. Whereas some jobs are available at the bachelor's level, others require a master's or PhD degree. More details on the various careers available to you in marriage and the family follow.

Family Life Education

Family life education (FLE) is an educational process that focuses on prevention and on strengthening and enriching individuals and families. The family life educator empowers family members by providing them with information that will help prevent problems and enrich their family well-being. There are different ways that this education may be offered to families: a newsletter, one on one, or through a class or workshop. Examples of family life education programs include parent education for parents of toddlers through a child care center, a brown-bag lunch series on balancing work and family in a local business, a premarital or marriage enrichment program at your local church, a class on sexuality education in a high school classroom, and a workshop on family finance and budgeting at a local community center. Your role as a family life educator would involve your making presentations in a variety of settings, including schools, churches, and even prisons. As a family life educator, you may also work with military families on military bases, within the business world with human resources or employee assistance programs, and within social service agencies or cooperative extension programs. Some family life educators develop their own business providing family life education workshops and presentations.

To become a family life educator, you need a minimum of a bachelor's degree in a family-related field such as family science, sociology, or social work. You can become a certified family life educator (CFLE) through the National Council on Family Relations (NCFR). The CFLE credential offers you credibility in the field and shows that you have competence in conducting programs in all areas of family life education. These areas are Families in Society, Internal Dynamics of the Family, Human Growth and Development, Interpersonal Relationships, Human Sexuality, Parent Education and Guidance, Family Resource Management, Family Law and Public Policy, and Ethics. In addition,

Appreciation is expressed to Sharon Ballard, PhD, CFLE, for the development of Special Topic 1. Dr. Ballard is an assistant professor of Child Development and Family Relations at East Carolina University. She is also a certified family life educator through the National Council on Family Relations.

you must show competence in planning, developing, and implementing family life education programs.

Your academic program at your college or university may be approved for provisional certification. In other words, if you follow a specified program of study at your school, you may be eligible for a provisional CFLE certification. Once you gain work experience, you can then apply for full certification.

Marriage and Family Therapy

Whereas family life educators help prevent the development of problems, marriage and family therapists help spouses, parents, and family members resolve existing interpersonal conflicts and problems. The range of problems they treat include communication, emotional and physical abuse, substance abuse, sexual dysfunctions, and parent-child relationships. They work in a variety of contexts, including mental health clinics, social service agencies, schools, and private practice. As of 2007 there are about 50,000 marriage and family therapists in the United States and Canada.

Currently, 48 states and the District of Columbia license or certify marriage and family therapists (only one province in Canada has passed license legislation). Although an undergraduate degree in sociology, family studies, or social work is a good basis for becoming a marriage and family therapist, a master's degree is required in one of these areas. Some universities offer accredited master's degree programs specific to marriage and family therapy; these involve courses in marriage and family relationships, family systems, and human sexuality, as well as numerous hours of clinical contact with couples and families under supervision.

Full certification involves clinical experience with 1000 hours of direct client/couple/family contact; 200 of these hours must be under the direction of a supervisor approved by the American Association of Marriage and Family Therapists (AAMFT). In addition, most states require a licensure examination. The AAMFT is the organization that certifies marriage and family therapists. A marriage and family therapist can be found at http://family-marriage-counseling.com/therapists-counselors.htm.

Child and Family Services

In addition to family life educators and marriage and family therapists, careers are available in agencies and organizations that work with families, often referred to as social service agencies. The job titles within these agencies include family interventionist, family specialist, and family services coordinator. Your job responsibilities in these roles might involve your helping clients over the telephone, coordinating services for families, conducting intake evaluations, performing home visits, facilitating a support group, or participating in grant writing activities. In addition, family life education is often a large component of child and family services. You may develop a monthly newsletter, conduct workshops or seminars on particular topics, or facilitate regular educational groups.

Some agencies or organizations focus on helping a particular group of people. If you are interested in working with children, youth, or adolescents, you might find a position with Head Start, youth development programs such as the Boys and Girls Club, after-school programs (e.g., pregnant or parenting teens), child-care resource or referral agencies, or early intervention services. Child-care resource/referral agencies assist parents in finding child care, provide training for child-care workers, and serve as a general resource for parents and for childcare providers. Early intervention services focus on children with special needs. If you work in this area, you might work directly with the children or you might work with the families and help to coordinate services for them.

Other agencies focus more on specific issues that confront adults or families as a whole. Domestic violence shelters, family crisis centers, and employee assistance programs are examples of employment opportunities. In many of these positions, you will function in multiple roles. For example, at a family crisis center, you might take calls on a crisis hotline, work one on one with clients to help them find resources and services, and offer classes on sexual assault or dating violence to high school students.

Another focus area in which jobs are available is aging. There are opportunities within residential facilities such as assisted living facilities or nursing homes, senior centers, organizations such as the Alzheimer's Association, or agencies such as Area Agencies on Aging. There is also a need for elder-care resource and referral, as more and more families are finding that they have caregiving responsibilities for an aging family member. These families have a need for resources, support, and assistance in finding residential facilities or other services for their aging family member. Many of the available positions with these types of agencies are open to individuals with bachelor's degrees. However, if you get your master's degree in a program emphasizing the elderly, you might have increased opportunity and will be in a position to compete for various administrative positions.

Family Mediation

In the chapter on divorce, we emphasized the value of divorce mediation. It is also known as family mediation and involves a neutral third party negotiating with divorcing spouses on the issues of child custody, child support, spousal support, and division of property. The purpose of mediation is not to reconcile the partners but to help the couple to make decisions about children, money, and property as amicably as possible. A mediator does not make decisions for the couple but supervises communication between the partners, offering possible solutions.

Although some family and divorce mediators are attorneys, family life professionals are becoming more common. Specific training is required that may include numerous workshops or a master's degree, offered at some universities (e.g., University of Maryland). Most practitioners conduct mediation in conjunction with their role as a family life educator, marriage and family therapist, or other professional. In effect, you would be in business for yourself as a family or divorce mediator.

Students interested in any of the above career paths can profit from getting initial experience in working with people through volunteer or internship agencies. Most communities have crisis centers, mediation centers, and domestic abuse centers that permit students to work for them and gain experience. Not only can you provide a service, but you can also assess your suitability for the "helping professions" as well as discover new interests. Talking with persons already in the profession you want to enter is also a good idea for new insights. Your teacher may already be in the marriage and family profession you would like to pursue or be able to refer you to someone who is.

CERTIFIED FAMILY LIFE EDUCATOR

Money isn't everything as long as you have enough.

Malcolm Forbes (1919–1990),
publisher of Forbes Magazine

Money and Debt Management in Marriage

Contents

One way of staying on a budget is keeping expenses low. These women are at a flea market looking for bargains.

G etting money and staying out of debt are top issues couples report as creating considerable stress in their relationship. In this Special Topics section we discuss budgeting, investing, life insurance, credit, and buying a house. The latter is one of the most significant decisions a couple makes and one that requires considerable negotiation and cooperation.

Budgeting

Developing a budget is a way of planning your spending. Since money spent on X cannot be spent on Y, budgeting requires conscious value choices about which bills should be paid, what items should be bought, and what expenditures should be delayed. It's a good idea for everyone to have a budget, but couples in particular need to develop a budget if they are always out of money long before their next paycheck, if they cannot save money, and if they cannot make partial payments or pay off existing bills but keep incurring new debts.

To develop a budget (see Table ST 2.1), list and add up all your monthly take-home (after-tax) income from all sources. This figure should represent the amount of money your family will actually have to spend each month. Next, list and add up all your fixed monthly expenses, such as rent, utilities, telephone, and car payment. Other fixed expenses include such items as life, health, and car insurance. You may not receive a bill for some of these expenses every month, so divide the yearly cost of each item by twelve so that you can budget the amount on a monthly basis. For example, if your annual life insurance premium is $240, you should budget $20 per month for that expense.

Set aside a minimum of 5 percent of your monthly income—more if possible—for savings, and include this sum in your fixed expenses. By putting a fixed amount in a savings account each month, you will not only have money available for large purchases, such as a car or a major home appliance, but also have an emergency fund to cover unforeseen expenses such as those caused by an extended illness or a long-distance move. The size of an emergency fund should be about twice your monthly income. Although you can personally set aside some of your monthly income for savings, an alternative is to instruct the bank to transfer a certain sum each month from your checking account to your savings account, or you can join a payroll savings plan. Under the latter arrangement, a portion of your monthly salary is automatically deposited in your savings account without ever passing through your hands.

After adding together all your fixed monthly expenses, including savings, subtract this amount from your monthly take-home income. What remains can be used for such day-to-day expenses as food (groceries and restaurant meals), clothes (including laundry, dry cleaning, alterations, and new clothes), personal care (barber and hairdresser, toiletries, cosmetics), and recreation (theater, movies, concerts, books, magazines).

If you come out even at the end of the month, you are living within your means. If you have money left over, you are living below your means. If you had to tap your savings or borrow money to pay your bills last

Table ST 2.1 Monthly Budget for Two-Earner Couple
(both spouses employed full-time)

Sources of Income	Fixed Expenses	Day-to-Day and Discretionary Expenses
Husband's take-home pay: $2,250	Rent: $650	Food: $425
Wife's take-home pay: $1,575	Utilities: $135	Clothes: $170
Interest earned on savings: $68	Telephone: $70	Personal care: $90
	Insurance: $85	Recreation: $120
	Car payments: $665	Miscellaneous: $110
	Cable/Internet: $50	
	Savings: $300	
TOTAL: $3,893	TOTAL: $1995	TOTAL: $915
	DIFFERENCE: $1938 (amount available for day-to-day expenses)	

This dual-earner couple should have $1023 in unspent money at the end of each month. This budget is of a model couple who have two solid incomes. The reality is that many couples have lower-paying jobs and can't or don't live within their incomes and go into debt each month.

month, you are living above your means. Knowing whether you are living within, below, or above your means depends on keeping accurate records.

And remember, you must pay taxes on what you earn. Most taxes (minimal at low incomes and around 40 percent for higher incomes) are paid through withholding and do not need to be considered in one's budget. However, self-employed people do need to include taxes as a budget item.

Investing

Saving and investing money should be a part of every budget. By allocating a specific amount of your monthly income to savings or an investment plan, you can accumulate money for both short-term goals (vacation, down payment on a house) and long-term goals (college education for children, retirement income).

By investing, you use money to make more money. All investments must be considered in terms of their risk and potential yield. In general, the higher the rate of return on an investment, the greater the financial risk.

There are several ways to invest your money to earn more money. A common method is to put money in a bank or savings and loan institution. Doing so is risk free, and your deposit is insured by the federal government. However, banks pay a relatively low rate of interest compared with other alternatives. Some individuals keep only the amount that they will need quick access to and invest the rest of their funds in alternatives that earn them a greater return.

One alternative to a savings account is to buy blue chip stock in companies such as AT&T. A complete list is available in the financial section of the newspaper in larger cities. Market crashes (e.g., 1929) and highs (the Dow Jones average hit over 12,000 in late 2006) remind investors that all stocks, even those of "blue chip" companies, can gain or lose considerable value. One strategy investment analysts recommend is to diversify your holdings—spread your investments over a lot of different stocks. However, all of these can go down together. Be careful. The stock market is called "white-collar gambling."

Do not invest in speculative stocks unless you can afford to lose the money. In addition to risk and return on investment, the liquidity—the ease with which your investment can be converted into cash—is an important consideration. Stocks and bonds can be sold quickly to provide cash in hand. In contrast, if you have invested in a building or land, you must find a buyer who is willing to pay the price that you are asking for your asset before you can convert it to cash.

The amount of your time required to make your investment grow is also important. A real estate investment can give you a considerable return on your money, but it may also demand a lot of your time—perusing the newspaper, arranging for loans, placing ads, and showing the property to prospective tenants, not to mention fixing leaky faucets, mowing grass, and painting rooms.

Also, consider the maturity date of your investment. For example, suppose you invest in a 10-year certificate at a savings bank. Although the bank will pay you, say, 5 percent interest and guarantee your investment, you can't get your principal (the money you deposited) or the interest until the 10 years is up unless you are willing to pay a substantial penalty. Regular savings accounts have no maturity date. You can withdraw any amount of your money from your savings account at the bank whenever you want it, but you may earn only, say, 1 percent interest on the money while it remains in the account.

A final investment consideration is taxation. Investment decisions should be made on the basis not of how much money you can make but how much money you can keep. Tax angles should be considered as carefully as risk and yield issues.

Although savings, life insurance, real estate, and stock investments are probably familiar forms of investment to you, the other types of investments may need further clarification.

Annuities provide a monthly income after age 65 (or earlier if desired), in exchange for your investment of monthly premiums during your working years. For example, a 65-year-old man might receive $200 per month as long as he lives (or a lump sum) if he has paid the insurance company $338 annually since age 30. Bonds are issued by corporations and federal, state, and local governments that need money. In exchange for your money, you get a piece of paper that entitles you to the return of the sum you lend at a specified date (up to 30 years) plus interest on the money. Although bonds are safer than stocks, you could lose all your money if the corporation you lend the money to goes bankrupt. United States savings bonds are safe but pay a comparatively low rate of interest.

Mutual funds offer a way of investing in a number of common stocks, corporate bonds, and government bonds at the same time. You invest your money in shares of the mutual fund, whose directors invest the fund's capital in various securities. If the securities they select increase in price, so does the value of your shares in the mutual fund. But they may also decrease in value (as might stocks, referred to earlier).

Treasury bills (T-bills) are issued by the Federal Reserve Bank. You pay a lower price for the bill than its cash value. For example, you might pay $900 for a T-bill that will be worth $1,000 on maturity. Maturity of the bill occurs at 3, 6, 9, or 12 months. The longer the wait, the more money paid on the investment.

Money market investments require a stockbroker who uses your money, together with the money others wish to invest, to purchase high-interest securities. Your money can be withdrawn in any amount at any time. As with all investments, you pay a fee to the broker or agent for investing your money. In the past several years, increasing numbers of people have put their money in money market funds.

Individual retirement accounts (IRAs) permit you to set aside up to $2,000 each year for your personal retirement fund. The money you put in your IRA is not automatically tax-free. (You must meet certain criteria.) Each spouse may open his or her own IRA.

Certificates of deposit (CDs) are insured deposits given to the bank that earn interest at a rate from one day to several years. These high-yield, low-risk investments have become extremely popular. A minimum investment of $500 or more is usually required.

One of the purposes of investing is to prepare for retirement. Yet, according to the Employee Benefit Research Institute, three-fourths of workers have not started saving for retirement and have no idea how much money they will need. Furthermore, they are unaware that inflation has eroded what savings they have.

A sound investment strategy emphasizes identifying your goals (having money for child's education, retirement, etc.), determining your appetite for risk (stocks are more risky than Treasury bills), and diversifying (not putting all your eggs in one basket). The best decisions are also informed and based on a lot of homework. If an investment sounds too good to be true, it probably is.

Life Insurance

In addition to saving and investing, one must be knowledgeable about life insurance. The major purpose of life insurance is to provide income for dependents when the primary wage earner dies. With dual-earner couples, life insurance is often necessary to prevent having to give up their home when one wage earner dies. Otherwise, the remaining wage earner may not be able to make the necessary mortgage payments.

Unmarried, childfree college students probably do not need life insurance. No one is dependent on them for economic support. However, the argument used by some insurance agents who sell campus policies is that college students should buy life insurance while they are young when the premiums are low and when insurability is guaranteed. Still, consumer advocates typically suggest that life insurance for unmarried, childfree college students is not necessary.

When considering income protection for dependents, there are two basic types of life insurance policies: (1) term insurance and (2) insurance plus investment. As the name implies, term insurance offers protection for a specific time period (usually 1, 5, 10, or 20 years). At the end of the time period, the protection stops. Although a term insurance policy offers the greatest amount of protection for the least cost, it does not build up cash value (money the insured would get upon surrendering the policy for cash).

Insurance-plus-investment policies are sold under various names. The first is straight life, ordinary life, or whole life, in which the individual pays a stated premium (based on age and health) as long as the individual lives. When the insured dies, the beneficiary is paid the face value of the policy (the amount of insurance originally purchased). During the life of the insured, the policy also builds up a cash value (which is tax free), which permits the insured to borrow money from the insurance company at a low rate of interest. A second type of life insurance is a limited payment policy, in which the premiums are paid up after a certain number of years (usually 20) or when the insured reaches a certain age (usually 60 or 65). As with straight life, ordinary life, or whole life policies, limited payment policies build up a cash value, and the face value of the policy is not paid until the insured dies. The third type of life insurance is endowment insurance, in which the premiums are paid up after a stated number of years and can be cashed in at a stated age.

Regardless of how they are sold, insurance-plus-investment policies divide the premium paid by the insured. Part pays for the actual life insurance, and part is invested for the insured, giving the policy a cash value. Unlike term insurance, insurance-plus-investment policies are not canceled at age 65. Which type of policy—term or insurance-plus-investment—should you buy? An insurance agent is likely to suggest the latter and point out the advantages of cash value, continued protection beyond age 65, and level premiums. But the agent has a personal incentive for your buying an insurance-plus-investment policy: the commission on this type of policy is much higher than it is on a term insurance policy.

A strong argument can be made for buying term insurance and investing the additional money that would be needed to pay for the more expensive insurance-plus-investment policy.

The annual premium for $50,000 worth of renewable term insurance at age 25 is about $175. The same coverage offered in an ordinary life policy—the most common insurance-plus-investment policy—costs $668 annually, so the difference is $493 per year. If you invested this money and were able to increase its value by 5 percent, at the end of 5 years you would have $2,860.32. In contrast, the cash value of an ordinary life policy after 5 years would be $2,350. But to get this money, you have to pay the insurance company interest to borrow it. If you don't want to pay the interest, the company will give you this amount but cancel your policy. In effect, you lose your insurance protection if you receive the cash value of your policy. With term insurance, you have the $2,860.32 for your trouble and you can use the money whenever you want. And you can do so without affecting your insurance program.

It should be clear that for term insurance to be cheaper, you must invest the money you would otherwise be paying for an ordinary life insurance policy. If you can't discipline yourself to save (and if your investments turn out to be unlucky ones), you might wish you had bought an insurance-plus-investment policy to ensure savings.

Finally, what about the fact that term insurance stops when you are 65, just as you are moving closer to death and needing the protection more? Again, by investing the money that you would otherwise have spent on an insurance-plus investment policy, you will have as much money for your beneficiary as your insurance-plus-investment policy would earn, or even more.

Whether you buy a term policy, an insurance-plus-investment policy, or both, there are three options to consider: guaranteed insurability, waiver of premiums, and double or triple indemnity. All are inexpensive and generally should be included in a life insurance policy.

Guaranteed insurability means that the company will sell you more insurance in the future regardless of your medical condition. For example, suppose you develop cancer after you buy a policy for $10,000. If the guaranteed insurability provision is in your contract, you can buy additional insurance. If not, the company can refuse you more insurance.

Waiver of premiums provides that your premiums will be paid by the company if you become disabled for 6 months or longer and are unable to earn an income.

Such an option ensures that your policy will stay in force because the premiums will be paid. Otherwise, the company will cancel your policy.

Double or triple indemnity (rent the video of the classic movie *Double Indemnity* to see how this works) means that if you die as the result of an accident, the company will pay your beneficiary twice or three times the face value of your policy.

An additional item you might consider adding to your life insurance policy is a disability income rider. If the wage earner becomes disabled and cannot work, the financial consequences for the family are the same as though the wage earner were dead. With disability insurance, the wage earner can continue to provide for the family up to a maximum of $3,500 per month or two-thirds of the individual's salary, whichever is smaller. If the wage earner is disabled by accidental injury, payments are made for life. If illness is the cause, payments may be made only to age 65. A 27-year-old spouse and parent who was paralyzed in an automobile accident said, "It was the biggest mistake of my life to think I needed only life insurance to protect my family. Disability insurance turned out to be more important."

In deciding to buy life insurance, it might be helpful to consider the following:

1. Decide whether you really need insurance. People with no dependents rarely need it.

2. Decide how much coverage you need. The cost of rearing a middle-class child from conception through college (public) is approximately $300,000.

3. Compare prices. Not all policies and prices are the same. In some cases, the higher premiums are for lower coverage.

4. Select your agent carefully. Only one in 10 life insurance agents stays in the business. The person you buy life insurance from today may be in the real estate business tomorrow. Choose an agent who has been selling life insurance for at least 10 years.

5. Select your company carefully. The big names in insurance, such as John Hancock, Prudential, and State Farm, may not always offer you the best protection at the most affordable prices.

6. Seek group rates. Group life insurance is the least expensive coverage. See whether your employer offers a group plan.

7. Proceed slowly. Don't rush into buying an insurance policy. Consult several agents, read *Consumer Reports,* and talk with friends to find out what they are doing about their insurance needs.

8. Select features carefully. Don't buy features you don't need. A waiver of premium, for which you are assessed a percentage of the yearly cost of the policy as a surcharge, continues to pay your premiums if you become permanently disabled. But you may be better off purchasing more disability insurance.

Although this section emphasizes life insurance, health coverage is equally important. With hospital costs of over $1,000 a day, life savings can be eliminated within a short time if you suffer a serious injury or illness. The least expensive health policies are group policies available through some employers. If such policies are not available, major medical policies with high deductibles are relatively inexpensive and will protect the insured in case of major illness.

Credit

College students might be mindful that their economic future can be wrecked by using credit unwisely. The "free" credit cards they receive in the mail are a Trojan horse and can plunge them into massive debt from which it will take years to recover.

Types of Credit Accounts

You use credit when you take an item (Ipod, Blackberry, high-definition television) home today and pay for it later. The amount you pay later will depend on the arrangement you make with the seller. Suppose you want to buy a small high-definition television set that costs $600. Unless you pay cash, the seller will set up one of three types of credit accounts with you: installment, revolving charge, or open charge.

Under the installment plan, you make a down payment and sign a contract to pay the rest of the money in monthly installments. You and the seller negotiate the period of time over which the payments will be spread and the amount you will pay each month. The seller adds a finance charge to the cash price of the television set and remains the legal owner of the set until you have made your last payment. Most department stores, appliance and furniture stores, and automobile dealers offer installment credit. The cost of buying the $600 high-definition TV is calculated in Table ST 2.2.

Instead of buying your $600 television set on the installment plan, you might want to buy it on the revolving charge plan. Most credit cards, such as Visa and MasterCard, represent revolving charge accounts that permit you to buy on credit up to a stated amount during each month. At the end of the month, you may pay the total amount you owe, any amount over the stated minimum payment due, or the minimum payment. If you choose to pay less than the full amount, the cost of the credit on the unpaid amount is approximately 1.5 percent per month, or 18 percent per year. For instance, if you pay $100 per month for your television for 6 months, you will still owe $31.62 to be paid the next

Table ST 2.2 Calculating the Cost of Installment Credit

Amount to be financed

Cash price	$600.00
− down payment (if any)	−50.00
Amount to be financed	$550.00

Amount to be paid

Monthly payments	$ 35.00
× number of payments	×18
Total amount repaid	$630.00

Cost of credit

Total amount repaid	$630.00
− amount financed	−550.00
Cost of credit	$ 80.00

Total cost of TV

Total amount repaid	$630.00
+ down payment (if any)	+50.00
Total cost of TV	$680.00

month, for a total cost (television plus finance charges) of $631.62. It is estimated that 70 percent of college students in 4-year universities have at least one credit card (Joo et al., 2003). In a sample of 242 undergraduates, about 10 percent paid only the minimum monthly balance each month. Almost half paid the balance on their cards in full at the end of each month. The average balance for those who held a credit card was $890.42; of those who paid only the monthly minimum, the average balance was $1,769.85 (Joo et al., 2003). In a sample of 1027 university students, 11 percent reported that they owed over $1,000 on one or more credit cards (Knox and Zusman, 2006).

It is important to protect your credit card numbers. Identity theft is the number one concern of consumers today (see the Identity Theft section below), because it can ruin your credit rating for years.

You can also purchase items on an open charge (30-day) account. Under this system you agree to pay in full within 30 days. Since there is no direct service charge or interest for this type of account, the television set would cost only the purchase price. For example, Sears and JCPenney offer open charge (30-day) accounts. If you do not pay the full amount in 30 days, a finance charge is placed on the remaining balance. The use of both revolving charge and open charge accounts is wise if you pay off the bill before finance charges begin. In deciding which type of credit account to use, remember that credit usually costs money; the longer you take to pay for an item, the more the item will cost you.

Credit Rating

When you apply for a loan, the lender will seek a credit report from credit bureaus such as Equifax, Trans Union, and Experian. In effect, you will have a "credit score"—also referred to as a NextGen score or a FICO (Fair Isaac Credit Organization) score—calculated as follows:

35% based on late payments, bankruptcies, judgments
30% based on current debts
15% based on how long accounts have been opened and established
10% based on type of credit (credit cards, loan for house/car)
10% based on applications for new credit or inquiries

As these percentages indicate, the way to improve your credit is to make payments on time and reduce your current debt.

Your score will range from 620 to 850; the higher your score, the lower your rate of interest. For example, on a $150,000 30-year fixed-rate mortgage, a score above 760 would result in a 5.5% interest rate with a monthly payment of $852. In contrast, a score of 639 and below would result in an interest rate of 7.09%, with a monthly payment of $1,007 (see http://www.myfico.com/ or type in "credit report" on www.google.com).

Three Cs of Credit

Whether you can get credit will depend on the rating you receive on the three Cs: character, capacity, and capital. Character refers to your honesty, sense of responsibility, soundness of judgment, and trustworthiness. Capacity refers to your ability to pay the bill when it is due. Such issues as the amount of money you earn and the length of time you have held a job will be considered in evaluating your capacity to pay. Capital refers to such assets as bank accounts, stocks, bonds, money market funds, and real estate.

It is particularly important that married individuals establish credit ratings in their own name in case they become widowed or divorced. Otherwise, their credit will depend on their spouse; if one spouse dies, the other spouse will have no credit of his or her own. Similarly, if a couple divorces, each ex-spouse will want to have established his or her own credit during the marriage.

It is also important that individuals not depend on credit to pay for necessities such as food, rent, and utilities. Continually spending more than one's income and taking all credit cards to the limit can lead to financial trouble and eventual bankruptcy.

Identity Theft

When someone poses as you and uses your credit history to buy goods and services, that person has stolen your identity. More than 10 million Americans are victims of **identity theft,** which can destroy your credit, plunge you into debt, and keep you awake at night with lawsuits from creditors. It happens when someone gets access to personal information such as your Social Security number, credit card number, or bank account number and goes online to pose as you to buy items or services. Identity theft is the number one fraud complaint in the United States.

Safeguards include (1) never giving the above information over the phone or online unless you initiate the contact and (2) shredding bank and credit card statements and preapproved credit card offers. (A shredder can cost as little as $20.) Also, don't pay your bills by putting an envelope in your mailbox with the flag up—use a locked box or the post office. Finally, check your credit reports, scrutinize your bank statements, and guard your personal identification number (PIN) at automatic teller machines (ATMs). If you use the Internet, protect your safety by installing firewall software.

Buying A Home

This text has been about choices. In reference to housing, the number of choices is seemingly endless. Such choices begin with what you can afford.

Figuring Out How Much You Can Afford

A prerequisite for owning a house or condominium is being able to afford one. The median price of a single-family home in the United States in 2006 was $217,900. The price of a single-family home varies by region. Median prices in 2006 by region were as follows: West, $344,000; Midwest, $158,800; South, $179,700; and Northeast, $285,200 (Bandy, 2006).

Lest we forget, individuals throughout the world live in a variety of housing structures. It is not the cost of the house but the value of the relationships between the individuals inside that make a house a home.

A financial officer in a bank or a real estate agent can determine the value of the dwelling you can afford. Such a determination is based on your net worth—what you own minus what you owe, how much you can afford as a down payment, and how much you can afford in monthly payments. In general, the more you pay down, the lower your monthly payments. Some loans involve a down payment of only 5 percent of the value of the dwelling, and others may require 10 percent or 20 percent.

Deciding What Type of Housing You Want

If you are buying a house, you must decide whether to buy a house that has already been built, buy one that is to be built identical to a model already available, or build your own custom-made dream house.

Many would-be homeowners believe that custom houses cost more than those already built or those based on a model. This is not necessarily so since custom houses carry no real estate agent fees. But be careful. The investment in a custom-built home is in terms of your own time (and frustration).

An alternative to buying a single-family house (old, new, custom-built) is to buy a condominium, a cooperative apartment, or a mobile home. A condominium involves your owning the space in which you live and sharing the ownership of common areas such as sidewalks, grounds, parking lot, and elevators. A cooperative apartment means that you own a share of the corporation that owns the apartments and your privilege to live there is based on your ownership. Mobile homes, some of which are referred to as double-wides, are rarely mobile, since 90 percent stay on their original site. Some individuals buy land on which they put their mobile home. Others lease the land on which they place their home. Such an arrangement makes the homeowner vulnerable, since the owner of the land can force you to move your home.

The type of housing that individuals select is based on what they can afford and what they need. The following are some of the factors to be considered:
Location: neighborhood and proximity to work, school, grocery stores, etc.
Space: number of bedrooms and baths; basement; workshop area; garage; etc.
Floor plan: one or two levels, separation of activities, kitchen placement, etc.
Other: need for fenced-in backyard, land for garden, etc.

The lot on which the house sits is particularly important for persons who enjoy gardening, who have small children, or who have a pet. The best house on an undesirable lot is not a house to buy. Other considerations are how much privacy you want and whether you will maintain the grounds yourself. Finally, identify your personal objections, such as small bathrooms, fake brick siding, heavy traffic, a lack of trees, or too little sunlight.

Deciding on a House

Having identified what you can afford and what you want, look at prospective houses advertised in the newspaper or recommended by a real estate agent or a buyer's broker. The latter is a person you hire at a fixed price to find the house you want at the least cost. The typical real estate agent gets a 6 percent commission on the house you buy. Hence, the more expensive the house you buy, the more money the agent makes. You can find the name of a buyer's broker from your local Board of Realtors. About 10 percent of homebuyers hire a buyer's broker.

As you begin to look at houses, look at a lot of them so that you become aware of the range

Buying a house or a condominium must include insuring your purchase against fire damage.

© David Knox

of possibilities. Once you have identified the house that you want, consider having it inspected by a person listed in the Yellow Pages under "Building Inspection Services" or "Real Estate Services." Such a person is knowledgeable about houses and will make a written report to you on issues such as water drainage, exterior walls, roofing, basement, electrical system, plumbing system, heating/cooling systems, and kitchen and bathrooms in terms of ventilation and flooring.

After the house passes inspection, have the house appraised. This involves hiring a professional residential appraiser who is not invested in selling you the house to tell you what the house is actually worth. This may prevent you from paying too much or make you aware that you have found a good deal. Once you have decided what you are willing to pay, make an offer in writing to the seller and identify all conditions, such as gutters to be fixed by seller within a certain period of time or taking action on a positive termite inspection report. Many asking prices have a good bit of padding built into them, to see if anyone will take the bait. "Don't feel you have to offer the full asking price, or even something close to it, just because that's what the owner is seeking," noted Knight Kiplinger, a financial adviser. If you are working with a real estate agent, this person can help you find a reputable home inspector and advise you about the listing price via a report on comparable sales in the area.

Finding the Right Mortgage

Once you have signed a contract for the house, find the best deal on a mortgage—the amount of money you will pay a lender for the house. Shop around— you are looking for a loan, not a lender. A difference of 0.5 percent on a 30-year $100,000 mortgage can mean thousands you save or pay unnecessarily. When comparing rates, ask for the true annual percentage rate (APR).

Also decide whether you want a fixed rate or an adjustable rate. The latter means that your monthly payment will go up or down depending on how the economy is doing. Conventional wisdom would have buyers lock in low interest rates with a fixed-rate loan when it appears that interest rates will be heading up over the next few years. On the other hand, if it looks as if rates will go down or stay about the same, an adjustable-rate mortgage might be better.

Regardless of the rate you decide on, be aware of the enormous price you are paying for the loan. The first payment on a $100,000 loan at 7 percent (in 2007 you should be able to get a loan for much less than 7 percent) for 30 years is $665.25. Only $50.00 of this amount is applied to the principal (the amount that actually goes toward your owning the house). The remainder ($615.25) is interest.

Over the 30-year period, you will pay $239,490 for the $100,000 loan. To shorten the number of payments, and therefore the total amount you must pay, consider making extra payments each month directed specifically toward the principal. The sooner the principal is paid off, the sooner all payments will stop.

KEY TERM

identity theft

WEB LINKS

Identity Theft: Federal Trade Commission
http://www.consumer.gov/idtheft/
Identity Theft: Prevention and Survival
http://www.identitytheft.org/
Ms.Money.com (budgeting documents)
http://www.msmoney.com
Countrywide Financial
http://www.countrywide.com

HomeGain
http://www.homegain.co
MBA Online (Mortgage Bankers Association)
http://www.mbaa.org
Quotesmith (Insure.com)
http://www.quotesmith.com

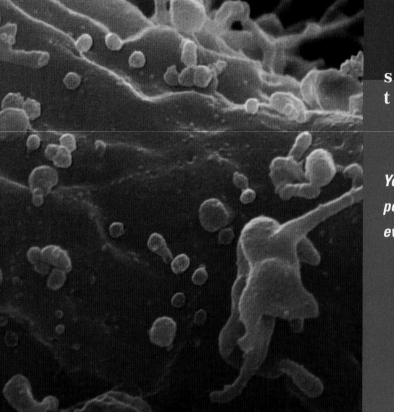

special
topic

3

You're not just sleeping with one person, you're sleeping with everyone they ever slept with.

Teresa Crenshaw, sex therapist

Human Immunodeficiency Virus and Other Sexually Transmitted Infections

Contents

Human Immunodeficiency Virus Infection

Other Sexually Transmissible Infections

Getting Help for HIV and Other STIs

T he focus of this Special Topic section on **sexually transmitted infections** (STIs) is that sexual encounters necessitate making choices about the level of infection risk one is willing to take. The wrong choice might lead to an early death.

Media information about the value of abstinence, safer sex, and the regular use of latex and polyurethane condoms may not translate into consistent condom use. Denial ("It won't happen to me") is the primary reason for failure to use condoms.

Human Immunodeficiency Virus Infection*

Human immunodeficiency virus (HIV) attacks the white blood cells (T-lymphocytes) in human blood, impairing the immune system and a person's ability to fight other diseases. Of all the diseases that can be transmitted sexually, HIV infection is the most life-threatening. Symptoms, if they occur at all, surface between 2 and 6 weeks after infection and are often dismissed because they are similar to symptoms of influenza. Antibodies may appear in the blood within 2 months but more often take 3 to 6 months before they reach reliable detectable levels.

International Data

Since the beginning of the HIV epidemic, more than 25 million people have died of acquired immunodeficiency syndrome (**AIDS**). Four million new cases are infected annually with about 40 million individuals living with AIDS currently (Worldwide HIV and AIDS Statistics, 2006). Barring some major medical breakthrough, many of these people worldwide will die within the next 10 years or so (U.S. Census Bureau, 2004). About three million die of AIDs annually (Worldwide HIV and AIDS Statistics, 2006).

Before HIV or its antibodies are detectable, infected individuals will test negative for the virus, making them silent carriers (this is called the "window period"). Although not all persons who have HIV get AIDS (half of those infected with HIV will develop AIDS within 10 years after the infection), all are infectious and are able to transmit the virus to others. Hence, even though your partner tested negative for HIV, she or he could still transmit the virus to you. If HIV progresses to AIDS, the person's body is vulnerable to opportunistic diseases that would be resisted if the immune system were not damaged. The two most common diseases associated with AIDS are a form of cancer called Kaposi's sarcoma (KS) and *Pneumocystis carinii* pneumonia (PCP), a rare form of pneumonia.

Seventy percent of all HIV-related deaths result from PCP. HIV can also invade the brain and nervous system, producing symptoms of neurological impairment and psychiatric illness.

Modes of Transmission of HIV Infection

In the United States, HIV infection was first noted in 1981 among men who had sex with men (or MSM) who had multiple sex partners. Estimates show that male-to-male sexual contact is still the leading transmission mode of HIV infection among U.S. males. The predominant mode of transmission of HIV for women is heterosexual contact (Stephenson, 2003). HIV is also transmitted through sharing needles during injection drug use and, more rarely, through perinatal transmission and workplace needle contact.

*Appreciation is expressed to Beth C. Burt, M.A.Ed., CHES, for updating Special Topic 3. She is a health education specialist with the Wake Area Health Education Center at Wake Med Hospital, Raleigh, North Carolina.

Prevalence of HIV/AIDS

More than 90 percent of people worldwide with HIV infection or AIDS live in developing countries (U.S. Census Bureau, 2004). At the end of 2003, an estimated 1.1 million persons in the United States were living with HIV infection or AIDS; 24 to 27 percent were unaware of their HIV infection because it had not been diagnosed (Glynn and Rhodes, 2005). About 80 percent of these infected persons are male, and 20 percent are female. According to the Centers for Disease Control and Prevention (CDC; Atlanta), AIDS affects nearly seven times more African-Americans and three times more Hispanics than whites (NIAID, 2005). *Risk group* is a term that identifies a certain demographic trait associated with a higher chance of having a certain condition, such as infection with HIV. However, anyone who is exposed can become infected.

Tests for HIV Infection

Early medical detection of HIV has decided benefits, including taking medications to reduce the growth of HIV and preventing the development of some life-threatening conditions. An example of the latter is pneumonia, which is more likely to develop when one's immune system has weakened.

HIV counselors recommend that individuals who answer yes to any of the following questions should definitely seek testing:

- If you are a man, have you had sex with other men?
- Have you had sex with someone you know or suspect was infected with HIV?
- Have you had an STI?
- Have you shared needles or syringes to inject drugs or steroids?
- Did you receive a blood transfusion or blood products between 1978 and 1985?
- Have you had sex with someone who would answer yes to any of these questions?

Additionally, counselors suggest that if you have had sex with someone whose sexual history you do not know or if you have had numerous sexual partners, your risk of HIV infection is increased, and you should seriously consider being tested.

Finally, if you plan to become pregnant or are pregnant, the American Medical Association recommends being tested. Many states, such as North Carolina, passed laws in the 1990s requiring health care providers to counsel pregnant women as early in pregnancy as possible and to offer HIV testing. However, one study found that only 58 percent of surveyed women reported that they were asked by their caregivers if they wanted to be tested for HIV, and only 49 percent reported receiving pretest counseling (Sengupta and Lo, 2004). The newest methods of HIV testing provide the fastest accurate results. In 1994, OraSure Technologies® released its new oral specimen collection device. This device allows for HIV testing with a shorter turnaround time for results, using a simple, 2-minute collection procedure that could be performed by trained health care providers. In November 2002, Orasure® received approval from the U. S. Food and Drug Adminstration for its OraQuick® Rapid HIV-1 Antibody Test. OraQuick became the first rapid, point-of-care test designed to detect antibodies to HIV-1 within approximately 20 minutes. This test requires taking a blood sample through a finger-prick. Finally, in 2004, OraQuick® *ADVANCE*™ Rapid HIV-1/2 Antibody Test, a rapid test that provides accurate results for both HIV-1 and HIV-2 in 20 minutes, was introduced; it uses oral fluid, finger-prick, or venipuncture whole blood or plasma specimens.

Other forms of testing may take up to 2 weeks for results. All three methods are 99 percent effective for detecting HIV antibodies when present.

Home HIV testing is available by calling 1-800-HIV-TEST (800-448-8378). The cost is $49. You mail in a sample of blood and call 7 days later to get anonymous test results.

Treatment for HIV and Opportunistic Diseases

While research to find a vaccine for HIV continues (Sternberg, 2003), several drugs (AZT, 3TC, indinavir, ritonavir, and saquinavir) used in various combinations have demonstrated the most efficacy in the treatment of HIV infection. We emphasize "the most efficacy" since some of the drugs have proven to be less potent and more toxic than previously thought. Patients are not cured, but the progress of the disease is slowed and the survival rate is increased.

About 25 other drugs are used to treat AIDS-related illnesses. Drug therapy for AIDS and associated illnesses is expensive. The cost for just the drugs used by one AIDS patient in 1 month is over $1,000.

Not all persons who become HIV-infected progress to AIDS. Indeed, about 3 percent are referred to as long-term nonprogressors—people who have not suffered any apparent damage to their immune system in 20 years.

Treatment for HIV and AIDS is not just medical. Human interaction networks, including spouses, parents, and siblings, are affected by the person who is HIV-positive. Getting the secret out and establishing a supportive network are essential in managing the psychological trauma of being diagnosed with HIV.

The following sections consider other STIs. The relationship of HIV infection to other STIs is being given increased attention by researchers. HIV infection leads to altered manifestations of other STIs and thereby probably promotes their spread. Genital and some herpes ulcers normally heal within 1 to 3 weeks, but they can persist for months as highly infectious ulcers in persons with HIV infection. Likewise, persons with an STI are at greater risk of contracting HIV than those without.

Other Sexually Transmissible Infections

There are numerous other STIs. Young women, formerly married women, persons with little education who are smokers, persons with a high number of sexual partners, and persons who do not use condoms regularly are more likely to be among those who contract HIV and other STIs (Lane and Althaus, 2002; Capaldi et al., 2003). Some of the more common ones include human papillomavirus (HPV) infection, chlamydia, genital herpes, gonorrhea, and syphilis.

Human Papillomavirus

HPV infection is the most common STI. The more sexual partners an individual has, the more likely the person is to contract HPV (Sellors et al., 2003).

There are more than 70 types of HPV. More than a dozen of these types can cause warts (called **genital warts,** or *condyloma*) or more subtle signs of infection in the genital tract. The virus infects the skin's top layers and can remain inactive for months or years before any obvious signs of infection appear. Often warts appear within 3 to 6 months after infection. However, some types of HPV produce no visible warts. Fewer than 1 to 2 percent of people who are infected with HPV develop symptoms. Any sexual partners of an infected individual should have a prompt medical examination.

HPV can be transmitted through vaginal or rectal intercourse, through fellatio and cunnilingus, and through other skin-to-skin contact. Genital warts are small bumps that are usually symptom-free, but they may itch. In women, genital warts most commonly develop on the vulva, in the vagina, or on the cervix. They can also appear on or near the anus. In men, the warts appear most often on the penis but can appear on the scrotum or anus or within the rectum. Incidence of infection radically increases as the number of sexual partners increases.

Health care providers disagree regarding the efficacy of treating HPV when there are no detectable warts. However, when the warts can be seen, either by visual inspection or by colposcope, providers do typically advise treatment. A number of treatment options are available. Choosing among them depends upon the number of warts and their location, availability of equipment, training of health care providers, and the preferences of the patient. Most of the treatments are at least moderately effective, but many are quite expensive. Treatments range from topical application of chemicals to laser surgery. Treatment of warts destroys infected cells, but not all of them, as HPV is present in a wider area of skin than just the precise wart location. To date, no therapy has proved effective in eradicating HPV, and relapse is common.

A danger for women exposed to certain strains of HPV is a higher risk for cervical cancer. The majority of cervical cancers (80 percent) are caused by just four types of HPV. Women who are diagnosed with HPV infection should carefully follow recommendations for cervical cancer screening and have Pap smears as directed by their health care providers.

New HPV vaccines could help reduce the risk of cervical cancer for women. Research shows that a vaccine for HPV has the potential to protect most women from cervical cancer for at least several years, but it doesn't protect completely against all strains of HPV. GlaxoSmithKline submitted Cervarix to the U.S. Food and Drug Administration (FDA) for approval in late 2006. Merck, which makes the other HPV vaccine, Gardasil, gained FDA approval for it in June 2006. On the basis of recommendations of a federal advisory panel (Harper, 2006), the vaccine is approved for females aged 9 to 26 and is administered in three shots over a 6-month period.

Chlamydia

Chlamydia trachomatis (CT) is a bacterium that can infect the genitals, eyes, and lungs. **Chlamydia** (clah-MID-ee-uh) is the most frequently occurring bacterial STI on college campuses. Indeed, it is estimated that 3 million new cases occur annually, and chlamydia is the most frequently reported infectious disease in the United States. Worldwide, chlamydial infections are even more extensive. Trachoma inclusion conjunctivitis, a chlamydial infection that occurs rarely in the United States, is the leading cause of blindness in Third World countries. In women, untreated infections can progress to involve the upper reproductive tract and may result in serious complications. Around 40 percent of women with untreated chlamydia infections develop pelvic inflammatory disease, and 20 percent of those become infertile, according to the CDC (Burnstein et al., 2005). Other complications include spontaneous abortion and premature birth.

CT is easily transmitted directly from person to person via sexual contact or by sharing sex toys. The microorganisms are most often found in the urethra of the man, in the cervix, uterus, and fallopian tubes of the woman, and in the rectum of either men or women.

Genital-to-eye transmission of the bacteria can also occur. If a person with a genital CT infection rubs his or her eye or the eye of a partner after touching infected genitals, the bacteria can be transferred to the eye, and vice versa. Finally, infants can get CT as they pass through the cervix of their infected mother during delivery. CT infection rarely shows obvious symptoms—up to 70 percent of women are asymptomatic (Shih et al., 2004)—which accounts for its being known as "the silent disease." Both women and men who are infected with CT usually do not know that they have the disease. The result is that they infect new partners unknowingly, who affect others unknowingly—unendingly.

Genital Herpes

Herpes refers to more than 50 viruses related by size, shape, internal composition, and structure. One such herpes is **genital herpes.** Although the disease has been known for at least 2000 years, media attention to genital herpes is relatively new.

Also known as **herpes simplex virus type 2** (HSV-2) infection, genital herpes is a viral infection that is almost always transmitted through sexual contact. Symptoms occur in the form of a cluster of small, painful blisters or sores at the point of infection, most often on the penis or around the anus in men. In women, the blisters usually appear around the vagina but can also develop inside the vagina, on the cervix, and sometimes on the anus. Pregnant women can transmit the herpes virus to their newborn infants, causing brain damage or death.

Another type of herpes originates in the mouth. **Herpes simplex virus type 1** (HSV-1) is a biologically different virus with which people are more familiar as cold sores on the lips. These sores can be transferred to the genitals by the fingers or by oral-genital contact. In the past, genital and oral herpes had site specificity: HSV-1 was always found on the lips or in the mouth, and HSV-2 was always found on the genitals. But because of the increase in fellatio and cunnilingus, HSV-1 can be found in the genitals and HSV-2 can be found on the lips.

Herpes symptoms range from none at all to painful ulcers or blisters. Other symptoms that may occur include discharge, itching, or a burning sensation during urination, back pain, leg pain, stiff neck, sore throat, headache, fever, aches, swollen glands, fatigue, and heightened sensitivity of the eyes to light. The outbreaks with herpes last about 10 to 14 days on average, although they can last for as long as 6 weeks if not treated. "I've got herpes," said one sufferer, "and it's a very uneven discomfort. Some days I'm okay, but other days I'm miserable."

As with syphilis, the sores associated with genital herpes subside (the sores dry up, scab over, and disappear), and the person feels good again. But the virus settles in the nerve cells in the spinal column and may cause repeated outbreaks of the symptoms in about one-third of those infected.

Stress, menstruation, sunburn, fatigue, and the presence of other infections seem to be related to the reappearance of herpes symptoms. Although such recurrences are usually milder and of shorter duration than the initial outbreak, the resurfacing of the symptoms can occur throughout the person's life. "It's not knowing when the thing is going to come back that's the bad part about herpes," said one woman.

The herpes virus is usually contagious during the time that a person has visible sores but not when the skin is healed. However, infected people may have a mild recurrence yet be unaware that they are contagious. Aside from visible sores, the itching, burning, or tingling sensations at the sore site also suggest that the person is contagious. Using a latex or polyurethane condom reduces the risk of transmitting or acquiring herpes, since the virus doesn't permeate the condom. However, if the condom doesn't cover the site of the sore, the virus may be spread through skin-to-skin contact.

At the time of this writing, there is no cure for herpes. Until recently, scientists have not fully understood how HSV enters human cells, making it hard to develop drugs that could treat the virus. Researchers from the University of Michigan Medical School have now identified a receptor that appears to function as one "lock" that HSV opens to allow it to enter cells, which could be a major breakthrough for developing more effective treatments (Fuller and Wald, 2005). Because it is a virus, herpes does not respond to antibiotics as do syphilis and gonorrhea. A few procedures that help to relieve the symptoms and promote healing of the sores include seeing a physician to look for and treat any other genital infections near the herpes sores, keeping the sores clean and dry, taking hot sitz baths three times a day, and wearing loose-fitting cotton underwear to enhance air circulation. Proper nutrition, adequate sleep and exercise, and avoiding physical or mental stress help people to cope better with recurrences.

Acyclovir, marketed as Zovirax, is an ointment that can be applied directly on the sores that helps to relieve pain, speed healing, and reduce the amount of time that live viruses are present in the sores. A more effective tablet form of acyclovir, which significantly reduces the rate of recurring episodes of genital herpes, is also available. Once acyclovir is stopped, the herpetic recurrences resume. Acyclovir seems to make the symptoms of first-episode genital herpes more manageable, but it is less effective during subsequent outbreaks. ImmuVir—an alternative to acyclovir—is primarily for use by persons who have frequent outbreaks of genital herpes (once a month or more). This ointment is designed to reduce pain, healing time, and number of outbreaks. The drug has no known side effects.

Coping with the psychological and emotional aspects of having genital herpes is often more difficult than coping with the physical aspects of the disease.

Gonorrhea

Also known as the clap, the whites, morning drop, and the drip, **gonorrhea** is a bacterial infection that is sexually transmissible. Individuals contract gonorrhea through having sexual contact with someone who is carrying *Neisseria gonorrhoeae* bacteria. The gonorrhea bacteria, and most other STIs, cannot live long outside the human body—outside mucous membranes—so they could not survive on a toilet seat unless fluid were present, and even then they would not survive long. These bacteria thrive in warm, moist cavities, including the urinary tract, cervix, rectum, mouth, and throat. A pregnant woman can transmit gonorrhea to her infant at birth, causing eye infection. Many medical experts recommend gonorrhea testing for all pregnant women and antibiotic eye drops for all newborns.

Although some infected men show no signs, 80 percent exhibit symptoms between 3 and 8 days after exposure. They may begin to discharge a thick, yellowish pus from the penis and to feel pain or discomfort during urination. They may also have swollen lymph glands in the groin. Women are more likely to show no signs (70% to 80% have no symptoms) of the infection, but when they do, the symptoms are sometimes a yellowish discharge from the vagina along with a burning sensation or spotting between periods or after sexual intercourse. More often, a woman becomes aware of gonorrhea only after she feels extreme discomfort, which results when the untreated infection travels up into her uterus and fallopian tubes, causing pelvic inflammatory disease (PID). Salpingitis (inflammation of the fallopian tube) occurs in 10 percent to 20 percent of infected women and can cause infertility or ectopic pregnancy.

Undetected and untreated, gonorrhea can do permanent damage. Not only can the infected person pass the disease on to the next partner, but other undesirable consequences can result. Untreated gonorrhea commonly causes long-term reproductive system complications, such as blocking and/or scarring of the urethra and possible infertility in men and PID (infection of the fallopian tubes), infertility, and damage to the uterus, fallopian tubes, and ovaries in women. In rare cases, the untreated bacteria can affect the brain, heart valves, and joints. Both men and women could develop endocarditis (an infection of the heart valves) or meningitis (inflammation of the tissues surrounding the brain and spinal column), arthritis, and sterility. Infected pregnant women could have a spontaneous abortion or a premature or stillborn infant.

A physician can detect gonorrhea by analyzing penile or cervical discharge under a microscope. A major problem with new cases of gonorrhea is the emergence of new strains of the bacteria that are resistant to penicillin. Because of high rates of resistance to penicillin and tetracycline, the current recommended treatment for gonorrhea is a single shot of ceftriaxone or a single dose of such oral medications as ofloxatin, cefixime, and ciprofloxacin.

Syphilis

Syphilis is caused by bacteria that can be transmitted through sexual contact with an infected individual. Syphilis can also be transmitted by an infected pregnant woman to her unborn baby. The national rate of primary and secondary syphilis, the early stages of the disease that indicate recent infection, has increased every year since an all-time low in 2000 (CDC, 2005). Although syphilis is less prevalent than gonorrhea, its effects are more devastating and include mental illness, blindness, heart disease—even death. The spirochete bacteria enter the body through mucous membranes that line various body openings. With your tongue, feel the inside of your cheek. This is a layer of mucous membrane—the substance in which spirochetes thrive. Similar membranes are in the vagina and urethra of the penis. If you kiss or have genital contact with someone harboring these bacteria, the bacteria can be absorbed into your mucous membranes and cause syphilitic infection.

Syphilis progresses through at least three stages, plus a latency stage before the final stage. In stage one (primary-stage syphilis), a small sore, or chancre, will appear at the site of the infection between 10 and 90 days after exposure. The chancre, which can show up anywhere on the man's penis, or in the labia, vaginal membranes, or cervix of the woman, or in either partner's mouth or rectum, neither hurts nor itches and, if left untreated, will disappear in 3 to 5 weeks. The disappearance leads infected people to believe that they are cured—one of the tricky aspects of syphilis. In reality, the disease is still present and doing great harm, even though there are no visible signs.

During the second stage (secondary-stage syphilis), beginning from 2 to 12 weeks after the chancre has disappeared, other signs of syphilis appear in the form of a rash all over the body or just on the hands or feet. Welts and sores can also occur, as well as fever, headaches, sore throat, and hair loss. Syphilis has been called the great imitator because it mimics so many other diseases (for example, infectious mononucleosis, cancer, and psoriasis). Whatever the symptoms, they, too, will disappear without treatment. The person may again be tricked into believing that nothing is wrong.

Following the secondary stage is the latency stage, during which there are no symptoms and the person is not infectious. However, the spirochetes are still in the body and can attack any organ at any time.

Tertiary syphilis—the third stage—can cause serious disability or even death. Heart disease, blindness, brain damage, loss of bowel and bladder control, difficulty in walking, and erectile dysfunction can result. Early detection and treatment are essential. Blood tests and examination of material from the infected site can help to verify the existence of syphilis. But such tests are not always accurate. Blood tests reveal the presence of antibodies, not spirochetes, and it sometimes takes 3 months before the body produces detectable antibodies. Sometimes there is no chancre anywhere on the person's body.

Treatment for syphilis is similar to that for gonorrhea. Penicillin or other antibiotics (for those allergic to penicillin) are effective. Infected persons treated in the early stages can be completely cured with no ill effects. If the syphilis has progressed into the later stages, any damage that has been done cannot be repaired.

Getting Help for HIV and Other STIs

If you are engaging in unprotected sex or are having symptoms, you should get tested for HIV and other STIs. There are different tests for different STIs.

Women must specifically ask for such tests from their gynecologist, since only a few doctors routinely perform them. If you do not know whom to call to get tested, call your local health department or the national STI hotline at 1-800-227-8922. You will not be asked to identify yourself but will be given the name and number of local STI clinics that offer confidential, free treatment. In addition, Duke University, in Durham, North Carolina, has opened an AIDS clinic to treat AIDS patients (919-684-2660). For the CDC National AIDS hotline, call 1-800-342-2437. Students can also obtain information and assistance about HIV and STIs from their student health care facility on campus.

KEY TERMS

AIDS	genital warts	herpes simplex virus type 2	sexually transmitted infection (STI)
chlamydia	gonorrhea	human immunodeficiency virus (HIV)	
genital herpes	herpes simplex virus type 1		syphilis

WEB LINKS

American Social Health Association
　http://www.ashastd.org
Centers for Disease Control and Prevention (HIV testing)
　http://www.cdc.gov/
　http://www.cdc.gov/hiv/rapid_testing
HIV Testing Home Access
　www.homeaccess.com

OraSure Technologies, Inc.
　http://www.orasure.com/products/default.asp?cid=16&subx=2&sec=3
Food and Drug Administration
　http:www.fda.gov/bbs/topics/NEWS/NEW00503.html

*For women the best
aphrodisiacs are words. The
G-spot is in the ears. He who
looks for it below there is
wasting his time.*

Isabel Allende

Sexual Anatomy and Physiology

Contents

f we think of the human body as a special type of machine, *anatomy* refers to that machine's part and *physiology* refers to how the parts work. This Special Topic reviews the sexual anatomy and physiology of women and men and the reproductive process.

Female External Anatomy and Physiology

The external female genitalia are collectively known as the *vulva* (VUHL-vuh), a Latin term meaning "covering." The vulva consists of the mons veneris, the labia, the clitoris, and the vaginal and urethral openings (Figure ST4.1). The female genitalia differ in size, shape, and color, resulting in considerable variability in appearance.

Terms for male genitalia (e.g., *penis, testicles*) are more commonly known than are terms for female genitalia. Some women do not even accurately name their genitals. At best, little girls are taught that they have a vagina, which becomes the word for everything "down there"; they rarely learn they also have a vulva, a clitoris, and labia.

Mons Veneris

The soft cushion of fatty tissue overlying the pubic bone is called the *mons veneris* (mahns vuh-NAIR-ihs), also known as the *mons pubis*. This area becomes covered with hair at puberty and has numerous nerve endings. The purpose of the mons is to protect the pubic region during sexual intercourse.

Labia

In the sexually unstimulated state, the urethral and vaginal openings are protected by the *labia majora* (LAY-bee-uh muh-JOR-uh), or outer lips—two elongated folds of fatty tissue that extend from the mons to the *perineum,* the area of skin between the opening of the vagina and the anus. Located between the labia majora are two additional hairless folds of skin, called the *labia minora* (muh-NOR-uh), or inner lips, that cover the urethral and vaginal openings and join at the top to form the hood of the clitoris. Some contend that the clitoral hood provides clitoral stimulation during intercourse. Both sets of labia—particularly the inner labia minora—have a rich supply of nerve endings that are sensitive to sexual stimulation.

Figure ST4.1
External Female Genitalia

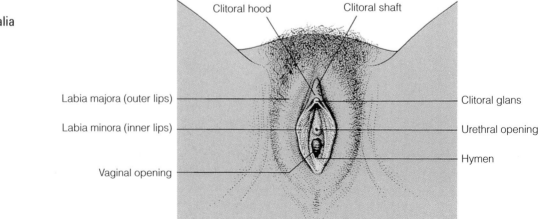

Special Topic 4 Sexual Anatomy and Physiology

Clitoris

At the top of the labia minora is the *clitoris* (KLIHT-uh-ruhs), which also has a rich supply of nerve endings. The clitoris is a very important site of sexual excitement and, like the penis, becomes erect during sexual excitation.

Vaginal Opening

The area between the labia minora is called the *vestibule*. This includes the urethral opening and the vaginal opening, or *introitus* (ihn-TROH-ih-tuhs), neither of which is visible unless the labia minora are parted. Like the anus, the vaginal opening is surrounded by a ring of sphincter muscles. Although the vaginal opening can expand to accommodate the passage of a baby at childbirth, under conditions of tension these muscles can involuntarily contract, making it difficult to insert an object, including a tampon, into the vagina. The vaginal opening is sometimes covered by a *hymen,* a thin membrane.

Probably no other body part has caused as much grief to so many women as the hymen, which has been regarded throughout history as proof of virginity. A newlywed woman who was thought to be without a hymen was often returned to her parents, disgraced by exile, or even tortured and killed. It has been a common practice in many societies to parade a bloody bed sheet after the wedding night as proof of the bride's virginity. The anxieties caused by the absence of a hymen persist even today; in Japan and other countries, sexually experienced women may have a plastic surgeon reconstruct a hymen before marriage. Yet the hymen is really a poor indicator of virtue. Some women are born without a hymen or with incomplete hymens. In others, the hymen is accidentally ruptured by vigorous physical activity or insertion of a tampon. In some women, the hymen may not tear but only stretch during sexual intercourse. Even most doctors cannot easily determine whether a woman is a virgin.

Urethral Opening

Just above the vaginal opening is the urethral opening, where urine passes from the body. A short tube, the *urethra,* connects the bladder (where urine collects) with the urethral opening. Because of the shorter length of the female urethra and its close proximity to the anus, women are more susceptible than men to cystitis, a bladder inflammation.

Female Internal Anatomy and Physiology

The internal sex organs of the female include the vagina, pubococcygeus muscle, uterus, and paired fallopian tubes and ovaries (Figure ST4.2).

Vagina

Some people erroneously believe that the *vagina* is a dirty part of the body. In fact, the vagina is a self-cleansing organ. The bacteria that are found naturally in the vagina help to destroy other potentially harmful bacteria. In addition, secretions from the vaginal walls help to maintain the vagina's normally acidic environment. The use of feminine hygiene sprays, as well as excessive douching, can cause irritation, allergic reactions, and, in some cases, vaginal infection by altering the normal chemical balance of the vagina.

Some researchers have reported that some women experience extreme sensitivity in an area in the front wall of the vagina one to two inches into the vaginal opening. The spot (or area) may swell during stimulation, and although a woman's initial response may be a need to urinate, continued stimulation generally leads to orgasm. The area was named the *Grafenberg spot,* or *G-spot,* for gy-

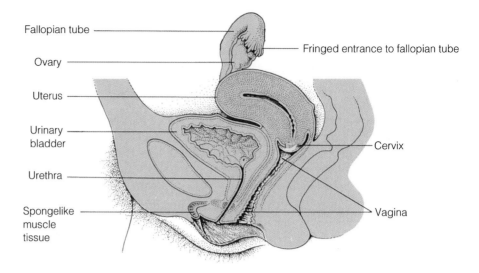

necologist Ernest Grafenberg, who first noticed the erotic sensitivity of this area more than 40 years ago. Not all women notice a G-spot (Figure ST4.3).

Pubococcygeus Muscle

Also called the *PC muscle*, the *pubococcygeus muscle* is one of the pelvic floor muscles that surround the vagina, the urethra, and the anus. To find her PC muscle, a woman is instructed to voluntarily stop the flow of urine after she has begun to urinate. The muscle that stops the flow is the PC muscle.

A woman can strengthen her PC muscle by performing Kegel exercises, named after the physician who devised them. The Kegel exercises involve contracting the PC muscle several times for several sessions per day. Kegel exercises are often recommended after childbirth to restore muscle tone to the PC muscle, which is stretched during the childbirth process, and help prevent involuntary loss of urine.

Figure ST4.3
The alleged "G" spot.

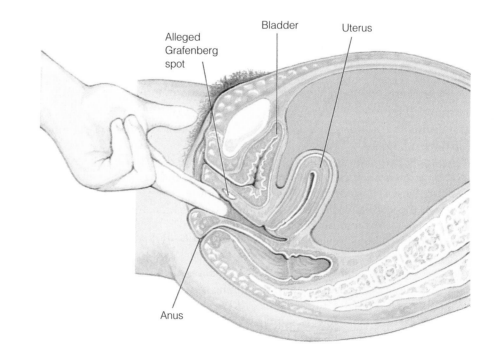

Uterus

The *uterus* (YOOT-uh-ruhs), or *womb*, resembles a small, inverted pear, which measures about 3 inches long and 3 inches wide at the top in women who have not given birth. A fertilized egg becomes implanted in the wall of the uterus and continues to grow and develop until delivery. At the lower end of the uterus is the *cervix*, an opening that leads into the vagina.

All adult women should have a pelvic examination, including a Pap test, each year. A Pap test is extremely important in the detection of cervical cancer. Cancer of the cervix and uterus is the second most common form of cancer in women.

Some women may neglect to get a Pap test because they feel embarrassed or anxious about it or because they think they are too young to worry about getting cancer. For all women over age 20, however, having annual Pap tests may mean the difference between life and death.

Fallopian Tubes

The *fallopian* (ful-LOH-pee-uhn) *tubes* extend about 4 inches laterally from either side of the uterus to the ovaries. Fertilization normally occurs in the fallopian tubes. The tubes transport the ovum, or egg, by means of *cilia* (hairlike structures) down the tube and into the uterus.

Ovaries

The *ovaries* (OH-vuhr-eez) are two almond-shaped structures, one on either side of the uterus. The ovaries produce eggs (ova) and the female hormones estrogen and progesterone. At birth, a woman has a number of immature ova in her ovaries. Each ovum is enclosed in a thin capsule forming a follicle. Some of the follicles begin to mature at puberty; only about 400 mature ova will be released in a woman's lifetime.

Male External Anatomy and Physiology

Although they differ in appearance, many structures of the male (Figure ST4.4) and female genitals develop from the same embryonic tissue (the penis and the clitoris, for example).

Penis

The *penis* (PEE-nihs) is the primary male sexual organ. In the unaroused state, the penis is soft and hangs between the legs. When sexually stimulated, the penis enlarges and becomes erect, enabling penetration of the vagina. The penis functions not only to deposit sperm in the female's vagina but also as a passageway from the male's bladder to eliminate urine. In cross-section, the penis can be seen to consist of three parallel cylinders of tissue containing many cavities, two *corpora cavernosa* (cavernous bodies), and a *corpus spongiosum* (spongy body) through which the urethra passes. The penis has numerous blood vessels; when stimulated, the arteries dilate and blood enters faster than it can leave. The cavities of the cavernous and spongy bodies fill with blood, and pressure against the fibrous membranes causes the penis to become erect. The head of the penis is called the glans. At birth, the glans is covered by *foreskin*.

Circumcision, the surgical procedure in which the foreskin of the male is pulled forward and cut off, has been practiced for at least 6000 years. About 80 percent of men in the United States have been circumcised.

Circumcision is a religious rite for members of the Jewish and Muslim faiths. To Jewish people, circumcision symbolizes the covenant between God and Abraham. In the United States, the procedure is generally done within the first few days after birth. Among non-Jewish people, circumcision first became popu-

Figure ST4.4
External Male Sexual Organs

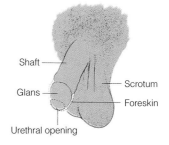

Shaft

Glans

Urethral opening

Scrotum

Foreskin

lar in the United States during the nineteenth century as a means of preventing masturbation. Such effectiveness has not been demonstrated.

Today, the primary reason for performing circumcision is to ensure proper hygiene and to maintain tradition. The smegma that can build up under the foreskin is a potential breeding ground for infection. But circumcision may be a rather drastic procedure merely to ensure proper hygiene, which can just as easily be accomplished by pulling back the foreskin and cleaning the glans during normal bathing. However, being circumcised is associated with having a lower risk of contracting penile HPV; similarly, female partners of men who engage in risky sexual behavior have a reduced likelihood of having cervical cancer if the man is circumcised (Lane, 2002). Circumcision is a relatively low-risk surgical procedure. And although the male does feel pain, it can be minimized by administering local anesthesia.

Scrotum

The *scrotum* (SCROH-tuhm) is the sac located below the penis that contains the *testes*. Beneath the skin covering the scrotum is a thin layer of muscle fibers that contract when it is cold, helping to draw the testes (testicles) closer to the body to keep the temperature of the sperm constant. Sperm can be produced only at a temperature several degrees lower than normal body temperature; any prolonged variation can result in sterility.

Male Internal Anatomy and Physiology

The male internal organs, often referred to as the reproductive organs, include the testes, where the sperm are produced; a duct system to transport and propel the sperm out of the body; and some additional structures that produce the seminal fluid in which the sperm are mixed before ejaculation (Figure ST4.5).

Testes

The male gonads—the paired testes, or testicles—develop from the same embryonic tissue as the female gonads (the ovaries). The two oval-shaped testicles are suspended in the scrotum by the *spermatic cord* and enclosed within a fibrous sheath. The function of the testes is to produce spermatozoa and male hormones, primarily testosterone.

Figure ST4.5
Internal and External Male
Sexual Organs

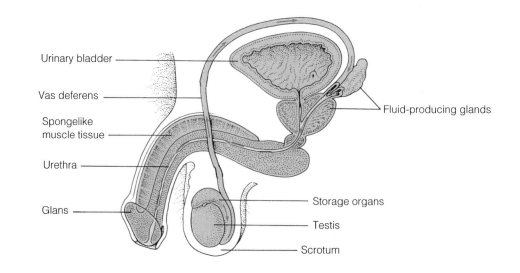

Special Topic 4 Sexual Anatomy and Physiology

Duct System

Several hundred *seminiferous tubules* come together to form a tube in each testicle called the *epididymis* (ehp-uh-DIHD-uh-muhs), the first part of the duct system that transports sperm. If uncoiled, each tube would be 20 feet long. Sperm spend from 2 to 6 weeks traveling through the epididymis as they mature and are reabsorbed by the body if ejaculation does not occur. One ejaculation contains an average of 360 million sperm cells.

The sperm leave the scrotum through the second part of the duct system, the *vas deferens* (vas-DEF-uh-renz). These 14- to 16-inch paired ducts transport the sperm from the epididymis up and over the bladder to the prostate gland. Rhythmic contractions during ejaculation force the sperm into the paired ejaculatory ducts that run through the prostate gland. The entire length of this portion of the duct system is less than 1 inch. It is here that the sperm mix with seminal fluid to form semen before being propelled to the outside through the urethra.

Seminal Vesicles and Prostate Gland

The *seminal vesicles* resemble two small sacs, each about 2 inches in length, located behind the bladder. These vesicles secrete their own fluids, which empty into the ejaculatory duct to mix with sperm and fluids from the prostate gland.

Most of the seminal fluid comes from the *prostate gland,* a chestnut-sized structure located below the bladder and in front of the rectum. The fluid is alkaline and serves to protect the sperm in the more acidic environments of the male urethra and female vagina. Males older than 45 should have a rectal examination annually to detect the presence of prostate cancer.

A small amount of clear, sticky fluid is also secreted into the urethra before ejaculation by two pea-sized *Cowper's,* or *bulbourethral,* glands located below the prostate gland. This protein-rich fluid alkalizes the urethral passage, which prolongs sperm life.

The fluid secreted by the Cowper's glands can often be noticed on the tip of the penis during sexual arousal. Since this fluid may contain sperm, withdrawal of the penis from the vagina before ejaculation is a risky method of birth control.

*Couples might consider
increased intimacy, rather than
sexual performance,
as the goal of sexual therapy.*

Clark Christensen, sex therapist

Sexual Dysfunctions

Contents

A team of researchers (Laumann et al., 2005) analyzed data from the Global Study of Sexual Attitudes and Behaviors on sex and relationships of 13,882 women and 13,618 men from 29 countries and concluded that sexual difficulties are common among mature adults throughout the world. Also referred to as sexual dysfunctions, these may be classified by time of onset, situations in which they occur, and cause. A **primary sexual dysfunction** is one that a person has always had. A **secondary sexual dysfunction** is one that a person is currently experiencing, after a period of satisfactory sexual functioning. For example, a woman who has never had an orgasm with any previous sexual partner has a primary dysfunction, whereas a woman who has been orgasmic with previous partners but not with a current partner has a secondary sexual dysfunction.

A situational dysfunction occurs in one context or setting and not in another, whereas a total dysfunction occurs in all contexts or settings. For example, a man who is unable to become erect with one partner but who can become erect with another has a situational dysfunction. Finally, a sexual dysfunction can be classified according to whether it is caused primarily by biological (organic) factors, such as insufficient hormones or physical illness, or by psychosocial or cultural factors, such as negative learning, guilt, anxiety, or an unhappy relationship.

Both women and men report experiencing sexual problems. In a study of 1,768 adults, about a third (34%) of the women and four in ten men (41%) reported having a current sexual problem (Dunn et al., 2000). Sexual problems often appear in tandem. Greenstein et al. (2006) assessed the degree to which 113 female partners of men with erectile dysfunction defined themselves as having a sexual dysfunction. Sixty-five percent reported having one or more sexual problem, including orgasmic disorder, low sexual desire, and low arousal. The authors emphasized the need to evaluate both partners when one of them has a sexual dysfunction.

Sexual problems become visible in societies where specialists (sexologists, sex therapists, marriage and family therapists) emphasize the importance of problem-free sexual relationships and offer (for a price) a therapeutic remedy. Minorities (African-American, Hispanic, and Asian-American), particularly women, may have a higher incidence of sexual dysfunctions than Caucasians due to cultural inducement of depression (typically associated with sexual dysfunction) (Dobkin et al., 2006).

In this Special Topics section we examine the sexual dysfunctions common to women and to men.

The national sex study reported by Dunn et al. (2000) identified the most frequent sexual problems creating dissatisfaction in women as arousal problems (51%), unpleasurable sex (47%), and inability to achieve orgasm (39%). A fourth of the women in a study by Bancroft et al. (2003) reported marked distress over their sexual relationship, with the greatest predictor of such distress being the woman's emotional and relationship distress. We will examine the potential causes and alternative solutions for the top three problems.

Sexual Dysfunctions of Women

In this section we discuss the most frequent sexual problems of women.

Arousal Problems

Also referred to as hypoactive sexual desire, lack of interest in sex is the most frequent sexual problem reported by women in the United States. Colson et al. (2006) also noted that diminution of sexual desire was the primary sexual complaint of 519 French women. Lack of interest and difficulty becoming aroused

(Quirk et al., 2005) may be caused by one or more factors, including restrictive upbringing, nonacceptance of one's sexual orientation, learning a passive sexual role, and physical factors such as stress, illness, drug use, and fatigue. Dennerstein (2006) also noted that menopause may be related. More often, lack of interest in sex and difficulty in becoming aroused can be explained by the woman's emotional relationship with her partner. In addition to a negative emotional context, there may be lower testosterone levels in women reporting low sexual desire.

The treatment for lack of interest in sex depends on the underlying cause or causes of the problem. The following are some of the ways in which lack of sexual desire can be treated.

1. Improve relationship satisfaction. Treating the relationship before treating the sexual problem is standard therapy in treating any sexual dysfunction, including lack of interest in sex. A prerequisite for being interested in sex with a partner, particularly from the viewpoint of a woman, is to be in love and feel comfortable and secure with the partner. Sammons (2003) emphasized the role of subjective mental factors in a woman's sexual arousal. Couple therapy focusing on a loving egalitarian relationship becomes the focus of therapy. Lau et al. (2006) also emphasized the importance of trust in one's partner as a prerequisite for positive sexual functioning.

2. Practice sensate focus. Sensate focus is a series of exercises developed by Masters and Johnson used to treat various sexual dysfunctions. Sensate focus may also be used by couples who are not experiencing sexual dysfunction but who want to enhance their sexual relationship. In carrying out the sensate focus exercise, the couple take turns pleasuring each other in nongenital ways, with each taking turns giving and receiving pleasure (while getting feedback from the partner about what is and is not pleasurable). On subsequent occasions, genital touching is allowed, but orgasm is not the goal. Indeed, the goal of sensate focus is to help the partners learn to give and receive pleasure by promoting trust and communication and by reducing anxiety related to sexual performance.

3. Be open to re-education. Re-education involves being open to examining and reevaluating the thoughts, feelings, and attitudes learned in childhood. The goal is to redefine sexual activity so that it is viewed as a positive, desirable, healthy, and pleasurable experience. A national study of Finnish women shows dramatic changes in their reported increases in sexual satisfaction. Women from unreserved and nonreligious homes who had high education and who were sexually assertive reported the greatest pleasure in sexual intercourse (Haavio-Mannila and Kontula, 1997). The authors suggested that increased emancipation of women would be associated with increases in sexual pleasure experienced by women.

4. Consider other treatments. Other treatments for lack of sexual desire include rest and relaxation. This is particularly indicated where the culprit is chronic fatigue syndrome (CFS), the symptoms of which are overwhelming fatigue, low-grade fever, and sore throat. Still other treatments for lack of sexual desire include hormone treatment and changing medications (if possible) in cases where medication interferes with sexual desire. In addition, sex therapists often recommend that people who are troubled by a low level of sexual desire engage in sexual fantasies and masturbation as a means of developing positive sexual images and feelings.

Unpleasurable Sex

Sex that is not pleasurable may be both painful and aversive. Pain during intercourse, or dyspareunia, occurs in about 10 percent of gynecological patients in the United States. Colson et al. (2006) noted that 15.5 percent of 519 French women identified dyspareunia as the primary sexual complaint. Dyspareunia may be caused by vaginal infection, lack of lubrication, a rigid hymen, or an improperly positioned uterus or ovary. Because the causes of dyspareunia are often medical, a physician should be consulted. Sometimes surgery is recommended to remove the hymen.

Dyspareunia may also be psychologically caused. Guilt, anxiety, or unresolved feelings about a previous trauma, such as rape or childhood molestation, may be operative. Therapy may be indicated.

Some women report that they find sex aversive. Sexual aversion, also known as sexual phobia and sexual panic disorder, is characterized by the individual's wanting nothing to do with genital contact with another person. The immediate cause of sexual aversion is an irrational fear of sex. Such fear may result from negative sexual attitudes acquired in childhood or sexual trauma such as rape or incest. Some cases of sexual aversion may be caused by fear of intimacy or hostility toward the other sex.

Treatment for sexual aversion involves providing insight into the possible ways in which the negative attitudes toward sexual activity developed, increasing the communication skills of the partners, and practicing sensate focus. Understanding the origins of the sexual aversion may enable the individual to view change as possible. Through communication with the partner and through sensate focus exercises, the individual may learn to associate more positive feelings with sexual behavior.

Inability to Achieve Orgasm

Orgasmic difficulty, also referred to as **inhibited female orgasm,** or orgasmic dysfunction, occurs when a woman is unable to achieve orgasm after a period of continuous stimulation. Colson et al. (2006) noted that 15.5 percent of 519 French women identified orgasm difficulty as the primary sexual complaint. Difficulty achieving orgasm can be primary, secondary, situational, or total. Situational orgasmic difficulties, in which the woman is able to experience orgasm under some circumstances but not others, are the most common. Many women are able to experience orgasm during manual or oral clitoral stimulation but are unable to experience orgasm during intercourse (i.e., in the absence of manual or oral stimulation).

Biological factors associated with orgasmic dysfunction can be related to fatigue, stress, alcohol, and some medications, such as antidepressants and antihypertensives. Diseases or tumors that affect the neurological system, diabetes, and radical pelvic surgery (e.g., for cancer) may also impair a woman's ability to experience orgasm.

Psychosocial and cultural factors associated with orgasmic dysfunction are similar to those related to lack of sexual desire. Causes of orgasm difficulties in women include restrictive childrearing and learning a passive female sexual role. Guilt, fear of intimacy, fear of losing control, ambivalence about commitment, and spectatoring may also interfere with the ability to experience orgasm. Other women may not achieve orgasm because of their belief in the myth that women are not supposed to enjoy sex.

Relationship factors, such as anger and lack of trust, can also produce orgasmic dysfunction. For some women, lack of information can result in orgasmic difficulties (e.g., some women do not know that clitoral stimulation is important for orgasm to occur). Some women might not achieve orgasm with their partners because they do not tell their partners what they want in terms of sexual stimulation out of shame and insecurity. Or, even in those cases where the woman is open about her sexual preferences, the partner may be unwilling to provide the necessary stimulation. Hence, a cooperative partner rather than a nonorgasmic woman should be the focus for resolution.

Kelly et al. (2006) found that inorgasmic women were more likely to be in marriages where there was poorer communication. Specifically, marriages where the wife had difficulty achieving an orgasm were characterized by greater blame and conflict.

Since the causes for primary and secondary orgasm difficulties vary, the treatment must be tailored to the particular woman. Treatment can include enhancing positive communication, rest and relaxation, testosterone injections, or limiting alcohol consumption prior to sexual activity. Sensate focus exercises might help a woman explore her sexual feelings and increase her comfort with her partner. Treatment can also involve improving relationship satisfaction and teaching the woman how to communicate her sexual needs. Teaching the woman how to masturbate is also a frequent therapeutic option. The rationale behind masturbation as a therapeutic technique for a nonorgasmic woman is that masturbation is the behavior that is most likely to produce orgasm and can enable her to show her partner what she needs. Masturbation gives the individual complete control of the stimulation, provides direct feedback to the woman of the type of stimulation she enjoys, and eliminates the distraction of a partner.

Sexual Dysfunctions of Men

Men also report sexual problems. The national sex study reported by Dunn et al. (2000) identified some of the most frequent sexual problems creating dissatisfaction in men as erectile problems (48%) and premature ejaculation (43%).

Erectile Dysfunction

Loss of erection, also referred to as **erectile dysfunction,** involves the man's inability to get and maintain an erection. Like other sexual dysfunctions, erectile dysfunction can be primary, secondary, situational, or total. Occasional, isolated episodes of the inability to attain or maintain an erection are not considered dysfunctional; these are regarded as normal occurrences. To be classified as an erectile dysfunction, the erection difficulty should last continuously for a period of at least 3 months.

Erectile dysfunction may be caused by physiological conditions. Such biological causes include blockage in the arteries, diabetes, neurological disorders, alcohol or other drug abuse, chronic disease (kidney or liver failure), pelvic surgery, and neurological disorders. Smoking is also related to erectile dysfunction; the more frequent the smoking, the more likely the erectile dysfunction.

Viagra (and similar medications Cialis and Levitra) are frequently recommended by therapists to restore erectile function.

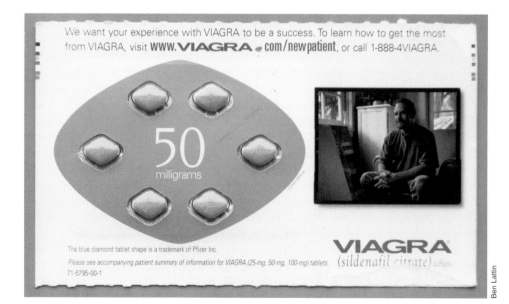

Special Topic 5 Sexual Dysfunctions

Psychosocial factors associated with erectile dysfunction include depression, fear (e.g., of unwanted pregnancy, intimacy, HIV infection, or other STIs), guilt, and relationship dissatisfaction. For example, the man who is having an extradyadic sexual relationship may feel guilty. This guilt may lead to difficulty in achieving or maintaining an erection in sexual interaction with the primary partner and/or the extradyadic partner.

Anxiety may also inhibit the man's ability to create and maintain an erection. One source of anxiety is performance pressure, which may be self-imposed or imposed by a partner. In self-imposed performance anxiety, the man constantly checks (mentally or visually) to see that he is erect. Such self-monitoring creates anxiety, since the man fears that he may not be erect.

Partner-imposed performance pressure involves the partner's communicating to the man that he must get and stay erect to be regarded as a good lover. Such pressure usually increases the man's anxiety, thus ensuring no erection. Whether self- or partner-imposed, the anxiety associated with performance pressure results in a vicious cycle—anxiety, erectile difficulty, embarrassment, followed by anxiety, erectile difficulty, and so on.

Performance anxiety may also be related to alcohol use. After consuming more than a few drinks, the man may initiate sex but may become anxious after failing to achieve an erection (too much alcohol will interfere with erection). Although alcohol may be responsible for his initial failure, his erection difficulties continue because of his anxiety.

Treatment of erectile dysfunction (like treatment of other sexual dysfunctions) depends on the cause(s) of the problem. When erection difficulties are caused by psychosocial factors, treatment may include improving the relationship with the partner and/or resolving the man's fear, guilt, or anxiety (i.e., performance pressure) about sexual activity. These goals may be accomplished through couple counseling, re-education, and sensate focus exercises.

A sex therapist would instruct the man and his partner to temporarily refrain from engaging in intercourse so as to remove the pressure to attain or maintain an erection. During this period, the man is encouraged to pleasure his partner in ways that do not require him to have an erection (e.g., cunnilingus or manual stimulation of partner). Once the man is relieved of the pressure to perform and learns alternative ways to satisfy his partner, his erection difficulties (if caused by psychosocial factors) often disappear. Most therapists bypass the use of these exercises in favor of Viagra, Cialis, or Levitra, discussed below.

Treatment for erectile dysfunction related to biological factors can include modification of the use of medication, alcohol, or other drugs. Increasingly, physicians are prescribing sildenafil (marketed as Viagra), which increases blood flow to the penis and results in an erection when the penis is stimulated. Though a physician should be consulted before the medication is taken, about 80 percent of men experiencing erectile dysfunction report restored potency as a consequence of Viagra (Sammons, 2003). Two new FDA-approved drugs are Cialis and Levitra. The newer medications can be taken 12 hours prior to sex and last for 24 to 36 hours. The man does not have a constant erection but may become erect with stimulation.

Viagra is having some unanticipated effects. Men in their 80s are beginning to take Viagra, which does produce an erection. But in some cases, the couple may not have had intercourse in 10 years, and the man's partner is no longer interested. A reshuffling of the sexual relationship of a couple in their later years may be indicated with medications such as Viagra, Levitra, or Cialis.

Men with low libido as well as erectile dysfunction may also benefit from testosterone replacement therapy (TRT). Once a low testosterone level (20% of men over age 60 have low levels) is confirmed by a blood test, testosterone supplements in the form of a gel, patch, or injection may be given. Potential risks include an increase in prostate size and changes in blood levels of cholesterol (AFUD, 2003).

Rapid Ejaculation

Also referred to as premature ejaculation, **rapid ejaculation** is defined as recurrent ejaculation with minimal sexual stimulation before, upon, or shortly after penetration and before the person wishes it. Whether a man ejaculates too soon is a matter of definition, depending on his and his partner's desires. Some partners define a rapid ejaculation in positive terms. One woman said she felt pleased that her partner was so excited by her that he "couldn't control himself." Another said, "The sooner he ejaculates, the sooner it's over with, and the sooner the better." Other women prefer that their partner delay ejaculation. Some women regard a pattern of rapid ejaculation as indicative of selfishness in their partner.

This feeling can lead to resentment and anger. Regardless of the definition, rapid ejaculation, or failing to control the timing of his ejaculation, is a man's most common sexual problem (Metz and Pryor, 2000) and can lead to other sexual dysfunctions, such as female inorgasmia, low sexual desire, and sexual aversion.

The cause of rapid ejaculation may be biological, psychogenic, or both (Metz and Pryor, 2000). Some men are thought to have a constitutionally hypersensitive sympathetic nervous system that predisposes them to rapid ejaculation. Psychogenic factors include psychological distress, such as shame, or psychological constitution, such as being obsessive-compulsive. Rapid ejaculation is less of a problem for the woman than for her male partner. And, while lower sexual satisfaction for both the woman and the man is associated with rapid ejaculation, relationship satisfaction is not affected (Byers and Grenier, 2003).

Treatment depends on the cause. If a hypersensitive neurological constitution seems to be the primary culprit, pharmacological intervention is used. For example, 50 mg of clomipramine hydrochloride taken 4 to 6 hours before sex has been found to delay orgasm in 30 percent of the cases. Where the cause is identified as psychogenic, cognitive-behavioral-sex therapy is indicated. This may include the stop-start technique or frequent ejaculations.

The pause, or stop-start, involves the man's stopping penile stimulation (or signaling his partner to stop stimulation) at the point that he begins to feel the urge to ejaculate. After the period of preejaculatory sensation subsides, stimulation resumes. This process may be repeated as often as desired by the partners.

Still another method of increasing the delay of ejaculation is for the man to ejaculate often. In general, the greater the number of ejaculations a man has in one 24-hour period, the longer he will be able to delay each subsequent ejaculation. The man's relationship with his partner is also important. Success of the respective treatments in reference to the cause awaits further research (Metz and Pryor, 2000).

The buyer should beware, and couples who decide to seek sex therapy should see only a credentialed therapist. The American Association of Sex Educators, Counselors, and Therapists (AASECT; www.aasect.org) maintains a list of certified sex therapists throughout the country.

KEY TERMS

erectile dysfunction	primary sexual dysfunction	rapid ejaculation	secondary sexual dysfunction
inhibited female orgasm			

WEB LINKS

FEMALE SEXUAL PROBLEMS

http://www.aamft.org/families/Consumer_Updates/FemaleSexualProblems.asp

MALE SEXUAL PROBLEMS

http://www.aamft.org/families/Consumer_Updates/MaleSexualProblems.asp

AMERICAN ASSOCIATION OF SEX EDUCATORS, COUNSELORS, AND THERAPISTS

http://www.aasect.org

Resources and Organizations

ABORTION—PRO CHOICE

Religious Coalition for Reproductive Choice
1025 Vermont Avenue NW, Suite 1130
Washington, DC 20005
Phone: 202-628-7700
www.rcrc.org

ABORTION—PRO LIFE

National Right to Life Committee
419 Seventh Street NW
Washington, DC 20004
Phone: 202-626-8800
www.nrlc.org/Unborn_Victims/

ADOPTION

Dave Thomas Foundation for Adoption
Phone: 800-275-3832
www.davethomasfoundationforadoption.org/

Evan B. Donaldson Adoption Institute
120 Wall Street, 20th Floor
New York, NY 10005
Phone: 212-269-5080
www.adoptioninstitute.org/

AL-ANON FAMILY GROUPS

1600 Corporate Landing Parkway
Virginia Beach, VA 23454
Fax: 757-563-1655
www.al-anon.org

CHILD ABUSE

Prevent Child Abuse
www.preventchildabuse.org/index.shtml

CHILDREN

National Association for the Education of Young Children
1509 Sixteenth Street NW
Washington, DC 20036-1426
Phone: 800-424-2460
www.naeyc.org

COMMUNES/INTENTIONAL COMMUNITIES

Intentional Communities
www.ic.org

Twin Oaks
138 Twin Oaks Road
Louisa, VA 23093
www.twinoaks.org/

DIVORCE- CUSTODY OF CHILDREN

Child Custody Evaluation Services of Philadelphia, Inc.
Dr. Ken Lewis, P. O. Box 202
Glenside, PA 19038
Phone: 215-576-0177

DIVORCE RECOVERY

The divorce room
   http://heartchoice.com/divorce/index.php

American Coalition for Fathers and Children
22994 El Toro Road, Suite 114
Lake Forest, CA 92630
Phone: 800-978-3237
   www.acfc.org

DOMESTIC VIOLENCE

National Coalition against Domestic Violence
P.O. Box 18749
1201 East Colfax 385
Denver, CO 80218
Phone: 303-839-1852
   www.ncadv.org

National Toll Free Number for Domestic Violence
Phone: 1-800-799-7233

FAMILY PLANNING

Planned Parenthood Federation of America
810 Seventh Avenue
New York, NY 10019
   www.plannedparenthood.org

GRANDPARENTING

The Foundation for Grandparenting
108 Farnham Road
Ojai, CA 93023
   www.grandparenting.org

AARP (grandparent information center)
601 E Street NW
Washington, DC 20049
   www.aarp.org/grandparents/

HEALTHY BABY

National Healthy Mothers, Healthy Babies Coalition
121 N Washington Street
Alexandria, VA 22314
Phone: 703-836-6110
   www.hmhb.org

HOMOSEXUALITY

National Gay and Lesbian Task Force
2320 Seventeenth Street NW
Washington, DC 20009
   www.ngltf.org

Parents, Families and Friends of Lesbians and Gays (PFLAG)
P.O. Box 96519
Washington, DC 20009-6519
   www.pflag.org

INFERTILITY

American Fertility Association
www.theafa.org/

MARRIAGE

National Marriage Project Rutgers, The State University of New Jersey
54 Joyce Kilmer Avenue, Lucy Stone Hall A347
Piscataway, NJ 08854-8045
Phone: 732-445-7922
http://marriage.rutgers.edu

Marriage: Keeping It Healthy and Happy
www.heartchoice.com/marriage/

Coalition for Marriage, Family and Couples Education
5310 Belt Road NW
Washington, DC 20015-1961
Phone: 202-362-3332
www.smartmarriages.com/

The Couples Place: Successful Marriages
www.couples-place.com/marriage.asp

MARRIAGE AND FAMILY THERAPY

American Association for Marriage and Family Therapy
1133 Fifteenth Street NW, Suite 300
Washington, DC 20005
Phone: 703-838-9808
www.aamft.org/index_nm.asp

MARRIAGE ENRICHMENT

ACME (Association for Couples in Marriage Enrichment)
56 Windsor Court
New Brighton, MN 55112
E-mail: hamb1001@tc.umn.edu
www.bettermarriages.org

MATE SELECTION

Right Mate
http://heartchoice.com/rightmate

MEN'S AWARENESS

American Men's Studies Association
www.mensstudies.org/

MOTHERHOOD

www.mothersoughttohaveequalrights.org/

PARENTAL ALIENATION SYNDROME

www.deltabravo.net/custody/pasarchive.php

PARENTING EDUCATION

www.familyeducation.com/home/

REPRODUCTIVE HEALTH

Association of Reproductive Health Professionals
2401 Pennsylvania Avenue, NW Suite
350 Washington, DC 20037-1718
Phone: 202-466 3825
    www.arhp.org

SEX ABUSE

VOICES in Action, Inc. (Victims of Incest Can Emerge Survivors)
P.O. Box 148309
Chicago, Illinois 60614
Phone: 800-7-VOICE-8
    www.voices-action.org

National Clearinghouse on Marital and Date Rape
2325 Oak Street
Berkeley, California 94708-1697
Phone: 510-524-1582
    www.ncmdr.org

SEX EDUCATION

Sexuality Information and Education Council of the United States
New York University
32 Washington Plaza
New York, New York 10003
    www.siecus.org

SEXUAL INTIMACY

    http://www.heartchoice.com/sex_intimacy/

SEXUALLY TRANSMISSIBLE DISEASES

American Social Health Association (Herpes Resource Center and HPV Support Program)
P.O. Box 13827
Research Triangle Park, NC 27709
Phone: 919-361-8400
    www.ashastd.org

National AIDS Hotline
Phone: 800-342-AIDS
    www.thebody.com/hotlines/national.html

National Herpes Hotline
Phone: 919-361-8488 or 800-230-6039
    www.ashastd.org/herpes/herpes_overview.cfm

STD Hotline
Phone: 800-227-8922
    www.ashastd.org/

Sex Therapy OnLine
    www.therapyoncall.com

STD/AIDS

Phone: 919-361-8400
    www.doe.state.in.us/sservices/hivaids_cdchotline.html

SINGLEHOOD

Alternatives to Marriage Project
P.O. Box 991010
Boston, MA 02199
Fax: 781-394-6625
www.unmarried.org/

SINGLE PARENTHOOD

Parents without Partners
8807 Colesville Road
Silver Spring, MD 20910
Phone: 301-588-9354
www.parentswithoutpartners.org

Single Mothers by Choice
1642 Gracie Square
Station New York, NY 10028
Phone: 212-988-0993
www.singlemothersbychoice.com

STEPFAMILIES

Stepfamily Association of America
650 J Street, Suite 205
Lincoln, NE 68508
www.saafamilies.org

Stepfamily Associates
1368 Beacon Street Suite 108
Brookline, MA 02146
Phone: 617-734-8831
www.stepfamilyboston.com

Stepfamily Foundation
333 West End Ave
New York, NY 10023
Phone: 800-759-7837
www.stepfamily.org

TRANSGENDER

Tri-Ess: The Society for the Second Self, Inc.
8880 Bellaire Blvd B2, PMB 104
Houston, TX 77036
www.tri-ess.org

WIDOWHOOD

Widowed Person's Service American Association of Retired Persons
609 E St. NW
Washington, DC 20049
http://seniors-site.com/widowm/wps.html

WOMEN'S AWARENESS

National Organization for Women
1000 Sixteenth Street NW, Suite 700
Washington, DC 20036
Phone: 202-331-0066
www.now.org

Bandy, J. 2006. Surplus of homes cools prices. Retrieved June 4, 2006, from http://www.bankrate.com/brm/news/real-estate/home-values1.asp.

Burnstein, G., M. Snyder, D. Conley, D. Newman, C. Walsh, G. Tao, and K. Irwin 2005. Chlamydia screening in a health plan before and after a national performance measure introduction. *Obstetrics and Gynecology* 106:327–34.

Byers, E. S. and G. Grenier. 2003. Premature or rapid ejaculation: Heterosexual couples' perceptions of men's ejaculatory behavior. *Archives of Sexual Behavior* 32:261–70.

Capaldi, D. M., J. Stoolmiller, S. Clark, and A. L. D. Owen. 2002. Heterosexual risk behaviors in at risk young men from early adolescence to young adulthood: Prevalence, prediction, and association with STD contraction. *Developmental Psychology* 38:394–406.

Centers for Disease Control and Prevention. National Center for HIV, STD and TB Prevention. (November 8, 2005). New CDC data show syphilis increasing in men. Retrieved April 27, 2006, from http://www.cdc.gov/od/oc/media/pressrel/r051108.htm.

Colson, M., A. Lemaire, P. Pinton, K. Hamidi, and P. Klein. 2006. Sexual behaviors and mental perception, satisfaction, and expectations of sex life in men and women in France. *The Journal of Sexual Medicine* 3: 121–31.

Dennerstein, L., P. Koochaki, I. Barton, and A. Graziottin. 2006. Hypoactive sexual desire disorder in menopausal women: A survey of western European women *Journal of Sexual Medicine* 3:212–22.

Dobkin, R. D., S. R. Leiblum, M. Menza, and H. Marin. 2006. Depression and sexual functioning in minority women: Current status and future directions. *Journal of Sex and Marital Therapy* 32:23–36.

Dunn, K. M., P. R. Croft, and G. I. Hackett. 2000. Satisfaction in the sex life of a general population sample. *Journal of Sex and Marital Therapy* 26:141–51.

Fuller, O. and A. Wald. 2005. Session 80V, Symposium: Virus Receptors, Congress of Virology, International Union of Microbiological Societies, San Francisco, CA. *Journal of Virology* June, 79(12):7419–37.

Glynn, M. and P. Rhodes. 2005. Estimated HIV prevalence in the United States at the end of 2003. Abstract 595, National HIV Prevention Conference, June, Atlanta.

Greenstein, A., L. Abramov, H. Matzkin and J. Chen. 2006. Sexual dysfunction in women partners of men with erectile dysfunction. *International Journal of Impotence Research* 18:44–46.

Haavio-Mannila, E., and O. Kontula. 1997. Correlates of increased sexual satisfaction. *Archives of Sexual Behavior* 26:399–420.

Harper, D. 2006. Sustained efficacy up to 4.5 years of bivalent L1 virus-like particle vaccine against human papillomavirus types 16 and 18: follow-up from a randomized control trial. Published online April 6 in *The Lancet*.

Joo, S., J. E. Grable, and D. C. Bagwell. 2003 Credit card attitudes and behaviors of college students. *College Student Journal* 37:405–19.

Kelly, M., D. Strassberg, and C. Turner. 2006. Behavioral assessment of couples' communication in female orgasmic disorder. *Journal of Sex and Marital Therapy* 32:81–95.

Knox, D. and Zusman, M. E. 2006. Relationship and sexual behaviors of a sample of 1027 university students. Unpublished data collected for this text. Department of Sociology, East Carolina University, Greenville, NC.

Lane, T. and F. Althus. 2002. Who engages in sexual risk behavior? *International Family Planning Perspectives* 28:185–86.

Lau, J.T.F., X. L. Yang, Q. S. Wang, Y. M. Cheng, H. Y. Tsui, L. W. H. Mui and J. H. Kim. 2006. Gender power and marital relationship as predictors of sexual dysfunction and sexual satisfaction among young married couples in rural China. *Urology* 67:579–85.

Laumann, E. O., A. Nicolosi, D. B. Glasser, A. Paik, C. Gingell, E. Moreira, and T. Wang. 2005. Sexual problems among women and men aged 40-80: prevalence and correlates identified in the Global Study of Sexual Attitudes and Behaviors. *International Journal of Impotence Research* 17:39–57.

Metz, M. F., and J. L. Pryor. 2000. Premature ejaculation: A psychophysiological approach for assessment and management. *Journal of Sex and Marital Therapy* 26:293–320.

National Institute for Allergies and Infectious Disease (NIAID). 2005. HIV infection and AIDS: an overview. AIDS Info. US Department of Health and Human Services, Rockville, MD.

Quirk, F., H. Scott and T. Symonds. 2005. The use of the sexual function questionnaire as a screening tool for women with sexual dysfunction. *The Journal of Sexual Medicine* 2:469–77.

Sammons, R. 2003. Personal communication. Dr. Sammons is a psychiatrist in private practice in Grand Junction, Colorado.

Sellors, J. W., T. L. Karwalajtys, J. Kaczorowski, J. B. Mathony, A. Lltwyn, S. Chong, J. Sparrow, A. Lorinez, for the survey of HPV in Ontario Women (SHOW group). 2003 Incidence, clearance and predictors of human papillomavirus infection in women. *Canadian Medical Association Journal* 168:421–25.

Sengupta, S. and B. Lo. 2004. U.S. pregnant women's perceptions of universal, routine prenatal care. *AIDS & Public Policy Journal* 18(3/4, Fall/Winter):82–96.

Shih, S., S. Scholle, K. Irwin, G. Tao, C. Walsh, and W. Tun. 2004. Chlamydia screening among sexually active young female enrollees of health plans. *MMWR Morbidity and Mortality Weekly Report* October 29. 53(42):983–85.

Stephenson, J. 2003. Growing, evolving HIV/AIDS pandemic is producing social and economic fallout. *Journal of American Medical Association* 289:31–33.

Sternberg, S. 2003. Vaccine for AIDS appears to work. *USA Today* 24 February:A1.

United States Agency for International Development (USAID). 2006. HIV/AIDS. Washington, D.C. Retrieved April 27, 2006 from http://www.usaid.gov/our_work/global_health/aids/News/aidsfaq.html#people.

United States Census Bureau. 2004. International Population Reports WP/02/2, The AIDS Pandemic in the 21st Century. U.S. Government Printing Office, Washington, D.C.

Worldwide HIV and AIDS Statistics, 2006. http://www.avert.org/worldstats.htm Retrieved Oct 24, 2006.

Individual Autobiography Outline

Your instructor may ask you to write a paper that reflects the individual choices you have made that have contributed to your becoming who you are. Check with your instructor to determine what credit (if any) is assigned to your completing this autobiography. Use the following outline to develop your paper. Some topics may be too personal, and you may choose to avoid writing about them. Your emotional comfort is important, so skip any questions you want and answer only those questions you feel comfortable responding to.

I. Choices: Free Will Versus Determinism
Specify the degree to which you feel that you are free to make your own interpersonal choices, versus the degree to which your choices are determined by social constraints and influences. Give an example of social influences being primarily responsible for a relationship choice and an example of your making a relationship choice where you acted contrary to the pressure you were getting from parents, peers, and so on.

II. Relationship Beginnings
   A. *Interpersonal context into which you were born.* How long had your parents been married before you were born? How many other children had been born into your family? How many followed your birth? Describe how these and other parental choices affected you before you were born and the choices you will make in regard to family planning that will affect the lives of your children.
   B. *Early relationships.* What was your relationship with your mother, father, and siblings when you were growing up? What is your relationship with each of them today? Who took care of you as a baby? If this person was other than your parents or siblings (e.g., a grandparent), what is your relationship with that person today? How often were you told by you mother/father that you were loved? How often did they embrace or hug you and tell you of their love for you? How has this closeness or distance influenced your pattern with others today? Give other examples of how your experiences in the family in which you were reared have influenced who you are and how you behave today. How easy or difficult is it for you to make decisions and how have your parents influenced this capacity?
   C. *Early self-concept.* How did you feel about yourself as a child, an adolescent, and a young adult? What significant experiences helped to shape your self-concept? How do you feel about yourself today? What choices have you made that have resulted in your feeling good about yourself? What choices have you made that have resulted in your feeling negatively about yourself?

III. Subsequent Relationships
   A. *First love.* When was your first love relationship with someone outside your family? What kind of love was it? Who initiated the relationship? How long did it last? How did it end? How did it affect you and your subsequent relationships? What choices did you make in this first love relationship you are glad you made? What choices did you make that you now feel were a mistake?
   B. *Subsequent love relationships.* What other significant love relationships (if any) have you had? How long did they last and how did they end? What choices did you make in these relationships you are glad you made? What choices did you make that you now feel were a mistake?

C. Subsequent relationship choices. What has been the best relationship choice you have made? What relationship choices have you regretted?

D. Lifestyle preferences. What are your preferences for remaining single, being married, or living with someone?

How would you feel about living in a commune? What do you believe is the ideal lifestyle? Why?

IV. Communication Issues

A. Parental models. Describe your parents' relationship and their manner of communicating with each other. How are your interpersonal communication patterns similar to and different from theirs?

B. Relationship communication. How comfortable do you feel talking about relationship issues with your partner? How comfortable do you feel telling your partner what you like and don't like about his or her behavior? To what degree have you told your partner your feelings for him or her? To what degree have you told your partner your desires for the future?

C. Sexual communication. How comfortable do you feel giving your partner feedback about how to please you sexually? How comfortable are you discussing the need to use a condom with a potential sex partner? How would you approach this topic?

D. Sexual past. How much have you disclosed to a partner about your previous relationships? How honest were you? Do you think you made the right decision to disclose or withhold? Why?

V. Sexual Choices

A. Sex education. What did you learn about sex from your parents, peers, and teachers (both academic and religious)? What choices did your parents, peers, and teachers make about your sex education that had positive consequences? What decisions did they make that had negative consequences for you?

B. Sexual experiences. What choices have you made about your sexual experiences that had positive outcomes? What choices have you made that had negative outcomes?

C. Sexual values. To what degree are your sexual values absolutist, legalist, or relativistic? How have your sexual values changed since you were an adolescent?

D. Safer sex. What is the riskiest choice you have made with regard to your sexual behavior? What is the safest choice you have made with regard to your sexual behavior? What is your policy about asking your partner about his or her previous sex history and requiring that both of you be tested for HIV and STIs before having sex? How comfortable are you buying and using condoms?

VI. Violence and Abuse Issues

A. Violent/abusive relationship. Have you been involved in a relationship in which your partner was violent or abusive toward you? Give examples of the violence or abuse (verbal and nonverbal) if you have been involved in such a relationship. How many times did you leave and return to the relationship before you left permanently?

What was the event that triggered your decision to leave the first and last time? Describe the context of your actually leaving (e.g., left when partner was at work). To what degree have you been violent or abusive toward a partner in a romantic relationship?

B. Family, sibling abuse. Have you been involved in a relationship where your parents or siblings were violent or abusive toward you? Give examples of any violence or abuse (verbal or nonverbal) if such experiences were part of your growing up. How have these experiences affected the relationship you have with the abuser today?

C. Forced sex. Have you been pressured or forced to participate in sexual activity against your will by a parent, sibling, partner, or stranger? How did you react at the time and how do you feel today? Have you pressured or forced others to participate in sexual experiences against their will?

VII. Reproductive Choices

A. Contraception. What is your choice for type of contraception? How comfortable do you feel discussing the need for contraception with a potential partner? In what percentage of your first-time intercourse experiences with a new partner did you use a condom? (If you have not had intercourse, this question will not apply.)

B. Children. How many children (if any) do you want and at what intervals? How important is it to you that your partner wants the same number of children as do you?

How do you feel about artificial insemination, sterilization, abortion, and adoption? How important is it to you that your partner feel the same way?

VIII. Childrearing Choices

A. Discipline. What are your preferences for your use of "time out" or "spanking" as a way of disciplining your children? How important is it to you that your partner feel as you do on this issue?

B. Day care. What are your preferences for whether your children grow up in day care or whether one parent will stay home with and rear the children? How important is it to you that your partner feel as do you on this issue?

C. Education. What are your preferences for whether your children attend public or private school or are "home schooled"? What are your preferences for whether your child attends a religious school? To what degree do you feel it is your responsibility as parents to pay for the college education of your children? How important is it to you that your partner feel as do you on these issues?

IX. Education/Career Choices

A. *Own educational/career choices.* What is your major in school? How important is it to you that you finish undergraduate school? How important is it to you that you earn a master's degree, Ph.D., M.D., or law degree? To what extent do you want to be a stay-at-home mom or stay-at-home dad? How important is it to you that your partner be completely supportive of your educational, career, and family aspirations?

B. *Expectations of partner.* How important is it to you that your partner have the same level of education that you have? To what degree are you willing to be supportive of your partner's educational and career aspirations?

Family Autobiography Outline

Your instructor may want you to write a paper that reflects the influence of your family on your development. Check with your instructor to determine what credit (if any) is assigned to your completing this family autobiography. Use the following outline to develop your paper. Some topics may be too personal, and you may choose to avoid writing about them. Your emotional comfort is important, so skip any questions you like and answer only those you feel comfortable responding to.

I. Family Background

A. Describe yourself, including age, gender, place of birth, and additional information that helps to identify you. On a scale of 0 to 10 (10 is highest), how happy are you? Explain this number in reference to the satisfaction you experience in the various roles you currently occupy (offspring, sibling, partner in a relationship, employee, student, parent, roommate, friend, etc.).

B. Identify your birth position; give the names and ages of children younger and older than you. How did you feel about your "place" in the family? How do you feel now?

C. What was your relationship with and how did you feel about each parent and sibling when you were growing up?

D. What is your relationship and how do you feel today about each of these family members?

E. Which parental figure or sibling are you most like? How? Why?

F. Who else lived in your family (e.g., grandparent, spouse of sibling) and how did they impact family living?

G. Discuss the choice you made before attending college that you regard as the wisest choice you made during this time period. Discuss the choice you made prior to college that you regret.

H. Discuss the one choice you made since you began college that you regard as the wisest choice you made during this time period. Discuss the one choice you made since you began college that you regret.

II. Religion and Values

A. In what religion were you socialized as a child? Discuss the impact of religion on yourself as a child and as an adult. To what degree will you choose to teach your own children similar religious values? Why?

B. Explain what you were taught in your family in regard to each of the following values: intercourse outside of marriage, need for economic independence, manners, honesty, importance of being married, qualities of a desirable spouse, children, alcohol and drugs, safety, elderly family members, persons of other races and religions, persons with disabilities, homosexuals, persons with less or more education, persons with and without "wealth," importance of having children, and occupational role (in regard to the latter, what occupational role were you encouraged to pursue?). To what degree will you choose to teach your own children similar values? Why?

C. What was the role relationship between your parents in terms of dominance, division of labor, communication, affection, etc.? How has your observation of the parent of the same sex and opposite sex influenced the role you display in your current relationships with intimate partners? To what degree will you choose to have a similar relationship with your own partner that your parents had with each other?

D. How close were your parents emotionally? How emotionally close were/are you with your parents? To what degree did you and your parents discuss feelings? On a scale of 0 to 10, how well do your parents know how you think and feel? To what degree will you choose to have a similar level of closeness or distance with your own partner and children?

E. Did your parents have a pet name for you? How did you feel about this?

F. How did your parents resolve conflict between themselves? How do you resolve conflict with partners in your own relationship?

III. Economics and Social Class

A. Identify your social class (lower, middle, upper), the education/jobs/careers of your respective parents, and the economic resources of your family. How did your social class and economic well-being affect you as a child? How has the economic situation in which you were reared influenced your own choices of what you want for yourself? To what degree are you economically self-sufficient?

B. How have the career choices of your parents influenced your own?

IV. Parental Plus and Minuses

A. What is the single most important thing your mother and father, respectively, said or did that has affected your life in a positive way?

B. What is the single biggest mistake your mother and father, respectively, made in rearing you? Discuss how this impacted you negatively. To what degree are you still affected by this choice?

V. Personal Crisis Events

A. Everyone experiences one or more crisis events that have a dramatic impact on his or her life. Identify and discuss each event or events you have experienced and your reaction and adjustment to them.

B. How did your parents react to this crisis event you were experiencing? To what degree did their reaction help or hinder your adjustment? What different choice or choices (if any) could they have made to assist you in ways you would have regarded as more beneficial?

VI. Family Crisis Events

Identify and discuss each crisis event your family has experienced. How did each member of your family react and adjust to each event? An example of a family crisis event would be unemployment of the primary breadwinner, prolonged illness of a family member, aging parent coming to live with the family, death of a sibling, or alcoholism.

VII. Future

Describe yourself 2, 5, and 10 years from now. What are your educational, occupational, marital, and family goals? How has the family in which you were reared influenced each of these goals? What choices might your parents make to assist you in achieving your goals?

Prenuptial Agreement of a Remarried Couple

Pam and Mark are of sound mind and body and have a clear understanding of the terms of this contract and of the binding nature of the agreements contained herein; they freely and in good faith choose to enter into the PRENUPTIAL AGREEMENT and MARRIAGE CONTRACT and fully intend it to be binding upon themselves.

Now, therefore, in consideration of their love and esteem for each other and in consideration of the mutual promises herein expressed, the sufficiency of which is hereby acknowledged, Pam and Mark agree as follows:

Names

Pam and Mark affirm their individuality and equality in this relationship. The parties believe in and accept the convention of the wife's accepting the husband's name, while rejecting any implied ownership.

Therefore, the parties agree that they will be known as husband and wife and will henceforth employ the titles of address Mr. and Mrs. Mark Stafford and will use the full names of Pam Hayes Stafford and Mark Robert Stafford.

Relationships with Others

Pam and Mark believe that their commitment to each other is strong enough that no restrictions are necessary with regard to relationships with others.

Therefore, the parties agree to allow each other freedom to choose and define their relationships outside this contract, and the parties further agree to maintain sexual fidelity each to the other.

Religion

Pam and Mark reaffirm their belief in God and recognize He is the source of their love. Each of the parties has his/her own religious beliefs.

Therefore, the parties agree to respect their individual preferences with respect to religion and to make no demands on each other to change such preferences.

Children

Pam and Mark both have children. Although no minor children will be involved, there are two (2) children still at home and in school and in need of financial and emotional support.

Therefore, the parties agree that they will maintain a home for and support these children as long as is needed and reasonable. They further agree that all children of both parties will be treated as one family unit, and each will be given emotional and financial support to the extent feasible and necessary as determined mutually by both parties.

Careers and Domicile

Pam and Mark value the importance and integrity of their respective careers and acknowledge the demands that their jobs place on them as individuals and on their partnership.

Both parties are well established in their respective careers and do not foresee any change or move in the future.

The parties agree, however, that if the need or desire for a move should arise, the decision to move shall be mutual and based on the following factors:

1. The overall advantage gained by one of the parties in pursuing a new opportunity shall be weighed against the disadvantages, economic and otherwise, incurred by the other.

2. The amount of income or other incentive derived from the move shall not be controlling.

3. Short-term separations as a result of such moves may be necessary.

Mark hereby waives whatever right he might have to solely determine the legal domicile of the parties.

Care and Use of Living Spaces

Pam and Mark recognize the need for autonomy and equality within the home in terms of the use of available space and allocation of household tasks. The parties reject the concept that the responsibility for housework rests with the woman in a marriage relationship while the duties of home maintenance and repair rest with the man.

Therefore, the parties agree to share equally in the performance of all household tasks, taking into consideration individual schedules, preferences, and abilities.

The parties agree that decisions about the use of living space in the home shall be mutually made, regardless of the parties' relative financial interests in the ownership or rental of the home, and the parties further agree to honor all requests for privacy from the other party.

Property, Debts, Living Expenses

Pam and Mark intend that the individual autonomy sought in the partnership shall be reflected in the ownership of existing and future-acquired property, in the characterization and control of income, and in the responsibility for living expenses. Pam and Mark also recognize the right of patrimony of children of their previous marriages.

Therefore, the parties agree that all things of value now held singly and/or acquired singly in the future shall be the property of the party making such acquisition. In the event that one party to this agreement shall predecease the other, property and/or other valuables shall be disposed of in accordance with an existing will or other instrument of disposal that reflects the intent of the deceased party.

Property or valuables acquired jointly shall be the property of the partnership and shall be divided, if necessary, according to the contribution of each party. If one party shall predecease the other, jointly owned property or valuables shall become the property of the surviving spouse.

Pam and Mark feel that each of the parties to this agreement should have access to monies that are not accountable to the partnership.

Therefore, the parties agree that each shall retain a mutually agreeable portion of their total income and the remainder shall be deposited in a mutually agreeable banking institution and shall be used to satisfy all jointly acquired expenses and debts.

The parties agree that beneficiaries of life insurance policies they now own shall remain as named on each policy. Future changes in beneficiaries shall be mutually agreed on after the dependency of the children of each party has been terminated. Any other benefits of any retirement plan or insurance benefits that accrue to a spouse only shall not be affected by the foregoing.

The parties recognize that in the absence of income by one of the parties, resulting from any reason, living expenses may become the sole responsibility of the employed party, and in such a situation, the employed party shall assume responsibility for the personal expenses of the other.

Both Pam and Mark intend their marriage to last as long as both shall live.

Therefore, the parties agree that should it become necessary, due to the death of either party, the surviving spouse shall assume any last expenses in the event that no insurance exists for that purpose.

Pam hereby waives whatever right she might have to rely on Mark to provide the sole economic support for the family unit.

Evaluation of the Partnership

Pam and Mark recognize the importance of change in their relationship and intend that this CONTRACT shall be a living document and a focus for periodic evaluations of the partnership.

The parties agree that either party can initiate a review of any article of the CONTRACT at any time for amendment to reflect changes in the relationship. The parties agree to honor such requests for review with negotiations and discussions at a mutually convenient time.

The parties agree that in any event, there shall be an annual reaffirmation of the CONTRACT on or about the anniversary date of the CONTRACT.

The parties agree that in the case of unresolved conflicts between them over any provisions of the CONTRACT, they will seek mediation, professional or otherwise, by a third party.

Termination of the Contract

Pam and Mark believe in the sanctity of marriage; however, in the unlikely event of a decision to terminate this CONTRACT, the parties agree that neither shall contest the application for a divorce decree or the entry of such decree in the county in which the parties are both residing at the time of such application.

In the event of termination of the CONTRACT and divorce of the parties, the provisions of this and the section on "Property, Debts, Living Expenses" of the CONTRACT as amended shall serve as the final property settlement agreement between the parties. In such event, this CONTRACT is intended to effect a complete settlement of any and all claims that either party may have against the other, and a complete settlement of their respective rights as to property rights, homestead rights, inheritance rights, and all other rights of property otherwise arising out of their partnership. The parties further agree that in the event of termination of this CONTRACT and divorce of the parties, neither party shall require the other to pay maintenance costs or alimony.

Decision Making

Pam and Mark share a commitment to a process of negotiations and compromise that will strengthen their equality in the partnership. Decisions will be made with respect for individual needs. The parties hope to maintain such mutual decision making so that the daily decisions affecting their lives will not become a struggle between the parties for power, authority, and dominance. The parties agree that such a process, while sometimes time-consuming and fatiguing, is a good investment in the future of their relationship and their continued esteem for each other.

Now, therefore, Pam and Mark make the following declarations:

1. They are responsible adults.

2. They freely adopt the spirit and the material terms of this prenuptial and marriage contract.

3. The marriage contract, entered into in conjunction with a marriage license of the State of Illinois, County of Wayne, on this 12th day of June 2004, hereby manifests their intent to define the rights and obligations of their marriage relationship as distinct from those rights and obligations defined by the laws of the State of Illinois, and affirms their right to do so.

4. They intend to be bound by this prenuptial and marriage contract and to uphold its provisions before any Court of Law in the Land.

Therefore, comes now, Pam Hayes Stafford, who applauds her development that allows her to enter into this partnership of trust, and she agrees to go forward with this marriage in the spirit of the foregoing PRENUPTIAL and MARRIAGE CONTRACT.

Therefore, comes now, Mark Robert Stafford, who celebrates his growth and independence with the signing of this contract, and he agrees to accept the responsibilities of this marriage as set forth in the foregoing PRENUPTIAL and MARRIAGE CONTRACT.

This CONTRACT AND COVENANT has been received and reviewed by the Reverend Ray Brannon, officiating.

Finally, come Vicki Oliver and Rodney Oliver, who certify that Pam and Mark did freely read and sign this MARRIAGE CONTRACT in their presence, on the occasion of their entry into a marriage relationship by the signing of a marriage license in the State of Illinois, County of Wayne, at which they acted as official witnesses. Further, they declare that the marriage license of the parties bears the date of the signing of this PRENUPTIAL and MARRIAGE CONTRACT.

Living Will

I, [Declarant], ("Declarant" herein), being of sound mind, and after careful consideration and thought, freely and intentionally make this revocable declaration to state that if I should become unable to make and communicate my own decisions on life-sustaining or life-support procedures, then my dying shall not be delayed, prolonged, or extended artificially by medical science or life-sustaining medical procedures, all according to the choices and decisions I have made and which are stated here in my Living Will.

It is my intent, hope, and request that my instructions be honored and carried out by my physicians, family, and friends, as my legal right.

If I am unable to make and communicate my own decisions regarding the use of medical life-sustaining or life support systems and/or procedures, and if I have a sickness, illness, disease, injury, or condition which has been diagnosed by two (2) licensed medical doctors or physicians who have personally examined me (or more than two (2) if required by applicable law), one of whom shall be my attending physician, as being either (1) terminal or incurable certified to be terminal, or (2) a condition from which there is no reasonable hope of my recovery to a meaningful quality of life, which may reasonably be referred to as hopeless, although not necessarily "terminal" in the medical sense, or (3) has rendered me in a persistent vegetative state, or (4) a condition of extreme mental deterioration, or (5) permanently unconscious, then in the absence of my revoking this Living Will, all medical life-sustaining or life-support systems and procedures shall be withdrawn, unless I state otherwise in the following provisions.

Unless otherwise provided in this Living Will, nothing herein shall prohibit the administering of pain-relieving drugs to me, or any other types of care purely for my comfort, even though such drugs or treatment may shorten my life, or be habit forming, or have other adverse side effects.

I am also stating the following additional instructions so that my Living Will is as clear as possible: Resuscitation (CPR)—I do not want to be resuscitated. Intravenous and Tube Feeding—I do not want to be kept alive via intravenous means; I do not want a feeding tube installed if I am unable to consume food/ liquids naturally. Life-Sustaining Surgery—I do not want to be kept alive by any new or old life-sustaining surgery. New Medical Developments—I do not want to become a participant in any new medical developments that would prolong my life. Home or Hospital—I want to die wherever my family chooses, including a hospital or nursing home.

In the event that any terms or provisions of my Living Will are not enforceable or are not valid under the laws of the state of my residence, or the laws of the state where I may be located at the time, then all other provisions which are enforceable or valid shall remain in full force and effect, and all terms and provisions herein are severable.

IN WITNESS WHEREOF, I have read and understand this Living Will, and I am freely and voluntarily signing it on this the day of (month), (year) in the presence of witnesses.

Signed: [Declarant]

Street Address:

County:

City and State:

Witness:

We, the undersigned witnesses, certify by our signatures below, that we are adult (at least 18 years old), mentally competent persons; that we are not related to the Declarant by blood, marriage, or adoption; that we do not stand to inherit anything from the Declarant by any means, including will, trust, operation of law or the laws of intestate succession, or by beneficiary designation, nor do we stand to benefit in any way from the death of the Declarant; that we are not directly responsible for the health or medical care, or general welfare of the Declarant; that neither of us signed the Declarant's signature on this document; and that the Declarant is known to us.

We hereby further certify that the Declarant is over the age of 18; that the Declarant signed this document freely and voluntarily, not under any duress or coercion; and that we were both present together, and in the presence of the Declarant to witness the signing of this Living Will on this the day of (month), (year).

Witness signature:

Residing at:

Witness signature:

Residing at:

Notary Acknowledgment

This instrument was acknowledged before me on this the day of (month), (year) by [Declarant], the Declarant herein, on oath stating that the Declarant is over the age of 18, has fully read and understands the above and foregoing Living Will, and that the Declarant's signing and execution of same is voluntary, without coercion, and is intentional.

Notary Public

My commission or appointment expires:

Durable Power of Attorney

KNOW ALL MEN BY THESE PRESENTS That I, _____ , as principal ("Principal"), a resident of the State and County aforesaid, have made, constituted, appointed and by these presents do make, constitute, and appoint and, either or both of them, as my true and lawful agent or attorney-in-fact ("Agent") to do and perform each and every act, deed, matter, and thing whatsoever in and about my estate, property, and affairs as fully and effectually to all intents and purposes as I might or could do in my own proper person, if personally present, including, without limiting the generality of the foregoing, the following specifically enumerated powers which are granted in aid and exemplification of the full, complete and general power herein granted and not in limitation or definition thereof:

1. To forgive, request, demand, sue for, recover, elect, receive, hold all sums of money, debts due, commercial paper, checks, drafts, accounts, deposits, legacies, bequests, devises, notes, interest, stock of deposit, annuities, pension, profit sharing, retirement, Social Security, insurance, and all other contractual benefits and proceeds, all documents of title, all property and all property rights, and demands whatsoever, liquidated or unliquidated, now or hereafter owned by me, or due, owing, payable, or belonging to me or in which I have or may hereafter acquire an interest, to have, use and take all lawful means and equitable and legal remedies and proceedings in my name for collection and recovery thereof, and to adjust, sell, compromise, and agree for the same, and to execute and deliver for me, on my behalf, and in my name all endorsements, releases, receipts or other sufficient discharges for the same.

2. To buy, receive, lease as lessor, accept, or otherwise acquire; to sell, convey, mortgage, grant options upon, hypothecate, pledge, transfer, exchange, quitclaim, or otherwise encumber or dispose of; or to contract or agree for the acquisition, disposal, or encumbrance of any property whatsoever or any custody, possession, interest, or right therein for cash or credit and upon such terms, considerations, and conditions as Agent shall think proper, and no person dealing with Agent shall be bound to see to the application of any monies paid.

3. To take, hold, possess, invest, or otherwise manage any or all of the property or any interest therein; to eject, remove, or relieve tenants or other persons from, and recover possession of, such property by all lawful means; and to maintain, protect, preserve, insure, remove, store, transport, repair, build on, raze, rebuild, alter, modify, or improve the same or any part thereof, and/or to lease any property for me or my benefit, as lessee without option to renew, to collect and receive any receipt for rents, issues, and profits of my property.

4. To invest and reinvest all or any part of my property in any property and undivided interest in property, wherever located, including bonds, debentures, notes secured or unsecured, stock of corporations regardless of class, interests in limited partnerships, real estate or any interest in real estate whether or not productive at the time of the investment, interest in trusts, investment trusts, whether of the open and/or closed funds types, and participation in common, collective, or pooled trust funds or annuity contracts without being limited by any statute or rule of law concerning investment by fiduciaries.

5. To make, receive, and endorse checks and drafts; deposit and withdraw funds; acquire and redeem certificates of deposit in banks, savings and loan associations, or other institutions; and execute or release such deeds of trust or other security agreements as may be necessary or proper in the exercise of the rights and powers herein granted.

6. To pay any and all indebtedness of mine in such manner and at such times as Agent may deem appropriate.

7. To borrow money for any purpose, with or without security or on mortgage or pledge of any property.

8. To conduct or participate in any lawful business of whatsoever nature for me and in my name; execute partnership agreements and amendments thereto; incorporate, reorganize, merge, consolidate, recapitalize, sell, liquidate, or dissolve any business; elect or employ officers, directors, and agents; carry out the provisions of any agreement for the sale of any business interest or stock therein; and exercise voting rights with respect to stock either in person or by proxy, and to exercise stock options.

9. To prepare, sign, and file joint or separate income tax returns or declarations of estimated tax for any year or years; to prepare, sign, and file gift tax returns with respect to gifts made by me for any year or years; to consent to any gift and to utilize any gift-splitting provision or other tax election; and to prepare, sign, and file any claims for refund of tax.

10. To have access at any time or times to any safe deposit box rented by me, wheresoever located, and to remove all or any part of the contents thereof, and to surrender or relinquish said safety deposit box in any institution in which such safety deposit box may be located shall not incur any liability to me or my estate as a result of permitting Agent to exercise this power.

11. To execute any and all contracts of every kind or nature.

As used herein, the term "property" includes any property, real or personal, tangible or intangible, wheresoever situated.

The execution and delivery by Agent of any conveyance paper instrument or document in my name and behalf shall be conclusive evidence of Agent's approval of the consideration therefore, and of the form and contents thereof, and that Agent deems the execution thereof in my behalf necessary or desirable.

Any person, firm, or corporation dealing with Agent under the authority of this instrument is authorized to deliver to Agent all considerations of every kind or character with respect to any transactions so entered into by Agent and shall be under no duty or obligation to see to or examine into the disposition thereof.

Third parties may rely upon the representation of Agent as to all matters relating to any power granted to Agent, and no person who may act in reliance upon the representation of Agent or the authority granted to Agent shall incur liability to me or my estate as a result of permitting Agent to exercise any power. Agent shall

be entitled to reimbursement for all reasonable costs and expenses incurred and paid by Agent on my behalf pursuant to any provisions of this durable power of attorney, but Agent shall not be entitled to compensation for services rendered hereunder.

Notwithstanding any provision herein to the contrary, Agent shall not satisfy any legal obligation of Agent out of any property subject to this power of attorney, nor may Agent exercise this power in favor of Agent, Agent's estate, Agent's creditors, or the creditors of Agent's estate.

Notwithstanding any provision hereto to the contrary, Agent shall have no power or authority whatever with respect to (a) any policy of insurance owned by me on the life of Agent, and (b) any trust created by Agent as to which I am Trustee.

When used herein, the singular shall include the plural and the masculine shall include the feminine.

This power of attorney shall become effective immediately upon the execution hereof.

This is a durable power of attorney made in accordance with and pursuant to Section (this section to be completed in reference to laws of the state in which the document is executed).

This power of attorney shall not be affected by disability, incompetency, or incapacity of the principal.

Principal may revoke this durable power of attorney at any time by written instrument delivered to Agent. The guardian of Principal may revoke this instrument by written instrument delivered to Agent.

IN WITNESS WHEREOF, I have executed this durable power of attorney in three (3) counterparts, and I have directed that photostatic copies of this power be made, which shall have the same force and effect as an original.

DATED THIS THE day of (month), (year).

WITNESS:

Name of Principal here

THE STATE OF and (County)

I, a Notary Public in and for said County, in said State, hereby certify that _____, whose name is signed to the foregoing durable power of attorney, and who is known to me, acknowledged before me on this day that being informed of the contents of the durable power, executed the same voluntarily on the day the same bears date.

GIVEN under my hand and seal this day of (month), (year).

NOTARY PUBLIC

My Commission Expires:

Glossary

A

Abortion rate—the number of abortions per 1000 women aged 15–44.

Abortion ratio—the number of abortions per 1000 live births.

Absolute poverty—the lack of resources that leads to hunger and physical deprivation.

Absolutism—sexual value system based on unconditional allegiance to the authority of science, law, tradition, or religion (e.g., sexual intercourse before marriage is wrong).

Abusive head trauma—nonaccidental head injury in infants and toddlers.

Accommodating style of conflict—conflict style in which the respective partners are not assertive in their positions but are cooperative. Each attempts to soothe the other and to seek a harmonious solution.

Acquaintance rape—nonconsensual sex between adults who know each other.

Advance directive—a legal document detailing the conditions under which a person wants life-support measures to be used for him or her in the event of a medical crisis.

Agape love style—love style characterized by a focus on the well-being of the love object, with little regard for reciprocation. The love of parents for their children is agape love.

Age—term defined chronologically, physiologically, sociologically, and culturally.

Age discrimination—discriminating against a person because of age (e.g., not hiring a person because he or she is old or hiring a person because he or she is young).

Ageism—systematic persecution and degradation of people because they are old.

AIDS—acquired immunodeficiency syndrome; the last stage of HIV infection, in which the immune system of a person's body is so weakened that it becomes vulnerable to disease and infection.

Al-Anon—an organization that provides support for family members and friends of alcohol abusers.

Alcohol-exposed pregnancy (AEP)—pregnancy resulting from binge drinking.

Androgyny—a blend of traits that are stereotypically associated with masculinity and femininity.

Annulment—a mechanism that returns the parties to their premarital status. An annulment states that no valid marriage contract ever existed. Annulments are both religious and civil.

Anodyspareunia—frequent and severe pain during receptive anal sex.

Antinatalism—opposition to having children.

Arbitration—a third party listens to both spouses and makes a decision about custody, division of property, etc.

Asceticism—the belief that giving in to carnal lusts is wrong and that one must rise above the pursuit of sensual pleasure to a life of self-discipline and self-denial.

Avoiding style of conflict—conflict style in which the partners are neither assertive nor cooperative.

B

Baby blues—transitory symptoms of depression 24 to 48 hours after a baby is born.

Barebacking—intentional unprotected anal sex.

Battered-woman syndrome—general pattern of battering that a woman is subjected to, defined in terms of the frequency, severity, and injury she experiences.

Battering rape—a marital rape that occurs in the context of a regular pattern of verbal and physical abuse.

Behavioral approach—an approach to childrearing based on the principle that behavior is learned through classical and operant conditioning.

Beliefs—definitions and explanations about what is thought to be true.

Bigamy—marriage to more than one person at the same time.

Binegativity—see *Biphobia*.

Binuclear family—family in which the members live in two households. Most often one parent and offspring live in one household, and the other parent lives in a separate household. They are still a family even though the members live in separate households.

Biofeedback—in reference to stress management, a process in which information that is fed back to the brain helps a person to experience the desired brain wave state.

Biphobia—also referred to as binegativity; refers to a parallel set of negative attitudes toward bisexuality and those identified as bisexual.

Biosocial—emphasizes the interaction of one's biological/genetic inheritance with one's social environment to explain and predict human behavior.

Biphobia—negative beliefs about and stigmatization of bisexuality and those identified as bisexuals.

Binegativity—see *Biphobia*.

Bisexuality—a sexual orientation that involves cognitive, emotional, and sexual attraction to members of both sexes.

Blended family—a family created when two individuals marry and at least one of them brings with him or her a child or children from a previous relationship or marriage. Also referred to as a stepfamily.

Blind marriage—practiced in traditional China, a marriage in which the bride and groom were prevented from seeing each other until their wedding day.

Blurred retirement—the process of retiring gradually so that the individual works part-time before completely retiring or takes a bridge job that provides a transition between a lifelong career and full retirement.

Brainstorming—suggesting as many alternatives as possible without evaluating them.

Branching—in communication, going out on different limbs of an issue rather than staying focused on the issue.

Bundling—a courtship custom practiced by the Puritans whereby the would-be groom slept in the future bride's bed. Both were fully clothed and had a board between them.

C

Certified family life educator (CFLE)—a credential offered through the National Council on Family Relations that provides credibility for one's competence in conducting programs in all areas of family life education.

Cervical cap—a contraceptive device that fits over the cervix.

Child sexual abuse—exploitative sexual contact or attempted sexual contact by an adult with a child before the victim is 18. Sexual contact or attempted sexual contact includes intercourse, fondling of the breasts and genitals, and oral sex.

Chlamydia—known as the silent disease, a sexually transmitted disease caused by bacteria that can be successfully treated with antibiotics.

Cialis—known as the "weekend Viagra" (effective for 36 hours), a medication that allows the aging male (when stimulated) to get and keep an erection as desired.

Civil union—a pair-bonded relationship given legal significance in terms of rights and privileges (more than a domestic relationship and less than a

marriage). Vermont recognizes civil unions of same-sex individuals.

Cohabitation—two unrelated adults (by blood or by law) involved in an emotional and sexual relationship who sleep in the same residence at least 4 nights a week.

Cohabitation effect—the research finding that couples who cohabit before marriage have greater marital instability than couples who do not cohabit.

Coitus—the sexual union of a man and woman by insertion of the penis into the vagina.

Coitus interruptus—the ineffective contraceptive practice whereby the man withdraws his penis from the vagina before he ejaculates.

Collaborative practice—a process that brings a team of professionals (lawyer, psychologist, social worker, financial counselor) together to help a couple separate and divorce in a humane and cost-effective way.

Collaborating style of conflict—conflict style in which the partners are both assertive and cooperative. Each expresses his or her view and cooperates to find a solution.

Coming out—being open and honest about one's sexual orientation and identity.

Commensality—eating with others. Most spouses eat together and negotiate who joins them.

Commitment—an intent to maintain a relationship.

Common-law marriage—a marriage by mutual agreement between a cohabiting man and woman without a marriage license or ceremony (recognized in some but not all, states). A common-law marriage may require a legal divorce if the couple breaks up.

Communication—the process of exchanging information and feelings between two people.

Commune—see *Intentional community*.

Compersion—the "opposite of jealousy"; feeling good about and being supportive of a partner's emotional and physical involvement with and enjoyment of another person.

Competing style of conflict—conflict style in which the partners are both assertive and uncooperative. Each tries to force his or her way on the other so that there is a winner and a loser.

Complementary-needs theory—mate selection theory that persons with complementary needs are attracted to each other. Also known as the "opposites attract" theory.

Compromising style of conflict—conflict style in which there is an intermediate solution so that both partners find a middle ground they can live with.

Conception—see *Fertilization*.

Conflict—the interaction that occurs when the behavior or desires of one person interfere with the behavior or desires of another.

Congruent message—one in which the verbal and nonverbal behaviors match: the message sent by what a person says is the same as that conveyed by what he or she does.

Conjugal love—the love between married people characterized by companionship, calmness, comfort, and security. This is in contrast to romantic love, which is characterized by excitement and passion.

Control group—group used to compare with the experimental group that is not exposed to the independent variable being studied.

Coolidge effect—term used to describe the waning of sexual excitement and the effect of novelty and variety on sexual arousal.

Corporal punishment—the use of physical force with the intention of causing a child to experience pain, but not injury, for the purpose of correction or control of the child's behavior.

Covenant marriage—type of marriage that permits divorce only under conditions of fault (such as abuse, adultery, or imprisonment for a felony).

Crisis—a sharp change for which typical patterns of coping are not adequate and new patterns must be developed.

Cross-dresser—a generic term for individuals who may dress or present themselves in the gender of the other sex (e.g., a heterosexual male will dress as a woman).

Crude divorce rate—the number of divorces that have occurred in a given year for every thousand people in the population.

Cunnilingus—the oral stimulation of a woman's genitals by her partner.

D

Date rape—nonconsensual sex (sexual intercourse, anal sex, oral sex) between two people who are dating or on a date.

Dating—a mechanism whereby some men and women pair off for recreational purposes, which may lead to exclusive, committed relationships for the reproduction, nurturing, and socialization of children.

December marriage—a marriage in which both spouses are elderly.

Defense mechanisms—unconscious techniques that function to protect individuals from anxiety and minimize emotional hurt.

Defense of Marriage Act (DOMA)—legislation passed by Congress denying federal recognition of homosexual marriage and allowing states to ignore same-sex marriages licensed elsewhere.

Demandingness—the degree to which parents place expectations on children and use discipline to enforce the demands.

Dementia—the mental disorder most associated with aging, whereby the normal cognitive functions are slowly lost.

Depo-Provera®—also referred to as "dep," a contraceptive shot taken by the woman every 6 months.

Desertion—a separation in which one spouse leaves the other and breaks off all contact with him or her.

Developmental-maturational training approach—an approach to childrearing that views what children do, think, and feel as being influenced by their genetic inheritance.

Developmental task—skills developed at one stage of life that are helpful at later stages.

Diaphragm—a barrier method of contraception that involves a rubber dome over the uterus to prevent sperm from moving into the uterus.

Discrimination—behavior that denies individuals or groups equality of treatment.

Disenchantment—the change in a relationship from a state of newness and high expectation to a state of mundaneness and boredom in the face of reality.

Displacement—shifting one's feelings, thoughts, or behaviors from the person who evokes them onto someone else who is a safer target.

Divorce—legal ending of a valid marriage contract.

Divorce mediation—process in which divorcing parties make agreements with a third party (mediator) about custody, visitation, child support, property settlement, and spousal support. Divorce mediation is quicker and less expensive than litigation.

Divorcism—the belief that divorce is a disaster.

Domestic partnership—a relationship in which individuals who live together are emotionally and financially interdependent and are given some kind of official recognition by a city or corporation so as to receive partner benefits (e.g., health insurance).

Double standard—the idea that there is one standard for women and another for men (e.g., having numerous sexual partners suggests promiscuity in a woman but manliness in a man).

Dowry (also called **trousseau**)—the amount of money or valuables the woman's father would pay the man's father for the man marrying the woman. It functioned to entice the man to marry the woman, because an unmarried daughter stigmatized the family of the woman's father.

Dual-career marriage—a marriage in which both spouses pursue careers and maintain a life together that may or may not include dependents.

Durable power of attorney—a legal document that gives one person the power to act on behalf of another.

E

Ecosystem—the interaction of families with their environment.

Emergency contraception—also referred to as postcoital contraception, the various types of morning-after pills that are used in three circumstances: when a woman has unprotected intercourse, when a contraceptive method fails (such as condom breakage), and when a woman is raped.

Emotional abuse—the denigration of an individual with the purpose of reducing the victim's status and increasing the victim's vulnerability so that he or she can be more easily controlled by the abuser. Also known as verbal abuse or symbolic aggression.

Endogamy—in mate selection, the cultural expectation to marry within one's own social group in terms of race, religion, and social class.

Erectile dysfunction—a man's inability to get and maintain an erection.

Eros love style—love style characterized by passion and romance.

Escapism—the simultaneous denial and withdrawal from a problem.

Euthanasia—derived from Greek words meaning "good death," or dying without suffering or pain. Euthanasia may be passive, where medical treatment is withdrawn and nothing is done to prolong the life of the patient, or active, which involves deliberate actions to end a person's life.

Exchange theory—theory that emphasizes that relations are formed and maintained between individuals offering the greatest rewards and least costs to each other.

Exogamy—in mate selection, the social expectation that individuals marry outside their family group (e.g., avoid sex and marriage with a sibling or other close relative).

Extended family—the nuclear family or parts of it plus other relatives such as parents, grandparents, aunts, uncles, cousins, and siblings.

Extradyadic relationship (involvement)—emotional or sexual involvement between a member of a pair and someone other than the partner. The term *extradyadic*, external to the couple, is broader than *extramarital*, which refers specifically to spouses.

Extramarital affair—emotional or sexual involvement of a spouse with someone other than his or her mate.

F

Familism—philosophy in which decisions are made in reference to what is best for the family as a collective unit.

Family—as defined by the U.S. Census Bureau, a group of two or more persons related by blood, marriage, or adoption. Broader definitions include individuals who live together who are emotionally and economically interdependent.

Family career—the various stages and events that occur within the family.

Family caregiving—activities to help the elderly, which may include daily bathing and feeding, balancing checkbooks, and transportation to the grocery store or physician.

Family life course development—the process through which families change over time.

Family of orientation—the family of origin into which a person is born.

Family of origin—the family into which an individual is born or reared, usually including a mother, father, and children.

Family of procreation—the family a person begins by getting married and having children.

Family relations doctrine—an emerging doctrine holding that even nonbiological parents may be awarded custody or visitation rights if they have been economically and emotionally involved in the life of the child.

Fellatio—the oral stimulation of a man's genitals by his partner.

Female condom—a condom that fits inside the woman's vagina to protect her from pregnancy, HIV infection, and other sexually transmitted infections.

Female genital alteration—sometimes referred to as genital cutting, mutilation, or circumcision, the cultural practice (e.g., in parts of Africa and some Middle Eastern countries) of cutting off the clitoris to reduce the woman's libido and make her marriageable.

Female genital mutilation—see *Female genital alteration*.

Female genital operations—see *Female genital alteration*.

Feminization of poverty—the idea that poverty is experienced disproportionately by women.

Feral—wild; not domesticated; refers to children who are thought to have been reared by animals.

Fertilization—also known as conception, the fusion of the egg and sperm.

Filial piety—love and respect toward one's parents.

Filial responsibility—feeling a sense of duty to take care of one's elderly parents.

Formal separation—a separation between spouses based on a legal agreement drawn up by an attorney and specifying the rights and responsibilities of the parties, including custody issues.

Foster parent—also known as a family caregiver, a person who at home, either alone or with a spouse, takes care of and fosters a child taken into custody.

Frail—elderly person who has difficulty with at least one personal care activity or other activity related to independent living (e.g., bathing, dressing, getting in and out of bed, grocery shopping, taking medications).

Friends with benefits—a relationship consisting of nonromantic friends who also have a sexual relationship.

Functionalists—structural functionalist theorists who view the family as an institution with values, norms, and activities meant to provide stability for the larger society.

G

Gay—homosexual (women or men).

Gender—the social and psychological behaviors that women and men are expected to display in society.

Gender dysphoria—the condition in which one's gender identity does not match one's biological sex.

Gender identity—the psychological state of viewing oneself as a girl or a boy, and later as a woman or a man.

Gender role ideology—the proper role relationships between women and men in a society.

Gender roles—behaviors assigned to women and men in a society.

Gender role transcendence—abandoning gender frameworks and looking at phenomena independent of traditional gender categories.

Genital herpes—also known as herpes simplex virus type 2 infection, a sexually transmissible viral infection. Can also be transmitted to a newborn during birth.

Genital warts—sexually transmitted lesions that commonly appear on the cervix, vulva, or penis, or in the vagina or rectum.

Gerontology—the study of aging.

Gerontophobia—fear or dread of the elderly.

GLBT—see *LGBT*.

Gonorrhea—also known as the clap, the whites, and morning drop, a bacterial infection that is sexually transmitted.

Granny dumping—a situation in which adult children or grandchildren who feel burdened with the care of their elderly parent or grandparent drive the elder to the entrance of a hospital and leave him or her there with no identification.

H

Hanging out—refers to going out in groups where the agenda is to meet others and have fun.

Hedonism—sexual value system emphasizing the pursuit of pleasure and the avoidance of pain.

HER/his career—wife's career given precedence over husband's career.

Hermaphrodites—(also **intersexed individuals**) persons with mixed or ambiguous genitals.

Herpes simplex virus type 1 infection—a viral infection that can cause blistering, typically of the lips and mouth; it can also infect the genitals.

Herpes simplex virus type 2 infection—see *Genital herpes*.

Heterosexism—the denigration and stigmatization of any behavior, person, or relationship that is not heterosexual.

Heterosexuality—the predominance of cognitive, emotional, and sexual attraction to those of the opposite sex.

HIS/her career—husband's career given precedence over wife's career.

HIS/HER career—husband's and wife's careers given equal precedence.

HIV—human immunodeficiency virus, which attacks the immune system and can lead to AIDS.

Homogamy—in mate selection, selecting someone with similar characteristics, such as interests, values, age, race, religion, and education.

Homonegativity—a construct that refers to antigay responses such as negative feelings (fear, disgust, anger), thoughts ("homosexuals are HIV carriers"), and behavior ("homosexuals deserve a beating").

Homophobia—used to refer to negative attitudes toward homosexuality.

Homosexuality—the predominance of cognitive, emotional, and sexual attraction to those of a person's own sex.

Hooking up—one-time sexual encounter where there is generally no expectation of seeing each other again. The nature of the sexual expression may be making out, oral sex, and/or sexual intercourse. The term is also used to denote getting together periodically for a sexual encounter—oral sex or sexual intercourse with no strings attached.

Human ecology—the study of ecosystems or the interaction of families with their environment.

Hysterectomy—removal of a woman's uterus.

I

Individualism—philosophy in which decisions are made on the basis of what is best for the individual as opposed to the family (familism).

Induced abortion—the deliberate termination of a pregnancy through chemical or surgical means.

Infatuation—a state of passion or attraction that is not based on reality.

Infertility—the inability to achieve a pregnancy after at least 1 year of regular sexual relations without birth control, or the inability to carry a pregnancy to a live birth.

Informal separation—similar to a formal separation except that no lawyer is involved in the separation agreement; the husband and wife settle issues of custody, visitation, alimony, and child support between themselves.

Inhibited female orgasm—inability to achieve an orgasm after a period of continuous stimulation.

Internalized homophobia—a sense of personal failure and self-hatred among lesbians and gay men resulting from social rejection and stigmatization; has been linked to increased risk for depression, substance abuse and addiction, anxiety, and suicidal thoughts.

Intentional community—group of people living together on the basis of shared values and worldview (e.g., Twin Oaks in Louisa, Virginia).

Internalized homophobia—a sense of personal failure and self-hatred among lesbians and gay men resulting from social rejection and stigmatization which has been linked to increased risk for depression, substance abuse and addiction, anxiety, and suicidal thoughts.

Intersexed individuals—see *Hermaphrodite.*

Intimate partner violence—an all-inclusive term that refers to crimes committed against current or former spouses, boyfriends, or girlfriends.

Intrafamilial child sexual abuse—sex by an adult or older family member with a child in the family.

Intrauterine device (IUD)—a small object that is inserted by a physician into a woman's uterus through the vagina and cervix for the purpose of preventing implantation of a fertilized egg in the uterine wall.

"I" statements—statements that focus on the feelings and thoughts of the communicator without making a judgment on others.

J

Jadelle®—a contraceptive consisting of rod-shaped silicone implants that are inserted under the skin in the upper inner arm and provide time-release progestin into a woman's system for contraception.

Jealousy—an emotional response to a perceived or real threat to an important or valued relationship.

L

Laparoscopy—a form of salpingectomy (tubal ligation) that involves a small incision through the woman's abdominal wall just below the navel.

Legal custody—decisional authority over major issues involving the child .

Leisure—the use of time to engage in freely chosen activity perceived as enjoyable and satisfying.

Lesbian—homosexual woman.

LesBiGays—collective term referring to lesbians, gays, and bisexuals.

Levitra—a medication taken by aging men to help them get and maintain an erection (an alternative to Viagra and Cialis).

LGBT—also known as GLBT, refers collectively to lesbians, gays, bisexuals, and transgendered individuals.

Litigation—a judge hears arguments from lawyers representing the respective spouses and decides issues of custody, child support, division of property, etc., which become legally binding.

Living together—see *Cohabitation.*

Living will—a legal document that identifies the wishes of an individual with regard to end-of-life care.

Lose-lose solution—a solution to a conflict in which neither partner benefits.

Ludic love style—love style in which love is viewed as a game whereby the love interest is one of several partners, is never seen too often, and is kept at an emotional distance.

M

Mania love style—an out-of-control love whereby the person "must have" the love object. Obsessive jealousy and controlling behavior are symptoms of manic love.

Marital rape—forcible rape (sexual intercourse, anal intercourse, oral sex) by one's spouse.

Marital success—relationship in which the partners have spent many years together and define themselves as happy and in love (hence, the factors of time and emotionality).

Marriage—a legal contract signed by a heterosexual couple with the state in which they reside that regulates their economic and sexual relationship. (The highest court in Massachusetts ruled in 2004 that homosexuals may marry.)

Masturbation—stimulating one's own body (usually genitals) with the goal of experiencing pleasurable sexual sensations.

Mating gradient—the tendency for husbands to marry wives who are younger and have less education and less occupational success.

May-December marriage—an age-discrepant marriage in which the younger woman is in the spring of her life (May) and he is in his later years (December).

Mediation—*see Divorce mediation.*

Medicaid—state welfare program for low-income individuals.

Medicare—federal insurance program for short-term acute hospital care and other medical benefits for persons over 65.

Megan's Law—Federal law requiring that convicted sex offenders register with local police when they move into a community.

Mexican-American—referring to people of Mexican origin or descent who are American citizens.

Mifepristone—also known as RU-486, a synthetic steroid that effectively inhibits implantation of a fertilized egg.

Miscarriage—see *Spontaneous abortion.*

Modern family—the dual-earner family, in which both spouses work outside the home.

Mommy track—stopping paid employment to spend time with young children.

Munchausen syndrome by proxy—rare form of child abuse whereby a parent (usually the mother) takes on the sick role indirectly (hence, by proxy) by inducing illness or sickness in her child so that she can gain attention and status as a caring parent.

N

Natural family planning—a method of contraception that involves refraining from sexual intercourse when the woman is thought to be fertile.

Negotiation—divorcing spouses discuss/resolve the issues themselves of custody, child support, division of property, and spouse maintenance.

No-fault divorce—a divorce that assumes that neither party is to blame.

Nuclear family—family consisting of an individual, his or her spouse, and his or her children, or of an individual and his or her parents and siblings.

NuvaRing®—a soft, flexible, transparent ring approximately 2 inches in diameter that is worn inside the vagina and provides month-long pregnancy protection.

O

Occupational sex segregation—the concentration of women in certain occupations and men in other occupations.

Oophorectomy—removal of a woman's ovaries.

Open relationship—a stable relationship in which the partners regard their own relationship as primary but agree that each may have emotional and physical relationships with others.

Ortho Evra®—is a contraceptive transdermal patch that delivers hormones to a woman's body through skin absorption.

Oxytocin—a hormone from the pituitary gland during the expulsive stage of labor that has been associated with the onset of maternal behavior in lower animals.

P

Palimony—a take-off on the word *alimony,* referring to the amount of money one "pal" who lives with another "pal" may have to pay if the partners terminate their relationship.

Palliative care—health care focused on the relief of pain and suffering of the individual who has a life-threatening illness and support for them and their loved ones.

Pantogamy—group marriage where each member of the group is married to each other.

Parallel style of conflict—style of conflict whereby both partners deny, ignore, and retreat from addressing a problem issue (e.g., "Don't talk about it, and it will go away").

Parent effectiveness training—a model of childrearing that focuses on trying to understand what a child is feeling and experiencing in the here and now.

Parental alienation syndrome—a disturbance in which children are obsessively preoccupied with deprecation or criticism of a parent (usually a noncustodial parent); the denigration is unjustified or exaggerated. Found in situations where one divorced parent has "brainwashed" the child to perceive the other parent in very negative ways.

Parental investment—any investment by a parent that increases the chance that the offspring will survive and thrive.

Parental status—the degree to which the stepparent is considered to be a parent to a stepchild.

Parenting—the provision by an adult or adults of physical care, emotional support, instruction, and protection from harm in an ongoing structural (home) and emotional context to one or more dependent children.

Patriarchy—"rule by the father," a norm designed to ensure that women are faithful to their husbands and remain in the home as economic assets, childbearers, and child-care workers.

Periodic abstinence—refraining from sexual intercourse during the 1 to 2 weeks each month when the woman is thought to be fertile.

Physical abuse—intentional infliction of physical harm in the form of hitting, slapping, pushing, kicking, or burning (e.g., with water or cigarettes) by one individual on another.

Physical custody—also called "visitation," refers to distribution of parenting time following divorce.

Polyamory—"many loves," whereby three or more men and women have a committed emotional and sexual relationship. The partners may rear children in this context.

Polyandry—a form of polygamy in which one wife has two or more husbands.

Polygamy—a generic term referring to a marriage in which there are more than two spouses.

Polygyny—a form of polygamy in which one husband has two or more wives.

Positive androgyny—a view of androgyny that is devoid of the negative traits associated with masculinity (aggression, being hard-hearted, indifferent, selfish, showing off, and vindictive) and femininity (being passive, submissive, temperamental, and fragile).

POSSLQ—an acronym used by the U.S. Census Bureau that stands for People of the Opposite Sex Sharing Living Quarters.

Postpartum depression—a reaction more severe than the "baby blues" to the birth of one's baby, characterized by crying, irritability, loss of appetite, and difficulty in sleeping.

Poverty—the lack of resources necessary for material well-being—most important, food and water, but also housing and health care.

Power—the ability to impose one's will on another and to avoid being influenced by the partner.

Postmodern family—a departure from traditional models such as lesbigay couples and single mothers by choice, which emphasizes that a healthy family need not be heterosexual or have two parents.

Pragma love style—love style that is logical and rational. The love partner is evaluated in terms of pluses and minuses and regarded as a good or bad "deal."

Pregnancy—a condition that begins 5 to 7 days after conception, when the fertilized egg is implanted (typically in the uterine wall).

Prejudice—negative attitudes toward others based on differences.

Prenuptial agreement—a contract between intended spouses specifying which assets will belong to whom and who will be responsible for paying what in the event of a divorce.

Primary group—small, intimate, informal group. A family is a primary group.

Primary sexual dysfunction—a dysfunction that the person has always had.

Principle of least interest—principle stating that the person who has the least interest in a relationship controls the relationship.

Projection—attributing one's own thoughts, feelings, and desires to some one else while avoiding recognition that these are one's own thoughts, feelings, and desires.

Pronatalism—view that encourages having children.

Psychological abuse—see *Emotional abuse.*

Q

Quality of life—typically associated with physical function, visual acuity, continence, being bedfast, depressed, etc.

R

Race—group of individuals who have physical characteristics that are identified and labeled as being socially significant.

Racism—the attitude that one group of people is inferior to another group on the basis of physical characteristics.

Random sample—sample in which each person in the population being studied has an equal chance of being included in the sample.

Rapid ejaculation—persistent or recurrent ejaculation with minimal sexual stimulation before, upon, or shortly after penetration and before the partner wishes it.

Rationalization—the cognitive justification for one's own behavior that unconsciously conceals one's true motives.

Refined divorce rate—the number of divorces/annulments in a given year, divided by the number of married women in the population, times 1000.

Reflective listening—paraphrasing or restating what a person has said to indicate that the listener understands.

Relativism—sexual value system whereby decisions are made in the context of the situation and the relationship (e.g., sexual intercourse is justified in a stable, caring, monogamous context).

Resiliency—the ability of a family to respond to a crisis in a positive way.

Responsiveness—the extent to which parents respond to and meet the needs of their children. Refers to such qualities as warmth, reciprocity, person-centered communication, and attachment.

Rite of passage—event that marks the transition from one status to another (e.g., a wedding marks the rite of passage of individuals from lovers to spouses).

Role—the behavior individuals in certain status positions are expected to engage in (e.g., spouses are expected to be faithful).

Role compartmentalization—separating the roles of work and home so that an individual does not dwell on the problems of one role while physically being at the place of the other role.

Role conflict—being confronted with incompatible role obligations (e.g., the wife is expected to work full-time and also to be the primary caretaker of children).

Role overload—the convergence of several aspects of one's role resulting in having neither time nor energy to meet the demands of that role. For example, the role of wife may involve coping with the demands of employee, parent, and spouse.

Role strain—the anxiety that results from not being able to take care of several role needs at once.

Romantic love—an intense love whereby the lover believes in love at first sight, only one true love, and that love conquers all.

Rophypnol—date rape drug that renders persons unconscious so that they have no memory of what happens when they are under the influence.

RU-486—see *Mifepristone.*

S

Salpingectomy—tubal ligation, or tying a woman's fallopian tubes, to prevent pregnancy.

Sandwich generation—individuals who attempt to meet the needs of their children and elderly parents at the same time.

Satiation—the state in which a stimulus loses its value with repeated exposure. Partners sometimes get tired of each other because they are around each other all the time.

Second-parent adoption—(also called co-parent adoption) a legal procedure that allows an individual to adopt his or her partner's biological or adoptive child without terminating the first parent's legal status as parent.

Second shift—the cooking, housework, and child care that employed women do when they return home from their jobs.

Secondary group—large or small group characterized by impersonal and formal interaction. A civic club is an example.

Secondary sexual dysfunction—a dysfunction that the person is currently experiencing following a period of satisfactory sexual functioning.

Secondary virginity—the conscious decision of a sexually active person to refrain from intimate encounters for a specified period of time.

Sensate focus—an exercise whereby the partners focus on pleasuring each other in nongenital ways.

Sex—the biological distinction between being female and being male (for example, having XX or XY chromosomes).

Sexism—an attitude, action, or institutional structure that subordinates or discriminates against an individual or group because of their biological sex (e.g., women are discriminated against as national television news anchors).

Sex roles—behaviors defined by biological constraints. Examples include wet nurse, sperm donor, and child-bearer.

Sexual double standard—see *Double standard.*

Sexual orientation—the aim and object of one's sexual interests—toward members of the same sex, the opposite sex, or both sexes.

Sexual script—shared interpretations and expected behaviors in sexual situations.

Sexual values—moral guidelines for sexual behavior.

Shaken baby syndrome—behavior in which the caretaker, most often the father, shakes the baby to the point of causing the child to experience brain or retinal hemorrhage.

Shared parenting dysfunction—the set of behaviors by both parents that are focused on hurting the other parent and are counterproductive for the child's well-being.

Shift work—having one parent work during the day and the other parent work at night so that one parent can always be with the children.

Sibling relationship aggression—Behavior of one sibling toward another intended to induce social harm or psychic pain in the sibling (e.g. excluding a sibling from an event, spreading a rumor).

Single-parent family—family in which there is only one parent—the other parent is completely out of the child's life through death, sperm donation, or complete abandonment, and no contact is ever made with the other parent.

Single-parent household—household in which one parent typically has primary custody of the child or children but the parent living out of the house is still a part of the child's family.

Social allergy—being annoyed and disgusted by a repeated behavior on the part of one's partner.

Socialization—the process through which we learn attitudes, values, beliefs, and behaviors appropriate to the social positions we occupy.

Sociobiology—a theory that emphasizes that there are biological explanations for social behavior.

Sociological imagination—the perspective of how powerful social structure and culture are in influencing personal decision-making.

Socioteleological approach—an approach to childrearing that explains children's behavior as resulting from the attempt to compensate for feelings of inferiority.

Sodomy—oral and anal sexual acts.

Spectatoring—mentally observing one's own and one's partner's sexual performance. Often interferes with performance because of the associated anxiety.

Spermicide—a chemical that kills sperm.

Spontaneous abortion—an unintended termination of a pregnancy.

Stalking—unwanted following or harassment that induces fear in the target person.

Status—a position a person occupies within a social group, such as parent, spouse, or child.

STD—sexually transmitted disease that can be transmitted through sociosexual contact. Now, more often referred to as sexually transmitted infection.

Stepfamily—a family in which at least one spouse brings at least one child into a remarriage. Also referred to as a blended family.

Step relationships—see *Stepfamily.*

Stepism—the assumption that stepfamilies are inferior to biological families. Stepfamilies are stigmatized.

Sterilization—a permanent surgical procedure that prevents reproduction.

Storge love style—a love consisting of friendship that is calm and nonsexual.

Stratification—the ranking of people according to socioeconomic status, usually indexed according to income, occupation, and educational attainment.

Stress—a nonspecific response of the body to demands made on it.

Supermom—a cultural label that allows the mother who is experiencing role overload to regard herself as particularly efficient, energetic, and confident.

Superwoman—see *Supermom*.

Symbolic aggression—see *Emotional abuse*.

Syphilis—a sexually transmissible infection caused by spirochete entering the mucous membranes that line various body openings. Can also be transmitted by a pregnant woman to her unborn child.

T

THEIR career—career shared by a couple; they travel and work together (e.g., journalists).

Theoretical framework—a set of interrelated principles designed to explain a particular phenomenon and to provide a point of view.

Therapeutic abortion—an abortion performed to protect the life or health of the woman.

Third shift—the emotional energy expended by a spouse/parent in dealing with various family issues. The job and housework are the first and second shifts, respectively.

Time-out—a discipline procedure whereby the child is removed from an enjoyable context to a nonreinforcing one and left there (alone) for a minute for every year of the child's age.

Traditional family—the two-parent nuclear family with the husband as breadwinner and wife as homemaker.

Transgender—a generic term for a person of one biological sex who displays characteristics of the other sex.

Transgenderist—an individual who lives in a gender role that does not match his or her biological sex, but has no desire to surgically alter his or her genitalia.

Transgendered—individuals whose gender identities do not conform to traditional notions of masculinity and femininity.

Transition to parenthood—the period of time from the beginning of pregnancy through the first few months after the birth of a baby.

Transracial adoption—the practice of parents adopting children of another race—for example, a white couple adopting a Korean or African-American child.

Transsexual—an individual who has the anatomical and genetic characteristics of one sex but the self-concept of the other.

Transvestite—a person who enjoys dressing in the clothes of the other sex.

U

Utilitarianism—the doctrine holding that individuals rationally weigh the rewards and costs associated with behavioral choices.

Uxoricide—the murder of a woman by her romantic partner.

V

Values—standards regarding what is good and bad, right and wrong, desirable and undesirable.

Vasectomy—male sterilization involving cutting out small portions of the vas deferens.

Verbal abuse—see *Emotional abuse*.

Viagra—a medication taken by aging men to help them get and maintain an erection (the first medication of its kind—before Levitra and Cialis).

Violence—the intentional infliction of physical harm by one individual toward another.

W

Win-lose solution—a solution to a conflict in which one partner benefits at the expense of the other.

Win-win relationship—a relationship in which conflict is resolved so that each partner derives benefits from the resolution.

Withdrawal—see *Coitus interruptus*.

Y

"You statements"—statements that tend to assign blame (e.g., "You made me angry") rather than "I statements" ("I got angry and blew up").

Youthhood—that period of time between adolescence and adulthood.

Photo Credits

This page constitutes an extension of the copyright page. We have made every effort to trace the ownership of all copyrighted material and to secure permission from copyright holders. In the event of any question arising as to the use of any material, we will be pleased to make the necessary corrections in future printings. Thanks are due to the following authors, publishers, and agents for permission to use the material indicated.

Chapter 1. **1:** Authors **2:** Authors **13:** © HBO/Photofest **16:** Authors **21:** Authors **30:** Authors

Chapter 2. **41:** © ABC/Photofest **43:** Authors **44:** Authors **56:** © Bloomberg News/Landov **65:** Authors

Chapter 3. **71:** Kelly Lewis **73:** Authors **77:** © Paramount/Photofest **78:** Authors **93:** Authors

Chapter 4. **99:** Authors **102:** Authors **110:** Kelly Lewis **112:** © Match Events, Inc./www.8minuteDating.com **120:** Authors

Chapter 5. **130:** Authors **138:** Authors **141:** Authors **146:** Authors **152:** Authors

Chapter 6. **161:** © AP/Wide World Photos **164:** Authors **170:** Courtesy Human Rights Campaign/www.hrc.org **173:** Authors **177:** © AP/Wide World Photos **183:** Authors **185:** © Focus Features/Photofest

Chapter 7. **193:** © Digital Vision/Getty Images **195:** © AP/Wide World Photos **196:** Authors **207:** © Universal/Photofest **218:** © AP/Wide World Photos

Chapter 8. **223:** Authors **229:** Authors **235:** Authors **239:** Authors **242:** Kelly Lewis **244:** Authors

Chapter 9. **252:** Kelly Lewis **253:** © Dwayne Newton/PhotoEdit **261:** Authors **263:** Authors **266:** Authors

Chapter 10. **274:** © bilderlounge media GmbH/Alamy **277:** top, © AP/Wide World Photos **277:** bottom, Authors **280:** Authors **284:** Authors **296:** Ben Lattin **304:** Katie Cushman

Name Index

Subject Index

Alfie, 77
Alimony, 467
Allen, Woody, 195
Alternative to Marriage Project, 556
Alzheimer's disease, 485
AMA. *See* American Medical
    Association (AMA)
Ambiguity, 264, 473–474
Ambivalence, 5
American Association for Marriage
    and Family Therapy, 389, 554
American Medical Association (AMA),
    491
American Social Health Association,
    555
Amish
    child rearing by, 320
    children, 18–19
    divorce, 19
    elderly, 19
    founder, 17
    marriage, 18
Amish, Jacob, 17
Amnesty International, 186
Anabaptist Movement, 17–19
Anal sex, 144
Androgyny, 64–65
Angelil, Rene, 243
Anger, 401
Announcements, marriage, 11
Annual percentage rate (APR), 527
Annuities, 520
Annulments, 447–448
Anodyspareunia, 144
Anonymity, 429
Anti-fraternization clauses, 81
Antigay marriage legislation, 177–178
Antinatalism, 280
Anxiety, jealousy and, 91
Anxiety, performance, 549
APMR. *See* Abbreviated progressive
    muscle relaxation (APMR)
APR. *See* Annual percentage rate
    (APR)
Arapesh people, 44
Arousal problems, 545–546
Arranged marriages, 80, 85
Artificial insemination (AIH), 286
*As Nature Made Him: The Boy Who Was
    Raised as a Girl*, 43
Asceticism, 134
Asians. *See also* Chinese; Japanese
    domestic violence, 400
    extended families of, 227
    group subordination, 19
Association for Couples in Marriage
    Enrichment (ACME), 389, 544
Atkins, Barbara, 411
Attachment
    love theory, 84
    mate selection and, 199
    parenting, 341–342
Attention deficit hyperactivity disorder
    (ADHD), 257
Attitude-behavior discrepancy, 35
Attitudes toward infidelity scale, 46,
    379

Attitudes toward parenthood scale,
    279
Attitudes toward transracial adoption
    scale, 290
Authoritarian parents, 327
Authoritative parents, 327
Autobiography outlines, family,
    561–562
Autobiography outlines, individual,
    558–560
Autoeroticism, 139
Avoidance style, 264, 266
Azidothymidine, 150

B

Baby blues, 320
Baby Think It Over (BTIO), 283–284
Bachelor, The, 112
Baha'i religion, 239
Band-Aid operation. *See* Laparoscopy
Barebacking, 144
Barkley, Charles, 239
Battered-woman syndrome, 396
Battered women, 410
Battering, 396
Behavior
    -attitude discrepancy, 35
    conflict, 263
    desired, identifying, 267
    inappropriate, 329
    in last years, 505
    negative, divorce and, 430
    rewarding, 328
Behavioral androgyny, 64–65
Behavioral child rearing approach,
    338–339
Beliefs, 8, 438
Bias
    antigay, 167, 170, 183–185
    language, 47
    research, 34
Big Love, 12, 13
Bigamy, 448
Binegativity. *See* Biphobia
Binuclear families, 17
Biochemical theory, 84
Biofeedback, 372–374
Biosocial theory, 28, 48
Biphobia, 168–169
Birth control pills. *See* Oral contracep-
    tive agents
Birth parents, 290
Birth rate, 276–277, 283
Bisexuals
    defined, 162
    prevalence, 164–165
    relationships of, 174
Blame, 401
Blended families, 17, 459–460
Blind Date, 112
Blind marriages, 116
Blood transfusions, 149–150
Blurred retirement, 493
BMI. *See* Body mass index (BMI)
Body images, 59
Body mass index (BMI), 85, 376
Body types, 59

Bombeck, Erma, 328
Bonds, 520
Boundary maintenance, 456–457
Bowles, Camilla Parker, 80
Brainstorming, 268
Branching, 255–256
Bride wealth, 227
Brokeback Mountain, 80, 185
BTIO. *See* Baby Think It Over (BTIO)
Buddhism, 13, 80
Budgeting, 518–519
Buffy the Vampire Slayer, 185
Bulbourethral glands, 543
Bundling, 117
Burns, George, 482
Business associates, 211–212

C

Cancers, cervical, 532
Cancers, prostate, 374–375
Careers
    child and family services, 515–516
    family life educators, 514–515
    marriage and family therapy, 515
Caregivers
    adult children as, 485
    parents as, 316
    support for, 486
    typical, 485
Carey, Drew, 101
Caribbean, 53–54
Carlin, George, 7
Carpenter, Kelli, 162
CASI. *See* Computer-administered self-
    interviewing (CASI)
Catholic Church
    divorce policy, 447
    gender role views, 51
    sex abuse scandal, 414
CDs. *See* Certificates of deposits (CDs)
Ceremonies, marriage, 11
Certificates of deposits (CDs), 520–521
Certified family life educator, 514–515
Cervarix, 532
Cervical cancers, 532
Cervical cap, 299
CFLE. *See* Certified family life
    educator
CFS. *See* Chronic fatigue syndrome
    (CFS)
Chaplin, Charlie, 242
Chaplin, Oona, 242
Cheney, Dick, 170
Cheney, Mary, 170
Child abuse
    contributing factors, 412–413
    defined, 411–412
    effects of, 413–414
    family of origin, 402
    neglect, 411
    physical, 411
    resources, 552
Child and Family Services, 515–516
Child custody
    deciding on, 439
    GLBT discrimination in, 181–183
    joint, choosing, 443–444

Government
Internet control by, 326
policies, birth rate and, 276
working family policies, 362
Grafenberg spot, 539–540
Grandparents
average age, 500
baby-sitting role, 499–500
benefits of, 501–502
child rearing role, 500
resources, 553
styles, 500–501
support children and, 476
visitation rights of, 501
Granny dumping, 418
Great Gatsby, The, 73
Greece, 54, 82
Groups
control, 33
defined, 6
experimental, 33
interpersonal choices and, 6–8
polyamory, 94
secondary, 7
subordination to, 19
support, 385–386
GSR. *See* Galvanic skin response (GSR) biofeedback
Guaranteed insurability, 522
Guardianship, 335

H

Habits, 9
Hadith, 236
Hanging out, 108–109
Harvard Business School, 352
Hatch, Orin, 50
Hate crimes, 183
Hawn, Goldie, 121
HCG. *See* Human chorionic gonadotropin (HCG)
Health. *See* Mental health; Physical health
Health insurance, 377
Health Marriage Resource Center, 248
Hebrews, 82
Hedonism, 136–137
Heinz-Kerry, Teresa, 211, 455
Held, Myka, 183
HER/his careers, 352
Here and now cohabitation, 120
Hermaphrodites, 43
Herpes simplex virus (HSV), 533–534
Heterosexual-Homosexual Rating Scale, 163
Heterosexuals
antigay bias and, 183–185
defined, 162
homosexuals contact with, 185
prevalence, 164–165
HGH. *See* Human growth hormone (HGH)
HHS. *See* Department of Health and Human Services (HHS)
Hill, Faith, 360
HIS/HER careers, 352
HIS/her careers, 352

Hispanic Americans
fertility rate, 276
gender roles, 50
HIV/AIDS incidence, 530
partner searches, 112
HIV/AIDS
characterization, 529
disclosure, 259
divorced dating and, 115
homosexual risks for, 173
opportunistic diseases and, 531
oral sex transmission of, 141
prevalence, 530
prevention, 150, 295, 297, 302
resources, 555
singles' risk of, 107
testing, 530, 535–536
transmission modes, 529
transmission of, 147–150
transsexuals and, 45
treatment, 531
window period, 529
women with, 59
Homogamy
age and, 197
defined, 195
education and, 197
race and, 196
remarriage, 455
theory of, 195–196
Homonegativity, 167–168
Homophobia, 167–168
Homosexuals. *See also* Same-sex couples
anal sex and, 144
attitudes toward, 167–168
casual sex among, 173
causes, 165–166
civil unions, 15, 175–176
cohabitation and, 115
coming out, 170–171
conflicts, 172–173
defined, 162
discrimination against, 183–185
equality struggle, 186
harassment of, 170
heterosexual contact with, 185
HIV/AIDS transmission, 529
imprisonment of, 186
liberation movement, 101, 139
long-term care needs, 488
marriages, 511–512
media depiction, 185–186
monogamy among, 173
national data, 165
prevalence of, 164–165
relationship satisfaction, 172
resources, 553
in straight marriages, 381
Honesty, 257
Honeymoons, 230
Hooking up, 109
Hope, Bob, 482
Hormonal contraception, 292–295
Hormone therapy, 286
Hostage, 396–397
Households
children moving between, 473
same-sex, 165

by social class, 29
Housing
costs of, 525
elderly concerns, 489–490
mortgages, 527
selecting, 526–527
types of, 526
value of, 526
Howard, Margo, 9
HPV. *See* Human papillomavirus (HPV)
HSV. *See* Herpes simplex virus (HSV)
Human chorionic gonadotropin (HCG), 286
Human ecology
defined, 27
familles, 27–28
marriage, 27–28
principles, 27
Human growth hormone (HGH), 484–485
Human papillomavirus (HPV), 531–532
Humor, 372
Hurricane Katrina, 369
Hwang Woo Suk, 34
Hypersensitivity, 204
Hysterectomy, 301

I

I statements, 255
Iacocca, Lee, 494
Iceland, 120, 262
Identity
gender, 44
gender roles and, 49
mixed-race children, 240
singlehood, 106
Identity theft, 525
Implanon, 293
Impulse control, 204
In vitro fertilization (IVF), 286–287
In-laws, 233–234
Inadequacy, 91
Incest taboos, 28, 195
Income
elderly's, 489
singles', 106
women's, 55, 58–59
India, 281
Individual retirement accounts (IRAs), 520
Individualism
divorce and, 429
interpersonal choices and, 5–6, 8
respect for, 327–328
Industrial Revolution
dating during, 117
divorce rate and, 428
families and, 20–21
Infatuation, 73, 75
Infertility
causes of, 285
remedies, 285–289
types of, 285
Infidelity. *See* Extramarital affairs

Suicide
  Dutch, 491
  loved one, surviving, 388
  patient-assisted, 491
  physician-assisted, 491
Summers, Larry, 48
Superperson strategy, 360
Superwomen, 360
Support groups, 385–387
Supportive communication scale, 257
Surrogate mothers, 286
Sweden
  cohabitation and, 121
  couples therapy, 389
  gender roles in, 54
Symbolic interactionism
  communication, 261–262
  family framework, 26
  stepfamilies, 261–262
Syphilis, 533, 534–535
Systems framework, 27

T

Taliban, 53
Tapestry Against Polygamy, 12
Tarrying. *See* Bundling
Tchambuli people, 44
Teachers, 317, 325
Temperature biofeedback, 373
Terrorism, 21
Tester cohabitation, 120
Testes, 542
Testosterone replacement therapy
  (TRT), 549–550
The Netherlands, 491
THEIR careers, 353
Thermal biofeedback, 373
Third shift, 360
Thompson, Dianna, 468
Thoreau, Henry David, 106
Thoughts, end of life, 504
Time
  giving, 328–329
  lags, 34
  leisure, 363–364
  long-term care considerations, 487
  management, 361
Time outs, 331
Timmendequas, Jesse, 416
Titanic, 28–29
Tradeoffs, 4–5
Traditional families, 17
Traditional fatherhood scale, 323
Traditional marriage, 226
Traditional motherhood scale, 321
Transcendence, 65
Transdermal contraception, 294–295
Transgenderism, 44, 162, 556
Transgenderists, 45
Transracial adoption, 289–290
Transrectal ultrasound (TRUS), 375
Transsexuals, 45
Transvestites, 44
Treasury bills, 520
Triads, 94
Triangular theory of love, 75–77
Trigametic IVF, 287–288
Triple indemnity, 522

Troxel v. Granville, 501
TRT. *See* Testosterone replacement
  therapy (TRT)
True Love Waits, 133
TRUS. *See* Transrectal ultrasound
  (TRUS)
Trust, 245
Turner, Ted, 199
Twin Oaks International Community,
  13, 107–108

U

Unemployment, 383–384
Unfulfilling love, 90
Unhappiness, 401
Uninvolved parents, 327
United Kingdom, 140, 276
Unrequited love, 90
Unsatisfactory relationships, 217–219
Urbanization, 20
Urethral opening, 539
Uterus, 541
Utilitarianism, 23
Uxoricide, 396

V

Vacations. *See* Leisure
Vagina, 539–540
Vaginal intercourse, 142
Vaginal opening, 539
Vaginal rings, 294
Vaginal spermicides, 297
Values
  conflicts over, 263
  divorce incidence and, 430–431
  interpersonal choices and, 8
  marriage's influence, 231
  Muslim, 236–237
  sexual (*See* Sexual values)
  stress and, 371
Vas deferens, 543
Vasectomies, 303
Viagra, 494
Video chatting, 111
Vietnam, 133
Vincent, Norah, 56–57
Violence. *See* Domestic violence
Virginity pledge, 133
Virginity, secondary, 133–134
Visitation, 181–183
Vows, 230
Vulva, 538

W

Waiver of premiums, 522
Wealth, bride, 227
Weddings, 229–231, 237
Welles, Orson, 26
When "I Love You" Turns Violent, 88
WHO. *See* World Health Organization
  (WHO)
Widowhood, 104, 458–459, 556
Wilbanks, Jennifer, 217
Will and Grace, 185
Williams, Charles Andrew, 184
Win-lose solutions, 268
Win-win solutions, 268–269
Winfrey, Oprah, 101, 224

Withdrawal, 299
Women. *See also* Mothers; Stepmothers
  Afghani, 53
  as property, 400
  beliefs about, 47
  body image, 59, 85
  child-bonding, 61
  condoms for, 296–297
  divorced, 456
  economic independence of,
    427–429
  emotional abuse by, 399–398
  ERT for, 495–496
  external anatomy, 538–539
  FWB relationships, 136
  genital alteration, 60
  gonorrhea symptoms, 534
  in age-discrepant marriages, 243
  income levels, 55, 58–59
  infertility and, 285
  internal anatomy, 539–541
  jealousy and, 92–93
  life expectancy, 61
  material satisfaction, 61
  parenthood impact on, 279
  partner's death, 503
  physiology, 538–541
  relationship choices, 64
  resources, 556
  sexism and, 60
  sexual behavior and, 144–145
  sexual changes, elderly, 495
  sexual dysfunction, 545–548
  sexuality, views of, 154
  STI risk, 59
  unemployment effects, 384
  views of, male, 45–47
  violence against, 396
  widows, 458–459
  work/family demands, 359–360
  working outside home, 117,
    351–354
  younger men and, 504
Women's movement
  emphasis of, 100–102
  mate selection and, 200
  sexual values and, 139
Woods, Tiger, 239
Work-family fit, 361–363
Working poor, 349
Workplace
  Danish, 364
  friendships, 380
  GLBT harassment at, 171
  love in, 81
  same-sex couples and, 176–177
  women in, 117
World Health Organization (WHO), 2
Wright, Frank Lloyd, 482

Y

Youthhood, 101
Yuzpe method, 300

Z

Zeta Jones, Catherine, 243
Zidovudine, 150
Zovirax, 533–534